Computer-aided Production Engineering

Conference Organizing Committee

Corresponding Committee Members

IMechE Conference Transactions

16th International Conference on

Computer–aided Production Engineering CAPE 2000

7–9 August 2000
The University of Edinburgh, UK

Edited by
Professor J A McGeough

Organized by
The University of Edinburgh

Supported by
Institution of Mechanical Engineers (IMechE)
Scottish Enterprise Edinburgh and Lothian (SEEL)

IMechE Conference Transactions 2000–5

Professional Engineering Publishing

Published by Professional Engineering Publishing Limited for The Institution of Mechanical Engineers, Bury St Edmunds and London, UK.

First Published 2000

ISSN 1356–1448
ISBN 1 86058 263 X

A CIP catalogue record for this book is available from the British Library.

Printed by The Cromwell Press, Trowbridge, Wiltshire, UK.

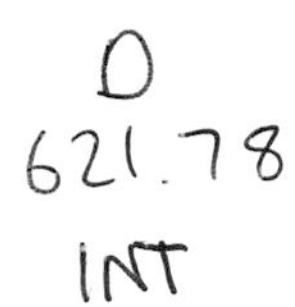

Related Titles of Interest

Title	Editor/Author	ISBN
IMechE Engineers' Data Book – Second Edition	C Matthews	1 86058 248 6
Advanced Manufacturing Processes, Systems, and Technologies (AMPST 99)	M K Khan, Y T Abdul-Hamid, C S Wright, and R Whalley	1 86058 230 3
Advances in Manufacturing Technology XIII	A N Bramley, A R Mileham, L B Newnes, and G W Owen	1 86058 227 3
Guide to Presenting Technical Information	C Matthews	1 86058 249 4
Software Quality Management V – The Quality Challenge	C Hawkins, M Ross, and G Staples	1 86058 110 2
Software Quality Management IV – Improving Quality	M Bray, M Ross, and G Staples	1 86058 031 9

For the full range of titles published by Professional Engineering Publishing contact:

Sales Department
Professional Engineering Publishing Limited
Northgate Avenue
Bury St Edmunds
Suffolk
IP32 6BW
UK

Tel: +44 (0)1284 724384
Fax: +44 (0)1284 718692

Contents

Manufacturing and Supply Chain Management

Concurrent Engineering and Design Manufacuture

About the Editor

Joseph McGeough is Regius Professor of Engineering at The University of Edinburgh, Fellow of the Institution of Mechanical Engineers, and Chairman of the Organizing Committee for CAPE 2000. His main field of research is in unconventional manufacturing in which he has published the books 'Principles of Electrochemical Machining', 'Advanced Methods of Machining' and is Editor of the new book 'Micromachining of Engineering Materials' to be published later this year. He has also published many papers in the learned journals.

Professor McGeough obtained his BSc and PhD at Glasgow University, his DSc at Aberdeen University, and is a fellow of the Royal Society of Edinburgh. He is an Honorary Professor of Nanking Aeronautical and Astronautical University and a visiting Professor to Glasgow Caledonian University.

Preface

On behalf of the Organizing Committee I have great pleasure in thanking you for attending the 16th International Conference on Computer-aided Production Engineering on the 7–9 August 2000. A programme of high quality research papers has been selected for presentation at the conference and I am grateful for the support of the Organizing Committee and the Corresponding Committee members for their advice in refereeing all the papers. I am appreciative of the help of the main sponsoring bodies, the Institution of Mechanical Engineers and Scottish Enterprise Edinburgh and Lothian (SEEL) in organization of the event especially Mr John Ling (IMechE), Mr Jim Reid (SEEL), and with particular thanks due to Miss Amy Middlemass (The University of Edinburgh).

A set of industrial presentations and visits have been arranged in conjunction with this years CAPE conference, as well as a social programme, which I hope will help to make the conference enjoyable for all our delegates and visitors to Edinburgh.

Professor J A McGeough
Chairman, Organizing Committee
Regius Professor of Engineering
School of Mechanical Engineering
The University of Edinburgh

Computer-aided Design and Manufacture

CAPE/048/2000

The construction of geometric models from point-cloud data

G SMITH and **T CLAUSTRE**
Faculty of Engineering, University of the West of England, Bristol, UK

SYNOPSIS

This paper is concerned with the utilisation of point-cloud data for reverse engineering activities and the development of geometric representations of free-form features, starting with data obtained from a digital scanning system. Consideration is given to the use of point data for the generation of CNC cutter paths and rapid prototype models, also to techniques for the construction of geometric surface representations. Particular consideration is given to the collection of point-cloud data, trimming of duplicated and redundant data, the definition of boundary curves and the establishment of appropriate continuity conditions at surface boundaries.

1 REVERSE ENGINEERING

Reverse engineering is concerned with the capture of data from existing 3D parts and utilising this to either construct geometric models or produce replicas. It is widely used in industries where it is commonplace for stylists and designers to use traditional techniques for the representation and modelling of their concepts, and for engineers to convert these models into manufactured products. Reverse engineering systems are to be found in industries as diverse as motor vehicle design/manufacture and foot wear design/manufacture.

1.1 Data acquisition

Co-ordinate measuring machines (CMM) equipped with touch-trigger probes are widely used to capture 3D geometric data, and a considerable amount of work has been done on the standardisation of programming formats and the operation of such machines. On-line and off-line programming facilities have been developed in an attempt to expedite the process of probe path planning and the generation of measurement data. Harris (1) discussed the development of one such system. Many co-ordinate measuring machines incorporate some facilities for surface data acquisition, but data acquisition rates are relatively low and this tends to limit their application to parts consisting of rational geometric features; such as cylinders, spheres and rectangular prisms. Whereas the trend is for engineering components to incorporate complex, irrational, features; such as free-form or sculptured surfaces.

Free-form features pose many unique problems for both design and manufacturing engineers. The techniques used to describe them are relatively complex; ranging from bi-parametric, bi-cubic, surface elements (e.g. cubic Bezier patches) to complete 3D surfaces (e.g. NURBS). There is no accepted standard for the measurement and representation of such features, and the problems of rebuilding geometric models from captured surface data have not been fully addressed. Data acquisition techniques for reverse engineering are usually concerned with the collection of masses of discrete position vectors (i.e. point-cloud data) in an attempt to capture all salient features on complex surfaces. Figure 2 shows the set of point-cloud data gathered from a small sculptured face. This is a very crude approximation consisting of just 2,000 vectors; many of the detailed features would be lost on such a representation. Figure 1 shows a rendered image based on the triangulation of a point-cloud data set of approximately 1.7 million vectors. This gives a much more complete representation of the sculptured face.

There are many systems used to capture shape data from surfaces, ranging from tactile probes to vision based systems. CMM with digital touch-trigger probes are generally considered to be the most accurate (with resolutions in the region of 0.005mm), but too slow when masses of data are required. Scanning systems, based on either laser triangulation or analogue contact probes, are considerably less accurate (resolutions in the region of 0.1mm) but much faster. Vision based systems are very rapid, but also very approximate (with typical resolutions in the region of 1.0mm). Consequently, vision systems are generally considered to be too approximate for the majority of reverse engineering applications.

Varady (2) presented a review of surface data acquisition techniques for reverse engineering. He concluded that tactile systems are generally both robust and accurate, and they result in nearly noise free data. The authors own work has also shown this to be the case. We have established (3) that laser scanning systems are generally faster than tactile ones, but the resulting data is often incomplete and may be very inaccurate, especially when scanning near-vertical faces. In contrast to this, tactile probing has consistently resulted in good quality point-cloud data. The work presented in this paper is based on point-cloud data acquired by a tactile, analogue, scanning system.

1.2 Point-cloud data

Point-cloud data simply consists of masses of 3D position vectors, hence it is inherently approximate and it inevitably gives an incomplete representation of scanned objects. It is analogous to the vertices in a B-Rep modelling system based on a face, edge vertex (FEV) schema. Consequently, such data gives a crude representation compared to CAD generated models. Many of the inherent advantages of CAD based modelling are therefore lost when parts are simply represented by position vectors, in particular the topological relationship between modelled features will be absent from such representations. However, point-cloud data does form a useful basis for several reverse engineering activities; such as CNC machining, rapid prototyping and the generation of surfaces for geometric modelling.

1.2.1 The generation of CNC cutter paths

CNC cutter paths for sculptured surfaces consist of many linear moves approximating the surface curvature. The degree of approximation being controlled by chordal deviation tolerances set at the tool path planning stage. Tactile systems for the collection of point-cloud data generally use chordal tolerances in a similar way. Consequently the format of point cloud data is well suited to the generation of cutter location data and CNC cutter paths. Lin (4) presented a description of a system which automatically generates three-axis cutter paths from such data. There are also several commercial systems available for the collection of surface data and the generation of CNC cutter paths. The CYCLONE™ scanning machine (figure 4) and TRACECUT™ software, from Renishaw plc, is one such example. This system utilises an analogue, tactile, scanning probe to collect point-cloud data, which then forms the basis for tool path planning in the TRACECUT™ software. Facilities such as chordal tolerancing, scaling, male to female translations and off-setting of the scanned data are facilitated. Off-setting is required to compensate for the radius of the spherical probe tip in order to generate actual surface geometry (rather than geometry relating to the centre of the probe tip). This can be accomplished within the software, but one pragmatic approach is to use a finishing cutter with the same tip radius as the scanning probe; thus obviating the need for probe tip compensation.

Cutter paths generated directly from point-cloud data generally produce very satisfactory results. Particularly if the cutter size is large compared to the pitch of the surface data; under these circumstances the cutter will track across the data and automatically fill in any gaps or voids in the point-cloud. It is commonplace for machine shops to generate CNC cutter paths on massive point-cloud data consisting of many millions of vectors.

1.2.2 Rapid prototypes and shading

Point-cloud data may be relatively crude and incomplete but it can form a very useful basis for the generation of shaded images and data for a wide range of rapid prototyping processes.

Using planar facets to link the vertices in a set of scanned data will result in a tessellated surface. The degree of facetting being dependent on the density of the point-cloud data and tolerance applied to the fitting algorithms. Triangular facets are often preferred for shading and rendering, and they are used extensively to produce images of geometric models. Applying triangulation and shading directly to a set of scanned data can give very acceptable results, as shown in figure 1. This tessellated representation is still far short of a complete geometric model, although triangular meshes are often employed as an intermediate stage in the process of curved surface reconstruction (5).

Rapid prototyping processes are almost exclusively based on the notion of representing 3D objects as sets of 2D cross-sections. A review of these processes is beyond the scope of this paper, but it is worth noting that the starting point for most rapid prototyping systems is a tessellated representation in the standard triangulation language (STL) format. Triangulation of point-cloud data is a relatively straightforward operation (see above) and there are commercial systems with the capability of generating STL files directly from such data. The TRACECUT™ software is one example of this. The conversion of point-cloud data into rapid prototype models is therefore considered to be relatively routine.

1.2.3 Data for transfer to CAD

Point-cloud data is not sufficient for the complete representation of geometric models, but it does form a useful basis for the generation of such models. The challenges inherent in progressing from a set of points to a complete and consistent CAD model are well documented by Varady (2) and Puntambekar (6). Essentially these consist of the creation a series of boundary curves and fitting surfaces to these and the point-cloud data, with the aim of producing a higher level representation of the scanned object. There are a number of commercial modelling systems designed specifically to facilitate the process of fitting curves and surfaces to point-cloud data. CopyCAD™, from Delcam International plc, and Metris Base™, from Metris N.V are examples of such systems. The transfer of data to these systems is facilitated by graphics exchange formats such as IGES, DXF or VDA.

2 BUILDING GEOMETRIC MODELS FROM POINT-CLOUD DATA

The basic process of building a surface model from point-cloud data consists of the following steps: trimming to remove excess data, tolerancing to remove unnecessary data, triangulation to create a tessellated model, segmentation to split the data into logical subsets, boundary creation and finally surface fitting. Boundary curves and fitted surfaces may take many forms, but Bezier and B-Splines are by far the most common for free-form modelling (2)(6)(7)(8)(9)(10).

2.1 Trimming, tolerancing and triangulation

Trimming of point-cloud data consists of removing any points lying outside the boundary of the intended model. Such points are often generated as a direct result of the data capture process; for example when a tactile probe tracks off the scanned part, or when a laser scanner scans over the edge of a part. The effect of modelling without trimming excess data is clearly illustrated in figure 3. Trimming may be done manually, by selecting and deleting individual points or groups of points. Alternatively trimming may involve removing all points lying outside some plane or boundary curve. The latter approach is generally preferred when large point-clouds are involved.

Tolerancing involves the removal of point data in areas of low curvature. The process is generally based on some form of chordal deviation tolerance, resulting in a high density of points in areas of the model with high curvature and lower densities in areas of low curvature. Figure 5 shows the effect of tolerancing a set of scanned data.

Triangulation converts the trimmed and toleranced point-cloud data set to a tessellated surface consisting of triangular facets, with the points forming vertices for adjacent triangles. The effect of tolerancing will result in a high density of small triangles in areas of high curvature, and a low density of larger triangles in flatter areas. The resulting tessellated surface forms the basis for subsequent curve and surface fitting processes.

2.2 Segmentation and boundary creation

The major challenge in model construction is the detection of logical boundary curves for the segmentation process, and the establishment of appropriate continuity conditions at these boundaries. Logical, or natural, boundaries are created when one type of surface intersects another (e.g. a cylinder intersecting a plane) or when some functional characteristic is required (e.g. the creation of a parting or joining line). Varady (2) discussed the problems associated with automatic surface segmentation and the creation of boundary curves. He observed that it is relatively easy for humans to recognise natural boundaries on complex objects, but it is generally very difficult for computers to perform this task without some form of a priori information.

Sarkar (8) describes a technique for segmentation and boundary detection based on the Laplacian Gaussian operator. This approach is effective for edge detection, but natural topology is not taken into consideration. Consequently, such edges may not coincide with logical or functional boundaries. Chivate (11) proposed an alternative approach based on surface extension. However, this also requires some knowledge of model topology in order to determine which surfaces to create and extend.

The automatic segmentation of point-cloud data may result in arbitrary boundary curves bearing little resemblance to the natural topology or functional requirements of the model. It is therefore common practice to have some form of human intervention in the process of boundary curve generation. This may be done at the data capture stage by manually tracing along a functional boundary, as shown in figure 4. Alternatively it may be done at the modelling stage, by selecting points which lie on the chosen boundary and then creating a spline through these. However, both of these approaches have inherent problems; manual tracing of complex curves can be both tedious and error prone, the selection of points along a boundary will also pose problems when large point-clouds are involved.

2.3 Surface fitting

The processes of fitting surfaces to data points and boundary curves are well established. They generally involve some form of least squares approximation to give the 'best fit' surface to a set of data points (8)(9). The resulting network of curves and patchwork of surfaces are combined to give a smooth, or 'fair', representation of the modelled part. Such representations are achieved by applying tangent plane ($\mathbf{G}^1$) or curvature ($\mathbf{G}^2$) continuity conditions over the entire surface (8).

However, many engineering components have functional requirements for surfaces meeting at 'sharp edges' and 'square corners'. Such features would be lost by global $\mathbf{G}^1$ or $\mathbf{G}^2$ continuity, since sharp edges and square corners are defined by position ($\mathbf{G}^0$) continuity. These features will only be modelled effectively when surfaces are constrained to meet at functional boundary curves, and appropriate continuity conditions are set at there curves. This clearly implies the definition of such curves before the generation of modelling surfaces.

3 SHARP EDGES AND BOUNDARIES

Many engineering components have functional features defined by 'sharp edges'. The example of parting lines on moulds and dies is one obvious case. Other instances are found on motor vehicle body panels, these are generally geometrically smooth (i.e. having $\mathbf{G}^1$ or $\mathbf{G}^2$ continuity), but they may also have 'sharp' folded edges where they join adjacent panels. There are also less well known examples such as the 'feather edges' on shoe lasts, these define the boundary between soles and uppers for manufacturing purposes.

There are many other examples of parts consisting of smooth areas bounded by 'sharp edges'. These edges may be very significant in terms of design and manufacturing activities. Hence it is often advantageous to have them defined explicitly at the geometric modelling stage. It is also advantageous to have explicitly defined boundary curves in order to trim point-cloud data before segmentation and surface construction (section 2.1).

The concept of a 'sharp edge' is somewhat ambiguous because a truly sharp edge cannot be achieved in practice. However, it is reasonable to classify an edge as sharp if has a very small radius. The problem of segmenting point-cloud data along sharp edges then becomes one of searching for points lying on small radii, identifying the apex of these radii and constructing boundary curves which interpolate the apexes. Surfaces may then be generated either side of these boundary curves and joined with $\mathbf{G}^1$ continuity to give sharp edged features.

3.1 Detecting edges

There are several established techniques for edge detection. Perhaps the most widely used being the Laplacian of Gaussian (12). However, this requires smoothing of data, using a Gaussian filter, determination of first and second derivatives and the comparison of zero crossings in second derivatives with peaks in first derivatives. This approach has been shown to be effective for the segmentation of point-cloud data (8), but it is clearly a complex process. It may also result in multiple edges when several position vectors lie around edge radii. This condition is not uncommon with dense point-cloud data.

The authors have developed an alternative approach, for the detection of edges, based on the use of vector dot products to determine angles between consecutive vectors along lines of point-cloud data. The comparison of consecutive angles will indicate changes in gradient and hence curvature. Changes above a predetermined threshold are presumed to indicate areas of high curvature corresponding to sharp edges. The search for these sharp edges is structured by analysing lines of point-cloud data coinciding with the scan lines used at the data capture stage.

This approach to edge detection will result in two points of significance where a scan line crosses an edge radius; one where the scan line meets the edge, and one where it leaves. This effect is shown in figure 6. However, the process of generating boundary curves requires the definition of just one point at the apex of each radius. This is achieved by simply extrapolating tangentially along the scan lines, and determining the intersection of these tangents. Thus one point, corresponding to the sharp edge, is generated where a scan line tracks around a small radius.

The final stage of segmentation involves sorting the edge points into logical order, and fitting curves to these points to defined sharp edge boundaries. Figure 7 shows a sharp edge boundary curve created by interpolating a series of edge points with a cubic B-Spline.

This process of sharp edge detection and boundary curve generation has proved to be effective for a range of 3D objects. However, it may give ambiguous results if there are several sharp edges or other areas of high curvature on a surface. Under such circumstances it has proved necessary to have human intervention to guide the search. For example, when searching for the sharp edge defining a parting line on a highly curved die the operator would inform the system that just one boundary was being sought, corresponding to the first and last sharp edges encountered as the scan path tracks across the die. This would result in all intermediate points lying on high curvature, i.e. those other than the first and last, being discarded. Thus, edge points will only be generated as the scan path crosses sharp edges on entry and exit from the die. This method may be extended to more than one sharp edge, as long as each edge forms a closed loop.

Problems will be encountered when one, or more, edges do not form closed loops. Under such conditions the generated boundary may interpolate some internal surface features. Manual editing will then be required to correct the sharp edge definition.

4 CONCLUSIONS

The use of point-cloud data to support a range of reverse engineering activities has been presented. A range of techniques for the acquisition of point data have been discussed and the view that tactile scanning systems, based on analogue contact probes, are well suited to the collection of such data was expressed. The limitations of point-cloud data for geometric modelling were considered, and examples of the use of this data for the generation of CNC cutter paths and as a starting point for rapid prototyping processes were given.

It has been shown that point-cloud data forms a useful basis for the development of higher level CAD models. The need to structure and segment this data to correspond with functional attributes, particularly the creation of logical boundary curves, was considered in some detail. A technique for the automatic detection of 'sharp edge' features in point-cloud data has been presented. This technique has been used to generate boundary curves corresponding to the 'sharp edges' on geometric models.

The proposed technique has proved effective in the detection of sharp edges and the creation of boundary curves. This approach to point-cloud data segmentation works well when boundaries are formed by a closed loop of data points. Open loop boundaries require manual editing, but it is apparent that this represents a considerable improvement on existing, manual, methods for edge and boundary definition.

REFERENCES

1, Harris J 'The Automation of Inspection Planning Using Feature Based Object-Oriented Models' PhD dissertation, UWE Bristol, 1999.
2, Varady T, Martin RR and Cox J 'Reveres Engineering of Geometric Models – an Introduction' Computer-Aided Design, Vol 29, No 4, 1997.
3, Smith G, Hill T, Rockliffe S and Harris J 'Geometric Assessment of Sculptured Features' Advances in manufacturing Technology, Vol viii, Taylor and Francis. 1994.
4, Lin AC and Liu HT 'Automatic Generation of NC Cutter Paths From Massive Data Points' Computer-Aided Design, Vol 30, No 4, 1998.
5, Guo B 'Surface Reconstruction from Points to Splines' Computer-Aided Design, Vol 29, No 4, 1997.
6, Puntambekar NV, Jablokow AG and Sommer HJ 'Unified Review of 3D Model Generation for Reverese Engineering' Computer Integrated Manufacturing Systems, Vol 7, No 4, 1994.
7, Tuohy ST, Maekawa T, Shen G and Patrikalakis NM 'Approximation of Measured Data With Interval B-Splines' Computer-Aided Design, Vol 29 No 11 1997.
8, Sarkar B and Menq CH 'Smooth-Surface Approximation and Reverse Engineering' Computer-Aided Design, Vol 23, No 9, 1991.
9, Milror MJ, Bradley C, Vickers GW and Weir DJ 'G^1 Continuity of B-Spline Surface patches in Reverse Engineering' Computer-Aided Design, Vol 27, No 6, 1995.
10, Ma W and Kruth JP 'Parameterization of randomly Measured Points for Least Squares Fitting of B-Spline Curves and Surfaces' Computer-Aided Design, Vol 27, No 9, 1995.
11, Chivate PN, Puntambekar NV and Jablokow AG 'Extending Surfaces for Reverse Engineering Solid Model Generation' Computers In Industry, Vol 38, 1999.
12, Jain R, Kasturi R and Schunck BG 'Machine Vision' McGraw-Hill, 1995.

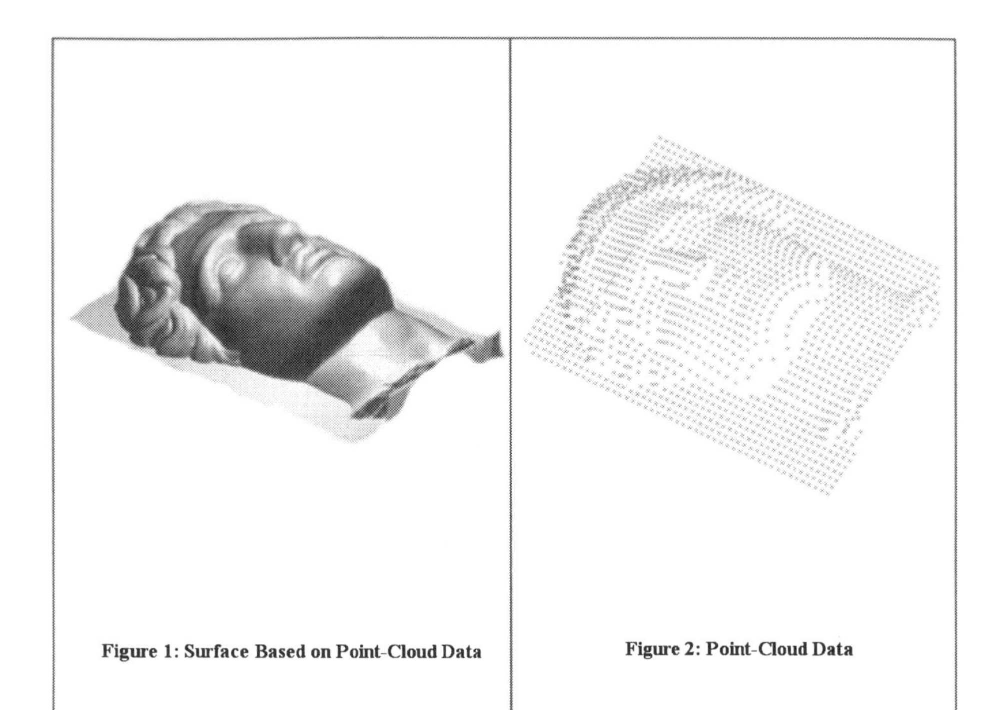

Figure 1: Surface Based on Point-Cloud Data

Figure 2: Point-Cloud Data

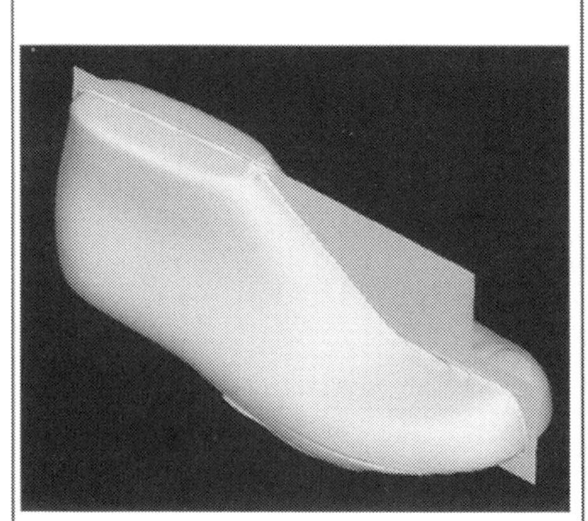

Figure 3: Geometric Model with Excess Data

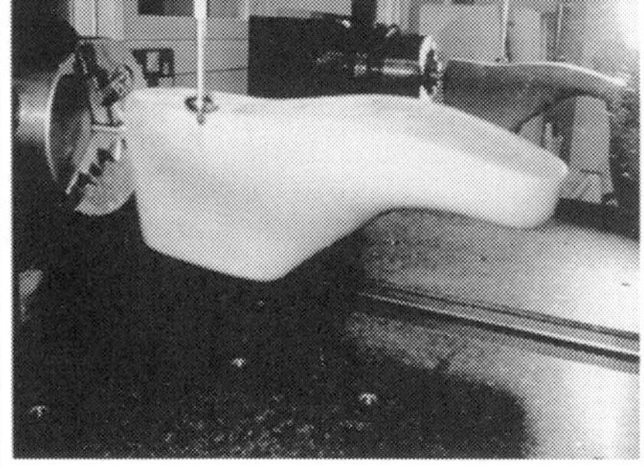

Figure 4: Manually Tracing a Boundary Curve

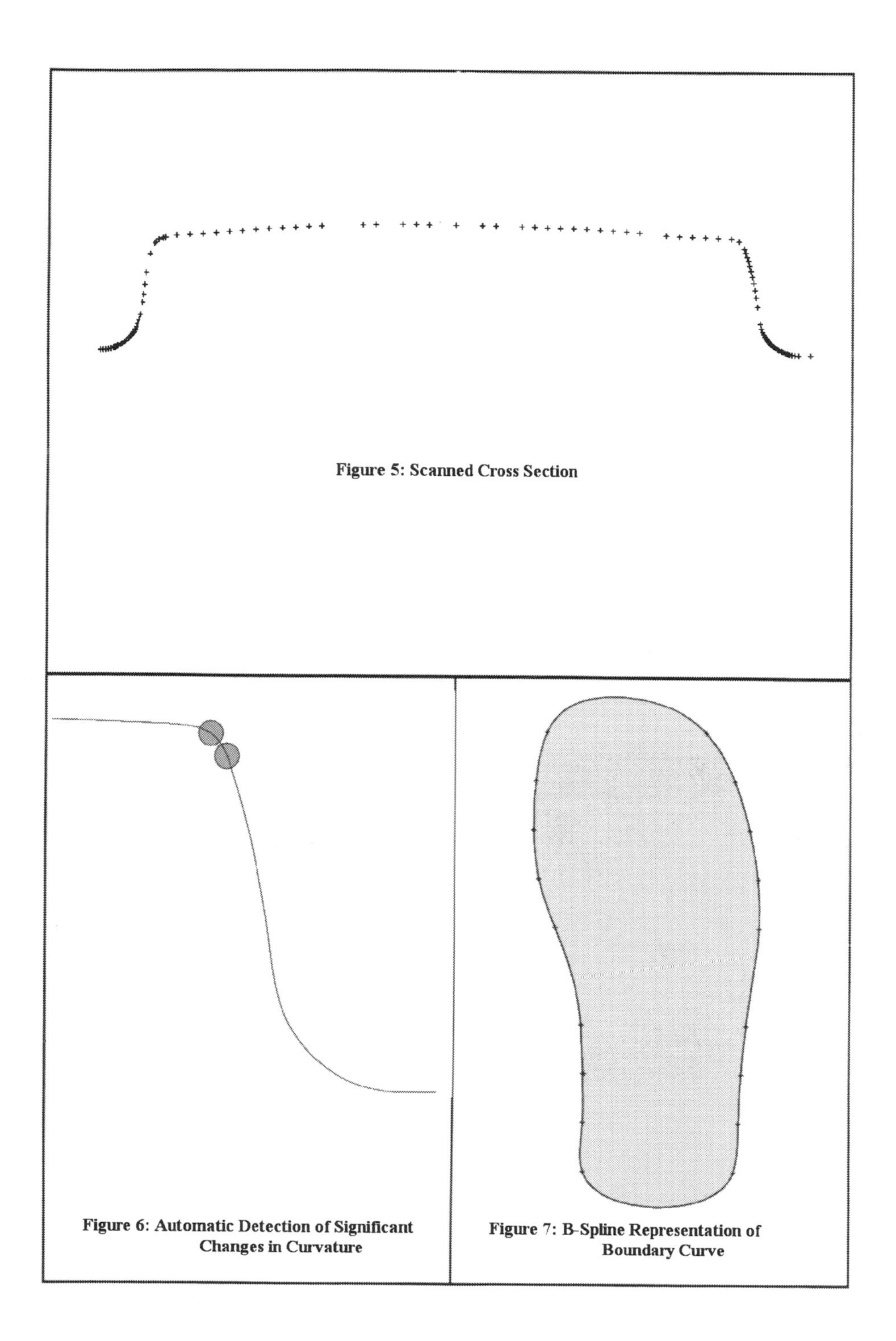

Figure 5: Scanned Cross Section

Figure 6: Automatic Detection of Significant Changes in Curvature

Figure 7: B-Spline Representation of Boundary Curve

CAPE/055/2000

A computer-based design decision logger

P CORBETT, J ATKINSON, and **S HINDUJA**
Department of Mechanical Engineering, University of Manchester Institute of Science and Technology, UK
P SHOLL
Advanced Manufacturing Technology Research Institute, Macclesfield, UK

ABSTRACT

With the need to become increasingly competitive and with the drive to achieve world class excellence, industry is becoming more aware of a requirement to accurately record design decisions. Such records constitute legacy information, which can be vital in subsequent, similar design work. They also provide traceability and facilitate design audits.

A methodology for logging design decisions, together with a computer-based prototype Design Decision Logger has been created and these are described in this paper. As an important step in structuring the design decision logging methodology, design decisions are classified in terms of constraining conditions. The approach taken permits logging of design decisions resulting from project meetings and other design activities and provides cross-referencing to areas of reference. Features of the prototype Design Decision Logger and its use are discussed. By means of an example, it is shown that the Design Decision Logger is of value to designers; and it also helps improve the efficiency of design teams, particularly within concurrent engineering environments, by developing techniques to identify and capture important design decisions during a project.

1 INTRODUCTION

Designers have, in the past, learnt the process of design through a combination of practical experience and exposure to increasingly complex design tasks. There have been many suggestions of specific routes through design activities in order to make the design process more efficient; in particular Pugh (1), and Pahl and Beitz (2). The process of design is difficult to model in the absence of complex interrelationships between design guidelines, reasoning and the decision-making process.

Pahl and Beitz (2) consider design as the conversion of information. Every conversion of information results in a comparison between specified requirements and conversion results. This can be considered as a performance ratio. Design iterations aim to achieve the necessary level of performance. These iterations are brought to a close when an acceptable level of performance is attained. However, the reasoning behind a decision to specify a particular level of performance is crucial in the realisation of design decision rationale.

Areas of Pugh's design core are used these include specification, concept, detail design, and the manufacturing phase of product design. The core model is useful as it is generic across homogeneous mass-produced products and specialised one-off products.

Decisions are made up of complex reasoning patterns and the use of abstract and concrete design models (3). There is a vast contrast in terms of effect and sometimes consequence, when making design decisions at different stages of the design activity. Starkey (4), from a decision theory point of view, states that initial project decisions have far more influence on a project than those made at a later date. It is suggested by Hurst (5) that "up to 70% of product cost could be accounted for once a concept has been selected". There is a disproportionate reduction in the level of decision-making opportunity as a project progresses. He goes on to suggest that "subsequent application of sophisticated design evaluation techniques cannot therefore have significant impact on overall profitability".

Typically, only the final design of a product is documented, and expressed in a variety of formats including CAD drawings and manufacturing instructions (3). Intermediate design activities frequently only exist in note form or in the memory of those involved in a particular project. The potential benefit of documenting intermediate design history is considered high. However, it is generally overlooked, and the focus is normally directed towards the provision of a final design. Historical, data used in further development of a similar product and in design reviews, are rarely considered until they are required. Furthermore, the regeneration of such data at this stage of a project requires re-allocating resources at a time when the opportunity to do so is small. A computer-based approach towards recording design histories was suggested by Malmqvist (3). This uses the "function-means tree" model of the design process and the "chromosome" model for product modelling. The chromosome model devised by Hubker and Eder (6) provides a good basis for detailing the design characteristics of a technical system. Malmqvist (3) suggests the introduction of function-means trees for the inclusion of design history information. A function-means tree considers the "co-evolution of function and form", it shows how the selection of a particular "means" for solving a certain function leads to a requirement on a secondary function, for which a means must be sought". The function-means tree supports the designer in thinking in an axiomatic way (7) which allows clearer visualisation of design concepts. Combining the chromosome model with the function-means tree allows design rationale to be embodied. Although the approach is valid, it has not been utilised in the design decision logging methodology. The approach would demand excessive time in its practical application and over-burden designers.

A more generic approach, suggested by Potts and Bruns (8) and based on Liskov and Guttag's (9) model for designing a text formatter, considered the design history as a network consisting of artefacts and deliberation nodes. Artefacts include design documentation and specifications, whilst deliberation nodes represent issues, alternatives or justifications. However, the approach could require excessive user interaction, and, although the design route optimisation stands to benefit, the trade-off is an excessive infringement of a designer's time. It has been shown that there is a requirement for a Design Decision Logger and that there is no known commercially available means of logging design decisions. The objectives of the Design Decision Logger methodology are included in the next section.

2 THE DESIGN DECISION LOGGER

The approach, in the present work, focuses on optimising routes through complete design activities (see Figure 1). A prototype application has been developed to reduce development costs on similar design activities by capturing and re-using historical design decision information. Subsequent, and similar, design activities can be carried out at reduced cost.

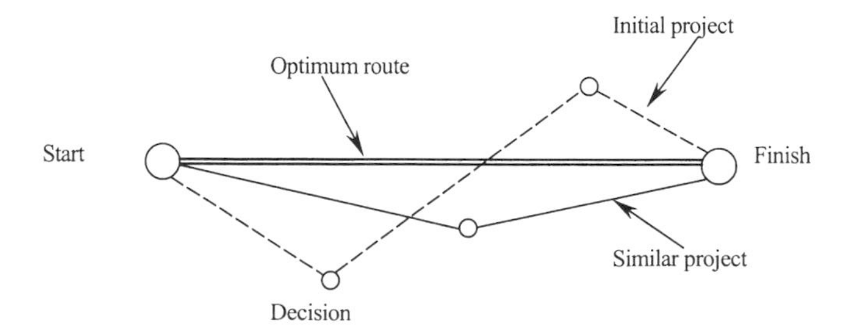

Figure: 1. Design route optimisation.

The Design Decision Logger captures and stores design decisions at critical phases, based on Pugh's (1) design core (see Figure 2) which occur throughout the life cycle of the design activity. The requirement to manage complex data in a clear and structured way led to the use of a relational database as the framework on which to build the Design Decision Logger. There was also a need to involve additional database criteria including details of people and their involvement in projects, companies and levels of authorisation.
The ultimate aim of developing a design decision logging application is to satisfy a commercial requirement for such a system. However, the initial objective is to provide a tool that justifies the value of logging design decisions and its application in optimising design routes.

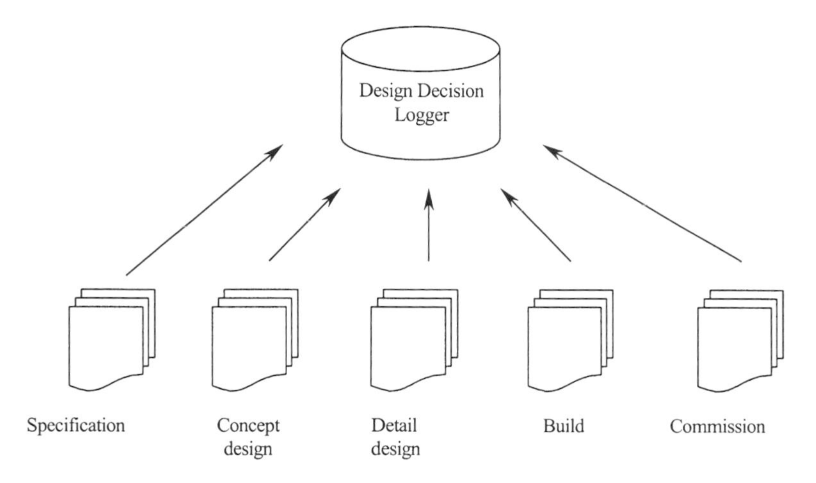

Figure: 2. Critical phases of the design process based on Pugh's (1) design core.

2.1 Methodology

The Design Decision Logger records design decisions and the reasons for those decisions. Decisions are recorded at designers' desks or at project meetings. The methodology forms part of the ISO 9001 quality system and benefits one-off and repeat projects. A relational database is used to efficiently manage the complex relationships that form the Design Decision Logger structure. The methodology is expected to aid project management with minor modifications to existing design procedures and the general working culture. Designers are encouraged to use the application by making them aware of the benefits to be gained.

3 APPLICATION DEVELOPMENT

This section reviews the approach in developing a design logging methodology using a requirement survey to generate a requirement specification and identify specific user groups.
The section also provides an insight into the significance of defining and classifying design decisions together with a method of introducing this into the design decision logging application.
Finally the interactive approach towards developing the design decision logging architecture together with the selection of suitable software and programming language is discussed.

3.1 Philosophy of the approach

Occurrences of design decisions are commonplace in design activities. However, it is understanding exactly when and why design decisions occur and their nature that provides the key to developing an application for logging design decisions. A survey of industrial requirements provided an initial platform in establishing the characteristics of a prototype design decision logger.

3.2 Requirement survey

A requirement survey to identify generic needs for the prototype Design Decision Logger was directed at a number of companies involved in design activities within the machine tool industry.
Personnel were interviewed in a bid to establish common ground on which to develop the design decision logging methodology. A range of design disciplines was encountered, including organisations with a wide range of similar products and small companies operating on a predominantly one-off basis. A requirement specification was written as a result of findings from the requirement survey.

3.3 Requirement specification

The requirement specification listed criteria that would need to be logged in order to explicitly document, and eventually provide, interrogation links to design decisions. The specification was divided into two phases. Phase 1 consisted of facilities to be implemented in the scope of the current work and Phase 2, the facilities to be implemented during further development. Commercial, system, management, operational and functional requirements were all considered in the generation of the requirement specification. Phase 1 requirements were quite broad; the five main requirements of the prototype Design Decision Logger were the ability to:

- capture design decisions as and when they are made;
- uniquely identify all recorded design decisions;
- be generic in order to suit different organisations and users;
- produce reports after interrogation activities; and to
- enable design decision data to be entered using tick, combo boxes, and drop down menus.

After defining these requirements there was a need to establish the contrast between different user requirements of the prototype application.

3.4 Users of the prototype Design Decision Logger

The design decision logging methodology covers the complete life cycle of the design process. This includes phases progressing from specification to commissioning. The prototype places restrictions on users by means of a password and authorisation levels. Different users will want to operate the software in different ways. Thus the prototype can be used by:

- design engineers who are primarily responsible for entering design decisions and performing some minor administrative and interrogation tasks;
- project managers to audit design decisions, produce minutes of a meeting and perform administrative and interrogation tasks; and
- installation and maintenance engineers to cross check the reasoning patterns and rationale of design features on a machine prior to making a modification and also to log reasons for a change or modification.

All users of the Design Decision Logger will require prior knowledge of simple fundamentals concerning operational methods to capture, identify and prioritise design decisions. These fundamentals are covered in the following sections and constitute the definition and classification of design decisions.

3.5 Definition and classification of design decisions

It is uneconomical to log every design decision. The definition and classification of design decisions concentrates effort on critical design decisions and minimises time spent on logging by:

- differentiating between importance levels of design decisions;
- providing a more structured route through storing and obtaining data; and
- encouraging users to realise and define why a decision was made and what the choices were.

Decision types were classified in terms of how they are influenced by various constraining conditions. The three classifications are as follows:

- Fully constrained: With a fully constrained decision, the designer is constrained to a specification. The designer may not make decisions that contradict specification requirements or which are physically impossible.
- Partially constrained: Here the specification allows the designer a degree of flexibility to make important choices. These could be in respect of manufacturing cost, time or engineering and technological constraints.
- Free: A free decision implies that the designer has freedom of choice. It must be realised that free decisions are still important and may have greater constraining influences at a later date.

3.5.1 Influences on decisions

The reasoning process during decision-making is influenced to varying degrees by experience, time, cost and attitude to risk. Advantages and disadvantages can be gained from an abundance, or lack, of experience. Experience can sometimes halt advancement and innovative progress. Here, inexperience could be advantageous. However, it must be pointed out that this is not always the case. People willing to take high risks will make different decisions to those who are more conservative. Typically, the best decision-makers take higher risks, but rarely do so without seeking relevant information. Prior to the decision being made a contingency plan is developed and the variance between high and low risk decisions is critically analysed. There have been many attempts to reduce decision-making risk by means of analytical methods. These are briefly discussed next.

3.5.2 Analytical models

Different types of analytical methods can be used to classify design decisions. These are as follows:

- Analytical Hierarchy Process or AHP by Saaty (10) which is a group decision-making process using a hierarchy of design factors.
- Goal Function Modelling or GFM (11) is a process consisting of two activities, firstly to specify customer needs and secondly to produce a system to solve the problem.
- Quality Function Deployment or QFD by Hales (12) is a process used to establish how a particular attribute will contribute to a system and its bearing on other attributes.

AHP, GFM and QFD are generally used to prioritise various design factors and analyse, and aid, decision-making. However, analytical methods were not used in the Design Decision Logger methodology. Analysing every design decision is time consuming and would unnecessarily hinder and overburden designers, which would have the effect of discouraging use.

3.6 Design Decision Logger architecture

The requirement specification was used as a guide in defining database tables, fields and relationships in the relational database. Database tables contain fields, which are classified by data types. The database contains a matrix of tables linked by relationships. The Design Decision Logger architecture is formed by relationships between database tables, which are ultimately important in establishing a realistic representation of reasoning patterns and the occurrence of design decisions. A conceptual version of the Design Decision Logger architecture and routines was initially developed. The first step in translating the conceptual version of the Design Decision Logger from paper to software was in constructing database tables and linking database tables using relationships in Microsoft Access (see section 3.7). Database forms were designed allowing users to enter the necessary data into the database tables. A data structure was developed which encapsulated all criteria in the requirement specification.

3.7 Development software

Microsoft Access (a relational database) requires a moderate level of programme familiarisation and provided a suitable means of rapidly developing the prototype. This enabled a high level of functionality to be achieved. Microsoft Access is compatible with other Microsoft programmes e.g. Word and Excel. This meant that establishing software links was made easier.

4 DESIGN DECISION LOGGER FACILITIES

This section discusses the main capabilities and facilities which have been incorporated into the prototype design decision logging application. The structure of the Design Decision Logger is depicted in Figure 3, showing 3 instances of use, Input, Administration and Interrogation.

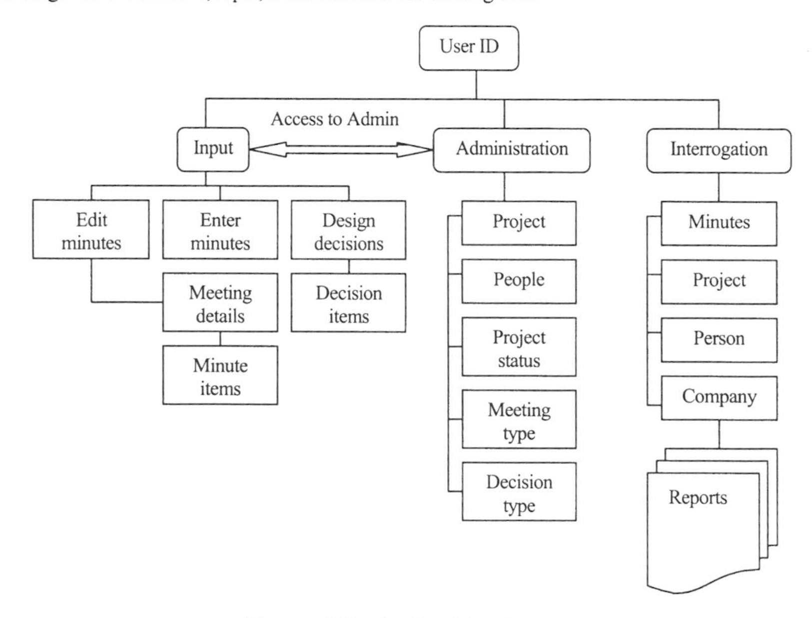

Figure: 3. Design Decision Logger structure

4.1 Software links

A fundamental requirement of the Design Decision Logger is that design decisions can be logged as they occur with minimum hindrance to the user. Software links aim to encourage use by allowing direct access from within other software programmes. Software links have been made to Microsoft Word and Excel with further links planned to Autodesk Mechanical Desktop.

4.2 ISO 9001 and concurrent engineering capabilities

The Design Decision Logger was created so that it would have the potential to improve the efficiency of design teams particularly within concurrent engineering environments by developing techniques to identify, capture and retrieve important design decisions and reasoning. In doing so the Logger assists in the implementation of the quality systems standard ISO 9001–Parts 4.4.4, 5,6,7 and 9 (13) which is a model for quality assurance in design, development, production, installation and servicing.

It is intended that utilising records of design decisions and their reasons and implications will streamline change management. Decisions that have resulted from a change, or will result in a change, can be identified. Related documentation can be used in decision-making, planning and modification. Reasons for any change can be logged and subsequently recalled using the interrogation interface.

Contract variations occur in most projects and are managed by the Design Decision Logger.

The Logger has been developed with a view to promoting concurrent engineering strategies and aiming to reduce the number of modifications and revisions during all stages of design activity. The inclusion of authorisation responsibilities and multi-user capabilities promotes additional communication between staff during projects that may otherwise not have occurred. A manager may query the Design Decision Logger to view all outstanding authorisations, then refer to the designer prior to progressing with the authorisation, after which the database is updated (see Figure 4).

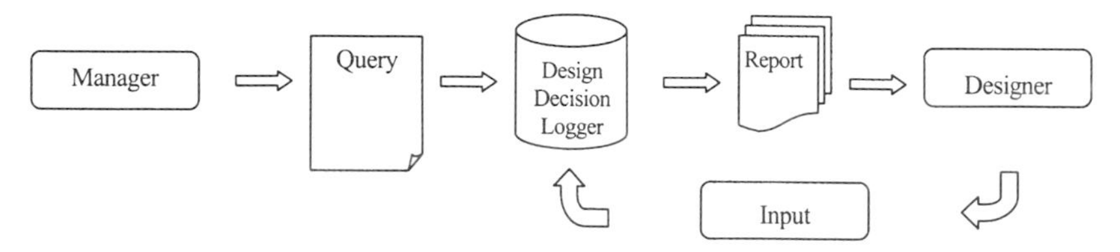

Figure: 4. A typical scenario of Design Decision Logger use.

4.3 Minutes-of-meeting facilities

A fundamental of capturing design decisions is to determine when they occur and to develop an approach to capture decisions at those moments in time. Minutes are a good medium for capturing decisions from within project meetings and the process of producing minutes has been automated and incorporated into the Design Decision Logger. Decisions are made by designers at their desks or at project meetings by various attendees. The Design Decision Logger treats a design decision as a special case of a meeting with only one item and only one attendee. It uses identical forms for entry of design decisions and entry of minute items.

4.4 Data input facility

Data input allows all design decisions, changes and minutes of meetings to be recorded.

A distinction between data input categories was made in order to relieve the user from having to enter irrelevant data and also to reduce the number of forms which the user has to navigate.

Data is entered through straightforward dialogue boxes with the ability to perform administration tasks at any stage of data input (see Figure 3).

4.5 Data interrogation facility

Data interrogation provides an interface that allows access to previously entered and stored information. The interrogation interface has been arranged into categories according to information requested. These are interrogations based on the following:

- Minutes of a meeting. A minutes report can be produced automatically by selecting a project and the relevant meeting description.
- Projects, including reports on design decisions and all project details.
- People, including reports on design decisions, and a person's involvement in a project.
- Companies, including reports on employees of a company and their contact details.

4.6 Administration facility

The administration interface provides a method of updating Design Decision Logger data. Data in need of update is related to a company, a person and a project. An example of this is that a person must have an involvement in a project in order to be selected. If the person does not have an involvement their name will not be seen in drop down boxes and must be added if they are to be henceforward involved. New projects are also initiated though the administration interface. This in turn may also require addition of new company data. The Design Decision Logger can be used across different organisations by allowing users to enter, Project, Person and Company details.

5 A CASE STUDY

The Design Decision Logger was used by Advanced Machine Tool Research Institute (AMTRI) as a tool to log design decisions during a project for a client company whose business involves extensive use of lathes. There were a number of lathes in need of a full refurbishment. The Design Decision Logger was employed to log changes specified by the customer after acceptance tests of the first lathe. An example of a change was the repositioning of a viewing window in the guarding enclosure surrounding the lathe spindle. Repositioning the viewing window would not involve any additional cost to AMTRI or its customer. However, there was an implication associated with a warranty supplied with the refurbished lathe. The customer suggested a change in the position of a viewing window. This would have required a retrofit to previous spindle guards. An implication associated with the ingress of coolant due to a poor seal around the viewing window would have been the result of retrofitting modified guards to previous lathes. It was agreed that the new position of the window was to be included only on the next and subsequent machines.
Here, the design change was entered into the Design Decision Logger and queried at a later date. The implications of changing the viewing window position were apparent and helped to avoid a potential future problem. Figure 5 shows the entry form used to log the design decision.

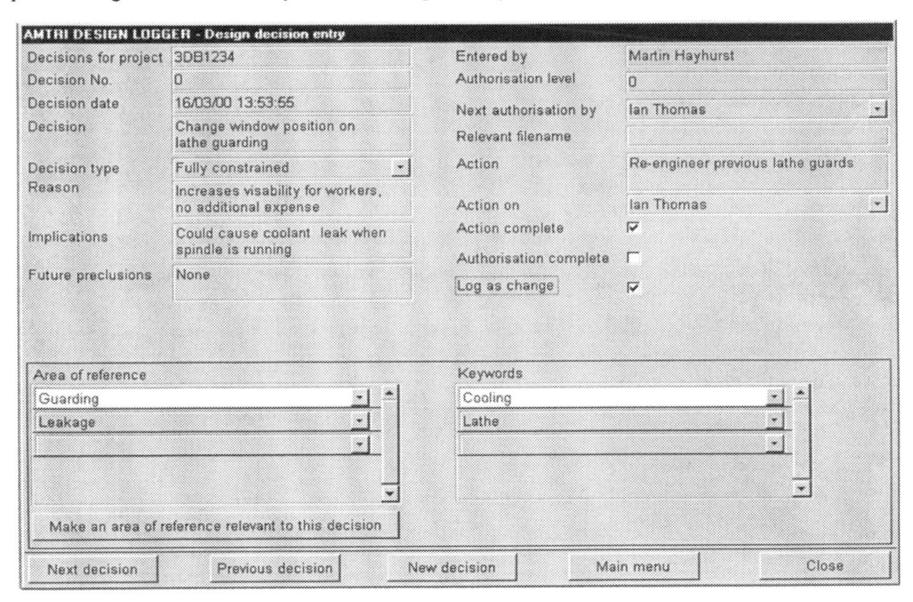

Figure: 5. Design decision entry form.

5.1 Main points from case study

The main objectives in using the Design Decision Logger during the lathe refurbishment project were to establish if the action of logging design decisions proved valuable to design projects and to see if the Logger was user-friendly. An assessment of the Design Decision Logger's suitability for logging and retrieving design decisions was also made. Logging design decisions was considered valuable for the scope of the refurbishment project by encouraging the designer to think of potential implications associated with a particular design decision.

5.2 Merits and limitations

The Design Decision Logger provided long-term benefits through its capability of closely monitoring design changes. Savings in both time and money were passed on to AMTRI and to the customer. The customer's confidence increased because the Design Decision Logger enhanced AMTRI's awareness of potential problems, which could otherwise have delayed the project. The case study highlights the importance of the interrogation interface's potential for locating design decision information. The main limitation of the Design Decision Logger is that it relies upon users to log design decisions, but it is expected that familiarisation and knowledge of benefits will encourage its use.

It can be seen that there is a difference in benefits to be gained between one-off projects and repeat or similar projects. These are explained in the following section.

6 CONCLUSIONS

The requirement survey and specification played invaluable roles in developing the prototype Design Decision Logger. Although these provided a good foundation for developing the initial version of the prototype some future enhancements are required. The importance of a reliable interrogation interface was highlighted by the case study, as was the need for access to design decision information to be simple, quick, reliable and easy. A combination of software links and the automation of minutes of a meeting encouraged design decisions to be logged. Project familiarisation times might be reduced if a graphical representation of the design route is included. This would include the predecessors and successors of design decisions. Such a facility is likely to aid change management by highlighting conflicting design characteristics prior to any modifications. A project archiving procedure is essential to make the Design Decision Logger fully acceptable.

The obvious benefit highlighted from the case study is in capturing design decisions. Normally only the final result of a design is documented, this leaves nothing to describe design iterations and design decision history. An historical record of design intent and rationale provides a key to understanding the global characteristics of a product. This would be particularly useful for organisations with a high labour turnover by reducing the time taken for design familiarisation. The Design Decision Logger is dynamic in terms of allowing design decisions to be logged against user defined areas of reference. When interrogation activities are being performed, the database restricts user-defined areas of reference to those related to the selected project. This approach is beneficial over existing procedures, which currently rely on organised filing systems. Filing systems are limited in use and functionality. This is due to the number of dividers or areas of reference one can fit into a filing system and the ease of finding requested data. Typical benefits of the Design Decision Logger are seen to be as follows.

- First-off projects: Improving communications between design team members and customers increases efficiency as a result of knowledge documentation made available to others as the design evolves. Reduced time for installation through design information being at hand is experienced with a reduction in cost through design re-use.
- Repeat or similar projects: A reduction in time-to-market and lead time can be experienced by reducing design time and errors through enabling access to the background of design decisions. Familiarisation time can be reduced by using design decision documentation, which leads to better design decisions based on the history of previous designs. Project schedules are less likely to overrun as potential pitfalls can be identified in the early stages of a project. The re-use of existing designs can be more reliable and there is greater potential to make technological advancements by eliminating the risk of regenerating design information that previously led to undesirable results.

7 ACKNOWLEDGEMENTS

The authors are grateful to AMTRI and to the Teaching Company Directorate for funding this research. They would also like to thank Dr R. Taylor (AMTRI) who provided valuable assistance in using Microsoft Access, and Mr. R. Thorn (AMTRI) who obtained reference materials and provided useful and constructive comments throughout the project.

8 REFERENCES

1. Pugh, S. Total design, integrated methods for successful product engineering. Addison-Wesley, 1997.
2. Pahl, G. and Beitz, W. Engineering Design, a systematic approach. The Design Council, 1988.
3. Malmqvist, J. A computer-based approach towards including design history information in product models and function means trees. DE-Vol. 83, Design Engineering Technical Conferences. Volume 2 ASME 1995.
4. Starkey, C. V. Engineering Design Decisions and Problem Solving. E. Arnold, 1992.
5. Hurst, K. An Expert System to Support the Concept Selection Phase of the Engineering Design. Engineering Designer. July/August 1991.
6. Hubka, V. and Eder, W. E. Theory of Technical Systems. Springer-Verlag, Berlin, Heidelberg. 1988.
7. Suh, N. P. The principles of design. Oxford University Press, New York. 1990.
8. Potts, C and Bruns, G. Recording reasons for design decisions. MCC Software Technology program IEEE, 1998.
9. Liskov, B and Guttag, J. Abstraction and Specification on Program Development, MIT Press, 1986.
10. Saaty, T. L. Decision making for leaders, the Analytical Hierarchy Process for decisions in a complex world. 2nd Edition, RWS publications, 1990.
11. Combined application of QFD and GFM. Symposium transactions, 11th symposium QFD., Novi-Michigan. 13th-16th June 1999.
12. Hales, R. F. Quality Function Deployment as a decision making tool. Pro-action Development Inc 1998.
13. ISO 9001 – Parts 4.4.4, 5,6,7 and 9 (Quality Systems Model for quality assurance in design, development, production, installation and servicing) 1994.

CAPE/015/2000

Feature-based progressive design and manufacturing planning in a PDM environment

R SHARMA and **J X GAO**
School of Industrial and Manufacturing Science, Cranfield University, UK

ABSTRACT

This paper reports the design and implementation of a Computer-Aided Conceptual Design system for single piece mechanical parts. The system provides an interface for input and editing of feature hierarchy trees. The feature tree can be visualised and after sufficient level of detail has been added, the feature tree can be evaluated for manufacturing and exported as an ACIS solid model. A detailed process plan can also be generated to view the routings if required. This system provides the designer with the complete control of not only creating the feature tree but also managing the associated Product Data Management (PDM) and Computer-Aided Process Planning (CAPP) activities. The system helps the designer in creating economically manufacturable components in the early design stages. Providing the designer with a tool to prevent unmanufacturable features reaching the detailed design stage ensures high quality and reduced design costs.

1. INTRODUCTION

The new millennium will be critical to the manufacturing enterprises driven by engineering, as they will have to operate in environments where customers demand more and more choices. One area of the product lifecycle which researchers have turned their attention to, is the product development process. Nearly 70 % of the cost of manufacture of a component is fixed during the design stage. Many of the important cost related design decisions are actually taken in the early design stages [1]. More often than not, designers have little or no knowledge of the manufacturing difficulties. Therefore, a design engineer typically designs components without any consideration for manufacturing difficulties. The result is a design, which meets the design specifications but may not be easily manufacturable with the available resources. The objective of this research is to develop a system that will guide the designer in creating economically manufacturable components in the early design stages, thus avoiding the problem of *'suitable, but difficult to manufacture'* end design.

Much research has been devoted to the theory of design. Extensive comparison of the various systematic modern design methods can be found in literature [2]. Product design begins with a need and ends with detailed drawings and all the accompanying information necessary for manufacturing the product. Product design is a complex and iterative process. Most studies have divided the design process into two, three or four stages. Phal and Beitz [2] described the four stages of design as – *Design Specification, Conceptual Design, Preliminary Design (embodiment design*), and *Detailed Design.* The transition of design from one stage to another is a gradual process and the design evolves from one stage to another rather than an abrupt jump from one stage to the next. This research groups the first three stages together as the *Early Design* stages. The evolution of design from one stage to the other in these early design stages is termed as *Progressive design.* A common characteristic of the *early design* stages is that the design options are still being explored, evaluated and compared. The *detailed* design stage is more about refining the design selected in the *Early Design* stages. Considering the first three design stages together as the *Early design* stages is especially helpful when developing computer support for the entire process. This is for two reasons. Firstly these three stages share a lot in common with each other but at the same time are very different from the detailed design stage. Secondly, because of the way design progresses in these early design stages it is difficult to draw a distinct boundary between them.

The early design stage is dominated by conceptual design. Conceptual design is perhaps the most important part of the overall product lifecycle although the time spent on it may be small when compared to the other stages. It should be noted that the conceptual design begins with the set of required functions, but the capture of *these* functions or specifications does not form part of conceptual design in context of this research. Conceptual design is responsible for transfer of product specifications or ideas into a practical product or a set of alternatives. Conceptual design is characterised by incomplete and ambiguous information. It is also the most creative stage of the product development life cycle. But at the same time the designer has to operate within the boundaries of goals and constraints. The culmination of this stage is a preliminary design configuration that is physically realisable although may not be the final design. In the subsequent design stages it becomes very difficult or even impossible to correct fundamental shortcomings in the design which makes this stage very critical. Although the main constraints and considerations in this stage are technical, other constraints like costs, manufacturing and assembly begins to emerge and may also be the major concern

2. MANUFACTURING EVALUATION

Traditionally, the transition of early design to detailed design to be manufactured is accomplished by iterations between the designers and the manufacturing engineers (or the process planners). Furthermore, transition of design from early design to detailed design involves a lot of rework as the two stages rarely used the same file standards. For the purpose of process planning, manufacturing evaluation and design analysis, most systems require the design to be represented in terms of manufacturing features [3]. This is usually achieved by form-features to manufacturing-features mapping, or feature recognition [4]. Feature mapping or feature recognition algorithms require a fully validated geometric model. *This primary requirement of having a fully validated model, is precisely the restriction in progressive design.*

It is hard to find a standard definition of *Manufacturability* in research literature. In general terms, *manufacturability* refers to how difficult or how easy it is to manufacture an object. In other words, it is a measure of the effort involved in manufacturing a product. This measure is relative and dependent on the type of object being manufactured and the resources available to manufacture it. This measure also depends to a large extent on the domain being considered. Another important issue to consider here is the scale of measurement. There is no standard scale for comparison of manufacturability. This is because different types of products and domains have different requirements.

2.1. Metrics for measuring manufacturability

Researchers have tackled the problem of measuring manufacturability in different ways. Many examples of different scales (or a combination of scales) to measure manufacturability can be found in literature [5]. The scales can be classified as below:

(1) Boolean: This is the most basic manufacturability rating. It is Boolean and simply reports if the part "can" or "cannot" be manufactured with the given resources. This can usually be achieved by generating a process plan.

(2) Cost: Cost estimation has been an important area to study both in manufacturing and marketing communities [6]. In today's highly competitive and globally distributed manufacturing environment, prompt estimation of cost with accuracy is crucial. The methodology to reduce manufacturing cost is also very important. Cost of manufacturing is a quantitative measure. Since all manufacturing operations have an associated cost, it is easy to arrive at a figure which represents the manufacturing cost of the product based on the manufacturing resources required to produce it. The cost of raw material (based on dimensions and weight) and its handling could also be included in the estimate. This figure may not be very accurate but is usually acceptable if it is within limits.

(3) Time: Manufacturing time is also a quantitative measure of manufacturing process. Every manufacturing process has an associated time. In most cases, it is the manufacturing time that is used to calculate the manufacturing costs. The quantitative measures of time and cost, may not be directly helpful in determining if the design is good or bad or if it meets the required specifications, but it is one of the most common constraints which a designers works in. The time and cost measures are also important inputs for management decisions associated with the product.

(4) Qualitative measure: Ishii [7] qualified design using an adjective qualifier: 'excellent', 'very good', 'good', 'poor', 'bad' and 'very bad'. These qualifiers were mapped on to a [0,1] measure for comparison. But such measures are hard to interpret and compare, especially if the rating come from different systems or designers.

(5) Abstract: This is similar to the qualitative measures discussed above, but involves each design attribute being assigned a manufacturability index or producability index (PI) instead of a qualifying adjective [8]. As with the qualitative measures, it can become difficult to interpret or compare designs.

2.2. Approaches to manufacturability measurement and evaluation

Given a set of manufacturing resources and geometric design, the problem of manufacturing evaluation is reduced to determining whether or not the design is manufacturable. If the design is found to be manufacturable, the next step is to determine one of the evaluation metrics discussed above. Based on the literature survey of existing manufacturing evaluation systems [5][9][10], the approaches to derive manufacturability metrics from design and resources as input can de broadly divided into two categories.

> ***(1) Direct or rule-based:*** In this approach, analysis is based on direct or rule based interrogation of the design description to assess manufacturability. Rule-based systems use a set of guidelines to evaluate the feasibility of a proposed design. They may also provide recommendations to designers for violated rules. Rule-based approaches do not usually provide quantitative metrics related to manufacturability that can provide a basis for comparison.
>
> ***(2) Plan based:*** This approach is based on generating the manufacturing process plan and then analysing this plan (or its rating) based on a pre-selected criterion. Rating-based approaches like this typically provide a quantitative evaluation based on scales like time or costs. The major advantage is that it can compare alternative designs and optimise a design based on a selected objective. Because Design-for-manufacturing (DFM) rules are not applied, this approach can not provide specific redesign recommendations.

2.3 Evaluation metrics and choice of approach of the proposed system

The system uses a combination of metrics and approaches to give the best possible evaluation of the design at the earliest possible stage. The system evaluates the design in two stages. In the first stage it attempts to generate a process plan for the given design selecting any or all of the resources available. This *plan-based* approach yields a *boolean* which indicates if the design is manufacturable or not. In case the design is manufacturable and there is at least one process plan which satisfies the design, then the design evaluated using a *rule-based* approach to arrive at manufacturing *time* and *costs*. Predicting the manufacturing time and costs, which are useful decision making inputs, during the early stages has obvious advantages.

3. CONCEPTUAL DESIGN AND THE PDM ENVIRONMENT

PDM applications have a common purpose of providing configuration management to engineering databases and the intent to start bridging the gaps between islands of automation. Configuration management in this sense represents a disciplined approach to define elements of product data, to control its change, and to track the changes made. For any CAD/ CAM / CAE application to be successfully integrated into an existing enterprise, it is mandatory that it is closely integrated with the PDM environment. This is for a number of reasons, such as document and workflow management. Integrating a computer-aided conceptual design system with a PDM environment would provide a suitable framework for data sharing as well as implementation of workflow. This is the best way to automate the clerical tasks associated with progressive design and transfer of design to the next stage. Implementing the application within the PDM environment would also avoid one very important pitfall – manual re-entering of data in the detailed design stage. Using the PDM system, it would be possible to carry design from the early design stages to the detailed design stage in a seamless manner. The detailed (conventional) CAD systems would take over where the early design system leaves.

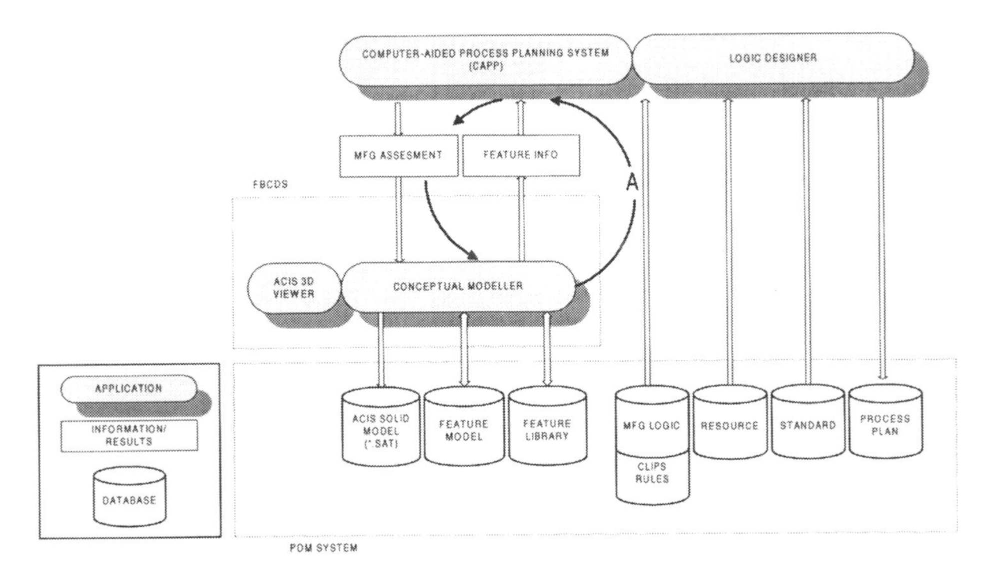

Figure 1. System architecture

4. SYSTEM ARCHITECTURE

The system layout is shown in figure 1. This system is based on the premise that the designers should be able to represent the preliminary design in terms of meaningful features which have manufacturing information associated with them. The representation of preliminary design as meaningful features allows close integration of design system with the CAPP system. This also eliminates any requirement of feature translation or feature recognition, allowing incomplete product model to be processed for manufacturing evaluation and process planning. Loop 'A' in figure 1, shows the information flow between the conceptual modeller and the process planning system.

The system can be divided into three distinct segments –

(1) The Feature-Based Conceptual Design System (FBCDS): This is basically a feature model editor that provides the user facility to build up the feature hierarchy tree using a simple interface. It also includes the feature library editor and solid model viewer.

(2) The Computer-Aided Process Planning System (CAPP): The process planning system used is a specially developed version of LOCAM process planing system [14]. A new set of enhanced commands has been included to allow calling of CLIPS routines and handle the conceptual model.

(3) Manufacturing Logic Designer: This is the database editor. It allows the user to build up the manufacturing evaluation logic graphically in the form a of a flow chart.

As discussed in section 2, there are various ways to measure manufacturability. This system measures the manufacturability in terms of manufacturing *time* and *costs*. It also determines if the design is manufacturable with the available resources. Instead of adopting a pure rule-based or plan-based approach to analysis and plan generation, a novel hybrid approach is used. This is achieved by combining the plan-based approach and the rule-based approach. CLIPS rules facilitate evaluation of partially completed model [11]. These rules can be called from within the analysis logic of the process planning and analysis system.

A view of the main application is shown in figure 2. The system provides the user access to the PDM System, feature library, part feature tree and the solid model in four separate panes. The solid model (or its wire-frame) shown is a visualisation only and the geometry cannot be directly edited. Any changes can be made only to the feature tree of the part being designed. The changes made to the feature tree are reflected in the solid model.

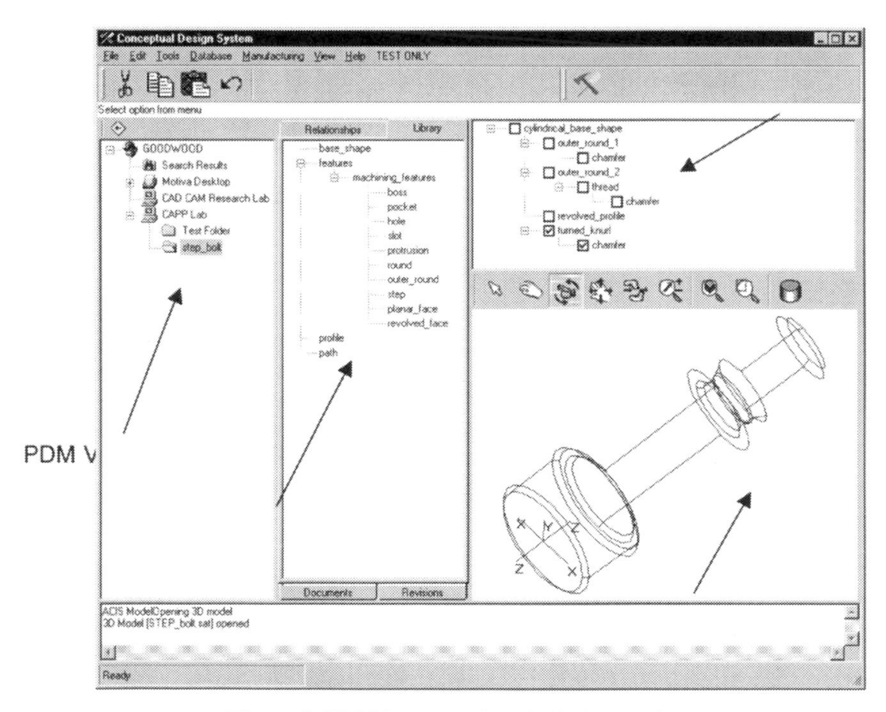

Figure 2. PDM integrated early design environment

The embedded feature model editor [12][13] is extremely user-friendly. Feature instances can be added to the part by simple drag-and-drop operations. The features can also be repositioned within a part using the drag-and-drop operation. The application also provides facilities for cut-and-paste operations. Each feature in the tree has a check-box that can be used to include or exclude the feature from the analysis. This check-box allows a quick evaluation of various product configurations. This can be exploited by the designer to decide if the cost and time overheads of a particular feature (like a chamfer or a knurl) are justified. Once a sufficient level of information has been added to the part model, it can be evaluated or exported as a SAT file to be later edited using a conventional detailed CAD system. The files and databases created are automatically checked-in the PDM system. The system can also set up PDM product structures for standard products based on predefined templates. This greatly speeds up the process of starting a new design project. The user manual is available on-line. The application also has extensive on-line context sensitive help

The process planning and manufacturing analysis logic is stored externally in a database and not built-in within the system. This program-data separation ensures that the system can be updated in the future when new manufacturing technologies and practices are adopted by the user. The logic and all its associated databases are edited using the *logic designer*. The logic designer represents a graphical view of the analysis flowchart as shown in figure 3. The analysis logic is developed using a rich vocabulary of commands provided by the system.

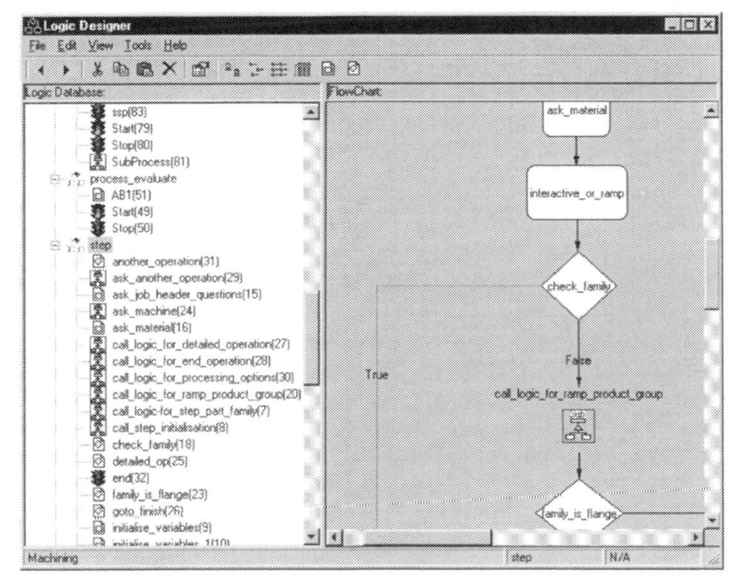

Figure 3. Logic designer

There are commands to extract data from the feature tree as well as from databases in which the data about resources and machining standards is stored. The command set is a sub-set of the commands provided by the LOCAM process planning system for automated analysis of STEP AP224 files. A few more commands have been especially added to cater for the conceptual model and calling of CLIPS rules.

4.1 Implementation of user-configurable library

If the features are defined-as-needed by the user, high level processing of user-defined features for viewing, solid modelling and process planning becomes a problem. Associating the parameters of the user-defined features with the compiled algorithm for processing of these features is extremely difficult. Using user-defined features, without recompiling the entire system every time a new feature is introduced in the library is difficult. User configurable feature library is implemented in this system based on a combination of procedural and Graphical User Interface (GUI) techniques. The data about the features, which includes all the attributes and parameters is stored in a plain ASCII file. This representation of data as a ASCII file is preferred over a proprietary database as it allows easier and quicker editing using a simple text editor. Each feature in the library is identified by a unique 'handle' which is the feature name. Implementation of feature visualisation without recompiling the complete software is done by compiling the visualisation routines of the user feature library into Component Object Module (COM) Dynamic Linked Library (DLL) server component. DLLs are compiled procedures which are linked to the calling program at run time. This late or run-time binding allows the DLL to be compiled separately of the main program. Thus adding new features to the library requires recompiling of the server component only and not the entire program.

A Visual Basic template for the COM server is provided to allow the user to add more features to the library. This could however be done using any high level language like C/ C++ or Ada which has the capabilities to compile COM DLLs. This choice of selecting the compiling tool would depend on the user's capabilities and requirements.

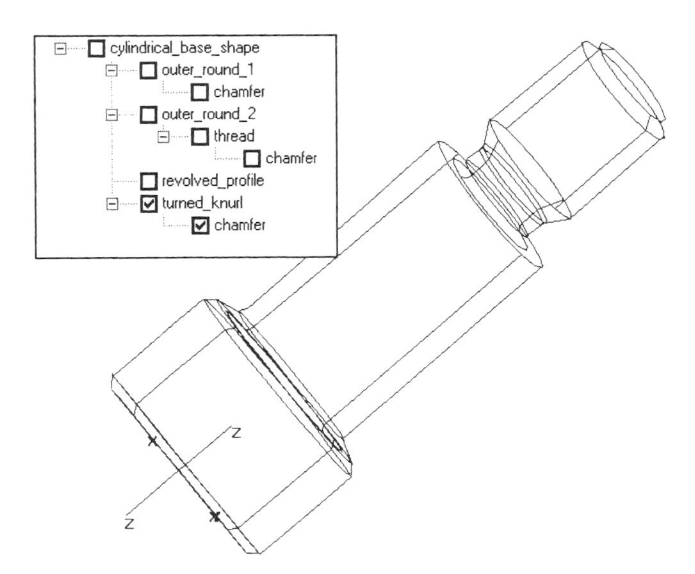

Figure 4. Feature tree and its visualisation

5. CASE STUDY

To understand the assistance provided by the system, it is necessary to consider the steps in building a complete model and evaluating it using the system. The example part used here is a simple stepped bolt shown in figure 4. The feature tree representing the stepped bolt is also shown in figure 6. The part is composed in terms of standard features provided by the AP224 STEP library [15]. All these features are available in a pre-defined library and there is no need to define any more features for this part.

5.1. Creating and evaluating the Part

The part feature tree is built by selecting the feature in the library and creating an instance of it in the part. A copy-and-paste operation or a drag-and-drop operation can do this. The hierarchy of the features in the part is not very important for analysis. But if a particular feature is excluded from the analysis by clicking on its inclusion check box, all features, which are children of this feature, are automatically excluded. Each feature has a set of attributes. The user needs to provide suitable values for these attributes. The attributes of the cylindrical_base_shape feature are shown in figure 5. Not all attributes need to be entered to analyse the design. Some of the attributes are derived from CLIPS rules during analysis if they have been left out. In case the user has entered a value, then this is checked against the value predicted using the rules.

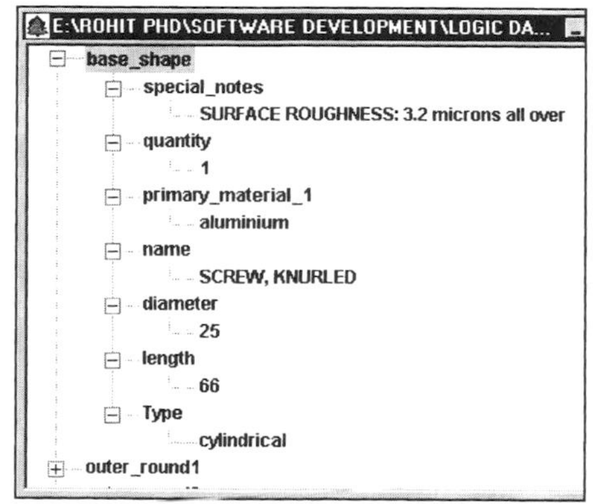

Figure 5. Feature attributes

Evaluation of the part gives a manufacturing cost and time estimate of the design. The user can then also generate a detailed process plan if required to view the routings information. The design can also be checked-in the PDM system SAT file format which Spatial Technology's non-proprietary geometric model file format. Further design activities can be done using a detailed design system based on Spatial Technology's ACIS® 3D CAD kernel.

6. FURTHER WORK

Further developments on the system are presently ongoing. The current implementation of the editor provides for a very limited validation and model consistency verification. This is not seen as a handicap at the initial stage as the model is intended for modelling only simple products. A trial copy of the system will be shipped to a few selected LOCAM customers. Feedback obtained would be critically analysed to decide on the future scope of work. The system and the analysis logic will be further developed to tackle more sophisticated machining examples.

7. CONCLUSIONS

The Computer-Aided Progressive Design system presented here can be used in the early design stages to improve the product quality from the manufacturing point of view. It is envisaged that this system will enable the designers to evaluate their design with respect to the available manufacturing processes and resources. It gives the planner complete freedom to create his own feature library dedicated to a particular application or a part. It also provides for quick evaluation of '*What-If*' scenarios by relocating a feature, changing its parameters or excluding it from the analysis logic. The proposed system also takes into account the other miscellaneous activities performed by the designer in the early design stages. This includes interaction with the PDM structure (to set up the product assembly structure) and transfer to design to the next stage through a solid model, usable by any detailed design CAD system Initial case studies reveal that the system is a highly desirable conceptual design and editing aid. It increases the efficiency by reducing the time taken to enter and edit the conceptual design data. Further studies will help to quantify the results.

Acknowledgement

The authors gratefully acknowledge the support of LSC, UK Ltd. during this research, especially Mr Alan Crawford, without whose support this would not have been possible.

REFERENCES

[1]. Jared G. E. M., Limage M. G. Sherrin I. J. Swift K. G. 1994. Geometric reasoning and design for manufacture. Computer-Aided Design 26, no. 7: 528-36.

[2] Pahl G., Beitz W. 1996. Engineering Design: A systematic approach. UK: Springer.

[3]. Case K., Gao J. X. 1993. Feature Technology : An overview. International Journal of Computer Integrated Manufacturing 6, no. 1-2: 2-12.

[4]. Salomons O.W., Houten F. J. A. M. Kals H. J. J. 1993. Review of research in feature based design. Journal of Manufacturing Systems 12, no. 2: 113-32.

[5]. Gupta S, Regli William C. Das D. Nau D. S. 1997. Automated Manufacturability Analysis: A Survey. Research in Engineering Design 9, no. 3: 168-90.

[6]. Schreve K., Schuster H. R. Basson A. H. 1999. Manufacturing cost estimation during design of fabricated parts. Proceedings of the Institution of Mechanical Engineers, Part B: Journal of Engineering Manufacture 213: 731-5.

[7]. Ishii K. 1993. Modeling of Concurrent Engineering Design. Concurrent Engineering: Automation, Tools and Techniques Kusiak AndrewJohn Wiley & Sons, Inc.

[8]. Priest John W., Sanchez Jose M. 1991. An empirical methodology for measuring producability in early product development. International Journal of Computer Integrated Manufacturing 4, no. 2: 114-20.

[9]. Mukherjee Amit, Liu C. R. 1997. Conceptual design, manufacturability evaluation and prelimnary process planning using function form relationshipd in stampted metal parts. Robotics and Computer -Integrated Manufacturing 13, no. 3: 253-70.

[10]. Pham D T, Ji C. 1999. A Concurrent Design System for machined Parts. Proceedings of the Institution of Mechanical Engineers, Part B: Journal of Engineering Manufacture 213: 841-6.

[11]. Bowland William, Gao J X. 1998. Embedded Knowledge based functionality for process planning software, MSc Thesis. Cranfield University.

[12]. Sharma R, Gao J. X. 1999. A feature model editor for process planning of sheet metal products. Proceedings of the 15th International Conference on Computer-Aided Production EngineeringUK: DMRG.

[13]. Gao J.X., Tang Y. S. Sharma R. (Accepted 1999). A Feature Model Editor and Process Planning System for Sheet Metal Parts. International Journal of Material Processing Technology.

[14]. LSC UK. 1997. LOCAM - System Development Manual, Version 10.9. UK.

[15]. Fowler Julian. 1995. STEP for Data management, Exchange and Sharing. Great Britain: Technology Appraisals.

AUTHORS

R Sharma is a PhD student at Cranfield University. He received an MSc in CAD/ CAM in 1998. His research interests include integration of CAPP, PDM and MRP systems. He has a strong background in the use of computers in these fields.

Dr Gao is a lecturer at Cranfield University since 1993. He leads a research team and a MSc course in CAD/CAM and published over 40 papers in this area. His current research interest is the integration of CAD, CAPP, CAAP and ERP systems through PDM.

CAPE/101/2000

Effects of material properties on stresses occurring in total hip replacement – a non-linear FE analysis

W SCHMOLZ, D R GORDON, A J SHIELDS, and **D KIRKWOOD**
Department of Engineering, Glasgow Caledonian University, UK
P GRIGORIS
Department of Orthopaedics, West Glasgow Hospital NHS Trust, UK

ABSTRACT

This study shows the influence of different prosthesis materials on stresses occurring in the interfaces of cemented total hip replacements. The 3-D solid geometry was reconstructed from measured point data. Four different prosthesis materials were investigated. Namely, titanium alloy and stainless steel, which are currently used in clinical applications, and alumina ceramic and a fibre-reinforced composite, which have a higher and lower elastic modulus than the materials currently used. Non-linear finite element analyses using friction contact elements to model the prosthesis-cement and the cement-bone interface were carried out. For a more flexible prosthesis stem the results show higher interface stresses and a lower risk of calcar bone resorption than for a stiffer prosthesis stem

1. INTRODUCTION

Total hip replacement (THR) is a very successful operation. THRs are looked on as being successful in that they do not produce discomfort, and are not likely to require revision surgery for around 15-20 years [1]. At the present time in the UK there are approximately 45 000 operations performed annually [2]. In the UK there are currently more than 60 different primary total hip replacements manufactured by over 19 different companies on the market [3]. They vary considerably in price, material, design and long-term clinical performance. Aseptic loosening of the femoral stem is often reported as the main cause of implant failure and is mainly caused by mechanical effects [4,5,6]. The mechanical failures include fracture of the prostheses, fracture of the cement and failure of the cement interfaces due to excessive stresses. Stress analysis may help to prevent the clinical application of mechanically unbalanced prosthesis designs and can help to improve the currently existing implants. Due to the complex shape and material properties involved in the structure of cemented total hip replacement the finite element method is the only efficient method for determining the stress levels in the prosthesis, cement and bone after joint replacement.

This study investigates the influence of the material used for hip prostheses on stresses occurring in the prosthesis and the cement. The specific prosthesis design analysed in this paper was chosen, because it was manufactured and clinically applied with two different materials, stainless steel and a titanium alloy. For both prostheses materials the prostheses have been reported to have a significantly higher failure rate in their clinical application [7,8]. To extend the study two hypothetical materials, alumina ceramic with a much higher elastic modulus and a fibre-reinforced composite material with a lower elastic modulus were simulated.

2. RECONSTRUCTION OF THE GEOMETRIES – REVERSE ENGINEERING

For this study a Capital stem (Capital, monoblock cemented hip system, manufactured by 3M) was used. The prosthesis was measured on a co-ordinate-measuring machine (TESA 3D Micro-MS 343). The point data obtained from the co-ordinate-measuring machine was imported into Pro/ENGINEER, where the three-dimensional solid geometry was reconstructed on the basis of the point data. Figure 1 shows the shaded solid model of the prosthesis in Pro/ENGINEER.

For the reconstruction of the femur, the point data from the "Standardised Femur" program was used. The "Standardised Femur" is a digitised geometrical model of a femoral bone analogue developed by the Laboratory for Biomaterials Technology of Insituti Orthopecici Rizzoli, Italy, within the frame of the Prometo Project. The femoral bone analogue is produced by Pacific Research Labs (Vashon Island, Washington, USA). The point data was obtained from a CT scan dataset, made of 99 slices non uniformly spaced. In PRO/Engineer the inner and outer surface of the cortical bone was reconstructed and solid models were generated. Figure 2 shows the shaded model of the surface of the femur reconstructed from the point data in PRO/Engineer [9].

The created solid models were assembled according to the instructions of an experienced orthopaedic surgeon using PRO/Assembly. Boolean operations were carried out to create the solid models of the cement mantle and the cancellous bone. The final assembly consisted of four different solid models with four different material properties. Figure 4 shows the cross-section of the created Assembly cut in the frontal plane. It also shows, in italics, the medical terms used to describe regions of the femur.

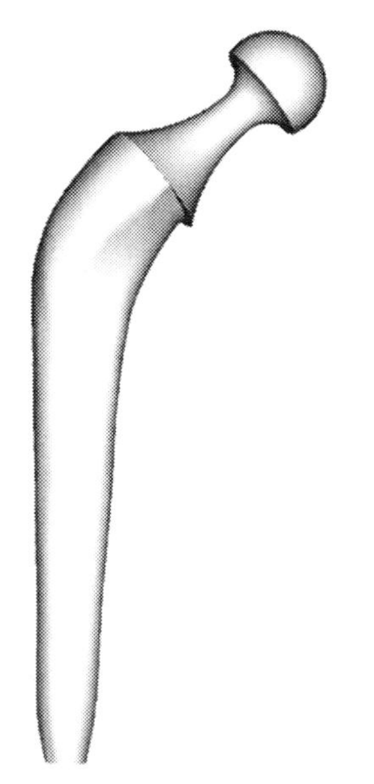

Fig. 1: Shaded model of prosthesis in PRO/Engineer

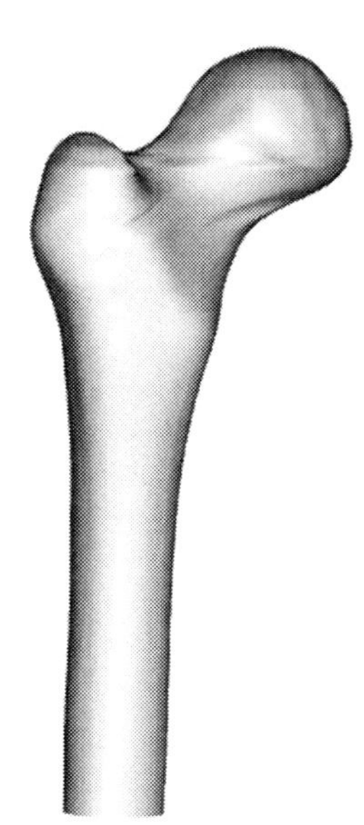

Fig. 2: Shaded final model of the proximal femur

3. FINITE ELEMENT MODELLING

The finite element models were generated using the Pro/MESH facility of Pro/ENGINEER (Parametric Technology Corporation). The solution and post-processing phase of the finite element analyses were carried out using the ANSYS 5.5 programme (Swanson Analysis Systems Incorporation). For the automatic mesh generation a 3-D 10-node tetrahedral structural solid (SOLID92) was used. The element has a quadratic displacement behaviour and is suitable for modelling of irregular meshes. It is defined by ten nodes having three degrees of freedom at each node.

The prosthesis–cement, cement–cortical bone and cement–cancellous bone interfaces were modelled with 6 noded triangular 3-D surface to surface contact elements (TARGE170 and CONTA174). The contacting surfaces were initially closed and a Coulomb frictional interface with a friction coefficient of 0.25 was applied [10,11]. The modelled interface is able to transfer shear stresses up to a set value (• $_{max}$) across the interface. After • $_{max}$ is exceeded, sliding between the interfaces will take place. Figure 3 shows the Coulomb friction model applied in this study. The cortical bone-cancellous bone interface was modelled as fully bonded and therefore all normal and shear stresses occurring at the contact surfaces are transmitted through the interface.

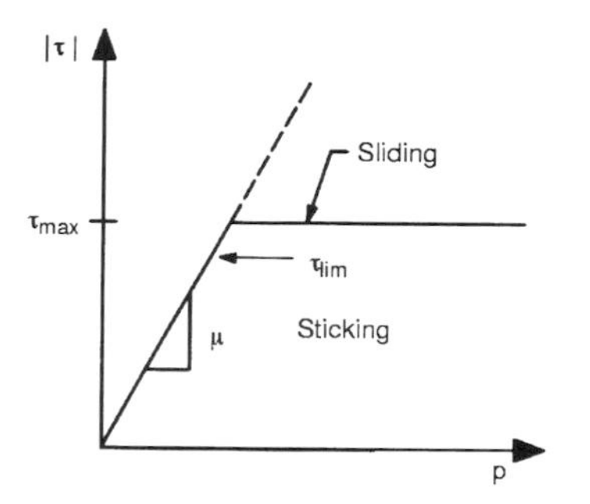

• $_{lim}$ = limit shear stress
• = equivalent shear stress
• = frictional coefficient
p = contact normal pressure

Fig. 3: Coulomb friction model applied to the interface

The materials for the prosthesis, cement, cancellous bone and cortical bone were assumed to be isotropic and homogenous. The Elastic modulus of cortical bone, cancellous bone and cement were taken as 14.4 GPa, 1GPa and 2.2 GPa, respectively. For the different prosthesis materials investigated, the following elastic moduli were taken: Alumina ceramic 400 GPa, stainless steel 200 GPa, titanium alloy 100 GPa, fibre reinforced composite 20 GPa. The Poisson ratio for all materials was taken as 0.3. The loading applied to the finite element model simulated the forces occurring during level walking. The resultant hip joint force was applied acting through the centre of the prosthesis head. The force was approximately four times an average body weight (2.9kN). The reaction force of the muscles attached to the femur was applied at the greater trochanter and the magnitude of the resultant force was 1.6 kN [12]. As a boundary condition, all nodes at the distal end of the femur were fixed. Figure 5 shows the meshed model in Pro/Mesh with the loading and the constraints applied to the finite element model. All the models consisted of approximately 16,500 SOLID92, 2,188 TARGE170 and 2,188 CONTA174 elements and of approximately 30,000 nodes.

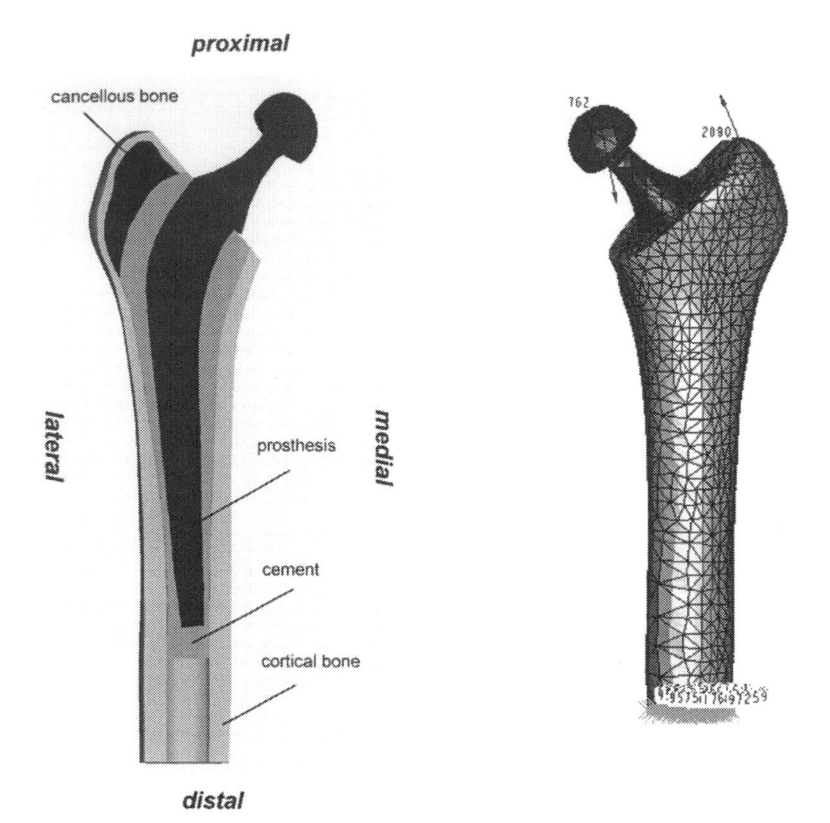

Fig. 4: Cross section of assembled model

Fig. 5: Meshed model with applied boundary conditions

4. RESULTS

The finite element models were cut in the frontal plane as shown in Figure 4 to obtain results from inside of the model. At the interfaces, paths were created and the von Mises stresses on each side of the interface was plotted along the paths. The x-axis in Figures 6-13 shows the distance from the start of the path in mm, whereby the zero indicates the start of the path at the distal end of the according interface and the y-axis shows the computed von Mises stresses in MPa. In all graphs ALU_SEQV, STS_SEQV, TI_SEQV and FRC_SEQV represent the stress plots for the alumina ceramic, stainless steel, titanium alloy and fibre reinforced composite, respectively.

Figures 6-9 show the stresses in the prosthesis-cement interface on the lateral and medial side. For the lateral side (Fig. 6 and 7), the stresses in the prosthesis increase with an increase of the elastic modulus and the stress peaks are close to the distal tip of the prosthesis. In the cement a more flexible prosthesis induces lower stresses in the distal and proximal part and higher stresses in the midsection then the more rigid prosthesis. At the medial side (Fig. 8 and 9) of the prosthesis the von Mises stresses in the midsection and distal region are increasing with an increase of the elastic modulus whereas the stresses in the proximal region are increased by decreasing the elastic modulus of the prosthesis. The stresses in the midsection and the distal region of the cement are slightly higher for a more rigid prosthesis, whereas in the proximal region the stresses are increased by a more flexible prosthesis.

Figures 10-13 show the stresses on the lateral and medial side of cement-bone interface. At the lateral side (Fig. 10 and 11) of the interface the stresses at the proximal end decrease, because the modelled interface is migrating from cortical to cancellous bone. The stresses in the distal part of the cement increase with a stiffer prosthesis. The stress peaks induced in the cement, in the area where the prosthesis tip is located, move proximal with decreasing stiffness of the prosthesis. Decreasing the stiffness also induces higher stresses in the midsection of the cement, with stress peaks close to the area of intersection of cancellous bone, cortical bone and cement. The stresses in the bone increase with a more flexible prosthesis and generate stress peaks in the proximal part of the midsection where the cancellous bone starts. At the medial side of the cement-bone interface (Fig. 12 and 13) the stresses in the distal region of the cement decrease with increasing prosthesis stiffness. A more flexible prosthesis induces higher stresses in the midsection and stress peaks at the proximal end of the cement. With increasing prosthesis stiffness, stress peaks are generated in the cement around the area of the prosthesis tip. The stresses at the medial side of the bone are increased with a more flexible prosthesis and stress peaks are generated at the proximal end of the bone.

Figure 14 and 15 show the von Mises stresses at the outer surface of the femur at the lateral and medial side. The results show higher stresses in the femur for a more flexible prosthesis. At the distal and proximal area of the paths, the differences in the stresses evoked by the different prosthesis materials are small. The more flexible prosthesis increases the stresses in the midsection of the path, which represents the area around the prosthesis tip up to the proximal end of the prosthesis.

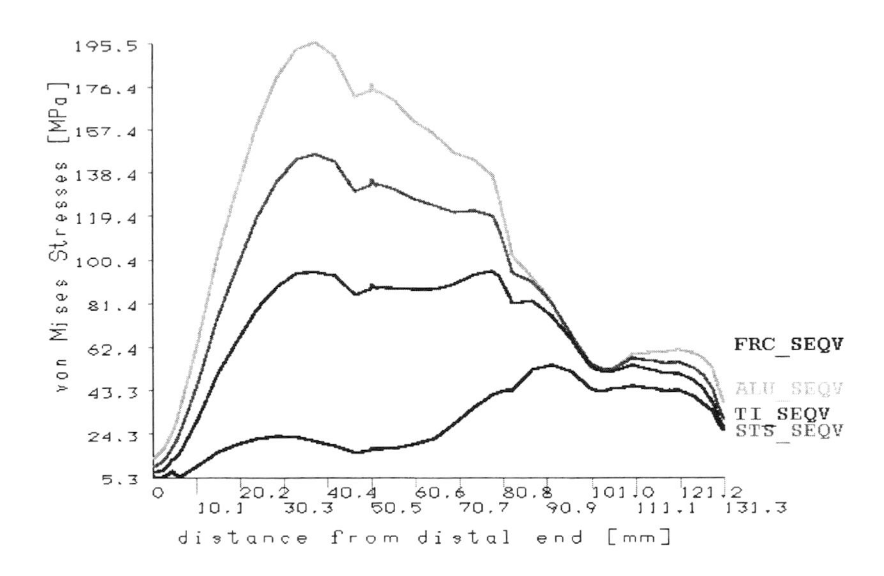

Fig. 6: Von Mises stresses in the prosthesis at the lateral side of the cement – prosthesis interface

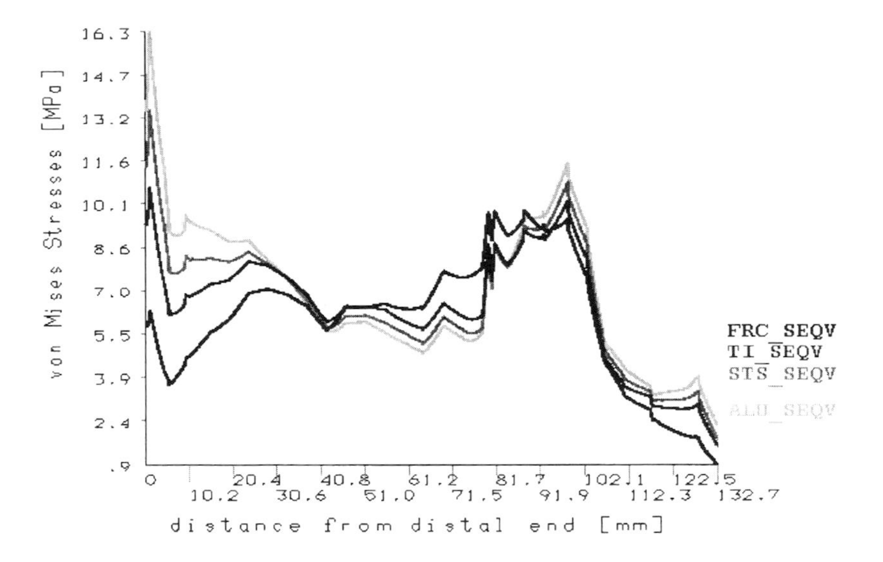

Fig. 7: Von Mises stresses in the cement at the lateral side of the cement – prosthesis interface

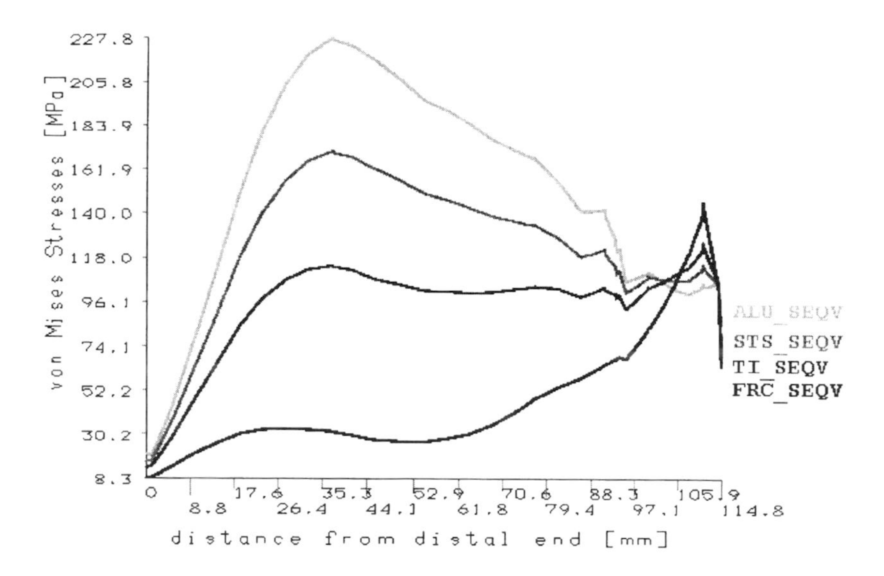

Fig. 8: Von Mises stresses in the prosthesis at the medial side of the cement – prosthesis interface

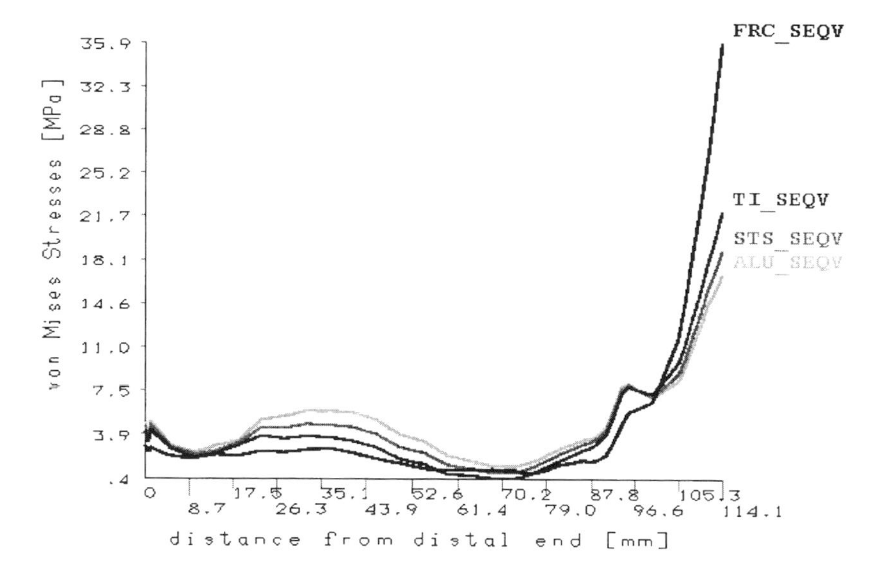

Fig. 9: Von Mises stresses in the cement at the medial side of the cement – prosthesis interface

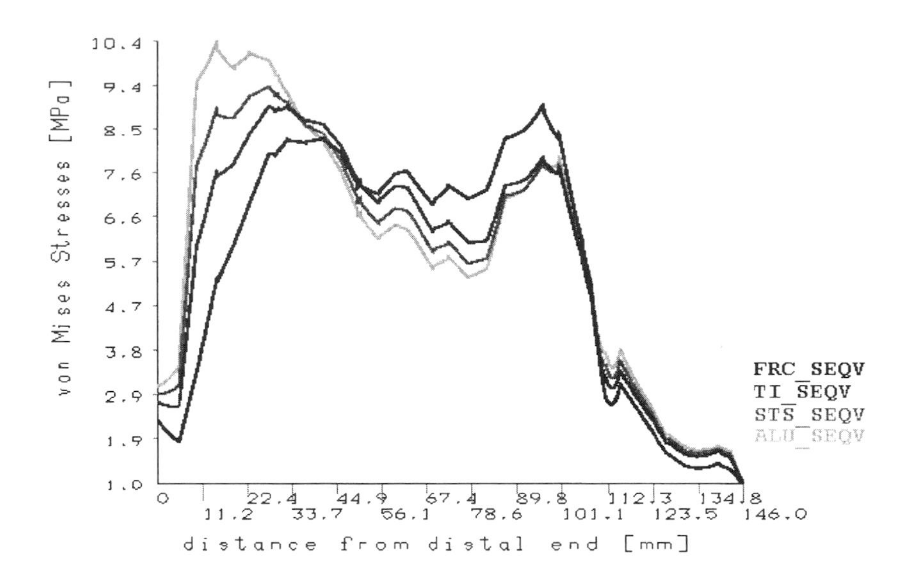

Fig. 10: Von Mises stresses in the cement at the lateral side of the cement – bone interface

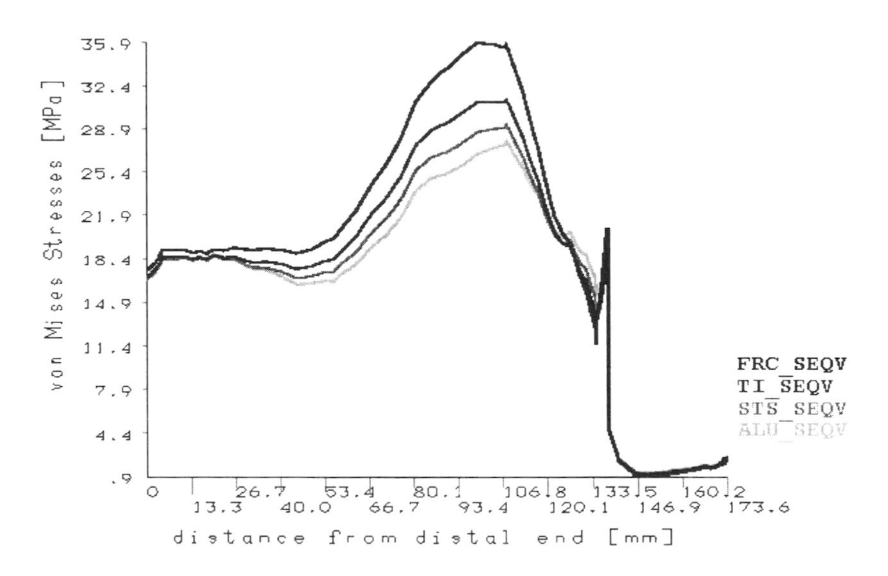

Fig. 11: Von Mises stresses in the bone at the lateral side of the cement – bone interface

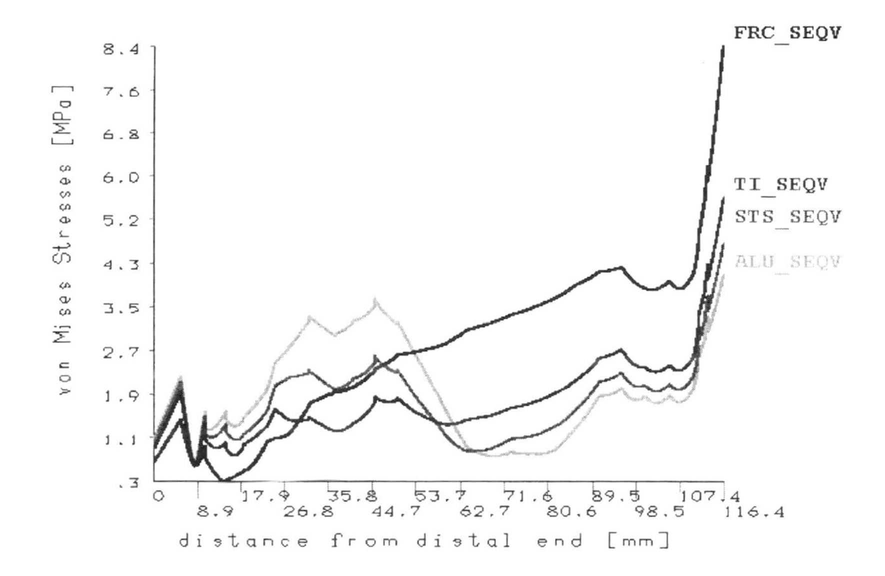

Fig. 12: Von Mises stresses in the cement at the medial side of the cement – bone interface

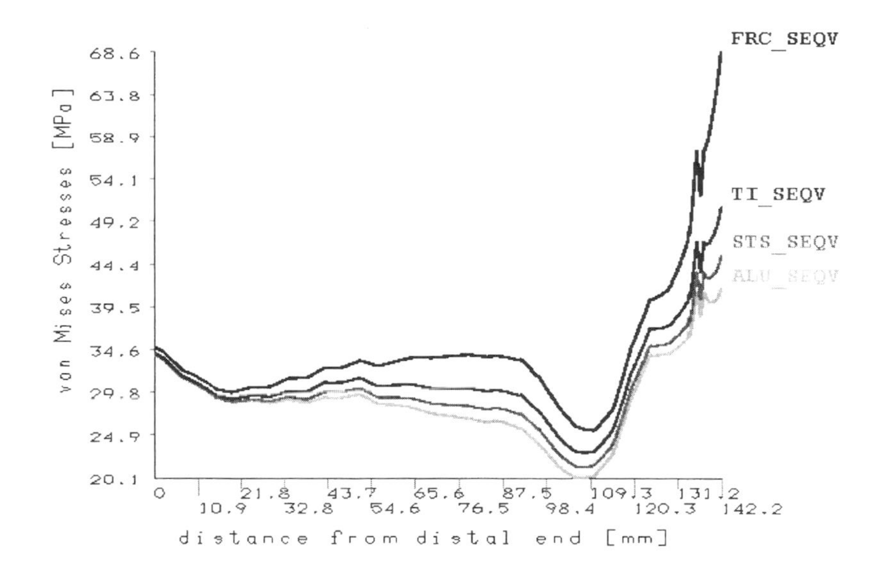

Fig. 13: Von Mises stresses in the bone at the medial side of the bone – cement interface

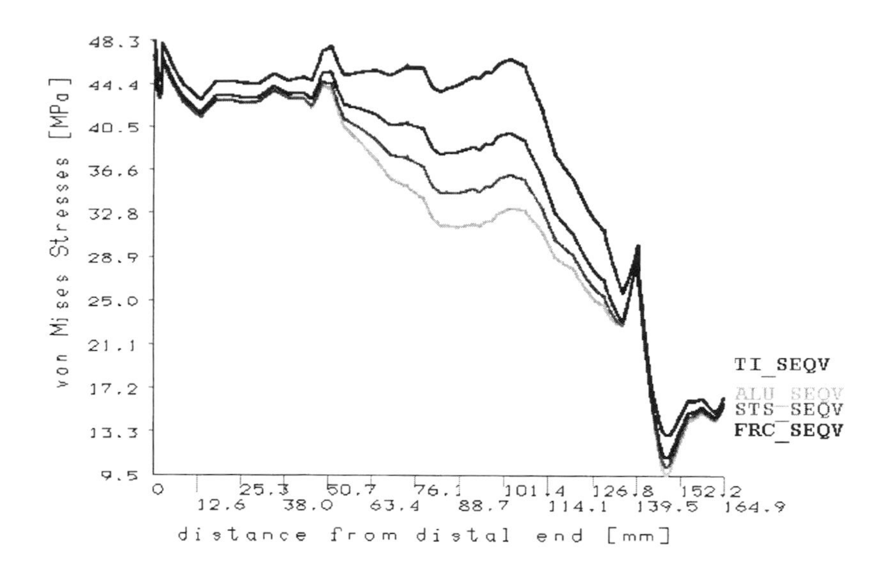

Fig. 14: Von Mises stresses at the lateral surface of the bone

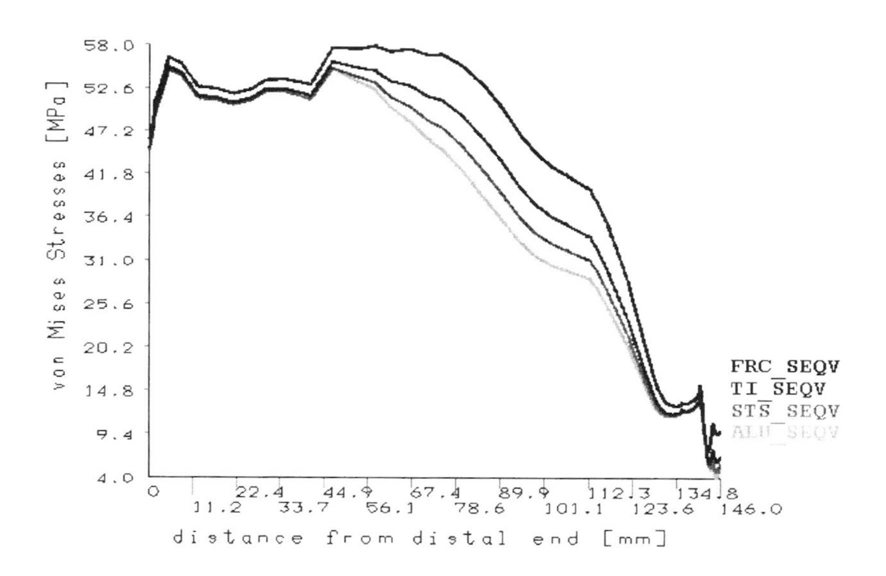

Fig. 15: Von Mises stresses at the medial surface of the bone

5. DISCUSSION AND CONCLUSION

The stiffness of a total hip prosthesis is dependent on two factors, its geometric shape and its material properties. The variation in the prosthesis geometry is limited by the size of the femur and the cement mantle thickness, which is recommended to be 2 to 5 mm, based on experimental, theoretical and long term clinical performance studies [13-16]. In this study one prosthesis design was used and the material properties assigned to the prosthesis were modified to assess the effects of the material properties on the occurring stresses.

The material properties assigned to the cortical and cancellous bone were assumed to be isotropic. There is a large inter-femur variability in geometry and material properties and in vivo both of these materials are anisotropic. The geometry and material properties used in this paper are within the in vivo range [17]. As the purpose of this study, was to compare the effects of the different material properties on occurring interface stresses and not to obtain absolute in vivo values, the simplifications made in the material properties of the femur do not affect the trends shown in this study.

In total hip arthroplasty a prosthesis that has a relative high stiffness compared to the bone is inserted in the femur. The load transfer and the stress pattern in the post-operative femur is therefore changed. The proximal femur is less stressed and more load is transferred through the intramedulary canal to the distal part of the femur. A more flexible prosthesis induces more stresses in the proximal part of the femur and the loading pattern of the femur is closer to the pre-operative condition. This indicates that calcar bone resorption, which is to some extent caused by the decrease in the femoral stresses after hip joint replacement, is less likely to occur with a more flexible prosthesis. The results of the finite element analyses support the outcome of long term clinical follow-ups of different hip prostheses, which has shown, that femoral components made of titanium alloy have a lower risk of calcar bone resorption [18]. On the other hand, the clinical statistics show that prostheses made of stainless steel have a lower risk of developing radiolucent lines, which indicate areas of local prosthesis debonding at the cement interfaces [18]. This clinical finding can be explained by the results of the finite element analyses carried out, which show higher stresses at the cement interfaces for a more flexible prosthesis. The peak stresses in the cement at the proximal medial area increase with a more flexible prosthesis. If these local stress peaks exceed the interface strength, prosthesis loosening will occur.

To assess the amount and progress of this localised interface loosening the current model has to be extended by additional gap elements, which are able simulate an interface bond and are deactivated once the strength of the interface is exceeded. The micro-motions occurring in the interfaces, the wear particles created by these micro-motions and the biological effects are not addressed in the paper.

The results of this study show that the addressed design goals are conflicting. With a more flexible stem, calcar bone resorption is less likely to occur but the cement interface stresses are increased and interface loosening is more likely to occur. For a stiffer stem, interface loosening is less likely to occur, but calcar bone resorption is more likely to occur. The current model shows only trends of the stress distribution within the prosthesis, cement, femur and its interfaces for one prosthesis design manufactured from different materials. To actually predict the clinical performance of the prostheses, the current model has to be further refined by the inclusion of additional contact elements, which will aid the prediction of the magnitude of interface loosening and whether this be stable and localised or unstable and progressing.

6. REFERENCES

[1] Neumann I., Knudd G.F., Sorenson K.H., Long term results of Charnley Total Hip replacements. Journal of Bone and Joint Surgery, 1994, Vol. 76-B, No. 2, pp. 245-251.

[2] The national audit of Primary Hip Replacement, R & D Priorities for Biomaterials and Implants, Publication June 1996.

[3] Murray A.W., Carr A.J., Bulstrode C.J., Which primary total hip replacement? Journal of Bone and Joint Surgery, 1995, Vol. 77-B, No. 4, pp. 520-527.

[4] Espehaug B., Havelin L.I., Engesaeter L.B., Vollset S. E., Langeland N., Early revision among 12,179 hip prostheses, Acta Orthopaedica Scandinavia, 1995, Vol. 66 No. 6, pp. 487-493.

[5] Sutherland C.J., Wilde A.H., Borde.n L.S., Marks K.E., A ten year follow-up of one hundred consecutive Mueller curved-stem total hip arthroplasties, Journal of Bone and Joint Surgery, 1982, Vol. 64-A, pp. 970-982.

[6] Rohlmann A., Moessner U., Bergmann G., Hess G., Koebel R., Effects of stem design and material properties on stresses in hip endoprostheses, Journal of Biomedical Engineering, 1987, Vol. 9, No. 1, pp. 77-83.

[7] Massoud S.N., Hunter J.B., Holdsworth B.J., Wallace W.A., Juliusson R., Early femoral loosening in one design of the cemented hip replacement, Journal of Bone and Joint Surgery, 1997, Vol. 79-B, No. 4, pp. 603-608.

[8] Ramamohan N., Grigoris P., Schmolz W., Chappel A.M., Hamblem D.L., Early failure of stainless steel 3M Capital femoral stems, Journal of Bone and Joint Surgery – British Supplements (British Hip Society meeting), in press.

[9] Schmolz W., Gordon D.R., Shields A.J., Kirkwood D., A three-dimensional investigation into the stress activities involved in the proximal end of a femur before and after total hip replacement, Proceedings of the Integrated Design & Process Technology Conference, July5-9, Berlin, Germany, 1998, Vol. 1, pp. 19-24.

[10] Verdonschot N., Huiskes R., The effect of cement stem-debonding in THA on the long term failure probability of cement, Journal of Biomechanics, 1995, Vol. 28, No. 8, pp. 795-802.

[11] Mann K.A., Bartel D.L., Wright T.M., Burstein A.H., Coulomb frictional interfaces in modelling cemented total hip replacements: A more realistic model, Journal of Biomechanics, 1995, Vol. 28, No. 9 pp. 1067-1078.

[12] Chang P.B., Mann K.A., Bartel D.L., Cemented femoral stem performance, Clinical Orthopaedics and Related Research, 1998, No. 355, pp. 103-112.

[13] Ebramzadeh E., Sarmiento A., McKellop H.A., Llinas A., Gogan W., The cement mantle in total hip arthroplasty, Journal of Bone and Joint Surgery, 1994, Vol. 76-A, No. 1, pp. 77-87.

[14] Fischer D.A., Tsang A.C., Paydar N., Milionis S., Turner C.H., Cement mantle thickness affects cement strains in total hip replacements, Journal of Biomechanics, 1997, Vol. 30, No. 11/12, pp. 1173-1177.

[15] Lee I.Y., Skinner H.B., Keyak J.H., Effects of variation of prosthesis size on cement stress at the tip of femoral implants, Journal of Biomedical Materials Research, 1994, Vol. 28 No. 9, pp. 1055-1060.

[16] Schmolz W., Gordon D.R., Shields A.J., Kirkwood D., Grigoris P., The effect of stem geometry on stresses within the distal cement mantle in total hip replacement, Journal of Technology and Healthcare, in press.

[17] Crisofolini L., Viceconti m. Capello A., Toni A., Mechanical Validation of whole bone composite femur models, 1996, Journal of Biomechanics, Vol. 29, No. 4, pp. 525-535

[18] Ebramzadeh E., On factors affecting the long term outcome of total hip replacements, 1995, PhD Thesis.

CAPE/082/2000

Computer-aid in rotary transfer machines (RTMs)

E GENTILI
Dip to di Ingegneria Meccanica, Università di Brescia, Italy
S MOLINARI
Dip to di Ingegneria Meccanica, Politecnico di Milano, Italy

ABSTRACT

In the past all RTMs were rigid and dedicated, able to work only big batches of pieces. Quick obsolescence and vulnerability to failure were critical problems. Now it has completely changed.
Some important technologies changed RTM: now all the productive system is governed by a computer, CNC reduced set-up time and improved precision.
We want to relate the production costs to the new kind of transfer machines. The investment is bigger than that of traditional RTMs, but the production costs can be decreased.
The paper presents an algorithm allowing the choice of the correct level of automation able to ensure the smallest production costs.

1. INTRODUCTION

Today modular flexible transfer with 3-8 units (equipped with 3 CNC axis) perform sequential machining and can work infinite shapes and dimensions within a 300mm cube. Static turning using 3 axis units is also allowed.
Some recent Italian transfers are able to machine hundreds of different and complex kinds of pieces with low set-up times (less than one hour) and high productivity (up to 10000 pcs/h).
Automatic load and unload carried out using a robot with artificial vision. This means a unmanned production with a computerised surveillance system and tool control. Tool or insert change using economical tool life are allowed in order to prevent tool breakdown and scrap and rework costs.
The American Hydromat designed a transfer machine able to work three different kinds of pieces without set-up time: automatically the computer changes the kind of pieces and the productive mix needed.
The computer gives user-friendly and self-learning programming and gives the possibility to prepare part programs off-line; all working parameters are stored in the PC supervisor.
Production data and machine failure statistics, availability of hardware and software for DNC, connections, for modem tele-assistance are now at hand.

The software includes self-diagnostic functions and is able to display alarm conditions in real time.
There is a great difference between modern rotary transfer machines and old transfers, so it is necessary explain the main differences before the describing of the criteria to utilise in choosing the right level of automation. The chief innovation has been made possible by means of numerical control and computers.

2. FUNCTIONING OF ROTARY TRANSFER MACHINES

In a rotary transfer machine the transfer of pieces is made by a rotary cylindrical table.
Rotary transfer machines make complex machining. To explain clearly and simply the functioning it is necessary to examine an elementary case [3]:

- the workpiece has a shape of a cube and the faces are called C1..C6
- the clamping devices are self centring jaws
- the machining axes are perpendicular to the faces (e.g. drilling, vertical milling, threading, etc)

When a workpiece is locked the faces C1 and C6 are parallel to the lateral area of the table. Face C1 is in front of the table and is near the jaws, so it is impossible to machine this face. Faces C4 and C5 are partially encumbered by the collet of the clamping devices. Faces C2 and C3 are completely free, and it is possible to machine all the area utilising the lateral units. Face C1 is completely free, and it is possible to machine all the areas utilising a radial unit. In this elementary case it is possible to machine simultaneously 3 faces (or 3 ways). Rotary transfer machines that machine simultaneously 3 faces are called 3-WAY.
To build a 3-WAY is easy and every constructor of rotary transfer machines can produce a good 3-way. The difficulty is to produce a good transfer that machines 4,5 faces of the cube in a single clamping. New technologies give the chance to machine 4,5 faces in a single clamping. The rotary jaws can work C1,C2,C3 completely and C4 and C5 partially: five faces in a single clamping, the rotary clamping devices can work C1,C2,C3,C6 completely: four faces in a single clamping.
Rotary jaws rotate the pieces around an axis radial to the rotary table. The first position (0°) machines simultaneously the faces C2,C3 on lateral units and C1 on radial unit, this situation is like a traditional 3-way transfer. The second position (+90°) machines on faces C4,C5 by lateral units and C1 by radial unit. There is a problem because the faces C4 and C5 are partially encumbered by the clamping device, so is impossible machine all the faces [6,8].
Rotary clamping devices allow the pieces to rotate around an axis tangent to the rotary table. The basic position (0°) is like a traditional 3-way transfer. The second position (+90°) machines faces C1,C6 by lateral units and C2 by radial unit. These devices allow to machine completely four faces [12,13,14].

3. OTHER NEW TECHNOLOGIES IN ROTARY TRANSFER MACHINES

Today many transfer machines can work different kinds of pieces. To change the piece it is necessary to stop the production and make a set-up: it is necessary to change the clamping devices or the collets, the tools, the charge and slot devices, and change the travel of the units. The numerical control improves set-up time because the time necessary to change the travels of the CNC units is zero.

One way to shorten change-over times for rotary transfer machines is to get more use from each of the tool stations. The idea here is to keep the drive unit in place and only change the head and cutters. Then by adjusting the offset, feed rate and speed, the station is ready for next workpiece. Specific tool stations are assigned a standardised operation, within certain parameters.

On a machine demo [1], CNC allows the operator to randomly select, from three different parts, which will be produced. For example tooling station two may be designated a rough turning operation. From job to job, that station makes rough turning operations with the CNC-controlled feed rate.

Tooling and set-up remain unchanged from part to part. The operator can make any combination of three parts in any quantities [1].

Revolver heads improve the times to change the tools because there are 6-8 tools on a head and the change is quick, only few seconds. Some important problems of revolver heads are: encumbrance, weight, inertia because the head is moving along X,Y,Z axis, eight is the maximum number of tools.

The constructors are searching for a new technology to try to make the change of tools with devices like the machining centre devices (for example chain devices). This technology can solve the problem of 8 tools and the inertia, but can't solve the problem of encumbrance which is perhaps more serious.

Some clamping devices have special patented systems to quickly unlock. These devices shorten change-over times.

Today the challenge is to develop a palletised system of clamping devices. The idea is like the palletised system in modern linear transfer machines, but they have a high level of encumbrance, so it is difficult use this device in rotary transfer machine.

This technology minimises the change-over time.

Another important innovation is a robot that charges and slots pieces automatically. The study is about recognising the kind of piece. Often the robot is too expensive, so another kind of charge and slot device is prefered.

The static turning units turn as a CNC lathe, so it is possible to carry out conical and profiled turning, single point treading. For static turning is possible to have two different tools, and it is possible, for example to carry out sequentially turning rough and turning finish with one unit in short times [14].

Recently in Ontario some screw machine shops compared a multispindles machine and a rotary transfer one and decided to change. The rotary transfer machine was better than multispindles [5,6,7].

4. A NEW KIND OF ROTARY TRANSFER MACHINE

In the past all rotary transfer machines were not flexible and quick obsolescence was a big problem. In 1995 Italian constructors presented a new kind of rotary transfer machine, the chief characteristic was: modular unit with 3 axis CNC.

This transfer can machine infinite shapes and configurations in a 300mm cube and can machine five faces in a single clamping. The new machines don't have the old problems and limits and is therefore a good solution for several industrial sectors. The first is automotive industry. The units have revolver heads with 6-8 tools and can perform quickly sequential machining with the same tool or different tools. The productivity of these machines is between traditional transfers and machining centres. Usually cycle times are some minutes. In traditional transfer usually cycle times are less than 1 minute for difficult pieces. In a transfer

machine the productivity is proportional to the number of units, the limit of this new machine is the number of units : 8 as maximum.
This limit is because the revolver head has big encumbrance and creates difficult problems of collision. Revolver heads are also heavy, for example 60kg for a 20kW unit [12].
This machine has high encumbrance for example 8400x4350x3970 mm for a transfer with 6 module 3 axis CNC and revolver heads with a 6 spindles [12].
This kind of rotary transfer machine is called by some constructors flexible modular cell

5 WHY CN AND PC ARE SO IMPORTANT

Numerical control gives various benefits. The most important are:
- Opportunity to obtain lower set-up time.
- Opportunity to have great precision during the machining.

The number of axis controlled by numerical control can be till 60, 70 axis per rotary transfer machine. Numerical control doesn't synchronise all the axis of the RTM, but there are groups of axis that need to be synchronised. The criteria used to create the groups are:
- The axis of a specific operating unit belongs to the same group, so they must be synchronised.
- Or the axis of a station (a station may have 1,2 or 3 operating units) belongs to the same group, so they mast be synchronised .

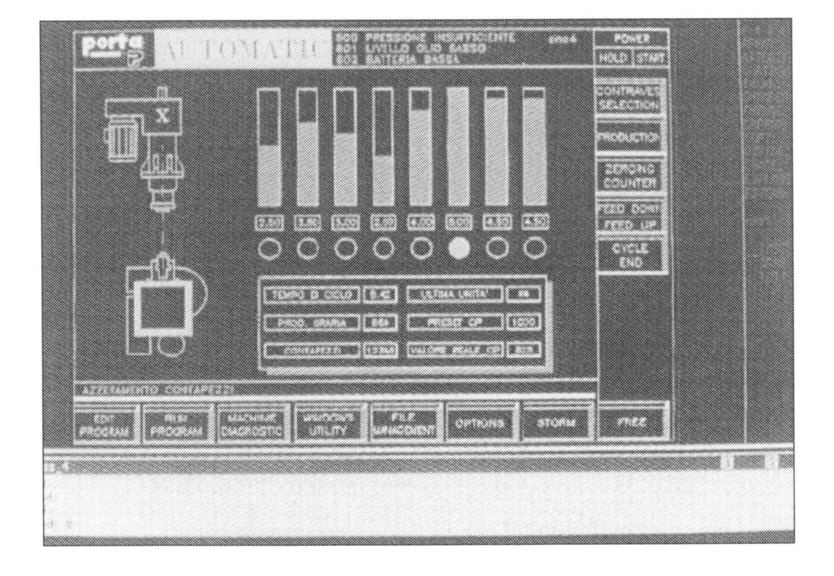

Fig. 1: The figure show a graphic interface in RTM (Courtesy of Porta Srl [16])

The reason of for such a great number of controlled axis in RTMs are:
- The great number of operating units in a transfer.

- The fact that the number of controlled axis per each operating unit is continuously increasing. While the traditional operating units have only the Z axe, or the spindle's axe synchronised with Z axe (max 2 axis), the recent operating units may have 5 or 6 axis that need to be synchronised by the numerical control.

Today there exists modern architecture that utilises only modular operating units with 3 CNC axis (these modular operating units are like machining centres) installed on rotary transfer machine. This architecture permits to work infinite shapes and configurations of workpieces. Numerical control is connected with a personal computer. The last one gives many benefits:

- Gives the possibility to insert the CN program, or simply to realise the program by means of user friendly graphic interfaces.
- Computer is a graphic interface (see fig. 1) and allows the operator to make sure that the machine functions correctly. Instead there are transducers placed in different parts of the machine connected with PC. In this way it is easy and instantaneous to see in witch zone of the machine the failure or problem has occurred.
- The network or modem connection gives the possibility to Tele monitor the right functioning of machine (in the case of unmanned production). The PC identifies automatically the kind of failure, or the failures and if the failure requires the participation of the man. Software gives help on line to solve the failure, and if eventually requires the participation of a technician. The technician can often modify software via Internet.
- Give the opportunity to realise efficiently control in process and automatically correction of the quotas of the workpiece.
- The PC supervisor stores working parameters and calculates efficiency measures, and stores the information coming from transducers and sensors and makes automatically control charts.

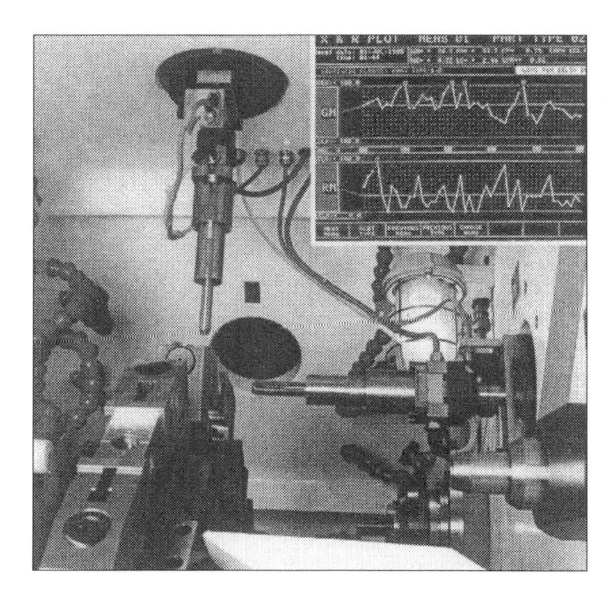

Fig. 2: The transducers and control in process(Courtesy of Porta Srl [16])

- The use of software is necessary to govern the robot (see fig. 3). One of the robots (two robots) is / are utilised for the automatic charge and discharge of the workpieces. The robots can keep the rough workpieces from palletised chains or conveyors. In the first case (palletised chains), the pieces have a specific position. In the second case (conveyors) the workpieces have random positions, so it is necessary to have an artificial vision system that identifies the orientation of the rough workpiece and allows orientation of the collet of robot in the right direction.

Fig. 3: The figure shows the robot charge and discharge (Courtesy of Porta Srl [16])

6. COSTS AND ADVANTAGE OF CN AND COMPUTER IN RTM

The modern numerical control RTMs with computer and control in process needs great investment, but this gives important advantages that guarantee lower costs during the production. These advantages are:

- The production is more reliable (the number of failure decreases), CNC and computer permit a greater efficiency for the productive system (productivity increases).

- Better cycle times, thanks to the efficient synchronisation of the axes given by CN. It is possible to optimise the little passive time as rapid approach, time to rotate and translate (along the rotation axes to unlock the Hirth system) the table or the RTM.
- More flexibility in the machine, so it is easy reconvert the RTM to new production. This flexibility allows to lengthen the working life of machine tool. Often to change the kind of parts machined by traditional it is necessary reconvert the transfer, and it is necessary high reconversion costs. Besides the production system has a high lost production time.
- Time and Costs to Market is lower.
- Set-up time is very really decreased.

The economic parameters are grouped in this class:

- Investment (cost to buy the RTM, and eventual future reconversions to change the production). The concept of working life of transfer is tightly correlated to the possibility to reconvert the machine to produce new kinds of parts. In the past there has been frequent cases of dismantlement of perfectly functioning transfer machine, but useless to market (in fact they couldn't machine the new pieces that the market needed).
- Maintenance costs
- Variable costs directly or indirectly chargeable to the batch.

Costs are summarised in the tables 1,2,3.

Table1

Investment	Opportunity to know the right value	Expensive charges
Costs of the productive system(RTM, robot for charge and discharge, etc..)	Known	Working life
Reconversion costs	Estimable	Working life
Costs of overhead exchanges of the lubricating oil treatment.	Known	Working life
Other starting investments	Known	Working life

Note, that we have made an approximate hourly cost considering the productive system investment.

Table 2

Cost items	Opportunity to know the right value
Routing maintenance	Known
Not-routing maintenance, in case of mechanic, electronic, CN, failure.	Estimable

Table 3

Costs items	Opportunity to know the right value	Expensive charges
Costs of pre-setting and balancing	Calculable	Directly to the batch
Manpower for set-up	Calculable	Directly to the batch
Other costs for set-up	Calculable	Directly to the batch
Tools cost	Calculable	Machining time
Direct manpower (during the machining)	Calculable	Machining time
Direct material (during the machining).	Calculable	Machining time
Electric power during the machining	Calculable	Machining time
Cooling by lubricating oil during the machining	Calculable	Machining time

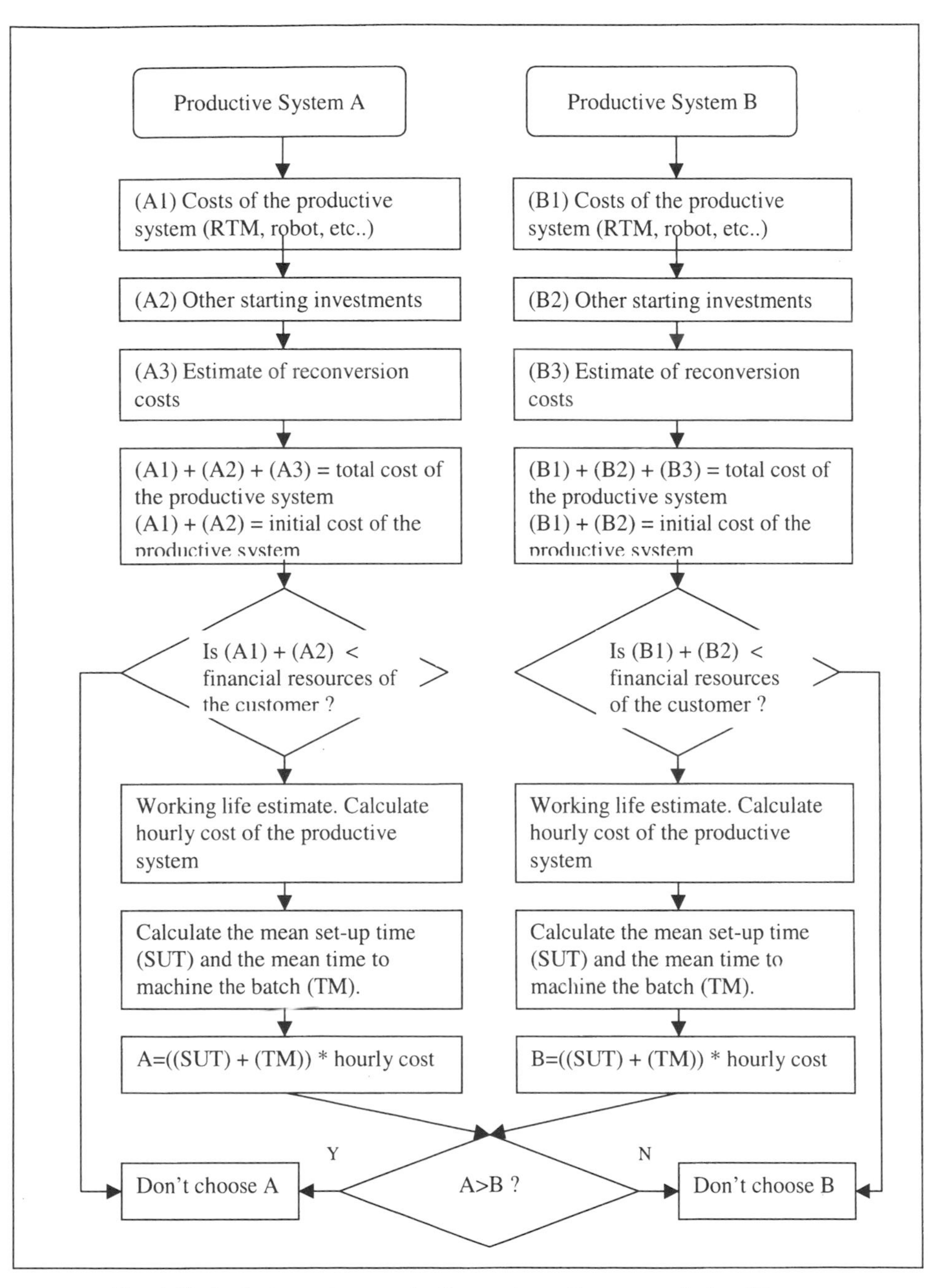

Fig. 4 Choice between two productive systems: the proposed algorithm

REFERENCES

(1) Koepfer C.,1993 *A new spin in rotary transfer machine.* Modern Machine Shop.
(2) Koepfer C., 1998, *Why not change. What you're doing?,* Modern Machine Shop.
(3) Molinari S., Gentili E., 1999, *New technology in rotary transfer machine.* Proceedings IV AITEM Conference.
(4) Consoli G., 1983, *Transfer di lavorazione*, Nuove Edizioni S. Rocco. (In Italian)
(5) Buffoli M., 1997, *Il transfer da barra,* RM. (In Italian)
(6) Agnesi F., 1996, *Tecnologia di qualità,* TM. (In Italian)
(7) Morelli R.,1998, *Il settore dei transfer,* TM. (In Italian)
(8) Costa P., 1998, *Italia in pole position,* TM. (In Italian)
(9) Giuliani *http://www.giulianico.com*
(10) Gnutti transfer *http://www.gnutti.com*
(11) Goss Deleeuw-Porta *http://www.goss-deleeuw.com*
(12) Riello macchine utensili *http://www.riellomu.it*
(13) Sinico S.p.a. *http://www.sinico.com*
(14) Buffoli Transfer S.p.a. *http:/www.buffoli.com*
(15) BTB Transfer *http://www.btb.com*
(16) Porta Srl catalogue

CAPE/088/2000

Three-dimensional modelling tools supports material handling devices operational quality

J SZPYTKO
Cracow University of Technology, Poland
J SCHAB
Univeristy of Mining and Metallurgy, Cracow, Poland

ABSTRACT

The material handling operation process generate numbers problems referring to modelling of qualitative operating characteristics, reconstruction of operating capacity of the device, and minimisation operating expenses, especially maintenance cost. The solid modelling is an effective tool supporting the operating process of the materials handling devices. The solid modelling makes possible support device maintenance and quality modernisation, as well s marketing process. The CAD solid modelling is moreover a usefully tool supporting the Business Process Reengineering and Total Productive / Preventive Maintenance.

1. INTRODUCTION

The known engineering software's are including following types:
- graphical (construction recording, solid modeling, image converting, chart drawing),
- database (documentation and information flow management),
- computing (solve a system of equation, characteristics calculate integral calculate),
- optimization (optimization and poli – optimization procedures),
- finite elements (rigid finite elements methods, edge finite elements methods),
- dynamic of discrete systems (movement equation compose and solving, graphical modeling, symbolic generate of movement equation),
- simulation (numerical solving of differential movement equation),
- artificial intelligence (experts systems, image recognition),
- results presentation (tables/ charts/ animations, spectral analysis, statistical processing).

The material handling devices (represented by large dimensional devices, e.g. cranes) during an exploitation generate numbers problems referring to: modelling of qualitative operating characteristics, reconstruction of operating capacity of the device, and minimisation operating expenses, especially maintenance cost (1). The computer software environment: aided design

CAD, aided manufacture CAM and aided engineering CAE, are supporting the device life phases integration, as well as exchange data necessary to modelling qualitative requirements of the device. The above software applications are based on the device structure and leading into the solid modelling process (2,3).

The CAD solid modelling is moreover a usefully tool supporting the Business Process Reengineering and Total Productive / Preventive Maintenance approaches.

2. DEVICE OPERATIONAL QUALITY FORMING

Each device must renew itself productivity and quality in response to competitive pressures, evaluation of technology and changing needs of its users (consumers). This renewal includes changing the device or changing the device quality. Reengineering and continuous improvement are complementary approaches to the device improvement. Stages of reengineering device project or modernization includes: vision – where you want to go, assessment – where you are now, concept – how you are going to get there, development – building the means to get there, implementation – getting there. Information system often plays a central role in reengineering efforts, especially material handling devices operational better quality.

Each device *DE* with the expected quality level *Q* with is consequence of acceptable tolerance of the device exploitation characteristics Φ, can be described through the set of exploitation parameters *a* with tolerances Δ and relationship between the device exploitation parameters *r*:

$$DE = \{a, \Delta a, r, \Phi\} \rightarrow Q$$

where:
a – set of exploitation parameters of the device; $a = \{a_1, a_2, ... a_n\}$, $a_i \in R^n$
r – set of relationship between the device exploitation parameters; $r = \{r_1, r_2, ... r_m\}$, $a_i \in R^m$
Δ – set of the device exploitation parameters tolerance zone; $\Delta = \{\Delta a_1, \Delta a_2, ... \Delta a_n\}$, $\Delta a_i \in R^n$
Φ - acceptable tolerance of the device exploitation characteristics
Q – exploitation quality of the device; $Q = \{ q_1, q_2, ... q_n\}$.

Quality of the device is mostly overlooked via prisms of safety, reliability and accuracy, as well as: functionality, durability, efficiency, lightly, unit cost and exploitation cost, material accessible, proper loads transmission, easy in exploitation, maintainability, ergonomic, surroundings protection, compatible with obligatory standards. The consequence of new user needs is better device quality Q* reach as the modernization result. Crane modernisation refers to an exercise in which is modified, refurbished or upgraded for the purpose to achieving improved operational quality (productivity, increased reliability and safety, enhanced maintainability or overcoming the difficulty in procuring obsolete spare parts). The device DE* which has been modernised is described by the following equation:

$$Q^* \rightarrow DE^* = \{a^*, \Delta a^*, r^*, \Phi^*\}$$

where (*) are modified sets.

3. SOLID MODELLING

Solid modeling helps to create 3D objects, which is possible to examine and change at the PC screen. The solid model is describing real physical property (geometry) of the device (outside look) and his units and elements (inside look), as an assembly system or separate units/ elements. Knowledge of the geometrical device property makes possible to calculate capacity and weight of the object, as well as his gravity centre and inertia moment's co-ordinates. The device structure is possible to design using bottom-up modelling, top-down modelling similar to the SADT method, as well as designs in context. Very usefully during the device solid modelling are *Revolution*, *Sweeping* with path, *Projection* and others modules.

All designed models of the device's units and elements have easy to modified structure which is very usefully during forming the exploitation parameters of the device and reconstruction his exploitation potential. *Feature Manager* similar to the SADT presentation could present the device structure. The *Feature Manager* makes easy and continuously possible review and supervision the device units' changes and verification established structure.

The solid model defines all geometrical links and relationships between all units/ elements of the device. Typical geometrical relationships are coincident, parallel, tangent concentric, perpendicular, distance, and angles. The solid modeling is offering also the automatic generation of the device documentation in 2D and 3D (using *Part* and *Assembly* modules), as well as presentation the device in VR environment.

The example of solid modelling of rope hoisting winch is presented at Figure 1. Figure 2 presents model of crane hook.

Fig.1. The rope hoisting winch

Fig.2. The crane hook model

4. CONCLUSION

The solid modelling is an effective tool supporting the operating process of the materials handling devices, as well as activities oriented to the Business Process Re-engineering and Total Productive/ Preventive Maintenance. The solid modelling makes possible:

- easily changing the device structure, as well as reliability relations describing,
- sequential device structure verification according to the certain quality criteria,
- the device visualisation for presentation, marketing, research, maintenance, e.g.: assembling or disassembling, personnel training in maintenance,
- decision making and device structure analysis
- visualisation in 3D joint co-operation of various devices operating into plant.

The above exercises might be leading simultaneously to the supervision process of the device exploitation parameters, as well as to reconstruction the exploitation potential of the device. The approach cut down the time and cost of the device modernisation.

REFERENCES

1. Szpytko, J. (1996). *Integrated Supervision System of the Chosen Exploitation Parameters of the Large-Dimensional Rails' Handling Device on the Automated Overhead Crane Example. UMM Press, Monographs*, **46**, Cracow
2. Winkler, T. (1997). *Komputerowy zapis konstrukcji.* WNT
3. SolidWorks 97Plus, (1997). *User's Guide.* SolidWorks Co.

CAPE/060/2000

Optimization algorithms for mapping three-dimensional woven composite preforms

S B SHARMA, P POTLURI, J ATKINSON, and **I PORAT**
Department of Mechanical Engineering, University of Manchester Institute of Science and Technology, UK

SYNOPSIS

Manufacturing of composite materials by moulding woven preforms into 3D shapes is getting increasingly popular due to the ability of a woven fabric to undergo complex deformations. However, there are limits to which a woven fabric can be deformed and this reflects in the form of wrinkles or fibre-rich areas which are unacceptable in the final product. The work presented in this paper deals with the algorithms used to find ways and means to remove such problems incurred during mapping. To start with, a strategic location of the starting point to map a woven fabric on the mould surface is found by analysing the surface curvature variation. The starting point and the fabric orientation are then varied to optimise the mapping quality defined in terms of an overall deformation factor or the fibre volume fraction. An algorithm are presented to investigate the fibre-rich areas and to apply local density variations of the yarns to make the fibre distribution more uniform.

NOTATIONS

$\mathbf{P}(u, v)$	Point in a parametric form
d_u, d_v	Spacing of yarns in the u- and v- direction
u, v	Parametric variables
u_c, v_c, α_c	Location and inclination of constrained yarns
v_f	Fibre volume fraction
Δ	Global deformation factor
$\theta_{i,j}$	Shear angle at point $\mathbf{P}_{i,j}$
δ	Local deformation factor
κ	Curvature

1 THE MAPPING ALGORITHM

The mapping algorithm is only briefly described here; a detailed discussion is presented in ref. (1). The fabric is represented as an assemblage of two sets of yarns, warp and weft, initially orthogonal to each other. Each yarn segment between the crossover points is assumed to be linear or piece-wise linear. The mould surface may be specified as:

- a set of connected parametric surface patches (2)(3), or
- in terms of the cross-section of the 2.5D preform geometry, and a sweep path (4) which characterises the deformation applied along the length of the woven preform.

Irrespective of the method of construction used and the number of patches in the super-set of the mould surface, the equation of the surface is reduced to the form

$$\mathbf{P}(u,v)=\{x(u,v)\}, \quad u_{min} \le u \le u_{max}, v_{min} \le v \le v_{max} \qquad \text{Eq. 1}$$

where u and v are the parametric variables and $\mathbf{P}(u, v)$ specifies a point on the surface for the given co-ordinate values u and v.

A 'closed' section is the one for which the cross-section is a closed curve. Draping of a closed section is more complex than the open sections. It involves fabric extension and local buckling and is not covered in detail in the present work. Some results to the same however are presented here because the optimisation and re-engineering strategies work equally well on them.

Fig. 1 illustrates the basic mapping algorithm applied to a curved open surface. Two "constrained yarns" on the fabric are initially specified to represent two orthogonal yarns on the fabric to align with curves on the mould surface. Depending upon their positions, they divide the surface into a maximum of four quadrants. The relative spacing of the crossover points on these yarns is fixed and is equal to the spacing of the yarns on the original fabric. These points are represented in a 2D array as $\mathbf{P}_{i,j}$, where i and j respectively define the indices of the point in the u- and v- directions. The mapping algorithm uses the fabric and mould surface properties to calculate the intersection points of the yarn segments with the parametric equation of the surface. Based on the known points on the constrained yarns, a sphere intersection model discussed in ref. (1) is used to locate the unknown points. An example of the application of this algorithm to an S-shaped curved surface is shown in Fig. 1(b); along with the 2D outline of the preform needed to cover it.

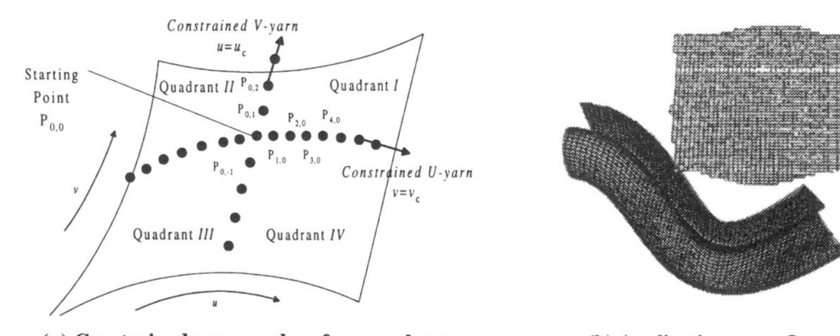

(a) Constrained yarns and surface-quadrants **(b) Application to an Open Section**

Fig. 1 Mapping Algorithm

2 MAPPING QUALITY: POINTERS TO IMPROVE MAPPING ALGORITHMS

Depending upon several parameters, e.g. constrained yarn locations and inclinations, fabric shear limit, and to some extent, the unit cell-size itself, different mappings would be obtained. There is a need to define the quality of mapping, so that relative comparisons can be made. The factors describing mapping quality can be categorised as follows:

- Qualitative
    1. Complete/partial cover of mould surface
    2. Presence of wrinkles
    3. Shear angle: its distribution
- Numerical
    4. Shear angle: maximum & minimum values
    5. Deformation factor
    6. Fibre volume fraction
    7. Area of fabric needed to cover surface

The qualitative factors do not give complete information about the mappings. There is a need for a parameter to define the overall fabric deformation. Also there should be a way to define the density of fibre distribution over the mould surface. The range of shear angles attained at the fabric cells provides a good numerical account of the fabric deformation, and therefore it has been used by previous researchers, e.g.(5). In addition, the following two factors are used to numerically define the quality of mapping: *deformation factor* and *fibre volume fraction.*

2.1 Shear angle distribution

In the mapping software developed during the course of present work, distribution of the shear angle is graphically plotted over the mould surface to obtain a feel for the extent of deformation. The program draws each of the points as a sphere, with its colour appropriately coded according to the shear angle attained by the yarns crossing over the point. When the points are drawn using these colour codes, the distribution of shear deformation within the fabric is visually very clear.

2.2 Deformation factor

The *deformation factor* relates the deformation to the closeness of yarns, or in other words, to the distortion of a unit cell. The *deformation factor* is cumulative, and is significant in local as well as global terms. In local terms, it can be used to locate the cell which has the maximum distortion. The distribution of deformation in a certain region can be used to highlight critical areas on the fabric which need re-engineering. Global deformation factor gives an average value of deformation that can be used to compare two mappings.

The area of a cell before shear deformation is given by $d_u \cdot d_v$. After shear deformation, it changes to $d_u \cdot d_v \cdot \sin(\theta_{i,j})$, where $\theta_{i,j}$ is the shear angle at cell $C_{i,j}$. The local ($\delta_{i,j}$) and global (Δ) deformation factors are given respectively by (n_c is the total number of cells in the fabric):

$$\delta_{i,j} = \frac{\text{Final Cell Area} - \text{Original Cell Area}}{\text{Original Cell Area}} = \sin(\theta_{i,j}) - 1 \qquad \text{Eq. 2}$$

$$\Delta = \frac{1}{n_c}\sum_{i,j}\delta_{i,j} = \frac{1}{n_c}\sum_{i,j}\sin(\theta_{i,j}) - 1 \qquad \text{Eq. 3}$$

Colour coding can again be used to plot the variation of deformation factor over the fabric region. The values of $\delta_{i,j}$ and Δ are always negative for an open section mapping, indicating a reduction in cell area. In the case of a closed section, the values may be positive or negative, depending upon the region being under extension or compression. The cell distortion in this case is more complex and the details for mapping and the method of determining the deformation factor are not covered here.

2.3 Fibre volume fraction (FVF)

Fibre volume fraction is another term which directly quantifies the density of fibre distribution. This factor too, in turn, depends upon the shear angle of individual cell for an open section mapping. Denoted by v_{f0}, its value lies between 0 and 1. Again the maximum and minimum values would be useful in comparing several types of fit. The expression for the fibre volume fraction for plain weaves is given in (6). For more complex weaves, expressions could be developed which measure the unit fibre volume fractions (FVF).

For a shear-distorted cell, the fibre volume fraction could be approximated to $v_{f0}/\sin(\theta_{i,j})$. There is a geometric limit to the shear angle which would also be reflected by the fact that beyond a certain angle, the fibre volume fraction might exceed unity. The FVF distribution is not much different from the deformation factor distribution, since it is in turn based on shear or extension/compression ratio. Fig. 2 shows the FVF distribution for an octant of a sphere.

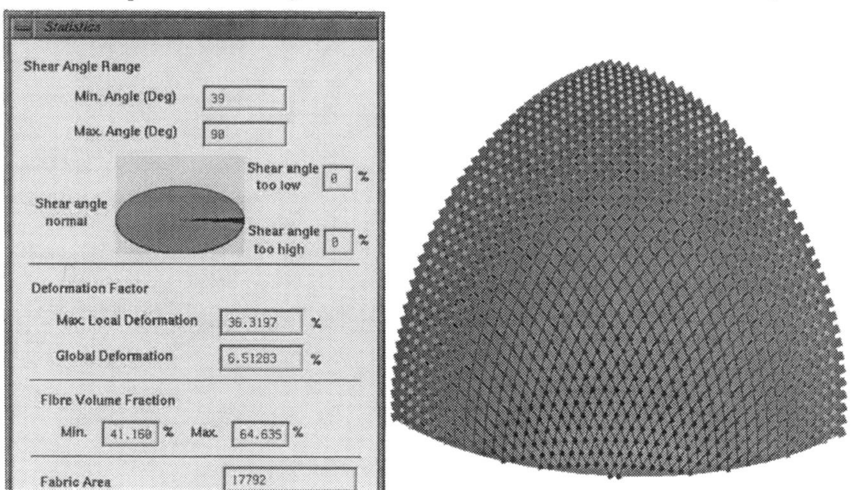

Fig. 2 FVF distribution for the octant of a sphere

2.4 Guidelines on using the mapping quality parameters

Shear deformation gives a good idea of the mapping over an open surface. The minimum and maximum values of shear, as well as its distribution too, give an indication of the extent of deformation which could be used to compare two mappings. The deformation factor and the fibre volume fraction distribution are two very useful factors for numerically expressing the quality of mapping. Deformation factor is a relative value, while FVF gives an absolute figure for the density of fibres. FVF directly affects the thickness of composite, so it is of high practical significance. Both these factors apply to open as well as closed sections. The main advantage of the deformation factor is that it expresses the values as a percentage, which makes it useful for comparison purposes. The local deformation factor is useful for plotting the distribution of the fabric distortion, while the global deformation factor gives an idea of the average distortion.

For the purpose of optimisation, global deformation factor offers a good way of comparing different mappings.

3 MAPPING OPTIMISATION

Based upon the above factors describing mapping quality, the following section presents strategies to optimise the mapping. To start with, a heuristic to predict the best starting point on the mould surface is presented. For most 2.5D mould surfaces, the goal point for optimum fit happens to be close to this point. Algorithms are then presented to search for the global optimum coordinates of the starting point and the orientations of the constrained yarns.

3.1 Good starting point for the mapping

It was observed that for the 2.5D types of preforms under study, constraining the yarns along the 'corner-lines' on the mould surfaces resulted in better mappings. These are the lines along which the cross-section curve and the path curve have sharp bends. Mathematically, these are the set of orthogonal curves on the mould surface patch along which the curvature is maximum, or the radius of curvature is minimum. These also happen to be the most logical locations to align the fabric for moulding.

So to achieve a better mapping, one has to locate the points of maximum curvature on the cross-section and the path curves. For example, curvature of the path curve $\mathbf{T}(u)$ is given by:

$$\kappa = \frac{\left\|\mathbf{T}^{u} \times \mathbf{T}^{uu}\right\|}{\left\|\mathbf{T}^{u}\right\|^{3}} \qquad \text{Eq. 4}$$

where $\mathbf{T}^{u}$ is the tangent and $\mathbf{T}^{uu}$ is the second derivative vector. Using standard numerical optimisation routines, one can determine the value of u for which curvature is the maximum.

Fig. 3(a) shows an example of an unsuccessful fit to a component of S-shaped cross-section. The same component, when mapped with the preferred constrained yarn locations, and by using the extended mapping algorithm, results in a better fit, as shown in Fig. 3(b).

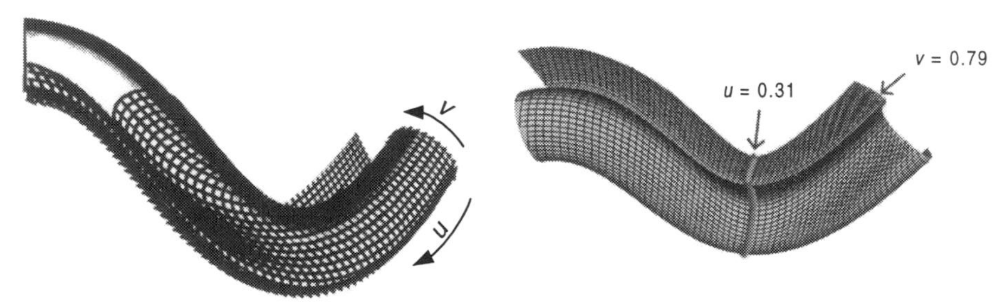

(a) Constrained yarns at the edges: $u_c = 0$, $v_c = 0$ **(b) Preferred locations: $u_c = 0.31$, $v_c = 0.79$**

Fig. 3 Choosing constrained yarn location based on local curvature variation

The results are also compared with respect to the mapping quality parameters in Table 1. One can see that the fitting in the second case does not result in any wrinkling, has a better shear distribution, and the values of local and global deformation factors are also lower.

Table 1 Comparison of mappings shown in Fig. 3

Parameter	*Case (a)*	*Case (b)*
Shear angle range (degrees)	0-162	39-144
δ_{max} (%)	99.8	41.5
Δ (%)	13.5	9.8
v_{fmax} (%)	100	50.7

3.2 Optimisation by varying the starting point and constraint angle

Once the starting point has been obtained using the above method, it is possible to iteratively seek a better solution by varying the starting conditions. As mentioned before, the most effective term to be optimised is the global deformation factor (Δ). It is the single term describing the cumulative distortions the cells of a fabric undergo when fitted over a doubly curved surface. The variables of the optimisation (with the assumption that warp follows the u-direction) are: constrained warp position (v_c), constrained weft position (u_c) and constraint angle (α_c). All these three variables are independent to each other. Keeping two of them fixed, one can vary the other variable and minimise Δ. The optimisation is thus sequential.

As an example, the S-shaped surface is considered again. The optimum mapping is obtained in three stages; Table 2 summarises these steps while Fig. 4 shows the final mapping.

Table 2 Steps in optimising the mapping over the S-shaped surface

Step	*Variable parameter*	*Fixed parameters*	*Optimum found at*	δ_{max}	Δ
1	u_c	v_c=0, α_c=0	u_c=0.31	40.29%	10.15%
2	v_c	u_c=0.31, α_c=0	v_c=0.05	40.31%	9.15%
3	α_c	u_c=0.31, v_c=0.05	α_c=0°	40.31%	9.15%

Step 1: Varying constrained weft location. u_c variable, $v_c = 0$, $\alpha_c = 0$.
The range of u_c is [0:1]. Optimum solution is found at $u_c = 0.31$, which is the same as the value predicted by the highest curvature based heuristic. The minimum value of Δ is 10.15%.
Step 2: Varying constrained warp location. $u_c = 0.31$, v_c variable, $\alpha_c = 0$.
The range of v_c is [0:1]. Optimum solution is found at $v_c = 0.05$. Although this is not the same as the value predicted by the highest curvature based heuristic, it still is one of the three points of high curvatures on the cross-section curve. The minimum value of Δ is 9.15%.

Step 3: Varying constraint angle. $u_c = 0.31$, $v_c = 0.05$, α_c variable.
The range of α_c is [-45°:+45°]. A simple linear search is performed to obtain the minima. The plot of global deformation factor (Δ) against α_c is shown in Fig. 4, and the optimum is found at $\alpha_c = 0°$. The minimum value of Δ is 9.15%.

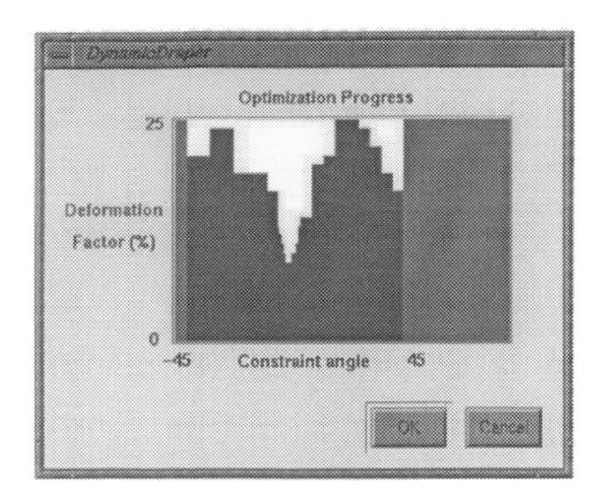

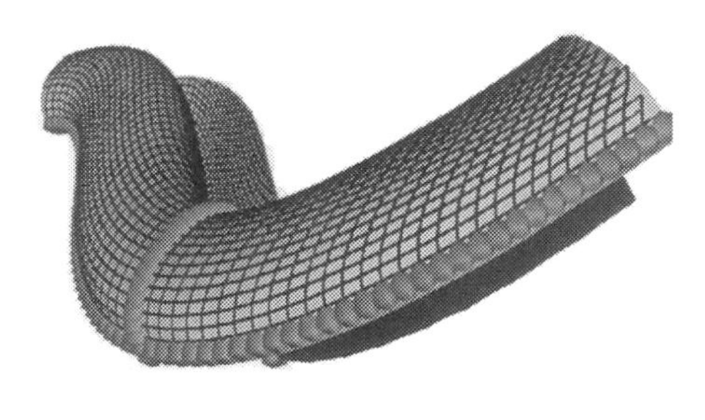

(a) Plot of Δ against α_c (b) Optimum fit

Fig. 4 Optimisation by varying the constraint angle

As another example, a vase surface is chosen. The values of u_c, v_c and α_c are sequentially determined and the optimum is found as shown in Fig. 5. The minimum value of Δ is found to be 4.97% (Table 3). Tilting the fabric over the vase surface results in a better distribution of the fibres in the four parts of the surface, and hence a better map.

Table 3 Steps in optimising the mapping over a vase

Step	*Variable parameter*	*Fixed parameters*	*Optimum found at*	δ_{max}	*Δ*
1	u_c	v_c=0, α_c=0	u_c=0.51	87.80%	13.81%
2	v_c	u_c=0.51, α_c=0	v_c=0.51	68.97%	9.85%
3	α_c	u_c=0.51, v_c=0.51	α_c=41°	28.83%	4.97%

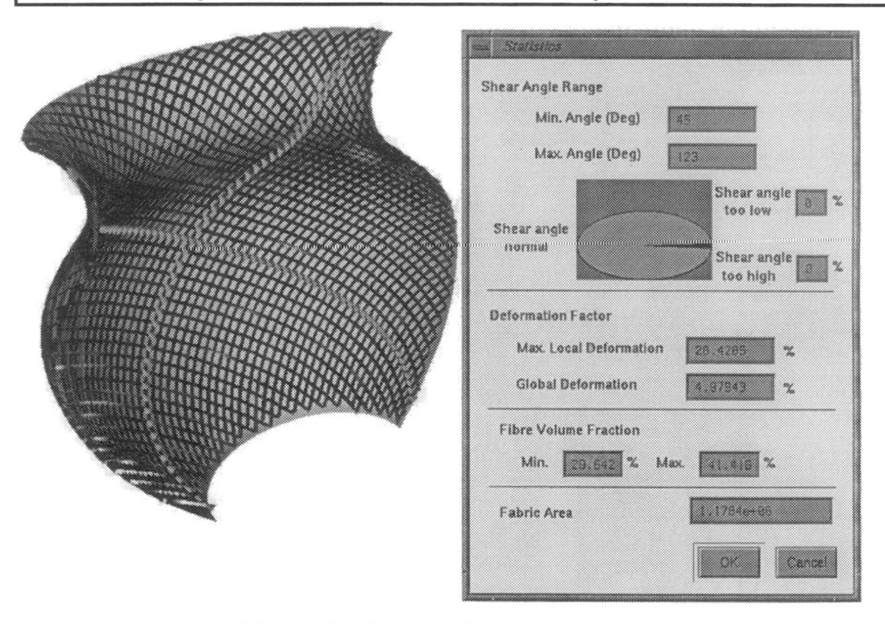

Fig. 5 Optimum fitting over a vase surface

4 FABRIC RE-ENGINEERING

Optimisation cannot always provide completely trouble-free mappings, and sometime wrinkled/fibre-rich areas cannot be completely avoided. In such a case the fabric needs to be modified or re-engineered. There are several ways in which this can be done. Aono *et al.* (7) suggested an algorithm of dart insertion in the fabric areas where the shear deformation is high. Approaches to modify the fabric structure or the yarn densities to overcome this problem have not been tried before and the aim of the present analysis is to do that.

The shear angle, the deformation factor and the fibre volume fraction all indicate the closeness of yarns, or the fibre density. Greater shear deformation corresponds to a higher deformation factor, and a higher fibre volume fraction. This leads to wrinkle formation, or, at the least, thickness increase. The thickness increases because the material volume remains constant, as the area of a unit cell decreases. Fibre rich areas are to be avoided; they prevent resin from providing sufficient bonding strength and as a result reduce the strength of the composite.

The fibre volume fraction happens to be the most crucial factor on which to base the re-engineering. This factor applies equally well to both open and closed sections.

4.1 Principle of re-engineering

Consider a set of four neighbouring fabric cells, each having length and width d, as shown in Fig. 6a. The FVF of a cell is directly proportional to the length of the fibres contained in it. In a given cell (either the smallest unit, or a combination of more cells), only half the volumes of the boundary fibres contribute to the FVF, while the full volumes of the inner ones contribute to the FVF. Let the fibre volume fraction of each of the cells be v_f.

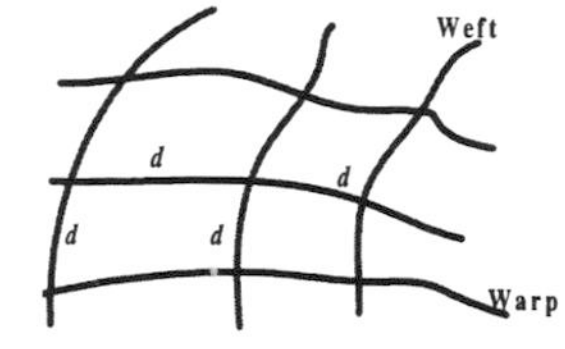

(a) Four cells, before re-engineering

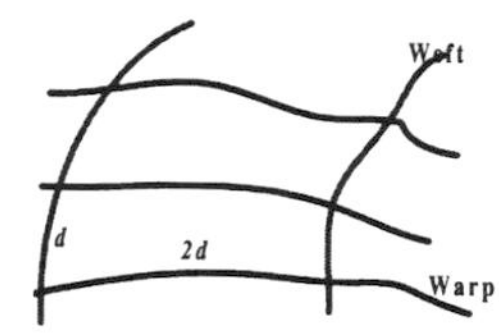

(b) Central weft removed

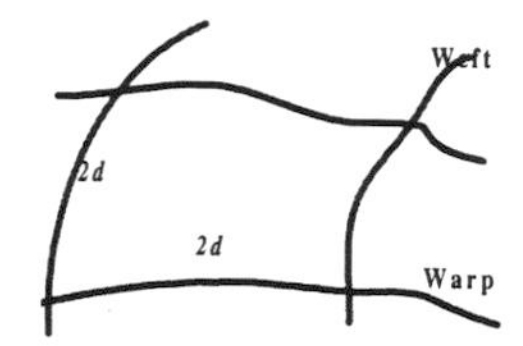

(c) Central weft and warp removed

Fig. 6 Principle of re-engineering

Now consider the central weft (the vertical thread) removed from the structure. Two cells will result (see Fig. 6b), the FVF of each of them will be (approximately):

$$v_f \frac{new\ perimeter\ of\ fibre\ lengths}{original\ perimeter} = v_f \frac{\frac{1}{2}(2d + d + 2d + d)}{4d} = v_f \frac{3}{4} \qquad \text{Eq. 5}$$

Finally, consider the central warp (the horizontal thread) too removed from the structure. Only one cell will result (see Fig. 6c), the FVF of which will be (approximately):

$$v_f \frac{new\ perimeter}{original\ perimeter} = v_f \frac{\frac{1}{2}(2d + 2d + 2d + 2d)}{8d} = v_f \frac{1}{2} \qquad \text{Eq. 6}$$

Thus, by reducing the yarn density in one direction to half the original value the effective FVF reduces to 75%, while by halving the yarn densities in both directions the FVF reduces to 50% of the original value. This is the basic principle upon which the re-engineering of fabric is based. The yarn density is reduced in the problem areas, bringing the effective FVF within acceptable limits.

The above principle is used to relax the yarn densities in the fibre-rich areas of fabric. The re-engineering is carried out once the mapping has been applied. The algorithm applies to both open and closed sections. Within the fabric, all the cells are scanned to locate the one with the highest FVF. A rectangular patch is outlined around this cell, containing cells which have high FVF. The yarn density in these cells is relaxed by removing every alternate weft from the patch. The process is repeated recursively for the fabric until all the fabric area is covered, and no cells with high FVF remain.

Two important points need to be noted here. First, the patch to be re-engineered must be rectangular in shape. This is due to limitations of weaving, where to modify the density over an arbitrarily selected patch is difficult. This constraint can be lifted from the patches that contain the edges of the fabric, since the excess fabric will be subsequently trimmed away. Second, it is easier to modify the weft density rather than the warp density, hence the mention of weft thread instead of the warp above.

4.2 Examples

The algorithm presented above applies equally well to open as well as closed section. One example of each of the two types are given in this section.

The first example is the octant of a hemisphere. As seen in Fig. 7, the maximum FVF has been reduced from 70.6% to 55% by re-engineering.

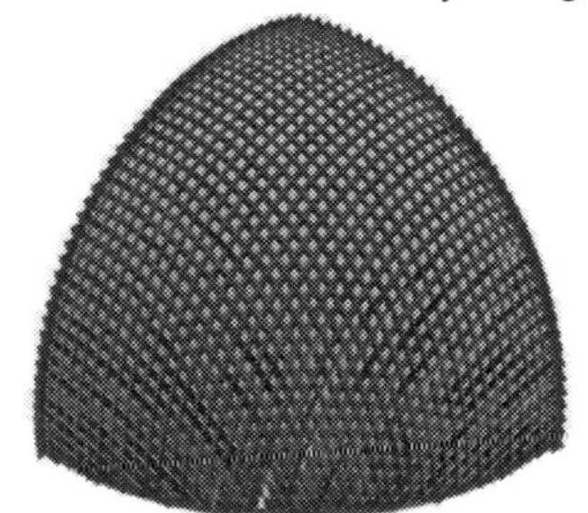

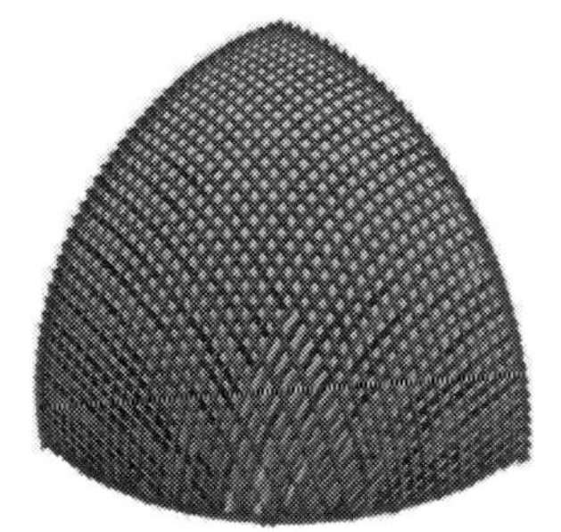

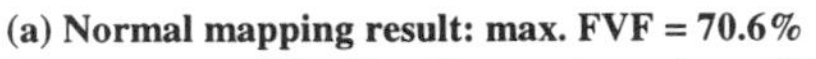

(a) Normal mapping result: max. FVF = 70.6% **(b) Re-engineered fabric: max. FVF = 55%**

Fig. 7 Re-engineering of fabric to fit an octant of a sphere

Another example is that of a circular bent tube. The normal mapped fabric is shown against the re-engineered one in Fig. 8; the maximum FVF in each case is 100% and 50% respectively.

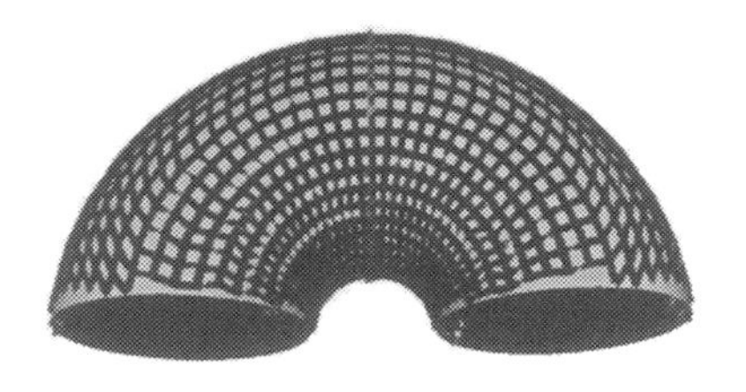

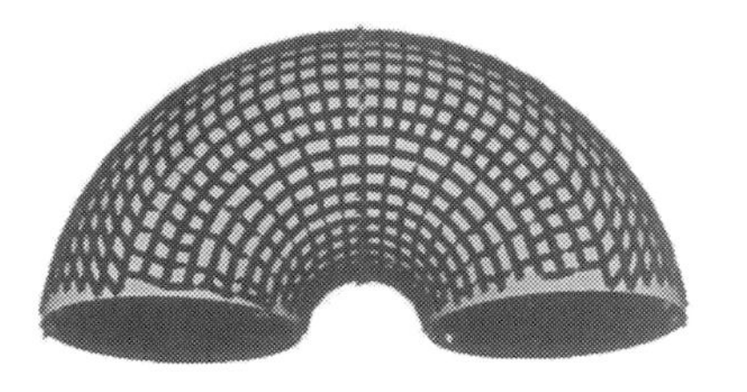

(a) Normal mapping result: max. FVF = 100% **(b) Re-engineered fabric: max. FVF = 50%**

Fig. 8 Re-engineering of fabric to fit a bent tube

CONCLUSIONS

This paper presented a way to utilise the results obtained by mapping algorithms developed in the past. Several factors describing the mapping quality were presented and their significance discussed. Optimisation of structures was achieved by linear search on the values of constrained yarn positions and orientations, and this was demonstrated using the S-shaped swept surface and a vase example. Re-engineering of fabric by modification of yarn densities presented at the end of this chapter could be very useful for automating the construction of performs having complex shapes. All the information provided in this paper has been implemented into a software. Using this software, one can engineer the weaving and moulding parameters to ensure good quality of the final product.

ACKNOWLEDGEMENTS

The authors wish to thank the Engineering and Physical Sciences Research Council (EPSRC), Fibrelite Composites Ltd, Carr Reinforcements Ltd, Bonas Machine Company Ltd, and S C Consultants for sponsoring the project on moulding analysis.

REFERENCES

(1) Sharma, S B, Potluri, P, and Porat, I. "Moulding Analysis of 3D Woven Composite Preforms: Mapping Algorithms", In: Proceedings of the twelfth International Conference on Composite Materials (ICCM-12), Paris, France, 5th-9th July 1999.

(2) Mortenson, M E. *Geometric Modelling*, 2nd Edition, Wiley Computer Publishing, New York, 1997.

(3) Farin, G. *Curves and Surfaces for computer Aided Geometric Design*, 3rd Edition, Academic Press, USA, 1992.

(4) Marhl M, Guid N, Oblonsek C, Horvat M. "Extensions of sweep surface constructions", *Computers and Graphics*, 1996, 20(6), pp893-903.

(5) Aono, M, Breen, D E, and Wozny, M J. "Fitting a Woven-Cloth Model to a Curved Surface: Mapping Algorithms", *Computer Aided Design*, 26, 6, 1994, 278-292.

(6) Gutowski, T G. *Advanced Composites Manufacturing*, Wiley-Interscience Publication, New York, 1997.

(7) Aono, M, Denty, P, Breen, D E, and Wozny, M J. "Fitting a Woven-Cloth Model to a Curved Surface: Dart Insertion". *IEEE Computer Graphics in Textiles and Apparel*, September, 1996, pp 60-69.

CAPE/075/2000

Development of a web-based system for collaborative product definition on the Internet

G Q HUANG, S W LEE, and **K L MAK**
Department of Industrial and Manufacturing Systems Engineering, University of Hong Kong, Hong Kong

SYNOPSIS

This paper presents a synchronised web application, called ProDefine, for early product definition. Three main components are provided to support the three main phases of early product definition. The Customer Requirements Explorer helps define and design objectives based on customer requirements. The Concept Generation Explorer generates feasible solutions that meet the design objectives. Alternative solutions are evaluated using the Concept Evaluation Explorer. Unique features for synchronisation and conflict resolution are particular useful for supporting project teams distributed in geographical and time terms.

1 INTRODUCTION

The product design and development process generally include four main stages, namely (1) planning and task clarification, (2) conceptual design, (3) embodiment design, and (4) detailing phases (Pahl and Beitz, 1984). In the context of this research work, the first two stages are considered as early phase of product definition and the last two stages as the detail phase of product definition. As the names imply, early product definition is mainly concerned with establishing key characteristics of a product under development, and detail product definition deals with concrete descriptions of the subject product. Majority of the design systems such as CAD and material selection software support and facilitate the detail product definition. Much of the STEP efforts are also dedicated to detail product definition. In contrast, early product definition has received relatively less attention from the research community although the practitioner community demands for better methods and computerised supports.

In order to address this gap, the research work reported in this paper focuses on the development of a software support for early product definition on the Internet and world wide web. This research aims to develop a generic framework for product definition at the concept stage of product design and development cycle, and demonstrate the methodology through a web-based system. There are four specific objectives expecting to be achieved: (1) To identify the key issues of product definition; (2) Establish an overall methodology for collaborative product definition; and (3) To implement the Web-Based Design Tool that supports the Early Product Definition.

There are a number of reasons that we concentrated on the early phases of the design process. First, the early product definition stage is considered to be an important stage in product design and development cycle since the rework or redesign of the product design in the subsequent process is expensive or even impossible (Pahl and Beitz, 1984).

Second, the early product definition stage is the most innovative, requiring considerable experience and skills from the designer or the development team. It consists of the complexity of the abstract process of conceptualisation. It is the phase that makes the greatest demands on the designer. Therefore, it is necessary to develop a tool that adequately supports the designers in the early design process (Hague and Bendiab, 1998).

Third, the early product definition stage still receives only a little support form the recent design tools because the abstract process is difficult to model in the conceptualisation process (Chakrabarti and Bligh, 1996). Thus, the success of product definition still relies heavily on designers' knowledge and experience.

Finally, the early product definition stage involves the collaborative nature. It comprises of people in different discipline and locations working together. It is therefore important to develop design tools which facilitate the collaborative nature. Web applications are one of the solutions that are gaining increasing popularity in these few years (Erkes *et al*, 1996). By implementing the design tools on the Internet, geographically distributed design teams can work together and different people can also participate to the design process through the web and give out ideas.

This paper will presents an effort made to develop an Internet-enabled support, called ProDefine, for early product definition collaboratively through a distributed project team. After brief review of the literature in Section 2, the system architecture, components and implementation issues are discussed in Section 3. Section 4 concentrates on explaining how synchronisation is achieved and how related issues such as conflict resolution are addressed. Some implications are drawn in the concluding section.

2 RELATED WORK

A number of researchers are designing electronic systems supporting various stages in the product design and development process. It is also noticed that computer tools is important in supporting and improving the product definition (Stauffer and Morris, 1991). Tseng and Jiao (1998) developed a database system to provide computerised environment for the Product Definition. A two-phase methodology is adopted i.e. pattern recognition phase and the pattern adoption phase. The pattern recognition phase is a preparatory phase in which functional requirement patterns are recognised from past designs. The recognised patterns are then used for new designs in the pattern adoption phase. This methodology mainly covers the elicitation of customer requirements and the definition of product specifications. It transfers customer requirements into functional specification by using the functional requirement pattern templates.

The main advantage of this methodology is that historical data from existing designs can be utilised when designing new products. It prevents the overlooking of relevant information about product requirements. However, some shortcomings are identified. First, the scope of the product definition is limited to the requirement analysis phase of the product design and development process. The later stages, such as the concept generation and evaluation stage, are not supported. Second, this approach is limited to similarly designed products. Innovative designs are not applicable by using this methodology.

While Tseng and Jiao (1998) focus on the methodology of the product definition, Tuikka and Salmela (1998) concentrate on the communication of designers and customers over the World Wide Web in the concept design domain. A computer system called WebShaman supports synchronous collaborative design work session with Internet technology. The designers and customers communication are facilitated by virtual prototyping technique with synchronous features over the Internet. As a result, geographically widespread design teams can communicate and introduce design choices to their customers. The virtual prototype in WebShaman is implemented by VRML and Java language. Thus, the main design focus is on the appearance of the products. The working mechanism of the products may be overlooked. Therefore, design by virtual prototype is suitable for small product with little variation in mechanical or electrical design.

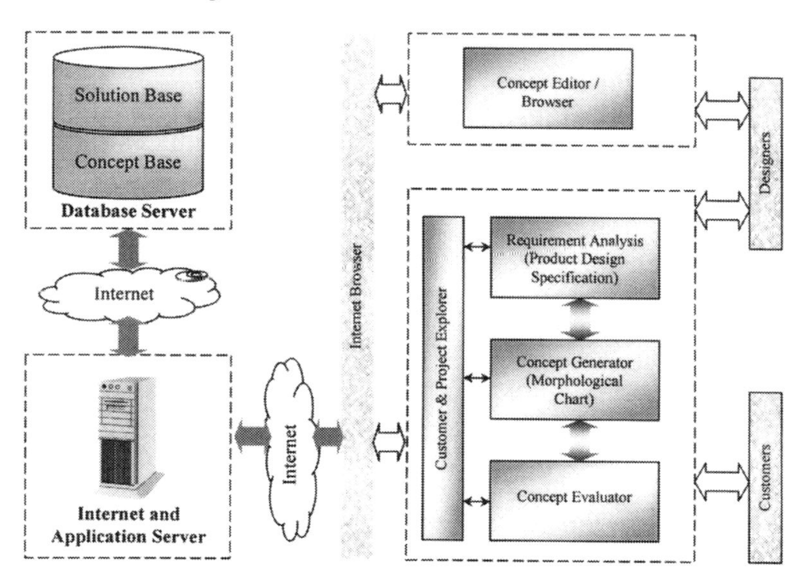

Figure 1 Web-based architecture of Collaborative Product Definition

3 *ProDefine*: A SYNCHRONOUS WEB APPLICATION FOR PRODUCT DEFINITION

Early product definition or conceptual product development is mainly concerned with establishing an appropriate description of the technology, working principles, and form of the product. It is generally accepted that three main stages are involved in early product definition. They are customer requirements or product design specification stage, concept generation stage and concept evaluation/selection stage. Requirement Analysis is responsible for defining the objectives or goals that the product under development aims to achieve. Concept generation is responsible to select different combinations of solutions based on the objectives in the Requirement Analysis. Concept evaluation is concerned with evaluating and comparing the solutions defined in concept generation. The outcome of this early product definition phase is a preliminary layout design reflecting the main working principles and features of the product. During this process, substantial amount of information is needed and generated.

Accordingly, the ProDefine system offers four main front-end components (or application clients in computational terms) and two main back-end databases, as shown in Figure 1. Main components of ProDefine include:

- The Solution and Concept Base – the solution and concept base is a repository for defining solutions and concepts.
- The Concept Editor/Browser – a tool for managing the Concept Base
- Customer & Project Explorer – the first module that allows designers to manage and organise different customers information and their respective projects in the database.
- Requirement Analysis Explorer – mainly concerned with establishing functional and other requirements of the product under development.
- Concept Generation Explorer – deals with identifying potential solutions that are able to meet the functional requirements established at the previous stage.
- Concept Evaluation Explorer – carries out comparative analysis of solutions and against different criteria.

These components are deployed and configured according to typical 3-tiered architecture of web and Internet applications (Jerke *et al*, 1998). The data source tier provides database management facilities for the concept and solution databases. The client tier includes a set of application clients of Project Explorer, Concept Browser/Editor, Concept Generation Explorer, and Concept Evaluation Explorer. Between the data source tier and the client tier is the middle tier. There are two types of servers in this middle tier. One is the web server from which application clients are automatically downloaded to the local client machines upon access and connection. The other is the application server shared by all the application clients. Its main function is to bridge the database source and the application clients. In addition, it maintains synchronisation and conflict resolution facilities. In the experiment, the application server and web server were deployed on the same computer, and the database server on a separate computer.

3.1 Solution Base and Concept Base

The two databases, Concept Base and Solution Base, are the most critical components in the system. The concept base is responsible for storing the concepts in different product domain and providing concepts in the concept generation process. Product domain is described as the collective term of the products. Examples of these are jig, fixture, extrusion machine, or printer etc. The concept base is built up using the Concept Editor/Browser. Once the concept base of a product domain is being built up, the design project of that domain can be initiated. The concept base primarily consists of different 'goals' and 'means' and their relationship in a particular domain of products. 'Goals' are the functional goals of the product domains and 'means' are the feasible ways to achieve goals. The design of the concept base is based on the notion of concepts. The relationship of 'goals' and 'means' are described as concepts. As a result, a concept can be depicted as "Goal is achieved by Means" or "Means achieve Goals". The mapping of 'goals' and 'means' can be one-to-one or one-to-many. A goal can be achieved by a number of means, on the other hand, a mean can also achieve a number of goals. Each mapping of goal and mean is referred to as a concept.

The solution base serves as a central repository for all the defined solutions. It is a database containing all the information of solutions. A solution is a combination of proposed concepts selected by designers or customers. The designers or customers can define different combinations of concepts among all the proposed concepts. The solutions are evaluated in the Concept Evaluation stage and the most feasible or outstanding solution will become the first preliminary layout of the new product design.

3.2 Concept Editor and Concept Browser

The two repositories, concept base and the solution base, works on the back-end of the product definition architecture. Designers and customers will have no direct interaction with these two databases. Instead, designers and customers will deal with the front-end part of the architecture. The Concept Editor/Browser is the front-end interface and is responsible for building up the content of the Concept Base. Experienced designers will use the Concept Editor/Browser to perform operations with the concept base. Different goals and means and their relationship are defined here. Customers and designers can then add, modify and evaluate solutions in the concept generation and concept evaluation stages.

3.3 Requirement Analysis Explorer

Requirement Analysis is the first stage in the product definition phase. It is responsible for defining objectives or goals that the product under development aims to achieve. It is the starting point of the product design and development process and also the product definition phase. In order to initiate a new product design task, the customers should have the general product idea or requirement. Moreover, the designers should have a thorough understanding to the customer and user needs, competitor information, technology risks and opportunities, regulatory or standards environment (Bacon et al, 1994) and all other parameters that affect the new product generation. Having all the necessary information all these affecting factors are used as an input to requirement analysis stage to build up a highly structured functional objective.

3.4 Concept Generation Explorer

The Concept Generation stage deals with identifying potential solutions that are able to meet the functional requirements established at the previous stage. The purpose is to conceive as many solutions to sub-functions and to produce as many feasible solution alternatives to the overall functional goal as possible. It converts the functional requirements as established in the preceding stage into the set of solution alternatives to be forwarded to the next stage for in-depth evaluation. The conversion makes full use of the information defined in the concept base. A morphological generation chart is used to help designers to define solution alternative combinations.

3.2.5 Concept Evaluation Explorer

In the Concept Evaluation stage, the solution alternatives are compared and evaluated against different criteria. The purpose is to narrow down the number of the feasible solution alternatives for further investigation. The basic mechanism used in concept evaluation stage is a morphological evaluation chart. A grade is given to each solution and the most promising solution among all the alternatives are deemed for further investigation. Also, Quality Function Deployment (QFD) is used to verify that the solution fully qualifies all the customer requirements or fulfils all the defined criteria.

4 *ProDefine* STRATEGIES FOR CONFLICT RESOLUTION

ProDefine aims to support collaborative product definition involving multiple members in the project team. Different members have different expertise from different disciplines. As a whole, they have the same project goal. Individually, their decisions naturally converge to their own expertise or disciplines, but not necessarily other members' domains. Ideally, individual decisions mutually support each other to constitute the best overall solution. In reality, decisions contradict, as well as complement, with each other. Conflict resolution, computationally or manually, is very complicated. This section outlines how ProDefine attempts to facilitate conflict resolution through some simple mechanisms.

4.1 Conflict Resolution in Collaborative Product Development

According to Klein, Lu and Baskin (1990), there are two categories of conflict situations. They are competitive conflict situations and co-operative conflict situations. In co-operative situations, different parties are co-operating with a common goal to achieve a globally optimal solution. In collaborative product development applications, share goal for the different expert domains does exist to improve the product design. This type of resolution involves the techniques for finding solutions with mutual benefits.

ProDefine does not provide any specific mechanisms for cooperative or proactive conflict resolution. Instead, conflicts are avoided before solutions are generated. This can be avoided by fully specifying the conditions or constraints upon which solutions are worked out. For example, the cost engineer specifies a constraint on the cost range and the reliability engineer specifies a reliability requirement. The two constraints or requirements are used as the conditions for ProDefine to generate conceptual solutions. The resulting concepts should meet both requirements and are therefore no longer contradictory to each other.

Conversely in competitive situations, different parties are working on their own interests or benefits only. They agree in one way and disagree in another, with each other's individual contributions. Therefore, there can be conflict of interest due to the interaction of the multiple diverse experts. For example, the marketing experts will put on the focus on the popularity and the functions of the developed product while the design experts put their focus on the feasibility of the system. The mechanical designers may have different viewpoint or idea with the electrical designers. Since they are working in a virtual space rather than the same place, this brings the necessary to provide a number of means to prevent or resolve the conflicts.

In order to deal with competitive conflicts, ProDefine provides four facilities, each supported by several strategies. The four facilities are as follows:

- Shared common workspace for synchronised maintenance of data and information integrity.
- Locking Mechanism for maintaining data integrity.
- Decision Fusion Explorer for resolving competitive alternatives in numeric terms.
- Vote Explorer for resolving competitive alternatives in qualitative terms.

4.2 Synchronisation Using Shared Workspace

Before we discuss how to deal with conflict resolution, it is necessary to address the issues of maintaining the data integrity between different users in order to avoid conflicts. While product development involves multiple functional perspectives, maintain a data consistency control mechanism can be a significant challenge on system development (Bentley *et. al.*, 1997; Klein, 1995). Fail to provide consistent information may have a direct impact on the product cost, quality and development time.

The workspace is a set of data objects. Each object represents certain portion of the decision space of the product development project. For example, the Customer Requirement Explorer maintains its own workspace through a data object such as a Treeview ActiveX control or an equivalent relational data table or resultset from an SQL query. Likewise, the Concept Generation Explorer maintains its workspace using the morphological charts or ActiveX grid control. The ProDefine application server also maintains its workspace using the same or similar data objects.

- Basically, each client maintains its own local workspace. Changes are first made to the local workspace and then sent to the ProDefine application server.
- The ProDefine application server updates its own workspace accordingly in response to the client's request.
- The ProDefine application server broadcasts a message to all the current clients requesting them to update their local workspaces.
- The ProDefine clients update their local workspaces accordingly.
- The process repeats itself with further changes in the status of the workspace.

4.3 Locking Mechanism

Locking mechanism is designed to protect the working data of designers. It prevents simultaneous editing of the same set of data by different users. It can also ensure the designers to be more aware of the current status of the product data (Bentley and Appelt, 1997, Sikkel, 1998, Dourish and Bellotti, 1992). Usually when the data is locked, no user can edit the same set of data. But it does not prevent the other designers from accessing the data.

In ProDefine, the workspace is shared among multiple users who may work on the same decisions. One typical problem is that one user is extending a decision while the other is deleting the decision from the workspace. To avoid this happening, locking facilities are provided in ProDefine.

A decision or contribution is automatically locked temporarily if it is under consideration by one user. All the other users are not allowed to change the status of this decision or contribution.

Permanent locking may be used to prevent any actions or events that might change the status of the decision concerned. One possible application of permanent locking is to lock design decisions that have been formally reviewed by the project team. That is, no changes should be made without the team consensus.

There are some general issues to be considered when designing the locking mechanism. First, the levels of locking should be designed carefully. For instance, in document writing system, it is necessary to decide whether the locking mechanism works on document-based, paragraph-based or sentence-based. Besides, it is common for the designers to analyse the product in a table format. Should one cell or the whole table be locked if it is being editing? This decision will depend on the nature of the design system. Second, the locking mechanism can be triggered automatically or manually. In some systems, the designers working context is locked automatically when the edit cursor is put on the system. This may be too 'sensitive' because some designers may use the edit cursor to point out some important figures to their colleagues. Alternatively, the designers need to do some extra task in order to trigger the lock event. Some typical way to achieve this is to have a locking button that is responsible to lock the working context.

4.4 Vote Explorer

ProDefine provides a Vote Explorer or voting mechanism for resolving differences among the team members about a decision. This voting mechanism is only activated if the team chooses a strategy that all or certain types of individual contributions must be voted before they can become the team decisions. The Vote Explorer would not work if the team decides to adopt the "first come first accepted" strategy. This agreement must be accepted or determined at the time of project planning.

When a user makes an individual contribution to the workspace, a record is made in the Vote Explorer. The contribution is marked as temporary decision in the workspace, pending the vote by the team members.

Individual team members may use the Vote Explorer to list all the individual contributions and choose some or all items to vote.

The voting is straightforward. If the member is in favour of the contribution, just click the "For" button to cast the vote. If the member is against the contribution, just click the "Against" button. If the member consider that the contribution does not him or her, then a "Don't care" button should be clicked. If the contribution receives all supports from all the members concerned, it becomes or is accepted as the team's decision and its status changes from "Temporary" to "Permanent" in the workspace. If the contribution is refuted by all the members, then it is removed completely from the workspace. However, these two situations of full support or full disagreement are unlikely in practice. Certain criterion is needed to determine the acceptance or rejection. Some possible simple criteria include "majority voting" and "50% votes". Again, the criterion must be determined by the team during the project planning stage.

Web-based collaborative design system usually involves group decision support. Voting is used in most system to represent group consensus. The product development system should allow group members to vote electronically with the selection of agree, disagree or abstain on the others' work (Hanneghan, Merabti and Colquhoun, 1996). The system is then responsible for collecting the voting results and displaying the outcome. This information together with the reasons behind each decision can be part of the rationale information. It should be retained within the repository for further review.

4.5 Decision Fusion Explorer

Very often, individual contributions made by team members are numerically represented. Take the example of concept evaluation. Each member is invited to evaluate the alternative concepts independent of other members. Their evaluation results are surely different, i.e. conflicts exist. ProDefine provides some methods for calculating numeric consensus.

One method is to calculate the average value from the individual evaluation results as the consensus value. Another method is to accept the maximum or minimum value among the individual evaluation results as the consensus value. There are more sophisticated algorithms for this purpose in the literature Which method should be used depends on the situation and the team's preference. The team must make up the mind during the project planning stage.

5 CONCLUDING DISCUSSION

This paper has presented a prototype system, ProDefine, for early product definition. The work contributes in a number of distinctive ways. First, we have not only further structured the methods involved in concept design, but also computerise them in an easy-to-use fashion. Without computer implementation, paper-based versions of these methods are extremely time-consuming to use. For example, with computerised systems, new items can be added and obsolete items can be removed at any time and the system automatically reconstructs the various charts used for product definition. This is impossible or extremely difficult in paper-based approach.

Moreover, the most important contribution is the use of web-based Internet/Intranet technology in computerising the system. Installation and maintenance are no longer necessary on the client side. Any changes made to the system will only be done on the server. As long as the users use the browser, he or she can have instant to the newest version of the system available on the Internet/Intranet.

Finally, the client-server architecture is a metaphor appropriate for collaborative product development, especially when the tools or teams are geographically distributed. In addition, ProDefine address the time dimension of the distribution through synchronisation features. Geographically dispersed design team is able to share the common workspace to maintain the consistency and integrity of design data and information. More importantly, they can share the latest design contributions made by different team members instantly. Strategies for resolving conflicts between team members are introduced in ProDefine. Individual contributions become team decisions only if they are agreed by the team members.

REFERENCES

1. Bacon G. *et al*, (1994), Managing Product Definition in High-Technology Industries: A Pilot Study, *California Management Review*, Spring, pp32-56
2. Chakrabarti, A., Bligh, T.P. (1996), An approach to functional synthesis of mechanical design concepts: Theory, applications, and emerging research issues, *Artificial Intelligence for Engineering Design, Analysis and Manufacturing*, Vol. 10, pp313-331
3. Cross, N. (1994), *Engineering Design Methods: Strategies for Product Design*, 2nd Ed
4. Erkes, J.W., Kenny, K.B., Lewis, J.W. (1996), "Implementing Shared Manufacturing Services on the World Wide Web", *Communications ACM*, Vol. 39, No. 2, pp 34-45
5. French, M.J. (1985), *Conceptual Design for Engineers*, London, Design Council
6. Gellersen and Gaedke (1999)
7. Hague, M.J., Bendiab, A. (1998), Tool for the Management of Concurrent Conceptual Engineering Design, *Concurrent Engineering: Research and Applications*, Vol. 6, No. 2, June, pp111-129
8. Huang, G.Q., Mak, K.L. (1999), "Web-Integrated Manufacturing: Recent Developments and Emerging Issues", IJ CIM on Web Integrated
9. Huang, G.Q., Mak, K.L. (1999), Web-Based Collaborative Conceptual Design, *Journal of Engineering Design*, Vol. 10, No. 2, pp183-194
10. Jerke, N., Szabo, G., Jung, D., Kiely, D. (1998), *Visual Basic 6 Client/Server How-To*, SAMS Publishing
11. Kim, N., Kim, C.Y., Kim, Y., Kang, S.H., O'Grady, P. (1996), "Collaborative Design using World Wide Web", *http://iil.ecn.uiowa.edu/Internetlab/Techrep/HTML/TR9702.htm*
12. Lewis, T. (1999), *VB COM*, Wrox Press Ltd.
13. Mills, A. (1998), *Collaborative Engineering and the Internet*, Society of Manufacturing Engineers
14. Pahl, G., Beitz, W. (1984), *Engineering Design*, London, Design Council
15. Stauffer, L.A., L.J. Morris (1991), The Product Realization Process: New Opportunities and Challenges, *Journal of Applied Manufacturing Systems*, Winter, pp13-19
16. Trevor, j. and Koch, T. (1997), "MetaWeb: bringing synchronous groupware to the WWW", Hughes, J. A., Prinz, W., Rodden, T. and Schmidt, K. (Eds), *ECSCW 97*, Kluwer Academic Publishers, Lancaster, pp65-80
17. Tseng, M.M., J. Jiao (1996), A Variant Approach to Product Definition by Recognizing Functional Requirement Patterns, *Proceedings of 1996 ICC&IC*, pp629-633
18. Tseng, M.M., Jiao, J. (1998), Computer-Aided Requirement Management for Product Definition: A Methodology and Implementation, *Concurrent Engineering: Research and Applications*, Vol. 6, No. 2, pp145-160
19. Tuikka, T., Salmela, M. (1998), Facilitating designer-customer communication in the World Wide Web, *Internet Research: Electronic Networking Applications and Policy*, Vol. 8, No. 5, pp442-451

CAPE/073/2000

Reverse engineering automotive, petrochemical and nuclear plants into three-dimensional viewers and CAD models

ABSTRACT

3 Dimensional (3-D) CAD is now widely used in many industrial applications for new design and also modification of existing design. It represents information in a format that can be visualised easily and which can be linked seamlessly to 3-D Simulation and Plant Information Management software packages.

The Automotive, Nuclear and Petrochemical engineering industries each use 3-D CAD to a significant degree. In the Automotive Industry many companies are producing new models on a 3 year cycle as new car sales are closely linked to the launches of new models. Increasingly this is resulting in the design and installation of new car lines within existing facilities. Where this is the case there is a paramount requirement for accurate information concerning the car plant facility. This needs to be used easily within the 3-D simulation package of the customers choice.

In a regulatory environment such as the Nuclear Industry the accuracy of plant data is especially important from a safety engineering perspective. It therefore needs to be closely monitored and updated regularly. Often original information takes the form of drawings which require updating and then conversion to 3-D CAD to be of maximum use for engineering project work. The conversion of drawing information to 3-D CAD is often a painfully time consuming process and one which is limited in its effect if the drawings are old and no longer truly reflect reality.

Engineering within the Petrochemical industry is often done on a large scale, both physically and financially. This means that the maxim "Time is money" is very true. Therefore careful engineering planning is done to ensure that the physical fitting of vessels and pipes into their plant environment both onshore and offshore goes smoothly without delay or incident. This engineering planning is often done using tools such as 3-D Plant Information Management and CAD packages. These software packages offer powerful capability to the user but are constrained by the accuracy and accessibility of the data input to their geometrical and information databases.

All three of these industries have a real need for a cost effective system which can capture 3-D data quickly and reproduce this data accurately into 3-D CAD or a similar 3-D viewable format. UK Robotics has recently introduced a laser scanning system called "Light Form Modeller" in conjunction with its partner Zoller & Frohlich in Germany. This system can capture millions of data points in literally seconds. UK Robotics have worked on the task of converting 3-D point data into 3-D CAD models since 1991. Conversion from 3-D point data into CAD objects is achieved by the application of powerful analysis algorithms which have been developed to facilitate swift points to primitives translation. They now provide this technology as a service to customers in the Automotive, Nuclear and Petrochemical industries.

This paper describes how the "Light Form Modeller" laser scanner and software works and then moves on to detail projects in these industry sectors which have benefited from this approach. It concludes with a discussion as to the future technical direction and commercial applicability of this technology in new market sectors.

1. BACKGROUND TO TECHNOLOGY

The "Light Form Modeller" (LFM) system comprises a 360 degree laser scanner and software which converts the collected 3-D data points into a 3-D visual database and 3-D CAD models. This system is used by UK Robotics to provide a 3-D laser scanning and modelling service to its clients in a number of industrial markets. The technology is UK Robotics own, in combination with its partners Z & F in Germany and Quantapoint in the U.S.

Data capture

The laser scanner (shown below) captures over 11 million points within a 2 minute scanning cycle.

Figure 1 - The LFM laser scanner

The complete system comprises the following. The laser scanning head uses Amplitude Modulated Continuous Wave (AMCW) technology(**3**). This is distinct from triangulation and direct time of flight techniques. The laser can pulse at up to 625,000 times per second but 125,000 pulses per second is found to be perfectly adequate in practice. The laser beam is directed at the environment via a nodding and rotating high grade optical mirror which rotates at 900 rpm providing a 360 degree field of view and nods through 70 degrees. The scanner is normally set to operate at a focal range of 12.5m for process and automotive applications (giving an operational diameter of 25m) but can be adjusted if necessary up to a 55m focal range for architectural and civil engineering applications (110m diameter).

The system operates from 110V or 240V industrial supply or through transformer or portable generator. The scanning head weighs about 20 kg and occupies a volume of about one foot cubed. This size and weight is compact for a unit of this type but further value engineering is being continued to make the unit smaller and widen the potential market for its use. The scanning head is connected to a PC interface via an umbilical cable which can extend up to 30m typically. Data can then reviewed on an industrial laptop (shown in picture) at the completion of each 2 minute scan. The scanning head is normally deployed on a pan and tilt platform mounted upon a tripod and dolly.

Data referencing

Rapid referencing of the data from each scan is effected through the LFM software which uses algorithms to interpret and compare large numbers of data points almost instantaneously. This process will be illustrated in the following industrial examples. Data capture at site is orders of magnitude faster than other surveying techniques. The quality of the survey is also improved significantly by the facility to immediately review the scanned data on screen at site. This is frequently found to be a difficulty with other survey techniques because the data can only be checked once the surveyors have returned to the office. This means that if any data is missing then a return trip to site is required. This is not always expedient if the site is far away or has resumed normal operation after a short shutdown (automotive industry).

Customer Deliverables

The customer can receive a range of deliverables dependent upon their engineering objectives:

1) A 3-D visual database of the scanned information showing the laser scanned information referenced together. This provides the customer with the facility to make many measurements of the plant from their computer screen in the process office many miles remote from the site of the work. This can be achieved within days of the site scanning, thereby saving significant project time by supplying accurate plant information much earlier in the revamp design cycle than has been hitherto possible. UK Robotics supply an "LFM viewer" software package to facilitate this.

2) Comparison of existing CAD models in the plant co-ordinate system with the scanned data. This saves process time if there are existing models of the plant which the customer wishes to test how up to date and valid they are.

3) Provision of CAD models into the customer CAD system. Output from LFM can be made into AutoCAD, Microstation, PDS and PDMS amongst others for the Process industry and CATIA, IDEAS and RobCAD amongst others for the automotive industry.

Accuracy
The scanned accuracy is dependent on the distance which the scanner is away from the environment to be scanned. There are two components of the accuracy; the linear accuracy of the laser and the angular accuracy of the pointing mechanism. The laser itself has a curved accuracy profile which improves relatively with distance whilst the angular accuracy is linear as one would expect being a component of the tangent.

Typically for process and automotive plant we would scan about 5-10m away from most plant detail but would go closer if a particular area requires significant detail e.g pipe intersections, welding guns. The resolution of the scanner is about 0.2mm at 10m. This means that the delivered CAD model can typically be specified up to an accuracy of about 3mm and that scanned information in a visual database can be more accurate than this. For process plant this meets the requirements of customers who require the visual database and CAD model data for full engineering review and Clash Detection work. Tie-in information requiring close millimetre accuracy can still be supported by theodolite references made by the customer in the normal manner.

2. AUTOMOTIVE INDUSTRY - EXAMPLE OF VOLVO

Background
At the time of this project Volvo wanted to know whether it would be possible to produce a variant of its S40&V40 model at their shared plant with Mitsubishi in Born, Holland. The Born Plant produced 280,000 cars per year (140,000 for Volvo Cars making a third of the total Volvo car output) and all the intended changes were to be made piecemeal while the factory continued to be fully operational **(1)**.

Volvo only had a two week shutdown in August in which to capture detailed 3-D geometrical information required for the decision making analysis. Data capture therefore had to be very quick. Traditional methodologies including standard EDM surveying and photogrammetry were simply not feasible owing to the limited amount of time available.

The approach taken by the Virtual Manufacturing Centre at Volvo Cars in Sweden was to prepare the launch of the replacement for the S40 and V40 by developing both the product and the production process simultaneously. This was done by cross-functional teams, where product designers and manufacturing engineers worked side by side to analyse the process using simulation throughout the project and do the final verification in the physical world shortly before the start of production. This approach shortens the lead-time for the development of a new car and is a very important step forward as the car market experiences increasing global consolidation and increasing need for regular new car model introduction.

To effect the concurrent process and product design Volvo therefore required detailed and accurate information for the car production line in an engineering format which they could use and communicate easily with internal engineers and engineering suppliers. Their preferred method to do this was within the CATIA CAD package and RobCAD simulation package which they used at a corporate level.

Figure 2 - Aerial view of the Born Factory

Therefore they required a rapid data capture process which provided data rapidly and cost effectively in a format compatible with their existing Cad packages. This led Volvo to consider using the LFM system to laser scan the car plant and reverse engineer the information into Cad format.

Project Detail

UK Robotics Scandinavian partners ATS understood Volvo's specific needs. Together they scanned the body and assembly shop, a total of 130 production areas, of which, over 50% were complex robot cells. About 120 robots were situated in the robot cells and their grippers and welding guns were scanned in detail.

The car plant scanned covered about 15 times the geographical size of a normal football pitch. Some 700 scans were completed within the two week period using two scanning teams comprising two people each working 10 hour shifts per day. The data collected exceeded 40Gbytes and filled over 70 CDs. The downloading of data onto JAZ drive and CD was completed each day in conjunction with data review to ensure that the data collected was correct and of suitable quality for modelling. The LFM software provided the facility for instantaneous data review. Although the laser scanner is calibrated at regular intervals within a controlled 360 degree calibration facility daily checks were also made to ensure that the scanner lay within calibration. This was achieved using a combination of simple quick surveying techniques and referencing of adjoining scans within the LFM software. Any problems experienced in registering the scans together would point towards a calibration problem and hence collection of unusable data could be avoided.

Figure 3 - Laser scan of robot cell

Although laser scanning using LFM is performed very rapidly through a 90 second scan cycle time the project still required careful planning and project management. ATS liaised carefully with Volvo to ensure that there were no superfluous scans taken and that specific areas of particular interest were clearly identified, which saved time. Each area was rated according to the customers engineering requirements and the number and locations of the scans were planned accordingly.

Use of laser technology is by no means new in the automotive industry but mass data capture of this type very definitely is. This meant that Volvo's needs had to be translated quickly into both the scanning and data processing and modelling stages of the project.

The key issues with regard to the laser scanning process were the following :

- collection of matt black surfaces such as welding gun tips
- collection of bright areas such as paint shop areas- ensuring the laser capture range agreed with accuracy specifications
- capturing sufficient detail to allow for occlusions and shadows
- speed of ingress to areas and portability and speed of scanner

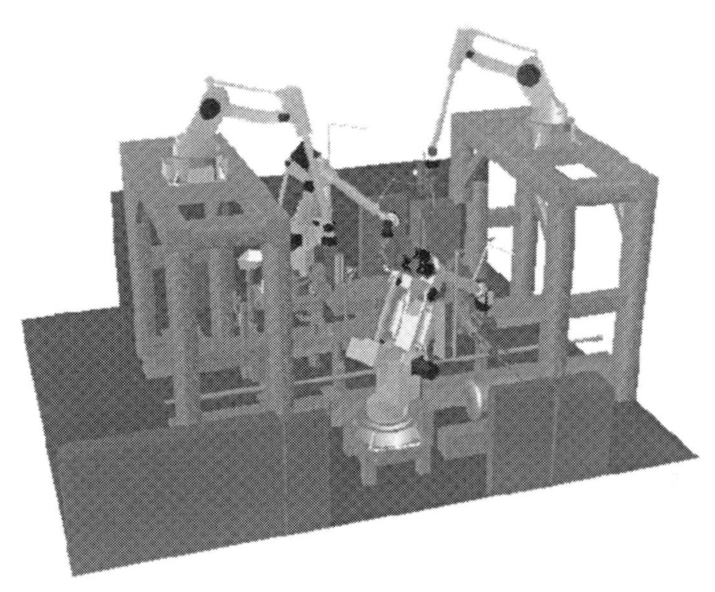

Figure 4 - Simulation of the robot cell

Expanding on these points briefly, laser scanning technology has evolved to a position where the range of subject matter which some laser systems can collect has increased dramatically to include almost all surfaces save extremely matt black and extremely bright surfaces. This is an important consideration in car plants because welding gun tips are usually very matt black but also of great interest to the automotive engineer. Likewise the case for paint shop areas where bright lighting conditions exist. The problem with matt black surfaces is that if there is no returned laser signal then measurement of the engineering data is not possible. We were aware of this situation prior to this project and tested some matt black surfaces to test measurement range and also developed other simple techniques to amplify the returned laser signal. This is an important point because the laser proximity to weld guns needs to be greater than to other automotive plant. The LFM system has a very high detectable laser light range and so this problem did not emerge during the project. However it remains an important general consideration for the employ of laser based data capture in the automotive environment.

Similarly every commercial laser scanning system has a defined accuracy capability which will vary linearly or non-linearly with distance. This accuracy will be a function of the ranging device and whatever pointing device is employed. It is therefore important to plan the data capture carefully to ensure that the data collected is accurate to the customers specifications. Over capture of data is sensible because you can then always throw more data away after the scanning operation but you cannot create data when you do not have it in the first place. This is particularly true of automotive plant environments where "time is literally money" and second chances to gain data are not usually granted.

Based on the scan data, over 50 detailed 3-D models describing complex production cells were delivered to Volvo. Several different point clouds have been added to a common co-ordinate system, and these served as the basis for data reduction to 3-D solid models. The final CATIA models form a 3-D CAD replication of parts of the Born Plant. This enables Volvo to perform their concurrent engineering process as previously described.

3. PETROCHEMICAL INDUSTRY - CHEVRON AND FLUOR DANIEL

Background

The Tengiz oilfield in West Kazakhstan has produced oil since 1991 **(2.)**. It is one of the largest oilfields in the world and is the largest discovered in the last 25 years. It has 24 billion barrels of oil with 6-9 billion barrels of this estimated to be recoverable. The Tengizchevroil (TCO) partnership, a Joint Venture of Chevron, Kazakhoil, Mobil and LukArco, has been operating the field since 1993. Since acquiring the facility, TCO has undertaken several capital investment programmes and expansions. This has increased crude oil production from 24,000 barrels per day (BPD) in 1993 to 220,000 BPD in 1999.

"Program 12" is a current project which is intended to raise production to 280,000 BPD. A key part of this project is to expand and upgrade the existing LPG and gas processing facilities at the plant.

The Republic of Kazakhstan, shown in Figure 5, is a large country which is about the same size as Western Europe. It is borders China, Russia, Kyrgyzstan, Turkmenistan, and Uzbekistan. The climate is arid with cold winters and hot summers. The country contains a lot of mineral wealth, including large oil reserves. Tengiz is located on the Northeast shore of the Caspian sea and experiences large very hot summers (44 degrees C) and very cold winters (-40 degrees C). The progression from the summer to the winter is also extremely rapid and can literally happen within a couple of weeks. This coupled with periodic sandstorms and a particularly remote location makes Tengiz a particularly tough practical test for the practicality and potential oflaser scanning.

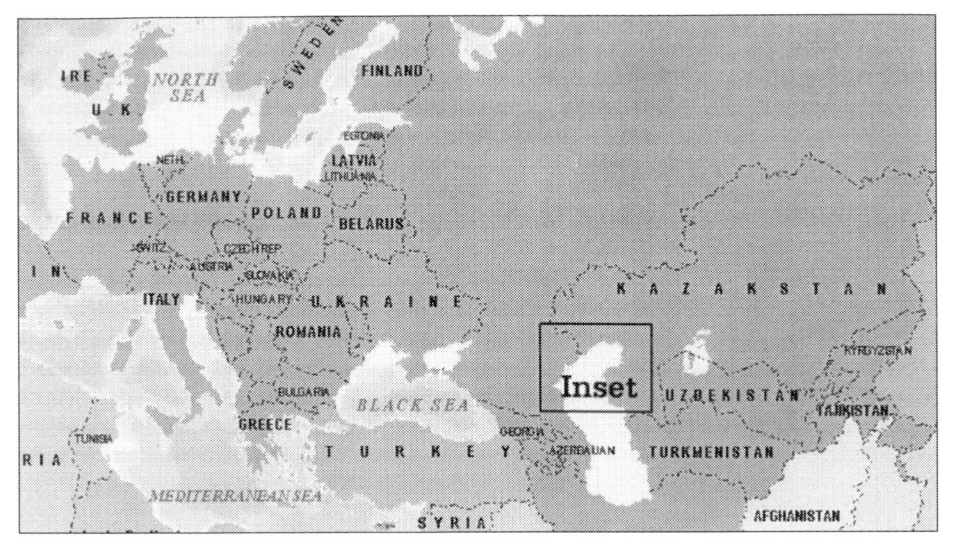

Figure 5 – Map of Europe and Kazakhstan

Project Detail

The overall project at Tengiz is considerable in size involving many process engineering companies working in partnership with TCO. The Tengiz plant is a "Brownfield" site in that the project is a revamp rather than a brand new build. In such a "Brownfield" project it is critical for the design engineers to have ready access to accurate data for the existing plant. Without this, retrofits and plant modifications cannot be efficiently executed. Although this need has remained constant over the years, what has varied is the capability of technology and industry to acquire the data. This capability has grown from simple use by field based designers of rulers and sketchpads to the application of more advanced technologies such as photogrammetry and laser scanning.

Fluor Daniel have significant experience in the application of both photogrammetry and laser scanning and the Fluor Daniel project management evaluated available survey techniques in order to select the most "fit for purpose" solution for the project. Key factors **(2)** in their evaluation included the fact that they had no existing drawings and that the financial cost and physical risk associated with placing engineers into the field were unacceptably high.The sheer speed of the project meant that manual surveying methods were simply not feasible and the extreme weather conditions meant that the surveying duration needed to be minimized as much as possible.

Photogrammetry is a technique which Fluor Daniel were very familiar with and had a high level of expertise with. However Fluor engineers calculated that in excess of 7000 photographs would be required which would take considerable time to gather and then post process to 3-D data.

Laser scanning techniques have now evolved to a point where they are economically viable for data capture on process plants provided that the data capture is sufficiently rapid for the task concerned. Fluor Daniel evaluated the LFM system, being attracted by the fact that 11 million data points could be captured in a 90 second scan, thus overcoming many of the time and cost issues associated with Plant data capture.

The process plant areas where accurate information was required were large. Data was required in five areas in total, the largest being some 500m by 250m, with the smallest being still 200m by 200m. Survey data was required up to an elevation of 15-20m, therefore elevated data capture was very important. Data was collected at these elevated sections either by using close by elevated vantage points or through use of a cherry picker. The use of a laser scanner set to a larger focal range does not obviate the need for an elevated vantage point because scanning from ground level still causes shadow areas to exist and occlusion of data is a real problem. Also as mentioned previously the further you are away from the area of interest the lower the positional accuracy of the data collected becomes. In total some 900 scans were taken altogether and some 150 scans were taken from a cherry picker. Of these 150 scans one scan was affected by wind and had to be repeated.

Given the challenges of the location, it was crucial to be able to view and analyze the collected data at site immediately after data collection. The modeling software enabled this to be carried out on completion of each 90 second scan and highlighted any deficiencies. In this way the scan affected by wind was immediately noticed and a repeat scan made which meant no return to site or a permitted area proved to be necessary. It would clearly be unacceptable to realise a problem existed with the captured data after the surveyors had returned to the UK.

The project would have taken a significant amount of time and been very labour intensive if either manual data capture or photogrammetry had been used. Using laser scanning, and in close liaison with the client's team in Tengiz, the data was captured within 5 weeks. A total of 80 Gigabytes of data were recorded. This clearly offered significant time and cost benefits to the project.

In addition the data captured was downloaded to CD and viewed back in the UK offices much more quickly than would have been possible using more standard technology. This was extremely important to the project because early engineering decisions could be taken. The project personnel at Fluor Daniel chose to use the 3-D visual database provided. This is a referenced and registered collection of the scans in an easily viewable and usable format. The project engineers took the view that because time was so important they could maximise the benefits of the rapid laser data capture provided by LFM through making measurements using the "3-D photobrowser" or 3-D visual database almost within days of the scanning being completed in Kazakhstan. The facility to ftp the information was also considered carefully and although proving extremely attractive from the project perspective the sheer amount of data would have used up all available bandwidth for this task alone. However as bandwidth technology obviously advances this will no doubt prove to be viable shortly.

A 3-D CAD model is not always required for initial decision making. The use of early 3-D data in an easily viewable format provided significant benefit and enabled modeling to be executed as a parallel function. This approach significantly increased flexibility in project execution and also permitted the project to request future CAD models as and when required thus avoiding costly abortive work.

4. NUCLEAR INDUSTRY - EXAMPLE OF BNFL

Nuclear applications need accurate documentation and up to date Plant CAD drawings. The pressure to achieve this is maintained both internally through top down management focus and externally via government appointed Inspectorate bodies. As nuclear plants get older there is an increasing need to ensure that records and drawings are accurate. This need is driven by refurbishment and decommissioning activities in addition to stringent regulatory requirements.

Two factors are of paramount importance when organising data collection and surveying of a nuclear plant installation. These are

1) The minimisation of dose uptake by all people associated with the survey
2) Reducing the survey time on site to a minimum and avoiding the need to revisit the site to capture information that was missed on the first occasion

Building B268 at Sellafield was built in the 1960s to do several operations. Our customer at Sellafield is primarily concerned with the Thermal Denitrification Plant. The customer required accurate CAD models for two main reasons

- To do refurbishment work, identifying lines to break to replace certain vessels
- Enable the identification of process lines which may require replacement

Because of the age of the plant original drawings and isometrics were extremely limited. Access time per day to the plant was a big concern. This varied according to the floor and location within the building. At its worst man access was limited to 10 minutes per day. Clearly traditional surveying techniques would result in significant dose uptake and would not demonstrate ALARP in most parts of the plant due to the limited access time.

The scanning of B268 was planned carefully. The primary issues considered were:

- lack of illumination present on certain floors
- limited time available due to dose uptake
- the ongoing processing requirements of the customer

Lack of illumination

The laser scanner was ideal for this application because it would operate in a wide range of light conditions ranging from total darkness to bright internal light conditions. The scarcity of light in B268 was therefore not a problem.

Time availability

This was a very important issue. Although a single scan could be completed within 2 minutes it was important to establish with the customer the detail required in the model. This in turn enabled us to plan the rough location of the tripod for each scan. Importantly, the LFM software registers or correlates information from different viewpoints so that various views can be “sewn” together accurately. This will be further discussed below. This is crucial in rapid scanning of a process plant both in terms of collecting all the data and also minimising the post data capture processing as the CAD model is generated. The scanning routine was therefore planned in advance and minimised the number of entries to B268 because of the dose access times.

Ongoing processing requirements

About 16 scans per floor were required to capture all the detail in B268. The level of detail generated included all pipes, vessels, tie-ins and valves in addition to the civil engineering information such as floors, beams, ceilings and walls.

Modelling

The customers specific requirements were as follows:

- A CAD model in AutoCAD format which could then be manipulated in AutoPLANT
- A "layered" model
- The facility to view all 5 floors of the building in continuation

Data collected from site was backed up on hard drive and CD. Each scan represented 75MB of information. The total information collected for the five floors of B268 represented 6 GBytes of information.

Figure 6 - Screen shot of LFM software in action

Modelling of the data was achieved using the LFM software. The software compares the data points collected in 3 dimensions with the geometrical description of recognised process industry equipment and shapes. This comparison is extremely rapid and powerful. For instance a single cylinder could represent over 10,000 data points. The user can set convergence limits within the software to control the conversion of data points into the 3-D model. This provides a "human governance" over the automatic process.

The way that the LFM software performs this task is that it uses registered points to compare certain points in space between each scan. These "registration" points may be naturally occurring e.g. flanges, beam supports etc or they may be artificial if the environment is extremely bland e.g. a large blank wall or floor. Where artificial points are required they can range from small spherical objects to typical surveying reference tools. The important issue is that the software assesses each individual point within the registration object in x,y,z space co-ordinates and then compares these with the x,y,z co-ordinates of a similar object in another scan which is suspected to be the same object. Typically three reference points would be used in any one scan to ensure accurate registration or overlay.

Clearly the advance of computer processing technology has enabled this technology to be viable and will further increase its advantage and cost effectiveness in years to come.

The LFM output to the customer was provided in the ACIS sat. format. This communicates directly with AutoCAD. Additional software provided by UK Robotics then enabled the customer to "layer" their plant. This meant that they could label vessels and pipes in an organised manner which enabled associated information such as connecting pipes, their sizes and materials to be clearly catalogued within the CAD model.

5. WHERE NEXT?

The three examples of industrial applications given in this paper comprise of applications where the forerunners of laser scanning have been used eg. Petrochemical, and those where traditional surveying methods have proved too slow eg. Automotive.

As the speed of data capture straight into a digital format accelerates, and the post processing of data improves, so the breadth of applications will multiply. What will dictate the direction of the technology development will be the requirements of the various industrial sectors. For example there are applications such as police and forensics where accurate and fast data capture is an imperative. However the technology has to be priced such that you can place one unit in every police vehicle in the country. It also has to be so straightforward to operate flawlessly that almost anyone can use one eg. Like a normal instamatic camera. In this example it is likely that the users will compromise on accuracy and move towards use of lower quality and lower priced data capture hardware supported by software downloaded or bought from the internet. This choice will be driven by cost and the need to have a common system widely available from all vehicles.

However in examples such as the petrochemical example it is clear that data capture needs are project based and accuracy is paramount because the cost of an inaccurate decision is very high, both in money and time. It is a similar case for automotive and nuclear work.

The interesting area is the civil engineering market. In this market 3-D CAD has been restricted in uptake due to the nature of the market eg. Small architect firms with limited funds and a liking for working with 2-D paper. As 3-D data capture becomes quicker and cheaper this will threaten the traditional 2-D world and is likely to cause segmentation in this market. There will be some work, probably a very significant part, which will only require total station surveying and 2-D drawings. However there will also be other areas where the ability to capture total 3-D data will cause the market to change, possibly influencing design methods, process control and the use of virtual reality tools.

6. REFERENCES

1) Kolk R, " Scan-do attitude PAYS OFF", Automotive Engineer May 2000, pp 51-55.

2) Connell, C; Ormiston B (Tengizchevroil) and Amott, N; Cullum,C (Fluor Daniel), " Challenges of a Complex Gas Plant Revamp in Kazakhstan", GPA Conference, Atlanta March 2000.

3) Dalton G, "Reverse engineering using laser metrology", Sensor Review Volume 18, Number 2, 1998 pp 92-96.

Virtual Reality Applications in Manufacturing

CAPE/034/2000

Intelligent manipulator work system using tracking vision hardware

N KAWARAZAKI, R KAJI, and **K NISHIHARA**
Department of System Design Engineering, Kanagawa Institute of Technology, Japan

Abstract

This paper provides an intelligent manipulator work system using tracking vision hardware. In our system, the hand gesture of human is recognized and the manipulator works based on it. We propose the new method called the block dividing method in order to detect and recognize the hand gesture rapidly. This method is that the template of the hand is divided into three blocks: base block, hand block and finger block. We can reduce the memory of templates in comparison with ordinary method. The effectiveness of our system is clarified by several experimental results.

1 Introduction

In the near future, robots are expected to cooperate with human in the same working space. The human friendly interface is necessary for the manipulator work system. A lot of research works concerning the interaction of the human and the robot has been proposed in recent years. The control algorithm for a service robot through the hand –over task has been proposed(1). A new user interface called "Active Interface" to interact with human beings has been presented(2). The use of human action (gesture) is useful for human-robot interaction(3). The method for recognizing head/hand gesture from a sequence of range images has been presented. Algorithms for real time visual recognition of human action sequences has been developed(4)(5). The human gesture recognition method using pattern space trajectory is presented(6). The "Interactive Sensing" to let the robot find a human in a complex background has been proposed(7). The new method uses multiple color extraction, stereo tracking and template matching for human pointing action has been developed(8). On the other hand, various types image processor for robot has been proposed(9). The high performance robot vision system has been implemented as a transputer-based vision system (10).

This paper provides an intelligent manipulator work system using tracking vision hardware. When the manipulator cooperates with the human in the same working space, it is necessary for the manipulator work system to recognize instructions of the human. The human can give many kinds of instructions for the manipulator through hand gestures. In our system, the hand gesture is recognized and the manipulator works based on it. We propose the new method called the block dividing method in order to detect and recognize the hand gesture rapidly. The memory of templates is reduced in comparison with ordinary method. This paper is organized as follows. The concept of our manipulator work system is provided in section 2. The detection of the target object is described in section 3, and recognition of the hand gesture is presented in section 4. Several experimental results are discussed in section 5. Conclusions are provided in section 6.

2 Manipulator Work System

Our manipulator work system is shown in Fig. 1. This system is composed of a manipulator, a CCD camera and a PC with a tracking vision hardware. The manipulator used here has six degrees of freedom of motion. Since the manipulator has to recognize the position and posture of the hand in real time, we use the tracking vision hardware with the advanced correlation processor. The tracking vision hardware is very useful for manipulator to recognize the environment. The goal of our system is that the manipulator works with the human in the same working space based on the hand gesture. For example, manipulator picks up an object and hands it over to the human according to the instruction of the hand gesture. Since our system has one camera because of simplicity of the condition, the motion of the manipulator is under restriction in the 3D space.

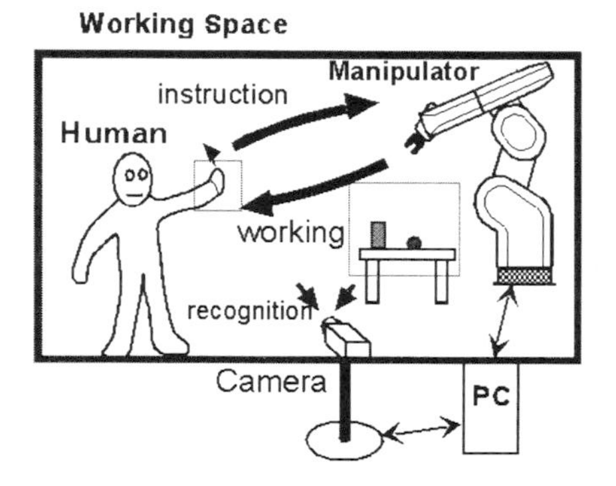

Fig. 1 Manipulator work system

3 Detection of the target object

At first, the system has to detect the target object in the working space. We prepare several templates of the target object in advance. In order to detect the target object in the search area, we use the pattern matching method. The correlation value between the template and the image of the search area is calculated. We consider that the area with the minimum correlation value is the target position. Since this compare process is very time-consuming, we separate this process into two steps: global search and local search. As shown in Fig. 2, the candidate area is detected roughly based on the global search. Moreover the detail position of the object is determined in the candidate area based on the local search. The size of the search window is 64×64 pixels and one of the candidate block is 48×48 pixels.

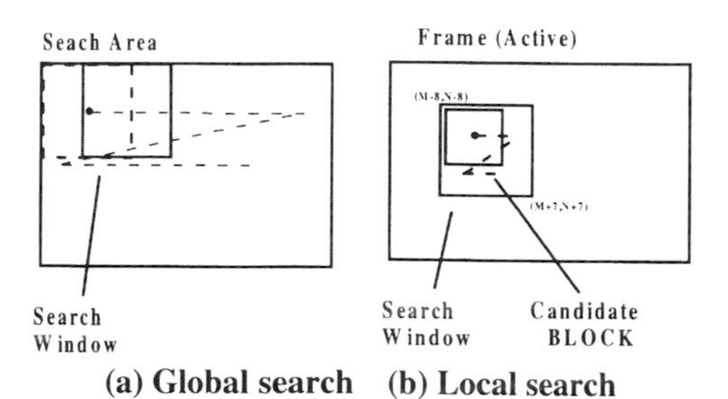

(a) Global search (b) Local search

Fig. 2 Detection of the target object

We define the correlation function *Cor* in order to detect the target object. The correlation function *Cor* is shown below.

$$Cor(u,v) = \frac{Dpp_{all} - Dpp(u,v)}{Dpp_{all}} \qquad (1)$$

where

D_{ppall} : total distortion value

D_{pp} : distortion value

$$Dpp(u,v) = \frac{1}{M \cdot N} \sum_{x=0}^{M-1} \sum_{y=0}^{N-1} |R(x,y) - C(x,y)| \qquad (2)$$

$$C(x,y) = S(x+u, y+v) \qquad (3)$$

$$(-8 \leq u, v \leq 7)$$

where

R: reference block C: candidate block

M,N : size of reference block

S : search window

This correlation function is the distance value between the distortion of one point in the current image and the average of total distortion value. We regard the position of the positive maximum value of this function as the position of the target object. The example of the correlation function is shown in Fig. 3. In our System, the peak position of the correlation function is regarded as the position of the target object.

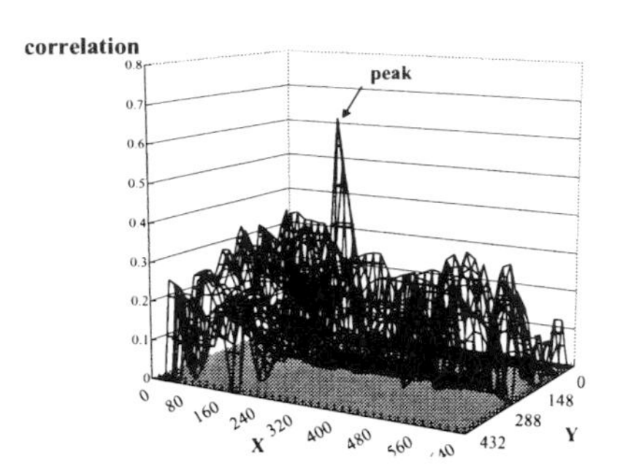

Fig. 3 Correlation function

4 Recognition of the hand gesture

When the shape of the hand is recognized by the pattern matching method, many kinds of templates are necessary. In order to reduce the memory of templates, we propose the block dividing method. This method is that the template of the hand is divided into three blocks: base block, hand block and finger block (Fig. 4).

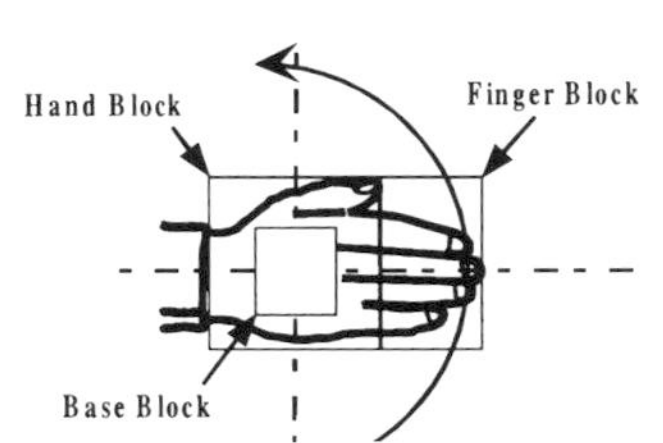

Fig. 4 Block dividing method

The base block is small area in the back of the hand. The area whose color is a flesh tint is regarded as the part of hand. Once the base block is detected, the position of the hand is recognized by the tracking of the base block. Since the image of the base block has the flat texture, there is little influence of the background. We define five templates for the base block.

The posture of the hand is determined based on the pattern matching of the hand block. The position of the finger is estimated based on the determined shape of the hand block (Fig. 5). After the definition of the hand position, the configuration of the finger is estimated based on the pattern matching of the finger block. The hand block with the minimum distortion value is selected in several hand templates. The finger block is selected as the same manner. The total configuration of the hand is estimated by the combination of the hand block and the finger block.

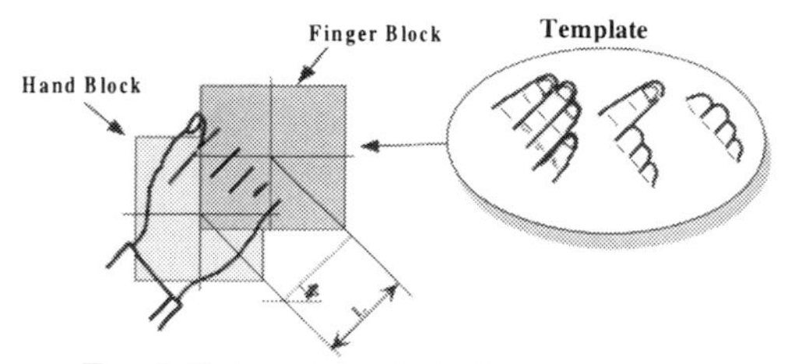

Fig. 5 Estimation of the finger position

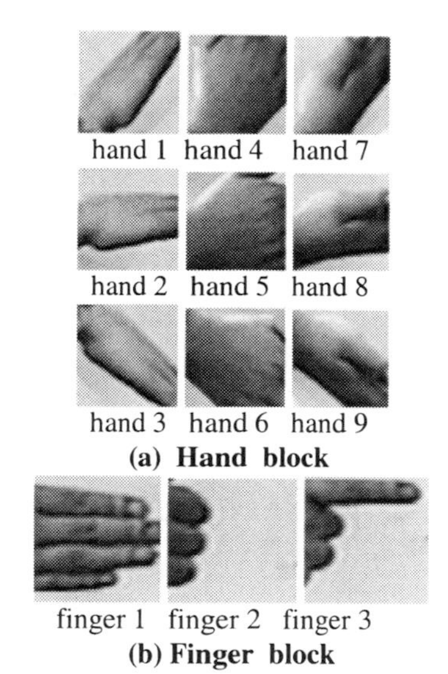

(a) Hand block

(b) Finger block

Fig. 6 Templates

As shown in Fig. 6, we define nine templates for the hand block and three templates for the finger block. Since we can define twenty-seven patterns of hand configuration from twelve templates, fifteen templates of the hand configuration is reduced. The feature of this block dividing method is that the memory of templates is reduced in comparison with the ordinary method.

5 Experimental Results

We made several experiments in order to clarify the effectiveness of our system. As shown in Fig. 7, we define several instructions using hand gestures. We make the manipulator move in accordance with the instruction of hand gesture. For example, when the hand opens and closes three times (Fig.7 (c)), manipulator hands the object over to the human.

(a)Stand by (b)Approach••• (c) Hand on the object

Fig. 7 Instruction patterns

The experimental results are shown in Fig.8. First, the target object is detected in the working space and manipulator grasps it (Fig. 8(a)). After the grasping the object, manipulator waits for the instruction of the hand gesture. As shown in Fig. 8(b), the hand is detected based on the pattern matching of the base block. Fig. 8(c) shows that the manipulator picks the object up and hand on it to the human. Manipulator can move according to the instruction of the hand gesture in real time.

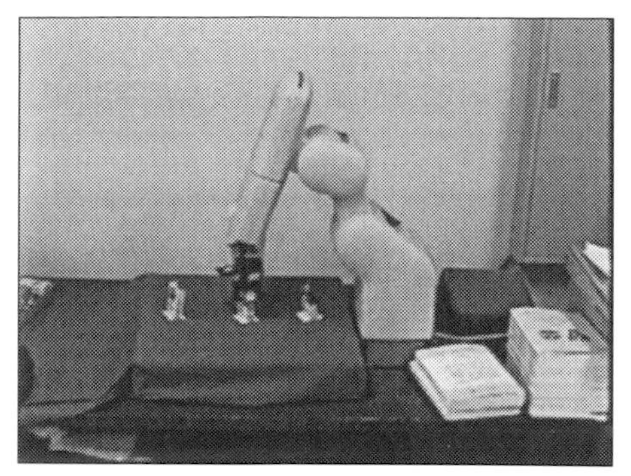

(a) Grasping an object

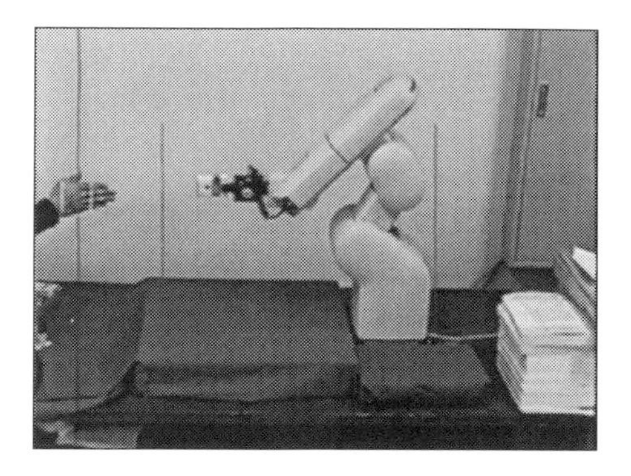

(b) Detection of the hand

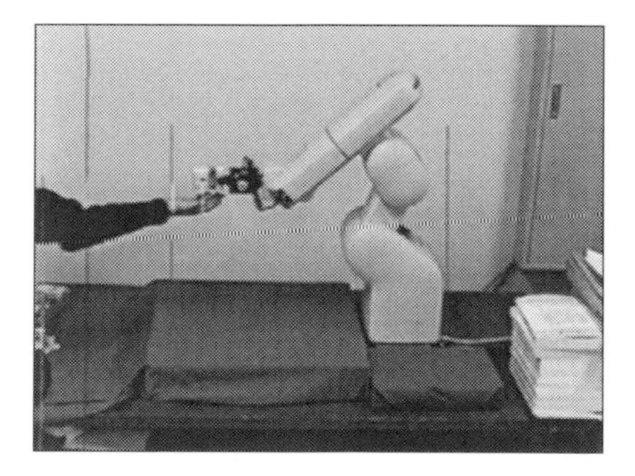

(c) Delivering the object to a human

Fig. 8 Experimental results

6 Conclusions

In this paper, we propose the intelligent manipulator work system using tracking vision hardware. In our system, the hand gesture is recognized and the manipulator works based on it. We propose the new method called block dividing method in order to detect and recognize the hand gesture rapidly. This method is that the template of the hand is divided into three blocks. We can reduce the memory of templates in comparison with ordinary method. The effectiveness of our system is clarified by several experimental results.

Future work of our system is expansion for the 3D working space. When the manipulator moves to whole direction in the 3D working space, it is necessary that the system has two cameras. We have to develop a new algorithm in order to adapt our method for the stereo image.

References

(1) A. Agah and K. Tanie, "Human Interaction with a Service Robot : Mobile-Manipulator Handing Over an Object to a Human", Proc. of the IEEE Int. Conf. on Robotics and Automation, pp.575-580, 1997.

(2) N. Yamasaki and Y. Anzai, "Active Interface for Human-Robot Interaction" : Proc. of the IEEE Int. Conf. on Robotics and Automation, pp.3103-3109, 1995.

(3) V. I. Pavlovic, R. Sharma and T. S. Huang, "Visual Interpretation of Hand Gestures for Human-Computer Interaction : A Review" : IEEE Trans. on Pattern Analysis and Machine Intelligence, vol.19, no.7, pp.677-695, 1997.

(4) Y. Kuniyoshi, M. Inaba and H. Inoue, "Seeing, Understanding and Doing Human Task" : Proc. of the IEEE Int. Conf. on Robotics and Automation, pp.2-9, 1992.

(5) C. R. Wren, A. Azarbayejani, T. Darrell and A. P. Pentland, "Pfinder : Real-Time Tracking of the Human Body" : IEEE Trans. on Pattern Analysis and Machine Intelligence, vol.19, no.7, pp.780-785, 1997.

(6) S. Nagaya, S. Seki and R. Oka, "Pattern Space Trajectory for Gesture Spotting Recognition", Proc of the 3rd Japan-France Congress on Mechtronics, pp.208-211, 1996.

(7) T. Inamura, M. Inaba and H. Inoue, "Finding Human based on the Interactive Sensing" : Proc. of Intelligent Autonomous Systems, pp.86-92, 1998.

(8) T. Mori, T. Yokokawa and T. Sato, "Recognition of Human Pointing Action Based on Color Extraction and Stereo Tracking" : Proc. of Intelligent Autonomous Systems, pp.93-100, 1998.

(9) H. Moribe, M. Nakano, T. Kuno and J. Hasegawa, "Image Preprocessor of Model-Based Vision System for Assembly Robots" : Proc. of the IEEE Int. Conf. on Robotics and Automation, pp.366-371, 1987.

(10)H. Inoue, T. Tachikawa and M. Inaba, "Robot Vision System with a Correlation Chip for Real-time Tracking, Optical Flow and Depth Map Generation" : Proc. of the IEEE Int. Conf. on Robotics and Automation, pp.1621-1626, 1992.

CAPE/098/2000

Implementation of four-dimensional machining simulation in virtual reality over the Internet

N AVGOUSTINOV
Insitute of Production Engineering/CAM, Saarland University, Germany

ABSTRACT

With the rapidly increasing use of Internet and virtual reality (VR) in engineering applications and in education, the need of respective software tools increases too. There is already demand for tools supporting tasks like remote diagnostics, remote consulting and support, co-operative work, sharing of complex multimedia data, etc. On the other hand, there is an obvious need to verify NC-programs before putting them in use, but (especially in the case of small and medium sized enterprises) the respective software tools are not always locally available. Although a couple of programs dealing with this problem are already offered on the market, none of them is well suited for use in a heterogeneous environment and over Internet.

The presented solution pursues a compact and reliable and at the same time exact model of the machining in order to allow efficient simulation and verification over Internet. The main problem is that some of the calculations supporting the simulation are repeatedly performed thousands and even millions of times. This requires clear concepts, optimized algorithms, parsimonious underlying structures and a well chosen compromise between the speed and the accuracy for achieving an acceptable (real-time) simulation over Internet.

This paper discuses how different algorithms and data structures affect the achieved speed and accuracy of simulation. Since the implementation of the proposed solution is based on `HTML`, `VRML` and `Java`, it is by design network-ready and portable without additional efforts. The safety of the method is guaranteed by a 4D machining simulation (i.e. 3D space and time) with a process and result visualization.

1 INTRODUCTION

In 1997 the international standardization organization (ISO) adopted the second version of the *virtual reality modeling language* (`VRML`) as an international standard. At that time the author of the present paper worked on a project for a verification of CLDATA-programs[1] and NC-programs and was looking for a (possibly quick and easy) way of visualizing the tool paths in these programs.

One of the experiments in the project aimed at determining the suitability of `VRML` for visualization of large amounts of 3D data. Although that experiment was performed with the version 1.0 of the language (and the respective support tools), the results were so exciting that they forced us to study in details the new features of `VRML2.0` (later on `VRML97`) and this, in turn, caused a major change in our concepts of CAD-CAM data exchange and verification of NC-programs.

2 CONCEPT

`VRML` allows a compact ant portable representation of 3D data, as well as its transfer over heterogeneous networks and consequent local visualization on the client's computer. It provides means for easy integration of 3D data in `HTML`-documents and, depending on the software environment, also in "normal" office application documents.

`VRML97` (cf. (8)) has in addition extended features like multimedia integration, parameterised objects, event processing, inter-object and over-network communication, animation, behaviour description. Interestingly enough, although almost every modern CAD-system offers an export of 3D models also as `VRML`, most of these exports either support only `VRML1.0` or do not exploit the extended features of `VRML97` and rely only on plain static geometry instead.

The idea is to employ, on the one hand, the available 3D geometry and, on the other hand, the extended functionality of `VRML97` in a *distributed web-based simulation system* for a rapid verification of NC-programs including a visualisation of the expected results. The main objective of this project is to develop software allowing a remote verification of NC-programs by employing only virtual models of the necessary real components: tools, NC-machine tools and materials.

A scenario for the use of the system is illustrated in Figure 1.

Each *request for machining simulation* (RMS) is assessed from the "Task distribution and control" server, and a *resource request* (RR) is broadcasted to all known simulation servers (DWBSS). When free resources are reported back (not represented in the figure!), the RMS is assigned to one or more of the free simulation servers so that a balanced load and minimal response time are achieved.

2.1 Requirements and desired features

The following requirements are posed on the developed system:

2.1.1 Portability

As the most important design rule for this system we consider achieving an independence from any hardware and software, and an ability both to run on low-end computers and to use the advantages of the high-end ones.

[1] CLDATA is the name of a standard for representation of tool positions (abbreviated from cutter location data; standardized in Germany under DIN66215).

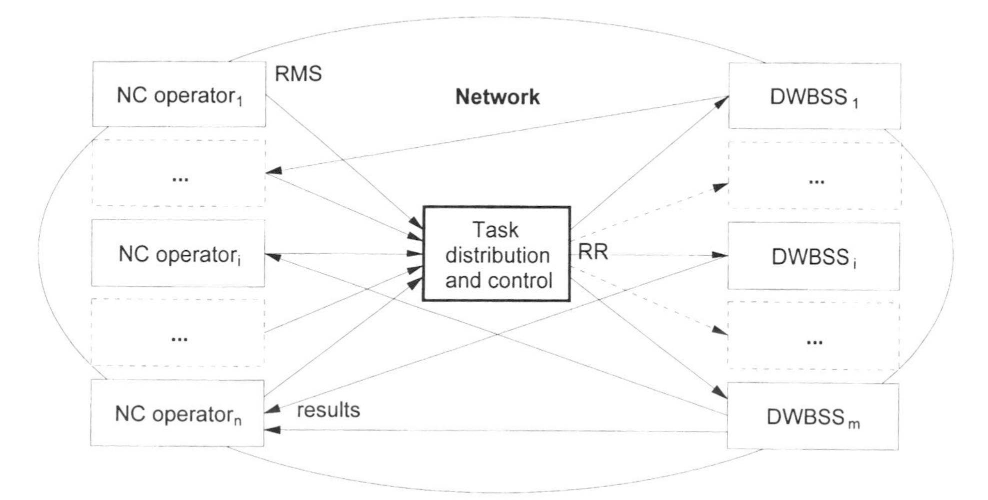

Figure 1: Scenario for use of the system

2.1.2 Simulation in 4D

Since the NC-machining is performed in 4D (3D space and time), the simulation has to support 3D geometry and be time-bound. **VRML97** offers a native support for both these properties which is considered in the concept and employed in the implementation.

2.1.3 Support for objects and features

An object-oriented design of the software is applied (as far as the languages involved allow) and at the same time the modules are developed in a way providing for feature-based simulation of the machining. This approach guarantees a more intuitive behaviour of the models, a simpler user interface of the system and a more natural simulation as a whole.

2.1.4 Ease of use

Another important design rule is the ease of use, which is achieved by means of:

- an intuitive **HTML**-based user interface with seamless integrated "menus", help and working samples;
- including 2D and 3D graphics and other multimedia elements for better expressive power and understandability;
- employing wizards to guide the inexperienced users through the system.

2.1.5 Extendibility

The architecture and the implementation of the separate modules are as open as possible. No modules, classes or others are intentionally prohibited from inheritance and improvement, except in the cases which require this for guaranteeing better stability or security.

2.1.6 Performance

Achieving high performance of the simulation is not a goal of the project, but it is required that the performance should not be a hindrance for use over network.

2.1.7 Support for cooperative work over network

The possibility of many users viewing or working simultaneously on the same model or simulation scene is considered not obligatory, but very important and desirable.

3 IMPLEMENTATION

3.1 Architecture

The simplified architecture of the system is presented in Figure 2, more details are given in (1).

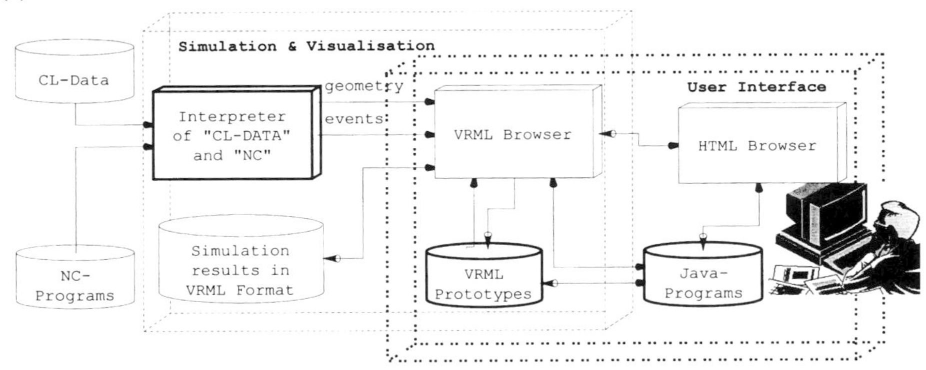

Figure 2: Architecture of a system for 4D machining simulation

3.2 Interface and structure

3.2.1 Basic objects

There are three basic objects in the system: a universal (or generalized) tool, a universal workpiece model and a universal tool path control object. The implementation of these objects is the absolute minimum needed, but even without additional objects fairly complicated simulations can be (fully) performed. A tool holder, fixtures and clamps, a machine table and other NC-machine elements are considered for implementation in the later stages of the project and are partially under development.

3.2.2 Inter-object communication

The basic objects communicate with each other by means of messages. **VRML** supports only message routing (between any two objects) and fan-out (from one object to a couple of objects). Since this is not always enough for the purposes of the machining simulation, an extension to this mechanism was developed in **Java**, which in addition allows to broadcast any message within a scene or to send a message to more than one potential receivers. A simplified representation of such message exchange is illustrated in Figure 3.

3.3 Technical portability

Since the portability was one of the main goals of the project, the product is implemented by means of platform-independent languages: **`HTML`** is used as an integration language, supported by **`ECMAScript`** (also known as **`JavaScript`**, cf. (7)). It is planned to switch to **`XML`** later on. For the 3D and 4D (i.e. 3D plus time) visualization **`VRML`** is used, which is originally designed as a language for 3D-data for the network and, hence, is expected to be 100% portable. The extension of the **`VRML`** and all computational and simulation algorithms are implemented in Java, which is also portable by design, so that the portability of the package is guaranteed. On the server-side there are a couple of scripts and support programs written in **`C++`**, **`PHP`**, **`AWK`** and **`Perl`**. Although they do not influence the portability, it is attempted to keep the number of the programs written in these languages low and to reduce it in the future as much as possible.

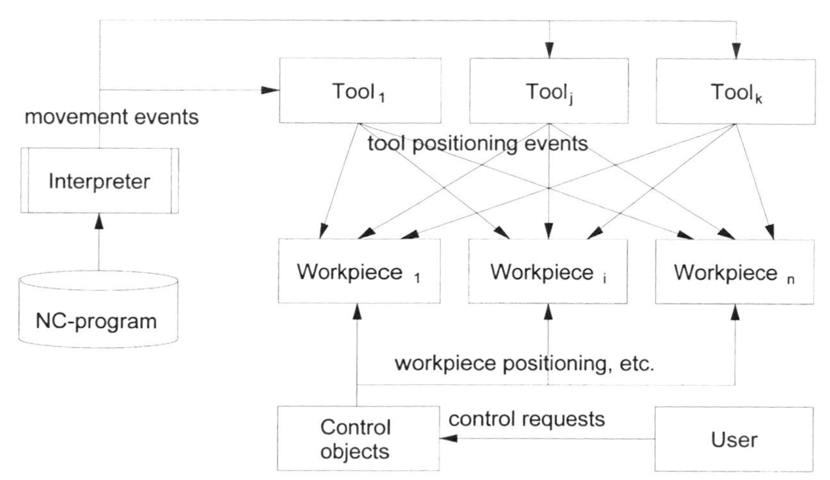

Figure 3: Communication within a simulation scene

3.5 Performance vs. accuracy

During the introductory phase of the project, the so-called **`ElevationGrid`** VRML-node (cf. (3)) was used as a basic element in implementing a prismatic workpiece model. It is a digital model, allowing to avoid complex Boolean operations. The problem of this model, as shown in (6), is that the computing effort **`CE`** is a cubic function of the accuracy, represented by the step[2] **`S`** of the elevation grid (e.g., **$CE=f(S^3)$**).

Again in (6) a method is presented for reducing CE to a quadratic dependency, but it has the following drawbacks.

1. It supports only three types of tools – cylindrical cutter, spherical cutter and torus cutter.
2. It is inapplicable or false for continuous simulation (i.e., animated simulation of the tool movement from its last position to the position requested by the current NC-

[2] *Step* is used instead of the VRML-term *spacing*. The node **`ElevationGrid`** has **`xSpacing`** and **`zSpacing`**, which are always the same in our model.

command).

3. It cannot calculate the magnitude and the direction of the cutting force acting on the tool.

Moreover, for supporting a 4D simulation, as required in section 2.1.2, a time scaling (i.e., not only simulating in real time, but also slowing down or speeding up) should be also possible. The discussed method had neither enough flexibility for covering large time scale nor enough performance for working even on low-end machines. If we also take into consideration that the run time performance of `Java` programs is worse than that of programs written in Fortran, `C/C++` and similar, it is clear that a radical improvement of the performance is needed. It can be achieved only with a complex of measures, including a small change in the concept for user interaction. Some of these measures are discussed below.

3.5.1 Support for arbitrary tool geometry

We have decided that the most natural way to support arbitrary tool geometry is to adopt the 7-parameter definition used in the CLDATA standard for our tool model. This definition is illustrated in Figure 4, it is compact and relatively easy to implement. A further motivation for this decision is the fact that this information is encoded in many NC-programs as a comment (when not, it can be easily added). The NC-program interpreter is modified to understand the encoded as a comment information by the known NC-generators and to be able to learn to understand the encoding of the same information from new sources of NC-programs. Some of the basic methods and techniques, used in this interpreter, are described in (2) and (5).

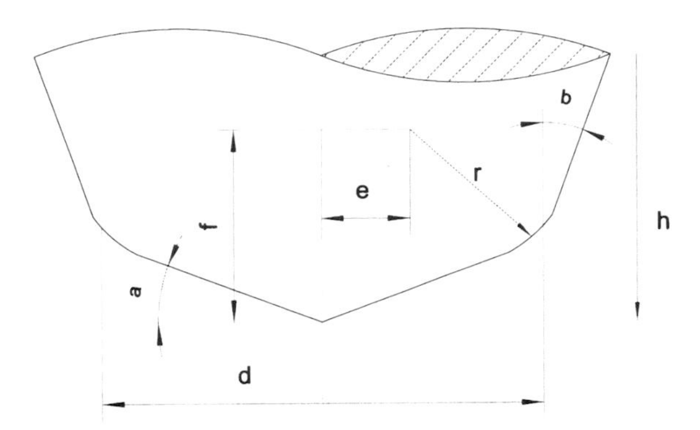

Figure 4: Tool geometry defined through 7 parameters (after DIN66215)

3.5.2 Performance scaling

Since the accuracy has a strong impact on the performance, its careful choice is a key to the successful simulation. The simulation accuracy has two aspects: computational (or internal) accuracy and visualization accuracy. Since the latter cannot and need not always be as exact as a NC-machine tool (cf. (4)), we have implemented an algorithm with a possibility of an automatic reduction of the visualization accuracy in two cases:

1. when it would be redundant (e.g., screen dimensions of an object are smaller than its

real dimensions);

2. the hardware cannot cope with the model complexity, caused by the initially requested accuracy.

It is important to mention that, if simulation results are used for further processing or optimization of the source NC-programs, the visualization accuracy does not influence the machining accuracy. Therefore, the values for both accuracies are kept in different parameters and controlled separately.

Our experience shows that it makes no sense to hardcode the accuracy (both computational and visualization) in the software, since depending on the specific case different requirements are posed on the simulation and different aspects of the simulation results are important. For this reason, the product proposes an initial accuracy to the user but leaves the final decision on him.

3.5.2.1 Discrete tool model

When an analogue model of the tool is used, the algorithm for cut-checking has to do for each new tool position the following:

1. select all points of the workpiece model which are under question;
2. calculate the "height" (or the distance from the machine-table) of the respective tool points;
3. compare the height of each selected point from the workpiece model with the height of the corresponding tool point;
4. change the height of the point from the workpiece model if it is higher than that of the corresponding tool point.

The main problem of this algorithm is step 2. For supporting arbitrary tool geometry, as mentioned in section 3.5.1, this step involves use of resource-hungry, piece-wise defined trigonometric functions. Since during one simulation this step is repeated thousand to million times, it is obvious that any optimisation of this calculation would have great impact.

An example of such optimisation is the introduction of a discrete model of the tool. This model is similar to the model of the workpiece, but its behaviour is different: when after a given movement a collision between the tool and the workpiece is detected, the points of the tool model do not "deform", contrary to those of the workpiece model. This model performs step 2 from the abovementioned algorithm only once per discrete tool point for each tool during one simulation, so it leads to significant improvement of the performance. A problem of the discrete model is its worse accuracy due to rounding errors. Our experience shows that the tool accuracy should be greater or equal to those of the workpiece. A very good ratio is 3, but acceptable results can be achieved also with 1.

3.5.2.2 Factored out calculations

The algorithm in the previous section and the frequency of its use convinced us to change most of the used algorithms so that intermediate results are saved in variables when this allows avoiding their re-calculation. To stay by the same example: instead of calculating repeatedly, say, the trigonometric functions for `a` and `b` in Figure 4, calculate them once and save (cache) for the consequent uses.

3.5.2.3 Different calculation algorithms

It is obvious, that no algorithm offers both good speed and sufficient accuracy. It is our strong conviction that the final decision about the compromise between speed and accuracy should belong to the end user of the product. For these reason, the implementation provides him with the possibility to choose himself the parameters and algorithms which are critical for the simulation.

3.5.2.4 Partial simulation

It is recommended to start each simulation with accuracy that is low enough to terminate in a reasonable time (rough simulation), and thereafter to increase the accuracy according to the needs and the hardware capabilities. If the desired accuracy makes the simulation brake down or is too slow, a conceivable solution is to reduce the dimensions of the workpiece model, to place it on a critical place recognised during the rough simulation, and to simulate again. We call this technique partial simulation.

4 CONCLUSION AND PROSPECTS

4.1 Advantages

The product offers a *NC-program verification through machining simulation* service over the network (application service provider, or in short, ASP). It is reachable for all end users of NC machine-tools, having a computer with network access, and allows them to use the knowledge and the skills of many experts, concentrated in the always latest version of the product on the server.

This approach avoids the transfer of NC-programs, which typically are extremely large, since their processing and simulation are performed locally at the user's site. Due to its network-based architecture, the product can be used from every country, having a network access to the project's web site. The only additional effort needed (after the development is completed) would be the translation of the menus in the user interface into the desired languages.

4.2 Examples of use

There are at least five main areas where the product presented here could be used: education, manufacturing, research, service and advertisement.

The product was used one semester for illustration of the machining process in the course "Information processing in design and manufacturing", taught at the Institute of Production Engineering/CAM at Saarland University. It is intended in the future to use experience, materials and results of the project in teaching students how to build simple network-oriented applications for exchange of design, manufacturing and other data, as well as how to visualize 3D data. It is also planned to use the product for (preliminary) training of NC operators and NC programmers.

The product can be used for verification of NC-programs and visualization of the expected results. Until now the commands for movement, speed control (start and stop, linear and rotational speed/direction), tool change and tool geometry can be interpreted, showing results from the interaction in (almost) real time. A module, which would be able to simulate and respectively visualize all important NC commands, is under development.

4.3 Future work

It is planned to focus the future work in two main directions. The first is to improve the support for using a digitally signed code and integrating user-designed extensions. The second is to adjust the architecture and the design of the separate modules for supporting Common Object Request Broker Architecture (CORBA), and thus to provide for using extensions over network as well as advanced co-operative work.

5 ACKNOWLEDGEMENTS

I am very indebted to Professor Helmut Bley, who provided me with excellent research conditions at the Institute of Production Engineering in Saarbrücken. I am grateful to Pavel Avgoustinov for implementing the cookie-based saving and loading of the user-based and session-based parameters.

6 REFERENCES

(1) Avgoustinov, N: "VRML as Means of Expressive 4D Illustration in CAM Education", to appear in a special issue of "Future Generation Computer Systems", Amsterdam, Elsevier publishing, 2000

(2) Avgoustinov, N., (1997), "Minimizing the Labour for Exchange of Product Definition Data Among *N* CAx-Systems", Ph.D. thesis, *Schriftenreihe Produktionstechnik, Band. 14, Universität des Saarlandes, ISBN 3-930429-43-8*

(3) Avgoustinov, N., "Virtual Shaping and Virtual Verification of NC-Programs", Proceedings, 31st *CIRP International Seminar on Manufacturing Systems*, Berkeley, 1998, pp 293-298

(4) Avgoustinov, N., "VRML as Means of Efficient 4D Machining Simulation in Intranet/Internet", Proceedings, *X Workshop – Innovative and Integrated Manufacturing*, Karpacz, 1999, pp 26-35

(5) Avgoustinov, N., Bley, H. "Acquisition and Organization of Conversion-Related Knowledge", Proceedings, *International Conference on Quality Manufacturing*, Stellenbosch, 1999, pp 150-156

(6) Enselmann, André: "HSC-Hartfräsen von Formen und Gesenken", Dissertation, Vulkan Verlag, Essen, 1999

(7) ISO/IEC DIS 16262, "ECMAScript: A general purpose, cross-platform programming language", http://www.ecma.ch/stand/ecma-262.htm.

(8) ISO/IEC 14772-1, (1997), "Information technology -- Computer graphics and image processing -- The Virtual Reality Modeling Language (VRML)", http://www.vrml.org/Specifications/VRML97/.

(9) Krause, F.-L., „Auf dem Weg zur virtuellen Produktentwicklung", VDI Berichte, 1998(?), pp. 17-33

(10) Weinert, K., Appelt, H., „Virtuelle Welten für die Produktionstechnik", Werkstattstechnik H.3, 1998, Seiten 112-116

(11) Weinert, K., Enselmann, A., Friedhoff, J., Milling Simulation for Process Optimization in the Field of Die and Mould manufacturing", Annals of the CIRP Vol. 46/1/1997, pp 325-328

CAPE/024/2000

Scene recognition for navigation of a five-legged walking robot 'Cepheus-2'

T HONDA, Y KUSHIHASHI, K FUJIWARA, K FUJITA, and **F NAKAHARA**
Department of Computer Science, Tokyo University of Agriculture and Technology, Japan

ABSTRACT

A method for scene analysis and recognition based on vertical edge segments extracted from the scene image has been developed and implemented as an integrated autonomic system for a five-legged walking robot Cephus-2. The system gets a scene image to be in the path of the walking robot through a single CCD camera mounted on the body of Cepheus-2, and processes the image to get contours or edge segments of the objects in the scene. Then only the vertical edge segments are detected from those contours and the area including no vertical edges is searched as the safe walkway for the robot.

The depth and width of the safety area in the scene are estimated by the calculation based on the x and y coordinates in the image plane. The relationship between the depth and width of the scene and the coordinates of the image plane has been calibrated in advance by experiments. These are processed by a Scene Analysis PC. The distance which Cepheus-2 can safely walk along is transmitted from the Scene Analysis PC to the Gait Control PC through RS-232C interface.

The experiments at the actual workshop in a laboratory room have shown the effectiveness of the system.

1. INTRODUCTION

The purpose of the study is to get a scene recognition system for navigating a walking robot in real time manner. As the first prototype system, a vision system to find no obstacle area on the floor including the surrounding objects in the scene and to calculate the distance allowable to go forward by walking has been developed. As the walking robot, a five-legged walking robot, Cepheus-2, with the height of 784 mm and the weight of 51 kgf, has been adopted. It is controlled by a PC which is an interface of a human operator, generates the trajectories for all of the axes following to the selected gait, and does servo control of all axes. For the scene recognition, a single CCD camera has been mounted on the center of the body. Another PC accepts the images from a CCD camera, performs scene analysis and recognition, calculates the safety area allowable to walk without collision, then sends the distance data to the gait control PC.

The experiments have been done in the actual laboratory room. The system has succeeded in detecting the walkway in allowable processing speed and positioning precision, and in controlling the robot transmitting the walking distance.

2. CEPHEUS-2: A FIVE-LEGGED WALKING ROBOT

2.1 Why five-legged ?

A five-legged walking robot, Cepheus-1(1), has been constructed in 1993 after the studies on four-legged walking robots (one of the authors has first constructed a four-legged walking mechanism in early 1960s). The five-legged mechanism was from the results of the experiences on the four-legged ones which was essentially apt to be unstable at the static walking. The legged walking mechanism fits to irregular surfaced floor or terrain, for example stepped floor, but has the more dangerous demerit of falling down. The more legs the locomotion robot has, the more stable it becomes, but the more inefficient and redundant it is. The five-legged mechanism is one solution for these problems. In 1996, Cepheus-2 of which mechanism and function of omni-directonal walking was reinforced has been developed. It has succeeded in walking on small stepped shaped terrain using touching-force sensors attached to all of the five lower legs and a gyro-scope on the body as the attitude sensor(2).

2.2 Features of Cepheus-2

The overall view of Cepheus-2 is shown in Fig.1. Several features of Cepheus-2 are in Table 1. A single CCD camera has been mounted on the body.

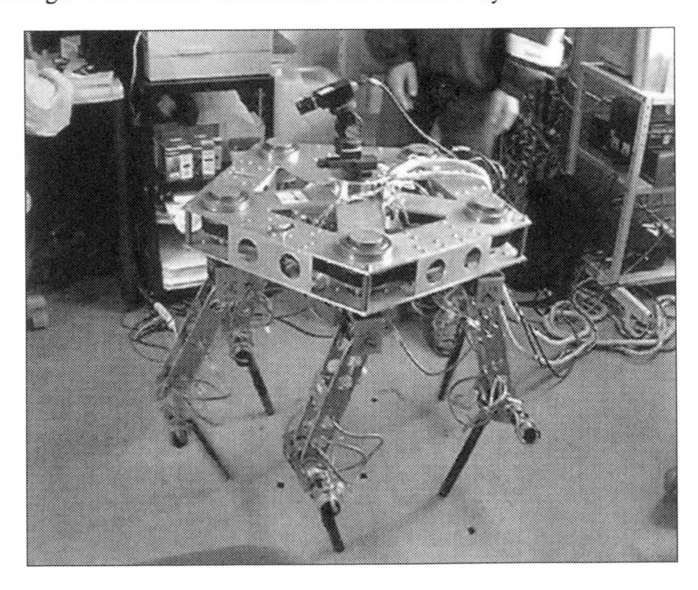

Fig.1 Overall view of Cepheus-2 with a CCD camera.

Table 1 Some features of Cepheus-2

one side of pentagonal body	485 mm
total height	784 mm
total weight	51 kgf
number of legs	5
number of joints per leg	2
yaw control axis	1

Cepheus is a platform for studying

(1) a new style of locomotion to convey materials,products,etc. as wheeled transportation vehicles,

(2) an autonomic-integrated architecture including a vision system for scene recognition.

The result in this paper concerns to the term (2) shown above.

2.3 Total system

The total system of Cepheus-2 is illustrated in Fig.2. Two PCs are utilized in the system and they are linked each other through RS-232C transmission interface. One of PCs is for controlling the mechanism of Cepheus-2. The other is for scene recognition and calculation of the distance to go forward. It accepts scene images from a single CCD camera, processes the images and sends the distance data to another PC for controlling.

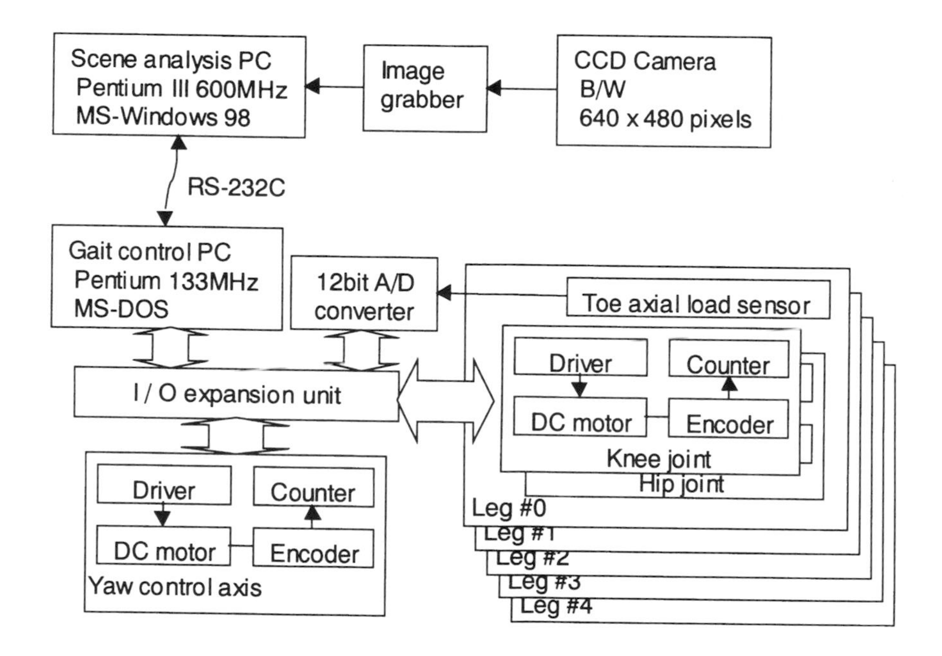

Fig.2 Total system of Cepheus-2

3. SCENE RECOGNITION SYSTEM

3.1 Total flow of processing

The accepted scene image from a CCD camera mounted on the body of Cepheus-2 is a bit map image. The scene recognition system processes the image as shown in Fig.3.

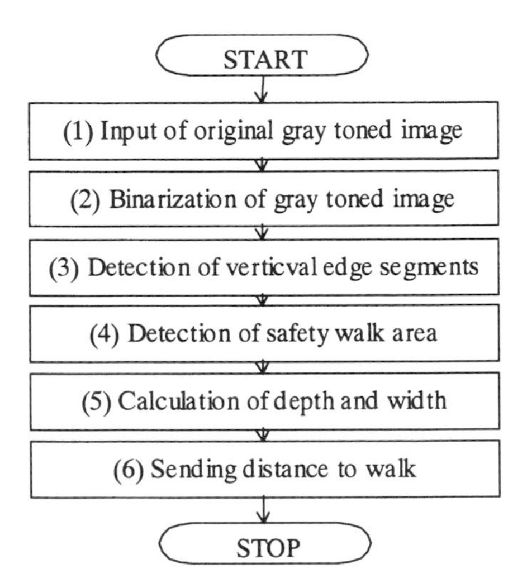

Fig.3 Flow diagram of scene recognition system

Because the results are utilized to control the walking robot in real-time manner, the algorithm has to be as simple as possible so that the processing may be carried out at least in several seconds. To satisfy the requirements, two simplifications have been applied, one is the use of a single CCD camera, and another of visual vertical edge segments.

The stereoscopic vision system is utilized in many robot visions, but it is often time consuming mainly because of the presence of the corresponding problem. In the system for Cepheus-2, a single CCD camera is used and the depth (the distance along the walkway) calculation is carred out based on the calibration data in advance.

Only the vertical edge segments in the image are noticed at the third step, because they are essential as the geometrical elements of contours for every things in our circumstances, and invariant even if the visual direction moves around the yew axis. As almost all things includes the vertical edge segments in their contours, no things, in other words no obstacle, are possively there in the area without vertical edge segments.

The safety area is a set of the horizontal scanning lines which do not collide the vertical edge segments.

3.2 Calculation of depth and width

After detecting the safety area to walk, the depth and width data of the necessary points are calculated. For the single camera method, the calibration has been done by corresponding the position of known distance to the coordinates (number of pixels) of the image plane.

The measured data have shown the relatioship between the actual depth and the number of pixels along the vertical axis on the image plane as

$$Y = 1.746\times10^{-2}\cdot x^2 + 1.169\cdot x + 1.446\times10^3 \qquad (1)$$

The relationship between the actual width at specified depth position and the number of pixels for one m along the horizontal axis on the image plane has been derived as

$$w = -1.576\times10^{-4}\cdot x^2 - 9.319\times10^{-1}\cdot x + 5.334\times10^2 \qquad (2)$$

These relationships are illustrated in Fig.4.

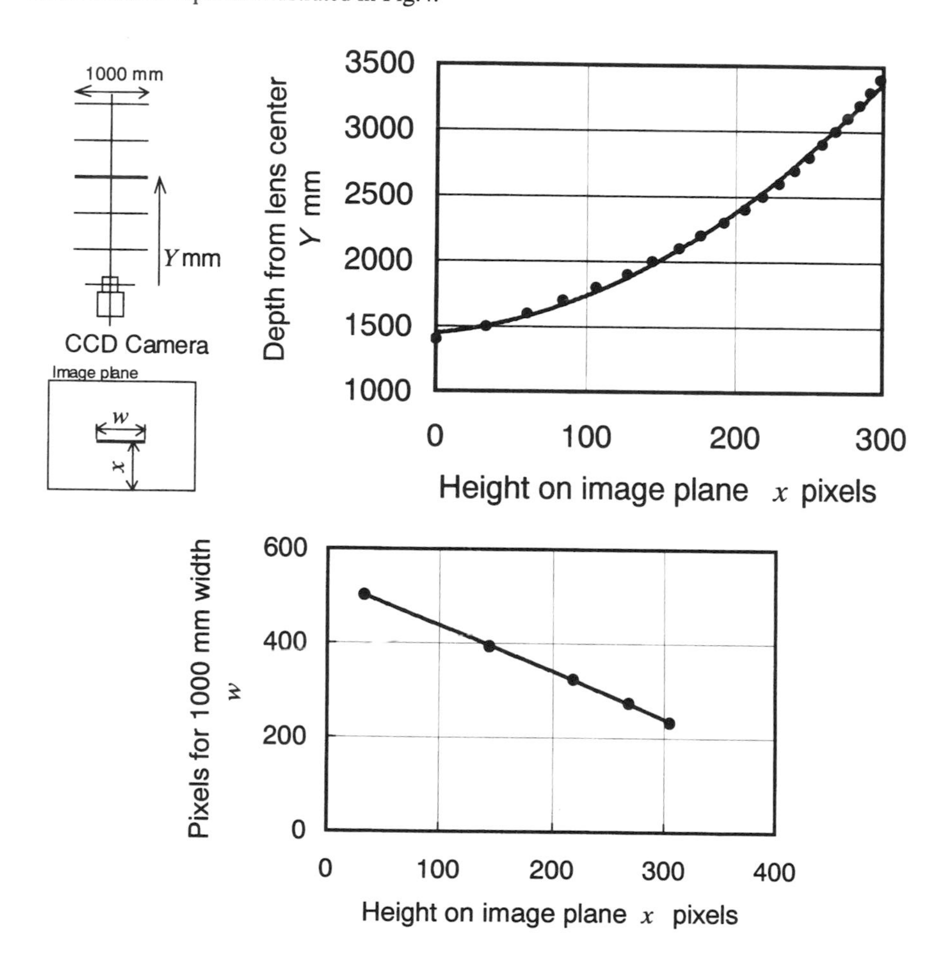

Fig.4 Relationships between depth and width in actual walkway to pixels in image plane

4. RESULTS OF EXPERIMENTS

4.1 Detection of obstacle and safety area

The techniques described above have been applied to operations in the actual workshop in the laboratory. Some object as an obstacle has been put on the floor and the CCD camera mounted on Cepheus-2 has taken the scene image. Then the system has processed the image and calculated the safety walk area.

Fig.5(a) is a result for a scene including a box on the floor, and Fig.5(b) for a human standing on the floor.

original image

edge segments

detected safety area

(a) a box on the floor

original image

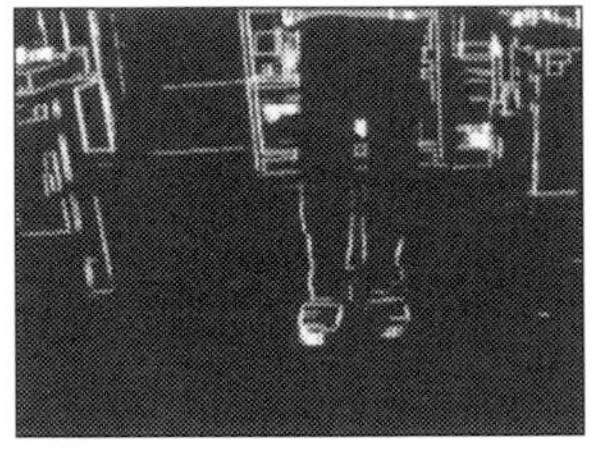

edge segments

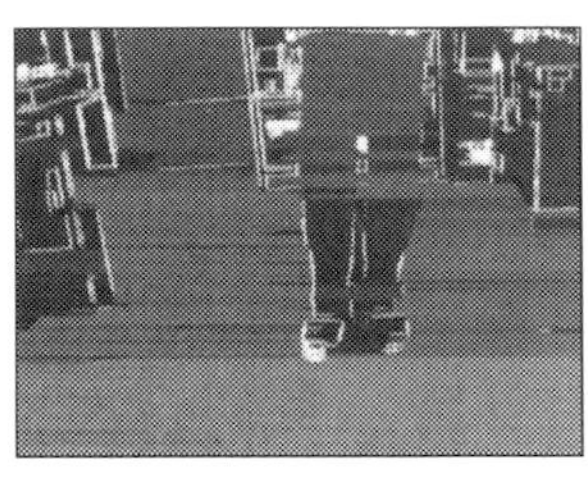

detected safety area

(b) a human standing on the floor

Fig.5 Examples of scene recognition

Table 2 shows the comparison between the actual distance to the object and the calculated depth of the safety walk area. The results show the practical precision of the method. The processings have been performed in a few seconds by the PC with Pentium III 600MHz CPU.

Table 2 Actual distance and calculated depth

object	actual distance	calculated depth
box	1,380 mm	1,316 mm
human	1,490 mm	1,416 mm

4.2 Application to autonomic walking

Since the effectiveness of the scene recognition system has been shown, the scene recognition system and the gait control system have been integrated for the autonomic walking of Cepheus-2.

Fig.6 and Table 4 show an example of the autonomic-integrated walking of Cepheus-2. Although the maximum error of the estimated distance in the safety area extends about 200 mm, the feasibility of the method has been certified.

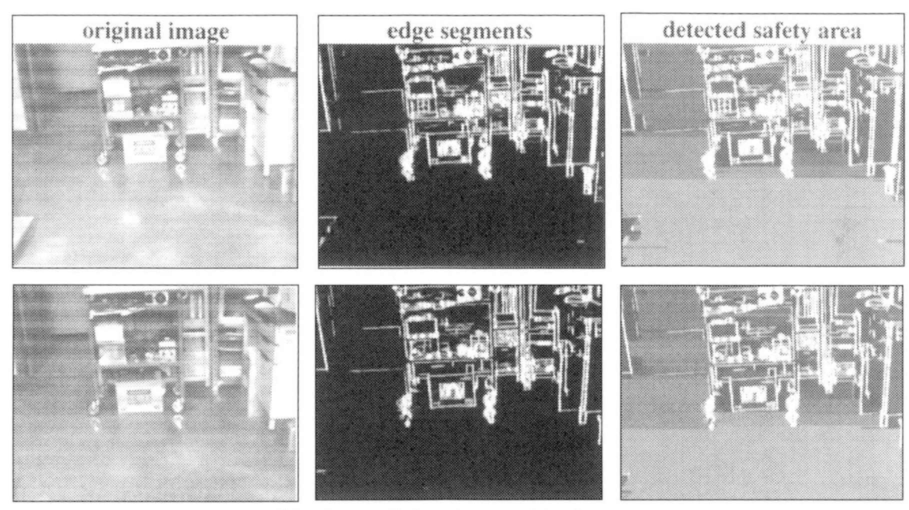

(b) after walking forward by 210 mm

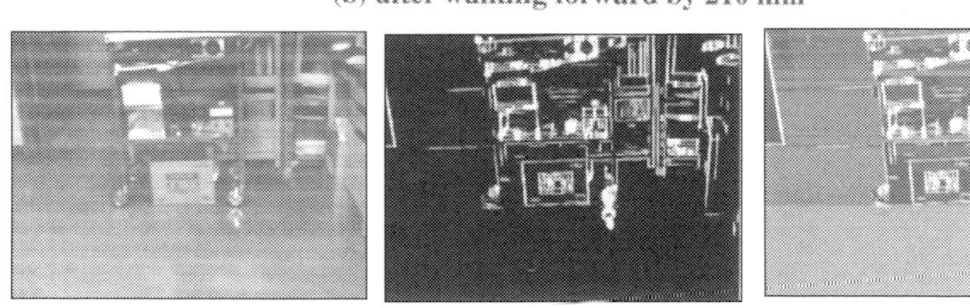

(c) after walking forward by 460 mm

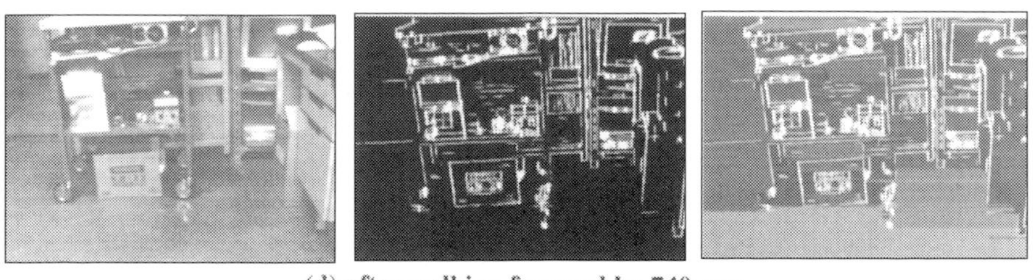

(d) after walking forward by 740 mm

Fig.6 Scene recognition images at autonomic-integrated walking

Table 3 Actual distance and calculated depth

actual distance	calculated depth
2,380 mm	2,115 mm
2,170 mm	1,988 mm
1,920 mm	1,757 mm
1,640 mm	1,443 mm

5. CONCLUSIONS

A scene recognition system to find the walkway for a five-legged walking robot has been developed. The results are collected as follows.

(1) By using a scene image through a single CCD camera, a simple depth calculation method has been realized.

(2) Utilization of the vertical edge segments in the scene image is effective to detect obstacles in the walkway and to find the safety area in front of the robot.

(3) In the experiments, the processings for total scene analysis have been done in a few seconds.

(4) Integration of the scene recognition system to the robot control system has shown a possibility of an autonomic navigation for a five-legged walking robot.

ACKNOWLEDGEMENT

The authors express their appreciations to the SECOM Science and Technology Foundation for his financial support.

REFERENCES

(1) T.Honda, S.Kaneko, S.Kan, et al. : Some Approaches for Intellectual Robot System with an Autonomic-Integrated Architecture, Proc.of the 10th Int. Conf. on Computer Aided Production Engineering, pp.435-440,Palermo, June 1994.

(2) T.Chiba, K.Fujiwara, Y.Kushihashi and T.Honda : Gait Control at Non-flat Floor for a Five-legged Walking Robot "Cepheus-2", pp.505-510,Tokyo, Sep. 1998.

CAPE/099/2000

A method for the reduction of machining time by the simulation of NC-programs in virtual reality

N AVGOUSTINOV and **H BLEY**
Institute of Production Engineering/CAM, Saarland University, Germany

ABSTRACT

There is an interesting technique, popular among the manufacturers of dies and moulds: instead of replacing a worn out mould by a new one each time the tolerances are lost, the geometry of the old one is refreshed by a shift along one of the axis (typically "Y") and machining it with the original NC-program once again. This leads to savings of material (due to the reuse), to reduced volume of chip – and, consequently, savings of energy and tools – and to time savings (due to skipping of the roughing). The main problem of this method is the fact, that if the same NC-program is used both for the first and for the following machining, the mill spends most of its time (up to 90%) in "cutting air". This useless work cannot be avoided, but it is possible to reduce dramatically the time necessary for performing it.
This article presents intermediate results of an in-progress research, which attempts to minimise the "air-cutting" time based on the NC-programs alone (i.e. without the help of any additional CAD, CAM or similar systems, and without having access to the rated or required geometry). Since the implementation of the proposed solution is based on HTML, VRML and Java, it is by design network-ready and portable without additional efforts. The safety of the method is guaranteed by 4D machining simulation with process and result visualisation.

1 BACKGROUND

1.1 State of the art

Several factors make the manufacturing of dies and moulds a very expensive process - the prices of the used tools, machine tools, materials and NC programs. For this reason different optimisation strategies, aimed at reducing the manufacturing costs, prolonging the life time of optimisation strategies, aimed at reducing the manufacturing costs, prolonging the life time of dies and moulds, and reuse of material and work have permanently been tried by the respective manufacturers. One of the often used techniques is known under the name "shift

Legend:
Required (or rated) geometry
Material to be removed (MTR):
during the first processing
during the re-machining

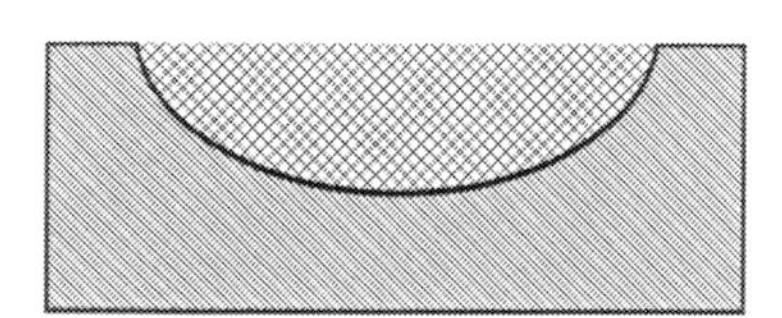

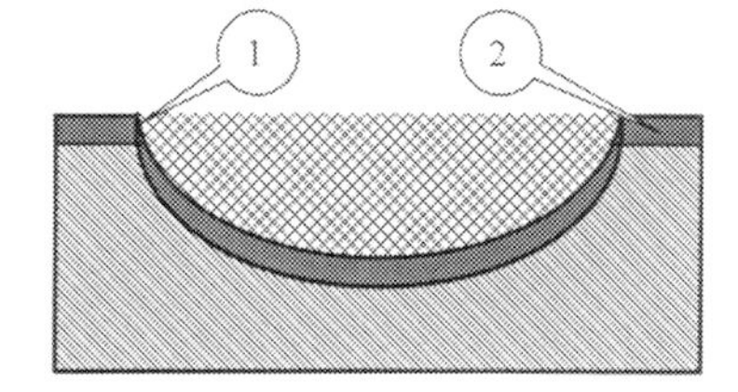

Figure 1: Shift and re-machine idea

and re-machine": instead of replacing a worn out mould by a newly produced one each time the tolerances are lost, the geometry of the old one is refreshed by a *shift*[1] along one of the axis (typically "Y") and machined by the original NC-program once again (*re-machining*). Re-machining could be applied repeatedly until there is no enough material left on the workpiece. This has the following advantages that are more or less mutually dependent:

1. Saving of material as a result of multiple reuse of each workpiece/mould;
2. Reduction of the *amount of the material to be removed* (in short MTR; cf. the figure on the right) during each consequent machining, which leads to energy saving, longer life of the used tools, reduced volume of chip and, hence, cheaper disposal;
3. Time saving due to skipping of the roughing;
4. Savings resulting from the reuse of the original NC-program for the re-machining instead of preparing a new one;
5. Neither CAD nor CAM or other CA-system is needed for the preparation of the NC-program for the re-machining. This is really important in cases of co-operation when the generation of the NC-program and its use happen to be in different enterprises.

Unfortunately, this technique has also some disadvantages.

1.2 Problems

As mentioned above, typically, the original NC-program is used for the consequent machining, with an empirically determined amount of the shift. There are two main problems when this method is applied:

1. When the same NC-program is used for both machining and re-machining, the mill spends most of its time (up to 90%) in "cutting air". This useless work cannot be avoided, but it is possible to reduce dramatically the time necessary for performing it;

[1] The value of the shift is determined empirically.

2. Due to different placement of the MTR for the first and for any consequent machining, the tool load during the re-machining differs from the rated tool load in the original NC-program. The point here is that the MTR is not smooth (cf. places marked by 1 and 2 in the figure above) and this could even cause dangerous for the tool jumps in the load.

Obviously, the shift and re-machine technique could be improved if there is a way to eliminate or at least to reduce the influence of these problems.

2 SOLUTION

A conceivable algorithm for striping all "air cut" movements from the original NC-program is presented in Figure 2. The main problem of this algorithm is how to determine whether a given movement of the tool would cause cutting or not (i.e., whether it would cause a "collision" with the workpiece). This problem is not mathematical, but rather technical: there is no easy way to determine and describe the geometry of the worn out part. The point is that the wear along the machined surface not regular. Theoretically, it is possible to perform a series of measurements on a coordinate measuring machine (CMM), then to create a 3D model of the part in a CAD (or CAM) system, to calculate there the differences between this part and the projected geometry and to generate a new NC-program. However, such an approach would be very expensive and time consuming. Consequently, both to prepare a new NC-program for the necessary refresh of the geometry and to strip all "air cut" movements from the original one are difficult tasks even if a (non-dedicated) CA-system is involved. And for suppliers who have no CA-systems, the "CMM-solution" is totally inappropriate.

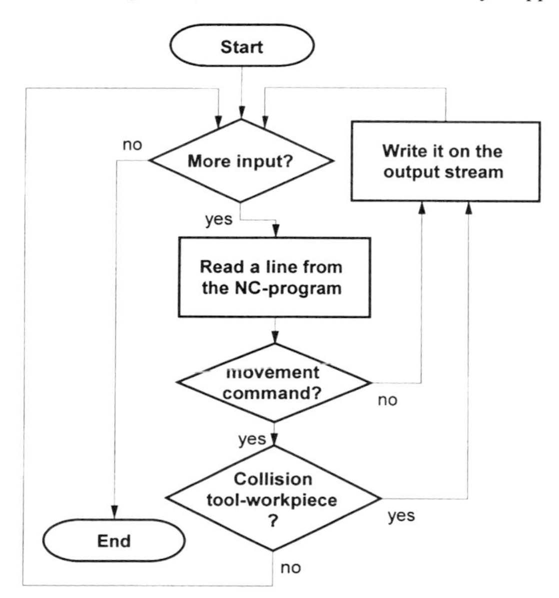

Figure 2: Algorithm for stripping out unneeded movements

2.1 Concept

In this context, a dedicated but at the same time inexpensive and easy-to-use software is desired. The easiest way to think of was extending the software package for verification of NC-programs and machining simulation, which was under development in our institute at the time we confronted the shift and re-machine challenge (cf. (1)). Since this package is able on the basis of a NC-program, a digital workpiece model and a tool model, to simulate the machining, it could be used for implementing the algorithm from Figure 2. Therefore, the first thing we did was to determine what kind of conceptual (if any) and technical changes to our package would be necessary for extending its functionality to support the shift and re-machine technique.

Thus we found that, on the one hand, we had the necessary software for fast determination whether a given movement would cause cutting or not, but on the other hand, the required geometry was not in form which could be processed by the developed software easily. A careful analysis of the situation, though, has shown that it is possible to achieve the desired support of the shift and re-machine technique even without conceptual changes.

The crucial question is how to produce the 3D data needed as a required geometry from arbitrary NC-program. We have found that this can be achieved by an extra (preliminary) machining simulation. The simulation is performed much faster than the real machining and running it once more does not consume much time. The result of the machining simulation is a digital model of the mould (expected from the real machining) and has the accuracy requested by the user.

2.2 Implementation

The scheme of the process is given in Figure 3.

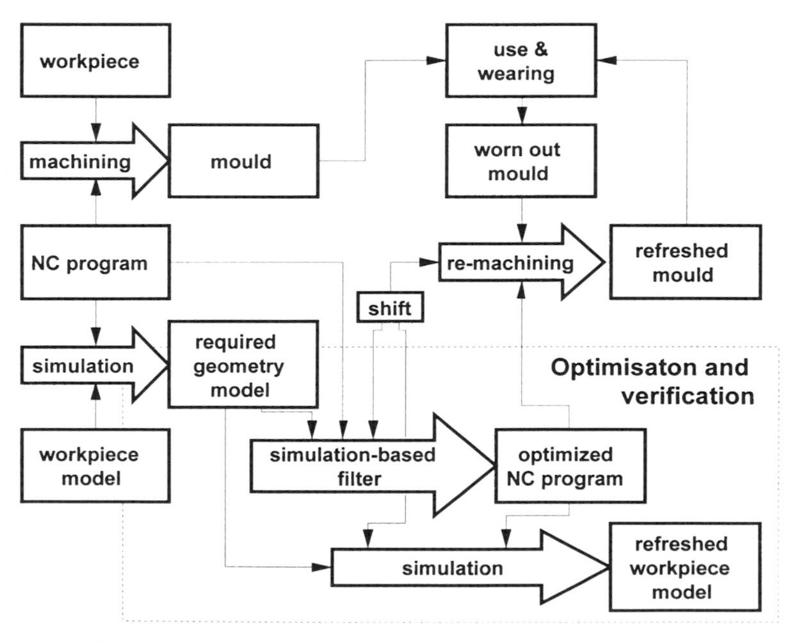

Figure 3: Scheme of the optimised shift and re-machine process

The simulation is performed from the software described in (1) after changing it a bit for supporting the shift and re-machine technique. This software converts the source NC-program to 3D geometry in VRML format and thereafter employs Java powered VRML extensions (workpiece model, tool model, etc.) for visualisation of the machining simulation in virtual reality. We call this kind of simulation virtual shaping or virtual machining (cf. also (3), (4)).

One of the major technical changes, needed in the software for extending its functionality to support the shift and re-machine technique, was to provide for any virtually processed digital model of the workpiece to be saved in order to allow its use as required geometry.

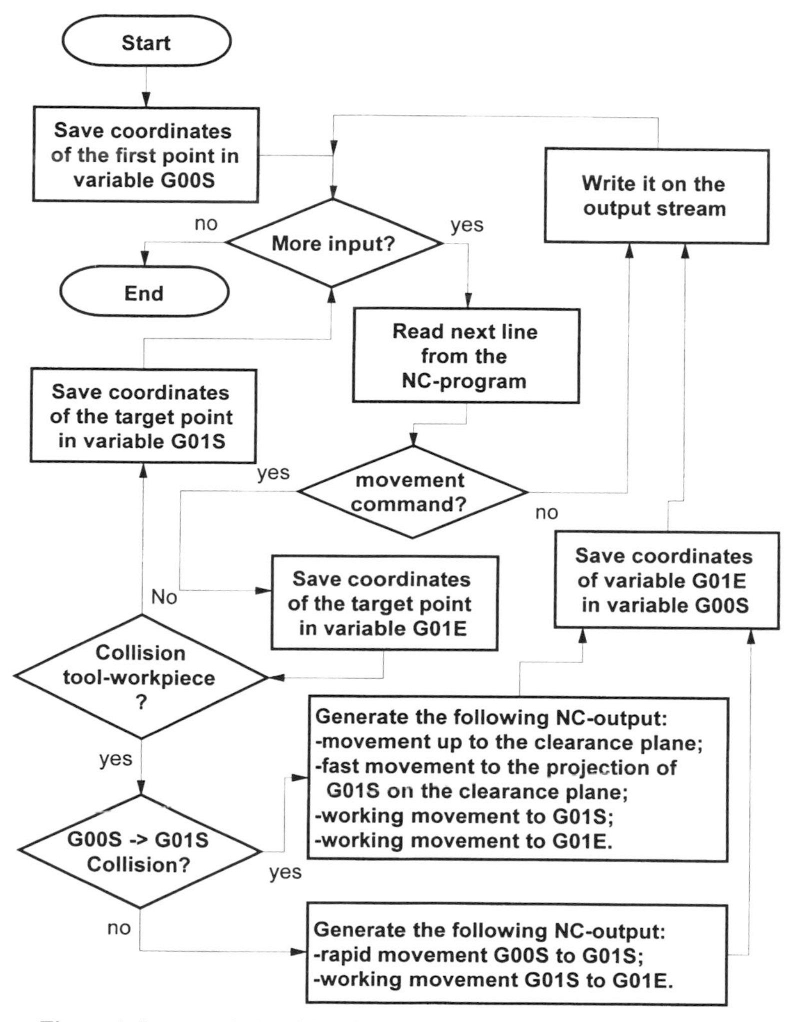

Figure 4: Improved algorithm for stripping out redundant movements

Another important change was necessary in the algorithm stripping out redundant movements in order to avoid damages in the geometry caused from stripped and replaced through NC-commands for rapid tool movements. The respectively modified and more detailed algorithm is illustrated in Figure 4.

There is one more modification of the software package which aims at employing its capability to calculate the volume of the material, cut at the moment, for adapting the feed rate to the MTR. The feed rate can be re-calculated for the newly expected MTR on a "per (movement-) NC-command" basis or on a "MTR within a given range" basis. The former method is easier to implement, but leads to inaccurate results if the thickness of the MTR is not the same along a given movement. The latter method gives more precise results but requires permanent calculation of the MTR during the tool movement along each segment of the tool path. Therefore, the simulation requires more computing resources and could be slower. A minor drawback of the method is that in order to have more than one feed rate along one movement-command of the original NC-program (i.e., between two points) it is necessary to split the respective movement in sub-movements and insert the respective NC-commands in the optimised NC-program. On the other hand, if a respective sub-movement causes no cutting, the software can stripped it out and thus achieve yet better grade of optimisation.

Since the implementation of the proposed solution is based on languages, which are hardware and software independent by design (HTML, VRML and Java), it is also portable by design and network-ready without any additional efforts. Moreover, the concept of the package and its architecture allow for co-operative (both consecutive and simultaneous) work of multiple users over any local or global network.

2.3 Correctness of the results

Digitalisation always implies some (rounding) errors. Due to the use of digital workpiece/mould models (cf. (5) or (3)) in the described method, it is important to guarantee that the re-machined with an optimised NC-program part is not of poorer quality than the original part. There are two different aspects:

1. to proof that striping parts from the original NC-program does not cause damages of the geometry by unforeseen collisions of the tool with the workpiece;
2. to check whether the geometry, achieved through the optimised shift and re-machine method, corresponds to the required geometry.

The proof against damages is performed by the operator by means of 3D visualisation of the simulation results. There is one special capability in addition to the traditional possibilities (changing the view point, the scale, the cross-section or the representing accuracy): it is possible to overlay the results of different simulations for comparing their differences visually. This is especially useful for comparing the result of an optimised NC-program and its (non-optimised) original, whereas different colours can be used for distinguishing the both results and different grades of transparency can be applied to them for estimating, e.g., their mutual penetration.

Interesting feature of the method is that it is not very sensible to digitalisation inaccuracies of the models since they compensate themselves during the second run of the simulation (cf. the simulation-based filter in Figure 3).

3 CONCLUSION AND FUTURE WORK

The shortening of the optimised NC-programs achieved by the described technique is case and NC-program-generator dependent, but can be up to 90%. As a result, the NC-machines using optimised with the proposed technique NC-programs can be used much more efficiently. A comparison between parts, produced by a given NC-program and its optimised variant will be available in the middle of this year (2000).

4 ACKNOWLEDGMENTS

We are very indebted to Krupp Gerlach GmbH for providing us with "real life" geometrical data for performing these experiments.

5 REFERENCES

(1) Avgoustinov, N: "On the Implementation of 4D Machining Simulation in VR over Internet", to appear in the proceedings of CAPE2000, Edinburgh, 2000
(2) Avgoustinov, N: "VRML as Means of Expressive 4D Illustration in CAM Education", to appear in a special issue of "Future Generation Computer Systems", Amsterdam, Elsevier publishing, 2000
(3) Avgoustinov, N., "Virtual Shaping and Virtual Verification of NC-Programs", Proceedings, 31st *CIRP International Seminar on Manufacturing Systems*, Berkeley, 1998, pp 293-298
(4) Avgoustinov, N., "VRML as Means of Efficient 4D Machining Simulation in Intranet/Internet", Proceedings, *X Workshop – Innovative and Integrated Manufacturing*, Karpacz, 1999, pp 26-35
(5) Enselmann, André: "HSC-Hartfräsen von Formen und Gesenken", Dissertation, Vulkan Verlag, Essen, 1999
(6) ISO/IEC 14772-1, (1997), "Information technology -- Computer graphics and image processing -- The Virtual Reality Modeling Language (VRML)", http://www.vrml.org/Specifications/VRML97/.

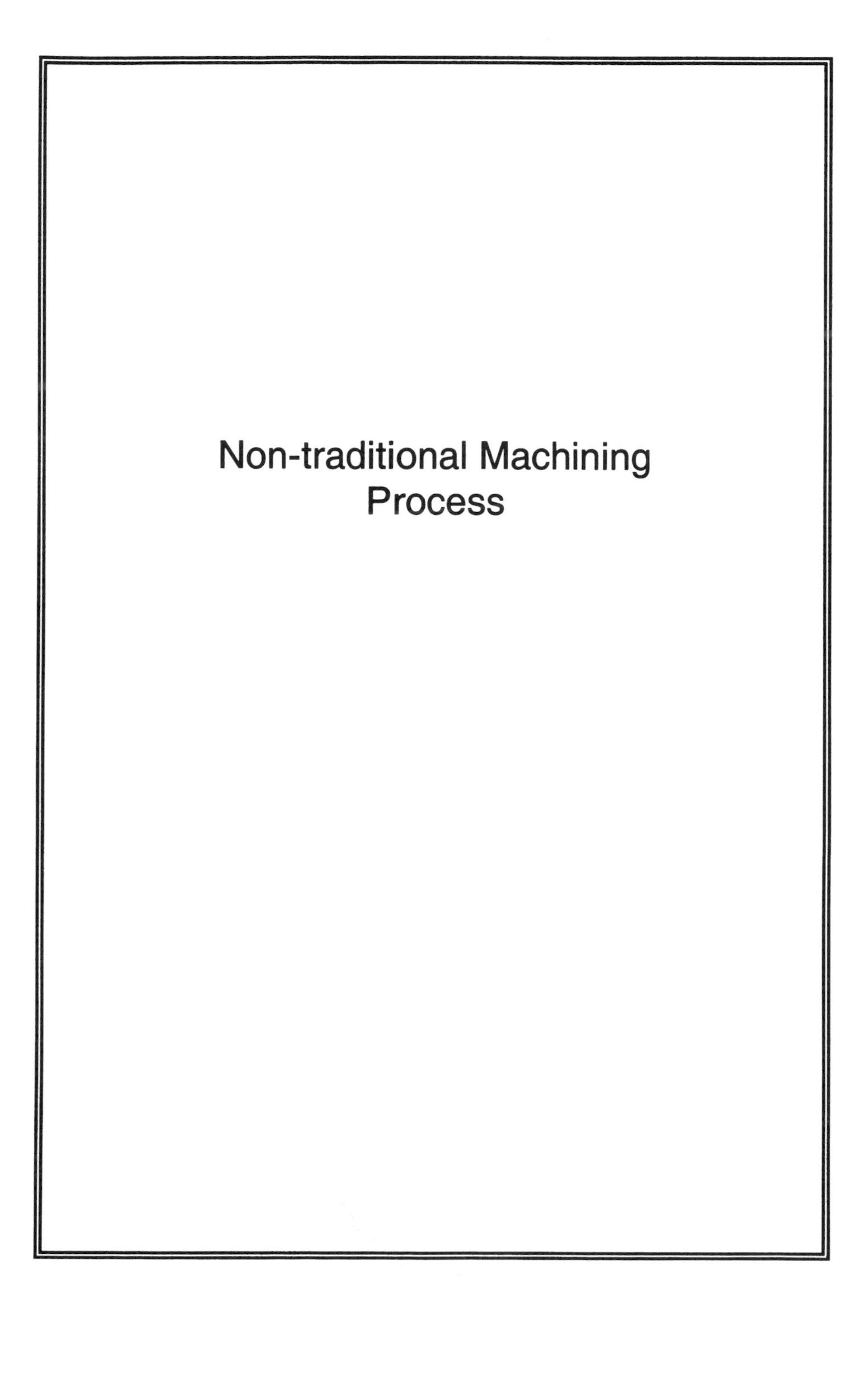

Non-traditional Machining Process

CAPE/059/2000

Accuracy of electrochemical forming of dies and molds

H EL-HOFY and **M A YOUNES**
Production Engineering Department, Alexandria University, Egypt

SYNOPSIS
Electrochemical machining (ECM) provides one of the best alternatives for producing complex shapes in advanced materials used in die and mold industries. This paper presents experimental assessment for the accuracy of shapes produced in high speed stationary electrochemical forming of holes into square shapes. Using specially designed tools, a test cell is prepared for experimental verification, where the effects of machining allowance, applied voltage, tool geometry, and forming time are investigated. Dimensions, straightness of sides and corner radii of produced workpieces are evaluated and compared with corresponding theoretical values. The degree of shape conformity that reflects the level of process accuracy is also determined.

1. INTRODUCTION

In electrochemical machining (ECM) difficult-to-cut materials and complex shapes are produced without distortion, scratches, burrs or stresses. The process provides an effective alternative for producing a wide range of components such as aircraft turbine parts, surgical implants, bearing cages, molds and dies and even micro-components [1]. Using EC-die sinking, the shape of the tool electrode can be transferred onto, or duplicated-in the workpiece. Forging dies combining horizontal and slopping planes, cylindrical, spherical and complex shaped surfaces are typical applications with tolerances in the range of 0.1-0.3 mm.

The main objective of ECM is to maintain the form errors as small as possible. This calls for the design of suitable tools, that generate the required form and dimensions of finished parts under specific machining conditions. The most general method for calculating the shape produced, by a tool, involves the solution of Laplace equation for the field using iterative numerical methods, complex numbers [2] and finite element methods [3].

The large expenditure of electrical energy and the use of considerable amount of electrolyte used for dissolving the entire machining allowance and removing the anodic products along the inter electrode gap impairs the level of accuracy obtained. Actually, some metal is dissolved from the adjacent areas of the workpiece by the stray machining action thus causing form errors [4]. Such undesirable effects can be reduced by using a tubular-section cathodic tools [5] or NC wire machines that utilize ECM, EDM or EEDM [6,7]. Loskutove [8] improved the accuracy of holes by adding sodium tungestinate to sodium nitrate thus reducing both the end and lateral gaps. ECM accuracy can also be improved by introducing pulsating voltage provided that the applied pulses are short enough and their duration is optimized to suit the gap size and electrolyte pressure [9]. Electropolishing and electrobrightening of holes using different feeding and rotating electrodes has been considered by Hocheng [10]. Sizing of electrochemically drilled holes have been experimented using a moving tool during the combined electrochemical sinking-broaching process [11-12]. In a further work [13], tools for correcting errors in hole shapes have been introduced. Tool profiles for stationary electrochemical machining of drilled holes that produce definite forms have been presented in reference [14]. Such tool forms are based on the distribution of the initial inter electrode gap if the shape and form of the finished part are specified. These tools have usually complicated forms and, therefore, can be made using NC wire EDM machines. In this work, the 4-wings tools of reference [14] are used to produce square shapes from pre-drilled holes. The accuracy of shapes produced under variable removal allowance, gap voltage, tool geometry as well as machining time is evaluated. Accuracy measures in terms of shape conformity factors are also determined. The results obtained can be useful for the design of tools used in machining dies of complicated shapes.

2. EXPERIMENTAL

Experiments were conducted using the setup shown in Figure 1. Hardened steel specimens 5 mm in height with initial hole of 13 mm diameter, and brass tools are used, Figure 2. The experimental program is devised to investigate the effects of the allowance removed, the gap voltage, the machining time and the tool geometry on the accuracy of machined parts. NaCl (170 g/liter) electrolyte is pumped under pressure to minimize the possible variation in its properties along the inter-electrode gap. During machining the weighed specimens were first located, concentric with the specified tool, and clamped in position. The electrolyte is then pumped through the pressure chamber and the power is switched on. Machining current was monitored using an ammeter (100 A range), and the weight loss and machining time were used to determine the volumetric removal rate and current efficiency.

In order to evaluate the accuracy of machined parts, the profile coordinates of both tools and workpieces were determined before and after machining, using a digital profile projector that measures the dimensions down to 1μm. for these measurements. The recorded coordinates were used to evaluate tool geometry including nose radius and wing width. Meanwhile, workpiece corner radii, out-of-straightens of sides and diagonal deviations were also estimated using appropriate algorithms. The shape conformity factors shown in Figure 3, are calculated in terms of; a radial factor (r_1/r_2), an angular factor (θ_1/θ_2), and a linear one (x_1/x_2). The same figure also shows the position where the machining allowance is measured with respect to the original and final workpiece shapes.

3. THEORETICAL MODELS AND ALGORITHMS

The profile of the tools necessary for producing square shapes from initially drilled cylindrical holes are presented in Figure 2. To calculate tool nose radius and the radius of curvature of workpiece corners a least squares approach was adopted [15,16]. If x_i, y_i are the profile coordinates at equiangular spacing from an assumed center, the sum of the squares of the deviations ($E_s = \Sigma E_i^2$) is given by, [15]:

$$E_S = \left\{\left[(x_i - a)^2 + (y_i - b)^2\right]^{\frac{1}{2}} - R\right\}^2 \quad (1)$$

$$E_S \approx (r_i - R - a\cos(\theta_i) - b\cos(\theta_i))^2 \quad (2)$$

Where:

R : radius of curvature.

a, b : estimated coordinates of the center of curvature.

$$r_i = (x_i^2 + y_i^2)^{\frac{1}{2}} \quad (3)$$

$$\theta_i = \tan^{-1}\left(\frac{y_i}{x_i}\right) \quad (4)$$

Solving for a, b and R, that give the minimum value for E_s results in [15,16],

$$a = \frac{2\sum x_i}{n} \quad (5-a)$$

$$b = \frac{2\sum y_i}{n} \quad (5-b)$$

$$R = \frac{\sum r}{n} \quad (5-c)$$

Where:

Σx_i : sum of all x_i values

Σy_i : sum of all y_i values

n : number of measurement points

Σr_i : sum of radial distances of measurement points from the least squares center (a, b).

On the other hand if x_i, y_i are the recorded coordinates for the workpiece sides, then the sum of the squares of the deviation of the side from a straight line is [17, 18],

$$L_S = \sum L_i^2$$

$$L_S = \sum (y_i - \beta_0 - \beta_1 x_i)^2 \quad (6)$$

Where:

β_o : intercept of the least squares line with the y-axis.

β_1 : slope of the least squares line.

Solving for β_o and β_1, gives the minimum value for L_s [17], hence,

$$\beta_0 = \bar{y} - \beta_1 \bar{x} \tag{7}$$

$$\beta_1 = \left[\sum x_i y_i - \frac{(\sum y_i)(\sum x_i)}{n}\right] / \left[\sum x_i^2 - \frac{(\sum x_i)^2}{n}\right] \tag{8}$$

Where:

$$\bar{y} = \frac{1}{n}\sum y_i \tag{9 - a}$$

$$\bar{x} = \frac{1}{n}\sum x_i \tag{9 - b}$$

The least square fitted line is therefore; $y = \beta_o + \beta_1 x$
The perpendicular signed deviation of each point from the fitted line is then calculated and, the difference between maximum and minimum deviations represents the out-of-straightness in the workpiece side.

4. RESULTS AND DISCUSSION

4.1 Material removal

Figure 4 shows the effect of the machining allowance and gap voltage on the volumetric removal rate. Accordingly, the experimenntal and theoretical removal rates decrease when removing larger machining allowances using smaller gap voltages. This trend is attributed mainly to the low electrolyzing current, shown in Figure 5. Under such conditions, as the machining process continues, the interelectrode gap becomes wider and consequently the average machining current decreases. The effect of the machining allowance and gap voltage on the machining time is shown in Figure 6. While the removal of larger machining allowances raises the volume to be removed, the machining time is prolonged which leads to a reduction of the metal removal rate. In contrast, for a given allowance, the increase of gap voltage drives more current and speeds up the machining process reducing the machining time with consequent increase in the removal rate.

The theoretical values of machining current and hence the removal rates are greater than the experimental ones. This behaviour can be directly related to the assumptions considered in the theortical model where effects of gass bubbles and Joule's heating are neglected [14]. Such deviations are reflected on the current efficiency values shown in Figure 7. The current efficiency is markedly affected at low voltage where it has the smallest value, irrespective of machining allowance. In addition, at larger voltages the results showed no clear trend with machining allowance. The same figure also indicates the increase of the current efficiency with machining voltage where it reaches 90% at 28 volts and 1.5 machining allowance. This rise can be attributed to Joule's heating that raises the electrolyte conductivity and hence more current is used in enhancing the metal removal process.

4.2 Accuracy and shape conformity

The machined profiles contain rounded corners, the radii of which depend on the machining conditions. Figure 8 depicts the decrease of corner radius at high machining allowance.

However, the same figure shows slight increase of the corner radius with gap voltage. The difference between the corner radii evaluated experimentally using the least squares model and those calculated using the theoretical model [14] are shown in Figure 9. Accordingly, smaller deviations are observed at smaller machining allowance as well as larger gap voltages. Similarly, Figure 10 shows the deviations between theoretically calculated and experimentally measured workpiece-width at different machining allowances and gap voltages. Accordingly these deviations reach minimum levels at small and large machining allowances. Moreover, smaller deviation from target values can be obtained at machining voltages of 21 volts probably due to the high current efficiency observed under such conditions, Figure 7. The effect of the machining allowance and gap voltage on the straightness error of the produced sides is shown in Figure 11. The out of straightness error increases with machining allowance and reaches maximum level at 1 mm. Such a maxima predominates at gap voltages greater than 14 volt.

The geometry of sizing tools, in terms of nose radius and wing width, Figure 2, affects the produced workpiece geometry. Their effect on both workpiece corner radius and straightness error of resulting sides is evaluated for different gap voltages. Figure 12 indicates that larger tool nose radius and greater gap voltage generate wider corner radii since greater cathodic tool arc length conducts more machining current to the anodic surface. Moreover, the deviation in straightness of workpiece sides increases with tool wing width as shown in Figure 13. Therefore, narrow tool wings having smaller nose radii are recommended for producing better shape accuracy

.

The conformity factors reflect the errors in the produced shape as a result of the formation of corner radius. In this regard, linear, radial, and angular factors are considered, Figure 3. These factors can be expressed in terms of each other using trigonometric relationships. Figure 14 shows the radial factor at different machining allowances as well as gap voltages. The greater the factor, the smaller the corner radius and the better the produced articles. The shape conformity factor becomes higher at larger machining allowances while no clear effect for the gap voltage is noticed.

Error in the diagonals of the square shapes is also evaluated and shown in Figure 15. It is therefore evident that, at a given gap voltage, such deviation rises at larger machining allowance. Additionally, at a given machining allowance, smaller deviaions can be obtained at greater machining voltages.

5. CONCLUSIONS

The analysis of the experimental observations highlight the following conclusions:-

1. Smaller machining times, higher removal rates and hence greater productivity can be obtained when removing smaller allowances using larger gap voltages.
2. For improved process accuracy, smaller gap voltages are recommended to remove larger allowances using small tool radii.
3. Out-of-straightness of workpiece sides and deviation in its diagonals decrease when narrow-wing tools are used to remove smaller machining allowances at larger gap voltages.
4. Deviation, from theoretical values, in workpiece corner radius and side width decreases at higher gap voltages as well as smaller machining allowances.

6. REFERENCES

(1) M. Kunieda and H.Yoshida, "Influence of Microindents Formed by Electrochemical Jet Machining on Rolling Bearing Fatigue life", Manufacturing Science and Engineering, ASME, PED-Vol. 64, pp. 693-699, (1993).

(2) V.K. Jain and P.C. Pandy, "Design and analysis of ECM tooling", Precision Engineering, Vol. 1, pp. 199-206, (1979).

(3) K.P. Rajurkar, D. Zhu and B. Wei, "Minimization of Machining Allowance in Electrochemical Machining", Annals of the CIRP, Vol. 47/1, pp. 165-168, (1998).

(4) A.F. Rashed, H.A.Youssef, and H.A. El-Hofy, "Effect of Some Process Parameters on the Side Machining During Electrolytic Sinking", Proceedings of the 3rd PEDAC Conference, Alexandria, pp.733-746, (1986).

(5) C.V. Kargin, "Electrochemical Shaping Using Tubular Section Cathode Tools", Russian Engg. J.,Vol. IV, No 4, pp. 73-74, (1975).

(6) H.A. El-Hofy, and J. A. McGeough, "Evaluation of an Apparatus for Electrochemcal arc Wire Cuttng", ASME, J. of Industry, Vol. 110, pp. 110-117, (1987).

(7) M. S. Amalnik, H. El-Hofy, and J. McGeough, "An Intelligent Knowledge-Based System for wire Electro-Erosion dissolution in concurrent engineering Environment", Journal of Material Processing Technology, Vol. 79, pp.155-162, (1998).

(8) A.I. Loskutove, "Increasing the Accuracy of Electrochemical Machining", Indust. Processing and Biology, Vol. 3, pp. 32-34, (1980).

(9) J.A. McGeogh, "Principles of Electrochemical Machining", Chapman and Hall , London , (1974).

(10) H. Hocheng, P. S. Pa. "Electropolishing and Electrobrigtening of Holes Using Different Feeding Electrodes" Journal of Material Processing Technology , 89-90, pp. 440-446. (1999)

(11) H.A. Youssef, and H.A. El-Hofy, "Computer Simulation Towards a Feed Function for EC-Broaching of Electrochemical Drilled Holes", Proceeedings of 2nd CAPE Conference, Edinburgh, pp. 371-375, (1987).

(12) H.A. Youssef, and H. A. El-Hofy "A Combined EC-Sinking- Broaching Process", Proc. of 6th Int. Conference on Prod. Engineering, Osaka, Japan, pp. 658-660, (1987).

(13) H.A. El-Hofy, "Computer Aided Desgin of Tool-Shape for EC-Sizing of Drilled Holes", Alexandria Engineering Journal Vol. 28, No 3, pp. 383-402, (1989).

(14) H.A. El-Hofy, "Towards Electrochemical Machining of Dies and Molds", Alexandria Engineering Journal Vol. 34 No 2, pp. 185-192, (1995).

(15) T.S.R. Murthy, "A Comparison of Different Algorithms for Circularity Evaluation", Prec. Eng., Vol. 8, No.1, pp. 19-23, (1986).

(16) International Standard, "Methods for the Assessment of Departure From Roundness Measurement of Variation in radius", ISO 4291, (1985).

(17) D.C. Montgomery, and G. C. Runger, "Applied Statistics and Probability for Engineers", John Wiley and Sons, Inc, New York, (1994).

(18) S.H. Cheraghi, H.S. Lim and S. Motavalli, "Straightness and Flatness Tolerance Evaluation: an Optimization Approach", Prec. Eng., Vol. 18, No. 1, pp. 30-37, (1996).

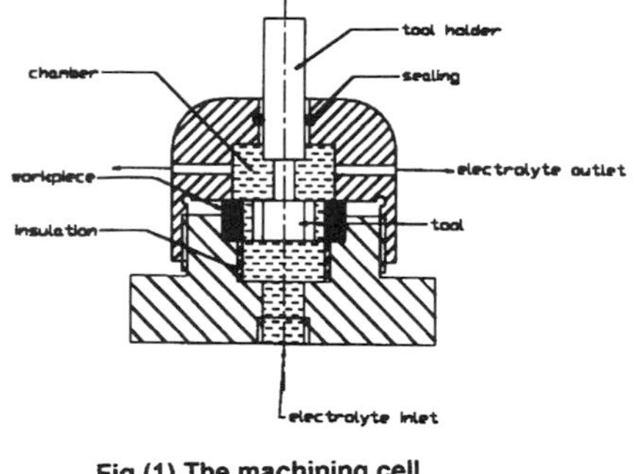

Fig (1) The machining cell

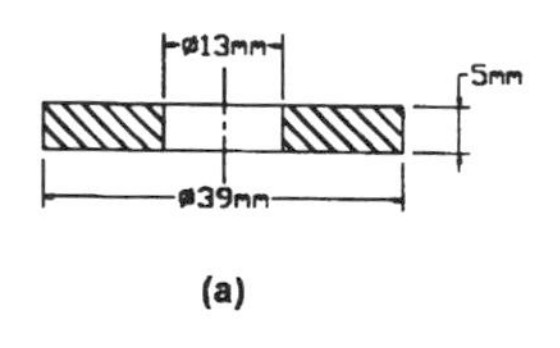

(a)

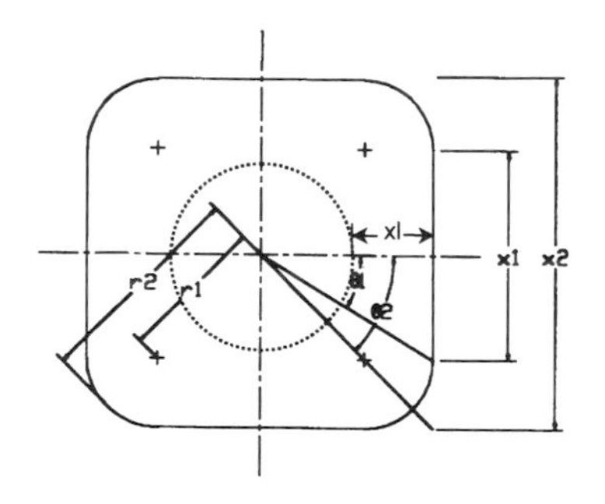

Fig (3) Main dimensions of machined hole

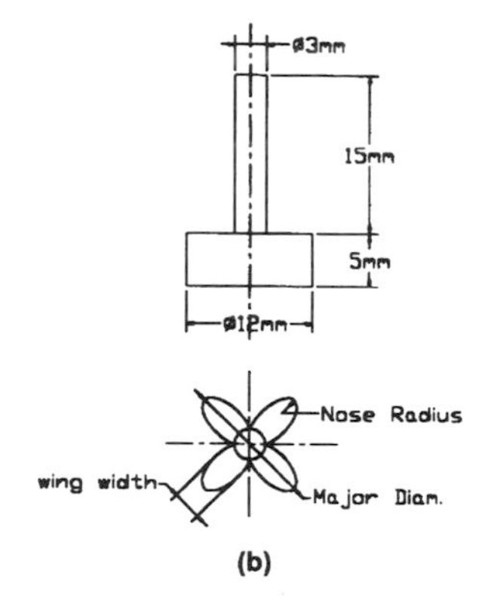

(b)

Fig (2) Shape and dimensions of, the workpiece (a), and tool (b)

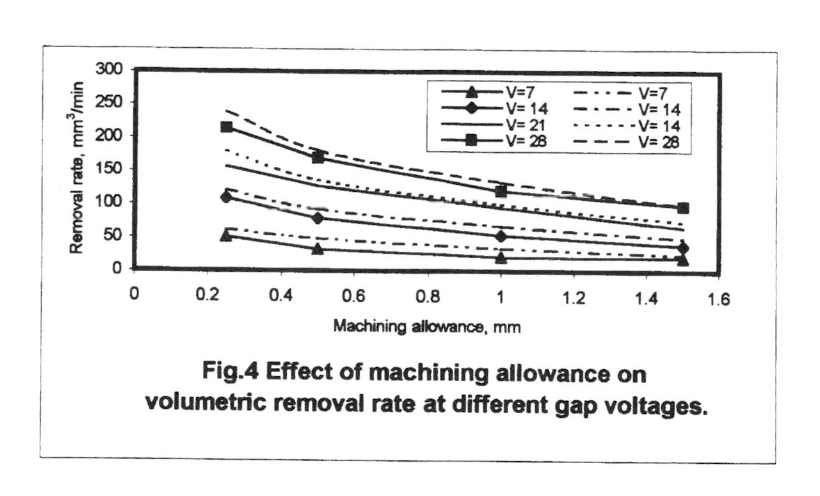

Fig.4 Effect of machining allowance on volumetric removal rate at different gap voltages.

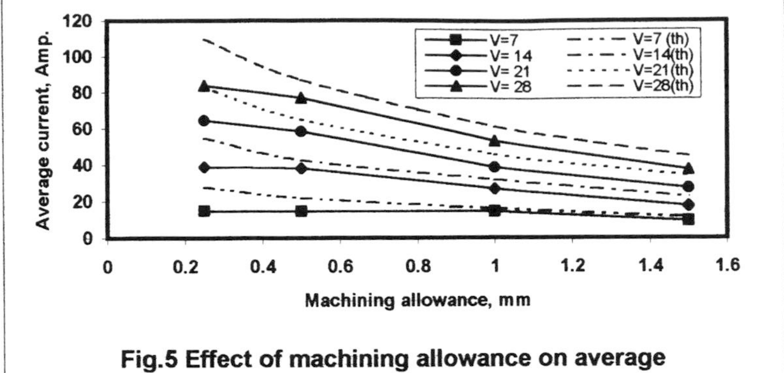

Fig.5 Effect of machining allowance on average current at different gap voltages.

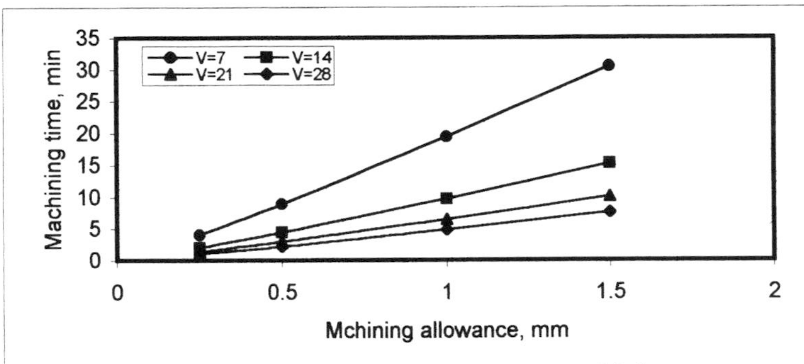

Fig.6 Effect of machining allowance on machining time at different gap voltages.

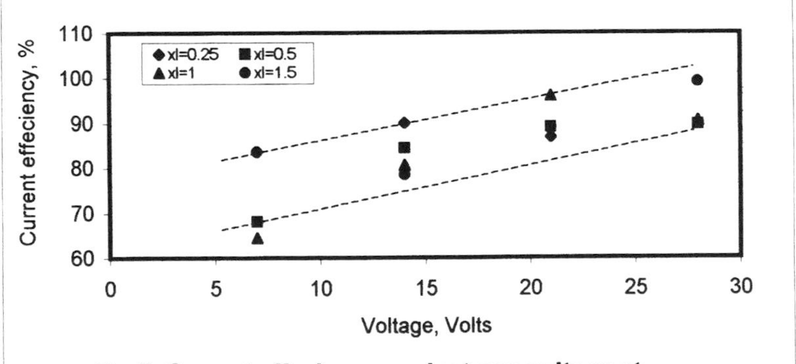

Fig.7 Currect effeciency against gap voltage at different machining allowances

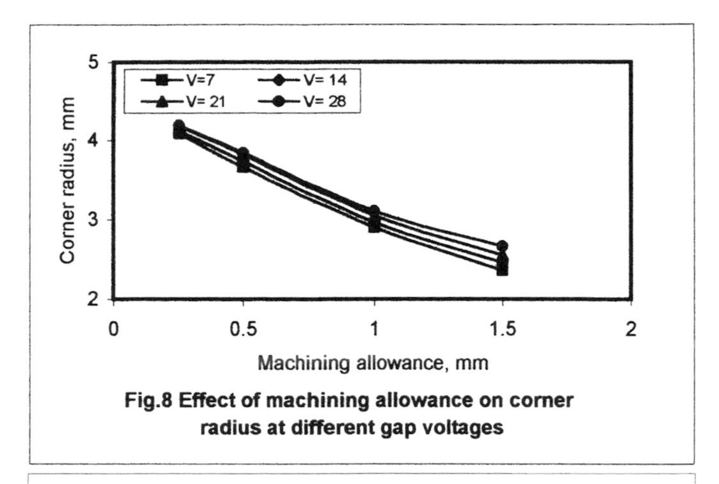

Fig.8 Effect of machining allowance on corner radius at different gap voltages

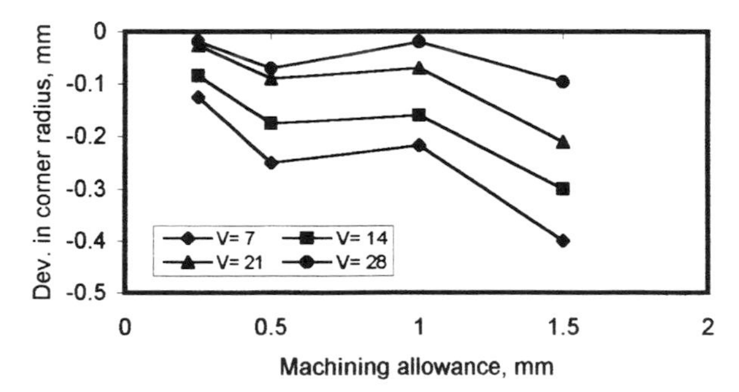

Fig.9 Deviation in corner radius from theoretical values at different machining allowances

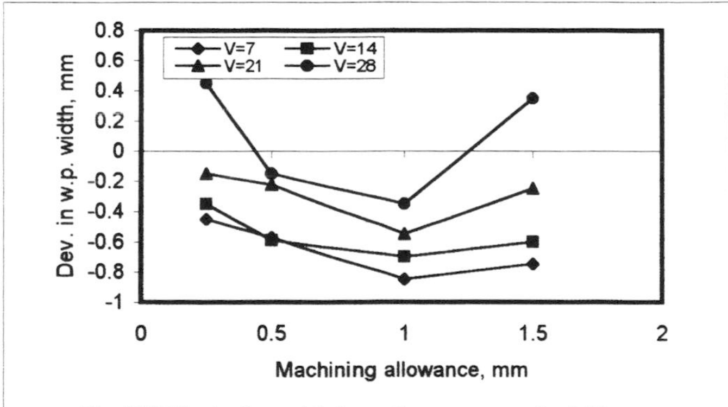

Fig.10 Effect of machining allowance on deviation in w.p. width from theoretical values

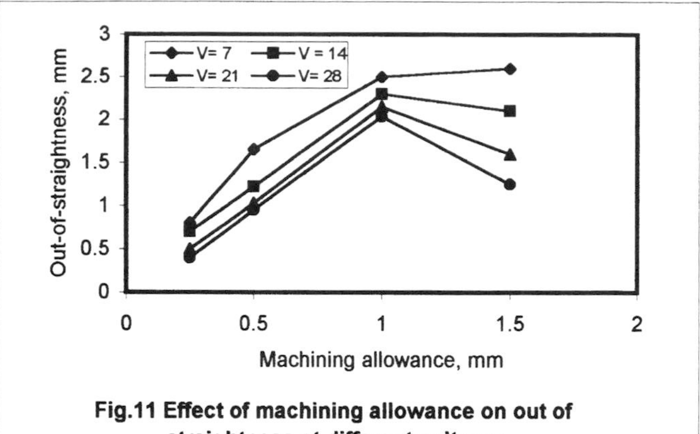

Fig.11 Effect of machining allowance on out of straightness at different voltages

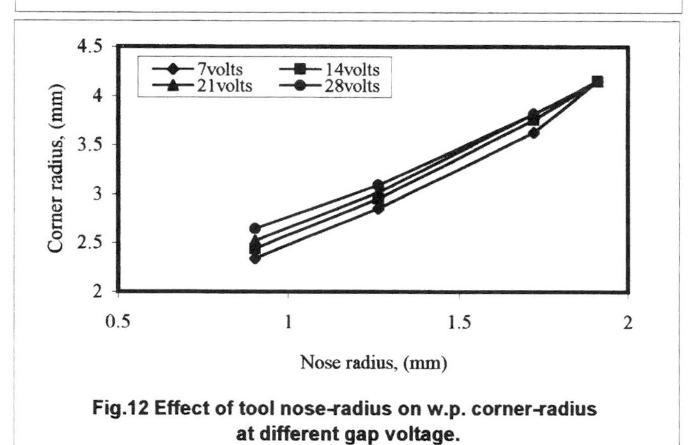

Fig.12 Effect of tool nose-radius on w.p. corner-radius at different gap voltage.

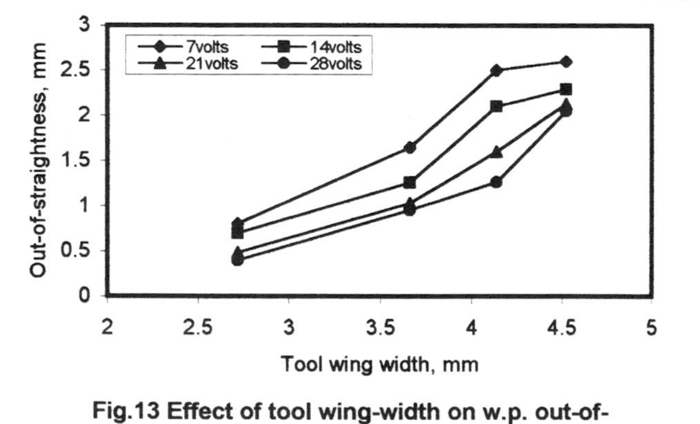

Fig.13 Effect of tool wing-width on w.p. out-of-straightness at different gap voltages

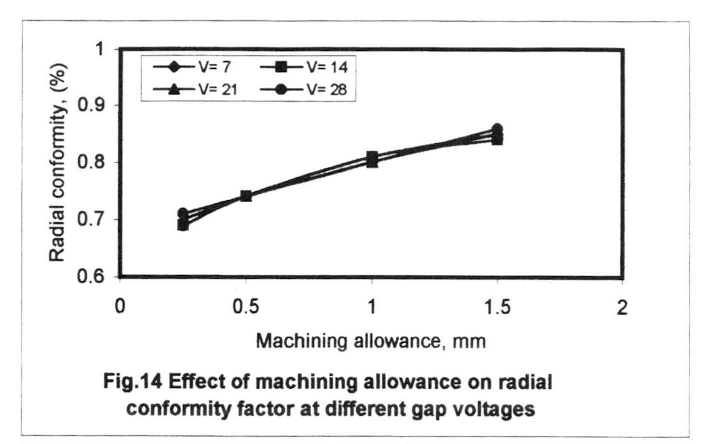

Fig.14 Effect of machining allowance on radial conformity factor at different gap voltages

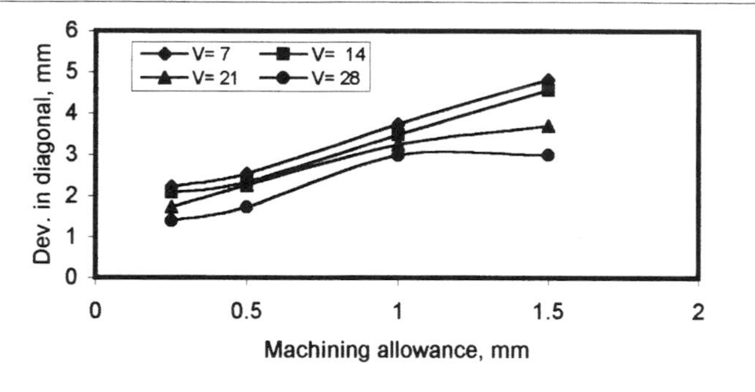

Fig.15 Effect of machining allowance on deviation in w.p. diagonal at different gap voltages

CAPE/096/2000

The laser drilling of multi-layer systems in jet engine components

A CORCORAN, L SEXTON, and **G BYRNE**
Department of Mechanical Engineering, University College Dublin, Ireland
B SEAMAN
SIFCO (Ireland) Limited, Cork, Ireland

ABSTRACT

Due to continuing advances in Jet Engine efficiency, components are exposed to ever-increasing combustion and exhaust gas temperatures. Air plasma sprayed Thermal Barrier Coatings (TBC's) protect Jet Engine components from direct exposure to the corrosive temperatures. The TBC is affixed to the superalloy substrate by an intermediate bond coat thus producing a multi-layer material system.
A laser drilling technique has been developed, to generate cooling holes in multi-layer systems, for use in the Aerospace industry. The cooling holes are required to conform to standards stipulated by the Original Engine Manufacturer (OEM). Two Nd:YAG lasers were used to study the parameters affecting hole generation in superalloy materials. This paper reports on the investigation into the negative effects of percussion laser drilling on material interfaces, bond strength, and the negative effects on the individual microstructures such as remelt-layers and microcracking.
Keywords: Laser, Percussion Drilling, Pulse Energy, Pulse Shape, Microcracking, Remelt Layer

1 INTRODUCTION

Laser drilling has become the accepted economical process for drilling thousands of closely spaced holes in structures such as aircraft wings and aircraft engine components. A specific application of modern laser technology is the drilling of cooling holes in aircraft engine 'hot-end components' such as Combustion Chambers, Nozzle Guide Vanes and Turbine Blades. The laser drilling of these holes is a contact-free optical machining process.

Yeo[1] describes hole drilling by a laser as 'a thermal process, where a high intensity beam is focussed to a spot of sufficiently high energy so as to vaporise almost any substance in its path'. Two types of laser drilling exist, Trepan and Percussion laser drilling. Trepan drilling involves cutting around the circumference of the hole to be generated, whereas Percussion drilling 'punches' directly through the workpiece material with no relative movement of the laser or workpiece – see Figure 1. Thus, an inherent advantage of the Percussion drilling process is the reduction in processing time from 4-10 seconds for Trepan Drilling to < 1 second. Considering that modern aero-engines contain up to 100,000 such cooling holes, percussion drilling represents a significant time and cost benefit to the manufacturer. Schematic representations of the Percussion and Trepan drilling techniques are illustrated in Figure 1.

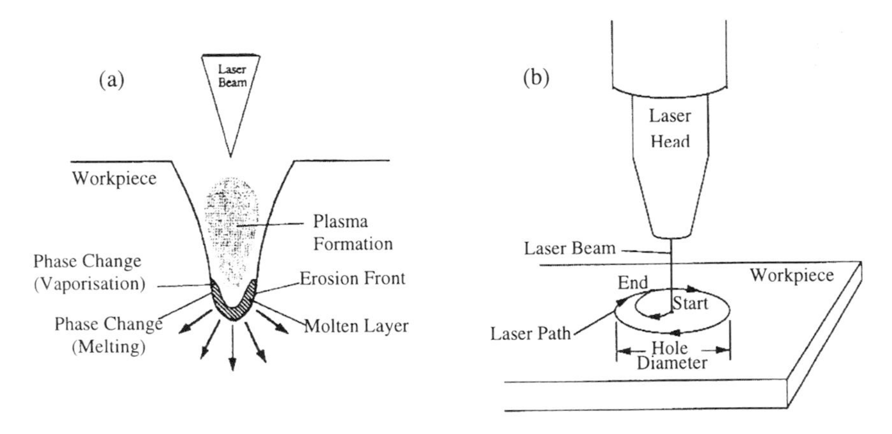

Figure 1: (a) Schematic of the Percussion drilling process
(b)Schematic of the Trepan drilling process

Laser drilled holes in aero-engine components must comply with strict quality standards that determine them suitable for in-service use. Much research into the suitability of percussion drilled holes for use in aerospace components has been carried out.

Research into the Percussion drilling process is primarily concerned with the geometrical and metallurgical quality of the hole. Some negative effects of the laser drilling process include microcracking, remaining remelt–material and taper of the drilled hole. Many parametric studies have been carried out on aerospace and other materials, to assess the relationship between various laser parameters and hole quality characteristics. French et al. [2], have used high speed filming and factorial experimental design to investigate the percussion drilling of uncoated aerospace alloys. Kamalu [4] also uses statistical design and analysis procedures to investigate the effects of the main parameters on the percussion drilling of 3mm thick Nickel based superalloy materials. Similar work is detailed in papers by Yilbas [5, 6] and Tam [7].

Investigation into the laser percussion drilling of coated aerospace materials has been carried out using both a Nd:YAG JK704 and Nd:YAG SD/C150 lasers. The high peak power of the solid-state laser is ideal to establish a high pressure in a keyhole and so efficiently blow melt out of the hole. The effect of laser parameters on the metallurgical quality of the percussion-drilled hole is investigated using statistical design of experiments.

The materials under test are aerospace superalloy materials, HastalloyX and Haynes188, coated with a Thermal Barrier Coating and intermediate plasma-sprayed bond coat. The Thermal Barrier Coating is an Yittria stabilised Zirconia and is bonded to the substrate material by means of a plasma sprayed MCrAlY bond coat.

2 EXERIMENTAL

2.1 Experimental Objective

The objective of this experiment is to identify the parameters that have the most significant effect on the metallurgical quality of laser drilled holes and thus optimise the process. The laser drilling process is complimented by a large number of process parameters. Figure 2 illustrates the parameters investigated.

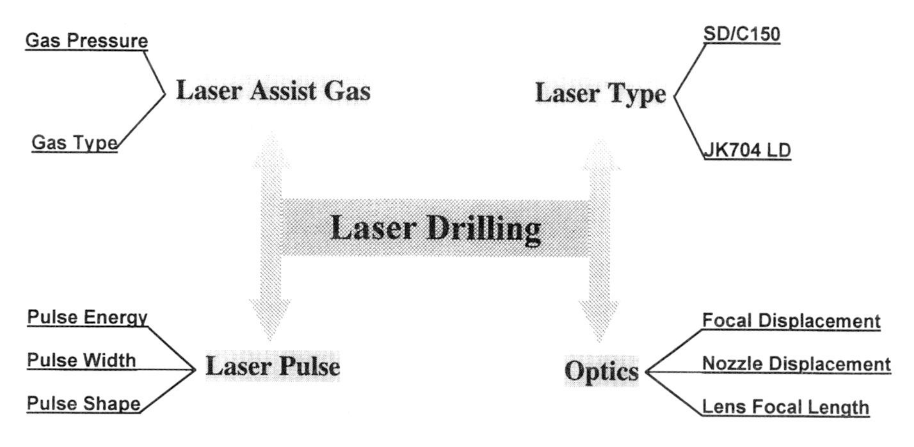

Figure 2: Laser Process Parameters Investigated

2.2 Experimental Procedure

2.2.1 Laser Systems

Two sets of tests were conducted using different laser systems, a JK704LD and a StarDrill/Cut 150. Specifications of each of the laser systems are outlined in Table 1 below:

Laser Specifications	Laser Type	
	JK704	SD/C150
Mode	LD2 (Drilling)	Drilling/Cutting
Peak Power (kW)	20	12
Average Power (W)	230W	150
Pulse Width range (ms)	0.3 – 5	0.05 – 20
Repitition Rate range (Hz)	5 – 150	1 – 2000
Pulse Energy (J)	0.3 – 50	15
Focal Lens used (mm)	120	80

Table 1: JK704 and SD/C150 Laser System specifications

The SD/C150 system incorporates a two-axis CNC system. Figure 3 illustrates a schematic of the system. The JK704 LD system features a five-axis CNC capability.

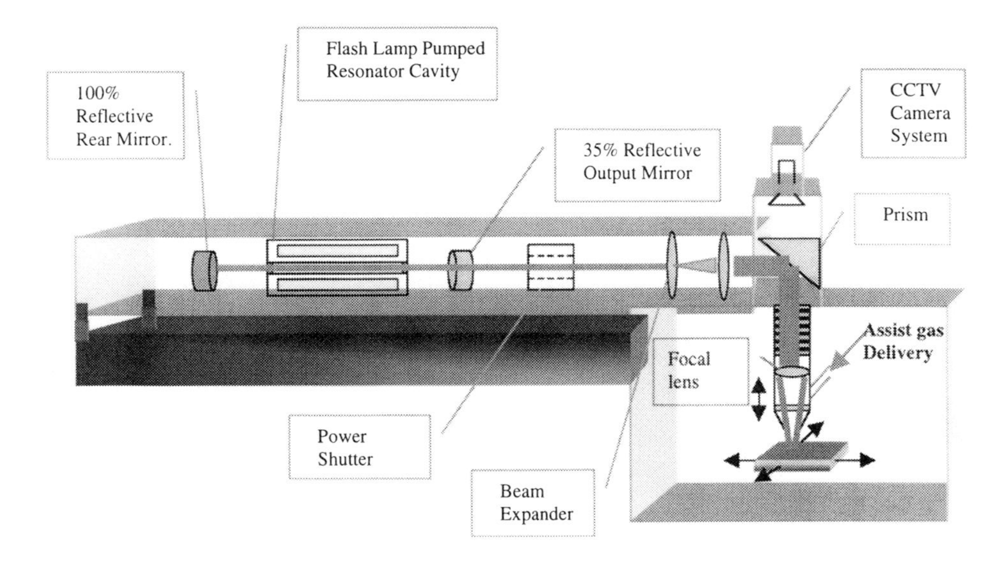

Figure 3: Schematic of StarDrill/Cut 150 Resonator and Workspace Cabinet

2.2.2 Design of Experiments

Figure 2 outlines the parameters investigated in the experiment. Traditionally, laser hole drilling experiments have been performed using a one-at-a-time approach. This method however, assumes that all other parameters remain constant other than the one being varied. Since interactions are not considered this method may lead to misleading results.

Thus, it was decided to use a factorial study that considered a number of parameters at a number of levels. The experimental design covers all the possible combinations of the parameters at the levels chosen.

Experimentation on the SD/C 150 used an 18 row orthogonal array to investigate the effect of each of the individual parameters and related levels outlined in Table 2. The orthogonality of the array means that the average effects of the factors can be estimated without the fear that the effects of other factors may interfere. The base material of the coated samples processed using the SD/C 150 is Haynes®188 alloy.

	Level 1	Level 2	Level 3
Assist Gas Type	Air	Argon	Oxygen
***Voltage (V)**	500	575	650
Pulse Duration (ms)	1	1.4	

*It should be noted that varying the voltage available to the flashlamps regulates the flashlamp power. The voltage available to the flashlamps determines the level of excitation of the Nd:YAG rod, and hence the energy of the laser pulses generated.

Table 2: Parameters and associated levels investigated using SD/C 150

Experimentation on the JK704 LD laser used a 12 row orthogonal array to investigate the effect of each of the parameters and related levels outlined in Table 3. The base material in this instance is HastalloyX, coated with a plasma sprayed bond coat and Thermal Barrier Coating.

	Level 1	Level 2	Level 3
Pulse Shape	Ramp-up	Square	Ramp-down
^% Height	95%	85%	
Pulse Duration (ms)	1.1	1.5	

^The percentage of available current that is made available to the rods is regulated by the %height function of the JK704.

Table 3: Parameters and Levels investigated using JK704

Figure 4 features an illustration of the pulse shapes outlined in Table 3.

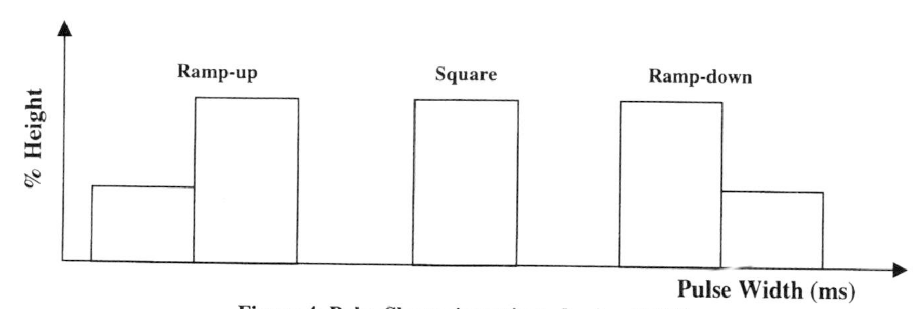

Figure 4: Pulse Shapes investigated using JK704

2.2.3 Procedure

Each of the experiments of the experimental arrays were carried out with two replications so as to minimise any experimental error. Mean diameters of the laser-drilled holes were recorded using a Profilometer. The samples were then sectioned approximately 1mm from the hole circumference using sufficient cooling lubricant to minimise negative thermal effects at the cut interface. A cold mounting technique was employed, thus eliminating errors arising from stresses and temperatures induced by hot mounting. Using a relatively coarse paper, the samples were ground until a suitable cross section of the hole was reached. Progressively finer grinding, polishing and etching completed the sample preparation.

3 RESULTS

3.1 StarDrill/Cut 150 Experimentation

The output responses recorded were:

- Mean Diameter
- Remelt-Layer Thickness
- Microcracking Depth
- Delamination of the Thermal Barrier Coating (Spalling)*
- Hole Diameter

*Delamination is measured prior to further testing of the drilled components at in-service conditions e.g. Thermal Cycling.

Remelt-layer thickness and microcracking measurements are used as examples of some of the results attained.
Figure 5 displays the direct results of experiments 1-18 in the Full Factorial Analysis.

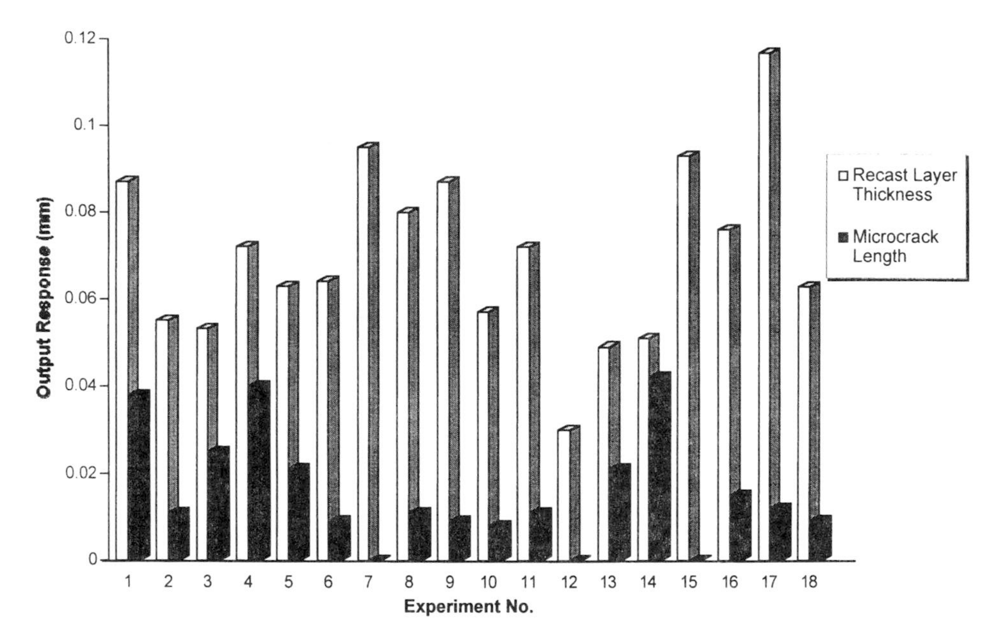

Figure 5: Direct Comparison of the Results attained for Experiments 1-18 conducted using the SD/C150

Using the results attained in Figure 5, parameter effects were calculated for each of the parameters investigated. This process involves calculating the average of each of the responses at each of the given levels and comparing these. The comparison allows one to see the factors that have the most significant effect on a given output response. A Parameter Effects Plot was generated using histograms– see Figure 6.

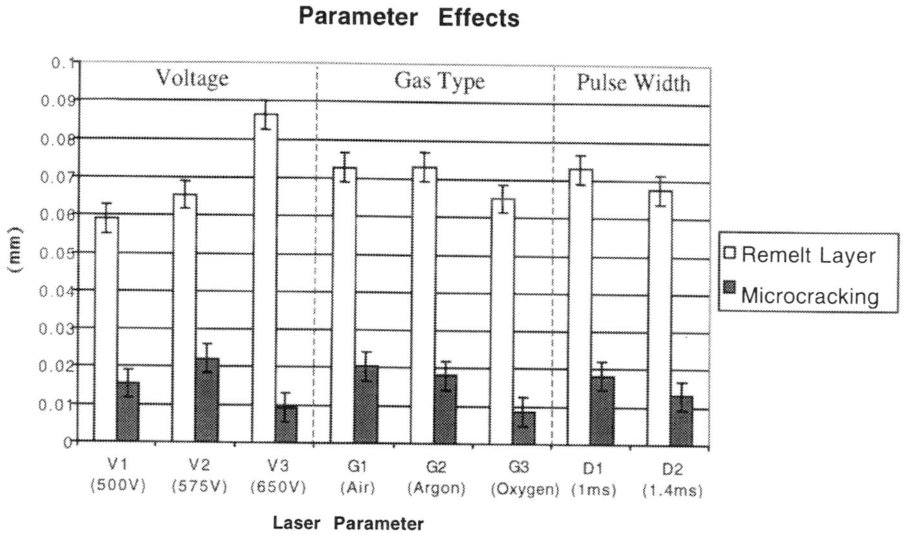

Parameter Effects

Voltage | Gas Type | Pulse Width

(mm)

0 0.01 0.02 0.03 0.04 0.05 0.06 0.07 0.08 0.09 0.1

V1 (500V) V2 (575V) V3 (650V) G1 (Air) G2 (Argon) G3 (Oxygen) D1 (1ms) D2 (1.4ms)

Laser Parameter

Remelt Layer

Microcracking

Figure 6: Parameter Effects Plot for SD/C Percussion Drilling Experiment

As a complete factorial analysis was carried out, it was possible to generate a surface map representing the results attained at each combination of the parameters under investigation. Figure 7 represents the surface generated by the microcracking results attained for each of the parameter combinations.

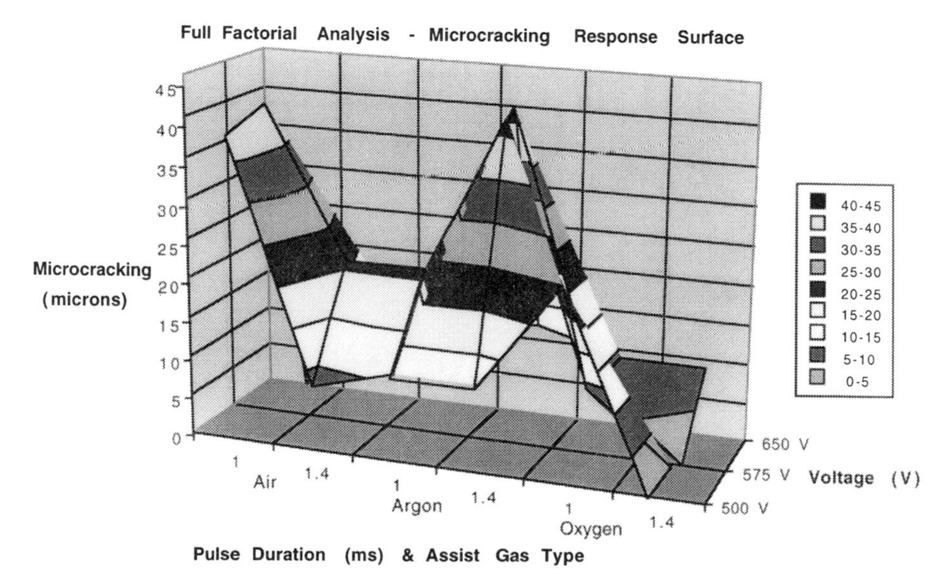

Figure 7: Parameter Map generated by complete factorial analysis

Using the parameter effects analysis and plot (Figure 6), the results suggested that to minimise microcracking a high voltage value coupled with long pulse duration and oxygen assist gas was to be used. This is also apparent from the surface plot illustrated in Figure 7. Confirmation experiments were carried out to verify this.

Figure 9(a), (b)&(c) show micrographs detailing the cross sections of laser drilled holes processed at the optimum configurations.

3.2 JK704 LD Experimentation

The output responses recorded were:

- Mean Diameter
- Remelt Layer Thickness
- Microcracking Depth
- Delamination of the Thermal Barrier Coating (Spalling)
- Taper

A Full Factorial Analysis was carried out on thermally coated Nickel-base superalloy, HastalloyX, using the JK704. As with the SD/C 150, a parameter effects plot was generated, see Figure 8. The average output responses for microcracking and remelt layer thickness were plotted for each of the parameter levels.

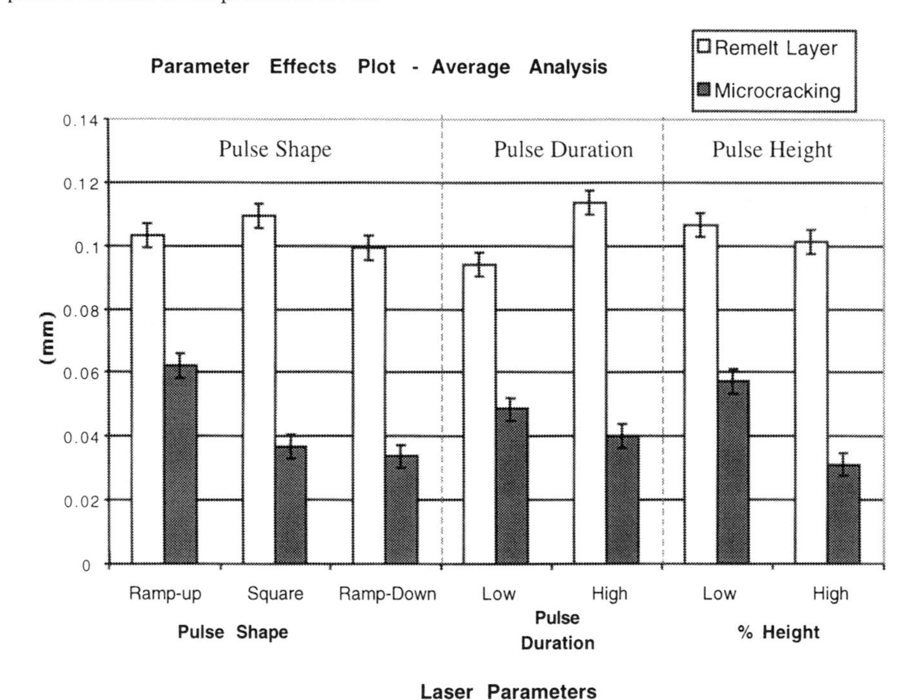

Figure 8: Parameter Effects Plot of JK704 percussion drilling experiment

Using the results attained, optimum parameters were derived and confirmation tests have been carried out to determine the validity of the experiment. The samples drilled at these parameters possess very little microcracking and adherent remelt material. The micrographs of Figure 9 illustrate examples of the optimum results attained.

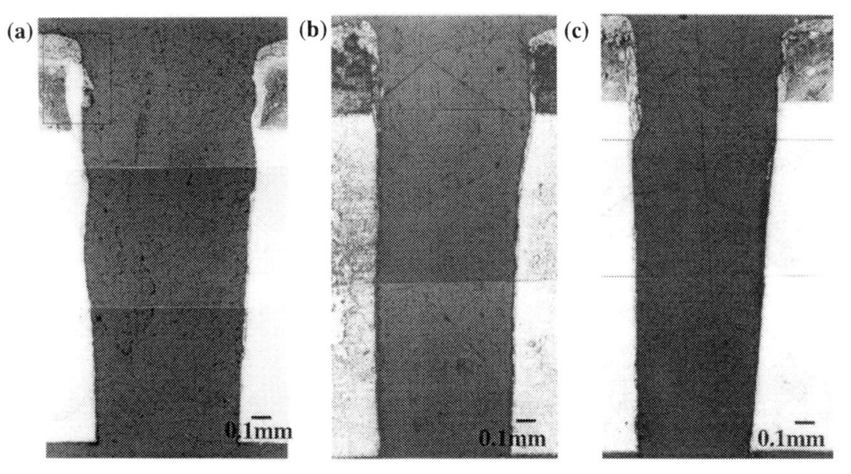

Figure 9: Micrographs of hole cross sections attained after drilling at optimum parameters using (a) JK704 and (b)&(c) using SD/C150

3.3 Discussion of results

A prevalent trend throughout the course of this experiment is summarised by Figure 10. Figure 10 compares microcracking and remelt thickness outputs for varying levels of pulse average power in the SD/C150 experiment. The results show clearly that increasing pulse average power decreases microcracking and increases the thickness of adherent remelt material.

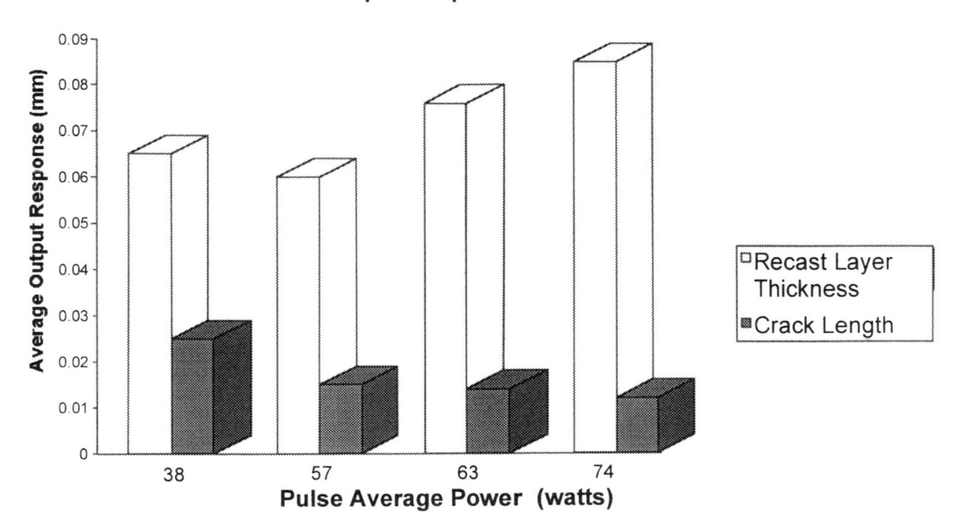

Figure 10: Pulse Average Power Vs Microcracking and Remelt Thickness for SD/C150 experiment

Other parameters and their effect on the quality of the laser-drilled hole have been investigated.

5 CONCLUSIONS

Results attained from the experiments carried out yield the following conclusions:

- Pulse Energy
  High pulse energy reduces the level of microcracking in the percussion laser-drilled hole. Low pulse energy however, reduces the level of adherent remelt material remaining on the side of the laser-drilled hole. Thus a conflict arises where one wants to minimise both these output responses.

- Assist Gas
  The results suggest that Oxygen is the most suitable assist gas for the drilling process.
  A limitation of using oxygen for this process is the dark metal oxide residue that remains on the bright component surface in the vicinity of the laser-drilled hole. The residue is metal oxide that did not completely escape the component surface during its evacuation by the assist gas.

- Pulse Width
  Results from both sets of experiments indicate that the longer pulse widths investigated reduce the depth of microcracking and the thickness of the remelt-layer of a laser drilled hole.

- Pulse Shape
  The experiments carried out would suggest that a ramp-down pulse is more suitable for minimising remelt and microcracking than a ramp-up or square pulse shape.

ACKNOWLEDGEMENTS

We gratefully acknowledge the funding and assistance from SIFCO (Ireland) Ltd. Carraigtohill, Cork, Ireland.

REFERENCES

(1) Yeo CY, *The Physics of Laser Processing*, J. of Matls. Proc. Tech. **42**, (1994), pp15-49

(2) French PW, DP Hand, C Peters, GJ Shannon, P Byrd & WM Steen, *Investigation of the Nd:YAG laser Percussion drilling process using high speed filming*, Sect. B, ICALEO '98

(3) Kamulu J, P Byrd, *Statistical Design of Laser Drilling Experiments*, Sect.B, ICALEO '98

(4) Yilbas BS, *Parametric Study to improve laser hole drilling process*, J. of Materials Processing Technology, **70**, (1997) 264 – 273.

(5) Yilbas BS, M Sami, *Study into the effect of beam waist position on hole formation in the laser drilling process*, Proc. ImechE **210**, 1996.

(6) Tam SC, CY Yeo, S Jana, Michael WS Lau, Lennie EN Lim, LJ Yang, and Yusoff Md Noor, *Optimisation of Laser deep hole drilling of Inconel 718 using the Taguchi Method*, J. of Materials Processing Technology, **37** (1993) pp.741 – 757.

CAPE/049/2000

An experimental study on molecular dynamics simulation in nanometer grinding

S YU, B LIN, and **Q GUAN,**
School of Mechanical Engineering, Tianjin University, China
K CHENG
School of Engineering, Leeds Metropolitan Univeristy, UK

ABSTRACT

Molecular dynamics method which is different to continuous linear mechanics is employed to survey the features of grinding energy dissipation, grinding forces, stress state and grinding temperature in the atomic space, and then explain the micro-scale mechanism of material removal and surface generation. The research shows that the atoms of the lattice reconstituting and some non-crystal layer are pilled up on the front of the abrasive grain, as a result of the continuous advancement of the abrasive grain, the material are removed and formed the grinding chips. The degenerating layer of machined surface are formed with the reconstituting of non-crystal atoms and fracture atomic bonds, it consists of outer non-crystal and inner lattice deformation layers.
Keywords: Nanometer grinding, Molecular dynamics simulation, Dislocation slip, Mechanism of material removal, Mechanism of surface generation

1 INTRUDUCTION

Ultra-precision machining, which enables us to produce optical, mechanical and electronic components with micrometer or sub-micrometer form accuracy and surface roughness to within a few tens nanometer, is one of the most successful development within precision engineering. Ductile range grinding of hard-brittle materials is an important newer research topic, researchers show that even in machining process of brittle solids, material can be removed by the action of plastic flow just as metal cutting. But there is not an efficient experiment and observation method for many physical phenomenon of occurring in micro-scale machining region, so many problems in ultra-precision and nanometer machining remains to be unsolved. The Molecular Dynamics method has spread from physics into the materials science, it is now entering the area of mechanical engineering, especially in ultra-precision and nanometer machining. One of the pioneering studied on MD simulation of nanometric cutting was initiated at LLNL in late 1980's, and proposed a MD model of the orthogonal cutting process. This was followed by the work of T. Inamura et. al. in Japan [1] who analyzed the 2-D nanometric cutting mechanism of copper with a diamond tool. Based on the nonlinear finite element formulation which regards atoms and atomic interaction as nodes and elements, the method proposed can handle discontinuous phenomenon due to instantaneous propagation of dislocation in a workpiece during cutting, and revealed that the process of chip formation as well as the stress distribution on the tool face during cutting is strongly dependent on the type of interaction energy. S. Shimida et. al. confirmed that very fine chips the uncut thickness of which is at the order of 1 nm can be stably removed in diamond turning of free machining workmaterials, and analyzed the behavior of an atomic solid model using MD analysis method [2]. Komandari et al. reported MD simulation studies on machining with large negative rake angle tools to simulate grinding process and compared them with the experimental results [3]. R. Rentsch et al. reported the MD simulation for abrasive process, they carried out the simulation of indentation tests and cutting process, and pointed out the difference between cutting and grinding process, especially the pile-up phenomenon in abrasive machining [4]. Overall, many of the challenging machining problems is yet to be investigated.

This paper will use MD method, which is different to continuous linear mechanics, to survey the features of grinding energy dissipation, grinding stress, strain state and grinding temperature in the atomic space, and explain the micro-scale mechanism of material removal and surface generation in nanometer grinding.

2 GRINDING MODEL AND POTENTIAL ENERGY

2.1 Grinding model

The simulation of nanometer grinding has been carried out using the model shown in Fig.1, where the workpiece and tool materials is assumed to be monocrystalline silicon with covalent bond and diamond, respectively. Workpiece dimension, grinding grain size and the depth of grinding are determined by means of the practical requirement of simulation experiments. For the purpose of simplicity, crystal orientation and grinding direction are (0, 1, 0) and (1, 0, 0). Tool atoms are assumed to be rigid, that is no deformation and wear. Simulation experiments are conducted in a quasi three-dimensional space, the calculation is mainly completed in x and y coordinates, but considered the effect z direction on x and y directions.

2.2 Potential energy

In order to carry out MD simulation, it is necessary to know correct interaction potentials of a system of many atoms, the general potential energy can be described by Microcanonical Ensemble contained N atoms as follows:

$$\Phi(R) = \sum_{i=1}^{N} \sum_{j>i}^{N} \phi(r_{ij}) \tag{1}$$

where i and j label the atoms of systems, r_{ij} is the distance between atoms i and j, $r_{ij} = |r_i - r_j|$, $\Phi(r_{i,j})$ is the interaction potential function of atoms i and j. The force F acted on atom i can be presented as follows:

$$F_i = \sum_{j \neq i}^{N} \nabla_i \Phi(r_{ij}) \qquad (j=1, 2, \ldots. N) \tag{2}$$

where $\nabla_i = \frac{\partial}{\partial x_i} i + \frac{\partial}{\partial y_i} j + \frac{\partial}{\partial z_i} k$.

Formula (2) represents effect other atoms on atom i in the system. In order to determine correct interaction potentials, it is necessary to know the proper electron ground state, but the calculating of electron ground state of condensed physics is a very complicated question of many body quantum system. According to the first principle of local densing functional, it needs to carry out electric structure calculations for $10^4 \sim 10^6$ sorts of atomic configuration due to its complexity. It is hardly to achieve statistic mechanics simulation for a long time, therefore an empirical potential is used to subsitude real action potential in MD simulation.

It is necessary to consider the mutual effect of multiple atomic bond for complicated multibody such as diamond crystal et al, therefore, Tersoff potential function suited to multibody system is employed to describe covalent bond crystals, silicon-silicon, carbon-carbon and silicon-carbon, in simulation experiments of nanometer grinding. The formula is given by:

$$\Phi = \sum_i \Phi_i = \frac{1}{2} \sum_{i \neq j} \Phi_{ij} \tag{3}$$

here $\Phi_{ij} = f_c(r_{ij})[f_R(r_{ij}) + b_{ij} f_A(r_{ij})]$ and

$$f_R(r_{ij}) = A_{ij} \exp(-\lambda_{ij} r_{ij})$$

$$f_A(r_{ij}) = -B_{ij} \exp(-\mu_{ij} r_{ij})$$

$$f_c(r_{ij}) = \begin{cases} 1 & r_{ij} < R_{ij} \\ \frac{1}{2} + \frac{1}{2}\cos[\pi(r_{ij} - R_{ij})/(S_{ij} - R_{ij})] & R_{ij} < r_{ij} < S_{ij} \\ 0 & r_{ij} > S_{ij} \end{cases}$$

$$b_{ij} = \chi_{ij}(1 + \beta_i^{n_i} \xi_{ij}^{n_i})^{-1/2n_i}$$

$$\xi_{ij} = \sum_{k \neq nij} f_c(r_{ik}) \omega_{ik} g(\theta_{ijk})$$

$$g(\theta_{ijk}) = 1 + C_i^2/d_i^2 - C_i^2/[d_i^2 + (h_i - \cos\theta_{ijk})^2]$$

$$\lambda_{ij} = (\lambda_i + \lambda_j)/2 \qquad R_{ij} = (R_i R_j)^{1/2}$$

$$\mu_{ij} = (\mu_i + \mu_j)/2 \qquad \delta_{ij} = (S_i S_j)^{1/2}$$

$A_{ij} = (A_i A_j)^{1/2} \qquad B_{ij} = (B_i B_j)^{1/2}$

In the above formulas *i, j, k* mean the under-marker of every atom, r_{ij} the bond length between two atoms, θ_{ijk} is the bond angle between atoms with the under-markers *i, j* and *k*, which also mean the angle formed by atom *i* and *j* with atom *k* at the range of truncation radius. χ_{ij} is a distinct revising coefficient which shows the strength of the hetropolar valence-bond. Besides, parameter ω_{ij} is also a revising coefficient adopted while there is more complex geometrical shape distribution of atoms and molecules, and let $\omega_{ij} = 1, \chi_{ij} = 1$ and $\chi_{ij} = \chi_{ji}$. The function $\phi_{ij}(R)$ in formula (3) express the energy relation of two atoms. $f_R(r_{ij})$ and $f_A(r_{ij})$ respectively represent the repulsion item and attraction item functions. In the truncation function, the truncation radius of atoms is at the range of the nearest atoms. Moreover, the affection of θ_{ijk} is also considered in the functions.

3 SIMULATION OF THE ABRASIVE PROCESS

After the definition of potential function, the Hummiton Quantity describing the interaction of N atoms can be acquired as follows:

$$H(p \cdot q) = \sum_{i=1}^{N} \sum_{j=1}^{N} \phi(r_{ij}) + \sum_{i=1}^{N} \frac{p^2{}_i}{2m_i} \tag{4}$$

here q_i, p_i and m_i are coordinate, momentum and mass of atom *j*. Then, the canonical ensemble motion equation describing the system will be expressed like this:

$$m_i \frac{d^2 q_i}{dt^2} = F_i = -\sum_{j \neq 1}^{N} \nabla_i \phi(r_{ij}) \tag{5}$$

From formula (5), we get the interaction force between atoms. The force acted from atom *i* to atom *j* is expressed by its three components on three coordinate axes. Summarized the above analysis and discussions, we can found the MD simulation algorithm as follows:

Step 1: Give the original position r of atoms.

Step 2: Give the original velocity v of atoms.
Under the corresponded temperature condition and from Maxwell Distributal Deviates get the original velocities of atoms needed to start the algorithm.

Step 3: Calculate the position of step n+1.

$$r_i^{n+1} = r_i^n + hv_i^n + \frac{1}{2m} h^2 F_i^n$$

Step 4: Calculate the force F_i^n of step n+1.

Step 5: Calculate the velocities of step n+1.

$$v_i^{n+1} = v_i^n + h(F_i^{n+1} + F_i^n)/2m$$

Step 6: Calculate the scale factor:

$$\beta = [T^{*}(N-1)/16\sum_{i} v_i^2]$$

Here T^{*} is reference temperature.
Step 7: Constraint the velocities of all atoms:

$$v_i^{n+1} = v_i^{n+1} * \beta$$

Step 8: Return to Step3 and repeat the algorithm through to the end.
Where scale factor β is set from the convention condition of system mentioned above. Periodic boundary condition is adopted to process the non-rigid boundary. Considering the efficiency of simulation and the capability of computer, we need to break in the potential, that is also to say while thinking of the case of simulation experiment, the value of truncation radius R_c should be selected reasonably. For the convenience of computer simulation, interaction forces between atoms can be represented by simple form.

4 RESULT AND ANALYSIS

Fig 2 expresses the transient state of an abrasive with the diameter of 2 nm at different time during grinding. The graph shows that following with the continuous advancement of the abrasive grain, the dislocation movements by slip of atoms appear on the front of the abrasive grain (refer to Fig 2 (a)). Then the dislocation pill up and form the plastic deformation of the crystal lattice (refer to Fig 2 (b, c)). The phenomenon is intensified with the continuous advancement of the abrasive grain, at last the atom crystal on the front of the abrasive grain are pilled up disorderly, which leads to the appearance of non-crystal layer. There is some features shown in this procedure:

(1) Non-crystal layer of atoms appears on the front of the abrasive grain, not like the one that appears under the indenter in the indentation test. The main reason is the action of tangential force in the grinding process.

(2) Because of the bigger negative rake angle, the abrasive resultant force points to the front and under the abrasive. As a result, not only under high compressive stress, the dislocation slip and deformation of the atom crystal lattice on the front and under the abrasive grain appear, but also for the action of shearing stress (abrasive shearing stress mainly distribute under the surface crystal lattice as shown in Fig 6) the atom bond break. Under the co-action of compressive stress and shearing stress, some crystal lattice will be reconstituted and the non-crystal layer formed due to atomic bond breakage. With the continuous advancement of the abrasive grain, some atoms of lattice reconstituted and a part of non-crystal pill up on the front of the abrasive grain (refer to Fig 2(b, c)). So the material is removed and the grinding chip forms.

(3) By the time the abrasive grain cut into the workpiece completely, the value of grinding force is basically stabilized to a certain level, increase and decrease about this level heavily. This kind of fluctuation will appear repeatedly while the abrasive grain advances continuously. This phenomenon is agree with the result of indentation test, that is the fluctuation of grinding

force is associated with the generation of dislocation (refer to Fig 3).

(4) The change of temperature in grinding process with single abrasive grain is larger than that in the indentation process. Moreover, the average temperature of the deformation zone is about 440K (refer to Fig 4). While considering the influence of abrasive grain number and dimension effect on it, the grinding temperature should not be neglected in the nanometer machining procedure.

(5) For the action of normal force and tangential force, in the grinding process, grinding compressive stress primarily distributes on the front and under the abrasive grain as shown in Fig 5. Therefore, crystal lattice deformation and non-crystal layer primarily centralize under the front of the abrasive grain while the distribution of deformation and non-crystal layer under the abrasive grain relatively decrease. From the result of simulation, the atom bond of crystal lattice is broken by the action of shearing stress. At the same time, the extrusion action of compressive stress of the front and under the abrasive grain leads to deformation, dislocation and reconstitution of crystal lattice. To transmission dislocation in covalent crystal, the high potential barrier should be overcome. Therefore, when the deformation, dislocation and crystal reconstitution can't satisfy the requirement for releasing all energy, non-crystal layer will be generated certainly in the front of the abrasive grain to release energy. As a result of the continuous advancement of abrasive grain, under the action of compressive stress, non-crystal atoms under the front of abrasive grain and the fracture atom bond of machined surface will combine together and reconstitute to form the degenerating layer of machined surface. The degenerating layer consists of outer non-crystal layer and inner lattice deformation layer which has been proofed in our ultra-precision grinding experiments of engineering ceramics, the details has proposed in reference [6].

5 CONCLUSION

By means of molecular dynamic simulation and from the view of micro-scale, problems such as the physical and mechanical features, the mechanism of material removing and surface generation in the microscopic zone near the action point of abrasive grain in grinding process is intensively studied. Conclusion is drown as follows:

(1) As the abrasive grain cut into the workpiece continuously, the value of grinding force increase gradually in a repeat fluctuation manner. When the abrasive grain cut into the workpiece completely, the average value of grinding force stabilizes on a certain level, but the peak value increases and decreases intensively. This kind of fluctuation should be regarded to have relationship to the generation of dislocation slip and the plastic deformation of the crystal lattice.

(2) The mechanism of microscopic material removing and surface generation is researched. The MD simulation experiment shows that atoms of crystal lattice reconstituted and part of the non-crystal atoms are pilled up on the front of the abrasive grain. As a result of the continuous advancement of the abrasive grain, the material is removed and grinding chip forms at last. Simultaneity, under the action of compressive stress, non-crystal atoms under the front of abrasive grain and the fracture atom bond of machined surface will combine together and reconstitute to form the degenerating layer of machined surface consisted of outer non-crystal layer and inner lattice deformation layer.

(3) For generation of dislocation and increase of the kinetic energy of atoms, the temperature of atomic crystal lattice also increases, which intensify the thermal vibration of atomic crystal lattice and help to the plastic deformation of material.

(4) The MD simulation method supplies a new and efficient way for the study of current theory about micro-scale grinding. However, there are also some problems occurring in the method such as the long time spent in computer, the definition of potential function, the correct description and definition of material default, and the ignorance of the wear of diamond abrasive grain, which should be noticed and solved.

REFERENCE

[1] Inamura, t. et al. 1992, “Cutting Experiments in a Computer Using Atomic Models of Copper Crystal and a Diamond Tool”, Int. J. Japan soc. Prec. Eng. , 25, No.4, 259

[2] Shimida, S. et al. 1992, “Molecular Dynamics Analysis as Compared with Experimental Results of Micro-machining”, Annals of the CIRP, 41, No.1, 117

[3] Komanduri, R. et al. 1998, “Some Aspect of Machining with Negative Rake Tools simulating Grinding, An MD Simulation Approach”, Philo. Mag. B 77, No.1, 7

[4] Rentsch, R. et al. 1994, “Molecular Dynamics simulation for Abrasive Process”, Annals of the CIRP, 43, No.1, 327

[5] Tersoff, J. 1989, “Modeling Solid-state Chemistry: Interaction Potentials for Multi-component Systems”, Physical Review, 39, No.8, 5566

[6] Namer, M, etal, 1999, “Ductile Grinding of Engineering Ceramics”, Journal of Tianjin University, 32, No.4, 486

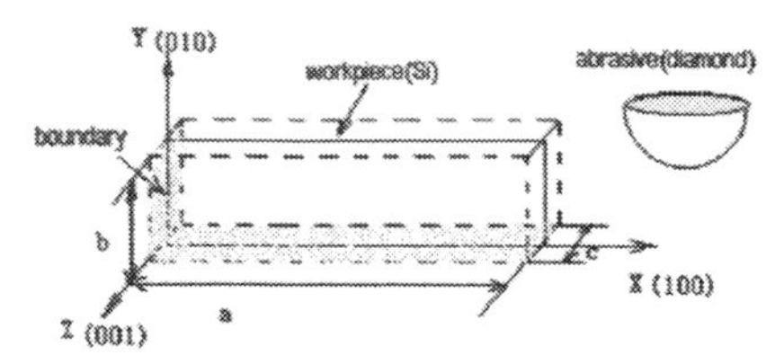

Fig 1 Grinding model for MD

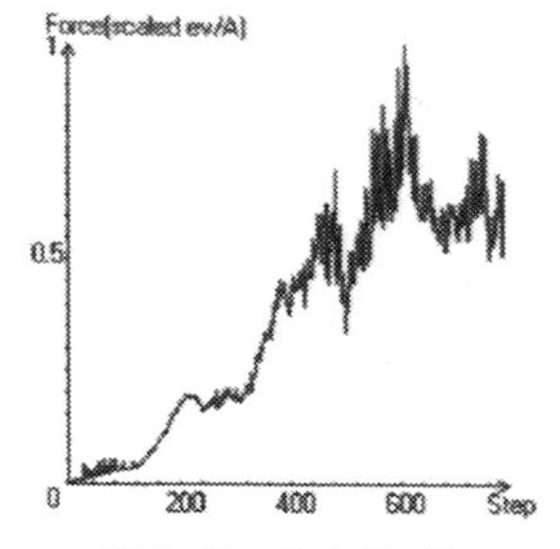

Fig 3 Cur of grinding force

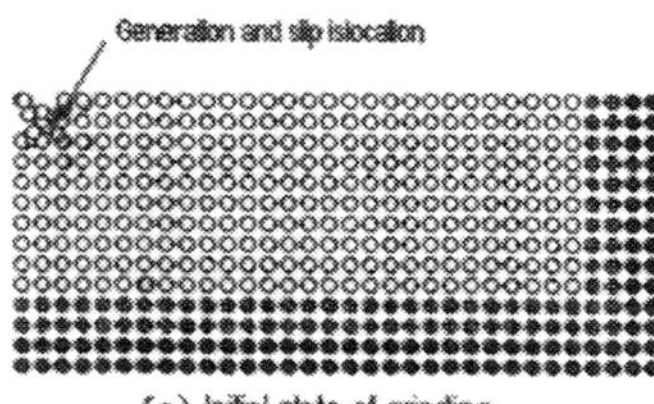

(a) Initial state of grinding

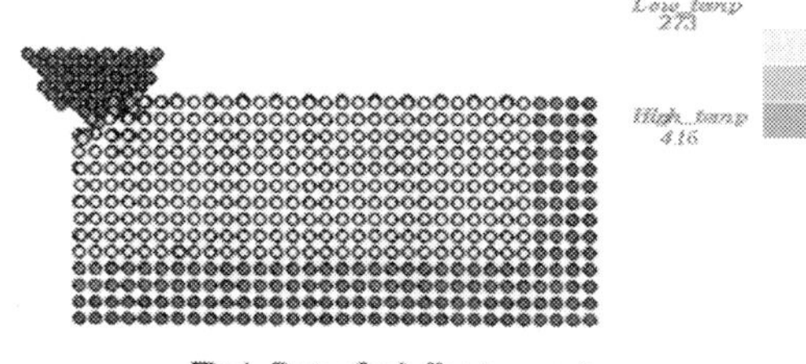

Fig 4 State of grinding temperature

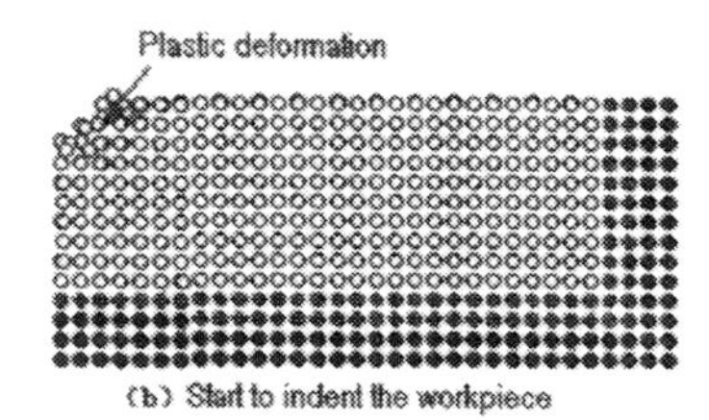

(b) Start to indent the workpiece

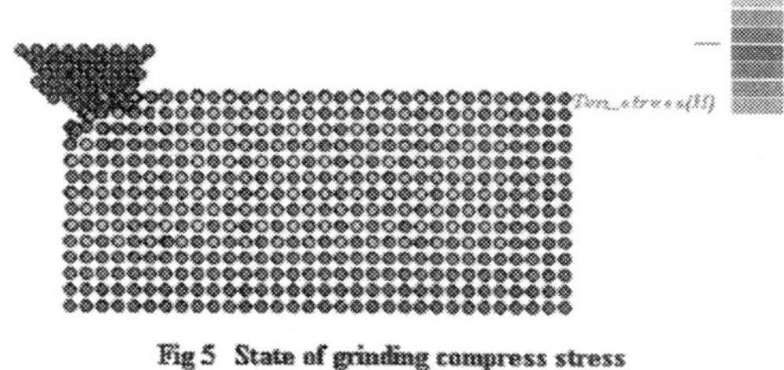

Fig 5 State of grinding compress stress

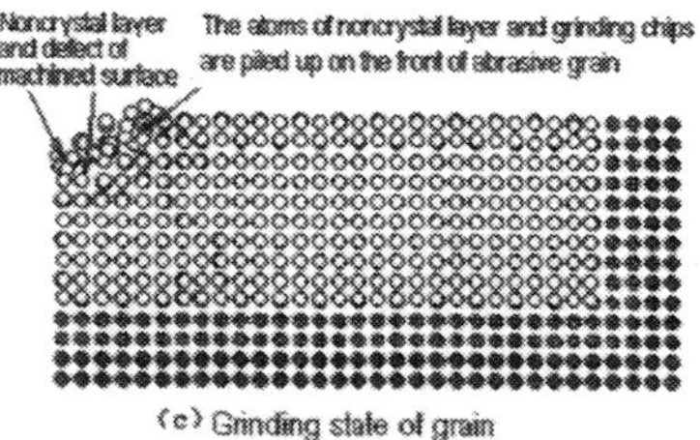

(c) Grinding state of grain

Fig 2 The abrasive processes of three different snap (small grain)

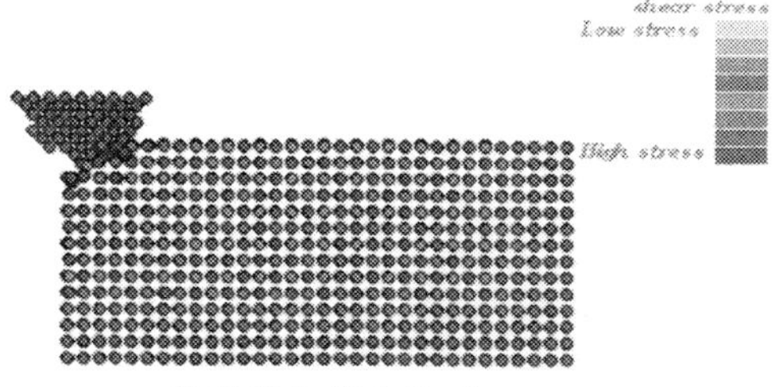

Fig 6 State of grinding shear stress

CAPE/083/2000

Investigations aiming to increase the rate of electrochemical dissolution process

M ZYBURA-SKRABALAK and **A RUSZAJ**
Department of Electrochemical Machining, The Institute of Metal Cutting, Cracow, Poland

ABSTRACT

The rate of electrochemical dissolution process depends mainly on the kinetics of electrode reactions and the activation energy of the electrode reaction . According to the Arrhenius law, it is possible to increase the rate of electrochemical dissolution by increasing the temperature in the interelectrode gap. The results of investigations have proved that together with temperature increase in the range from 20 to 44 ^{0}C also current density and intensity of dissolution process increase (from 0.6 to 1.8 mm^3/A*min). One of the efficient way for increasing the temperature in the interelectrode gap is to apply for machined surface heating the laser beam. The laser beam helps to increase the temperature on the machined surface and localisation of the ECM process.

1 INTRODUCTION

The electrochemical machining does not change significantly the properties of machined surface. It results from the fact that there is not a mechanical contact between electrode and machined surface. In electrochemical machining the material allowance is removed as a result of electrochemical dissolution process which is carried out usually in the temperature 20 to 80 ^{0} C. During machining process into interelectrode area there are not mechanical or thermal influence on machined surface. The surface layer quality is the same as in the core of material.

The run and kinetic of the electrochemical reactions during electrochemical machining depend mainly on electrolyte properties and machined material constitution, temperature, current density and velocity of dissolution product evacuation out of machined area [Davydov, Engelgardt 1988; Datta and other 1989].

In order to eliminate the indirect reactions and increase the velocity of the main reaction it is necessary:

a) to increase the current density:

$$j = \chi \frac{\Delta E_{om}}{h} \quad (1)$$

where: $\Delta E_{om} = U - (E_a^o - E_k^o + \Delta E_a - \Delta E_k)$;

b) to decrease the electrodes polarisation by:
- decreasing the thickness of diffusion layers adjacent to electrodes,
- decreasing the dissolution product concentration in interelectrode gap;

c) to increase the temperature in machining area, what influences also on surface passivation phenomenon.

In electrochemical machining the transportation process of heat and reaction products from machined area is carried out as a result of diffusion [Delui, Tribolle 1993; Landolt 1995]. The distribution of product concentration in the diffusion layer is described by relationship (3):

$$\frac{\partial c}{\partial t} + \vartheta_y \frac{\partial c}{\partial y} = D \frac{\partial^2 c}{\partial y^2} \ . \quad (2)$$

The values of diffusion coefficient, viscosity, solution density, transference number on the border line of electrode are different from those inside electrolyte solution. So, it is very important to work out the relationships, which will describe the mass and heat transportation process from these areas. On the border surface: machined material – electrolyte, the stream of mass in direction perpendicular to the electrode surface is created.

2 ELECTROCHEMICAL REACTIONS VELOCITY IN CONDITION OF DISSOLUTION PROCESS RUN

In case of high current density the transportation processes in direction to electrode limit the electrochemical reaction rate. This rate depends mainly on coefficient of ions diffusion from electrode.

When the concentrations of reagents near the electrode are constant and not dependent on current density and time, the concentration polarisation does not occur. It means that the rate of transportation and additional reactions is significantly higher than rate of main electrochemical reaction.

On assumption that in electrode process only one reaction occur (for instance reaction of metal ions transfer)

$$O^{+z} + ze^- \rightarrow R ,$$

it is possible that created products are involved in chemical reaction without electrical charge exchange with electrode. The measured total current density amounts:

$$j = j_+ - |j_-| \ . \quad (3)$$

According to the chemical kinetic principles, the reaction rate is proportional to concentration c_o of oxidised or reduced substance. For this reaction the energy of activation (difference between thermodynamic potentials of substrates and active complex) are ΔG_- or ΔG_+ given by (eq. 4):

$$\begin{aligned} -\frac{dn_o}{dt} &= k_- c_{oO} \exp\left[-\frac{\Delta G_-}{RT}\right] \\ -\frac{dn_R}{dt} &= k_+ c_{oR} \exp\left[-\frac{\Delta G_+}{RT}\right] \end{aligned} \ . \quad (4)$$

3 THE SIMPLIFIED MODEL OF THE PROCESS

The simplified model of the process makes, that it is possible to evaluate the relationship between constant of dissolution rate and temperature. The dissolution process intensity for reaction rate is connected with interaction between mass, heat, electrical discharge transportation and hydrodynamic parameters.
One of characteristic properties of electrochemical reaction carried out with high current density and high overpotential is the surface heat generation [Dikusar and other 1986]. The difference between surface and inside temperature can be calculated from the heat balance equation (5):

$$Q_s = q \tag{5}$$

where: Q_s - surface density of heat source,
$q = \alpha \Delta T$ - heat stream density from interphase border inside electrolyte solution,
α - coefficient of heat transformation for the system: electrode-electrolyte solution (in stationary conditions when electrolyte flow is intensive).

The reaction temperature influences on the ion diffusion process in electrolyte solution and is estimated by relation (6):

$$D = \lambda RT/F^2 \ , \tag{6}$$

where: λ - equivalent conductivity, which decides about dissolution reaction rate.
The relationship between constant of reaction rate K and activation energy W and temperature T is given by Arrhenius equation:

$$K = e^{-W/RT} \tag{7}$$

where: K - constant of reaction rate,
W - activation energy,
R - gas constant.
After finding the logarithm of reaction rate constant K, it was obtained:

$$ln\,K = ln\,z - \frac{W}{RT} \tag{8}$$

or

$$lg\,K = lg\,z - \frac{W}{2.3RT} \ . \tag{9}$$

Assuming that:

$$lg\,z = B \tag{10}$$

$$A = \frac{W}{2.3R} \ , \tag{11}$$

it is possible to obtain:

$$\lg K = B - \frac{A}{T} \ . \tag{12}$$

In case of anodic dissolution process K is dependent on current density $K \sim j$, so then the equation can be transformed in relationship (13):

$$lg\,j = B - \frac{A}{T} \ . \tag{13}$$

The chart $lg\,j - \frac{1}{T}$ is the straight line, which is characterised by A – angle of the line slope to axis of abscissa X and B - value in cross point of the line with axis of ordinates Y.
So, energy of activation can be calculated from the chart using the relation (14):

$$A = \frac{W}{2.3R} \Rightarrow W = 2.3AR \ . \tag{14}$$

4 EXPERIMENTAL INVESTIGATIONS

The aim of primary investigations was to find the relation between temperature and dissolution reaction rate. The experiments have been carried out for the electrochemical machining in the different temperatures. The tests were realised on electrochemical machine tool EOCA 40 [Ruszaj and others 1995]. The electrode potential E_a has been measured during experiments. The principle of E_a measurement and calculation of ηk_v have been presented in [Chuchro, Ruszaj, Zybura – Skrabalak 1995].

The following parameters have been taken into account:

<u>Investigated factors:</u>
C_e - electrolyte concentration; in range 5÷25%,
T - electrolyte temperature; in range 20 - 44°C,
U - interelectrode voltage; in range 8 -20 V,
v_f - electrode feed rate; in range 0,2 - 0,6 mm/min.

<u>Constant factors:</u>
- sort of electrolyte - $NaNO_3$ water solution,
- workpiece material - NC6 i 3H13,
- initial interelectrode gap s_o = 0,7 mm,
- thickness of removed material h = 4 mm.

<u>Resulting factors:</u>
j - current density,
E_a - electrode potential,
ηk_V - coefficient of electrochemical machinability,
R_a - roughness parameter.

5 EXPERIMENTAL TESTS RESULTS

The results of experiments are presented in the form of relations with neural nets application (Figs 1 to 6).
From Figs 1 to 6 it results that together with process parameters j and E_a change significantly. In investigated case current density can be calculated from relationship (15):

$$j = \frac{v_f}{\eta k_v} \qquad (15)$$

From relation (16) results that by changing temperature T and electrolyte concentration C_e coefficient ηk_v is also changed and the same current density j. The electrode potential E_a also depends on T and C_e. So, changing for instance temperature T it is possible to change E_a and the same kind and velocity of main reaction. More detail explanations need further investigations, which are carrying out now.

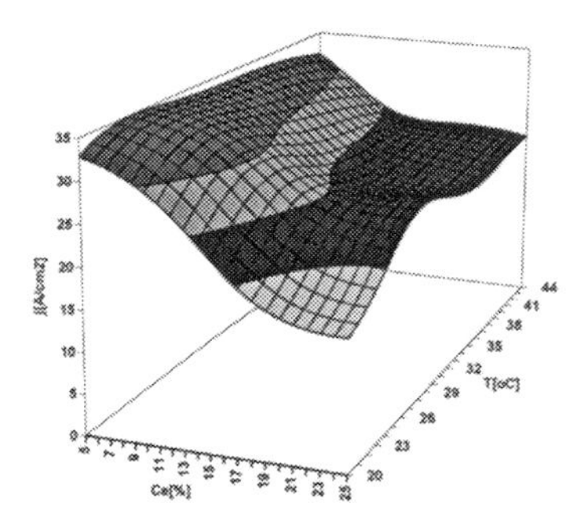

Fig. 1 Relation $j = f(C_e, T)$ for electrochemical machining NC6 steel with machining parameters: $U = 14V$, $v_f = 0.4$ mm/min

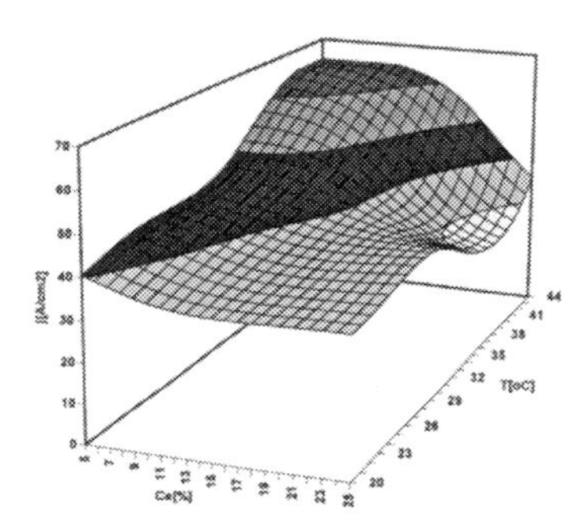

Fig. 2 Relation $j = f(C_e, T)$ for electrochemical machining 3H13 steel with machining parameters: $U = 14V$, $v_f = 0.4$ mm/min

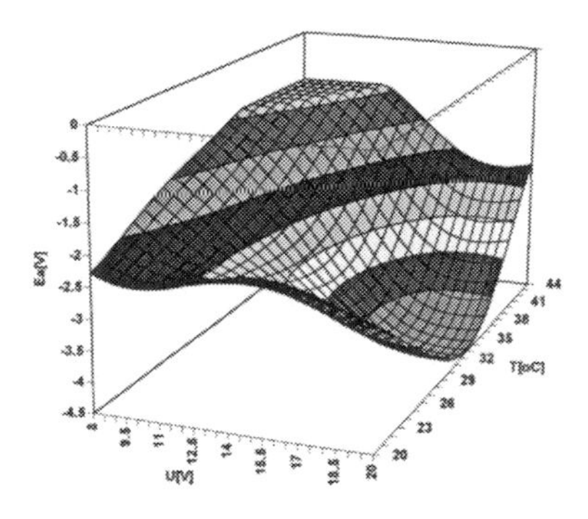

Fig. 3 Relation $E_a = f(T, U)$ for electrochemical machining NC6 steel with machining parameters $C_e = 15$ %, $v_f = 0.4$ mm/min

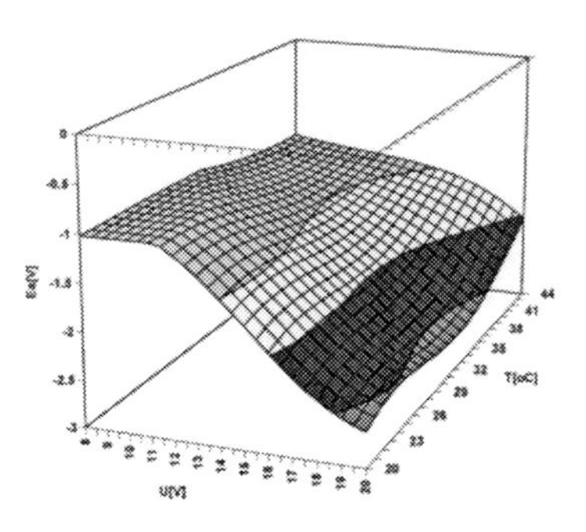

Fig. 4 Relation $E_a = f(T, U)$ for electrochemical machining 3H13 steel with machining parameters: C_e = 15 %, v_f = 0.4 mm/min

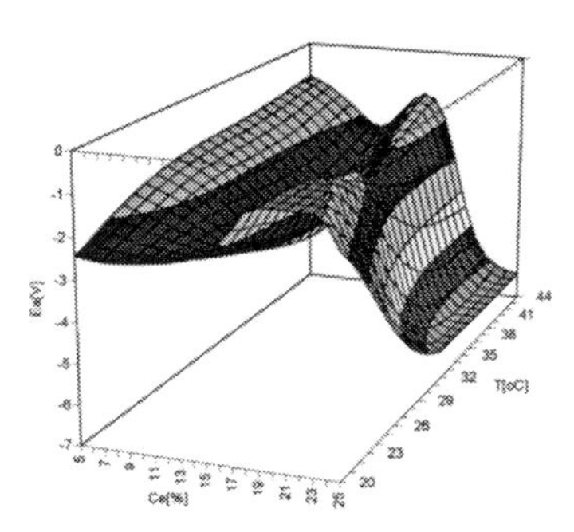

Fig. 5 Relation $E_a = f(C_e, T)$ for electrochemical machining NC6 steel with machining parameters U = 14V, v_f = 0.4 mm/min

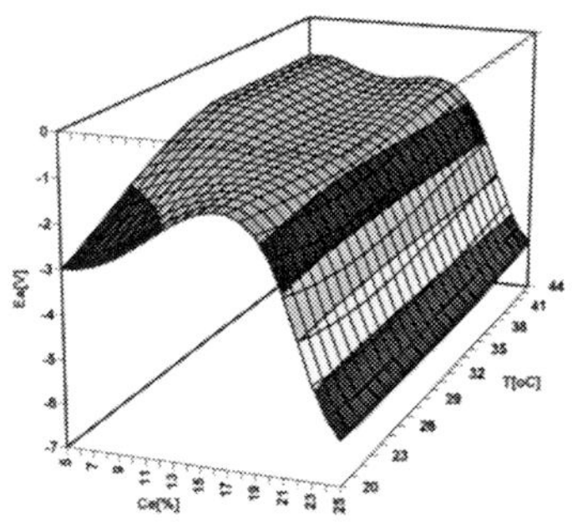

Fig. 6 Relation $E_a = f(C_e, T)$ for electrochemical machining 3H13 steel with machining parameters: U = 14V, v_f = 0.4 mm/min

The results of test give the possibility for estimating the activation energy. The activation energy W has been estimated taken into account the results presented in Figs 7 and 8. Using these figures it was possible to find out the value A which is necessary for evaluation W according to the equation (14).

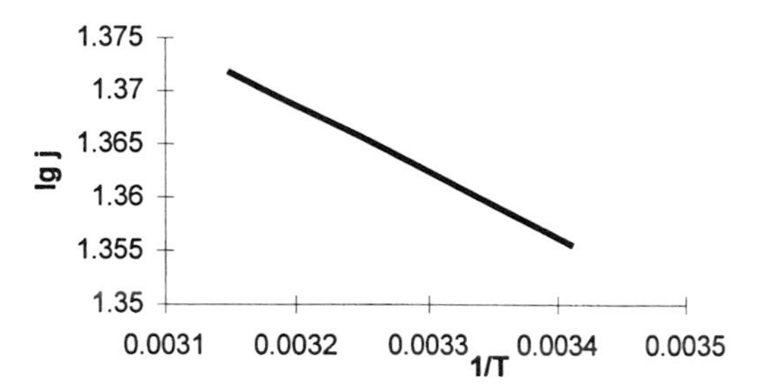

Fig. 7 Relation *lg (j) =f(1/T)* for evaluation *W* for dissolution process of NC6 steel for machining parameters: $U = 14$ V, $v_f = 0.4$ mm/min, $C_e = 15$ %

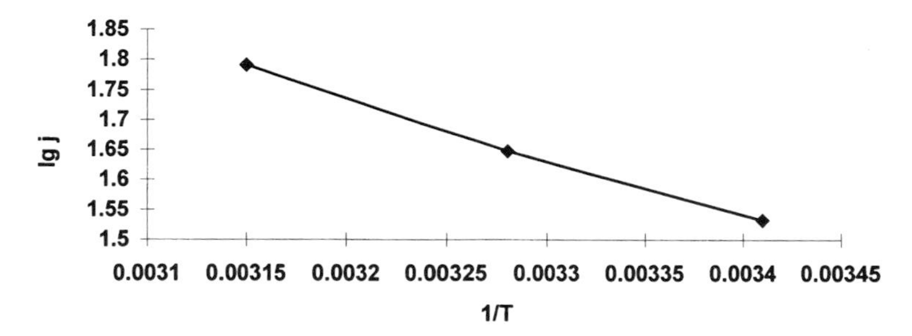

Fig. 8 Relation *lg (j) =f(1/T)* for evaluation *W* for dissolution process of 3H13 steel for machining parameters: $U = 14$ V, $v_f = 0.4$ mm/min, $C_e = 15$ %

The estimated activation energy W for NC6 i 3H13 steels was in range 24 – 45 kcal/mol. It should be very interesting to compare the results of solely electrochemical machining with the results of electrochemical machining assisted with laser beam.

The primary investigations have proved that the electrolyte temperature decides mainly about run of dissolution process. However, other technological parameters, such as: interelectrode voltage, electrode feed rate, are also responsible for the run of dissolution process in active, passive or transpassive state. The run of dissolution process depends also on the structure and composition of the machined material. During the investigations two different material: NC6 steel with austenite structure and content about 1.4 % of carbon and 1.4 % of chromium, and 3H13 steel with martensite structure and content about 0.26 to 0.35 % of carbon and 0.6 % of nickel have been tested.

It is expected that during electrochemical machining with proper assistance of laser radiation should be possible to increase the dissolution rate and the dissolution process localisation.

6 POSSIBILITIES OF ELECTROCHEMICAL DISSOLUTION ASSISTED WITH LASER BEAM

During last years the significant development in the field of electrochemical machining with assistance of laser beam took place. In this way of machining the area of electrochemical dissolution process is under the influence of laser beam [Davydov 1994]. This method is investigated now with high intensity, however it's practical application in industry are not very wide. It is expected that its prospective application will take place in shaping of small elements (5 – 500 μm) with high accuracy (1- 10 μm) for microelectronic and space industry, especially when they are made of special materials (alloys, composites, ceramics) which are difficult to cut [Davydov 1994, Ruszaj 1997].

As a result of laser beam influence many physical and physic-chemical phenomena occur on the machined surface and in surface layers of the material. These transformations are the result of heat flow into machined material. Laser radiation can change physical properties of machined material as well as kinetic of electrochemical reactions.

In case when there is no additional thermal focus the temperature of electrochemical reactions during electrochemical machining depends only on the activation energy of the electrode reactions and electrolyte temperature. The example of such relations is presented in Fig. 9.

In case when the electrode surface is heated by the additional heat source, for instance by the laser beam, the additional influence is also given by the heat transport. In electrochemical machining process supported by laser radiation it is very important to evaluate the interaction between electrode reaction kinetic and heat transportation.

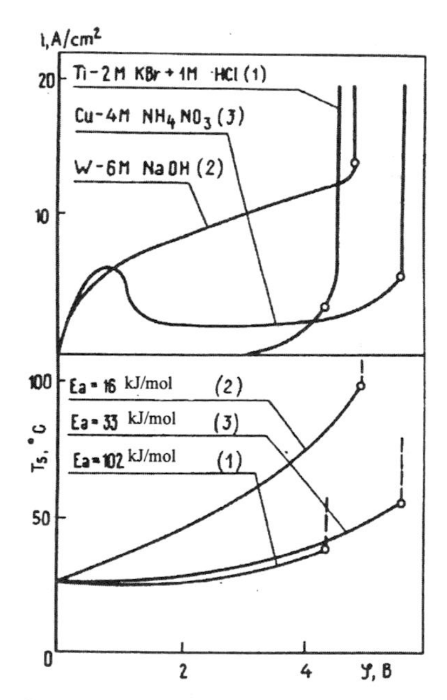

Fig. 9 Relations between the current density and surface temperature and the electrode potential for the different electrochemical systems [Dikusar and others 1986].

In the research carried out at the Warsaw Technical University the mathematical model of shaping process in Laser Electrochemical Machining (LECM) has been developed and experimental investigations of effects of laser beam on anodic dissolution of steel were performed. Experiments and computer simulations have demonstrated that increasing of metal removal rate and localisation of anodic dissolution can be reached by use a laser assistance of ECM [Kozak 1998, Kozak and others 1999].
Similar results has been presented by Prof. McGeough [McGeough 1999].

7 RECAPITULATION

From presented research results that temperature of machined area has significant influence on activation energy and main electrochemical reaction. These results will be taken into account in investigated now case when machined surface is heated by laser beam.
The electrochemical laser machining can be controlled by changing the following parameters: interelectrode voltage, initial thickness of interelectrode gap, composition of electrolyte, hydrodynamic conditions, power of laser beam, and length of laser waves.
The laser beam can give the increase of the velocity of dissolution process during electrochemical machining and reach the higher level of dissolution process localisation and the same the machining accuracy.
The model for electrochemical machining assisted with laser beam should also include the kinetics of electrode reaction and the heat existing during their runs.

8 AKNOWLEDGEMENTS

The Authors would like to thank the European Commission for supporting the cooperation between the Edinburgh University, Glasgow Caledonian University, Warsaw Technical University, PHILIPS, KALTEX and the Institute of Metal Cutting by funding the Concerted Action Contract No ERBIC15 CT98 0801 on the project: „Research on clean hybrid micromachining (HMM)".

REFERENCES

1. Chuchro M., Ruszaj A., Zybura-Skrabalak M.: *The influence of electrochemical dissolution process conditions on machined surface geometry.* Int. Symp. for Electromach., ISEM XI, Lausanne, Switzerland, 1995, p.521-531.
2. Datta M., Romankiw L.T., Vigliotti D.R., Gutfeld R.J.: *Jet and laser - jet electrochemical micromachining of nickel and steel.* J. Electrochem. Soc., 1989, Vol. 136, No 8, p.2251-2256.
3. Davydov A.D.: *Lazerno-elektrochimiczeskaja obrabotka metallov.* Elektrochimija, 1994, Vol. 30, No 8, p.965 - 976.
4. Davydow A.D., Engelgardt G.R.: *Metody intensifikacii nekotorych elektrochimiczeskich processov.* Elektrochimija, 1988, Vol. XXIV, No 1, p.3-17.
5. Delui K., Tribolle B.: *Elektrogidrodinamiczeskij impedans: sposob izuczenija elektrodnych processov.* Elektrochimija, 1993, Vol. 29, No 1, p.84-88.
6. Dikusar A. I., Engelgardt G. R., Molin A. N., Domente G. S.: *Vzaimnoje vlijanie perenosa i sinergiczeskije effekty pri elektrochimiczeskich i kombinitrovannych metodach obrabotki.* Proceed. Intern. Symp. for Electromach., ISEM 8, Moskva, 1986, p. 152 – 155.

7. Kozak J., Rozenek M., Dąbrowski L., Zawora J.: *Badania wpływu promieniowania laserowego na proces kształtowania elektrochemicznego.* Wyd. Pol. Warsz., Program Priorytetowy, Nowe Technologie, Prace naukowe, 1999, Zeszyt 2, p.233-242.
8. Kozak J.: *Mikrokształtowanie elektrochemiczne z laserową aktywacją procesów elektrodowych.* Wyd. Pol. Warsz., Program Priorytetowy, Nowe Technologie, Prace naukowe, 1998, Zeszyt 1, p.25 -33.
9. Landolt D.: *Processy massoperenosa pri anodnom rastvorenii metallov.* Elektrochimija, 1995, Vol. 31, No 3, p.228-234.
10. McGeough, Tang Y., De Silva A.: *Hybrid laser – electrochemical micromachining.* Report for INCO - COPERNICUS Programme, Contract No IC15-CT98-0801, 1999, pp.15.
11. Ruszaj A., Dziedzic J., Czekaj J., Krehlik M.: *Electrochemical machining with electrode displacement controlled in three axes.* Proceed. Int. Symp. for Electromach., ISEM XI, Lausanne, Switzerland, 1995, p.553-563.
12. Ruszaj A.: *Kierunki badań z zakresu technologii elektroerozyjnej (EDM) i elektrochemicznej (ECM).* Postępy Technologii Maszyn i Urządzeń, 1997, Vol. 21, No 2, p.83-101.

CAPE/079/2000

Study of electrochemical machining utilizing a vibrating tool electrode

J KOZAK, K P RAJURKAR, and **S MALICKI**
University of Nebraska-Lincoln, USA

ABSTRACT

Application in electrochemical machining (ECM) of vibration of the electrodes is one the effective methods for improving of the accuracy and reducing the correction of the tool electrode. This paper presents modeling and analysis of results of numerical calculations of the ECM process with vibrating electrode. The effect of input parameters and machining conditions on effectiveness of vibration during ECM has been investigated using computer simulation. An appropriate range of input parameters for the optimal effectiveness of machining has been identified.

INTRODUCTION

Electrochemical Machining (ECM) is an important manufacturing technology in machining difficult-to-cut materials and to shape complicated contours and profiles with high material removal rate without tool wear and without inducing residual stress. Industrial practices in ECM have revealed some problems impeding its further development and wider acceptance by industrial users. Among them, prediction and control of the local interelectrode gap distribution (and hence, the control of dimensional accuracy), along with the design of tool electrodes for complex WP shapes and optimization of process, are the major problems encountered by ECM users (1).

An improvement of accuracy and reducing of the tool electrode correction can be achieved by vibrating of the tool or the workpiece (2). One such version is the ECM with vibrating TE and with the pulse voltage applied in synchronism with the vibration, as shown in Fig.1 (3).

Thus, according to Morozov (4), the TE, which vibrates sinusoidally, touches the WP surface in every cycle and is removed from this surface for a distance, as determined by the vibration amplitude. The pulses are applied, when the electrodes are drawing together. At the moment of the electrode contact, the voltage is zero.
The application of vibration movement with frequencies from ten's Hz to hundred's kHz (for example, at constant working voltage), increases the slope of the curve of current density vs. the gap size. Finally it leads to an increased anodic localization and uniformity of the gap size distribution, and improved quality of the machined surface (3-6).

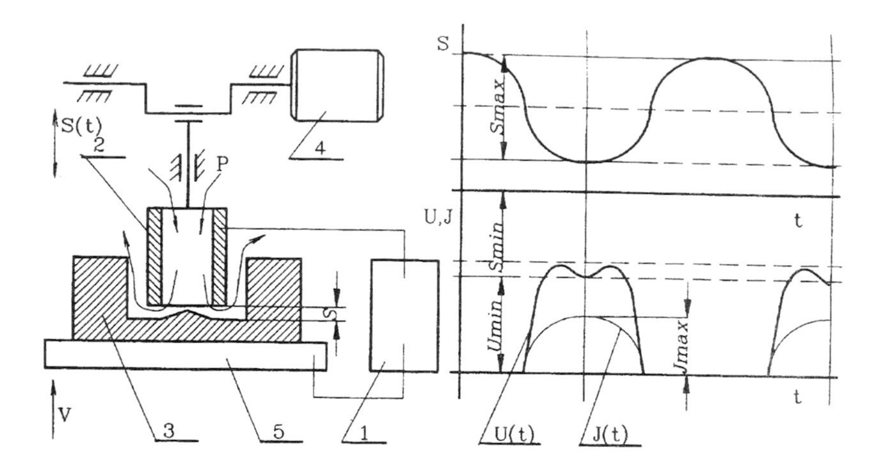

Fig. 1. Diagram of mode of contact-less vibrating PECM: 1 - pulse power supply, 2 - vibrating tool electrode (TE), 3 – workpiece (WP), 4 - vibrating unit, 5 - table (3)

This paper presents an analysis of basic characteristics of the ECM with vibrating electrodes.

ANALYSIS OF ECM WITH VIBRATING FLAT TOOL ELECTRODES

In some ECM and PECM operations, a vibrating tool electrode is used. In one type of the machines operating on this principle, constant feed rate alternates with vibrations.

Let us consider the case of ECM with plane parallel electrodes, one of which (for example cathode-tool, ER) oscillates periodically with amplitude A and the circular frequency ω, and simultaneously is moving towards anode with a constant feed rate V_f as shown in Fig.2.

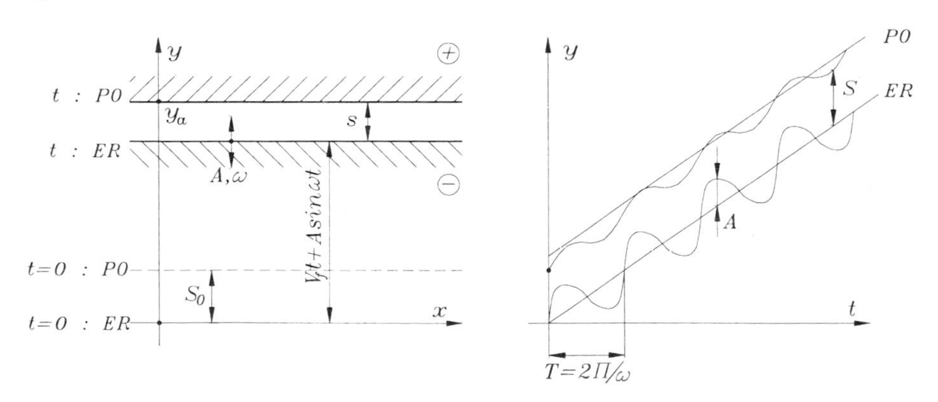

Fig.2 ECM with constant feed of vibrating tool-electrode (7)

The motion of tool - electrode can be described as follows:

$$s = V_f t + A\sin\omega t \tag{1}$$

The position of the anode surface y_a, in a fixed coordinate system (x, y), during electrochemical dissolution process is described by following equation (5,7):

$$\frac{dy_a}{dt} = \kappa \cdot K_v \frac{U - \Delta U}{y_a - V_f t - A\sin\omega t} \tag{2}$$

with initial condition for $t = 0$, $y_a = S_0$, where κ is electrical conductivity of electrolyte, K_V is the coefficient electrochemical machinability, which is defined as the volume of material dissolved per unit electrical charge, U is working voltage, ΔU is total overpotential of electrode processes,

The gap is given by,

$$S = y_a - V_f \cdot t \tag{3}$$

In these coordinates, the kinematics equation of the ECM gap (2), can be written as,

$$\frac{dS}{dt} = \frac{D}{S - A \cdot \sin\omega \cdot t} - V_f \tag{4}$$

where $D = \kappa K_V (U - \Delta U)$.

After substituting the equation (4) in dimensionless form in moving coordinates system, equation (4) can be written as:

$$\frac{d\zeta}{d\tau} = \frac{1}{\zeta - \overline{A}\sin\overline{\omega}\tau} - 1 \tag{5}$$

where: $\zeta = \frac{y_a}{S_f}$, $\tau = \frac{V_f t}{S_f}$, $\overline{A} = \frac{A}{S_f}$, $\overline{\omega} = \frac{S_f}{V_f}\omega$ are dimensionless variables.

This differential equation cannot be integrated in closed form; therefore the numerical methods should be employed. The solution of this problem can attained by application higher order Runge-Kutta method.

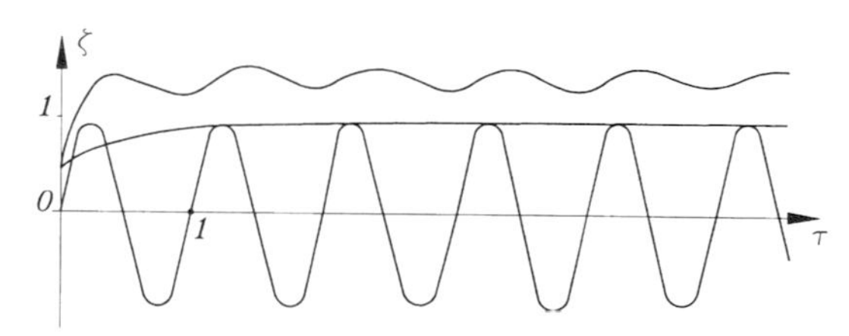

Fig.3 Changes distance S, the position of vibration tool-electrode and the gap size at ECM without vibration, in moving coordinate system: ζ_0=0.5, $\overline{A}$=1 and $\overline{\omega}$=2π

Result of numerical solution of Eqn.5 and the curves of vibration of the tool, and changes of gap size during ECM with constant feed rate are shown in Fig.3. For comparison, the solution for ECM with constant feed rate ($\overline{A}$=0), is also shown.

From the curve $\zeta(\tau)$ shows that the process reaches to quasi – steady state, a position of the anodic surface is described by periodic functions:

$$\zeta(\tau) = \zeta(\tau + n \cdot \overline{T}) \qquad n = 1,2,......, \tag{6}$$

Under the quasi – steady state, the gap size is changing from a value S_{min} to S_{max}. In the Fig.3, the value S_{min} = 0,29 S_f at "top" position of the tool-electrode, and is equal S_{max} = 2,52 S_f at the bottom position of the tool-electrode.

The Fig.4, illustrates the curve $\zeta(\tau)$ and quasi – steady state for various initial values of $\zeta(0)$. The steady state of ECM without vibration is presented by straight line, $\zeta = 1$.

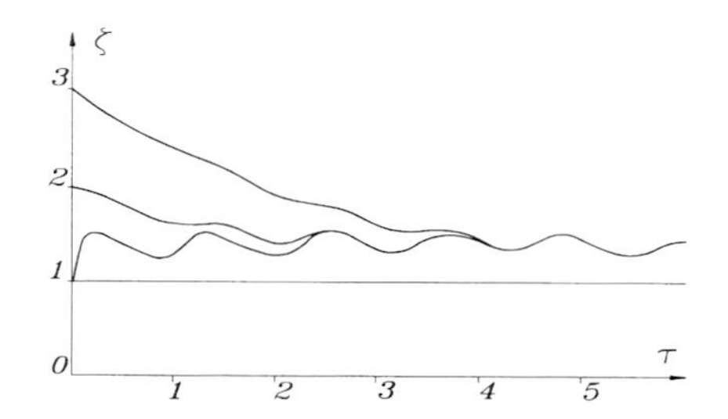

Fig. 4 Changes of anode position $\zeta(\tau)$ in time at different initial conditions Parameters: $\bar{A} = 1$, $\bar{\omega} = 2\pi$

The effect of the amplitude and the frequency of the vibration can be seen in Fig. 5 and 6, respectively.

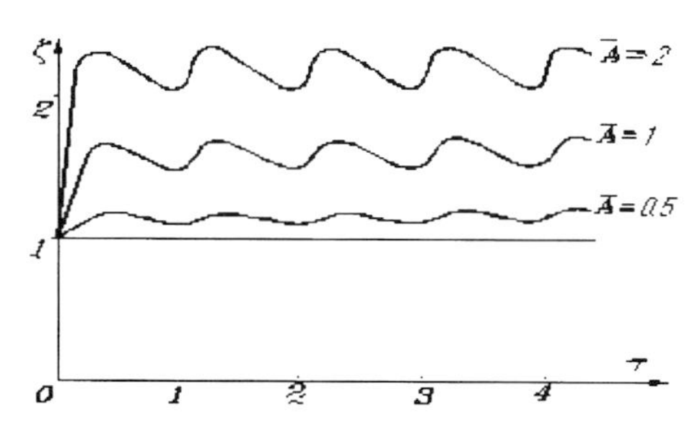

Fig. 5 Effect of the vibration amplitude of tool-electrode on the $\zeta(\tau)$

Amplitude of changes of the position anode-workpiece $a = ½ (\zeta_{max} - \zeta_{min})$, is less than amplitude of vibration $\bar{A}$ and depends on of the frequency $\bar{\omega}$ (Fig.6 and 7).

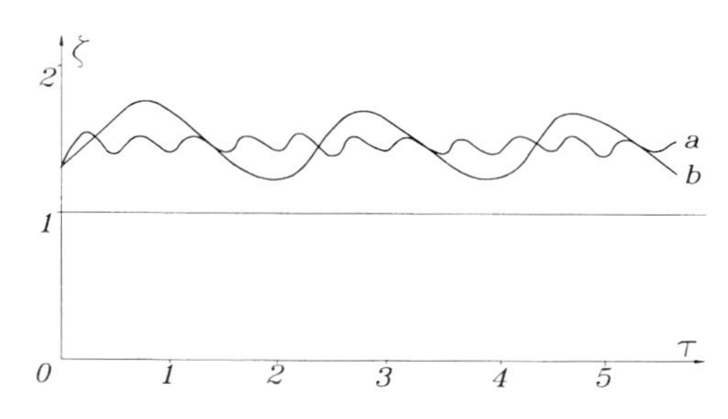

Fig.6 Effect of vibration frequency of tool-electrode on $\zeta(\tau)$ at $\bar{A} = 1$: a) $\bar{\omega} = \pi$, b) $\bar{\omega} = 4\pi$

To achieve machining accuracy it is necessary to use higher frequency, with reduced changes in position anodic surface relative to mean position of the tool electrode i.e. the value of a.

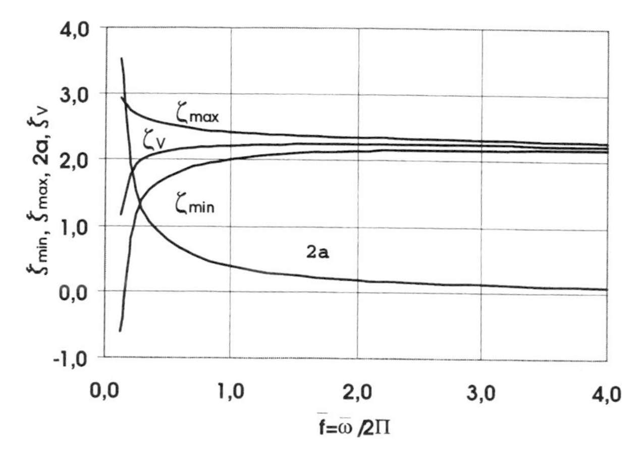

Fig.7 Effects of the frequency of vibration on 2a, ζ_{min}, ζ_{max} , at amplitude $\overline{A} = 2$ **(7)**

For decreasing of deviation of the anode surface position i.e. values amplitude, a, the dimensionless frequency $\overline{\omega}$ >40, or dimensional frequency can be recommended as,

$$f > 6.4 \cdot \frac{V_f}{S_f} \quad \text{or} \quad f > 6.4 \cdot \frac{V_f^2}{\kappa \cdot K_V \cdot (U - \Delta U)} \tag{7}$$

For example, for V_f = 1 mm/min and S_f = 0,2 mm, f > 0.5Hz, therefore, the condition (7) is obtained at relative low frequency of the tool-electrode.

The periodical change of gap persists as long as time iterations are carried out. It can be said a quasi steady state of ECM with vibration is attained after some transient stage of machining. In quasi-steady state, the gape changing around some mean value of ζ_v and solution of Eq. (5) can be described as:

$$\zeta = \zeta_v + \overline{\Delta}(\tau) \tag{8}$$

where: $\overline{\Delta}(\tau)$ is some periodical component with inharmonic character as a result of non-linearity of Eq. (5). With an increase of the frequency of vibration, amplitude of periodical component $\overline{\Delta}(\tau)$ decreases to zero. The reduction in frequency leads to an increase of amplitude of $\Delta(\tau)$, and its value equals the amplitude of vibration tool-electrode $\overline{A}$.

In the dimensional coordinate system, a solution of Eqn.2 can be rearranged as:

$$y_a = V_f t + S_v + \Delta(t) \tag{9}$$

where: $S_v = \zeta_v S_f$ is characteristic gap which is analogue to the equilibrium gap S_f,, and $\Delta(t) = \overline{\Delta}(\tau) S_f$ is periodic component of motion of anode.

For determination the value of S_v, let us consider the case of ECM when the effect of the periodic components $\Delta(t)$ on current is not significant and can be neglected (5, 7). Substituting Eq. (9) into Eq. (2), we have:

$$V_f + \frac{d\Delta}{dt} = \kappa k_v \frac{U - \Delta U}{S_v + \Delta(t) - A\sin\omega t} \tag{10}$$

Integrating both sides of Eq. (10) for one cycle length, we have:

$$V_f T + \Delta(T) - \Delta(0) = \int_0^T \frac{D}{S_v + \Delta(t) - A\sin\omega t} dt$$

Under assumption of disregarding of $\Delta(t)$,

$$V_f T + \Delta(T) - \Delta(0) = \frac{D}{\sqrt{S_v^2 - A^2}} T$$

where: $\Delta(T)-\Delta(0) = 0$.

This equation can be expressed in terms of S_f as:

$$S_v = \sqrt{S_f^2 + A^2} \tag{11}$$

This solution indicates that S_v is larger than the equilibrium gap size S_f and the amplitude A.

In dimensionless variables, the Eq. (11) can be written as,

$$\zeta_V = \sqrt{1 + \overline{A}^2} \tag{12}$$

The accuracy of the derived expression (Eq.11) is increasing with increase of frequency of vibration.

ELECTROCHEMICAL MACHINING WITH VIBRATING CONTOURED TOOL ELECTRODES

Let us consider of the effect of vibration electrode on quasi – steady state distribution of the gap size at curvilinear electrodes (Fig.7).

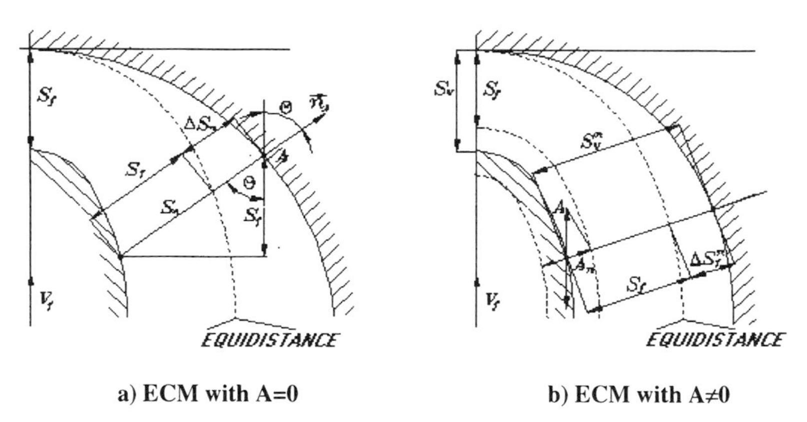

a) ECM with A=0 **b) ECM with A≠0**

Fig.8 Electrochemical shaping in: a). Steady state ECM, and b) quasi – steady state ECM with vibrating tool-electrode

The changes of gap size in steady state contoured ECM without vibration, by using approximation widely used in practice is described by:

$$S_n = \frac{S_f}{\cos\theta} \tag{13}$$

The value of tool-electrode correction (as shown in Figure 8) is,

$$e = \Delta S = S_n - S_0$$

Hence:

$$\Delta S = \frac{1-\cos\theta}{\cos\theta} \cdot S_f \qquad (14)$$

In the ECM with vibrating tool-electrode, the quasi – equilibrium frontal gap size, i.e. at $\theta=0$ (Fig.8 b), is approximated by the expression (11):

$$S_v = \sqrt{S_f^{\,2} + A^2} \qquad (15)$$

The amplitude of tool-electrode vibration in normal direction is given by:

$$A_n = A \cdot \cos\theta_E \qquad (16)$$

where θ_E is angle between a normal direction to the tool-electrode and the direction of feed rate (and vibration).

The quasi – equilibrium gap size along the normal to the anode is approximately equal:

$$S_v^n = \sqrt{S_n^2 + (A \cdot \cos\theta_E)^2} \qquad (17)$$

For $\theta<70^\circ$, a difference between θ_E and θ for corresponding points on the anode and cathode surface is not significant, therefore θ_E can be replaced by θ in Eq. (17).
Substitution of Eq. (13) to Eq. (17), gives the local gap size in quasi – steady state ECM with vibrating tool-electrode

$$S_v^n = \sqrt{\frac{S_f^2}{\cos^2\theta} + A^2 \cdot \cos^2\theta} \qquad (18)$$

The value of tool-electrode correction $\Delta S_V^n = S_V^n - S_V$ (as shown in Figure 7) is

$$\Delta S_v^n = \left(\frac{\sqrt{1+\overline{A}^2 \cdot \cos^4\theta}}{\cos\theta} - \sqrt{1+\overline{A}^2} \right) \cdot S_f \qquad (19)$$

The calculation using Eq. (18) and (19) for typical conditions of ECM shows, that the values $\Delta S_v^n < \Delta S$ for the same values S_f. Therefore, application of vibrating tool-electrode results in more uniform gap size distribution than in the ECM without vibration.

For example, at $\theta=60^\circ$:

- In ECM without vibration $\Delta S = S_f$,
- In ECM with vibrating tool-electrode at the amplitude $A=S_f$, the correction is equal $\Delta S_v^n = 0.647 \cdot S_f$, i.e. 35% less than in previous case.

Increasing the amplitude to $2S_f$, reduction of ΔS_v^n to zero, but using the amplitude $A>S_f$ can lead to increasing probability of short circuit and damage of the electrodes by electrical discharges in practical ECM operations.

Results of computer simulation of ECM with vibrating contoured tool electrodes confirmed above conclusions. For example, the evolution of shape of machining surface during ECM with vibration of the tool electrode is shown in Fig.8. After some time, quasi-steady state is reached and changes of the gap size distribution are periodical as illustrated in Fig. 8, by the two lines at tool electrode. The distributions of the correction values along x-axis for ECM without vibration (ΔS^n) and with vibrating tool (ΔS_v^n) are shown in Fig. 10. The simulation results shows that the maximum value ΔS_v^n is less then maximum value of ΔS^n about 42%.

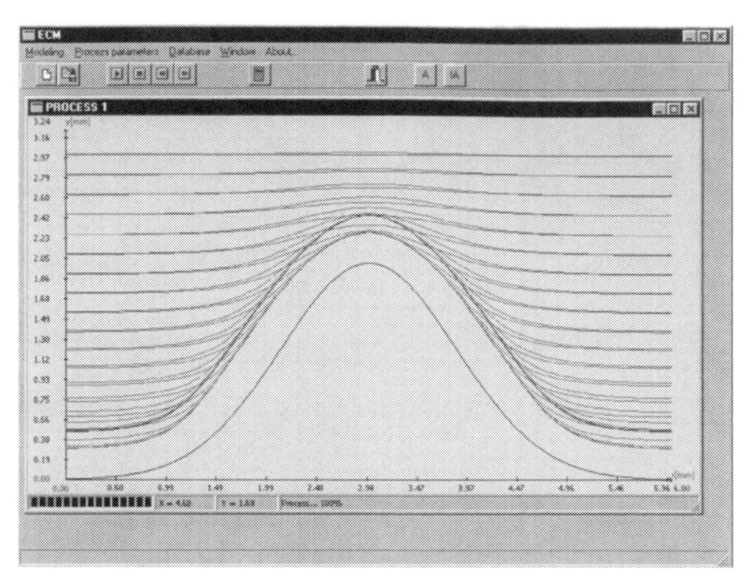

Fig.9 Changes of workpiece shape during ECM with vibrating tool electrode. Parameters ECM: U=10 [V], V_f=1 [mm/min], κ=10 [A/Vm], K_V=2 [mm^3/Amin], f=2.7 [Hz].

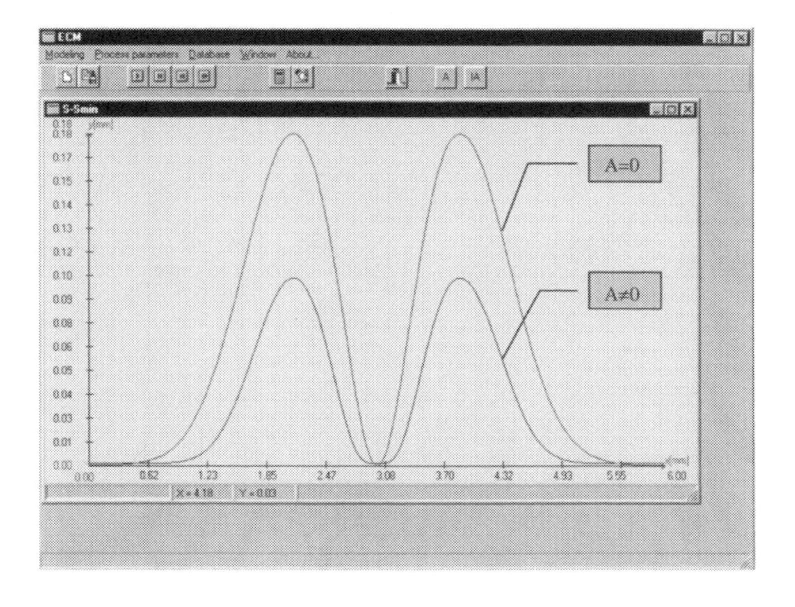

Fig.10 Distribution of correction values ΔS^n and ΔS_v^n at A=0 [mm] and A=0.35 [mm], respectively. Parameters of ECM: U=10 [V], V_f=1 [mm/min], κ=10 [A/Vm], K_V=2 [mm^3/Amin], f=2.7 [Hz]

Additional positive effects of vibration are related to changing of physical conditions in the gap, which leads to limiting the reduction in current density in ECM process, increasing of uniformity of interelectrode medium properties and increasing of localization of anodic dissolution (3, 4, 6).

These positive results are obtained with vibration of tool-electrode in continuous ECM. These advantages of vibrating ECM, we get amplified with electrochemical machining with using pulse current, i.e. in PECM process (1).

In practice a variant of vibrating tool PECM without touching TE and WP during machining is also used. In machining of large and heavy parts or stiff part like large airfoils such as PECM system is used. Depending on the specifically technical performance this variant includes several schemes, such as:

- PECM with vibration TE and constant feed rate WP,
- PECM with constant feed rate of vibration TE,
- PECM with constant feed rate of the tool electrode and with vibration WP.

Typical PECM with vibration TE equipment for machining turbine and compressor blades are M-71, EKU-250 and EKhS-10A (made in of Russia). Specifications of M-71 are: pulse voltage Up = 10-12 [V], pulse current I = 6300 [A] at blades with length to 250 [mm], minimal gap size S_{min} from 0.05 [mm] to 0.2 [mm], frequency of vibration from 20 to 50 [Hz], and inlet pressure p_{in} = 0.5 [MPa]. In EXS-10 [A]: U_p = 6-13 [V], I = to 6000 [A] at blades with length from 80 to 200 [mm], frequency from 20 to 50 [Hz], amplitude from 0.005 to 0.020 [mm] and p_{in} = 0.5-0.6 [MPa]. Blades are machined from both sides, simultaneously.

For comparison, machining of parts by continuous ECM without vibrating, has been performed with a feed rate of 1 [mm/min], and it was found that the accuracy was only 0.4 [mm], i.e. 5-8 times less in comparison with PECM with vibration TE. Simultaneously many hydrodynamics defects on machined surface were observed, when ECM without vibrating tool was used (6).

Acknowledgement

The support of the State of Nebraska (Research Initiative Fund) is gratefully acknowledged.

REFERENCES

1. Rajurkar K., P, McGeough J., A., Kozak J., De Silva A., New Developments in Electro-Chemical Macining. Annals of the CIRP Vol.48/2, 1999, pp. 567-580.
2. US Patent 3271283, 1966; USSR Inventor's Certificate no. 194510, 1967; USSR Inventor's Certificate no. 670410, 1979; UK Patent 1577766, 1980
3. Zhitnikov V., P., Zajtsev A., N.,*Mathematical Modeling of ECM*. Publ. Aeronautical University of Ufa, Ufa 1996 (in Russian)
4. Morozov B.,I.: Electrochemical machining with vibrating tool electrode, Elektronnaja Obrabotka Marerialov (Electron Treatment of Materials), no.6, 1974, pp.26-28
5. Davydow A.D., Kozak, J.: *High Rate Electrochemical Shaping*, Nauka, Moscow, 1990 (in Russian)
6. Atanasiantz A., G., *Electrochemical Manufacturing of Nuclear Reactors Parts.* EnergoAtom, Moscow, 1987 (in Russian)
7. Kozak J.: Analysis of Electrochemical Machining Process with Vibration of Tool Electrode. Proceed. on Int. Symposium "Electromachining '97", Bydgoszcz 1997 (in Polish), pp.204-212.

CAPE/072/2000

Process parameter analysis for an optimal feed rate control of EDM

G WOLLENBERG, H-P SCHULZE, and **T PAPE**
Institute for Fundamental Electrical Engineering and Electromagnetic Compatibility, Otto-von-Guericke-University, Magdeburg, Germany

ABSTRACT
The optimisation of the feed rate control is an important factor for maintaining the process stability while spark erosion machining. The parameters of the feed rate control are a part of process analysis and must satisfy the goals of feed rate control. A control to an constant gap occurs because as a result a high accuracy and a firm offset for the CNC are guaranteed.

In the presented paper relevant process parameters are investigated and assessed concerning their suitability to different applications.

1 FEED RATE CONTROL AND THE PROCESS PARAMETERS

The spark erosion should occur with an as constant as possible gap width to guarantee the high accuracy of the machining process. The quality of feed rate control determines the stability of the gap width the machining process.

The following four process variables can be used for feed control:

- gap voltage,
- ignition delay time,
- voltage breakdown,
- high-frequency parts of the burning voltage.

In contrast to the process analysis the gap width control needs a rating which corresponds to the time constant of the feed moving system. The simplest method is the averaging the process variable over this time constant (Fig. 1). The modern and comprehensive method of connection between process parameters is based on Fuzzy systems.

In the investigations, goal was to be found, which process variables are best suitable for feed-rate control, and which inserted signals are to be faded out from this control.

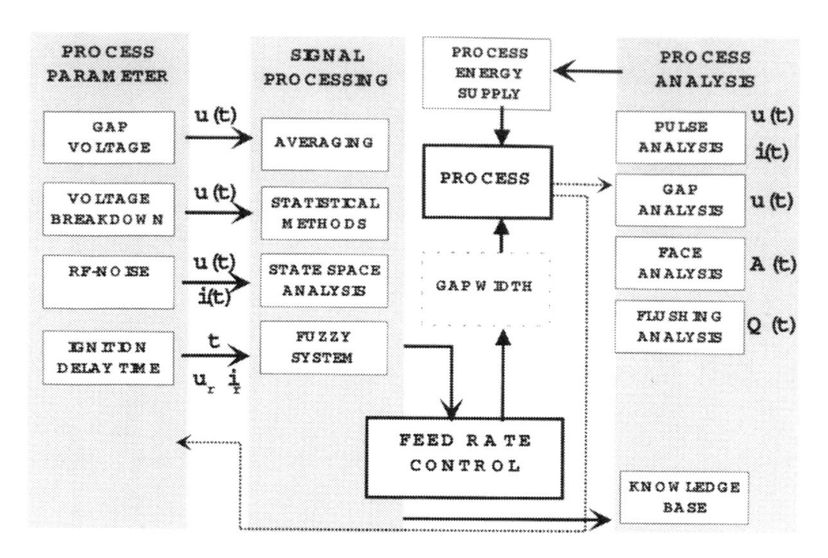

Fig. 1: Process parameters for the Feed Rate Control

2 THE PROCESS VARIABLES IN DETAIL

2.1 The Middle Gap Voltage

This process characteristic is a average value of the gap voltage. This mean value depending on the different burning voltage levels of spark and arc as well as on the ignition delay time characterises the gap stage. If the gap voltage inclines to the higher spark voltage level/open circuit, the working gap width has to be reduced. In the other way, the gap width control should enlarge the gap only in critical processing stages.

The determination of the middle gap voltage and its technical implementation for gap width control is relatively easy in comparison with other methods. As long as generators were applied with voltage source characteristic, the gap voltage was directly proportional to the working gap width. However, due to the use of generators with current source characteristic and the modular design (ignition and power units), this simple relation between middle gap voltage and the gap width is no more valid.

2.2 The Ignition Delay Time

The ignition delay time characterises the conditions for the voltage breakdown of the gap. The theoretical consideration, that a small gap offers better conditions for a breakdown than a large gap leads therefore to shorter ignition delay times confirmed in (1) and (5). In this way, the ignition delay time can be used as a controlled variable for the feed rate control.
The gap conditions at spark erosion are ideal if the voltage breakdown occurs after a short ignition delay. Longer ignition delays are uncritical for the processing result, however, they decrease the removal rate. At shorter ignition delays, the danger for occurring arc discharges arises.

The measurement of the ignition delay of the dielectric breakdown has the advantage compared to other procedures that critical states of the gap before the actual power pulse can be found. In addition, the disturbance afflicted current and voltage measurements are replaced by time measurement. The start of measurement is given by the generator control. The time interval ends with reaching a reference voltage or a reference current.

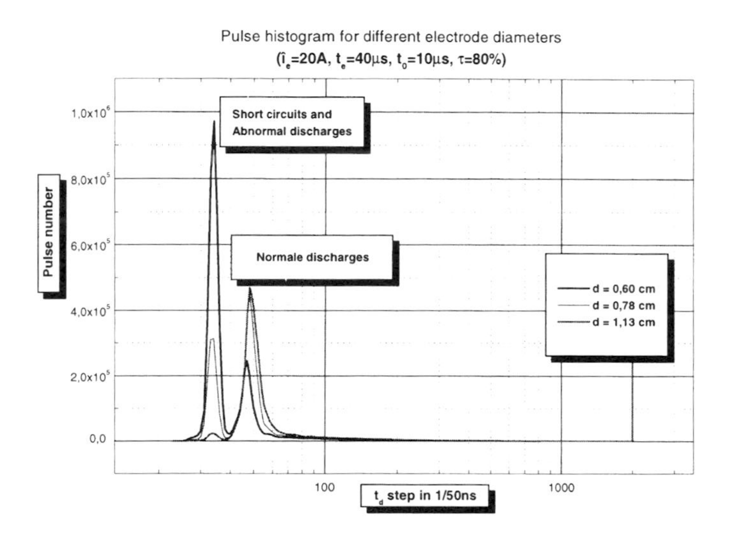

Fig. 2: t_d-Histogram for different tool diameters (2)

Fig. 2 shows the influence of current density on the distribution of normal ($t_d > t_{d\ crit}$) and abnormal discharges ($t_d < t_{d\ crit}$).

2.3 Voltage Breakdown

The voltage breakdown is characteristic of subsequent discharge because the slope of this breakdown is determined by the dominating mechanisms of conducting (combination and recombination of the charge carriers). (9)

2.4 The High Frequency Part

The practical investigations show that the spark voltage has a high-frequency part while the arc burning voltage does not have this one. Using this, process analyses and feed rate controls achieving a sufficiently good machining result were developed.

Investigations at the University of Magdeburg showed that the high-frequency part can be measured directly from the gap current, how it is to be recognised in **Fig. 3**. The influences of the process in the form of its electrical discharges represents itself in a frequency range of above 100 MHz. This field is independent from the pulse shape and the wiring system.

In other investigations, the measurement of the high-frequency part occurred via a receiving antenna. The main radiation source is the gap loop forming a loop antenna and excited by the erosion current.

The pulse shape is in particular reflected in the frequency range of 30 to 100 MHz. Below 30 MHz, no process relevant information can be determined. The spectral functions are only determined by the pulse shape. (3), (4)

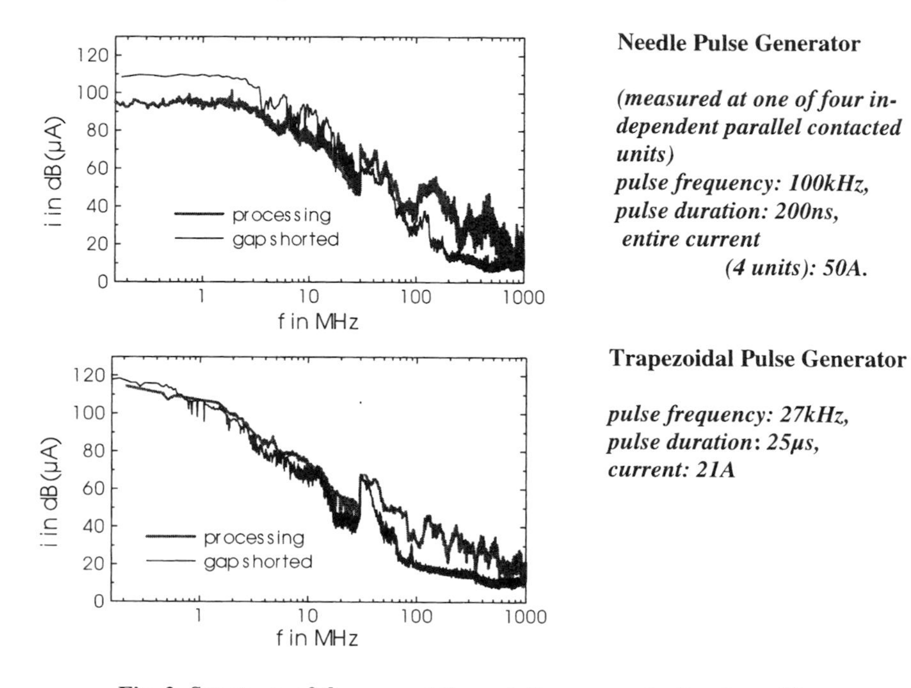

Fig. 3: Spectrum of the current through the gap contacting-loop (3)

Of course, for correct measuring high-frequency process characteristics are needed measuring systems suitable for this frequency range. Geometrical changes in wiring structures, for instance, can result in a wrong evaluation of the process. Are fades out too many pulses the process efficiency goes down. Practical investigations showed even a decreasing of the efficiency below 20%.

3 CRITICAL LIMITS OF PROCESS PARAMETER

3.1 Problems

Today the feed rate control is carried out via the analysis of the above-mentioned process parameters. The simplest method is the averaging. More extensive are the determination of statistical ratings or Fuzzy control. For the pulse parameter areas of standard Sinking and Wire Erosion, there are hardly problems if a sufficiently good process analysis determines the "critical" process states and eliminates them via a process control.

Since in modern eroding plants the pulse parameters can be adapted to the current processing situation, the new feed rate controls must be adapted to these variable conditions at shortest times. The fastest variant is the adaption via Fuzzy Control of the feed control parameters.(5)

For Fuzzy systems the question must be set up whether the process parameters sufficiently fulfil the demand concerning process stability and distinction in “desired” and “undesired” discharges. In the following the separation in "normal" and "abnormal" discharges will be considered.

3.2 Analysis Periods

The selection what the kind of feed rate control is to use in the first place depends on the hardware solution of feed system and the signal processing in the CNC. The hypothesis, simple feed systems could be corrected essentially by high-quality analysis and software systems, is only conditionally correct. In particular cases an improvement can occur, but a true optimisation of the processing problem can not be reached.

The analysis periods depend on several factors.

- computation time for the evaluating algorithm;
- time constant of the hardware system;
- time parameter of pulses.

Because of interdependencies between these factors a combinatorial analysis is reasonable. The computation time for the evaluating algorithms can be determined very simply and can be reduced by μ-Processors working in parallel. The time constants of the feed systems are in the millisecond range, i.e. the time consumption for calculations is uncritical.

For several reasons the order of magnitude of time parameters of pulses has an influence on the feed rate control. For series eroding plants the pulse duration is in a range from 2 to 100 μs, i.e. about 100 pulses come in evaluation. With these 100 pulses, the demands of statistics and the time constant of the feed system are met. No essential decrease in the removal rate is achieved by the extraction of "unwanted" discharges.

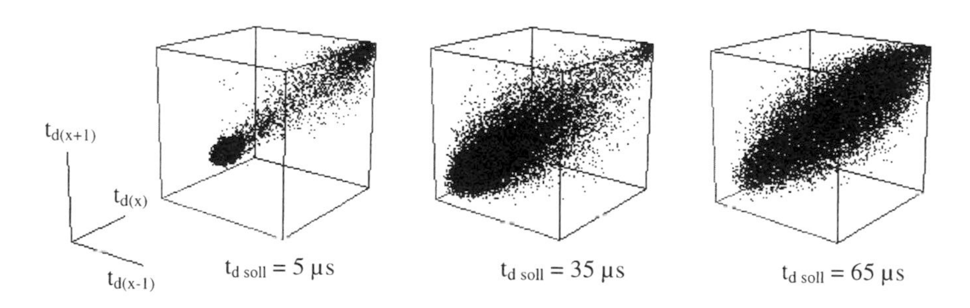

Fig. 4: Ignition Delay Time in the State Space – Pulse Sequence Analysis

In Fig. 4 distributions of the ignition delay times averaged over the time constant of the feet system at different given control parameters $t_{d\ soll}$ are shown. A too small controlled variable shows accumulations in the field of abnormal discharges, a too big controlled variable causes a cone-shaped distribution around the space diagonal. For a wanted concentration of discharges with low ignition delay times (near to arcs and short circuits) the points in the state space should lie near, but outside of a “forbidden” sphere with the centre in the origin of coordinates.

Especially, in the field of high current erosion with larger pulse trains over 100 A and pulse durations above 100 µs this evaluating strategy is critical to achieve a stable erosion process. A statistically necessary number of pulses leads to evaluation times being a multiple one of the time constant of the hardware system. The extraction of pulses unwanted leads to a drastic reducing of the removal rate, in spite of the larger removal volume of individual discharges.

3.3 Discharge Classification

The main problem of spark erosion is the clear definition of the spark as the "wanted" discharge and the wrong discharge as the "unwanted" discharge. For gap width control, this knowledge is only conditionally necessary because the controlled variables as middle gap voltage or middle ignition delay do not require this separation. For stable process courses, the detection of "unwanted" discharges must occur via a separate process analysis.

The feed rate control demands the correct evaluation of discharges directly or indirectly via an additional process analysis independent of using statistical evaluations or Fuzzy rules. Fig. 5 shows, that e.g. the decision between sparks and arcs basing on the ignition delay times and according the knowledge can not explain the pulse train represented. That is, because specific pulse and gap conditions are not kept.

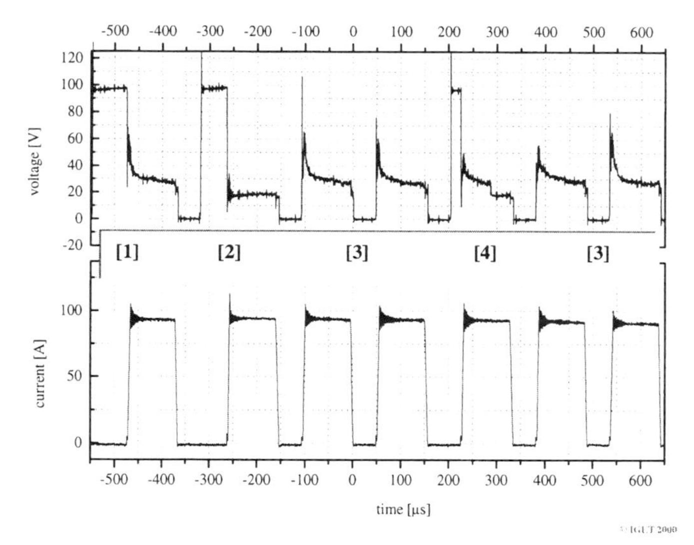

Fig. 5: Typical electrical discharge for High-Current-Erosion (6)

The pulse [1] corresponds to normal conditions, i.e. low gap pollution or a sufficiently large gap. The burning voltage is 25-30 V corresponding to a spark. The pulse [2] following directly, after normal ignition delay unexpected achieves only the arc burning voltage. The effect of this arc can only be analysed in a limited manner. Regardless of the very short ignition delay times the pulses [3] are spark discharges because of their voltage breakdown and the level of burning voltage. In the subsequent pulse [4] the spark discharge (ignition delay, 1st burning voltage stage) passes suddenly to the lower arc burning voltage.

As a result of investigation of process stability (2) the classification of discharges in only sparks and arcs is not sufficient. An essential role plays the localisation of the discharges. "Uncritical arc discharges" deliver an essential contribution for increasing the removal rate. However, the wear behaviour is another in comparison with the spark discharges. (7)

At process energy supplies generating very short needle pulses strong interferences due to fast transients have influence on up to 80% of the entire pulse duration. Burning voltage levels and process-induced RF parts can be analyzed no more. The middle gap voltage is almost only determined by ignition delay. The discharges can be analysed via the voltage breakdown. Because of the less pulse energies and the higher pulse repetition frequency, single arc discharges are uncritical.

4 CONCLUSIONS

For controlling the feed rate while spark erosion machining the four process variables middle gap voltage, ignition delay time, high-frequency part of the gap voltage and voltage breakdown are suitable as controlled variables .

Due to the relatively big time constant the feed system can not react to individual impulses. Therefore, it appears sufficient to use only one process variable or its mean value for the feed control. However, for precise trend statements the knowledge over every individual discharge as well as its classification is important. The exact assignment of single discharges to specific categories can only occur by consideration of several process variables. Therefore, for more powerful control strategies a combination of process variables has to be used. Fuzzy control systems, for instance, are a reasonable way for this.

REFERENCES:

(1) Dehmer, J. M.: Prozessführung beim funkenerosiven Senken durch adaptive Spaltweitenregelung und Steuerung der Erodierimpulse, Fortschrittsberichte VDI Reihe 2: Fertigungstechnik Nr. 244, VDI-Verlag Düsseldorf 1992.

(2) Schulze, H.-P.; Wollenberg, G.; Steinmetz, Th.: Stability of the EDM-Sinking Process. 12th International Symposium for Electromachining (ISEM) Aachen 11-13, 1998. VDI BERICHTE 1405 , pp. 215-223.

(3) Wollenberg, G.; Luhn, F.: RF-Noise Generation and Emission by Electrical Discharge Machines. International Symposium on Electromagnetic Compatibility October 5-7, 1999, Magdeburg, Germany. Otto-von-Guericke-University Magdeburg. Symposium Record. Pp.95-100.

(4) Luhn, F., Wollenberg, G., Giebel, St.: Modellbildung und Analyse der Abstrahlung von Funkenerosionsanlagen. EMV 2000, 22.-24. Februar 2000 in Düsseldorf. VDE-Verlag Berlin Offenbach, S.511-518.

(5) Pape, Th.; Steffen, Th.: Prozessführung mittels Fuzzy-Control. Unveröffentlichter Forschungsbericht. Otto-von-Guericke-Universität Magdeburg 1999.

(6) Wollenberg,G., Schulze, H.-P., Pape, Th., Läuter, M. : Scientific Report to EC-Project ECOPROD. April 2000.

(7) Kojima, H., Kunieda, N., Nishiwaki,N.: Understanding Discharge Location Movements During EDM. ISEM X, 6-8 May 1992 Magdeburg (Germany) Proceedingsof the 10th Intern. Symposium. S.144-149.

(8) De Silva, A., McGeough, J.A.: Process Monitoring of Electrochemical Micromachining. Int. Conf. On Machining and Measurements of Sculptured Surface. Krakow, 1997. P. 243-252.

(9) Rajukar, K.P., Wang, W.M.: A New Model Reference Adaptive Control of EDM. Annals of the CIRP Vol. 38/1/1989, pp. 183-186.

Production and Control

CAPE/004/2000

Flexible laser-based measuring cells in car-body assembly for dimensional management systems

A KALDOS and **A BOYLE**
School of Engineering, Liverpool John Moores University, UK
F STRONK
Perceptron (Europe) B V Rotterdam, The Netherlands

ABSTRACT

Car body manufacture and assembly involves a complex set of operations involving a major capital investment for the production tooling. This applies particularly to the manufacture and assembly of body panels in the "body-in-white" phase. Computer controlled co-ordinate measuring machines (CMM) are traditionally used to determine the dimensional accuracy of both the sub-assemblies and the complete car body. The paper considers the need for the replacement of CMM based methodologies with laser based optical measuring systems. The experiments reported conclude that laser based optical measuring systems provide the required speed of data capture and data processing capacity to control the production process and to assure the necessary dimensional accuracy of the car bodies.

1 INTRODUCTION

The need to reduce manufacturing costs, to increase assembly flexibility and to obtain high and consistent quality is an ever-present challenge to the automotive industry. This applies particularly to high quality volume car manufacturers where world wide competition is particularly fierce (1, 2). It is generally accepted the car body accounts for the greatest part of the value of the car structure and the assembly process represents 40% to 60% of the total manufacturing cost. Car bodies are assembled from a large number of individual parts and sub-assemblies.

Quality control in car body manufacturing and assembly is of paramount importance. Off-line dedicated fixture based measurement systems are widely used to check both sub-assemblies and completed car bodies. The rate at which the sub-assemblies and assemblies can be measured is limited by the measuring and inspection technology used. Further limiting effects are associated with the manual transport system employed for collecting and returning assemblies between the production line and the measurement area, the degree of manual setting up and adjustment required and the number of gauging or measuring points involved. The consequence of this is that a very low measurement sample rate is achieved compared with the production rate of the sub-assemblies. Coordinate measuring machines (CMM) are widely used to check both major sub-assemblies and complete car bodies. Although the accuracy of the system is excellent with data processing software and hardware support, the sample rate is still very low, typically in the region of 1%-2%, depending on the size and complexity of the structure to be measured.

The paper reports on the measurement and inspection being decisions made by one volume car manufacturer resulting from having available only one skilled inspector supported by one further semi skilled inspector to operate the CMM. Increasing production volume demands and greater quality demands are forcing the company to reconsider the performance of the CMM system to satisfy anticipated new inspection requirements. Failure to assess the dimensional integrity of the framed bodies is unacceptable to the senior management of the company. In addition they recognise any deterioration in the dimensional aspects of the body can result in increased repair costs, increased scrap and quality deterioration of the fitment of the doors, liftgates, boot lids and trim parts to the framed body. Any deterioration in quality standards is unacceptable as it is recognised this can result in loss of sales and loss of customer satisfaction. The need to continue to measure the framed bodies under conditions of increased volume of production and reduced operator levels is of critical importance.

The paper also examines the potential for the use of off-line measuring cell employing a fixturing system complete with a laser based measuring system and data feedback facilities. The strategy includes moving a laser sensor-measuring head to a set of predefined measuring positions. The geometry and size of the part under investigation determine the required measuring positions. The measuring system is provided with data processing software. The use of the proposed off-line flexible measuring cell substantially increases the measurement sample rate and it is envisaged that this type of cell can be incorporated into a car body assembly line. The accuracy meets requirements, the process is fast allowing a 100% sample rate to be achieved (3). Non-contact laser based measuring systems can be used to provide in-line measurement and quality control. The present quality assurance systems in part production, sub-assembly and assembly production are typically based on statistical process control (SPC) methods. The problem at present with these on-line laser based quality measurement systems is the substantial up front capital investment cost requirements.

2 MEASURING CAR BODY ASSEMBLIES

Car bodies, individual body panels and sub-assemblies are often geometrically complex and large in size. Due to the inherent mechanical flexibility of body shells and sub-assemblies, it is quite difficult to handle the car body parts and sub-assemblies, which makes dimensional management a particularly difficult task. Many of the major problems in car body production are associated with single parts such as the large openings for the boot space or the doors. The boot-space opening in particular is subject to errors caused by torsion as it is a three dimensional opening and is not confined to a single plane. These openings are often the most unstable part of the car body (4). The liftgate/door, which is designed to close the opening of the boot, is a three-dimensional curved surface, it is flexibly unstable and subject to dimensional errors in three dimensions. The complexity of the problem is indicated by the number of location points necessary to determine both its dimensional accuracy and its geometric form. Table 1 presents the tolerance information associated with each measuring point of a typical liftgate and Figure 1. shows the location of the required measuring points.

Statistical data can be of two types, namely Attribute data or Variables data (5). Attribute data consists of measuring the parameters of a product on the basis of either passing or failing a predefined specification while variables data consists of taking a measurement against a continuous scale, such as size. Statistical Process Control (SPC) can be applied to both sets of data. From the values given in Table 1 and using a 5-day week and a 46-hour working week, approximately six five-door units are measured each day. In terms of the measuring performance, if the production is 0.1% defective, then by measuring six five-door units and considering the data as attributes there is only a 6% chance of finding a defective unit (6). In addition, it takes three hours to measure the units using CMM methodology. This further indicates the need to improve the measuring system for framed bodies. One difficulty when measuring several parameters around a liftgate aperture and using Control Charts to determine whether the process is in control for each parameter, is that the Charts take some time to analyse and do not easily give rise to Management Reports.

The annual production of the body plant by volume of model type is given in Table 2 along with the number of bodies measured by the CMM system. The table shows that during a year

Table 1. Tolerance and features checked at measuring points on liftgate

Check no.	2	3	4	5	6	9	10	11	12	13	14
Feature	flush	flush	flush	flush	gap	flush	flush	flush	gap	twist	twist
Tolerance [mm]	+1.00 -1.25	+1.00 -2.00	+1.00 -2.25	+0.50 -2.00	+1.00 -1.25	+1.0 -1.50	+0.50 -1.75	+1.60 -1.50	+1.00 -1.75	Ok / N ok	Ok / N ok

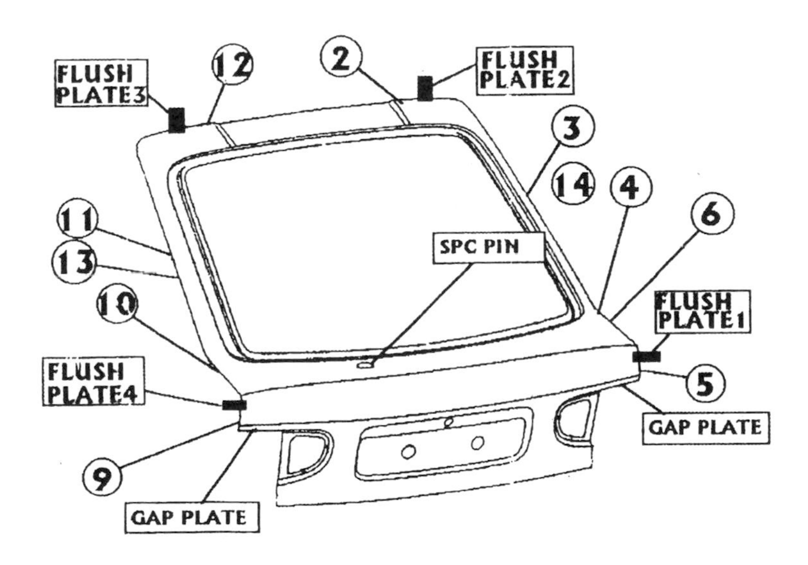

Figure 1. Measuring points on a typical liftgate

Table 2 Plant production by volume (number of units)

Model	3-Door	4-Door	5-Door	Van	All Models
Production	12675	6303	118403	30609	167990
Percent	7.5	3.8	70.5	18.2	100
Measured	179	229	1316	360	2084
Percent	1.4	3.6	1.1	1.2	1.2

the five-door model accounts for some 70.5% of the total production volume but only 1.1% of this volume is measured. Consideration clearly needs to be given to increasing the number of measurements made. There is a need to increase substantially the measurement and inspection sample rate, ultimately to 100%, by means of a relatively low cost but high quality measuring technique.

The production rate achieved for the manufacture of the liftgate is typically 100 parts per hour. This is a high volume part production requirement and in order to achieve total process control a significantly higher measurement and inspection sample rate than the 0.65% currently achieved is needed. This low sample rate results from the use of a dedicated fixture based off-line measuring station (7). The liftgate parts are loaded and unloaded manually, clamping is also manual and the wide use of manually based data collection and analysis system is typical. The ideal situation is to achieve a 100% sample rate but this cannot be achieved under present measuring methodologies used in off-line measuring arrangements.

3 ALTERNATIVE MEASURING STRATEGY

The purposes of the work is to examine the measuring method used for framed bodies and if possible recommend a system which is potentially an improvement on the existing CMM arrangement. Improvements are required from the viewpoint of operating cost, an increase in the number of bodies measured, improved statistical output and the tooling of the framed bodies. The CMM is also considered to be an inadequate measuring system to use for Statistical Process Control, because the amount of time taken to measure a framed body provides an inefficient sample size. Several measuring systems are used in body manufacture in the automobile industry including traditional dedicated gauging, automatic gauging, multiple CMM systems and laser camera systems (8). The wide scale introduction of laser based measuring systems into fully integrated in-line assembly is currently prohibited by the high capital costs and associated retraining requirements.

Discussions concerning the requirement for measuring body components, in particular liftgate assemblies, have raised the possibility of using a laser camera system. The suitability of the system and its viability for carrying out a large measuring frequency and its use in quickly assessing changes made to production tooling has been determined (8). The company has successfully manufactured a locating device to hold framed bodies in fixed tooling so that the laser units can carry out the measurements. For trial purposes the decision was taken to monitor the 3-door and 5-door model types and store the measured data in separate files. The rear aperture of the framed body has been selected for measurement because the rear opening is the most difficult opening in the body to control for dimensional stability. This is from a manufacturing viewpoint and from a product design viewpoint, as in general it is the largest aperture in the body. The two models selected for the tests have a rear aperture of common design dimensions.

It was agreed the process parameters should be controlled because then the quality parameters are also controlled. It was finally agreed eight indirect process parameter-measuring points should be employed along with two quality parameters. The parameters selected for measurement are two points on each of the two liftgate side match up lines, two points on the roof and the position of each of the bodyside master tooling locator holes. The two other points selected are on each side of the bodyside Rear Lamp Fin areas. These quality parameters are selected because this is the area of the body where the liftgate is set flush in the body aperture. Thus ten points are selected for measurement on the rear aperture, requiring one laser camera unit for each selected point (4). Figure 1 shows the selected measuring points on the aperture and the master tooling locator holes.

4 LASER CAMERA MEASURING

The measuring system is based on triangulation. A plane of laser light is emitted from the opening of the unit. When this plane strikes a surface it forms a contour line which is reflected back from the target surface to strike a mirror located above the laser which directs the returning beam to a CCD camera. Because the two fixed points, the laser diode and the camera, are located at a known angle the location of the laser line can be determined and hence the contour of the surface. Surface sensors combine laser-structured light and triangulation with feature image capabilities to measure in three dimensions. In the first step of the measuring process LED light is used to flood the surface to locate the feature in (X, Y) space then a laser plane is used to determine the Z range. The arrangement is typically used to identify the lateral and up/down motion of the centroid of a hole or slot (two-dimensional measurement) and its distance from the sensor (three-dimensional measurement). The arrangement is generally as shown in Figure 2.

The measurement programme consists of a hierarchy of models and zones. All measurement features on the liftgate are organised to one zone that represents logical groupings of measurements. The model, which can contain different zones, is a separate and distinct measurement programme. In this instance the designed measurement programme consists of one model, called Liftgate, and one zone called Gapflush. The programme measurement routine algorithms are predefined tools in the set-up section of the system software. In each algorithm the parameters required for the special part of the measurement programme is defined. Each sensor is calibrated and is provided with its own individual rectification data, which provides the system with an accurate calibration table of the camera and optics that are entered in to the system. The logical sensor set-up routine defines the measurement programme assigned to a physical sensor. Each physical sensor represents a single measurement point and is assigned its own algorithm in order to process the measurement. To determine the position of the measured part relative to the measurement system, a set of co-ordinate systems are defined and programmed as follows: -

- Body co-ordinate system. This defines the position of the car body. In the interests of standardisation the body co-ordinate system is defined as being the same as that used by the CMM body co-ordinate system.

- Model angles system. This defines the location of the part relative to the car position (body co-ordinate system) if the part is out of the original position during measurement.

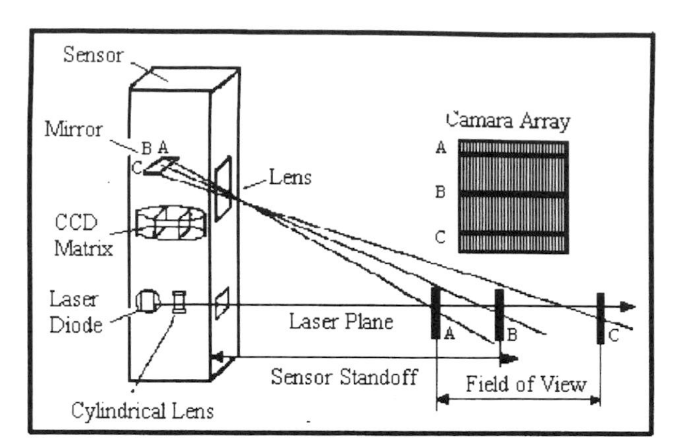

Fig 2 Sensor operation

- Sensor co-ordinate system. It is necessary to define the orientation of each sensor axis relative to the body co-ordinate axis of the part in the fixture in order to report dimensional deviation.

- Sensor sweep face angles are used to describe the orientation of the sensor, relative to the fore/aft (X-axis).

Sensor calibration is the process of setting up a nominal reference against which all measurements are to be compared. The set up is carried out using the master part of the production process. The master part is measured 100 times to allow average, minimum and maximum measurements to be calculated and displayed. In addition a standard deviation or sigma value representing the difference over the course of those number of measurements is also displayed, so that the sensor's repeatability can be evaluated.

The use of a manual measurement mode, for the production of n-line measurements, means the measurement cycle is under the control of the plant line controller. Different measurement cycles are carried out to prove the installed system in order to evaluate the accuracy and repeatability as well as to determine out of tolerance events. The time to carry out one measurement cycle is approximately one second. In order to check the dimensional quality of the liftgate, the deviations of Gap and Flush between part and fixture are measured. To calculate the desired measurement, the results in absolute co-ordinates are subtracted from the calibrated model. The resulting sensor co-ordinate deviation vector is then transformed into the body co-ordinate system via a transformation matrix.

5 OFF-LINE FLEXIBLE MEASURING CELL

A third possible strategy is based on the use of a robotised measuring cell involving sensory locating, clamping and measuring devices. The measurement task is to design a fast operation off-line measuring cell capable of significantly increasing the current sampling rate for the measurement of vehicle body assemblies and sub-assemblies. One solution is to move to an intermediate situation between an off-line measuring system and a true on-line measuring system, by providing the off-line measuring system with a high degree of automation. The progressive development of such a measuring cell arrangement from a standard off-line system to a fully integrated on-line system has been reported previously (8). The starting point of the development is a sensor based non-contacting measuring cell employing manual loading and unloading of the parts to be measured. The part loading and part unloading of the fixture can be further automated by the use of an appropriate robot and the has the potential to be completely automated with a suitable conveyor system for moving parts to and from the cell. The planning function for the measuring process requires a number of procedural steps in order to identify the important part location points, the part clamping points, the part clamping devices and any relative displacement of the part with respect to the mating parts. The selection of several location points on the surface is necessary to establish the presence of any possible surface orientation irregularities due to mechanical distortion.

In general the laser measuring system is designed to operate in a static arrangement with the part to be measured located in a measuring fixture and one point on the surface measured with one sensor. With this arrangement the static measurement fixture for the tailgate needs eight laser measuring sensors to measure both the gap and the flush conditions. To overcome the need and capital investment of having eight laser measuring sensors in the system the off-line measuring cell combines a six-axis PUMA762 industrial robot with a single laser-measuring sensor to generate a flexible three-dimensional measuring cell. This is achieved by attaching the laser-measuring sensor to the end plate of the robot arm. The robot is then programmed to deliver the laser-measuring sensor to the necessary measuring positions for any number of measuring points and for any part geometry that allow access to the measuring laser beam. In designing the arrangement of the measuring cell a number of decisions have to be made including: -

- How the different parts to be measured by the laser measuring system can be located.
- The accuracy of location of the parts to be measured so the measuring errors are due to variations in the manufactured parts and not to errors of location.
- The feasibility of using a single contour laser measuring sensor to measure a number of different points on the surface of the part under examination.
- The accuracy and repeatability of the robot system to move the laser-measuring sensor from one measuring point on the component to the next required measuring point.

For measuring purposes the part to be measured is located generally horizontally on a working or measuring fixture designed to provide a rigid, accurate and repeatable location for parts during the measuring cycle. The fixture is designed to hold the part in its required measurement position at a set of known three-dimensional coordinate positions. In addition the fixture provides high repeatability of location, it has a high resistance to vibration and allows the part to be simply located in the necessary position for measuring. For cell development and for determining the system accuracy and repeatability experiments involving five measuring points on a liftgate have been undertaken. During a single measurement cycle the coordinate values of each measurement point in three dimensions are determined. The measuring point locations are given in Table 3. A set of experiments have also been conducted to determine the inherent variation in the values recorded by a single sensor when collecting data from a number of different points. Prior use of the system has

Table 3. Coordinates of measuring points on liftgate

Measuring Axis	Measuring Point P1	Measuring Point P2	Measuring Point P3	Measuring Point P4	Measuring Point P5
X (mm)	-134.56	-556.06	-591.88	224.00	213.31
Y (mm)	719.95	830.19	1004.80	993.63	841.75
Z (mm)	691.13	311.25	332.06	311.44	301.31

indicated that readings can vary at a measuring position even though the sensor is held in a constant position.

The first experiment consists of visiting each of the five measuring points on a liftgate and while the sensor is held in the same place by the robot arm the coordinates of the point are measured from 5 to 100 times without moving the robot arm. The experiment results in a range of measured values for each of the five measuring points on the liftgate producing a standard deviation of 0.03mm. The distributions of measurement values at any given measuring position given by three standard deviations of the mean is +/- 0.09mm, which is an acceptable measuring error for the part being measured.

Repeatability and accuracy are meaningful only within a certain range of the independent variables of robot velocity and payload. The measuring cell requires the robot to make a defined movement and orientation to bring the sensor to the desired measuring point on the liftgate. So the repeatability of the robot to return again and again to the same point that has previously taught within set of fixed point is also to be considered. To reduce errors that stem from the momentum of the robot as it travels to each fixture point, a delay of five seconds is included to allow the robot ample settling time to activate the measurement operation.

In the experiments to determine the repeatability and accuracy of the robot, the robot is programmed to move to the required measuring positions at six different velocity values varying from a minimum value of 150 mm s^{-1} to a maximum value of 1500 mm s^{-1}. Each time the robot approaches the required measurement position, the X-Y-Z coordinates of the robot position are recorded. The mean deviation of the recorded position is shown to increase slightly as the robot velocity is increased. The repeatability of the robot is influenced by the speed, but the deviation is extremely low. Even though the higher speeds lead to shorter measuring cycle times and higher productivity it is recommended the robot is operated at the intermediate range of 1000 mm s^{-1}. This corresponds to an actual measuring cycle time, for moving to the five required measuring points and having a five second delay at each measuring point to take the measurements, of 29.5 seconds. The single sensor concept to measure a series of points on the tailgate is found to be applicable.

If a manual part handling time for loading and unloading the measuring cell is included the floor to floor time for measuring five different points on a liftgate can be reduced to an average value of six minutes. This increases the sampling rate for liftgates to ten per hour from the previous sample rate using CMM methodology of typically six per day. As the production rate for liftgates is typically 100 per hour the sampling rate is considerable increased to 10% from the previous value of 0.65%. A sample size of 10% is an acceptable level of data input for statistical process control systems.

The measuring cell is currently being expanded to incorporate two further development programmes. The first is to increase the size of the measuring table to accommodate two or more body panels to provide for multiple measuring positions on several parts within the same measuring cycle by the robot. This allows one body panel to be loaded or removed while a second or subsequent body panel is being measured. In this way further flexibility is incorporated into the measuring cell. The second development programme is to replace the laser-measuring sensor on the robot arm by a laser-scanning sensor. This will allow surface scanning to take place while the robot arm is programmed to move in a prescribed path to suit the particular geometry of the part under investigation.

6 CONCLUSIONS

The system has proved to be a very capable non-contact measuring system, which will become more important in the next few years. It will remain a supplement of the current measurement system in selected areas of process measurement. The system offers the potential advantage eventually achieving of 100% in-line measurement, monitoring and control of the process that is unattainable by traditional measurement techniques such as CMM's and checking fixtures.

The system introduces a new process control philosophy, which allows problems via suitable error identification up to the sub-assemblies stage to be found. The system offers the required flexibility, being able to accommodate a large number of different parts. The system has an excellent SPC software package and because of the capabilities of the system, it will be used in more non-automotive applications in the future which provides scope for further investigations. With the right application of the system, it can be an excellent contribution to quality improvements strategies.

References

1 Kaldos, A., Boyle, A., Newham, C. Influence of an Automatic Non Contacting Measuring System on the Assembly Strategy for the Front Frame of a Car Underbody. Proceedings of 12 International Colloquium on Advanced Manufacturing and Repair Technologies in Vehicle Industry, Balatonfured, Hungary, pp 114-119, 1995.

2 Golding R. How More can mean Less. Automotive World, February, p21, 2000.

3 Perceptron 1000, Optical Coordinate Measuring System for Quality Control. Perceptron (Europe) B V Rotterdam, 1994.

4 Dodd, I. A. Full Automation of a Non-Contact Measuring System, B.Eng. Report, School of Engineering, Liverpool John Moores University, 1994

5 Murphy, J A. Quality in Practice. Gill and MacMillan, Dublin, 1986

6 Scherkenbach, W. W. The Deming Route to Quality and Productivity – Road Maps and Road Blocks. Mercury. pp111-112, 1991.

7 Kaldos, A., Binger, G., Bingel, G. Development of hybrid, flexible measuring cells in car body assembly. Innovations-Forum, Dresden'96, Technical University of Dresden, 9-10. October 1996.

8 Kaldos, A., Boyle, A., Takacs, J. Development of Laser Based off-line Measuring Cells for Dimensional Management in Car Body Assembly. Proceedings International Conference on Laser Assisted Net Shape Engineering (LANE'97), Erlangen, Germany, pp287-296, 1997.

9 Ainsworth, K. The Perceptron – An Alternative to Co-ordinate Measuring Machines for Measuring Automobile Body Dimensions. B.Eng. Report, School of Engineering, Liverpool John Moores University, 1997

CAPE/050/2000

On-line production control via discrete event simulation – an industrial application

M DASSISTI and **L M GALANTUCCI**
Dipartimento di Progettazione e Porduzione Industriale, Politecnico di Bari, Italy
G CICIRELLI
GETRAG spa, Bari, Italy

ABSTRACT

A different perspective on the use of discrete event simulation is proposed to respond on-line to daily practical production-control problems. A simulation model has been designed and implemented, for a real manufacturing system, with the aim of supporting real-time operative decisions concerning contingencies in production control. Two benchmarking cases of applications have been addressed, with the flow production environment running in a clearly transient condition. The applications provide some interesting insights into future development of the application.

1. INTRODUCTION

1.1 Use of Discrete Event Simulation

Discrete Event Simulation (DES) is widely adopted for several purposes in manufacturing: manufacturing system design, layout design, production planning; production scheduling. Despite the increasing adoption in the recent time, in the literature it is still considered to be rather immature technology and its use is limited in today's industry [1].

Usually, the horizon of analysis is long-term both because of equilibrium analysis (for statistical collection reasons) and the limitation to update models when even slight changes occur in the manufacturing systems [2]. This has resulted in wide use of this powerful tool as an off-line support to engineer decisions, despite several authors in recent years suggesting on-line applications of DES, particularly for production planning and control purposes of flexible manufacturing systems (see, for instance, [3], [4], [5]). Most of the effort in this field has been devoted to Flexible Manufacturing Systems, for profitability reasons due to their capital-intensive features [6].

Nevertheless, in our opinion DES can be used in the field of integration with other types of production settings, with a lower level of flexibility and a low degree of unmanned operations such as the system referred to in this paper. In this kind of system, a different perspective of the use of DES has to be considered, namely as a decision support tool for line personnel; this new view requires an appropriate strategy for data updating of simulation models to overcome the present limitation in production planning and control due to its poor 'adherence' to actual operating conditions.

By lowering to the minimum the degree of abstraction, simulation model can be made a 'virtual copy' of the real working conditions of a manufacturing system. This strategy might turn simulation into an extremely useful tool for driving real-time decisions about problems related to production control, even within a complex manufacturing environment.

1.2 On-line production control

Quite often in the literature, production planning and control is addressed as an operation research problem, to be solved using several approaches (see, e.g., [7]). This might be true if dealing with a totally unmanned factory, seldom present in the current scenario. The most of the time, operators are responsible for implementing production plans and also for facing contingencies.

Production planning is based on process planning and long-term production scheduling. Process planning establishes the sequence of manufacturing processes to be performed in order to convert a part from an initial to a final form. It answers the question of how the product is going to be made and results in the description of the processes and their parameters, as well as the equipment and the machine tools to be used for production [8].

The production scheduling activity integrates time-varying information of the facility on a long-term basis to generate the Master Production Schedule (MPS). It answers the question of how to send the parts through the system to reach a specific production volume in a specified time span [9] given constraints on personnel, equipment and facilities.

Production control deals with the execution of the Master Production Schedule.

Operation scheduling (or process scheduling) starts from the MPS and adapts to the actual situation at the moment of execution. The time horizon might vary from days to hours or minutes. At the lowest level, there is real-time control, which is responsible for the execution and monitoring of all the activities necessary for production, from the scheduled time to their completion. The time horizon here is in minutes or even seconds. [9]

It is evident, from the above, that controlling production activities, particularly in real-time, may be a critical as well as complex task, because of the importance of correcting decisions taken at this stage, where dynamically changing conditions in a complex system make it necessary to change scheduling solutions to adapt the MPS to suddenly changing situations.

Without significant reshaping of the production organisation or of the PPC approaches [10], this might become a very significant problem, particularly if a decision has to be taken by not very experienced personnel or without a high skill profile, resulting in a very poor effect on the whole production.

2. INDUSTRIAL CASE ANALYSED

In order to derive sound indications about the use of simulation as an on-line tool to support decisions in production control, an existing manufacturing system has been benchmarked for the present study. The choice was in order to test the DES model with reference to physical and logical operating conditions of a real system.
The company studied is the GETRAG S.p.A. plant in Bari, which is one of the major mechanical transmission manufacturers for the automotive industry in the South of Italy (it produces about 20 transmission variants), with 350 machines, about 800 permanent employees and 27,000[m^2] of manufacturing area, with a target rate of 600,000 transmission/year over 4 shifts. GETRAG production setting is organised in 4 autonomous 'channels', as shown in figure 1 (one each per major sub-component of the final product); each channel consists of one or more autonomous production sub-units, which manufacture a complete part of a sub-component; the ATG22 line modelled in the present paper produces the output shaft of the transmission (German "abtriebswelle") which also has 9 variants.

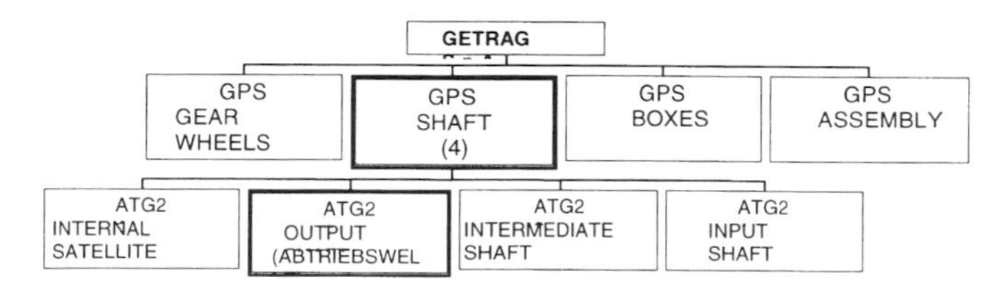

Figure 1. Organisation of GETRAG S.p.A.: 4 production channels

The processing sequence for the 'abtriebswelle' is shown in figure 2, which also defines the physical layout of the ATG22 line. It consists of 22 stations plus one dedicated heat-treatment oven (shared with other lines), 16 on-board buffers and 10 interoperation buffers. The processing technologies are quite standard, while production machines are of the new generation without use of lubricants to reduce the environmental impact of machining.
The ATG22 is an independent line, being interfaced with the raw part warehouse and the assembly channel via another warehouse; the production pace is set by the 'internal customer' (assembly channel). This latter is driven by a production plan with a monthly horizon.
The 9 variants of the output shaft have almost the same processing sequence, with the same processing time (including transportation). No difference exists between tool life amongst variants, which is controlled for preventive replacement via a PLC on board each machine. Tool change time, as well as set-up time between variants are dependent on the ability of the operator and have been accurately recorded for simulation purposes. The company runs a preventive maintenance policy, so as to minimise WIP and improve efficiency of production flow. To this aim a maintenance plan is prepared at scheduled intervals and is implemented

under the responsibility of each production line, with direct involvement of all the personnel in the line, according to a Total Quality Management approach.

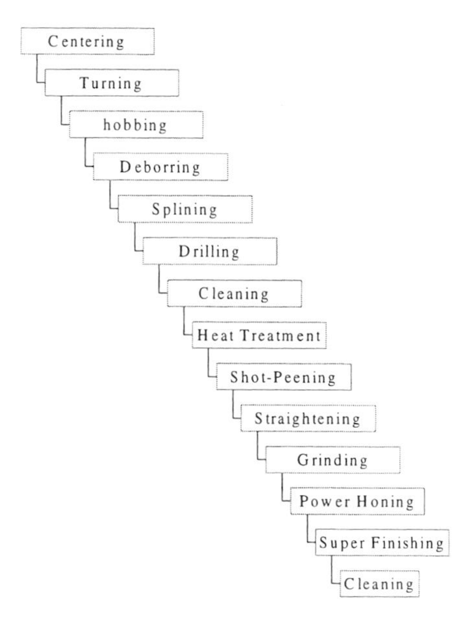

Figure 2. Processing sequence of the 'abtriebswelle' component.

3. THE SIMULATION APPROACH

Simulation offers more advanced potential than ordinary tools available for production control: apart from statistical features, it allows the capture and representation of the complexity of information and production flows [11]. At the present time, the increase in calculation power and software tools for information management reveal to a different perspective for the use of simulation.

As stated above, the target of a simulation model is to support real-time operative decisions concerning contingencies in production control, where the horizon of 1 to 4 shifts is addressed (transient conditions). To this aim, special attention has been devoted to the expert judgements on failure or breakdown risk of machines, more than using statistical data which rely more on stationary conditions.

Simulation modelling did not differ from the traditional approach (see [2]). The model was built after an accurate analysis of the critical variables (bottlenecks; critical buffers; technological criticality of the processes). This was also done to reduce the number of variables and parameters in order to simplify model updating and modification. In this sense, the model can be considered a 'macro model' with, according to [12], a low level of detail and encompassing a narrowly defined system).

Two special criteria have been adopted in modelling as follows. The first criteria was to reduce simplifications to a minimum, such as having the 'virtual copy' of the logical

operational of the system. A second criteria, allowed by the simulation environment chosen [13], was a flexible design of the model to simplify data updating at the beginning of the simulation runs (*model initialisation*): material in the line, processing times, possible reduction of the functionality of part of the system; priorities between codes. Furthermore, the model design has been made flexible, so as to allow the analysis of a general number of codes with a general processing time and set-up time.
The initialisation strategy adopted was chosen to completely allow automatic update of the state variables recognised (which here are represented as a decisional set of numbers, i.e. the module of a decisional vector). A set of matrices has been used to this aim (processing times; set-up times, WIP state; technological risk on the machines), which are input for each code automatically, using 20 routines. At present, the true problem remains to retrieve and collect the more than 200 data from the factory. Each time a simulation is needed, the initialisation procedure is requested by the simulation model, so as to start in the closest conditions to the real state of the production system.
Much effort is required in improving this stage, to exploit the data management system SAP/R3 implemented at GETRAG. A research is under way at present to address this problem.

3.1 The simulation model

In this section the simulation model developed for the ATG22 line is briefly described (see figure 3). The simulation environment adopted to model the production setting is SIMPLE++ [13] for its object-oriented features and its friendliness in modelling the real manufacturing setting. The model consists of approximately 1000 elementary objects and there are 100 entities contemporary present during simulation.
The demand pattern coincides with the real production plan scheduled (it is input using the table file).
The product processing sequence of Figure 2 also traces the physical layout of the line. Deterministic processing times have been set according to those corresponding to the ideal production cycle. Machines have been considered to be subject to deterministic breakdown, the duration and occurrence of which is provided according to expert judgement. Machines also produces scraps, which have been assumed to be lost, as in the real case given the tight tolerances of the operations that occurs according to historical data.
Set-up times were also assumed to be deterministic, and include the first setting of the machine for the variant as well as the 1st piece conformity check.
For the sake of brevity, the numerical details of the simulation model will be omitted here.

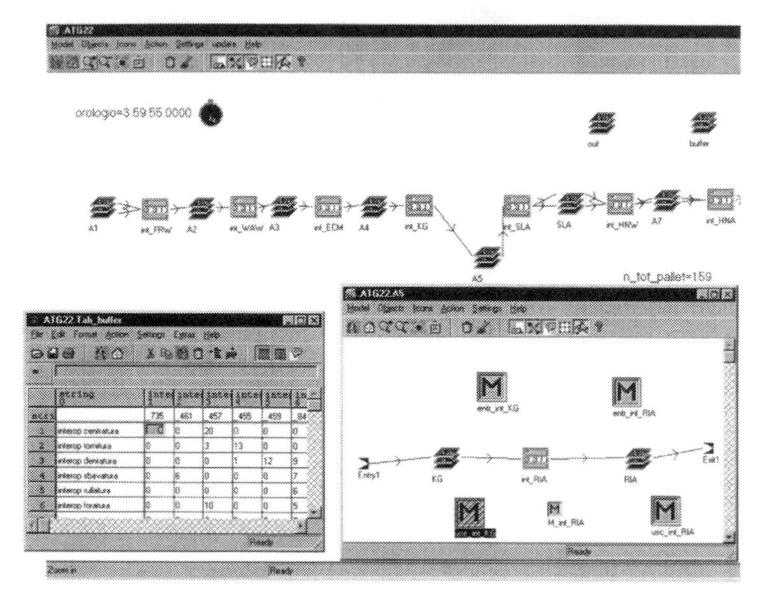

Figure 3. Simulation model of the ATG22 line.

As stated above, in order to have the 'virtual copy' of the system, no hypothesis have been made on availability of raw parts (raw part warehouse). Furthermore the physical place for finite products (output warehouse) was assumed to be the real one.
At the present, the only departures from a virtual copy are the following assumptions: 1) a complete availability of human resources, tools and spare parts has been assumed, as it is in reality; 2) no modelling of human operator efficiency has been made so far; 3) intersection of the current operations with other lines has been modelled using an availability coefficient of the shared resources (thermal treatments; honing).
The transportation system has not been taken into account here, due to the true state of the line (pallets move during processing time).
The model has been validated by running a deterministic simulation and comparing the manufacturing lead time and throughput per code either with the analytic calculation in ideal conditions or with output of real production over several days.
The output performance measures derived form the model are quite standard ones: throughput, work-in-process (WIP), job tardiness, lead time, total amount of parts produced per shift, scraps, machine utilisation so as to allow use of appropriate decision rules to be adopted.
Simulation CPU time for 1 shift simulation is less than one minute.

4. BENCHMARKING TESTS

In this section, two test cases are presented to benchmark the usefulness of the model developed. The operating conditions of these cases were clearly transient (start-up phase), where the system did not have all the 22 machines running (mostly redundant, parallel

machines were down). Furthermore, with statistical knowledge of machine behaviour was available. This choice was voluntary, given the scope of the approach proposed.
The updating criteria of the line state are here addressed , since this is a critical design factor of the approach from a forecasting accuracy point of view [3]. The effect of the updating horizon has thus been analysed by considering three different conditions: no updating of line state during simulation (NO case); updating of line state (only WIP) every 1 day (1D case) and, finally, every 2 days (2D case)
The first test case concerns the analysis of a 3-week run period. Three replicate simulation runs were made for each condition analysed, recording the system performance (namely Work-in-process) at the end of each horizon (apart from the no updating case, where daily recording was made). WIP performance measure has been considered here because in the transient condition analysed it was the most critical target for the production (this index is a direct consequence of throughput and scraps).
Table I shows the true WIP recorded for the indicated days. Dates have been altered for to respect privacy of the Company; figures are the real ones for the three different updating horizon. Table I also shows the percentage of relative errors (with respective sign) on WIP and the average of their absolute value (reckoned with a 2-day step for coherence with data recorded in the 2D case).

Table I. Simulation output for WIP for the three-week test.

WIP		NO (no updating)		1D (1 day updating)		2D (2 day updating)	
Date	WIP real	Absolute err.	% error	Absolute err.	% error	Absolute err.	% error
15/xx/9x	1343	-515	-38,3	-515	-38,3	-515	-38,3
16/xx/9x	1128	-772	68,4	-314	-27,8	--	--
17/xx/9x	1062	-770	-72,5	-35	-3,3	421	39,6
19/xx/9x	1285	-519	-40,4	102	7,9	--	--
20/xx/xx	2049	-100	-4,9	759	37,0	881	42,9
21/xx/ xx	1805	-393	-21,8	-45	-2,5		--
22/xx/ xx	1667	-581	-34,9	104	6,2	-445	-26,9
23/xx/ xx	1853	-166	-9,0	82	4,4	--	--
24/xx/ xx	1611	-416	-25,8	-278	-17,3	-261	-16,2
26/xx/ xx	2093	-290	-13,9	590	28,2	--	--
27/xx/ xx	2283	-50	-2,2	150	6,6	108	4,7
28/xx/ xx	2445	-96	-3,9	-108	-4,4	--	--
29/xx/ xx	2480	-268	-10,8	-22	-0,9	80	3,2
30/xx/ xx	2714	-171	-6,3	540	19,9	--	--
31/xx/ xx	2605	-474	-18,2	-208	-8,0	34	1,3
ABSOLUTE AVERAGE	VALUES	(2 DAY STEP)	(13,84)	---	(7,84)	---	(11,54)

It is worth noting, to avoid misunderstandings, how apparent convergence of relative errors on WIP is not due to the updating criteria adopted but to the actual line working conditions. This is in fact confirmed by all the three criteria having the same convergence. What it is significant to say is that updating significantly reduces the errors per day, as it is also

possible to note from the average values presented. Using this latter parameter, it can be seen that the 1D updating strategy results the best.
Another test has been run to check discrete event simulation as an on-line decisional support, using the model for driving a decision over a personnel assignment problem, again during the start-up phase. In this case, only 15 of the 22 machines were working; 2 operating shifts out of 4 were held and an expected throughput of 450 parts/day was fixed (instead of 2400 for the regime state). The operation problem to solve was as follows: to decide whether or not to allocate human resources on a critical bottleneck (honing machine), which was causing a significant reduction of the throughput (and, therefore, an increase in the job tardiness). Given the line state, the two alternative solution were i) to decide whether maintain 2 shifts or ii) to assign 2 operators for 8 hours overtime for the rest of the week. The average cost of this latter solution was estimated about 1250 Euro.
Table II reports the WIP level observed and evaluated by the simulation model for the week concerned; no initialisation has been done here.

Table 2. Simulation output for WIP for the three-week test.

DAILY THROUGHPUT			
	(TRUE)	(SIMUL.): 2 SHIFTS	(SIMUL.): 3 SHIFTS
Monday	361	(361)	(361)
Tuesday	68	(68)	(68)
Wednesday	457	(457)	(457)
Thursday	(475)	360	368
Friday	(499)	450	690
Saturday	(445)	455	600
TOTALS (target 2700)	2305	2151	2544

As can be noted, the 3 shift solution was the best one as concerns the WIP target. Simulation revealed how this solution was only a temporary one, because it came out that that in the next week the honing machines would be starving (extra processing capacity).
In addition, the real decision taken at that time by the production engineer was to increase partially the number of extra-shift hours for both the employees, thus resulting in the least cost but not in a good agreement with the production targets.
The effect of updating the breakdown risks on the machines was also tested, in the same cases shown above, by changing the risk of breakdown occurrence according to real events. No significant influence was recorded on WIP for either test (a slight influence was recorded on the throughput) for the criteria adopted. For this reason we made the assumption of neglecting the outcome of these trials and do not present here the numerical results. Nevertheless, this outcome does not invalidate the idea, which needs more cases to be proved.

5. DISCUSSION

The paper presents an approach of using discrete event simulation as a decision support tool for production control. It does not try to optimise either the control strategy or the scheduling

rules to be adopted. Decisions driven by the test cases analysed derive from well-experienced line personnel.
The results of benchmarking cases discussed in paragraph 4, despite encouraging (for the error figures on the performance parameter), have to be considered as partial proof of the effectiveness of the approach proposed. Indeed, these do not provide any statistically significant evidence of it.
The applications allow to draw some interesting hints on future development of the application proposed.
It is worth noting how important it is to improve the accuracy of initialisation of the model, which may also prove to be very expensive in terms of human hours. The initialisation may also give the chance to implement expert opinion on the breakdown risk. In this way it would be possible to have a more accurate and reliable technological state of the system but the only statistical one (information about what can be called the 'technological state').
As a concluding remark, what is important to stress here is the necessity for a new perspective on the use of simulation tools for real-time application also for other systems than FMS. This fact brings a different order of problem: it is not possible to rely only on statistical information, due to the non steady state condition of analysis! Statistics, in fact, neglect the 'technological state' of the system, which belongs to the instantaneous conditions. These, in turn, forces planner to take control decisions on unpredictable events which dynamically vary the history of the production environment.

6. ACKNOWLEDGEMENTS

The Authors wish to thank Mr. Tobias Hagenmeyer (President) , Dieter Schlenkermann and Herbert Roth - board of directors - for their encouragement to operate in the factory and Mr. Franco Modeo – GPS2 leader - of GETRAG S.p.A., Bari, for his personal support of the research activities.
This research was developed under the research program supported by the Italian MU.R.S.T. "Evaluation cost models for management of manufacturing systems under uncertain operating conditions" - ex.40% - 1998 - co-ordinated by Prof. A. Villa.

7. REFERENCES

[1] KLINGSTAM, P., 1999, "A methodology for supporting manufacturing system development: success factors for integrating simulation in the engineering process", Proc. 32nd Int. Sem. on Manufacturing Systems, Leuwen, Belgium, 359-367.
[2] SCHROER, B. Y , TSENG, F.T., 1989, "An intelligent assistant for manufacturing system simulation", Int.. J. of Prod. Res., Vol. 27, No.10, 1665-1683
[3] WU, S-Y D., 1989, "An application of discrete-event simulation to on-line control and scheduling in flexible manufacturing", Int.. J. of Prod. Res., Vol. 27, No.9, 1603-1623
[4] ALTING, L., BILBERG, A., LARSEN, N.E., 1988, "Extended applications of simulation in manufacturing systems", Annals of the CIRP, Vol, 37/1, 417-420.
[5] BENGU, G, 1994, "A simulation-based scheduler for flexible flow lines", Int.. J. of Prod. Res., Vol. 32, No. 2, 321-344.
[6] HEBRON, A, 1998, "On-line production control of a flexible multi-cell manufacturing system operating in a highly dynamic environment", Int.. J. of Prod. Res., Vol. 36, No.10, 2771-2791.

[7] GUPTA, M.C., GUPTA, Y.P., EVANS, G.W, 1993, "Operation planning and scheduling problems in advanced manufacturing system ", Int.. J. of Prod. Res., Vol. 31, No.4, 869-900.
[8] Chryssolouris, G., 1992, "Manufacturing systems (theory and practice)", Springer – Verlag, NewYork, Inc.
[9] VAN BRUSSEL, H., 1990, "Planning and scheduling of Assembly systems", Annals of the CIRP, Vol,39/2, 637-644.
[10] ENGELBERT, W., SIHN, W., May 1999, "Order Management by means of the 3-L-PP concept", Proc. 32nd Int. Sem. on Manufacturing Systems, Leuwen, Belgium, 521-525.
[11] Law, A.M., Kelton, W. D, 1991, "Simulation modelling and analysis", McGraw Hill, New York.
[12] WILLIAMS, E.J., AHITOV, I, 1996, "Scheduling analysis using discrete event simulation"; Proc. 29th Annual Simulation Symposium, 148-154.
[13] Aesop GmbH, "SIMPLE++ reference manual", Stuttgart.

CAPE/071/2000

Visual modelling for CORBA-based negotiation scheduling of holonic manufacturing systems

K H P TAM, S T R FUNG, W H R YEUNG, and **H M E CHEUG**
Department of Manufacturing Engineering and Engineering Management, City University of Hong Kong, Hong Kong

Abstract – One of our recent research projects focuses on the use of Object Oriented Technology (OOT) and Unified Modelling Language (UML) in the design of a new class of advanced manufacturing system, called the Holonic Manufacturing System (HMS). HMS builds on a modular mix of standardised, autonomous and co-operative agents called holons. A negotiation control model for HMS, which advocates a dynamic structure, has been devised. It enables messages passing to be carried out for scheduling, operation planning and physical execution of tasks in the system. The Common Object Request Broker Architecture (CORBA) services are incorporated in our model to provide point-to-point, and point-to-multi-points connections in a distributed environment. This paper attempts to describe this software system using UML. In particular, it illustrates the interaction between various components of the system by using activity diagrams, sequence diagrams and so forth. It therefore aims to provide, at this stage, a snapshot of the behaviour of the system.

1. INTRODUCTION

The major problem facing the manufacturing industry nowadays is the rapid change of market demands and the high cost of changing the manufacturing environment to meet the demands. It is essential in today's flexible manufacturing systems (FMS) to incorporate the *agility, configurability, extendibility and scalability* attributes in its architectural design and development [1-2]. This is necessary in order to ensure and enhance the growth of productivity and maintain the industry's competitiveness in the market.

The concept of holonic manufacturing system (HMS) was introduced in the early 1990s as a new approach for the design of a FMS [3-7]. HMS builds on a modular mix of standardised, autonomous and co-operative agents called holons [8]. Under the holonic shop floor control framework, shop-floor elements such as machine tools, AGVs, parts and so on, are now modelled as distributed and autonomous holons [9-11]. Negotiation-based communication protocol is used between holons for message passing. With this feature built in a HMS, it is expected that scheduling and control activities carried out at the shop-floor will be integrated in a highly configurable, extendible and co-operative manner.

Common Object Request Broker Architecture (CORBA) [12-14] has been proposed by Object Management Object (OMG) in 1991 [15-18] to provide high flexibility and interoperability in developing distributed software systems. It enables us to develop a highly distributed and interoperable holonic negotiation protocol which is agile, configurable and extendible to meet the needs of a modern HMS.

Having developed a HMS model, a representation method is required. The Unified Modelling Language (UML) [19-24], which is evolved from merging Jacobson's Object Oriented Software Engineering (OOSE), Rumbaugh's Object Modelling Technology (OMT) and Booch's methodologies, is a computer aided software engineering (CASE) tool for object-oriented modelling and design of systems. With the use of various diagrams and relationship, it shows and illustrates how a software system model is built, how its functionality is conceived by end users and its internal interaction.

In this paper, UML is used to illustrate the software system which includes the CORBA based communication services such as event service, and naming service for the negotiation scheduling.

2. SYSTEM ARCHITECTURE

In our CORBA-based negotiation scheduling of software system, hereafter referred to as the system, the end users are the "Job Request" and "Machine". They are attached to the system for negotiation and scheduling as shown in figure 1. The entire system is mainly divided into five parts, namely Naming Server, Event Server, CORBA based system, part holon and machine holon. The Component of the system is shown in figure 2.

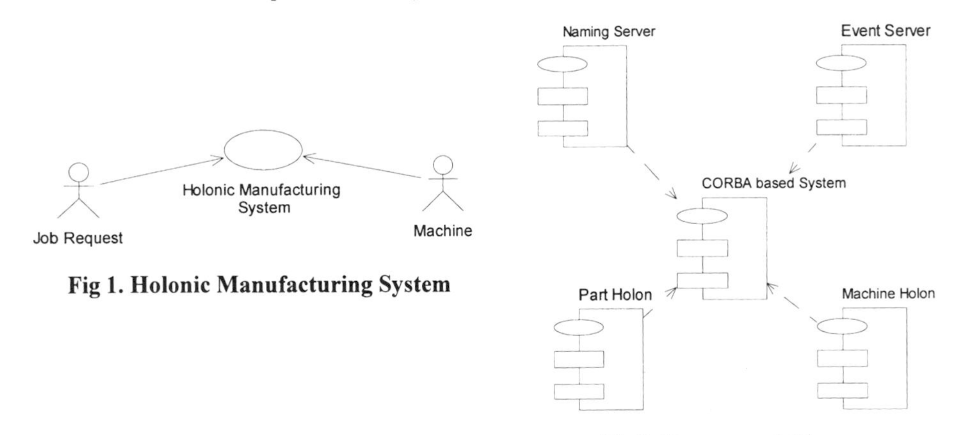

Fig 1. Holonic Manufacturing System

Fig 2. Component Diagram

A machine holon models a machine tool found in typical HMS. It waits for job request from a part holon and gives the reply. A part holon represents jobs from customers, usually from high level production planners. It sends job requests to the machine holons and analyses the response from machine holons to select the best machine which will satisfy the job request with best schedule. In our system, Event Service, Naming Service and Callback are implemented to facilitate and support the development of a truly distributed version of our system.

The Event Services provide different kinds of event channels for machine holons and part holons to attach to. Each event channel logically represents a capability pipeline where only machine and part holons having the same capability, such as drilling, milling, boring, etc, will connect to the channel. In this way, the messages between part and machine holons are more targeted and reduced to a manageable number. The Naming Services is used to store the information of different kind of event channel such that the machine holons and part holons will retrieve the channel information and connect to the related channels irrespective to the physical location of the event server. Callback functions break the traditional concept of client/server programming model. With the callback functions, not only the client can send data to the server, the server can also send data back to the client effectively.

2.1 Use Cases of the system

Use case diagrams are a kind of diagrams employed in the UML for modelling the dynamic aspects of system and are central to modelling the behaviour of a system, a subsystem, or a class. Each one of them shows a set of use cases and actors and their relationships. As shown in figure 3, the behaviour of the CORBA-based negotiation-scheduling model is documented in a use case model using a number of use cases to illustrates the system's intended functions (use cases), its surroundings (actors), and relationships between the uses cases and actors (use cases diagram).

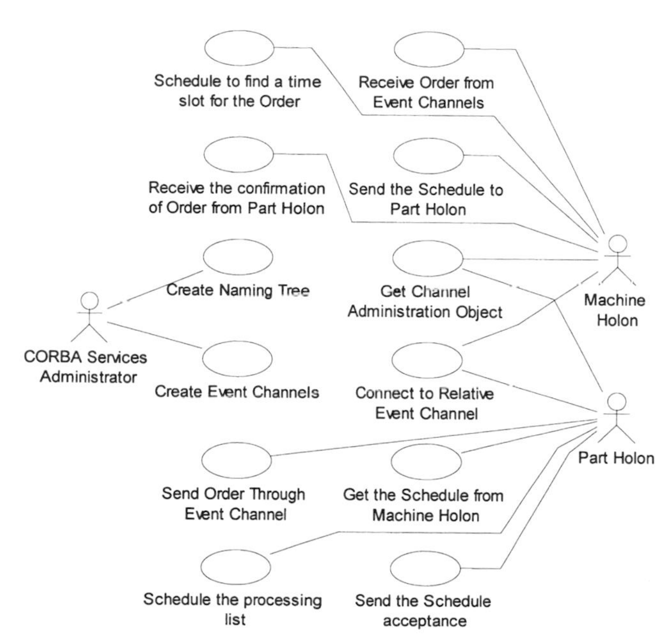

Fig 3. Use Case Diagram

In the figure, there are all together 3 actors and 12 use cases identified and they can be grouped into the following 3 categories: -

2.1.1 *The behaviour of the CORBA Service Administrator:*
The CORBA Service Administrator is responsible to initialise the CORBA services for the system including the Naming Tree and Event Channels. The Naming Tree will store the information of the event channel. Different Event Channels are created for different type of machine capabilities support in the HMS.

List of CORBA Service Administrator behaviour:

- Create Event Channels
  - Event Channels are created based on the machining capability, such as drilling, milling, boring and so forth. Part holon then sends job request to machine holons which are attached to the event channel with the same capability as those required for the job.
- Create Naming Tree
  - The Naming Tree is used to store the information of different event channels for the holon retrieving and connecting to related event channels.

2.1.2 *The behaviour of the machine holon:*
The machine holon represents the capabilities of the machine tool in a typical HMS. It receives the job requests from part holons and find a time slot for the task if possible.

List of machine holon behaviour:

- Get Channel Administration Object from Naming Tree
  - Before the machine connects to the related event channels, it must retrieve the channel information from Naming Tree for channel connection.
- Connect to Related Event Channels
  - In order to receive the job requests (orders) from part holons through the event channel, the machine holon must connect to related event channel. For machine holon with both drilling and boring capabilities, it will connect to both the 'drilling' channel and 'boring' channel at the same time.
- Receive Job requests from Part Holon through Event Channel
  - After the connection to event channel, the machine holon waits for the job requests from part holon and store the job requests when they are received locally.
- Schedule to find a time slot for the job request
  - Having the information of job requests, the machine holon will base on its own subsequent process scheduling rule to find an available time slot for the job request, if available.
- Send the schedule to part holon through Object Request Broker (ORB)
  - After the scheduling, the scheduled result will be sent back to part holon for verification in part holon.
- Receive the confirmation of Job requests from Part Holon by callback function
  - When the part holon accepted the schedule from a machine holon, the machine holon will receive an acknowledge for confirmation of job request. Then the machine holon will allocate the time slot for the job request.

2.1.3 *The behaviour of the part holon:*
The part holon represents returned job request to manufacture a part, which include information such as part ID, batch size, deadline represent and machine capabilities requirements and so forth. It will send the jobs to machine holons through appropriate event channels and analyse the schedule from machine holons. Finally, it will decide on implementing which machine holon's schedule.

List of part holon behaviour:

- Get Channel Administration Object from Naming Tree
  - Before the part holon connects to the related event channel, it must retrieve the channel information from Naming Tree for channel connection.
- Connect to Relative Event Channel
  - In order to send job requests (orders) to machine holons through the event channels, the part holon must connect to the related event channel such as drilling channel.
- Send Job requests through Event Channel
  - After connecting the event channel, the part holon will send job requests to machine holons which are connected to the event channels.
- Get the Schedule from Machine holon
  - After the part holon sends the job requests to machine, it will wait for the proposed schedule from machine holons.
- Schedule the processing list
  - After it receives the schedule from machine holons, the part holon will base on its internal scheduling rule, select a suitable machine holon for the job request.
- Send the schedule acceptance
  - After the machine holon is selected, the part holon will send an acknowledgement of confirmation to it for completing a job later according to the agreed schedule.

2.2. Design Pattern
Figure 4 below shows the Class Diagram with Stereotype Display of the system and their relationship.

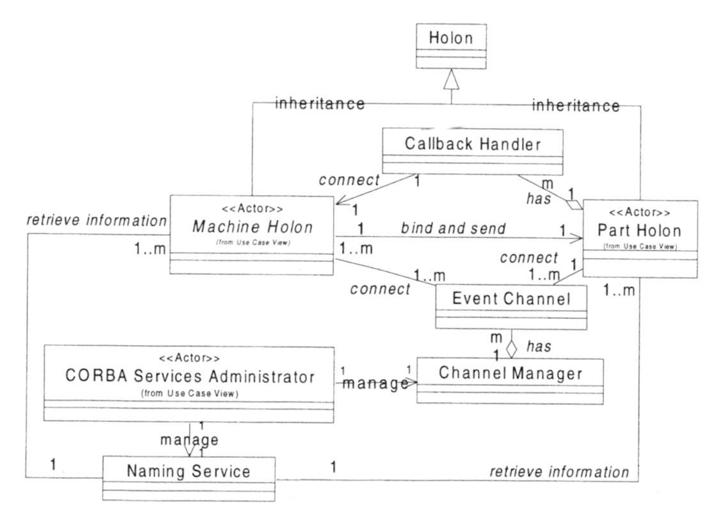

Fig 4 Class Diagram of CORBA based HMS

As shown in the figure, **Machine holon** and **Part holon** are the children of the **Holon** class. They inherit the features of a holon. Part Holon and Machine Holon make use of the naming tree from **Naming Service** to retrieve channel information. Naming Service and Channel Manager are managed by the **CORBA Service Administrator** using navigable association relationship.

Channel Manager contains the class of **Event Channel** which makes an association relationship between itself and the holon for job requests sent from part holon to machine holon. Moreover, machine holon will send the schedule to part holon by one-to-one communication so that navigable association relationship is formed between them. The part holon also contains the class of **Callback Handler** which handles the message directly passing from part holon to machine holon. There are three communication paths shown on the CORBA based HMS. Therefore, this design pattern shows three different communication paths for the system.

3. DYNAMIC BEHAVIOUR OF THE SOFTWARE SYSTEM

This section makes use of a number of sequence and activity diagrams to illustrate the basic operation of the system, how the part and machine holon attach to the event channels, how the job requests are sent to machine holons and how the machine holons send back their proposed schedule and so forth.

3.1 *Activity Diagram*

Figure 5 below shows the activity diagram of the entire system. It can be divided into 3 different phases of operation, namely the initialisation phase, the scheduling phase and the negotiation phase.

In the initialisation phase, the system will prepare the necessary naming trees, event channels and make the machine and part holons connected to the appropriate channels.

Phrase 2 is the negotiation phrase. After the part holon sends the job request to machine holons through event channels, the part holon will wait for the reply from machine holons. If no reply is received within a given period of time, it will re-announce request. When the part holon receives the schedule from machine holons, it will go into the third phase to decide which schedule to select and send back an acknowledgement to the machine holon. In the machine holon, it waits for the job request to arrive. If the request is received, it will go into phase 3 to find a time slot for the job request. If the machine holon successfully finds out a time slot for the job request, it then waits for the confirmation. If no confirmation is received, it will release the time slot. Otherwise, it will allocate the time slot for the job request.

It then is followed by the scheduling phrase. The machine and part holons will base on their own scheduling rule to schedule. Machine holon will determine the best schedule to fulfil the job request while the part holon will decide which schedule from machine holon is most suitable for the job request.

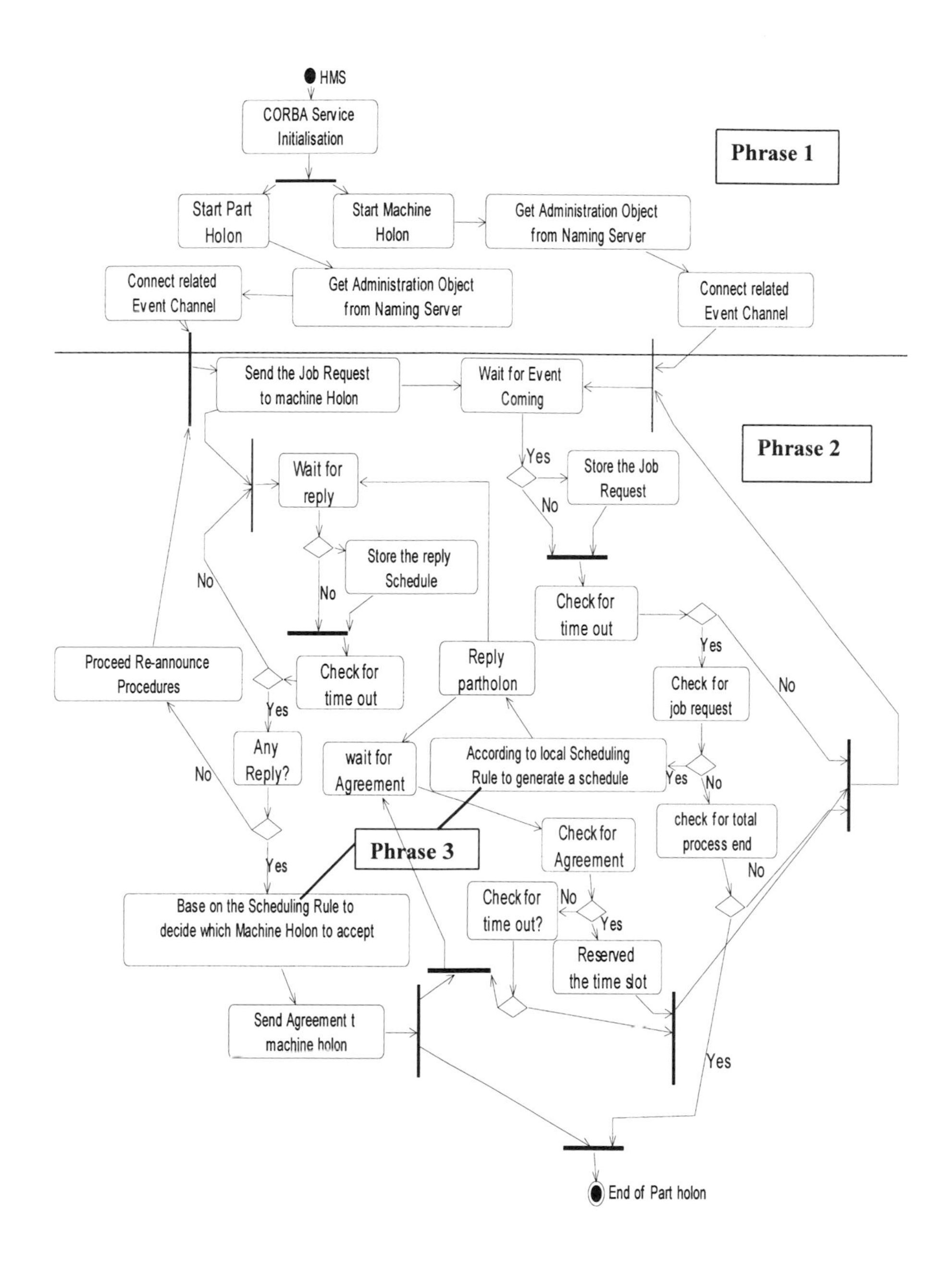

Fig. 5 Activity Diagram of CORBA based HMS

3.2 *Establishment of Event Channels*

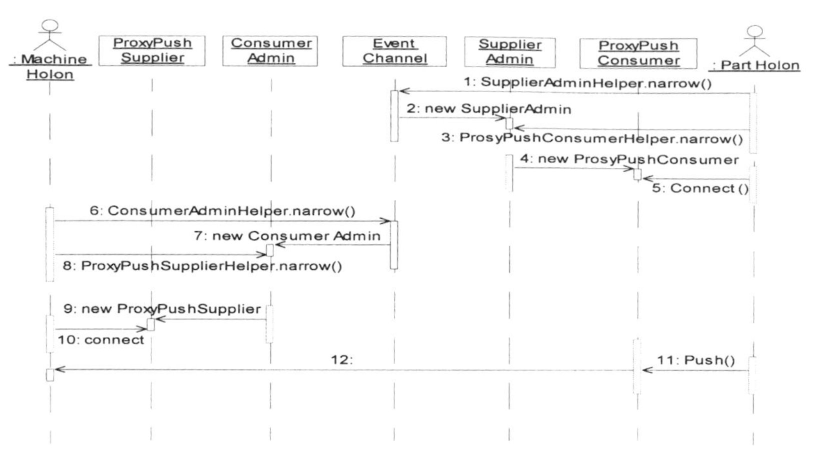

Fig. 6 Sequence Diagram of Event Channel Handling

Figure 6 above shows the sequence diagram which illustrates the typical sequences involved for part and machine holons to connect to event channels. First of all, the part and machine holons will connect the Event Channel. (Procedure 1 and 6) to obtain the **SupplierAdmin** and **ConsumerAdmin** (Procedure 2 & 7) Objects respectively. Then the holon will contact the SupplierAdmin and ConsumerAdmin Objects (Procedure 3 & 8) to obtain the **ProxyPushSupplier & ProxyPushConsumer** objects (Procedure 4 & 9). Finally, part and machine holons will register with the event channel on ProxyPushSupplier and ProxyPushConsumer objects. After the set-up is completed, the part holon can send messages to machine holon (procedure 11 & 12) through the channels.

3.3 *Interaction between the part holon and machine holon*

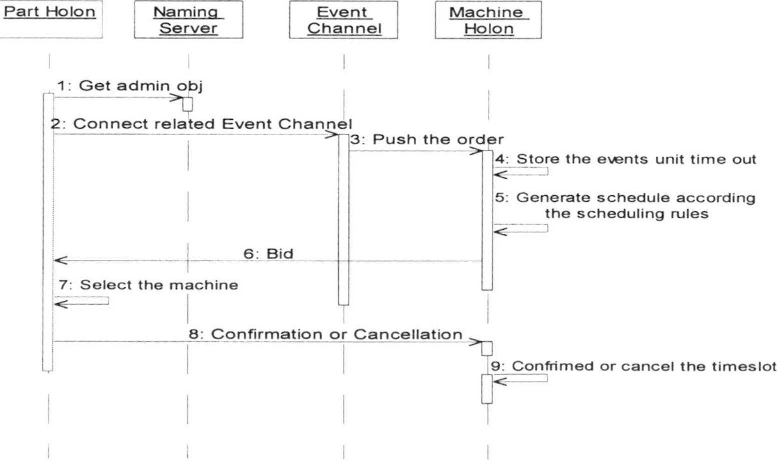

Fig. 7 A typical sequence diagram for interaction between part and machine holon

Figure 7 represents one of the cases in our HMS. Part holon obtains the Event Channel Administration Object for connecting with related Event Channel (details are shown in the sequence diagram in figure 6). It will send the job request through the event channel to machine holon. The machine holon will store the job request until time out. After this, the machine holon will generate a schedule according to its own scheduling rules and send it back to the part holon. When the part holon receives the schedules from machine holons, it will select a suitable one also according to its own scheduling rules. Moreover, the part holon will send the confirmation or cancellation of job request to the related machine using callback technique. Finally, the machine holon will either cancel or reserve the time slot depending whether the confirmation is positive or not.

4. CONCLUSION:

In this paper, we describe the use of UML to illustrate an object-oriented architecture for a CORBA-compliant Holonic Manufacturing System. While UML provides us unified tools to model and design our system, CORBA provides a very good and robust distributed environment for developing the system with high flexibility and reconfigurability. We are now at the stage to finishing the design of the entire system, and the implementation of the system is being proceeded in stages.

5. ACKNOWLEDGEMENT

The authors wish to thank the Research Grant Committee of the Hong Kong SAR Government for funding the research project.

6. REFERENCES:

[1] P. Valckenaers & H. Van Brussel, *IMS TC5: Holonic Manufacturing Systems*, National Center for Manufacturing Sciences, Inc., Dec 1994.

[2] Cheung H.M.E., Yeung W.H.R. Yeung, Ng H.C.A., & Fung S.T.R., "HSCF: a holonic shop floor control framework for flexible manufacturing systems", *International Journal of Computer Integrated Manufacturing,* Vol.13, No.2, 2000, pp 121-138.

[3] M. Winkler & M. Mey, "Holonic Manufacturing Systems", *Europe Production Engineering* 18 (1994) 3-4, pp 10-12.

[4] Dipl.-Ing. M. Höpf, "Holonic Manufacturing Systems: The basic concept and a report of IMS Test Case 5", National Center for Manufacturing Sciences, Inc., Feb 1995.

[5] P.P. Groumpos, "The challenge of Intelligent Manufacturing Systems (IMS): the European IMS", *Journal of Intelligent Manufacturing* (1995) 6, pp 67-77.

[6] S. M. Deen, *Cooperation issues in Holonic Manufacturing Systems, Information Infrastructure Systems for Manufacturing* (B-14), H. Yoshikawa and J. Goossenaerts (Editors), Elsevier Science B.V. (North Holland), 1993 IFIP. pp 401-412.

[7] Tharumarajah A., Wells J. & Nemes L, "Comparison of bionic, fractal and holonic manufacturing system concepts", *International Journal of Computer Integrated Manufacturing*, Vol.9, No.3, pp.217-226.

[8] Koestler A., *The Ghost in the Machine*, London, 1967.

[9] Ng H.C.A., Yeung W.H.R. & Cheung H.M.E., "New Paradigm for flexible manufacturing systems: a holonic approach", *Proceedings of The 3rd International Conference on Manufacturing Technology in Hong Kong*, Dec. 1995, pp.555-560.

[10] Ng H.C.A., Yeung W.H.R. & Cheung H.M.E., "An object-oriented information model for the control of holonic manufacturing systems", *Proceedings of The 12th International Conference on CAD/CAM, Robotics and Factories of the future*, London, Aug. 1996, pp 8-14..

[11] Ng H.C.A., Yeung W.H.R. & Cheung H.M.E., "HSCS - the design of holonic shop floor control system", *Proceedings of the 5th IEEE International Conference on Emerging Technologies and Factory Automation (ETFA '96),* Kaùai, Hawaii, Nov. 1996, pp.179-185

[12] The OMG Homepage, Object Management Group,http://www.omg.org.

[13] *CORBAservices: Common Object Services Specification*, OMG document 1998, http://www.omg.org/cgi-bin/doc?formal/98-07-05.

[14] *The Common Object Request Broker: Architecture and Specification, Revision 2.2*, OMG document 1998, http://www.omg.org/cgi-bin/doc?formal/98-07-01.

[15] "*OrbixWeb Programmer's Guide and Reference*", IONA Technologies, 1998.

[16] "*OrbixNotification Programmer's Guide and Reference*", IONA Technologies, 1998.

[17] Robert Orfali, Dan Harkey, & Jeri Edwards, *Instant CORBA*, John Wiley & Sons, pp 3,7,8 1997.

[18] Robert Orfali & Dan Harkey, *Client/Server Programming with JAVA and CORBA Second Edition,* John Wiley & Sons, 1998. ISBN 0-471-24578-X

[19] Unified Modelling Language Resource Centre. http://www.rational.com/uml/index.jtmpl

[20] *"Rational Rose 2000 – Using Rose",* Rational Software Corporation, 2000.

[21] "*Rational Rose 2000 – Using Rose CORBA*", Rational Software Corporation, 2000.

[22] Grady Booch, James Rumbaugh, & Ivar Jacobson. *The Unified Modeling Language User Guide*, Addison Wesley, 1998. ISBN 0-201-57168-4

[23] Perdita Stevens with Rob Pooley, *Using UML – Software Engineering with Objects and Components,Updated Edition,* Addison Wesley, 2000. ISBN 0-201-64860-1

[24] Terry Quatrani *"Visual Modeling with Rational Rose and UML",* Addison Wesley, 1998. ISBN 0-201-31016-3

A process analysis methodology for enterprise resource planning (ERP) implementation

T BILGE
Department of Industrial Engineering, Marmara University, Turkey
M OZBAYRAK
Department of Systems Engineering, Brunel University, Uxbridge, UK

ABSTRACT

With ERP through computers and a common database, the work in many different functions of the enterprise can be better coordinated, and volumes of common information can be shared. ERP has sought sophistication but achieved complexity instead. Therefore implementation of ERP applications has become a more complex and expensive issue. In this study, a process analysis methodology for ERP implementation is presented. This methodology consists of the steps necessary to select and define processes on which an ERP implementation will be carried out and the steps to identify the gaps between these processes and the standard processes of the ERP application. The major aim of this methodology is to determine the degree of functional suitability of an ERP application to the business requirements of an enterprise by determining the gaps systematically.

1 INTRODUCTION

There are two options in an information technology (IT) project to automate the processes. The first one is to develop the software system from the beginning. This approach can be taken if the enterprise has special business requirements. But it takes too long to develop a software, and this fact causes major problems in many IT projects. The delays caused by slow systems development have been the second most common cause of failures in IT projects (1).

The second option in IT projects which is an increasingly common reaction to the time frame dilemma is the use of the packaged client/server ERP applications. These applications can meet 80% to 85% of the business process requirements. That means only 15% to 20% of the software needs to be customised by writing new code. For the moment this option - purchase and modify - still seems to be the likeliest choice for large corporations that consider installing an integrated client/server manufacturing system. More and more corporations are buying their business applications and modifying them. No one is writing human resources or general

ledger applications from scratch. Rather they are purchasing them and then adapting them with tools (1).
Implementing ERP applications requires extensive effort and care. There are two questions that must be answered before the implementation. The first one is "will the business processes be improved and redesigned?" If the answer is "no" then the ERP package will be modified to a great extend to adapt to the current processes. That will both increase the cost and duration of the implementation. But if the answer is "yes" the implementation will be much easier with the redesigned processes although some minor modifications still can be done in the package.
The second question is "when will the business processes be improved and redesigned?" Before the ERP implementation or after the ERP application has been selected? There are two opposing schools of thought related to this problem (2). One school believes that implementation should be considered only after a process has been improved. The other school believes that IT should be considered before a process has been improved. This school of thought defends that ignoring advances in IT can limit the validity of the models that are defined by the project team. But it is also stated that applying information technology before process improvement tends to put the process into the hands of the IT people. It is said that when adverse situations occur in the implementation, process participants will not take the ownership of the process rather, they will very quickly blame IT people. Others have said that project teams consider only software rather than to truly understand the process (2).

Considering the advantages and the disadvantages of both thinking third way is could be followed: seeking improvement and redesign opportunities for the processes while considering the limits of the IT which is the ERP package in our case. This is the main approach in this study. This approach can also be formulated by a linear programming problem:

Max {improvement and redesign}

St.

ERP application.

This study consists of four parts. The first part is introduction. The second part is the core of this study. It includes the process analysis methodology for ERP implementation. The methodology consists of four steps as can be seen in Figure 1. Processes are selected and mapped in steps 1 and 2. In step 3 the mapped processes are matched with the processes in the ERP application. Finally in step 4, the gaps between the mapped processes and the ERP processes are identified. This methodology ensures that the processes are analysed and the right ERP application is selected before the ERP implementation. The third part of the study includes an application of this methodology for a selected process in a manufacturing firm. Finally part four is the conclusion.

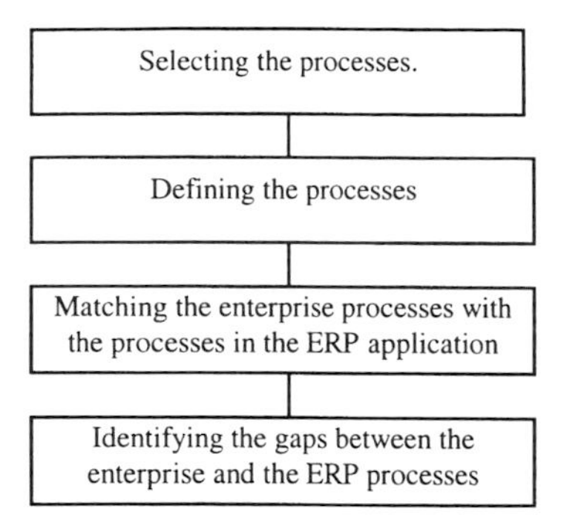

Fig. 1 The four step process analysis methodology

2 A PROCESS ANALYSIS METHODOLOGY FOR ERP IMPLEMENTATION

2.1 Selecting the processes

2.1.1 Determining the altitude of the process analysis

The altitude of the process analysis must be determined before selecting and analysing the process (3). Altitude is critical because it establishes the vision and perspective in both selecting and documenting a process. Selecting and documenting a process for an international corporation will require a much bigger vision than selecting and documenting a process for a single local company.

There are several questions that should be asked for determining the altitude of the process analysis for ERP:

- What is the type of the organisation on which the process analysis study will be carried?
  It can be a department, a manufacturing unit of a company, a sales organisation, a single local company, a group of local companies or an international conglomerate.

- Is the organisation a legal entity or a strategic business unit?
  If the organisation is a legal entity it has some legal requirements like paying tax, preparing balance sheet and income statement.

- What are the company's primary and secondary markets?

- Does the company have immediate plan to expand into new markets?

- What are the customer base and customer groups?

- What are the products and services from which revenue is derived?

- What will be the scope of the ERP implementation? Which modules are to be implemented? Will it comprehend all the processes, or a partial implementation will be carried on?

2.1.2 Determining the level of detail of process analysis

Level of detail of the analysis determines the resources that must be allocated for selecting and documenting the processes (3). Lots of detail results in tight control but there is only little freedom. Little detail results in limited control but lots of freedom. There is a trade off between control and flexibility. Once the desired level of detail is determined, we do not spend time for unnecessary activities in our analysis.

The questions that should be asked for determining the level of detail of the process analysis are as follows:

- What is the intent of documenting the process?
  In our study the intend is to make a comparison between enterprise processes and ERP functions and to determine the gaps.
- How much detail is necessary to document the process?
  In our analysis, there is no need to prepare data dictionaries since we are not building a database in our study.

2.1.3 Selecting the processes that are in the scope of the ERP implementation

After determining the altitude and the level of detail of the analysis the processes can now be selected. The resources allocated for defining the altitude and the level of detail of the analysis will pay big dividends for selecting the right processes (3).

It is recommended to focus on the following questions when selecting a process (3):

- What is the purpose of the process? What do you want the process to achieve?

- Do you need the process?

- Why do you need the process?

- What are the alternatives?

- What are the consequences of no action?

- Who are the intended customers/users?

2.1.4 Defining the process boundaries.

Once the processes are listed and a convention for the altitude and detail agreed on, most of the boundaries should be reasonably clear. Before process mapping, the start-point, end-point, and boundaries of the process with other processes should be identified (4). Where the process boundary lines are drawn? Where does another process precede, supplement, or follow? If this study is done for a department, the boundaries get smaller. If it is for an international corporation, the boundaries are much larger. The analyst should focused on the following questions when determining the boundaries of a process (4):

- Where do other processes stop?

- Where do other processes begin?

- How do my customers/users influence and are influenced by the process?

The answers to the questions above should help us determine the boundaries of the process. Once the boundaries are determined, definite starting and end points of the process and the framework are established for the process.

2.2 Defining the processes

After selecting the processes that are in the scope of the ERP implementation, the next step is to define them. Defining the existing processes requires an extensive understanding of the activities that constitute the business processes and the other processes that support them, in terms of their purpose, inputs, outputs and their boundaries.

There are several points to consider when defining the processes (2):

- The aim of the defining the processes should be to understand them in order to create enhanced customer satisfaction and improved business performance, not only to document them or standardise them.

- No solutions should be suggested when defining the processes. A common error when defining the business processes is the tendency to describe what should be done according to a standard operating procedure instead of what is actually done. It is difficult enough to capture the true flow of a process since people have different views in the process. Adding solutions just add confusion to the situation.

- Individuals tend to focus on a process flow that occurred seldom. The approach should be to capture what usually happens, 80 - 90 percent of the time. Too many processes have been documented and designed with an emphasis on the exceptions rather than usual.

There are two steps for defining business processes: collecting data and mapping the processes.

2.2.1 Collecting data

Data must be collected to define the processes. The following data is useful for a complete definition of the process (5):

- Output of the process
- Customers of the output
- Process participants

- Process owner
- Stakeholders
- Process boundaries
- Inputs and their suppliers

There are several methods for collecting data about the processes. Three of them can be used for our analysis: interviewing, questionnaire and observing behaviour & environment.

2.2.2 Process mapping

After necessary data is collected the processes should be mapped to be defined and improved. Process mapping is defining the process in a compact manner, as a means of better understanding and improving it.
Before starting to map the processes a convention to describe the hierarchy of process levels should be declared. A typical three level hierarchy includes "process", "activity" and "task" (6).

- Process: A top level description of a major business activity, e.g. "Sales".

- Activity: This is the next level expansion of each step in the process, e.g. "Processing Customer Orders".

- Task: A task is a step of the activity in the process hierarchy. For example, "Printing Customer Invoices".

There are several techniques for mapping processes. "Flowcharting" is one of them. Another technique for mapping processes is "process table" which will be used primarily in our analysis. It is a tabular way of mapping processes. This technique is explained in detail below.

Process table

Process table is a tabular way of mapping a process. The process table consists of rows and columns. All the information that can be stored in the flowchart symbols can also be stored in rows and columns of the process table.

The basic methodology for preparing process tables is:

- Decomposing the process into activities and activities into tasks.

- Arranging the process table by placing the activities and their tasks in rows and their attributes in columns of the process table.

Each task of an activity is represented by a row. The number of rows in the table is equal to the number of tasks in the process. Each column in the process table stores one attribute of the activity task. The fields that must be located in the process table are number and definition of process, activity and task. Additional information can be recorded on a process table by adding other fields.

Table 1. Process table for internal purchasing process.

	A	B	C	D	E	F	G	H	I	J	K	L	M
1	Process no.	Definition of process	Activity no.	Definition of activity	Task no.	Definition of task	Person/Dept that carry out the work	Task type	Input data	Output data	Media of input data	Media of output data	Problems, proposals for improvement&exceptions
2	1	Internal purchasing	1	Purchase requsition	1	Filling purchase requsition form	The dept.that make the requsition	Operation	Material, short text, quantity requested,unit of measure, person and dept.who make the request	Request form		Paper form	The request forms that were not filled completely causes interruptions, delays in next process steps.
3	1	Internal purchasing	1	Purchase requsition	2	1th approval	The head of the dept.that make the requsition	Approval	Info in the request form	Signature in the request form	Paper form	Paper form	The request form ought to be filled in a computer file format and transferred for approvals electronically to speed up the process.
4	1	Internal purchasing	1	Purchase requsition	3	2nd approval	General manager	Approval	Info in the request form	Signature in the request form	Paper form	Paper form	The request form should reach to the director of accounting before the approval of general manager for fixed asset purchases.
5	1	Internal purchasing	1	Purchase requsition	4	Budget control for fixed assets	Budget plannnig dept.	Operation	Budget plan, info in the request form	Request form	Paper file	Paper form	
6	1	Internal purchasing	1	Purchase requsition	5	Decision of purchase of fixed assets that is unnplanned in budget	General manager	Decision	Info in the request form	Decision in the request form	Paper form	Paper form	
7	1	Internal purchasing	1	Purchase requsition	6	Signing purchasing form	Accounting manager	Approval	Info in the request form	Signature in the request form	Paper form	Paper form	
8	1	Internal purchasing	1	Purchase requsition	7	Recording purchasing form into the Excel purchase monitoring file	Purchasing dept.	Operation	Info in the request form	A record in the excel purchase request file	Paper form	Excel file	If the purchasing form is filled and processed in a computer file instead of in paper form then there is no need record it to a PC file.
9	1	Internal purchasing	1	Purchase requsition	8	Filing purchase requsition form	The dept.that make the requsition	Operation					If the purchasing form is filled and processed in a computer file instead of in paper form then there is no need file it.
10	1	Internal purchasing	2	Purchasing	1	Collecting vendor proposals	Purchasing dept.	Operation	Info in the purchase request in the Excel file	Price, terms of payment, delivery date	Excel file	Paper form	The data like price, payment terms, discount conditions for each vendor should be available in the computer system.
11	1	Internal purchasing	2	Purchasing	2	Evaluating vendor proposals	Purchasing.,Acc.and demanding depts.	Decision	Price, terms of payment, delivery date of each vendor proposal.	The winnning vendor that is recorded in the Excel purchase request file.	Paper form	Excel file	Accounting manager attends the evaluation meeting if an only if a fixed asset is being purchased.
12	1	Internal purchasing	2	Purchasing	3	Giving an order by telephone or fax	Purchasing dept.	Operation	Material, short text, quantity requested,unit of measure.	Date of order	Excel file	Excel file	Purchase orders are not monitered in the system as a separate entity. They are a part of purchase requisitions.
13	1	Internal purchasing	3	Stock monitoring	1	Good's receipt	Warehouse	Operation	Bill of laiding number, material, quantity receipt, unit of measure, name of vendor	Location number	Paper form	Paper form	
14	1	Internal purchasing	3	Stock monitoring	2	Recording good's receipt	Stock planning dept.	Operation	Bill of laiding number, material, quantity receipt, unit of measure, name of vendor, location number	A record in the stock file	Paper form	Text file	
15	1	Internal purchasing	3	Stock monitoring	3	Signing bill of laiding for material that is under stock control	Stock planning dept.	Approval	Info in bill of laiding	Signature in bill of laiding	Paper form	Paper form	
16	1	Internal purchasing	4	Posting invoice	1	Revision of payment plan	Accounting dept.	Operation	Payment amount, due date for payment, vendor name, vendor no., vendor bank account no.in vendor master data.	A payment plan record in the excel file	Paper form, text file	Excel file	The invoice that was posted should be output in the payment list and this list should be in an electronic data format and sent to the bank by e-mail.
17	1	Internal purchasing	4	Posting invoice	2	Accounting posting of vendor invoice	Accounting dept.	Operation	Pym.amount, due date for pym., invoice date, vendor no. in vendor list, G/L accounts to be posted in accounts list.	Accounting document no.	Paper form, text file	Text file	

A process table has two advantages over process flowcharts:

1. The type of data that can be stored in the table is extremely large since we can add as many rows and columns to the table as are required. But the analyst is limited to several types of symbols when using flowcharts.

2. A process table is a database once it is stored in a computer file. Therefore it can be processed by using the capabilities of computers.

A sample process table prepared for documenting "internal procurement process" of a paper manufacturer firm is given in Table 1. The properties of this table are as follows:

1. The fields that are located in the table are "process number", "definition of process", "activity number", "definition of activity", "task number", "definition of task", "person/dept. that carry out the work", "task type", "input data", "output data", "media of input data", "media of output data" and "problems, proposals for improvement and exceptions". Other fields can be added to the table or some fields can be dropped.

2. The table was prepared in an Excel™ spreadsheet. Therefore the data in this table can easily be stored, retrieved, filtered, summarised, printed and distributed by using computers. For example, the table was filtered by "task type" to select approval tasks as can be seen in Table 2 by using 'filter' function of Excel™. Also the process table was summarized by using the pivot table function of Excel™ and this summary table is given in Table 3.

Table 2 The filtered process table to select tasks which are 'approval'

	A	B	C	D	E	F	G	H
1	Process no.	Definition of process	Activity no.	Definition of activity	Task no.	Definition of task	Person/Dept that carry out the work	Task type
3	1	Internal purchasing	1	Purchase requsition	2	1th approval	The head of the dept.that make the requsition	Approval
4	1	Internal purchasing	1	Purchase requsition	3	2nd approval	General manager	Approval
7	1	Internal purchasing	1	Purchase requsition	6	Signing purchasing form	Accounting manager	Approval
15	1	Internal purchasing	3	Stock monitoring	3	Signing bill of laiding for material that is under stock control	Stock planning dept.	Approval

Table 3 A summary of process table - the distribution of tasks by activity.

SayýTask type	Definition of activity				
Task type	Posting invoice	Purchase requisition	Purchasing	Stock monitoring	GenelToplam
Approval	0	3	0	1	4
Decision	0	1	1	0	2
Operation	2	4	2	2	10
GenelToplam	2	8	3	3	16

3. The last field of the table is "problems, proposals for improvement and exceptions". It is used for storing the problems, improvement opportunities and exceptions of the process task if there are any.

4. The field which is called "task type" is used for storing the type of work. A task can be an operation, a decision, a transportation etc.

5. There are four fields in the table for modelling data flow in the process:

- Input data. It stores the data/information required for the process task.
- Output data. It stores the data/information that is produced in the activity task.
- Media of input data. It is the media in which the input data/information is carried to our system.
- Media of output data. It is the media in which the output data/information is stored.

2.3 Matching the documented processes with the processes in the ERP application

At the end of step 1 we have selected the processes that will be implemented and in step 2 we have defined them. Before identifying the gaps between the ERP application and requirements of the enterprise, the enterprise's documented processes should be matched with the processes in the ERP package.

There are several sources from which ERP processes can be extracted:

1. The ERP application's own documentation. Detailed information about ERP application processes and functions are available in help documents of the ERP applications.

2. Functional consultants that have worked in the implementation of the modules of the ERP applications. They know well the processes in the application and they are capable of identifying which enterprise process step(s) coincides with which function(s) of the ERP package. Therefore functional consultants of the ERP application should attend the process analysis study together with domain experts from the enterprise.

In this step, documented processes are simply matched with business processes defined in the ERP package. There may be one to one, one to many and many to one relationship between these two sets of processes. Examples of three types of match can be seen in Figure 2.

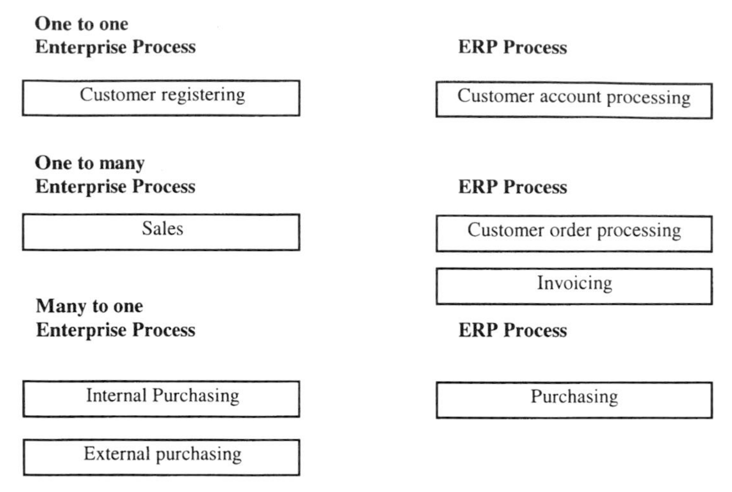

Fig. 2 Match of processes

2.4 Identifying the gaps between the analysed enterprise processes and the ERP processes

After processes are matched they can be compared with each other to identify gaps. The three types of gaps between enterprise and ERP processes and the ways to identify them are as follows:

1. A missed function in the ERP application. A function that is required for an enterprise process can be missed in the ERP application. For each process step the corresponding function(s) are searched in the application. If there is a missed function it can easily be identified.

2. An inadequacy in an ERP function. This is the most frequent case. A function is available but is inadequate to meet the requirements. For example, the requirement of a firm with about 40,000 customers may be entering customer incoming payments very fast. But if the screens for posting incoming payments function are not user friendly then entering data with this function will be a very difficult task.

Functionality determines the capability of the application to track, manage and report information in a timely, flexible, and proactive manner (6). An ERP application function can be modeled as a mathematical function, ie. $y=f(x \mid a)$. "x" is the input variable to the function. "f " itself denotes the function. It is a mapping or transformation. "a" is called the parameter. A parameter is a constant that is variable. It can take virtually any value. "y" is called output variable or depended variable of the function.

Determining the inadequacy of a function is a more elaborate task than simply searching for a missed function. The steps for determining an inadequacy are as follows:

i. The function should be matched with the process step or steps.

ii. The inputs and outputs of the process step should be matched with the inputs and outputs of the function. Input and output can be a data entity, a physical object or a service. Inputs and outputs of a process step are documented in process tables. If there is a mismatch between the inputs and outputs then it is probable that the function is inadequate to meet the process requirements.

iii. The ERP function itself should be compared with the process step. An ERP function can be configured to a certain extend to adapt to the requirements of process steps. The ERP function contain parameters. An ERP function can be configured with the help of these parameters. This process is called "configuration" in ERP terminology. For example interest rate is usually a parameter in interest calculation programs so that users can calculate interest with different interest rates. But sometimes some attributes in an ERP function can not be configured because they are not parametric. In that case the source code of the function can be changed with programming. But this is usually the least desired practice in ERP implementations. Changing source code of programs is costly and alters the standards in the ERP application.

3. An inconsistency between the flow of ERP and enterprise process. Determining the inconsistencies between the enterprise and ERP process flows requires a broader perspective than evaluating functions. A process in the ERP application mostly includes several functions in more than one module of the application. Therefore being an expert in one module is absolutely not enough to compare process flows. A stock movement in the logistics module causes an accounting document to be posted in the finance module. Usually functional consultants of the application with over-modules perspective are valuable sources of information to compare the logical flow of enterprise processes and ERP processes.

3 A MANUFACTURING FIRM CASE STUDY

3.1 The analysis

The process analysis methodology that is presented in this study is applied to a manufacturing firm before the implementation of ERP application which is the SAP R/3™. The purchasing process of a paper manufacturing company was selected for the analysis. The reason for selecting the purchasing process is that this process is related with both logistics and finance departments of the firm and materials management (MM) and finance (FI) modules of SAP R/3™. The process was first mapped. Then, the process was matched with the purchasing process of the ERP application. Finally the gaps between the enterprise and ERP process were identified. The aim was to determine whether the purchasing functions of the package were suitable for the enterprise requirements.

The analysis for identifying the gaps has taken place in six steps.

1. The process which is to be analyzed and compared was selected
It is the internal purchasing process.

2. The formal and informal data about the process was collected by applying a questionnaire

The questionnaire consisted of three parts which are also the three activities of the process: purchase requisition, purchase order and good's receipt. The questions were directed to the people from purchasing department and the warehouse in the factory. Specific data were collected with closed questions and informal and undocumented data were collected with open-ended questions.

3. The process was mapped by following the guidelines in the second step of the process analysis methodology.

The tool that is used for mapping the process is process table. The internal purchasing process was mapped by a process table which is given in Table 1.

4. Internal purchasing process was matched with the purchasing process of the material management (MM) module of SAP R/3 package

The internal purchasing process of the enterprise coincides with the purchasing process of the ERP. There is a one to one relationship in this match which is also illustrated in Figure 3.

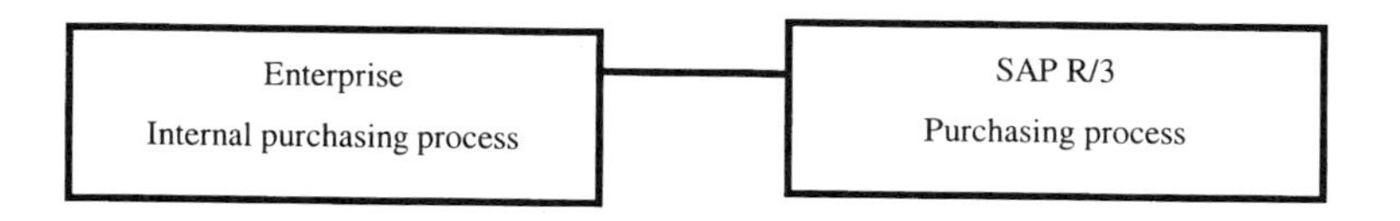

Fig. 3 Match of ERP and enterprise processes

5. Detailed information about the flow and functions of purchasing process of ERP application was collected.

Figure 4 gives an overall view for the purchasing process.

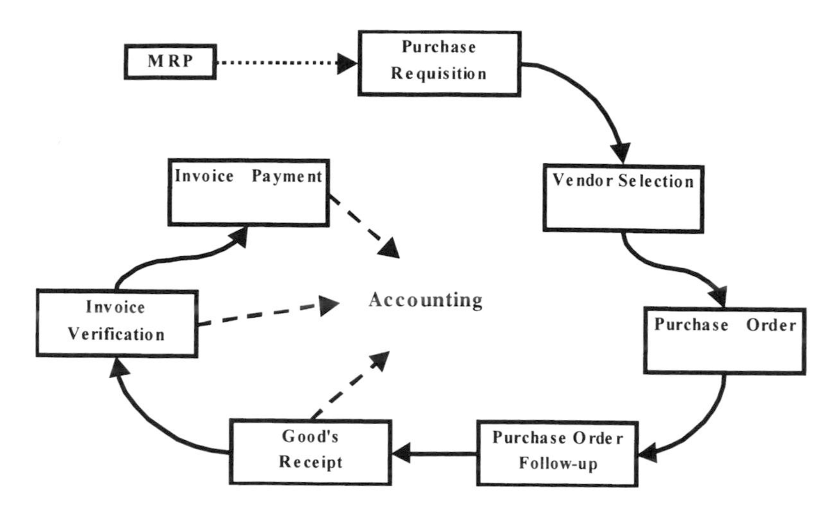

Fig. 4 Purchasing process overview (SAP R/3 documentation)

6. The steps of the documented process were compared with the purchasing functions in the ERP application.

The detailed information that is the outcome of the previous step for the internal procurement process and its functions was used together with the process table that is outcome in the third step.

Additionally one logistics and one finance consultant of the application have helped in matching and evaluating the functions for each task in the table. The process table was revised to include corresponding ERP functions for each task and is given in Table 4. A function in SAP R/3™ application is represented by a four-letter transaction code. The tasks that has no corresponding functions or has corresponding functions that are inadequate are represented by dark rows in this table.

3.2 The results of the analysis

Three gaps were identified between the enterprise and ERP processes. They are as follows:

1. There is an inconsistency between the flow of processes. Vendor invoices can only be entered in the ERP by taking reference from purchase order. Therefore using purchase order is a must in the application. But in purchasing process of the company, purchase order is nothing but a purchase requisition as can be seen in Table 1. Purchase requisitions are used to make and monitor orders. A comparison of flows can be seen in Figure 5.

Table 4. Revised process table including matched ERP functions for each task in internal purchasing process.

	A	B	C	D	E	F	N
1	Process no.	Definition of process	Activity no.	Definition of activity	Task no.	Definition of task	ERP application function
2	1	Internal purchasing	1	Purchase requsition	1	Filling purchase requsition form	Creating purchase requisition-ME51
3	1	Internal purchasing	1	Purchase requsition	2	1th approval	Changing purchase requisition-ME52
4	1	Internal purchasing	1	Purchase requsition	3	2nd approval	Changing purchase requisition-ME52
5	1	Internal purchasing	1	Purchase requsition	4	Budget control for fixed assets	Changing purchase requisition-ME52
6	1	Internal purchasing	1	Purchase requsition	5	Decision of purchase of fixed assets that is unplanned in budget	Missed
7	1	Internal purchasing	1	Purchase requsition	6	Signing purchasing form	Changing purchase requisition-ME52
8	1	Internal purchasing	1	Purchase requsition	7	Recording purchasing form into the Excel purchase monitoring file	No need any more.
9	1	Internal purchasing	1	Purchase requsition	8	Filing purchase requsition form	No need any more.
10	1	Internal purchasing	2	Purchasing	1	Collecting vendor proposals	Missed
11	1	Internal purchasing	2	Purchasing	2	Evaluating vendor proposals	Missed
12	1	Internal purchasing	2	Purchasing	3	Giving an order by telephone or fax	Creating purchase order - ME21 (Inadequate)
13	1	Internal purchasing	3	Stock monitoring	1	Good's receipt	Good's receipt-MB01
14	1	Internal purchasing	3	Stock monitoring	2	Recording good's receipt	Good's receipt-MB01
15	1	Internal purchasing	3	Stock monitoring	3	Signing bill of laiding for material that is under stock control	Changing material document-MB02
16	1	Internal purchasing	4	Posting invoice	1	Revision of payment plan	Payment program-F110
17	1	Internal purchasing	4	Posting invoice	2	Accounting posting of vendor invoice	Entering vendor invoice-F043

2. There is an inconsistency between the inputs of the purchase order requisition task and inputs of the purchase order function. Account assignment data (vendor account number and material number) should be available during the purchase order in the application. Otherwise purchase order cannot be created. This is a limitation of the purchase order function. But account assignment data is only available with the receipt of invoice in the enterprise.

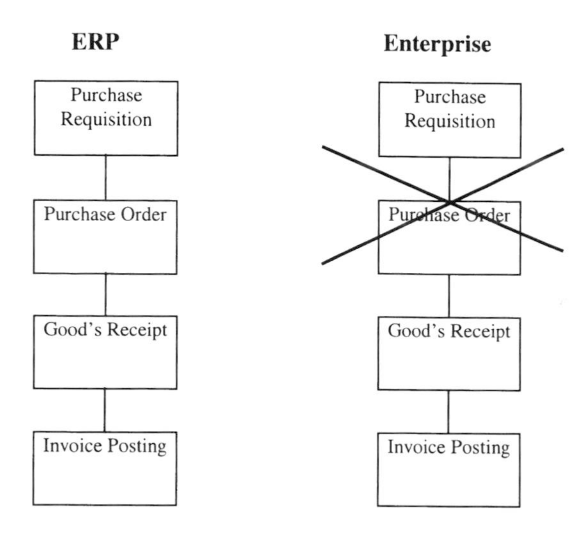

Fig. 5 A comparison of ERP and enterprise internal purchasing processes

3. There are three tasks that have no correspondent ERP functions in the purchasing process as can be seen in Table 4. They are the tasks named "decision of purchase of fixed assets that is unplanned in budget", "collecting vendor proposals" and "evaluating vendor proposals". These functions can not be performed in the selected ERP application.

In short, there are sixteen tasks in the internal purchasing process. Out of these 16, three of them have no correspondent functions in the ERP application. Also two functions were found to be inadequate for the two tasks of the process. Therefore five tasks out of 16 can not be met by the applications functionality. In other words the application is 68 % ((11/16)*100) suitable for that specific process if we only look at the process steps. Of course other types of gaps that were also explained in this study (mismatches in process flows, user friendliness of functions' screens etc.) should also be incorporated to that percentage figure which is an very difficult task.

3.3 The decision

By looking at the results of the analysis of purchasing process of manufacturing firm, the following decisions was taken by the manager who would sponsor the project:

1. The ERP application which is the SAP R/3™ will be implemented with three modules, namely materials management, sales and distribution and finance.

2. Purchasing process of the company will be revised to comply with the R/3 functionality:

i. Purchase orders will be opened by the purchasing department. The firm was only using purchase requisition for purchasing stock material. This modification will eliminate the first gap determined in the analysis.

ii. A temporary account assignment will be entered in the purchase order. But this temporary account assignment will be corrected before the invoice receipt by the accounting deparment. This decision will eliminate the second gap which was determined in the analysis.

4 CONCLUSION

What the process analysis methodology for ERP implementation aims to achieve can be summarised as follows:

1. Being process oriented in ERP implementations. Making a process analysis study for an ERP implementation ensures that processes will be taken into consideration during the implementation; not just single activities and tasks.

2. Selecting the right processes before the ERP implementation.

3. Defining the selected business processes at a detail required for ERP implementations.

4. Determining the gaps between the enterprise processes and ERP application processes.

Functional fit of ERP applications with the company's business processes is still the most important selection and success criteria (27 %) according to a research by Gartner Group in 1998 (6). With the help of this study, valuable information about the degree of functional suitability of an ERP application will be determined. This information will alert both decision makers and implementers about the problems that may arise in the implementation and help them in taking necessary actions against these problems. Therefore, by applying this methodology the risk of failure of the ERP project will probably be minimised and duration of the implementation will probably be shortened which is a crucial improvement for all IT projects.

REFERENCES

1. Semich, J.W. 1995. C/S Manufacturing: Build, Buy, or Reengineer? Plugin Datamation. (No.9, 1995) http://www.datamation.com/Plugin/issues/1995/sept15/09bev100.html.

2. Stickler, M.J. 1997. Reengineering the Front Office: Cashing in on the Cash Cow. 1997 APICS International Conference Proceedings.

3. Klement, R.E. and Richardson G.D. 1997. Business Process Mapping Techniques for ISO 9001 and 14001 Certifications. 1997 APICS International Conference Proceedings.

4. TQM International. 1994. Business Process Improvement, An Approach to Implementation. Chesire: TQM International Ltd.

5. Galloway D. 1994. Mapping Work Processes. Wisconsin: American Society for Quality Control (ASQC) Quality Press.

6. Levine P. 1998. How to make smart IT decisions, Gartner Group Symposium ITxpo98, Cannes, France.

Machining Processes

CAPE/001/2000

Computer-aided design of high-performance grinding tools

M J JACKSON and **K K B HON**
Department of Engineering, University of Liverpool, UK
N BARLOW
Liverpool John Moores University, UK

SYNOPSIS

This paper considers computer-aided design of high performance cylindrical grinding wheels. The finite element method has been used as the numerical tool to examine the influence of geometric shape and size of abrasive elements on the safe design of high-speed grinding tools. Conventional cylindrical grinding wheels are limited to low peripheral speeds due to the strength of the abrasive body. By replacing the centre section of the wheel with a stronger material and bonding abrasive segments to it, it is shown that a substantial increase in maximum operating speed is possible while maintaining the required margin of safety. Optimisation of the number and size of abrasive segments is considered as well as the material used for the reinforcing centre section. A new concept wheel is also presented that is used to increase the maximum operating speed even further.

1. CONVENTIONAL GRINDING WHEELS

Conventional cylindrical grinding wheels are limited in maximum operating speed because of the strength of the abrasive body and the levels of in-service stresses. It is usual to apply a factor of safety of four on the predicted 'bursting' speed of the grinding wheel. The most significant loading imparted to the grinding wheel that contributes to the limiting stress condition is due to centripetal loading. Stress levels associated with other in-service loading such as clamping have been discussed by Barlow, Jackson, Mills, and Rowe (1). For plain-sided cylindrical grinding wheels, centripetal loading can be predicted using the linear theory of elasticity as verified by Munnich (2). The levels of stress due to centripetal loading are a function of wheel speed, wheel geometry, and density of the wheel. If similar levels of stress are to be maintained with an increase in the maximum operating speed then changes in wheel geometry and wheel density need to be addressed. An increase in wheel speed by a factor of two will lead to a fourfold increase in the maximum stress due to centripetal loading, assuming that the geometry of the wheel and its density remains constant. The distribution of stresses across a plain (parallel-sided) cylindrical grinding wheel is shown in Figure 1. The stress factor is defined as the operating stress divided by the product of density and the square of the peripheral velocity, i.e. $\bar{\sigma} = \frac{\sigma}{\rho v^2}$. The grinding wheel considered in this analysis is a vitrified-bonded white alumina wheel (WA60L5V) having an outside diameter of 500mm, bore diameter of 300mm, and a thickness of 100mm. The density of the abrasive was 2235 kg/m^3. The modulus of elasticity was measured as 57.2 GN/m^2, and Poisson's ratio was 0.2. The maximum stress occurs at the bore of the grinding wheel and is tensile in the circumferential, or hoop, direction. The models presented in this paper assume that the grinding wheel is freely rotating at the bore. This allows mechanical designers to err on the side of caution and assumes that maximum stresses are tensile and occur at the bore in the hoop direction.

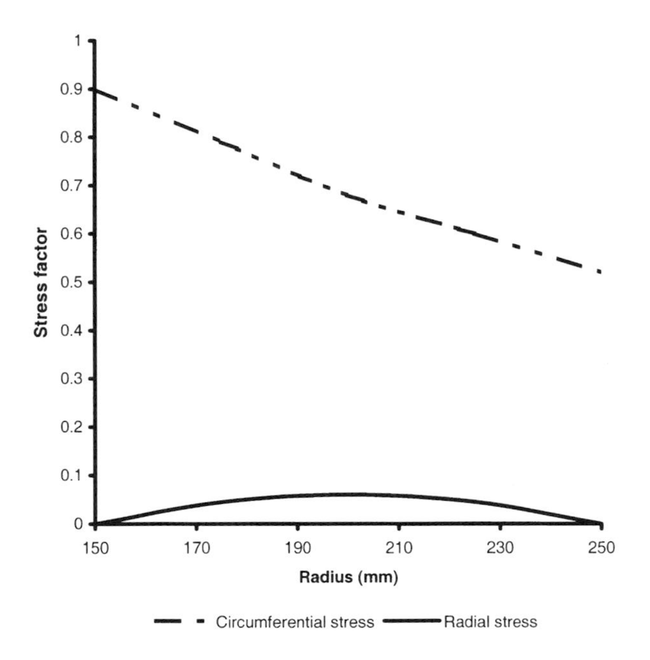

Figure 1. Stress distributions across a plain cylindrical grinding wheel

2. EFFECT OF SHAPE ON MAXIMUM STRESS

The cross-sectional shape of the grinding wheel also has an influence on the levels of stress due to centripetal loading. As a general rule, a geometry that allows the mass of the grinding wheel to be concentrated towards the centre of the grinding wheel leads to a reduction in the level of stress due to centripetal loading. For example, the use of a trapezoidal shaped cross section that has a width at the bore which is two times that at the periphery, leads to a 9% reduction in the maximum stress across the wheel due to centripetal loading.

3. REINFORCED GRINDING WHEELS

Reductions in the magnitude of maximum stress can be achieved by replacing the central, or non-grinding, portion of the grinding wheel with a stronger, less dense, rigid material. The use of a material with a higher modulus of elasticity at the centre of a grinding wheel restricts the radial movement of the abrasive, leading to a reduction in the circumferential stress due to centripetal loading. Figures 2-4 illustrate the effect of using a number of materials on the circumferential stress distribution across a cylindrical grinding wheel according to the specifications stated in section 1. The properties of the reinforcing centre section are given in Table 1. Two depths of abrasive layer are considered: 75mm and 25mm, respectively. It is shown that a 70% reduction in the magnitude of maximum stress in the abrasive layer can be achieved using a carbon fibre reinforced centre section with an abrasive layer that has a 25mm abrasive layer depth.

Reinforcing Material	Density (kg/m^3)	Modulus of Elasticity (GN/m^2)	Poisson's Ratio
Steel (En 24T)	7850	200	0.27
Aluminium Alloy	2752	70	0.33
Carbon Fibre Reinforced Composite	1500	189	0.3

Table 1. Properties of reinforcing materials

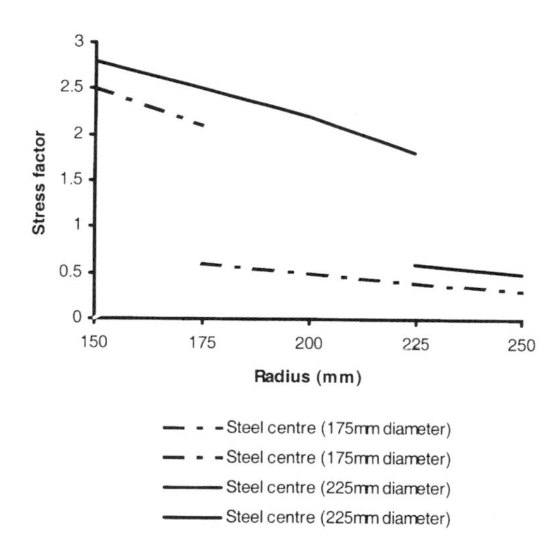

Figure 2. Circumferential stress distribution across a grinding wheel with a steel reinforcing centre section

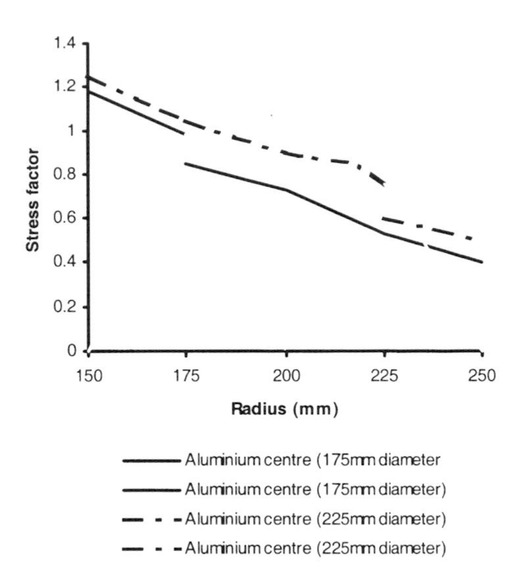

Figure 3. Circumferential stress distribution across a grinding wheel with an aluminium alloy reinforcing centre section

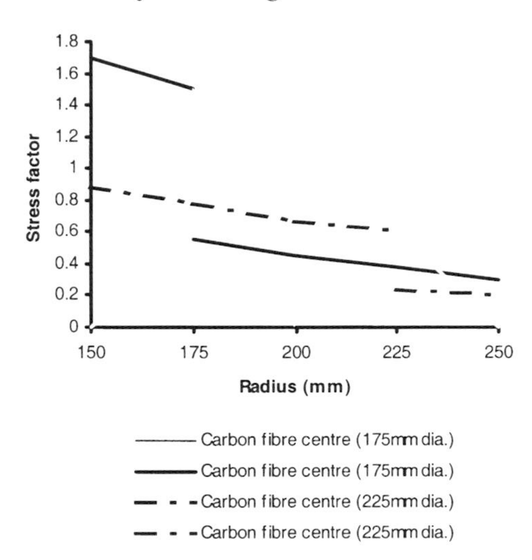

Figure 4. Circumferential stress distribution across a grinding wheel with a carbon fibre reinforcing centre section

4. SEGMENTED GRINDING WHEELS

The magnitude of the circumferential stress due to centripetal loading is a direct result of radial expansion that leads to 'circumferential stretching' of the abrasive layer. The magnitude of circumferential stretching can be reduced by replacing the continuous layer of abrasive, which is attached to the outer surface of the centre section, with a set of discontinuous segments. Although the linear theory of elasticity can be used to predict the level of stress due to centripetal loading in a continuous grinding wheel, a numerical technique such as the finite element method is required to predict stress levels in a segmented grinding wheel (3).

4.1 Finite element model

In order to examine the influence of the number and the size of the abrasive segment on the levels of stress, a finite element model of a segmented grinding wheel was created. Figure 5 shows one-quarter of the segmented grinding wheel due to symmetry about the Y- and Z-axis. The grinding wheel is now 350mm in diameter, and has a bore diameter of 150mm. The abrasive material is a vitreous-bonded cubic boron nitride that has a density of 2270 kg/m^3, a modulus of elasticity of 8.6 GN/m^2, and a Poisson's ratio of 0.2. The material used for the reinforcing centre section was En 24T steel. The segments were bonded to the steel centre section using an adhesive with a density of 1700 kg/m^3, modulus of elasticity of 1.5GN/m^2, and a Poisson's ratio of 0.4. The edges of the model shown in Figure 5 were constrained from movement in the 'Y' and 'Z' directions in order to simulate connectivity with the remaining parts of the wheel. All other nodes were allowed to move freely, and the size and number of segments were varied.

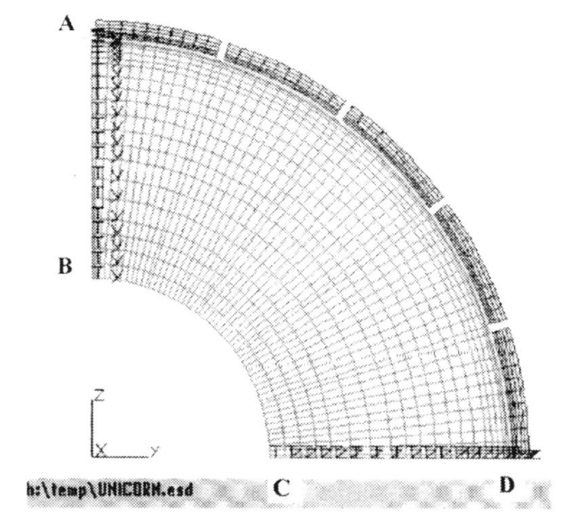

Figure 5. Finite element model of a segmented grinding wheel with a steel reinforcing centre section

4.2 Results of analysis

Figure 6 shows the influence of the depth of the abrasive segments on the circumferential and radial stresses in the abrasive part of the grinding wheel. The number of segments was constant at this stage of the analysis. The number of segments used in industrial practice for this particular diameter was 60. Figure 7 shows the influence of the number of segments on the circumferential and radial stresses in the abrasive part of the wheel. The depth of the abrasive segment was constant at 4mm.

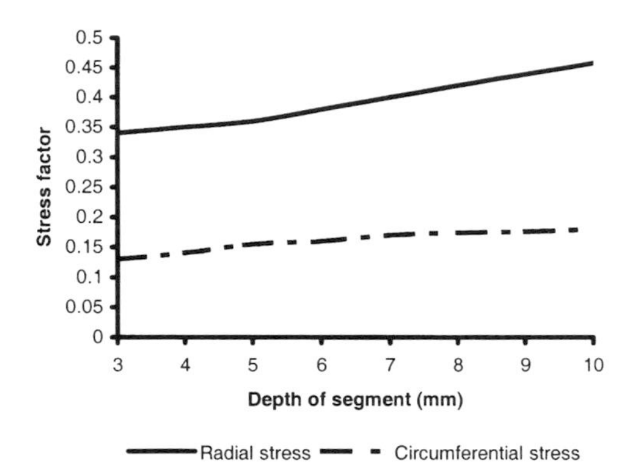

Figure 6. Effect of depth of the segments on the stress levels in the abrasive part of the grinding wheel

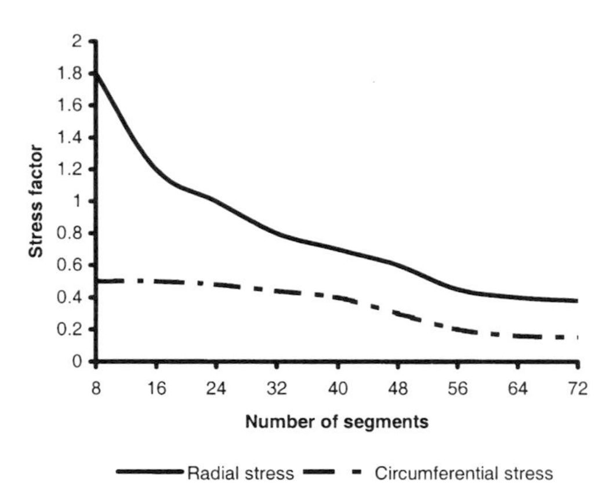

Figure 7. Effect of the number of segments on the stress levels in the abrasive part of the grinding wheel

For the grinding wheel considered here, it can be seen that the smaller the depth of the abrasive segment and the higher the number of segments leads to a reduction in the level of stress in the abrasive part of the grinding wheel. The greatest effect on the magnitude of stresses in segmented grinding wheels is clearly the number of segments. It can be shown that the maximum stress in the abrasive segments is in the radial direction, whereas the maximum stress in continuous layer of abrasive is in the circumferential direction.

5. ULTRA HIGH-SPEED GRINDING WHEEL

Ultra high grinding wheel speeds can be achieved if more exotic designs are used that effectively concentrate mass at, or near to, the bore of the wheel (4). Figure 8 shows a wheel design that was developed to reduce the mass of the reinforced centre section. This design allows a wider segment to be used that allows direct plunge-grinding operations to be undertaken. Hyperbolic and trapezoidal shaped grinding wheels use narrow abrasive segments, which means that plunge grinding operations are accompanied by traverse movements in order to remove the same stock of material compared to the ultra high speed grinding wheel shown in Figure 8. The model represents one half of the wheel that has symmetry about the 'Y' axis. The corners of the wheel are constrained from moving in the 'Y' direction to simulate connectivity. The properties of the grinding wheel are the same as those stated in section 4.1, in order to compare with the segmented grinding wheel.

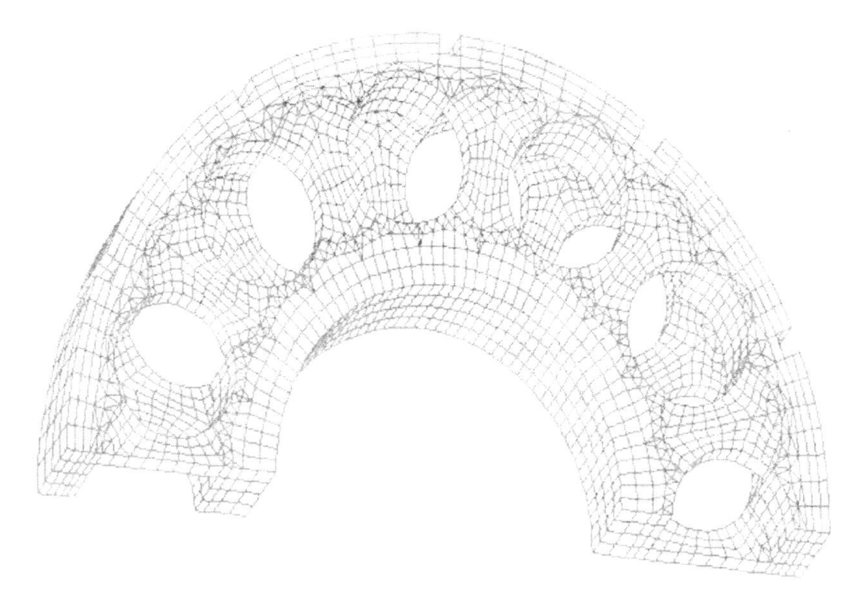

Figure 8. Finite element model of the ultra high-speed grinding wheel with 'porous' steel reinforced centre

Figure 9 shows the results of the finite element analysis carried out on this particular geometry. The results are compared with the results of the segmented grinding wheel. Figure 9 shows the effect of the number of segments on the circumferential and radial stress in the abrasive part of the grinding wheel shown in Figures 5 and 8. The depth of the abrasive segments was constant at 4mm.

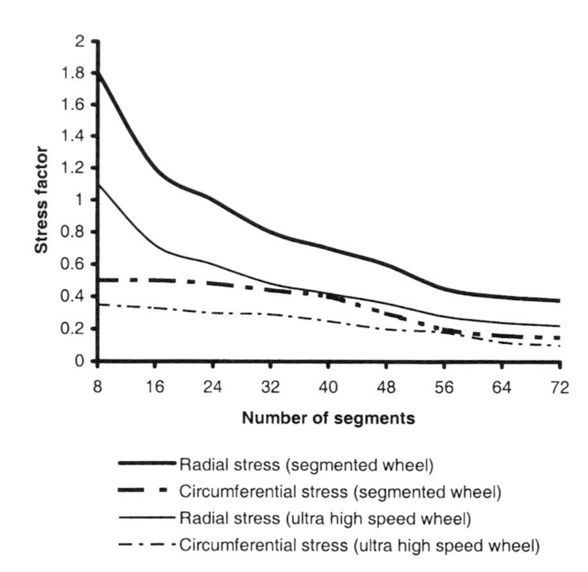

Figure 9. Influence of number of segments on stress levels in the abrasive for segmented and ultra high speed grinding wheels

It can be shown from Figure 9 that the ultra high-speed grinding wheel with the 'porous' reinforced steel centre exhibits a lower stress factor due to centripetal loading. For the same level of safety experienced by segmented grinding wheels, higher wheel speeds can be achieved using the novel design shown in Figure 8.

6. CONCLUSIONS

The levels of stress in a cylindrical grinding wheel due to centripetal loading can be reduced by using a reinforced centre made from a material that is rigid, lighter, and is stronger than the abrasive material.

Replacing the continuous layer of abrasive with a layer of discontinuous segments can reduce the levels of stress due to centripetal loading further.

The depth and number of segments has a significant influence on the levels of stress in a rotating grinding wheel. As a general rule, increasing the number of segments and reducing the depth of the abrasive segments can reduce stress levels.

The maximum stress in an abrasive segment due to centripetal loading is in the radial direction.

Stress factors in high-speed segmented grinding wheels can be reduced further by reducing the mass of the wheel by redesigning the centre section of the wheel using the new 'porous' design.

ACKNOWLEDGEMENTS

The principal author acknowledges support from Unicorn International and Saint-Gobain Abrasives.

REFERENCES

1. N. Barlow, M.J.Jackson, B.Mills, and W.B.Rowe, Optimum Clamping of CBN and Conventional Vitreous-bonded Cylindrical Grinding Wheels, *International Journal of Machine Tools and Manufacture,* 1995, **35**, 119-132.

2. H. Munnich, Doctoral thesis, Technische Hochschule Hanover, Germany, 1956.

3. M.J.Jackson, N.Barlow, B.Mills, and W.B.Rowe, Mechanical Design Safety of Vitreous-bonded Cylindrical Grinding Wheels, *British Ceramic Transactions,* 1995, **94**, 221-229.

4. W. Koenig, and F. Ferlemann, CBN Grinding at 500 m/s, *In 'Ultrahard Materials in Industry – Grinding Metals',* De Beers Industrial Diamond Company, U.K., 1991, 58-65.

Iterative learning control strategy for information model-based precision leadscrew grinding process

H BIN and **S CHEN**
School of Mechanical Science and Engineering, Huazhong University of Science and Technology, Peoples Republic of China

ABSTRACT This paper puts forward a new compensation strategy for the generalized kinematic errors (GKEs) using off-line measuring information of the machined workpiece. An iterative learning control algorithm (ILCA) supported by the information model of machining complex is adopted to realize the strategy when the GKEs is produced by error motion of two relative motions, e.g. the helical errors of a lead screw. Experimental results show that the proposed GKEs compensation strategy and the information model based ILCA is effective for the precision leadscrew grinding process control.

KEYWORDS generalized kinematic error; error compensation; iterative learning control; information model; uncertainty

Generalized kinematic errors (GKEs) are defined as including the conventional kinematic errors and the form errors of workpiece in machining process. Error compensation methods based on information technology and artificial intelligence are important techniques for the reduction of GKEs. To achieve good compensation results, acquiring process information by sensors reliably is as a key factor. However, accurate sensors, especially long linear scales, can hardly work reliably online in precision and ultra-precision machining processes. Because the GKEs of the machined workpiece reflect comprehensively the error motions of the machining complex, the authors put forward a new GKEs compensation strategy which considers the machined workpiece as the carrier of GKEs and an off-line measurement of the machined workpiece is carried out after each operating stroke, then the error information of the machined workpiece is mapped to be as the error map which will be used to control the machining process for the next operating stroke. Therefore, the off-line measuring loop and machining complex loop are integrated into the compensation control loop, which is called large-scale loop. In this system, computer network is adopted to transmit the involved information among computers in different location. The key problem is to find the information mapping relationship, $u(t)=f[e(t)]$, between errors $e(t)$ of the machined workpiece and the error map to be compensated $u(t)$ for the online industrial PC. This paper concentrates on the realization of the information mapping by developing an information model based iterative learning control algorithm (ILCA) for the precision leadscrew grinding process in which the GKEs is produced by the error motions of two relative motions.

1. INFORMATION SYSTEM BASED MODELING OF GKEs COMPENSATION SYSTEM

Machining complex that is composed of machine tool, cutting tool, workpiece, jig and fixture is a typical nonlinear time-varying system. Due to the complexity of machining complex, developing a general reliable model to predict the machining errors accurately for all machining complex is very difficult and a lot of barriers still have to be overcome[1]. Because the depth of cut in precision machining is small, machining process of a specific machining complex under specific machining conditions shows weak non-linearity and can be taken as a linear dynamic system[2,3]. Moreover, the disturbances acting on the machining process, such as the kinematic errors of machine tool, thermally induced deformation of machine tool and errors caused by the static deformations of the machining complex due to force, may be considered to possess good repeatability. Comparatively, the stochastic disturbances are much weaker. All these incite the authors to model the machining process in two steps. The first step is to develop an information model to map from the machine-fixture-workpiece-tool structural information to the information specific to the components of machining complex, such as the dynamic model of the machining process, geometric model of the workpiece, kinematic model of the machine tool, self-modeling algorithm etc. and related historical machining information which includes historical machining parameters, machining errors,

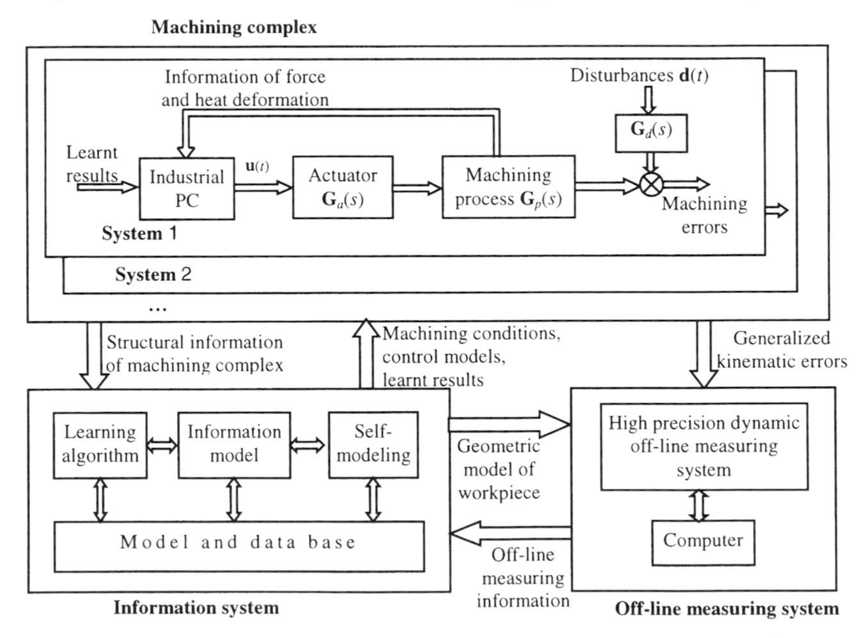

Fig. 1 GKEs compensation system using Off-line measuring information.

learnt error map to be compensated and so on. The second step is to derive the model of the repeatable disturbances, $\mathbf{d}(t)$, and the specific machining process, a linear dynamic system, $\mathbf{G}_p(s)$, with unknown structure and parameters by self-learning algorithm. Then we can use the information model based linear dynamic system model to control the output of the specific machining process to trace an arbitrary reference trajectory with high precision. Figure 1 shows the diagram of GKEs compensation system using off-line measuring information.

Deformation caused by heat and force can be neglected when the workpiece is measured off-line in the measuring room. This is very different from that of the workpiece machined online under shopfloor. To improve the precision of error mapping, it is necessary to real time measure and compensate the workpiece deformation induced by the variation of temperature and force during machining. Therefore the models for compensating heat and force deformation of the workpiece are also necessary to be contained in the information system. From Figure 1 we know information system receives machining information including off-line measuring information from measuring room through computer networks and activates the learner to perform learning, such as self-modeling of the generalized dynamic system, self-tuning and calculation of the learnt results of the error map. As an extension of information system, the online industrial PC sends structural information of the machining complex to and receives models and the error map from the information system before each machining process. The industrial PC also performs real-time measurement of the heat and force deformation of the workpiece. The manipulation of the compensation for the deformation is calculated and is synthesized with the error map, thus a resultant values are used to control the machining process.

Figure 2 shows the flow chart of the GKEs compensation system based on the off-line measuring information of the machined workpiece. In the Figure, $\mathbf{u}_{th}(t)$ and $\mathbf{u}_f(t)$ mean the online calculated values by the industrial PC for the compensation of the heat and force deformation of workpiece respectively, $\mathbf{u}_{iter}(t)$ means the off-line learnt error map derived by the ILCA from the off-line measuring information of the machined workpiece.

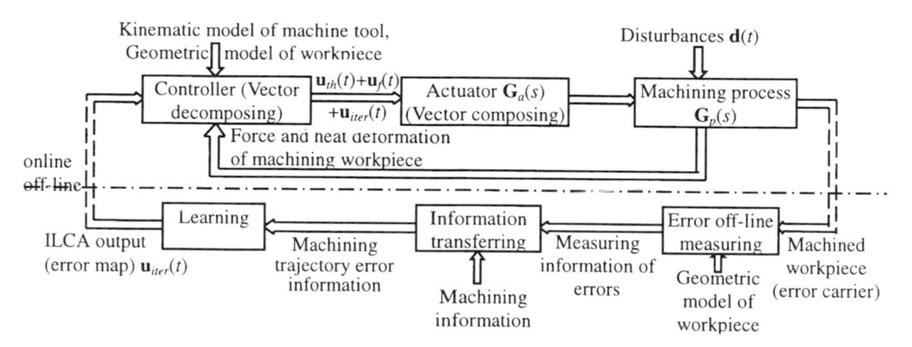

Fig. 2 Flow chart of GKEs compensation system

For the precision leadscrew grinding process, the GKEs are produced by error motions of two relative motions, so the GKEs can be denoted by a signed variable, $e(t)$. So in the remainder of the paper, the generalized dynamic system, $\mathbf{G}(s)=\mathbf{G}_a(s)\mathbf{G}_p(s)$, the input vector, $\mathbf{u}_f(t)\bullet\ \mathbf{u}_{iter}(t)\bullet\ \mathbf{u}_{th}(t)$, and the disturbances, $\mathbf{d}(t)$, are substituted by variables, $G(s)$, $u_f(t)$, $u_{iter}(t)$, $u_{th}(t)$ and $d(t)$, respectively.

For general control algorithm, it is necessary to model the generalized dynamic system, $G(s)$, and disturbances, $d(t)$, with high precision to fulfill the compensation of the GKEs. However, for different machining complex, the complexity of $G(s)$ and $d(t)$ varies greatly. This makes it very difficult to develop a general accurate mathematical model for all machining complex. Many artificial neural networks have been used to deal with such control problems and have obtained good experimental results. But these schemes are not friendly to meet industrial requirements because the variation of the complexity may cause different initial value choice, different training method and even different topology of the neural network.

This paper adopts iterative learning control algorithm (ILCA) for the precision leadscrew grinding process to avoid precise modeling of the process and the disturbances. The ILCA is a performance-based control algorithm, which attempts to improve the control performance through repetition[4] by changing the latest input sequences, which is the error map in the paper, according to the latest machining errors to get a more reasonable error map. By acquiring more information through repetitive operation, the ILCA gradually decreases the influences of repeatable disturbances and model errors of the dynamic system. Eventually high precision tracing of arbitrary trajectory can be achieved.

ILCA does not need precise model of the dynamic process and only the low frequency range of the dynamic system is of importance to the parameter determination of the ILCA. So the authors try to derive a 3rd order model to describe the dynamic system approximately by using the impulse response method [5]. Supposing the heat and force deformation of the workpiece can be compensated completely by online measurement and control, the input and output needed to model the generalized dynamic system is constructed as following. For

$$\begin{aligned}\Delta E(s,k) &= E(s,k) - E(s,k-1) = G(s)[U_{iter}(s,k) - U_{iter}(s,k-1)] + G_d(s)[D(s,k) - D(s,k-1)] \\ &= G(s)\cdot \Delta U_{iter}(s,k) + G_d(s)\cdot \Delta D(s,k), \qquad (1)\end{aligned}$$

Where, $U_{iter}(s,k) = L[u_{iter}(t,k)]$ denotes the k-th learnt element of the manipulation in the frequency domain and $E(s,k) = L[e(t,k)]$ is the Laplace transform of the output errors. The L denotes the operator of Laplace transform. Supposing the disturbance, $d(t)$, is the same in the continuous operations, i.e. $\Delta D(s,k) = 0$• then from equation (1) we can obtain

$$\Delta E(s,k) = G(s)\cdot \Delta U_{iter}(s,k). \qquad (2)$$

So, $u_{iter}(t,k) - u_{iter}(t,k-1)$ and $e(t,k) - e(t,k-1)$ can be taken as the input and output of the generalized dynamic system respectively. Then the unit impulse response $g(t)$ can be obtained by learning algorithm and the 3rd order transfer function $\frac{B(s)}{A(s)} = \frac{K_p\cdot(b_1 s^2 + b_2 s + 1)}{a_0 s^3 + a_1 s^2 + a_2 s + 1}$ of the dynamic system can be derived by Hankel matrix method [5].

2. ITERATIVE LEARNING CONTROL FOR PRECISION LEADSCREW GRINDING PROCESS

It is shown that the manipulation of the compensation system can be calculated as

$$u(i,k) = u_{th}(i,k) + u_f(i,k) + u_{iter}(i,k), \tag{3}$$

Where $u_{iter}(i,k)$ denotes the learnt element of the manipulation produced by the ILCA, $u_{th}(i,k)$ and $u_f(i,k)$ denote control action for heat and force deformation compensation respectively at the i-th sampling point of the k-th operating stroke.

The learning algorithm can only be an open loop control algorithm because the machining errors of the k-th operating stroke are unknown when the k-th learning is proceeding. Therefore, the manipulation obtained by the ILCA will be as following.

$$u_{iter}(i,k) = L_u u_{iter}(i,k-1) + K_c[1 + \beta \frac{d}{dt}] \cdot e(i,k-1). \tag{4}$$

Where, $e(i,k-1)$ is the off-line error measuring information of the GKEs of the machined workpiece after the (k-1)th operating stroke, L_u, K_c and β are iterative operator, L_u is set to 1 while K_c and β forms a PD-type ILCA. Then we have

$$E(s,k) = [1 - \frac{K_p \cdot (b_1 s^2 + b_2 s + 1) \cdot K_c(\beta s + 1)}{a_0 s^3 + a_1 s^2 + a_2 s + 1}] E(s,k-1) = G_{iter}(s) E(s,k-1). \tag{5}$$

If K_c is set to $1/K_p$ and β is set to a_2/b_2, the $|G_{iter}(s)|$ will be less than 1 in the whole frequency rage and $G_{iter}(s)$ shows better filtering effect in low frequency range. This ensures the convergence and rapid reduction of the machining errors.

When the first learning with k=1 has been carrying through, $u_{iter}(i,0)$ and $e(i,0)$ are not existent and set to zero. Consequently the $u_{iter}(i,1)$ must be zero. So during the machining of the first operating stroke of a workpiece, only the compensation of heat and fore deformation of the workpiece is carried out. When k=2, the process model is still not available. So the ILCA is simplified as a P-type ILC which means

$$\Delta u_{iter}(i,2) = -K_d \cdot e(i+1,1). \tag{6}$$

If Δ is defined to be as the compensation value per pulse input to the actuator. Reference[4] points out that the learning of the ILCA is convergent when $0 < K_d < 2/\Delta$. Here $K_d = 1/\Delta$. Figure 3 shows the block diagram of predictive learning control for GKEs compensation system using off-line measuring information of the workpiece.

In ILC algorithm, it is assumed that the initial state value of the process is equal to that of the desired trajectory for perfect tracing. Theoretically, an impulse input can be used to change the state value of the process suddenly to the desired trajectory at the start point. But this is not practical for physical system. Reference[6] pointed out that the tracing error can be controlled and decreases exponentially as time increases if the PD-type ILC algorithm is adopted with the initial state value keeping the same in every repeated operation. Because the initial state value is zero at the start point of every operating stroke, the initial state error only effects on the tracing accuracy of the process output at the beginning segment.

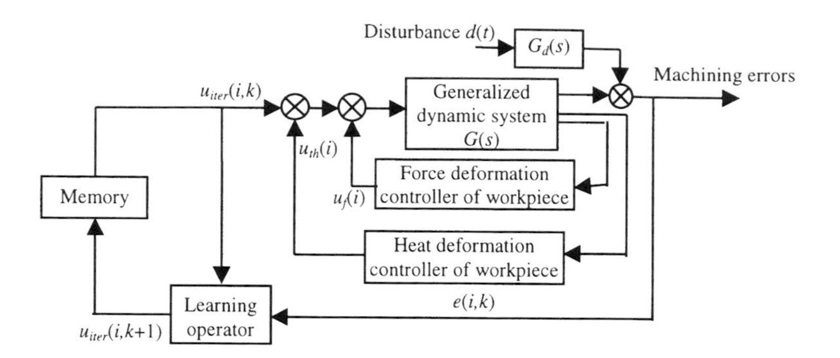

Fig. 3 Block diagram of iterative learning control compensation scheme

To perform perfect tracing, the disturbances must keep repeatable very well in the time domain. If the disturbance, $d(t)$, acts on the system with a time delay τ, in the next repeated operation, for

$$d(t-\tau)-d(t)=L^{-1}[D(s)e^{-\tau s}-D(s)]=-L^{-1}\{D(s)\sum_{n=0}^{\infty}[(\tau s)^n/n!]\}\approx -L^{-1}[\tau sD(s)]=-\tau\dot{d}(t),$$

then it is equivalent to adding another new disturbance, $-\tau\dot{d}(t)$. The Longer the time delay the greater the new disturbance will be. In machining processes, the demand for repeatability of the disturbance in the time domain is actually that of the space point. So high positioning precision for zero-point of the machine coordinate is required. To improve the precision, the actuator must be reset to the zero-point firstly. Then the closed loop control technique should be adopted to control the machine to be settled at the position where the measurement value is corresponding to the zero signal of the sensor and this point is set to be as the zero-point.

The average times of operating strokes can be decreased if the learnt error map which is derived by ILCA for a specific machining process is used to control the same machining process with a new workpiece of the same geometry and material. The learning result of ILCA is a satisfying error map which implies the dynamic behavior of the generalized dynamic system $G(s)$ and the disturbances $d(t)$, Moreover, the dynamic behavior of the disturbance and the generalized dynamic system are mainly determined by the structure of machining complex and the applied machining conditions. Consequently, When applying the same machining conditions to a machining complex, the model of the disturbance and the generalized dynamic system keep almost the same if only the workpiece is changed into a new one. To machining a new workpiece, there is little uncertainty to be learned for the learner if the learner has already derived perfect learnt results for a workpiece of the same type. So the application of the learnt results may decrease the learning iterative for each workpiece and thus decrease the average operating stroke times. In this case, the first learning for a new workpiece can be carried through assuming the $u_{iter}(i,0)$ equals to the latest learnt results $u_{iter}(i,terminal)$ specific to the qualified workpiece and $e(i,0)$ is set to zero.

3. EXPERIMENTS

Experiments are conducted to verify the proposed GKEs compensation scheme and the ILCA. An experimental device is developed and shown in Figure 4 schematically. The lead screw is driven by a stepping motor and forces the table to move along the roll slide way. The data acquisition and control board that is inserted in computer receives the signals from encoder and linear grating and sends them to the computer. The computer processes the signals and finds the GKEs information, which corresponds to the off-line measuring information of the machining errors derived in the measuring room. The ILCA learns from the GKEs information to get the error map to be compensated. The compensation stepping motor outputs the control actions and drives the differential nut through reducer to fulfill the compensation actions for the GKEs compensation in the next stroke. Since the error map to be compensated has been obtained from measuring information and the ILCA, it is unnecessary to offer error information acquired from online control process, say, under workshop conditions. In order to carry out the synchronization between the measuring time and the control time, the starting point should be the same, which may be achieved by zero-signal of the encoder and the linear grating.

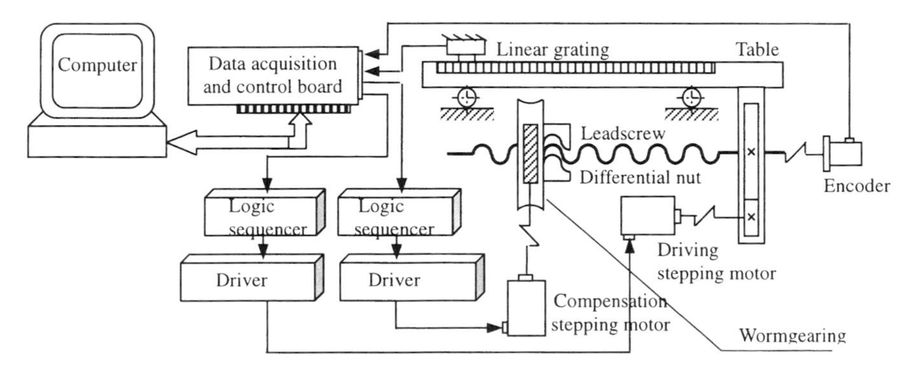

Fig. 4 Schematic drawing of the experimental device

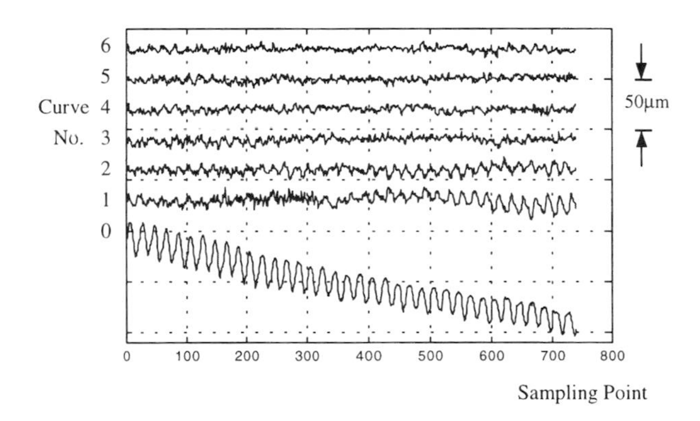

Fig. 5 Experimental results of ILCA for GKEs compensation system

Table 1 Statistical results of GKEs

NO. of curve	0	1	2	3	4	5	6
Errors of effective stroke(• *m*)	112.0	30.5	22.5	17.5	19.5	18.5	17.5
Cycle error(• *m*)	37.5	20.5	18.0	15.0	16.5	14.0	15.0
cumulative lead error(• *m*)	-85.3	-10.3	-3.75	3.45	-1.47	-0.44	1.58
Standard deviation(• *m*)	26.2	5.4	3.6	3.1	2.6	2.6	2.4

Figure 5 shows one of the typical experimental compensation results. Table 1 gives the statistical results. Among these curves, the first curve indicates the measuring results without compensation because of $u_{iter}(i,0)=e(i,0)=0$.Table 1 shows that the ILCA can compensate most of the cumulative lead errors of the lead screw for the first repetition and learning. The second to the fourth repetition also does much on GKEs compensation of the lead screw to decrease the cumulative lead errors by more than 98% and the cycle errors by about 60%. The experimental results show that machining accuracy comes to stable for the first four learning and further learning seems to be needless. The results also show that the algorithm of information model based predictive learning control using off-line measuring information of the machined workpiece is effective in compensating GKEs.

The compensation strategy proposed in this paper uses the learnt results derived from the off-line measuring information of the errors of the machined workpiece to find a practical scheme for the comprehensive compensation of the errors of the machining process. This scheme does not need to model the errors and can renew the error map after every operating stroke. So that, this compensation method is an advanced and shop floor friendliness error control method. This scheme can be applied in a wide area such as the precision machining of lead screw, gear and worm wheel, cylinder and so on.

REFERENCES

1. C.A. van Luttervelt, T.H.C. Childs, I.S. Jawahir et al. Present situation and future trends in modeling of machining operations. Progress report of the CIRP working group `Modeling of Machining Operations'. Annals of the CIRP, 1998,47(2): 587-626
2. J.Vinolas,J.Biera,J.Nieto et al. The use of an efficient and intuitive tool or the dynamic modeling of grinding processes. Annals of the CIRP,1997,46(1):239-242
3. M.D.Tsai,S.TakataM.Inui et al. Prediction of chatter by means of a mode-based cutting simulation system. Annals of the CIRP, 1990,39(1):447-450
4. Wang, Danwei.Convergence and robustness of discrete time nonlinear systems with iterative learning control. Automatica,1998, 34(11):1445-1448
5. Chongzhi Fang, Deyun Xiao. Process identification. Beijing, Tsinghua university press, 1988,92-100(in Chinese)
6. Kwang-Hyun Park, Zeungnam Bien, Dong-Hwan Hwang. A study on the robustness of a PID-type iterative learning controller against initial state error. International Journal of Systems Science,1999,30(1):49~59

Single set-up machining of moulds and dies

D I LEGGE
Division of Manufacturing Engineering, Lulea University of Technology, Sweden

ABSTRACT

Multi-axis and high speed machining can allow moulds and dies to be manufactured using fewer set-ups and with improved quality and lower cost than is possible using conventional manufacturing methods.

Two case studies are presented which highlight the benefits and problems associated with multi-axis and high speed machining of fully hardened steels and aluminium. The conclusions from this work are that there are significant opportunities to reduce the number of set-ups and overall machining time through the use of multi-axis machine tools or HSM. The work also highlighted the fact that there remains considerable scope for development of functionality in CAM software.

1. BACKGROUND

A major pressure on manufacturing companies today is to reduce overall manufacturing lead time. This is especially true of mould and die manufacture which often falls on the critical path for product development and introduction. Reduction in the lead time for mould and die development has a significant effect on new product introduction. (1) Whilst new techniques such as rapid tooling offer considerable possibilities for prototype and short series tooling, manufacture of moulds and dies in traditional materials, i.e. hardened steel is still required.

Swedish toolmakers have been under increasing threat in recent years from foreign suppliers of moulds and dies. Current demand for tooling in Sweden is ca. 6,750 MSEK of which only about 1/3 (2,200 MSEK) is supplied by Swedish companies; the majority of tooling being sourced outside Sweden and Scandinavia. (2)

Of the ca. 220-240 tool making companies in Sweden, 25% have five or fewer employees whilst only 5% have more than 50 employees. (ibid.) The small size of Swedish toolmakers makes new capital equipment a major investment. (3) This is reflected in the fact that only a handful of Swedish toolmakers have 5 axis machines; these being exclusively for manufacture of tooling for plastic products. (2)

The current work is part of two ongoing projects at the Department of Manufacturing Engineering at Luleå University of Technology. The original work on single-set-up machining was within a project known as "*Complete Manufacture*" which was carried out for *Pressoform*[1]. The aim of this project was to help raise awareness of good manufacturing practice within Swedish mould and die industry by highlighting opportunities offered by multi-axis machine tools.

A follow on project, *RAMOULDIE*[2] has similar objectives to Pressoform's Complete Manufacture and aims to demonstrate and quantify the benefits of using 5-axis machine tools over 3-axis machines. The project also indirectly compares the capabilities of a serial machine tool structure, Luleå University's Leichti Turbomill 1200, with a parallel kinematic machine, Nottingham University's Variax (4). Both machine tools are equipped with state-of-the-art controllers and spindles and can thus machine complex geometries in materials in a hardened state.

The RAMOULDIE project centres on a number of test cases, which covers a range of products which are typical for the tooling industry. The project is very much concerned with traditional machining processes, but using state of the are equipment and tooling. It is hoped to be able to demonstrate that using this kind of equipment can be economically justifiable, and to offer additional benefits such as reduction in overall lead times. Specific expectations for the RAMOULDIE test cases are to demonstrate:

- Reduction of machine set-ups by up to 60%
- Reduction of machining times by up to 50%
- Reduction of EDM machining requirements by up to 80%
- Reduction of manual polishing times by up to 80%
- Reduction of overall manufacturing cost by up to 15%

---

[1] Press och formverktyg i Luleå AB (Pressoform) was founded in 1995 as a limited liability company whose shareholders are companies in the tool and die industry in Sweden. The research and development, market analyses, competence development and technical demonstrator projects carried out Pressoform are instigated in response to requirements from industry.

[2] RAMOULDIE = Rapid and Accurate Manufacturing of Moulds and Dies. This is a CRAFT project with partners in Sweden, Germany and the UK.

2. SCOPE OF THE PROBLEM

There is a close link between the geometry of the final product (and hence mould), the machining process and the machine tool itself. This relationship is shown below in Figure 1.

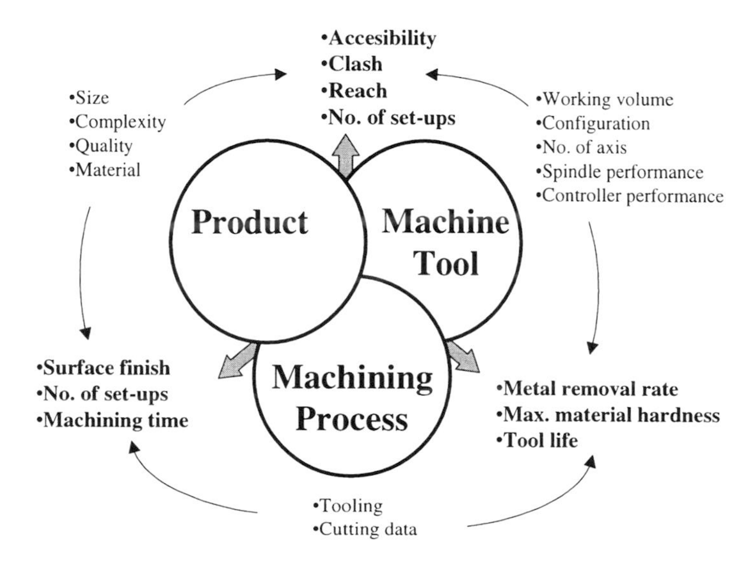

Figure 1 Relationship between product, machine and manufacturing process.

The reduction of set-up and overall machining time can be achieved in three ways; by careful design of products, through the design and optimisation of manufacturing processes and through the use of advanced machine tools, or by a combination of these methods. (5, 6, 7)

2.1. Set-up reduction in mould and die manufacture.

The machined features in a mould or die can be divided into two groups; the mould cavity and ancillary features such as cooling channels, sprues, runners, etc.

A generic manufacturing process would start with rough and semi-finish machining of the mould cavity, and many of the ancillary features is carried out in the soft state. This is followed by hardening after which the mould cavity is finish machined to achieve final size / form and surface finish. Electro-discharge machining (EDM) is traditionally used, although milling is becoming a viable alternative. In either case, hand grinding and / or polishing follows to produce the final form and to give the required surface finish. This generic process plan is shown in Figure 2 below.

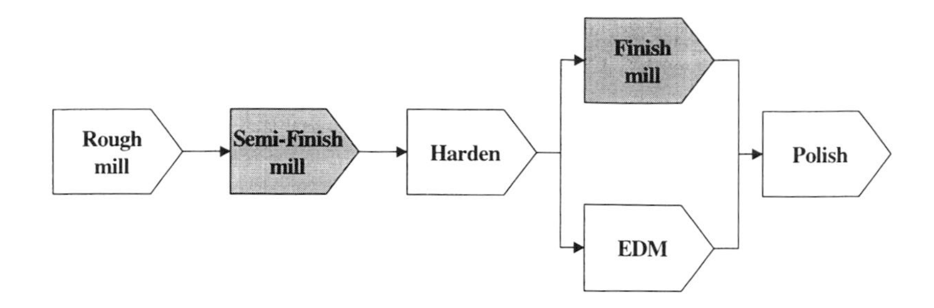

Figure 2 Generic manufacturing method for moulds and dies

The need for hardening prior to finish machining, and for polishing following finish machining, imply that the greatest scope for improvement or rationalisation in the manufacturing process lies with reducing the time required for individual manufacturing stages. Whilst this is true (and will be dealt with later) there are other possibilities open to reduce the number of set-ups or manufacturing operations.

2.1.1. Eliminating set-ups during machining in the soft state.

Multi-axis machines offer the possibility to machine from several different directions which can lead to a reduction in the number of discrete set-ups required during . Careful design of the mould or die can help reduce the number of directions from which machining must be carried out. This is especially the case for ancillary features.

2.1.2. Finishing in the soft state.

There are two avenues open to eliminate post-hardening finishing operation. Firstly, to manufacture the mould or die in a material which is used in an un-hardened condition. This option is commonly used where limited production runs are expected or for products which are expected to change following their initial market introduction and in which case the cost of a fully hardened tool with a life of several hundreds of thousands of cycles is unnecessary.

The other option is to use material with very predictable deformation characteristics during hardening. This allows the final form, with allowance for changes in dimensions, to be machined, with only polishing required after hardening. This option requires accurate prediction of the changes that will occur.

2.1.3. Machining from solid in the fully hardened state.

For materials which are through hardened, the possibility exists for carrying out semi-finish and finishing operations in the hardened condition. By eliminating the need for manufacture of EDM tools, and with the possibility to hold stocks of pre-hardened blanks in standard sizes, this option allows the fast turn round of durable tooling.

2.1.4. Elimination of EDM.

Even if hardening is carried out after a semi-finish operation, milling can still be is becoming a viable alternative to EDM in many cases due to improvements in machine spindle performance, machine stability and the use of cubic boron nitride (CBN) or even coated carbide tooling which are capable of cutting materials of 58 HRC or harder. Clearly, there are situations where the geometry to be machined is such that EDM represents the only realistic alternative; typically narrow sections where accessibility would be a problem for cutting tools, very small internal radii, macro textured surfaces etc. Again, all these can be positively affected by thoughtful product design which in turn reflects on the design of the mould cavity.

2.2. Reduction of cut times using "advanced machining."

In addition to the elimination of manufacturing operations, advanced machine tools, allow machining to be carried out in less time. There are two principle reasons for this; the ability to achieve better tool : workpiece orientation and better performance of spindles and tooling.

2.2.1. Improved cutting conditions.

When using ball end mills for machining freeform surfaces, the effective diameter of the cutter is often far less than its radius. This is especially true when shallow depths of cut are being taken which is often the case when machining moulds and dies. By tilting the tool with respect to the machined surface, which is possible with a 5-axis machine tool, the effective contact point can be moved to a larger radius on the tool and thus reduced the spindle speed required to achieve a given cutting speed. Another opportunity offered by 5-axis machines is the possibility to use flat end cutters. (8) This is, however, not without its problems.

Modern machine tools are being offered with spindles with far better performance in terms of maximum rpm. and torque characteristics than were available just a few years ago. This in turn has sparked development in terms of cutting tool design and allows higher metal removal rates to be achieved.

2.2.2. Gouging and clash **problems.**

Gouging of the tool into machined or to be machined surface(s) or *clash* between non-cutting elements of the tool and the workpiece are a potential source of problem. Advances in the functionality of CAM software have significantly reduced the problem of gouging, although not eliminated it completely. Multi-axis machine tools can help avoid gouging problems, but at the same time, can also cause gouging problems which would not have occurred with simple 3-axis machines. (9)

CAM software is usually very passive in that it solves the problem by generating tool paths that avoid gouging, but that do not necessarily generate the desired surface. A better, and not unrealistic solution, would be for the CAM software to suggest a tool that can generate the required surface without clash. Clash problems can also be detected by most leading CAM packages, but again, active solutions to eliminate the problem do not currently exist. (10)

3. CASE STUDIES

3.1. The Filter Basket Case

The filter basket mould is one of the test pieces being manufactured as part of the RAMOULDIE project. This case aims to demonstrate complete machining of fully hardened workpieces. The workpiece material used was Uddeholm's ORVAR, hardened to 56/57 HRC. Roughing / semi-finish machining of the mould cavity taking 0.5 mm waterline cuts was followed by finish machining using either a 25mm end mill (with 10mm diameter inserts) or a 12 mm diameter solid carbide ball mill. The mould cavity is relatively shallow and symmetrical about two axis. For this reason it has been possible to divide the surface into four quadrants and to apply a different cutting strategy to each. The results of the finish machining are given below in Table 1. In all cases a depth of cut of 0.25 mm and feed per tooth of 0.2 mm were used. Surface cutting speed was 100m/min for the end mill and 150 mm / min for the ball mill.

Table 1 Machining times for the Filter Basket case

	Tooling	**3axis**	**5axis**
		Actual time (min)	Actual time (min)
Quarter 1	25 mm end mill	48	
Quarter 2	Ditto		19
Quarter 3	12 mm carbide ball nose	35	
Quarter 4	Ditto		53

In both cases, a very small tilt angle of between 4° and 6° was used for the five axis strategies. Since this slight tilt makes only a minor difference in the effective cutting diameter of the tool, the differences in machining time observed are down to the efficiency of the CAM generated tool path. Due to the fact that the cutting path consisted of long sweeping passes along the mould surface and that relatively slow feed rates were used, there was no appreciable difference between the times indicated by the CAM software and the actual machining times.

3.2. The Sandia Case

Although designed for evaluation of parallel kinematic machines, the Sandia test piece (11) was one of the first 5-axis components to be machined on the department's Liechti Turbomill. The primary reason for machining this part was to obtain data about the geometric accuracy of the machine tool. However, since the material used, Uddeholm's Alumec, is free cutting, this component also proved a good test of the machine at high spindle speeds and feed rates as well as giving experience of programming 5-axis machine tools.

Of interest in the present work is the finish machining of the four spherical cavities which are the principle feature of the Sandia testpiece. This operation used a 20 mm ball nose tool, one testpiece being cut using a three axis strategy for all four cavities and one cut using a five axis strategy for all cavities. Whilst this may seem like unnecessary repetition, it must be remembered that the primary aim of the tests was to assess geometric accuracy and systematic errors were expected to be seen between the four cavities.

The machining times indicated by the CAM software and the actual machining times are shown below in Table 2. The slight variation in actual machining time is due to the fact that manual feed / speed over-ride was used during the initial entry into the workpiece

Table 2 Machining times for finish machining the Sandia testpiece spherical cavities

	Tooling	3axis		5axis	
		Actual time (min)	CAM time (min)	Actual time (min)	CAM time (min)
Sphere 1	Ball nose, 20mm	15'00"	8'23"	8'38"	3'25"
Sphere 2	Ditto	14'18"	8'23"	9'34"	3'25"
Sphere 3	Ditto	15'00"	8'23"	9'59"	3'24"
Sphere 4	Ditto	14'19"	8'23"	9'48"	3'24"

These results show that for surfaces with a significant degree of curvature such as the spherical cavities, a five axis machine tool offers significant reductions in finish machining times. However, irrespective of machine tool configuration, the times generated by CAM software for machining non-planar surfaces cannot be trusted if the CAM software takes no account of the controller block cycle time and acceleration limits of the machine tool axis when high feed rates and short path segments are used which is often the case with free form surfaces.

4. CONCLUSIONS

The aim of single set-up manufacture of moulds and dies is unlikely to be reached. However, today's business pressures are not about achieving single set-up manufacture, but rather to reduce the overall manufacturing lead time. In the case of moulds and dies, this can be achieved in one of two ways; set-up reduction and reduction of cut times. In addition, careful thought and the application of design for manufacture rules can ease the demands put on manufacturing. A number of guidelines are given below:

Mould and die design rules for reducing manufacturing lead time include

- Careful design of the mould with the aim of machining ancillary features from so few directions as possible. (DFM)

- Specification and use of mould and die materials with predictable deformation characteristics.

Or

- Specification and use of mould and die materials that do not require hardening, but rather can be used for limited manufacturing runs.

Manufacturing guidelines include

- Acquisition of machine tools and tooling which allow machining of materials in their fully hardened state. Although metal removal rates may be slow, the elimination of EDM and the possibility to hold pre-hardened blanks in standard sizes can offer significant lead time reductions.

- If free form or complex surfaces are to be machined, acquire 5-axis machine tools which allow improvements in tool : workpiece orientation with the aim of increasing metal removal rates.

Product design rules.

- Avoid features which require non-planar parting lines or other complex solutions.

- Avoid features which would require the use of EDM because of their absolute width or depth : width ratio or

- Avoid sharp external corners, these also require EDM tooling.

Organisational guidelines.

- Use the principles of concurrent engineering and get the tool maker involved in the product design process as early as possible. This will hopefully ensure that the final product geometry can easily be transformed into a mould or die cavity and allow rational series production.

5. ACKNOWLEDGEMENTS

The author is indebted to Stephanie Temme who completed much of the initial background work on the Complete Manufacture project. Thanks also to Markus Klaissle, Joachim Greinke and Mikko Mäkivuoti for providing the case study material and to other members of staff at the Department of Manufacturing Engineering for discussions related to this work; notably Charlotta Johansson, Mikael Beckström and Kjell Rask.

REFERENCES

1. Charlotta, J., Legge, D. and Wäppling, A. (1999) ***A Case Study of Concurrent Engineering in the Supply Chain***, Proc. CE 99, Bath 1-3 September 1999, ISBN 1-56676-790-3, pp 51-56

2. Bergwall Analys, (2000), ***Svensk verktygsindustri och verktygsmarknad***. Report commisioned by Pressoform.

3. Greis, N.P. (1995) ***Technology Adoption, Product Design, and Process Change: A Case Study in the Machine Tool Industry***. IEEE Transactions on Engineering Management, vol. 42, no. 2, pp. 192-202

4. Chrisp, A.G. & Gindy, N.N.Z. (1998) ***Parallel Link Machine Tools: Simulation, Workspace Analysis and Component Positioning.*** Proc. 1st European-American Forum on Parallel Kinematic Machines.

5. Breuer, E., (1993), ***Complete Machining on a 5-axis Machining Centre***, Zeitschrift-fuer-Wirtschaftliche-Fertigung-und-Automatisierung, vol. 88, no. 7-8, pp. 23-27

6. Feng, C.X., Kusiak, A. & Huang, C.C. (1997) ***Scheduling Models for Setup Reduction***. Journal of Manufacturing Science and Engineering, vol. 119, no. 11, pp. 571-579

7. Sarma, S.E. & Wright, P.K. (1996) ***Algorithms for the Minimisation of Setups and Tool Changes in Simply Fixturable Components in Milling***. Journal of Manufacturing Systems, vol. 15, no. 2, pp. 95-112

8. Li, S.X., & Jerard, R.B. (1994) ***5-axis Machining of Sculptured Surfaces with a Flat-end Cutter***. Computer-Aided Design, vol. 26, no.3, pp. 165-178

9. Elber, G. (1994) ***Accessibility in 5-axis Milling Environment***. Computer-Aided Design, vol. 26, no. 11, pp.796-802

10. Lauwers, B. (1998) ***State-of-the-art on CAD/CAM Technology***. ROBTOOL Technical Report, Brite-Euram.

11. Lagerfelt, T. (1999) ***Benchmarking Tests for Parallel Kinematics***. Technical report from the Brite/Euram project ROBTOOL - CEC Proposal BE97-4177

CAPE/081/2000

Modelling cutting surface of new type of the segmental abrasive wheel and simulation of the grinding process

L DABROWSKI and **M MARCINIAK**
Warsaw University of Technology, Poland

ABSTRACT

The estimation of kinematics system realizing planetary motion of new type the segmental abrasive wheel and complemented additionally by transportation, is presented in the paper. Computer method of grains trajectories visualization enables the theoretical analysis of their distribution on the machined surface. It corresponds to the working conditions of a wheel head with rotatable abrasive segments.
The computer method of such complex machining motion modeling enables determining the kinematic parameters in order to obtain constant intensity of allowance removal or minimize the errors of surface shape.

1 INTRODUCTION

The increase in the efficiency of grinding during the last half century was mainly stimulated by the implementation of new abrasive materials, the improvement of grinding wheels manufacturing technology and the increase of the grinding speed. In the forecasts for the future it is expected that a further increase in the efficiency of grinding will occur due to grinding tools with abrasive grains of the complex geometry and the definite edge shape, preservation of constant pressure of the tool on the work-piece as well as maintaining the properties of the machined grains constant. Aiming to guarantee the stable properties of the machined abrasive grains, it is necessary to solve many problems resulting from the specific character of contact between the grain and the machined material.
The grinding process can be compared with a micromilling process where the tools remove material in the form of microchips. The basic difference in this type of comparison is that in the grinding process the cutting tools or individual grits are not geometrically defined in terms of orientation, rake angle or individual depth of cut. This is due to the nature of production of the grits themselves and the fact that they are held in a random nature within a bond material [1]. Computer simulation is a modern and powerful method of investigation. It has been used

to investigate different methods of machining. The diagram of dynamics of grinding system is shown in Fig. 1. This paper presents the portion of it, known as "topography generation". Nontraditional surface grinding as a well established technology is basically limited by the kinematic parameters. The aim of the computer simulation was to asses the qualitative results of nontraditional grinding. The results depend on the shape and distribution of abrasive grain traces concentration on the machined surface (Fig. 2).

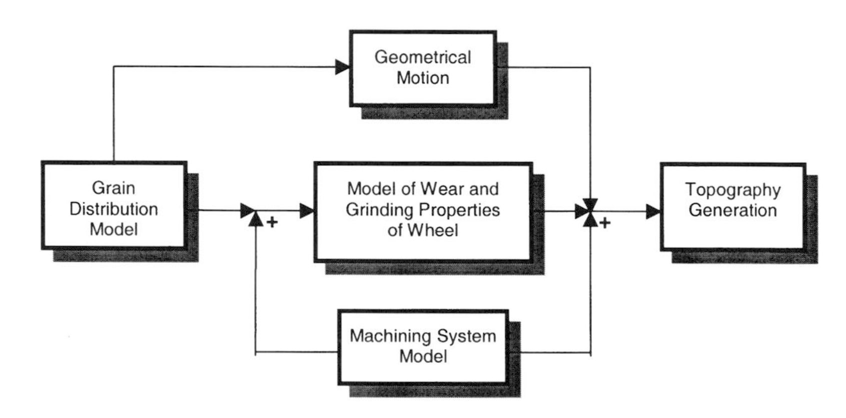

Fig. 1 Block diagram for computer modeling of surface topography after nontraditional grinding

Fig. 2 shows an example of nontraditional grinding process, which has variable direction of grinding wheel motion [2]. However, in the nontraditional surface grinding, the workpiece is attacked on many sides of the each active abrasive grain.

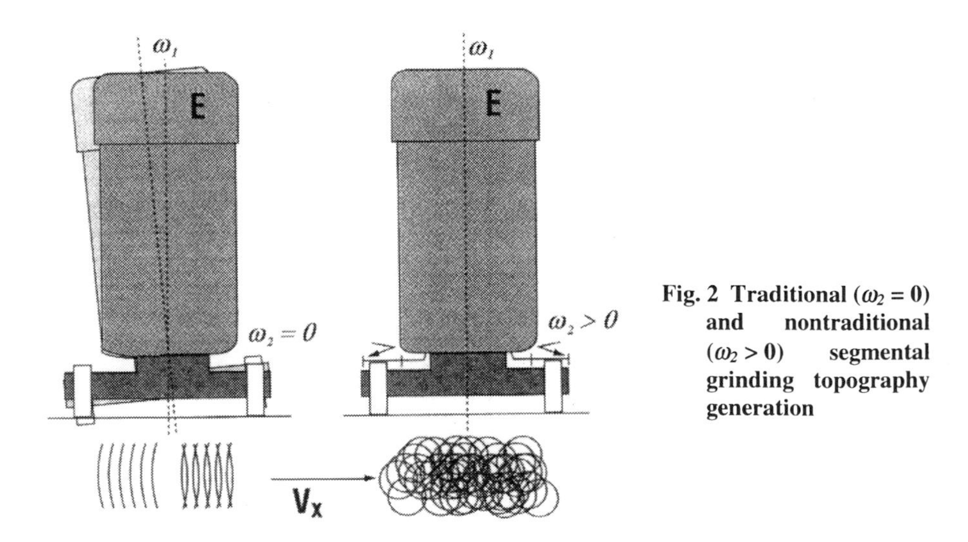

Fig. 2 Traditional ($\omega_2 = 0$) and nontraditional ($\omega_2 > 0$) segmental grinding topography generation

The segmental grinding wheel is equipped with abrasive segments, rotating around their own axes ($\omega_2 > 0$). The segmental grinding wheel is driven by engine E and abrasive segments by their own engine. The machined workpiece performs a longitudinal motion with feed rate v_x.

2 NEW TYPE OF THE ABRASIVE SEGMENT CHARACTERISTICS

In recent years brazing superabrasive grits to a steel substrate in a monolayer configuration with a suitable material has found application in manufacturing high performance wheels which can outperform their conventional galvanically bonded single layer counterparts [3]. Better grit retention better bond uniformity and higher crystal exposure are claimed to be advantageous features of such wheels. Claims are made that such types of superabrasive tools can efficiently solve abrasive machining problems involving loading and heat generation.

Fig. 3 Surface topography of brazed bonded diamond wheels

But in practice, a brazed tool does not always offer high crystal exposure because of non-uniform bond formation during brazing. From the surface topography of the commercial brazed diamond tool shown in Fig. 3, one can see that the relative height between the mean bond base and the tips of several crystals is fairly large. But the feature which draws immediate attention is the high bond level around the grits in the center of the picture. Irregular grit distribution as well as initial non-uniform thickness of the braze alloy layer will adversely influence the topography of the single layer diamond tool after brazing. It appears however, that under the right conditions, excellent bond uniformity can be maintained, the grits are distributed in a regular pattern and the thickness of the bonding alloy layer is kept uniform.

The cutting edges of a grinding wheel are geometrically undefined in location and shape. As a result, generalizations or estimations are difficult to make with any degree of accuracy. The wheels designed using open grain concept utilize the adjustable parameter described as the average distance between the grains. The average distance between the grains $l > 3a$ (a – the average grain size) arc characteristic for the new type of the abrasive segments intended for nontraditional segmental grinding. The rotary motion of the segments relative to their own axes makes the distance l_x between paths of two neighboring grains smaller within the range $0 \le l_x \le l$ (Fig. 4). The real distribution of the abrasive grains trajectory on the machined surface may be controlled using computer simulation method [4].

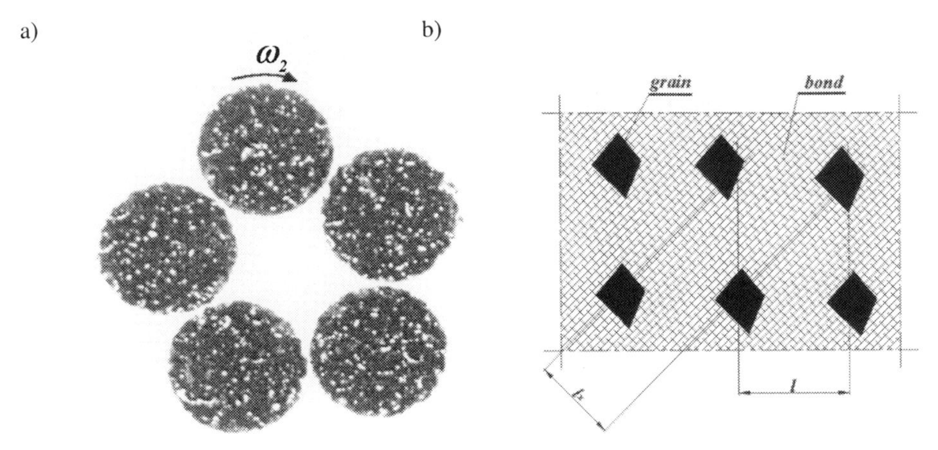

Fig. 4 New type of the abrasive segments (a) and variable distance l_x between path of two neighboring grains for $\omega_2 > 0$ (b)

3 COMPUTER SIMULATION OF GRAIN TRAJECTORIES

The wheel topography and the grinding parameters affect the kinematic interactions between the abrasive grains and the workpiece. Trajectory of the single grain of the rotary segments can be described with a set of two equations:

$$\begin{aligned} x &= R\sin(\omega_1 t + \beta_0) + r\sin(\omega_2 t + \beta_0) + v_x t \\ y &= R\cos(\omega_1 t + \beta_0) + r\cos(\omega_2 t + \beta_0) \end{aligned} \quad (1)$$

Fig. 5 shows logic flow of the program consisting of particular information concerning tasks to be realized. Concentration of trajectories in model testing can be increased by describing trajectories of many grains or by applying the scale factor v_x/x for feed rate v_x. In model testing, where grain trajectories distribution on the machined surface is performed, computer visualization method has been applied. We have made use of mechanism of image forming in the computer display screen, which uses the pixels. Sum of pixels on the horizontal (x) and vertical (y) levels y_j and x_k, represents concentration of grain trajectories on this levels. To determine variability of trajectories concentration in x direction, an index SX is accepted as well as index SY for y direction, respectively. They describe mean deviations from mean (per cent) value of sum of pixels equal to YS %.

Computer drawing of grain trajectories, are shown in Fig. 6. Basing on analysis of the cutting edge trajectories obtained for a given area of a geometric surface structure by numerical calculation, the following results have been obtained:

- mean (per cent) value of quantity of pixels *YS* %,
- mean deviation from mean value (along *y* direction) *SY* %,

$$SY = \frac{1}{n}\sum_{i=1}^{n}\left|s_{yi} - \bar{s}_y\right|; \quad \bar{s}_y = \frac{1}{n}\sum_{i=1}^{n}s_{yi}, \tag{2}$$

- mean deviation from mean value (along *x* direction) *SX* %.

$$SX = \frac{1}{n}\sum_{i=1}^{n}\left|s_{xi} - \bar{s}_x\right|; \quad \bar{s}_x = \frac{1}{n}\sum_{i=1}^{n}s_{xi}, \tag{3}$$

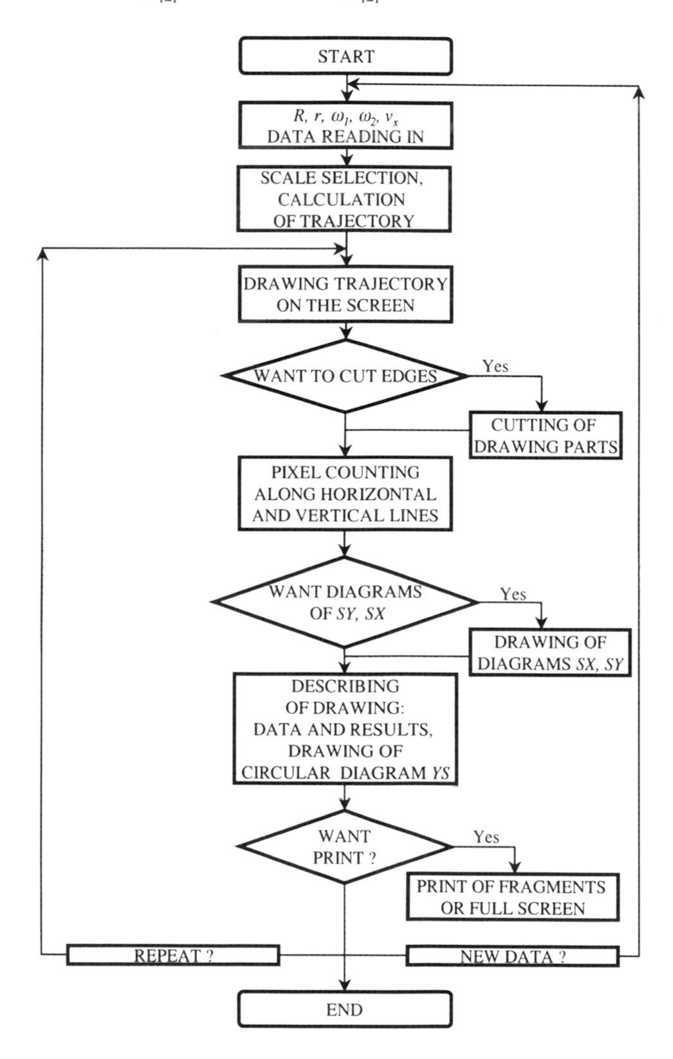

Fig. 5 Logic flow of computer simulation

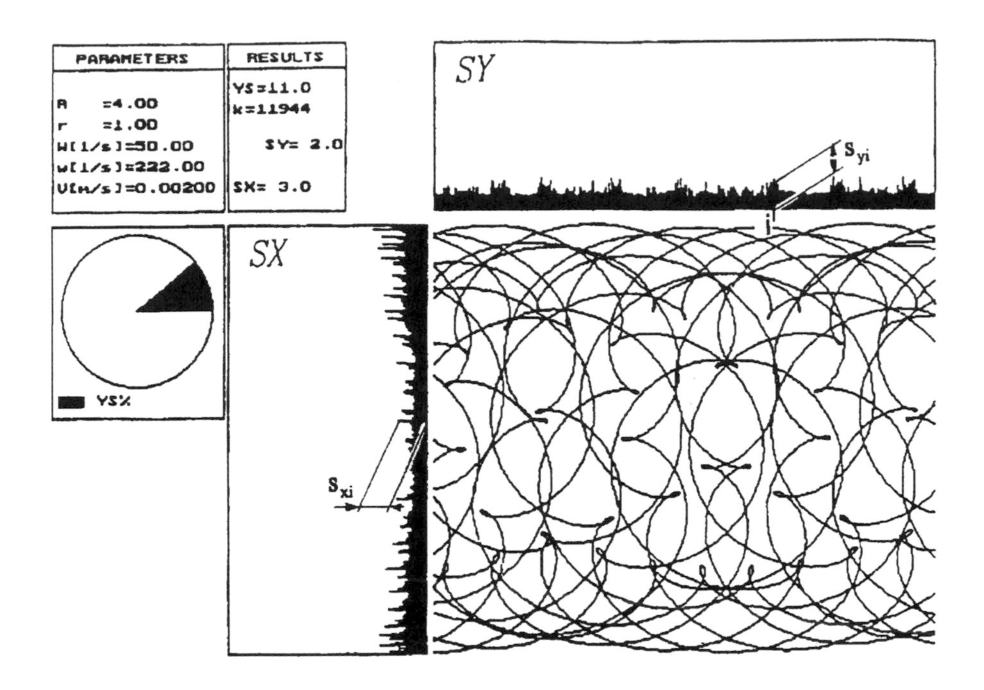

Fig. 6 Computer drawing of grain trajectories and trace concentration index *SX, SY*

The program user can get immediate general assessment of trace concentration distribution on the machined surface for the given machining parameters. It is possible to check visually how the changes of working parameters influence the uniformity of spots concentration on the machined surface. The quantitative estimation of the trace concentration distribution along *x*, *y* axes is illustrated by plots and described by the coefficients *SX* and *SY*. One can conclude, that the obtained data present a confirmation of the conclusions following from simulation tests, that the surface roughness correlates with indicators SX and SY. This serves as a basis for forecasting the possibilities of utilization of computer simulation to optimization of kinematic parameters of nontraditional grinding (Table 1). During trajectory analysis in the area of stable structure we have obtained maximum value the following: item 3, 7, 11.

4 RESULTS OF THE MACHINING TESTS

The results of modeling investigations with reference to the quantitative and qualitative efficiency of microcutting have been verified in the machining tests (Fig. 7).

The quantitative efficiency of nontraditional grinding has been estimated during the machining of aluminum alloy PA7 on face new abrasive segments. PA7 is the material of poor grindability, prone to built-up edge formation and to the gumming up effect on the cutting surface of the grinding wheel. Thus, the application of the nonconventional grinding enabled to observe the favorable influence of abrasive grains displacement in the tool-in-use system on their self-cleaning as well as on their self-sharpening.

Table 1. Results of computer modeling calculation for R/r = 0.4

Item	$\omega_1 R$	$\omega_2 r$	$v_x/x \cdot 10^6$	*YS* %	*SY* %	*SX* %
1	140	77	10	38.6	4.3	11.7
2	140	77	15	28.2	3.6	9.3
3	140	257	10	74.5	4.4	12.6
4	140	257	15	61.1	4.7	12.7
5	70	77	10	35.0	5.9	12.6
6	70	77	15	24.3	4.0	8.5
7	70	257	10	73.8	4.1	12.9
8	70	257	15	60.7	4.1	11.2
9	6	77	10	35.2	4.5	8.8
10	6	77	15	24.3	2.3	5.6
11	6	257	10	71.5	6.9	13.3
12	6	257	15	54.3	3.4	11.9

Fig. 7 The segmental grinding with rotary segments

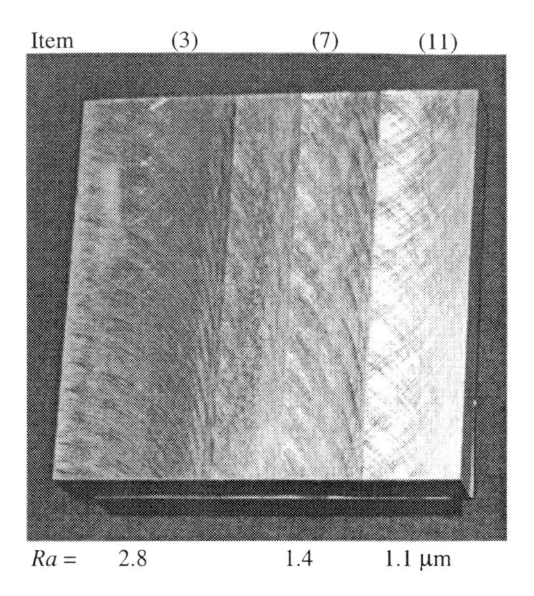

Fig. 8 Roughness of surface machined on new abrasive segments according to parameter item 3, 7, 11 provided in Table 1.

Fig. 8 illustrates the comparison of *Ra* measurements of surface after the nonconventional grinding, at varying ratio $K = \omega_1/\omega_2$ and feed v_x. From Fig. 8 it follows that the quotient K has an essential influence upon the roughness of the machined surface. The obtained data confirm, besides, the conclusions following from simulation tests that the surface roughness correlates with indicator *YS %*.
The investigation results prove, that it is possible to design the cutting surface for a new type of the abrasive segment and to predict surface roughness basing on theoretical analysis of trace concentration distribution obtained by computer visualization method.

REFERENCES

1. Blunt L., Ebdon S.: The application of free-dimensional surface measurement techniques to characterizing grinding wheel topography. Int. J. Mach. Tools Manufact., vol. 36, No. 11, pp. 1207-1226, 1996.
2. Marciniak M.: Patent No. 157746, Int. Cl. B24D5/06, 1993.
3. Inasaki J., Tonshoff H. K., Howes T. D.: Abrasive Machining in the Future Annals of the CIRP, vol. 42/2/1993, pp. 723-732.
4. Dabrowski L., Marciniak M.: Tribological Aspects of the Grinding Process. Advances in Manufacturing Science and Technology. Quarterly, vol. 25, No. 2, Warsaw 2000.

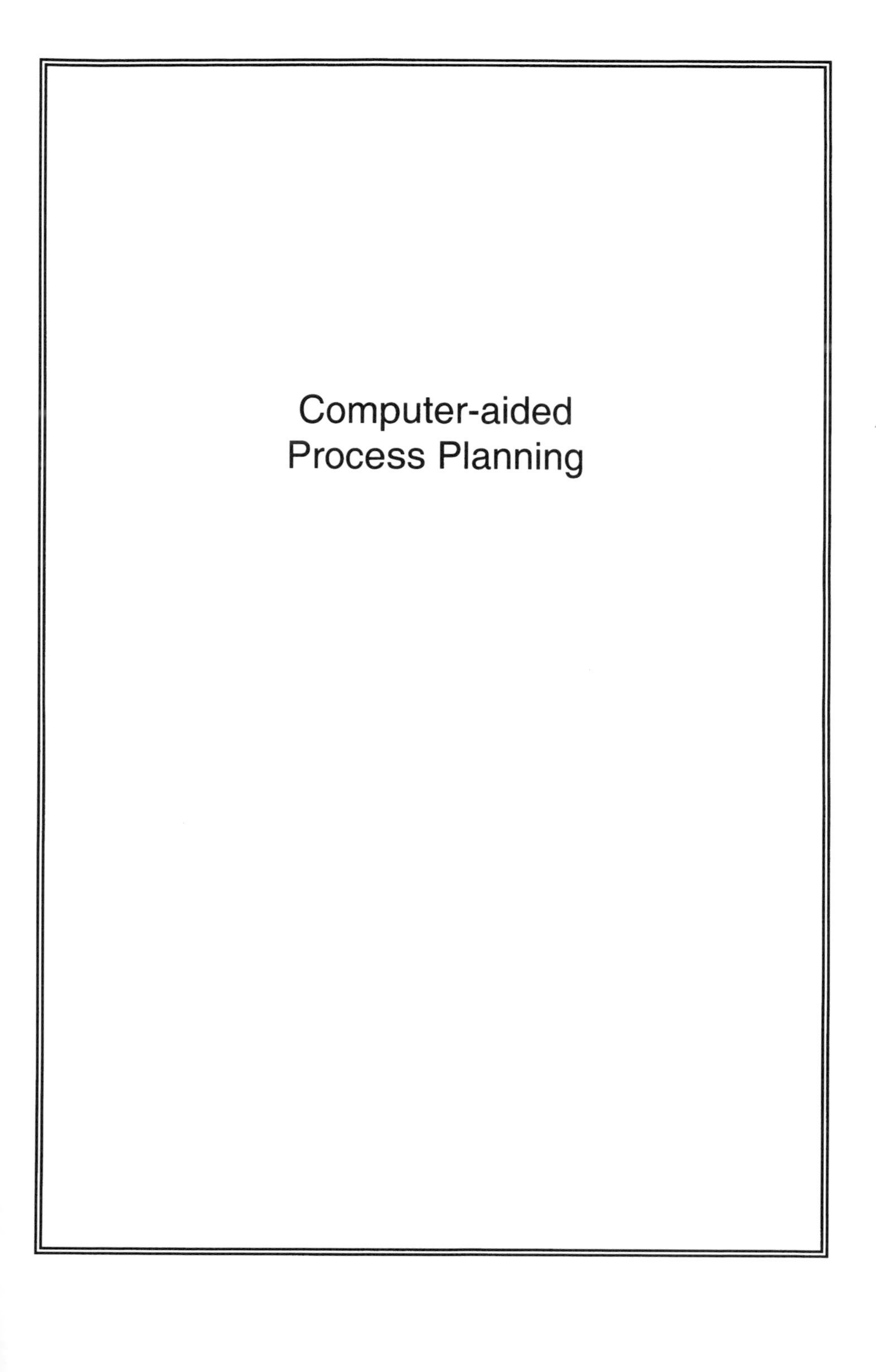

Computer-aided
Process Planning

CAPE/031/2000

Cost justification of AMT by comparison of normalized process plans

D I LEGGE
Division of Manufacturing Engineering, Lulea University of Technology, Sweden

ABSTRACT

The European CRAFT project "RAMOULDIE" is concerned with evaluating conventional machining technologies against 5-axis and high speed machining (HSM) for tool and die manufacture. Since the cost of these new technologies remains high, they can be difficult to justify economically. A model has been developed where costed generic process plans are used for the basis of economic comparison / justification. Process plans for a given mould or die are 'normalised' to fit into the generic model. Comparing generic plans eliminates the need for carrying out detailed planning for a machine or process that may be unfamiliar as would be the case if a company were in the process of evaluating new technologies. Instead, subjective expectations are used to tailor the generic plan for the new technology.

The method framework has been realised in a simple spreadsheet application but not as yet tested against real manufacturing data. This will be carried out once the machining tests in the RAMOULDIE project are completed.

1. BACKGROUND

The current work is part of a series of ongoing projects at the Department of Manufacturing Engineering at Luleå University of Technology related to mould and die manufacture. Much of this work has been financed by *Pressoform*[1], who been working actively for several years to raise standards of awareness of best practice amongst Swedish tool and die makers.

---

[1] Press och formverktyg i Luleå AB was founded in 1995 as a limited liability company whose shareholders are companies in the tool and die industry in Sweden.

The *RAMOULDIE*[2] project aims to demonstrate and quantify the benefits of using 5-axis / HSM machine tools over 3-axis machines. The project also indirectly compares the capabilities of a serial machine tool structure, Luleå University's Leichti Turbomill 1200, with a parallel kinematic machine, Nottingham University's Variax (1). Both machine tools are equipped with state-of-the-art controllers and spindles and can thus machine complex geometries in materials in a hardened state.

Central to the project are a number of case studies which were suggested by the industrial partners associated with RAMOULDIE, from which nine cases were initially selected. Since the capability of the two machine tools available to the project consortium are comparable, it was decided that the majority of cases would be machined on one or the other machine tool. However, some items were selected to be machined on both machines. This would provide some, although limited, scope to compare a parallel kinematic machine with a serial structure.

The project hopes to be able to demonstrate that using this kind of equipment can be economically justifiable, and to offer additional benefits such as reduction in overall lead times. The purpose of the test cases is to demonstrate and quantify these improvements. The expectations of the project group, which have been expressed as targets for the project, are:

- Reduction of machine set-ups by up to 60%
- Reduction of machining times by up to 50%
- Reduction of EDM machining requirements by up to 80%
- Reduction of manual polishing times by up to 80%
- Reduction of overall manufacturing cost by up to 15%'

2. ASSESSMENT METHOD

2.1. Comparison of costed process plans

The most obvious way of documenting the performance improvements expected by using advanced manufacturing technology (AMT) is to take actual process planning / cost details for an item manufactured using conventional techniques and to compare this with details for the manufacture of the same item but using high speed / 5-axis machining processes (2).

Whilst such a comparison clearly should be made, it could be argued that the particular results found by the RAMOULDIE team are specific to the products selected, and that the data cannot necessarily be taken at face value by another company considering investing in these technologies. Such an argument is reinforced by the fact that the RAMOULDIE project only has time/resources to look at a limited number of test pieces which have been selected for a variety of reasons, not necessarily because they are typical, if such a thing exists, in terms of complexity, manufacturing method etc.

---

[2] RAMOULDIE = Rapid and Accurate Manufacturing of Moulds and Dies. This project has partners in Sweden, Germany and the UK.

In order to make the results of the RAMOULDIE project more general, and hence applicable in a wider range of situations, a more generic method of assessing the benefits of high speed / 5-axis processes is proposed. This method aims to provide a framework for developing representative data for cost justification purposes, without the need to carry out detailed planning for a machine or process that may be unfamiliar as would be the case if a company were in the process of evaluating new technologies. Instead, subjective expectations as far as the performance of the new technology are used to tailor the content of a generic plan for the AMT. Whilst clearly not as accurate as a good costed process plan, it should be possible to get ball park figures which will help in any justification / selection work. It is important to remember that, if investment in AMT is made, that it invariably takes a while for staff to become competent in its use, and to iron out any implementation related problems.

2.2. Outline of the methodology

In the RAMOULDIE project, it has been possible to compare actual processing times and hence costs, for a number of case study components manufactured using traditional methods and also using AMT. Unfortunately, companies investing in a particular AMT seldom have this opportunity (3). Reference date from the existing process are invariably available but, short of actually having a component machined on the machine tool of interest, obtaining reliable data about the performance characteristics is problematical. What is required is a methodology to allow comparison of machining times and associated manufacturing costs for the current manufacturing method with the method possible with AMT.

The principle of the method developed, shown in Figure 1, is that an actual or reference process plan is "normalised" to remove unwanted detail but retain the substance of the process in the form of a synthetic / generic plan. The manufacturing cost, or other performance criteria based upon the original plan, should be the same as that from the synthetic / generic plan which is the result of the normalisation process.

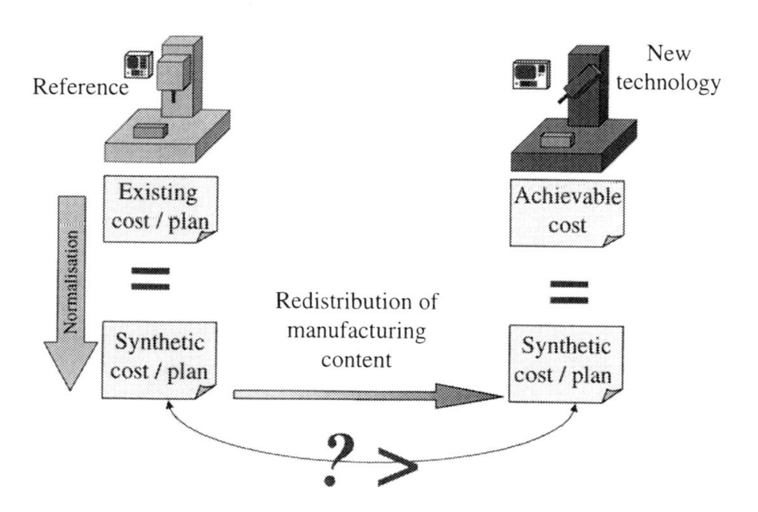

Figure 1 Concept of comparison of costed normalised process plans.

An equivalent synthetic / generic plan for the new technology is developed by re-distributing the content of the synthetic / generic process plan for the reference technology. This redistribution is to account for changes in the way in which a product is manufactured using the 'new' technology, which must also be reflected in the structure of the synthetic plan for the new technology. The redistribution is based on both objective criteria but also subjective expectations as far as the performance of the new equipment is concerned.

Assuming that the synthetic / generic plan for the new technology is indicative of the achievable performance, direct comparison of the performance criteria generated from the data in the synthetic / generic plans should be sufficient to indicate whether the new technology can be justified.

2.3. Normalised process plans

The data applied to the synthetic / generic plan is based upon company specific planning, but, in order to eliminate unnecessary detail the "normalised" process plan contains only the key stages associated with the manufacture of a particular type of mould/die using a particular technology. Normalising existing process plans occurs at two levels, firstly simplifying them to the level of a generic plan for the manufacture of a given type of product (Figure 2a).and secondly, for each operation standardising the cost / performance elements to be compared (Figure 2b). It is these elements that will, in the RAMOULDIE project, help indicate the cost / time advantages and possible disadvantages.

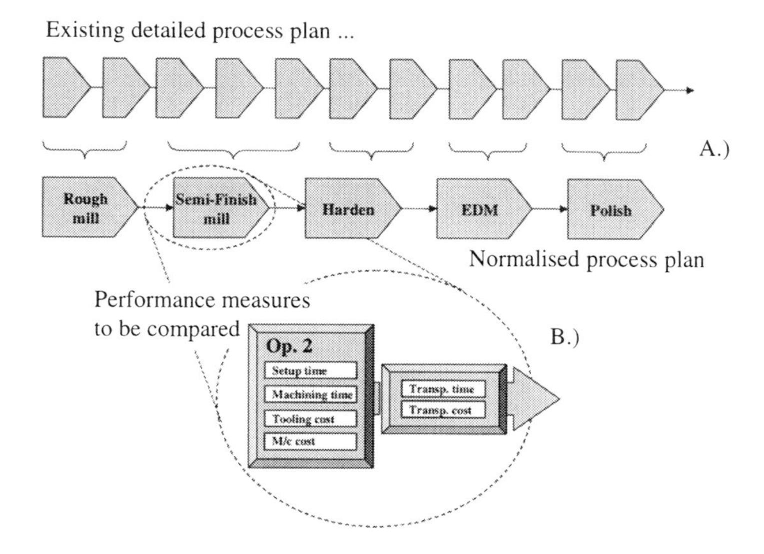

Figure 2 a.) Mapping of detailed process plans onto a normalised framework and b.) The costing elements associated with each normalised operation.

The cost / performance elements can be summed for all the operations in the normalised plan and hence provide the data required for performance comparison and justification. In order to provide a mechanism for comparison, generic process plans are required for each method of manufacture to be compared, i.e. 'conventional' manufacturing processes and those related to high speed / 5-axis machining or other AMT.

Machine tools with high speed spindles or multi-axis capability offers the possibility to eliminate set-ups and even operations or to reduce machining time compared to conventional machine tools. Possible changes include:

- Replacing EDM with milling in the hardened state.
- carrying out all machining in the hardened condition.

These options are shown as normalised / generic process plans in Figure 3.

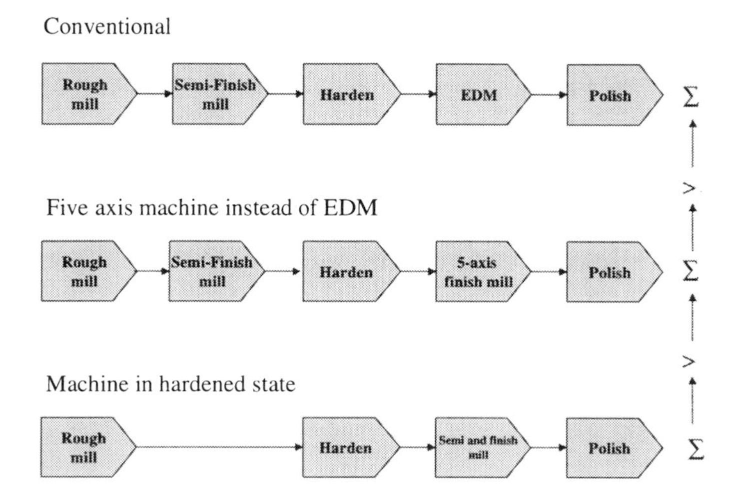

Figure 3 Possible alternatives to conventionl machining; a.) replacing EDM with 5-axis milling and b.) semi-finish and finish in the hardened condition.

Since the total amount of work required to manufacture a mould / die is the same, irrespective of the method of manufacture (The same volume of material must be removed and the same tolerances etc achieved.), the content, or rather the cost / performance elements, of the known or reference process must be "redistributed" amongst the operations in the generic process plan for the advanced manufacturing method(s). This redistribution can be objective and/or subjective. Redistribution of manufacturing content at this detailed level allows the particular bias towards specific operations associated with a company's different products to be catered for and also reflect the expectations of the staff carrying out the assessment / justification.

In the model developed, redistribution is carried out for each cost / performance element in turn. In the current model, element times / costs must be mapped onto equivalent elements and cannot be split between two locations.

3. DISCUSSION AND CONCLUSIONS

The costing model described above has been implemented in a spread sheet and will be tested using real manufacturing data once this is available. It is hoped that a number of generic plans will be developed that reflect the rationalisation opportunities offered by applying multi-axis machining and HSM to mould and die manufacture. In addition, it is hoped to identify general rules regarding how the work content of manufacturing moulds and dies should be re-distributed. If this is achieved, it will be very easy to take any component that fits one or more of the generic plans for conventional machining and to quickly document the benefits to be had from using AMT.

Whilst the model is good at handling machining times, actual costs are more difficult to assign since machine rates, labour costs and cost allocation methods which are used in a particular company can have a significant effect on the cost benefit that is produced. This problem is outside the scope of the present research, but indicates the difficulties associated with justifying incremental technological advances in industry.

5. ACKNOWLEDGEMENTS

The author is indebted to Achim Schulz who implemented the concept of comparison of normalised process plans in software. Thanks also to members of staff at the Department of Manufacturing Engineering for discussions related to this work; notably Kjell Lindfors, Mikael Beckström and Kjell Rask.

REFERENCES

1. Chrisp, A.G. & Gindy, N.N.Z. (1998) ***Parallel Link Machine Tools: Simulation, Workspace Analysis and Component Positioning.*** Proc. 1s[t] European-American Forum on Parallel Kinematic Machines.

2. Xirouchakis, D. & Persson, J.-G., (1998) ***A Petri-net Technique for Process Planning Cost Estimation,*** Annals of the CIRP 47/1.

3. Greis, N.P. (1995) ***Technology Adoption, Product Design, and Process Change: A Case Study in the Machine Tool Industry***. IEEE Transactions on Engineering Management, vol. 42, no. 2, pp. 192-202

A computer-aided process planning system based on branch and bound algorithm

P Y GAN, K S LEE, Y F ZHANG, and **C Y HUNG**
Department of Mechanical and Production Engineering, National University of Singapore, Singapore

ABSTRACT

With an increasing number of customized mold bases, mold base companies are looking into computer aided process planning (CAPP) as a way to ensure high quality process plans. This research will focus on developing a CAPP system to optimize the process plan of customized mold bases. Manufacturing information of a given part is extracted from its CAD model. Based on the information of available machining capacity (machines and cutters), the entire solution space of process plan is obtained. A search algorithm based on branch and bound is then used to search for the optimal plan. The problem formulation and mechanisms of the branch and bound algorithm used in this study will be described. As the industries are most concerned with the delivery time of mold bases, a performance measure of total time will be used.

1 INTRODUCTION

Computer-Aided Process Planning (CAPP) is receiving much attention in recent years. This is mainly due to increasing factory automation and greater emphasis on manufacturing efficiency. CAPP is essential in integrating Computer-Aided Design (CAD) and Computer-Aided Manufacturing (CAM). To reduce labor, the computer-integrated manufacturing (CIM) concept involves in linking the processes of an entire factory via computers to make it fully automated. This makes the development of CAPP vital in ensuring a fully functional CIM concept.

This research looks into process planning involved in mold base industries. The program is built on IMOLD®*. Currently, the process planning for mold bases are mostly done manually. The need for a CAPP system is justified by the following reasons:

a) A standard mold base normally has a standard process plan, which has already been optimized to the amount of machining required on it. Customization of the mold base disrupts the original process plan and new operations have to be added in manually. As new operations are added and changed continuously, the process planner is unable to keep up with the optimization of the changes.
b) Process planning should also consider the overall shop floor condition such as machine availability. CAPP would allow such conditions to be considered before generating an optimized process plan.

The ultimate goal of this research is to create a process planning module that can be used on IMOLD® to generate process plans of mold bases. Information of holes or slots to be machined is extracted from the model part file. Other information of machine and tools types is stored in a database, which will be considered in the process plan. The planner can determine the precedence constraints between each operations as well as add in information of customized operations. By considering all the above necessary information a process plan can then be generated using some form of artificial intelligence.

In this paper, emphasis will be placed on the intelligence in obtaining an optimal solution for a mold base using a branch and bound method. The background will be provided in the next section to describe artificial intelligence in process planning and some literature on the uses of the branch and bound algorithm. The problem formulation and some details of the algorithm used will be provided in the subsequent section. Lastly, an example and discussion will illustrate the potentiality of a branch and bound CAPP.

2 BACKGROUND

Process planning is about determining a set of instructions to produce or manufacture a particular part. In this study, the part is a mold base that can be simplified as a prismatic block with holes or slots. Each hole or slot has to be created in different ways, which might require more than one operation. For example, countersunk holes require drilling first before doing the countersunk. This kind of constraint is to be considered as a precedence constraint, which will be taken into account to prevent any clashes or contradiction in the final process plan.

Up to now, many CAPP systems have been reported, but few have been able to provide globally optimized process plans (1). As a result, there has been an increasing use of artificial intelligence in CAPP to address that problem (2,3). Presently, there is a range of artificial intelligence algorithms available, comprising mainly of heuristical methods. Many of these reported methods only involve sequencing features to be processed without details of its operations (4,5). The details are important as the required shop floor resources have to be allocated so that the operations can be performed smoothly.

* IMOLD® (Intelligent Mold Design) is a knowledge-based application software developed by Department of Mechanical and Production Engineering, NUS to facilitate plastic injection mold design

The performance measure or objective function is the value to be maximized or minimized in all optimization problems. Objective functions in the form of cost or time (6,7) are mostly used. As delivery time is most critical in mold base manufacturing industries, overall machining time for all the features on the mold base plate will be taken as the performance measure in this study.

Among many different ways of optimization, the branch and bound technique stands out as one of the most unique and powerful means of optimization. A handful of researches have been carried out using the branch and bound technique for a variety of planning tasks, ranging from tool selections (7) to sequencing of sheet metal bending (8,9). Our research aims to process plan all operations on all available machines, considering all tool access directions with available tools on a mold base using a branch and bound algorithm. To the best of our knowledge, such level of consideration has not been taken into account for other researches using the branch and bound technique for process planning.

3 PROBLEM FORMULATION

In this research, the information required for optimization is extracted from mold bases modeled using IMOLD®. A routine was written to extract all features like holes and slots from each mold base plate. The routine will also state the operations, precedence constraints, machines, tool types and corresponding processing time needed to create each feature from a database. This database will be read together with machines' availability during process planning.

3.1 Requirements for the process plan

The performance measure or objective function is taken to be the total machining time and the generated process plan has to satisfy the following conditions:

a) The features of the mold base plate are recognized with operations assigned to it. The operations assigned should yield the desired shape, dimension, tolerances and finishes.
b) The sequence of operations obtained from the process plans should not violate any of the precedence constraints governing the operations.
c) Operations can only be done on available machines with the available tools, which are capable of machining the particular feature.
d) The total machining time for the process plan will be calculated such that only one operation can be done at any one time.

The generated process plan should include the number of operations to be carried out, the sequence of these operations and the machines, machining direction and tools to be used on those operations. Such details are necessary so that time can be saved when the same set-up can be used on the same machine for two or more operations. For example, a blind hole is to be machined in the +x direction and a through hole is to be done in the x direction and it does not matter if it is drilled from the +x or –x direction. The process plan should strive to perform both operations on the same machine in the +x direction such that only one set-up is necessary for two operations.

3.2 The objective function

To account for all the machining time for the process plan, we use a similar calculation framework used by Zhang et al (10). There are altogether 3 areas, which contribute to the calculation of the objective function, the overall machining time (OMT). OMT is calculated for each successive sequence of process plans and the minimum will be taken as the final process plan.

3.2.1 Machine set-up time

Machine set-up time (MST) is to be considered when there is a change in machine between two operations. It is the time required to move between machines and the time to set-up the mold base plate onto the machine in a particular direction. It is defined for all n operations as,

$$MST = \sum_{i=1}^{n-1} \left(\Omega\,(M_{i+1}, M_i) \times MSTI_{i+1} \right) \qquad [1]$$

where $$\Omega\,(x, y) = \begin{cases} 1 \text{ if } i = 1 \\ 1 \text{ if } x \neq y, i > 1 \\ 0 \text{ if } x = y, i > 1 \end{cases} \qquad [2]$$

M_i refers to the machine selected to process operation i and $MSTI_i$ refers to the machine set-up time index for the machine used in operation i and n is the number of operations selected for the whole series of operations identified from its features.

3.2.2 Machining direction set-up time

Machining direction set-up time (MDST) refers to the time required to change the orientation of the mold base plate on the same machine. MDST is only calculated when there is a change in machining direction but no change in machine between 2 operations. It is defined as,

$$MDST = \sum_{i=1}^{n-1} \left(\Omega\,(MD_{i+1}, MD_i) \times [1 - \Omega(M_{i+1}, M_i)] \times MDSTI_{i+1} \right) \qquad [3]$$

MD_i is the machining direction selected to process operation i and $MDSTI_i$ is the machining direction set-up time index for the machine used in operation i. $MDSTI_i$ and $MSTI_i$ are related by the difference in time to move the part between the old and new machine.

$$MSTI_i = MDSTI_i + (\text{Time to move part between machines}) \qquad [4]$$

The case study used in this paper assumes $MDSTI_i$ and $MSTI_i$ to be the same. This estimate is based on the assumption that the difference between the two is in moving the job from one machine to the other and the bulk of $MSTI_i$ is in $MDSTI_i$.

3.2.3 Machining time

Machining time MT is the time required to perform all the machining operations on the assigned machines with their respective tools.

$$MT = \sum_{i=1}^{n} (\, MT_{M_i, T_i})_i \qquad [5]$$

The machining time for a single operation (i) can vary according to the assigned machine (M_i) and tool (T_i) selected. Therefore there is a number of possible MT_i for a particular operation.

3.2.4 Overall Machining time

Overall machining time is the total of all machine change set-up time, machining direction change set-up time and all machining times.

$$OMT_{min} = (MST + MDST + MT)_{optmised\ sequence} \quad [6]$$

The objective is simply to produce a sequence of operations that will require the least OMT.

4 THE BRANCH AND BOUND ALGORITHM

The branch and bound technique is chosen as a means of artificial intelligence due to its robust and enumerative nature. The computation load for any branch and bound is inherently high, but effective heuristics and efficient lower bound calculations would decrease the search space considerably.

4.1 The implemented branch and bound algorithm

The algorithm will start by sequencing one of the available operations and this is called branching the node. A node will contain information of all the operations sequenced so far. Operations that have not been sequenced and have their precedence constraints satisfied are considered as available operations. After one of the operations is sequenced, the potential of this node is estimated by its lower bound value. Each successive node's lower bound value is compared to the upper bound value, which can be obtained through a heuristic scheduler. The next node to be branched is the one with the best lower bound value as it is deemed to have the best potential. As the nodes are branched, more and more operations will be sequenced and the upper bound value will get smaller and smaller. It will reach a time where the upper bound value is smaller than all lower bound values and this means that the best solution is the upper bound solution.

To better illustrate the branch and bound algorithm, we use the conventions of the A* algorithm. A mathematical representation of the lower bound function $f(S_a)$ for sequence S_a is defined as

$$f(S_a) = g(S_a) + h(S_a) \quad [7]$$

where $g(S_a)$ is the cost incurred to reach S_a and $h(S_a)$ is a function that calculates the estimated cost of reaching the final schedule. The algorithm can then be briefly summarized as follows,

begin
Step 1. $S_0 \leftarrow$ Initial situation (No operation sequenced)
    Open $\leftarrow S_0$
while Open $\neq \varnothing$
Step 2. Choose in Open, a node with sequence S_a, which has the best lower bound
    Open $\leftarrow$ Open $- S_a$
Step 3. Sequence all the possibilities starting from S_a
    for each possibility S_k

Step 4. Use a heuristic to schedule the remaining operations $S_{k(final)}$
Since $h(S_{k(final)}) = 0$, the objective function can then be calculated as $f(S_{k(final)}) = g(S_{k(final)})$

Step 5. Update the upper bound value
if $f(S_{k(final)})$ is better
then upper bound = $f(S_{k(final)})$

Step 6. Calculated the lower bound
for each possibility k,
$f(S_k) = g(S_k) + h(S_k)$

Step 7. Include branched node into search space
if $f(S_k)$ better then upper bound
then Open $\leftarrow$ Open + S_k

end

Step 8. Discard all nodes in Open with f(S) worse than or equal to upper bound

end while

end

4.2 Constant machine and machining direction heuristic

A heuristic is critical in generating good solutions as early as possible. When good solutions are generated early, the upper bound value is going to reject many unpromising nodes, which will reduce the search space accordingly.

Normally, when a process planner plans for a particular job, operations that require both the same machine and machining direction will be done together. In view of that, the heuristic will try to keep looking for the subsequent operation that has the same machine and machining direction as the previous one. The heuristic can be briefly summarized as,

do

Step 1. To determine operation i+1,
for all other operations, which can be processed using, machine M_i
choose the operation with a corresponding machine and machining direction which will yield the lowest added time ($AT = MDST_{i+1} + MST_{i+1} + MT_{i+1}$)

Step 2. **if** no operation is chosen, i.e. none of the remaining available operations can be processed by machine M_i, choose the operation with the lowest AT

Step 3. **while** there are still unassigned operations, i = i + 1

end

The heuristic will choose the available operation that has the same machine and machining direction as the previous operation if the time is shorter. In cases where time is saved by doing that operation on a faster machine, the faster machine is chosen. By having a heuristic like this, the process plan will not always force operations to be done on the same machine and machining direction but rather allow operations to be done on different machines if time can be saved.

4.3 Lower bound calculation

The lower bound value is an estimate of the best possible solution that can arise from the current sequence of a node. A simple way of calculating lower bound would be to add up the minimum process times for all remaining operations. However, that will underestimate the lower bound value of that node as no set-up time is included in the estimate. Underestimating lower bound values causes many unpromising nodes to be branched and hence waste time. To better grasp the true best possible solution, we include the set-up when necessary into the lower bound calculations. It is found that by including set-up times, search space is reduced significantly especially for large problems.

5 RESULTS AND DISCUSSION

The algorithm is programmed in C language and ran on a workstation. A simplified standard mold base plate is used for the benefit of this exercise and the algorithm will be used to plan for all the operations for the simplified plate.

5.1 Case study

The information of the operations, machines and machining parameters are added into the CAPP system through a series of databases. The operations required and machines involved can be summarized in Table 1 that shows part of the operations required to manufacture a real industrial standard mold base plate.

Table 1 Operations selection results for a simplified core plate

Features	Operations (OPR)	Predecessors	Machines (M)	Tool Types (T)	Directions (MD)
F1 Cavity	1 Milling	-	2, 3	3	+z
F2 Support Pin Holes (x 4)	2 U-Drilling	7	1,2,3	6	+z, -z
	3 Reaming	2	2, 3	7	+z, -z
	4 Boring	3	2, 3	8	+z, -z
F3 Guide Pin Holes (x 4)	5 U-Drilling	-	2, 3	6	+z, -z
	6 Reaming	1,5	2, 3	7	+z, -z
	7 Boring	6	2, 3	8	+z, -z
F4 Counter boring of Hole	8 Counter bore Drilling	5	2, 3	3	-z
F5 Eye Bolt Hole	9 Spot Drilling	-	1,2,3	4	+x
	10 Drilling	9	1,2,3	6	+x
F6 Ejector Pin Holes (x 4)	**11 Drilling**	**1,4**	**1,2,3**	**6**	**+z, -z**

The last feature, ejector pins holes is a customized feature, which is added onto the standard mold base. For each operation, there are precedence constraints, machines, tools and machining direction. Precedence relationships are clearly shown in the table for example, reaming cannot commence before drilling the hole. All such operations are fixed, as they are required for a standard mold base. When customization of new holes or slots are added, the new operations have to be added manually as there are many ways to create a feature. For example, a hole can be drilled first, bore next, then reamed to the desired finish.

The available type of machines has to be determined and fed into the process planning system. Each machine will have a set of tools and machine set-up time. For this case study, we take the machines of a local industrial manufacturing shop as our available machines. Should the machines become unavailable, the process planner will have to omit those machines from the database. The details of the machines are described in Table 2. Note that the set-up times for machine change and machining direction change are taken to be the same in this case study. Tool types for the machines are also included with their corresponding codes. This is done so that for each operation that requires a certain tool type, all available machines for that operation can be easily recognized.

Table 2 Machines, MSTI, MDSTI and types of tools

	Machines (M)	MSTI, MDSTI (mins)	Suitable Tool Types (T)
1	Radial Drilling Machine	1, 1	4,6
2	MORI SEIKI MV65-50 Vertical CNC Milling	1.5, 1.5	1,3,4,6,7,8,9
3	MAKINO MC98 Vertical CNC Milling	2, 2	1,3,4,6,7,8,9
1: Face mill Cutter 2: Grinding Wheel 3: End mill 4: NC Spot Drill 5: Edge-Grinding Wheel		6: Drill 7: Reamer 8: Boring Tool 9: Tap Drill	

5.2 Branch and bound solution

The case study attempts to test the effectiveness of the branch and bound algorithm in several circumstances. The few scenarios are

1) When all machines are available and all standard operations must be completed
2) When there is an added customized operation to be completed
3) When machine 2 is unavailable and all standard operations must be completed

The presented process plan shows the sequence of operations to be performed and the corresponding machine, machining direction and tool selected for each operation. The algorithm was executed for all three situations and the following results were obtained.

Table 3 Process plan found for core plate in scenario 1

No.	1	2	3	4	5	6	7	8	9	10
OPR	9	10	1	5	8	6	7	2	3	4
M	1	1	3	3	2	2	2	2	2	2
MD	1	1	5	5	6	6	6	6	6	6
T	4	6	3	6	3	7	8	6	7	8
Computational time = 50 secs (Number of nodes = 3532) No. of M changes = 3 No. of MD changes = 0 OMT = 45.5 mins										

Table 4 Process plan found for core plate in scenario 2

No.	1	2	3	4	5	6	7	8	9	10	11
OPR	9	10	1	5	8	6	7	2	3	4	11
M	1	1	3	3	2	2	2	2	2	2	2
MD	1	1	5	5	6	6	6	6	6	6	6
T	4	6	3	6	3	7	8	6	7	8	6
Computational time = 55 secs (Number of nodes = 4171) No. of M changes = 3 No. of MD changes = 0 OMT = 52.5 mins											

Table 5 Process plan found for core plate in scenario 3

No.	1	2	3	4	5	6	7	8	9	10
OPR	5	8	9	10	1	6	7	2	3	4
M	3	3	1	1	3	3	3	3	3	3
MD	6	6	1	1	5	5	5	5	5	5
T	6	3	4	6	3	7	8	6	7	8
Computational time = 14 secs (Number of nodes = 934) No. of M changes = 3 No. of MD changes = 0 OMT = 47.0 mins										

Tables 3, 4 and 5 show the process plan for the three respective scenarios. From the solutions it can be clearly seen that the generated process plans try to maintain same machine and machining direction where it is possible to save machining time. When an extra operation is added, the process plan includes and fits the operation in the best possible way such that total number of machine and machining direction change is minimized as shown in Table 4. In an event where machine 2 is unavailable, the process plan will continue without the use of this machine. It can also be seen that as number of available machines decrease, search space also reduces considerably as shown in Table 5.

The objective is to minimize total machining time. For the same operation, the difference in machining time on different machines is small for an average mould base plate. This means that process plans with minimal machine and machining direction change are optimal. From Tables 3, 4 and 5, it can clearly be seen that the number of machine and machining direction change are minimized to the total number of machining direction required. Hence the process plans obtained are optimal or near optimal.

6 CONCLUSIONS

This paper illustrates the methodology and potential of a branch and bound based computer aided process planning system. It considers operation sequencing, machine and machining direction selection. It also has a flexible structure and can easily account for tool change time and other customization when necessary to accommodate different environments. The developed system offers a comprehensive shop floor consideration in its optimization of overall machining time.

The case study has achieved valid and good process plans that are capable of re-adjusting operation sequences in view of any shop floor changes. The module offers an approach to suit different dynamic changes and is more realistic as compared to approaches, which assume a fixed shop floor environment. The CAPP system will be able to provide quantitative comparisons between different generated process plans and helps to better evaluate manufacturing processes. Further research is conducted to explore other performance measures like costs and also the usage of this module on other more complex parts.

REFERENCES

1. Kayacan, MC, Filiz IH, Sönmez AI, Baykasoglu A, dereli T., "OPPS-ROT: An optimised process planning system for rotational parts", Computers in Industry, Vol. 32, pp. 181-95, 1996.
2. Pham D.T., P.T.N. Pham, "Artificial intelligence in engineering", International Journal of Machine Tools & Manufacture, Vol. 39, pp.937-949, 1999.
3. Leung Horris C., "Annotated Bibliography on Computer-Aided Process Planning", The International Journal of Manufacturing Technology", Vol.12, pp. 309-329, 1996.
4. József Váncza and András Márkus, "Experiments with the integration of reasoning, optimisation and generalisation in process planning", Advances in Engineering Software, Vol. 25, pp. 29-39, 1996.
5. Philip Husbands, Frank Mill and Stephen Warrington, "Generating optimal process plans from first principles", Expert Systems for management and Engineering, Chapter 8,1990.
6. Kiritsis, D. Neuendorf, K -P. Xirouchakis, P. "Petri net techniques for process planning cost estimation", *Advances in Engineering Software. v 30 n 6 Jun 1999. p 375-387*
7. Kyoung, Y M. Cho, K K. Jun, C S. ,"Optimal tool selection for pocket machining in process planning" *Computers & Industrial Engineering. v 33 n 3-4 Dec 1997. p 505-508*
8. Duflou, J. Kruth, J -P. Van Oudheusden, D. "Algorithms for the design verification and automatic process planning for bent sheet metal parts", *Cirp Annals. v 48 n 1 1999. p 405-408*
9. Duflou, J R. Van Oudheusden, D. Kruth, J -P. Cattrysse, D. ,"Methods for the sequencing of sheet metal bending operations", *International Journal of Production Research. v 37 n 14 1999. p 3185-3202*
10. F. Zhang, Y.F. Zhang, A.Y.C. Nee, "Using Genetic Algorithms in Process Planning for Job Shop Machining", IEEE Transactions on Evolutionary Computation, Vol.1 No.4, Nov. 1997.

CAPE/021/2000

A structured assembly planning tool using Intranet technology

H J REA, D LODGE, J L MURRAY, J E L SIMMONS
Dept of Mechanical and Chemical Engineering, Heriot-Watt University, UK

SYNOPSIS

Instructions on how to assemble a product, known as assembly plans, are traditionally presented to operators in the form of a paper document. The quality, clarity and format of this documentation rely largely on planning engineers' experience and expertise. Inconsistencies can occur from planner to planner or across project departments which effect product quality. This paper describes a computer assisted assembly plan generation system based on a novel structured approach capturing best practice. The system has been developed for initial application in the small batch, high technology, precision products sector (eg avionics, mechatronics, electronics). In this system, assembly plans are produced for Intranet display. On-site evaluations have taken place at three UK aerospace firms. The design and initial results of these evaluations are presented here.

1 INTRODUCTION

For manual assembly the assembly plan is the definitive means of communication between the planning engineer and individual operators on the shop floor. In small batch production, used in high precision engineering firms such as the avionics and aerospace industries, operators do not have the opportunity to repeat assembly tasks frequently, and hence learn the processes. Previously operators in these industries were highly skilled and specialised. Due to modern market variability, such specialisation is rarely viable and multiskilled work forces are now employed. In these situations it is important that the assembly instructions are clear, consistent and correct. To address these requirements a system to create and display assembly plans as Intranet documents has been developed (1). This paper focuses on the planner's interface to the system known as the GAP (Generate Assembly Plan) module.

The GAP module is based on the product tree structure (2). This interface promotes a top down approach to plan creation. A survey (3) has shown that the best assembly planning is done by experienced planners who: use a top-down approach; have a good understanding of the product hierarchy, and alternate between global and local detail. Research has concentrated on producing automated Computer Aided Assembly Planning (CAAP) tools (4-8), but it has been shown (3) that these approaches do not follow the best practice of experienced planners, and so may be of limited assistance.

It is a highly integrated system, based on the design product model and directly producing the assembly plan documentation on line. This integration will minimise the cost of updating assembly plans. The system is adapted for use by novice planners, by providing a strong initial structure, however the requirements of flexibility demanded by an experienced planner is also met. Prototype software of this system has been developed and evaluated on-site at the three participating manufacturing firms. The objectives and design of this evaluation, in addition to initial qualitative results obtained, are presented here.

2 A STRUCTURED APPROACH TO ASSEMBLY PLANNING

The specification for the overall Adaptive Planning and Process Learning Environment for Assembly (APPLE A) system has been described by Rea et al (1). The plan generation system described here forms the planning engineers' interface to the APPLE A system known as the Generate Assembly Plan (GAP) module.

The GAP module allows a planner to generate the plan structure and documentation simultaneously. It is recognised that in future the assembly plan structure may be provided by some other source such as a CAAP package. The assembly plan structure is a tree hierarchy based on the product hierarchy. The GAP module also acts as an integration package providing the planner with on-line access to: the bill of materials; product drawings; 3D models; component, tool and skills databases; text and image editing packages; standardised work instructions; and health and safety documentation. In addition to the actual assembly planning documentation, the final assembly plan structure holds extensive information about each assembly operation such as: the skills required; associated documentation; items lists; and tooling requirements. This information can then be compiled and accessed at a later date for, among other things, automated maintaining of the plans; resource planning; and skills matching.

2.1 Building an Assembly Plan Structure

The GAP module is a Windows based software package. To create a new assembly plan structure a planner must first access the product's bill of materials (BOM). This is displayed as a text list in the BOM Window as seen in Figure 1. A second window provides a canvas on which the planner can build a graphical hierarchy of the components and processes. This graphical hierarchy is the assembly plan structure. An example of a completed assembly plan structure is shown in Figure 2. On the canvas the planner is presented with an icon ( ) which represents the completed assembly, encouraging the planner to think top-down.

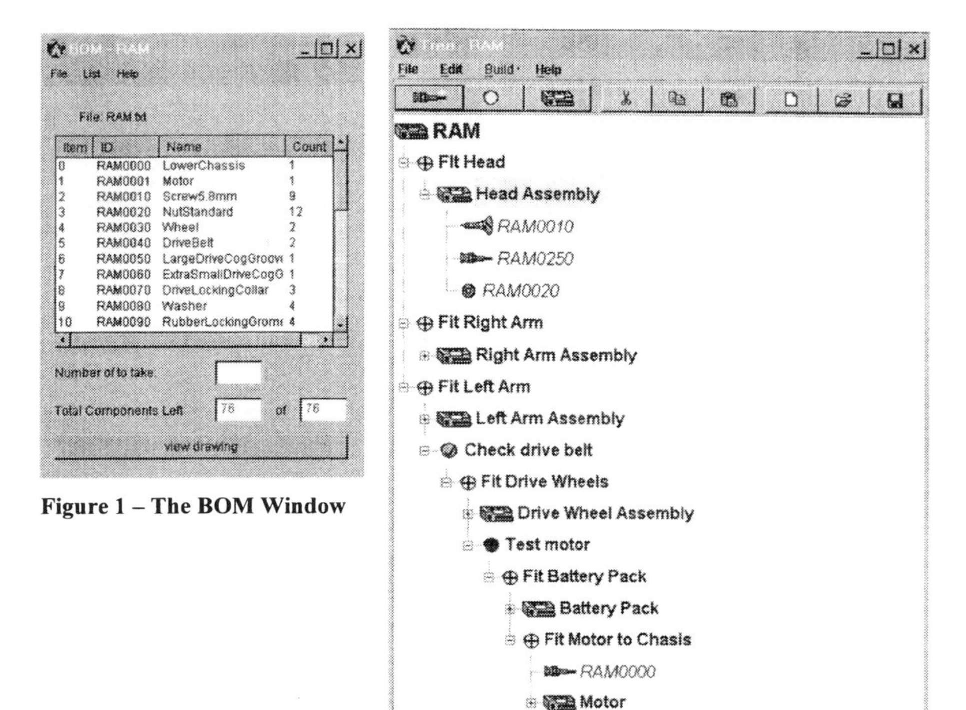

Figure 1 – The BOM Window

Figure 2 – Example Assembly Plan Structure

The planner can pick components in the BOM window and place them on the canvas. The components appear as icons (eg. , ,) with associated part numbers on the canvas, and are removed from the BOM list. Each component is represented as an object with certain features including links to drawings and models. In addition features such as relative cost (eg A, B or C items), weight, hazards, can be highlighted through a change in the icon colour. Any feature could be denoted this way to ensure the planner applies a company specific heuristic such as adding high cost or hazardous components as late as possible in the assembly sequence.

The planner can manipulate (drag and drop/cut and paste) the components on the canvas into operational subassemblies or manufacturing stages (also symbolised by). The next phase involves adding operations icons (eg. ⊕, ,) which represent the tasks required to incorporate the components into the assembly. Operations are used to link one or more components to a subassembly.

Operations are software objects that also have features, which include: sequence number; process description; process sheet (generally in the form of a web page); skills and tooling requirements; pertinent health and safety documents; and graphics. These features can be accessed, edited and in special circumstances created by the planner at any point during the plan development through the Operation Properties Window (see Figure 3). A library of company standard operations will be maintained, and here the planner may be required to add little or no information to these to create a process sheet. Depending on the application, it may be desirable to add automatically certain operations when an assembly plan structure is initially created, eg final test and quality assurance operations which may be required as standard.

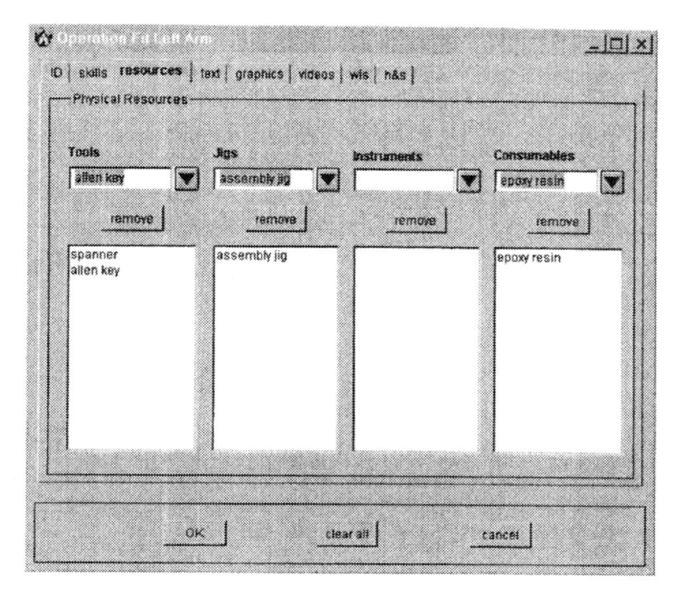

Figure 3 – A View of the Operation Properties Window

2.2 An Integrated Adaptive System

When the assembly plan structure has been completed it is a simple operation to publish the plan to an on-line document. The publishing tool allows version control to be maintained and also means that planners do not need to be trained in Intranet publishing technology.

The planning interface is closely integrated with the design model and the assembly plan documentation. Such integration will mean changes in design will only require minimum-planning time to update plans. Integration is further achieved by providing the planner with seamless interfaces to company data such as manufacturing resources (eg tooling, fixtures, machinery and skills base).

The product tree structure interface can be used to emphasise company specific heuristics making this an ideal tool for novice planners. However, the planner will have complete control over the plan layout, allowing experienced planners the freedom they may require.

3 ON-SITE EVALUATIONS

In order to validate the work of the project on-site evaluations were required to:

- make comparisons with the existing systems, particularly in terms of cost and quality;
- identify interface and functionality problems; and
- prove the system in a working environment with current and meaningful assemblies.

The evaluations were to be conducted at the three UK aerospace manufacturing firms collaborating with this project. In order to maintain commercial confidentiality these firms will be referred to as Companies A, B and C. The focus for this research has been based on the needs of the advanced, highly competitive, avionics and aerospace industry as exemplified by these three companies. The firms produce high specification, low volume, military and civil products. There is some variation in the nature of the products manufactured by the three firms. The assemblies at one firm (Company C) largely consist of high specification PCB's (printed circuit boards) which are assembled by a mix of automated, semi-automated and manual procedures. PCB's are inherently restricted to two dimensions, reducing the complexity of the task. The products at the other companies (A and B) are complex three-dimensional electromechanical/optical items (eg: ring-laser and mechanical gyroscopes; thermal, infrared and laser imaging devices) which typically can take several man-months to complete.

Data was to be gathered by allowing potential users (known as "participants") to interact with the APPLE A system. Although there was to be an emphasis on quantitative measures, qualitative data was also to be recorded. The evaluation test plan that was used is based on the test plan theory described by Rubin (9). The following sections detail the approach adopted for these evaluations.

3.1 Objective

The main objective of the evaluation was to determine what benefits the APPLE A system would provide to manual assembly planning, particularly in terms of cost and quality. Table 1 summarises these cost and quality issues for the APPLE A GAP module.

Table 1 – Cost and Quality Issues for the GAP module

	Measures
Cost	Training capabilities
	Reuse and adaptiveness of plans
	Time to produce plans
	Cost of paper used
	Time spent updating plans
Quality	Consistency (same plans produced each time)
	Clarity of plans

From this table the following questions about the APPLE A system were obtained. It is expected that the evaluations will provide answers to all of them.

- How does the APPLE A system compare in terms of learnabilty (i.e. time taken to reach a chosen level of proficiency) to the existing systems?
- How long does it take to produce plans/assemblies compared to the existing systems?
- Can time and therefore money be saved in the updating of existing plans?
- Are there differences in consistency, clarity and quality of the final plan?
- What level of cost savings will result with the use of less paper?
- Also, are there further cost savings to be made as a result of the reusability and adaptive features of the APPLE A system?
- How do users feel about the overall look and feel of the APPLE A interface, and what improvements can be made?
- Are there any areas of functionality missing from the APPLE A modules, which were not identified during previous evaluations or original specifications?

In addition to the above questions there is also the more general issue of what potential training capabilities might APPLE A provide.

3.2 Measures

The measurements that were collected during the evaluations can be split into two categories: performance measures and preference measures.

The performance measures include:

- time taken to reach a chosen level of proficiency during the training session;
- time taken to produce a plan;
- consistency of plans produced (number of differences between plans);
- number of sheets of paper used;
- time taken to update a published plan;
- number of sheets of paper used in updated plan; and
- time spent creating plans by using iteration and evolutionary methods,

Preference measures are:

- ease of building and modifying the tree;
- ease of working with the BOM;
- ease of operation data input; and
- clarity of terminology.

These measures were to be obtained by a number of means. Video recording of the users interaction with both their current system and the APPLE A system was to provide the time measures. Questionnaires and audio recordings were to be used to determine the preference measures.

3.3 Examples

Each of the three manufacturing companies kindly provided examples of typical product builds to test the APPLE A system. The choice of example was limited due to commercial sensitivity and the time available for the evaluation. The examples supplied varied considerably.

Company A was able to provide a live build example. This was a relatively uncomplicated piece, nevertheless some complications were encountered, as the drawings used were not complete.

Company B provided an archived example. This build was originally planned using hand drawn sketches. It was not ideal as no electronic form of the drawings or model were available for use in the APPLE A system. In addition, the time lapse meant part numbers had changed a number of times and cross-referencing proved difficult.

Company C also provided an archived example. It was more recent than company B's example, but more complex. Due to time constraints, a number of representative operations were created on both systems rather then a complete plan.

In order to familiarise the participants with the GAP module, training was to be provided using an example based on a Meccanno Robot Construction Set. The training sessions were limited to one and a half hours.

3.4 Participants

The system users were all experienced planners. Companies A and C provided two planners for the evaluation, each created a plan on both systems. Company B also provided two planners for the evaluation, however, due to the length of the example, one planner created a plan using the in-house system and the other created a plan with the APPLE A system. Each participant was asked to complete a background questionnaire, to provide information on their experience in planning and using windows software environment.

4 RESULTS

4.1 Performance Measures

The evaluation produced approximately 60 hours of video footage for analysis. Initial results show that in Company A, the time to create the plan on the APPLE A system is of the same order as their current system. A detailed analysis of the time and cost measures will be presented in future papers.

4.2 Preference Measures

In all cases the participants initially found the tree structure confusing. They all stated that they tend to think bottom-up, creating the assembly documentation in sequential order. The GAP module did allow them the freedom to build the plan in this manner, and none attempted a different approach. It was noted that the top-down approach was perhaps required earlier on, when designing for assembly. In all cases the participants indicated that the tree-structure was "understandable and easy to use once the logic was worked out".

All the participants found the BOM window "easy" or "very easy" to use, and all listed it as the most useful aspect of the APPLE A system. In Company A's example, the BOM window highlighted errors on the issued drawing, and reduced errors in the items list[†] provided for each operation.

Inputing data about each operation was considered to be relatively easy, although a view of the final plan page was desirable. This was not provided for on the prototype system, however could be included in a subsequent version. Some participants indicated that a wizard or standard path option would be useful for novice planners.

The terminology had been standardised for the project, and the participants found some of this confusing, as each company uses operation, process and stage to refer to different things.

In general, the participants found the system easy to use, except for the obvious limitations of the prototype system. Of the five participants who used the APPLE A system when asked if they would prefer to use the APPLE A system over their current system the response was:

- two disagreed;
- one was neutral;
- one agreed; and
- one strongly agreed.

The two that disagreed openly admitted their bias towards their current system due to their length of experience with it.

5 CONCLUSIONS AND FUTURE WORK

Considering the short training time (1.5 hours) and the limitations of the prototype system, the initial indication is that the cost of producing plans using the APPLE A system will be of the same order as the current systems used. It is anticipated that the assembly planning documents created by the APPLE A will also be considered to be of the same quality, or better, than those currently used. An additional output is obtained with the APPLE A system in the form of the assembly planning structure and operation properties within it. This capture and organisation of information is a valuable asset particularly for assembly plan maintenance. In addition, the APPLE A system, being an integrating tool, has the capability of dealing with evolving forms of multimedia data and software products.

The APPLE A system was well received by the participants of the evaluation. Although none of the participants created a plan top-down, this was due to their prevailing work practice of producing the plans bottom-up (or sequentially). It is foreseeable that the new system will prove increasingly valuable as users develop new ways of working to exploit the capabilities now available. A good parallel is the way in which word-processors now provide writers the freedom to create documents in any order they choose, and to work on different parts of the text at the same time. Furthermore sectors of text can easily be imported from other documents, moved around within the new document and modified at will. All this is in

[†] The items list is a table of each part used in an operation and cross references the part numbers to a localised item number, for Company A this would normally have to be typed in by hand

radical contrast to the sequential creation of documents and the difficulty of making amendments that applied when only typewriters were available.

The most widely received aspect of the GAP module was the BOM window, which actively reduced errors in:

- missed components; and
- miss-typed part numbers.

In summary initial analysis of the evaluation results indicate that APPLE A GAP module:

- promotes a top down approach to plan creation as it based on a tree structure;
- promotes clarity by use of multimedia information such as 2D and 3D graphics, animations and videos;
- promotes consistency by providing structured data presentation, and templates; and
- reduces errors by providing seamless interfaces with design and manufacturing databases.

Further work will include:

- a detailed analysis of the performance measures;
- further investigation into the benefits of using a top-down tree structure to prepare assembly plan documents; and finally
- the development of a commercial release of the APPLE A package.

ACKNOWLEDGEMENTS

This work has been developed as part of the Adaptive Process Planning and Learning Environment for Assembly (APPLE A) project. This project is funded by the UK Engineering and Physical Science Research Council (Grant GR/K92825) with support from: BAe Systems; TRW - Lucas Aerospace; Matra BAe Dynamics and Talkback Training.

REFERENCES

1. Rea, H.J., R.A. Falconer, J.L. Murray, and J.E.L. Simmons. *The requirement for an adaptive planning and process learning environment for assembly (APPLE A)*. in *The European Conference on Integration in Manufacturing*. 1997. Dresden Germany: Selbstverlag der Technischen Universitat.

2. Rea, H.J., R.A. Falconer, J.L. Murray, and J.E.L. Simmons. *A Structured Approach for Assembly Planning*. in *International Mechanical Engineering Congress and Exposition - Symposium on Assembly Modeling and Assembly Systems*. 1998. Anaheim, CA: American Society of Mechanical Engineers.

3. Ye, N. and D. Urzi, *Heuristic rules and strategies of assembly planning: experiment and implications in the design of assembly decision support system*. International Journal of Production Research, 1996. **34**(8): p. 2211-2228.

4. Delachambre, A., *Computer-aided Assembly Planning*. First ed. 1992, London: Chapman and Hall.

5. Laperriere, L. and H. ElMaraghy, *GAPP, A Generative Assembly Process Planner*. Journal of Manufacturing Systems, 1996. **15**(4): p. 282-293.

6. Liao, T., X. Wu, S. Zheng, and S. Li, *A computer-aided aircraft frame assembly planner*. Computers in Industry, 1995. **27**: p. 259-272.

7. Kim, G., S. Lee, and G. Bekey, *Interleaving Assembly Planning and Design*. IEEE Transactions on Robotics and Automation, 1996. **12**(2): p. 246-251.

8. Wright State University, *Computer Aided Assembly Planning System*, . 1998: Available: http://www.cs.wright.edu/research/caap/default.htm.

9. Rubin, J., *Handbook of Usability Testing: How to Plan, Design, and Conduct Effective Tests*. Wiley Technical Communication Library, ed. J.T. Hackos, W. Horton, and J. Redish. 1994, New York: John Wiley & Sons, Inc. 330.

CAPE/016/2000

PDM-integrated computer-aided assembly process planning

W BOWLAND and **J GAO**
School of Industrial and Manufacturing Science, Cranfield University, UK

ABSTRACT

Despite considerable research effort over the last two decades, assembly planning has not become widely accepted as a mature technology. This paper proposes a new approach to Computer-Aided Assembly Process Planning (CAAPP) which addresses some of the issues that have restricted industrial implementation of CAAPP. Primarily, the approach is based on the integration of CAAPP with a Product Data Management (PDM) system to provide a powerful data interface and a simple assembly model as input to the planning system. The idea of this is to simplify the operation of the planning system and provide a more integrated approach to CAAPP.

1. INTRODUCTION

The past ten or so years have seen a number of assembly planning systems developed as research projects. These have addressed some of the low-level problems of the assembly planning domain, such as assembly modelling and sequence planning. Several shortcomings are apparent from the research material generated over this time.

Firstly, the systems are developed as standalone applications. Integration with data management systems and downstream applications is not supported. One of the main drawbacks of this is that engineers involved with the assembly planning process still have to do the unnecessary and tedious work of preparing data for input to the planning system. Once planning is complete, the output from the system must also be stored and plan renditions disseminated to the relevant parties throughout the enterprise. In a contemporary engineering environment, levels of integration between data management systems and mature Computer-Aided Engineering (CAE) applications are so high that this kind of work is not acceptable to engineers.

Secondly, the assembly planning problem is recognised to be extremely complex. For this reason, simplifications that can be made without adversely affecting accuracy or productivity are highly desirable. The two main problems of complexity in the assembly planning domain can be considered as:

- The generation and verification of fully general vector trajectories.
- The simultaneous generation of all possible part configurations and the associated number of assembly paths through configuration graph; so-called combinatorial explosion.

One of the most common simplifications is the consideration of assembly trajectories only along the major axes. This simplification is very effective in allowing useful systems to be developed more easily. However, very few systems attempt to simplify the assembly planning problem by tackling the issue of combinatorial explosion.

Finally, most systems are overly sequence-driven in that they do not generate process or resource requirement information. An assembly plan cannot be considered merely as a chronological sequence of part names; this is at too lower level to be of any real benefit to industrial users. Process and resource selection in not only essential for full plan generation, it can also radically influence operation sequence. Therefore, such information must be considered before the sequence planning stage. Furthermore, if process and resource requirements are not derived as part of assembly planning, this information is unavailable for downstream applications. This implies that automation of assembly-related tasks (such as robot task planning) is not facilitated.

These problems have been limiting factors in the industrial acceptance of CAAPP as a mature technology. An approach limited by these issues is insufficient for contemporary engineering requirements:

- It lacks the formalism required for consistent and reliable operation in an integrated environment.
- There is little opportunity for the automation of data retrieval or storage routines.
- There is no easy method to manage the complexity of the assembly planning problem.
- Process and resource selection is incomplete.
- There is little opportunity for the automation of clerical or technical tasks downstream.

Assembly planning will remain a largely fringe engineering application domain if these problems are not addressed with simple yet effective solutions.

This work addresses these issues with the intention of developing the next generation in assembly planning systems. This is done primarily through a novel approach based on the integration of CAAPP with data management technology. This provides a consistent and intuitive data interface for all planning data transactions and allows formal security and tracing methodologies to be employed. Data management tools also provide a simple and effective method by which engineers can manage the complexity of the planning problem. This is done by using a simple, PDM-based hierarchical model to divide an assembly into configurations. To address issues of extensibility and generality, wherever possible the system should utilise external, user-configurable methods for reasoning in order to extend the life of the system under development. Any external resources, however, must be subject to PDM control to ensure the accuracy and repeatability of the planning process. Complementing these methods of formalisation and simplification with a CAAPP methodology that generates process-driven sequences of assembly offers considerable advantages over the current state-of-the-art in assembly planning.

RELATED WORK

Assembly planning, like many CAE domains, is a broad topic. A solution to the CAAPP problem is likely to embrace several technologies. The following section provides an overview of some important recent work in the CAAPP field. Pham and Dimov (1) describe a method for the extraction of feature-based assembly information from CAD models. This information is transferred to an object-oriented database for knowledge-based and geometrical analysis. The class-and-object structure of the system is simple, classifying the objects main assembly, subassemblies, parts and features. Library information about mating features is stored to characterise various families of feature type. The authors claim that the extraction and storage of the above data facilitates downstream planning and control applications.

Wang and Kim (2) discuss an approach for the extraction of mating relations between a set of polyhedral parts based on the Alternating Sum of Volumes with Partitioning (ASVP) method of feature decomposition. ASVP is a complex but powerful method of feature extraction from the boundary faces of solid models. This work considers only containment mating, where a positive feature is contained either partially or wholly by a negative feature. Considering liaisons of this type only allows the system to infer the mating features existing between a pair of related components. The system then finds all of the possible configurations given by the mating conditions of the components.

Ong and Wong (3) concentrate on the identification of subassemblies from a product model. These subassemblies are then treated as components in order to reduce the complexity of the assembly planning problem. They employ a multiple matrix representation to provide an assembly model where the system utilises an interference matrix and a connectivity matrix (the methodology of Dini and Santochi (4)). The interference matrix encodes which components block extraction of a particular part along possible assembly trajectories. The connectivity matrix simply encodes whether a pair of components is in physical liaison. Identification of possible subassemblies allows contraction of the matrices in order to simplify the sequence planning process.

You and Chiu (5) discuss a method of extending the methodology of Lee and Gossard (6) by providing knowledge-based support and utilisation of standard parts. Expert modules implemented by You and Chiu include a tolerance checker, a bolt analyst and a key analyst. The bolt analyst is particularly interesting, representing standard bolts as features rather than components. An expert system ensures that the bolt is used correctly and any necessary nuts or washers are present in the structure. Interference checks are conducted using a bounding block approach. Dimensional errors in the structure, such as a shaft dimension not matched to its associated hole, are also highlighted.

Tran and Grewal (7) take a different approach than many, opting for interactive generation of a sequence of work element rather than automated sequence generation. Their system provides task planning rather than simple sequence planning. An object-oriented model of assembly entities provides the core of the application, characterising objects such as assembly, part, task sequence, task and assembly machine and tool. The modelling of enterprise resources has been a major ingredient of Computer-Aided Process Planning (CAPP) since its conception, yet is widely ignored in CAAPP. The application assists engineers in the interactive creation of a task sequence of assembly, considering aspects such as handling analysis and time estimation.

3. SYSTEM PROPOSAL

This section details the methods by which the proposed extensions to traditional CAAPP technology are implemented. The workings of the CAAPP system as a whole are also discussed, such as kinematic pair and mating surface definition. The system runs under Windows NT and is facilitated by component-based integration of CAAPP with a PDM system (Motiva DesignGroup), a solid modelling tool (ACIS BuildingBlox) and diagram-authoring tool (AddFlow). Knowledge-based support is provided by the utilisation of the CLIPS expert system. Relational database technology is facilitated through integration with Access. The system itself is implemented in Visual Basic and C\C++. These tools provide the necessary power and flexibility to implement the advanced methodologies discussed in the Introduction.

3.1 Integration of CAAPP and PDM technologies

A seamless integration between CAAPP and PDM offers the formal and consistent data control methods that are a requirement in the contemporary engineering environment. Firstly, providing access to the PDM data vault directly from the CAAPP application will allow data transactions to be conducted in a controlled and traceable fashion. The benefits offered by this integration can be summarised as:

- The PDM configuration is set-up to maintain the integrity of the assembly model under incomplete definition or redefinition.
- The CAAPP application runs in partnership with the PDM system, therefore, administrative tasks such as launching workflow processes can be automatically conducted upon specific document actions or on demand.
- The PDM system can be used to specify what happens to document types (hence, different types of entity used in assembly planning) and its references under certain conditions, such as check-out and release. This can be used to help user retrieve data correctly.

These advantages are inherent in PDM usage; however, a fully integrated system becomes in many ways like part of the PDM system. The deeper the integration the easier the subsystems become to operate and implement. This improves the accuracy and repeatability of the planning process.

Secondly, because of the type of integration utilised (i.e. component-based integration), the data access methods used within the CAAPP application are actually part of the PDM system development kit. Therefore, the interface presented during data retrieval and storage is consistent with other applications in an integrated environment. These data retrieval methods also provide search routines (simple, property and contents searching) to facilitate location of the required documents.

3.2 Development of PDM-based assembly model for CAAPP support

An enterprise typically uses a PDM system as the centre of its information technology implementation. Therefore, the PDM system itself offers the possibility of supporting the CAAPP process by providing a suitable assembly model. This model is not intended to facilitate assembly planning but is used as an input the planning system proper.

DesignGroup allows the creation of a suitable assembly model using document types (to characterise the various entities used in assembly planning) and relationships (to define a simple hierarchical assembly structure for further analysis). Using this model as input to the CAAPP application, rather than a list of the constituent components, provides an improved starting point for planning and offers a simple and intuitive method of complexity management.

Document types are used to characterise the various types of entity used in assembly planning. Three document types are utilised:

- A **configuration** document is used to divide an assembly into smaller configurations. Each configuration has an individual plan. The configuration level plans are then merged to produce a plan for the entire assembly. A configuration document is associated to a database file in order to facilitate saving of incomplete structure definitions. The database file contains the information necessary to re-spawn the object model created during planning.

- A **component** document type is associated to the solid model of each part in the assembly. The solid model is used to define surface relations between kinematic pairs. Surface mating features are saved as attributes to geometrical entities in the solid model file.

- A **fastener** document type is used to characterise the operational parameters of the fasteners (both mechanical and non-mechanical) present in an assembly. Many CAAPP systems do not differentiate between components and fasteners. This approach is flawed in several reasons. Firstly, detailed component solid model data is useful in determining precedence relations , whereas such information concerning fasteners is not required. For example, a screw presents rather complex geometry that does not aid reasoning and can dramatically effect the speed of system algorithms. Secondly, enterprises primarily concerned with assembly are unlikely to manufacture their own fasteners, implying that solid model data may not be readily available. A fastener can be more effectively characterised by its parameters, such as thread classification or maximum tightening torque, rather than by their geometry. Therefore, providing suitable data for calculation or knowledge-based processing rather than solid model data is a more satisfactory method.

Relationships are then set-up to associate these different document types to each other, i.e. a configuration to a lower-level configuration, a component and a fastener. Different relationships are required as the handling method for each data type is different. For example, mating feature information is saved to the solid model files so a component document must be checked-out. Fastener information, however, is standard and should not be modified during planning. This information should be opened in read-only mode (copied-out), therefore.

As shown in figure 1, the document types and relationships discussed are sufficient to create an hierarchical model of an assembly. Using this method allows the CAAPP system to collect the required data in the correct manner to satisfy planning data needs simply by the user checking-out a configuration document. The extensions made to the assembly model (which could be integrated with minimum disruption to an existing PDM configuration) provide an advanced input to the CAAPP application itself.

3.3 Development of liaison definition tool

Liaison definition is an essential aspect of assembly planning. Without the data created at this stage, the planning activity would be impossible. The liaison definition tool should take the assembly model from the PDM system as its input and aid the user in definition of kinematic pairs, fastener details and surface contact data. Note that in this work, "fastener" means not only mechanical fasteners such as nuts, bolts and associated items but also non-mechanical methods of fastening such as adhesive or sealant.

Firstly, the user is presented with a sketchpad (actually an AddFlow window) and a toolbox containing all of the sub-configurations, parts and fasteners that a configuration holds. User-friendly interface dynamics aid the creation of a graph-based assembly structure showing components as nodes and liaisons between components as arcs. Fastener details may then be added to a liaison, shown as a block on the liaison arc. Ordered positions of mechanical fasteners

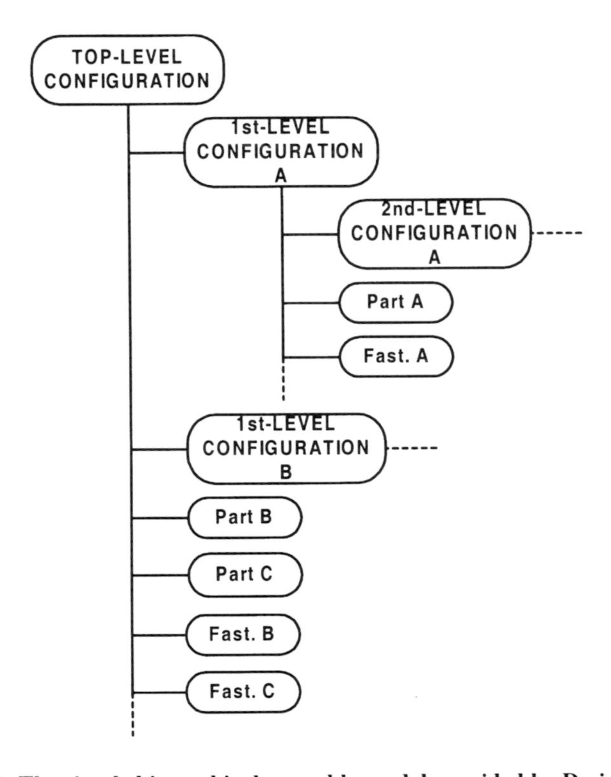

Figure 1: The simple hierarchical assembly model provided by DesignGroup.

(washers or spacers, for example) and ordered or delayed tightening of arrays of fasteners are supported to obtain accurate and repeatable plans. It should be noted that the structure graph is actually a representation of the underlying data structure (see 4. System Operation) in its current state.

Secondly, surface contact data should be specified using solid models of the components in a kinematic pair. The system should use the data generated at this stage to derive the local freedom matrices of the parts under positive and negative translations (of user-defined magnitude, δ) along and rotations (of user-defined angularity, ω) about the major axes (12 half-degrees of freedom). Thus the local freedom matrix, $\mathbf{F}_{1,2}$, is given by:

$$\mathbf{F}_{1,2} = \begin{bmatrix} \delta x & \overline{\delta x} & \omega x & \overline{\omega x} \\ \delta y & \overline{\delta y} & \omega y & \overline{\omega y} \\ \delta z & \overline{\delta z} & \omega z & \overline{\omega z} \end{bmatrix}$$

It is clear that the matrix, $\mathbf{F}_{2,1}$, can easily be derived from $\mathbf{F}_{1,2}$. One matrix of this form is generated and stored for each kinematic pair in the assembly structure. Each element in the matrix is given a binary value to indicate freedom (1) or constraint (0). The local freedom of a component, with respect to all of its relations, can be derived by utilising simple Boolean algebra on the respective matrices:

$$\mathbf{F}_1 = \mathbf{F}_{1,a} \wedge \mathbf{F}_{1,b} \wedge \ldots \wedge \mathbf{F}_{1,n}$$

$\mathbf{F}_1$ is termed the Resultant Local Freedom Matrix (RLFM). This resource of matrix data can be used to define precedence relations among the parts of an assembly during sequence planning. Simple reasoning methods, powered by geometrical manipulation of the components in the solid modeller, are used to fill each matrix. The manipulations and Boolean volumetric operations required do not take excessive amounts of time (as they may have done in the past) and so provide an effective and simple method to generate the necessary data.

If it is to be removed from an assembly, a component is also required to be globally free. A component is globally free if there exists a clear and feasible path from its initial position to a point well clear of the remaining parts. The system batch processes the structure for global freedom once the assembly model is fully defined by taking each component in turn and checking along its locally free trajectories. Any parts that are found to block motion along a locally free trajectory are stored. This resource can then be checked during sequence generation (which uses the assembly-by-disassembly strategy) to see if blocking parts are currently present in the structure. If a blocking part is present, it is clear that the component under consideration is not globally free.

4. SYSTEM OPERATION

Once the PDM-based model has been defined, the first stage of planning is to retrieve the required information from the PDM system's data vault. This can be done from within either the PDM system or the CAAPP application. The system uses the simple model created in the PDM system to retrieve the configuration references in either read-only or modify modes, depending on the data type. The retrieved data is used to build a toolbox containing the relevant parts, fasteners and sub-configurations used in the configuration. The toolbox is the main portal to the functional aspects of the system but also allows users to preview component solid models or view fastener data.

The core of the application is an object model (as shown in figure 2) which stores the assembly as a collection of component objects and a collection of link objects. Adding a component to the assembly model via the toolbox is equivalent to adding a component object to the object model. The component class holds information such as the DesignGroup identification string of the associated document, the transformation required to position it correctly in the 3D-space of the structure and its current RLFM. The link class holds information such as linked components, fastener specifications and the local freedom matrix given by the liaison. The classes to which the component and fastener objects belong handle any operations required, such as drawing a node or arc in the structure graph, showing the solid model or dumping themselves to relational a database. This keeps the code of the application itself relatively simple.

Once a liaison is completely defined (i.e. a link, fastener details and the surface relations have been entered), the liaison manager is used to derive information required for assembly planning.

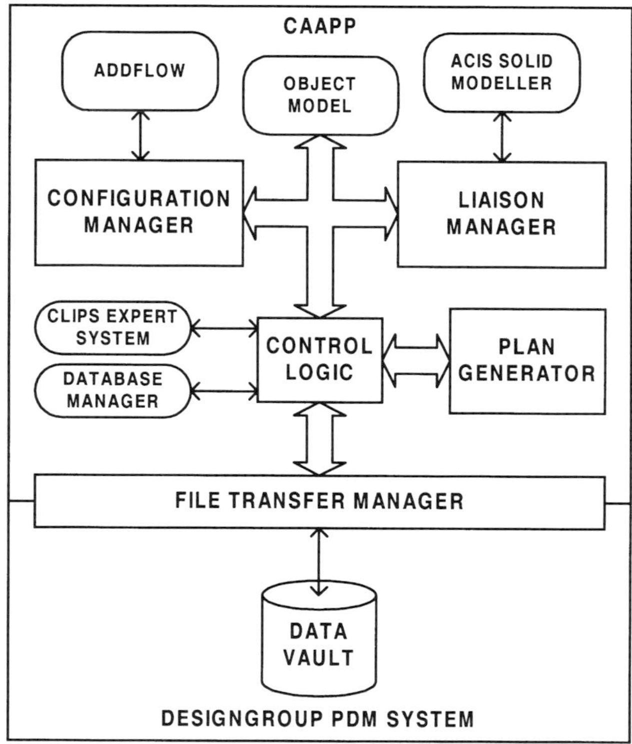

Figure 2: An overview of the system.

The mating expert uses geometrical reasoning and a combination of editable logic flowcharts and expert systems (8, 9) to extract the local freedom matrix, process requirements and the special information (such as sensitivity to physical conditions or stability under gravity). This data is added to the object model.

Once all the liaisons in a configuration have been defined, the sequence generator can batch process the global freedom of each part and begin to derive a process sequence from the assembly model (see figure 3). This is done by using the assembly-by-disassembly strategy, whereby the system examines the structure in its assembled configuration and attempts to remove components one by one. At each step, the system re-evaluates the RLFM of each component. A list of candidate components is compiled based on non-zero elements in the RLFM. The links broken by the removal of a candidate part are compiled and the associated tasks analysed by an expert system for feasibility and saliency. The global freedom of candidate components is also checked. Unsuitable candidates are discarded. Candidates that survive the review process are considered suitable for disassembly. The core of the plan is a list of links removed at each stage of disassembly. Multiple candidate selection is supported by the use of multiple lists. The system considers each list sequence separately, in a recursive manner. When one sequence is complete, the system backtracks to consider the other possible candidates for removal and refills the assembly model to the correct state.

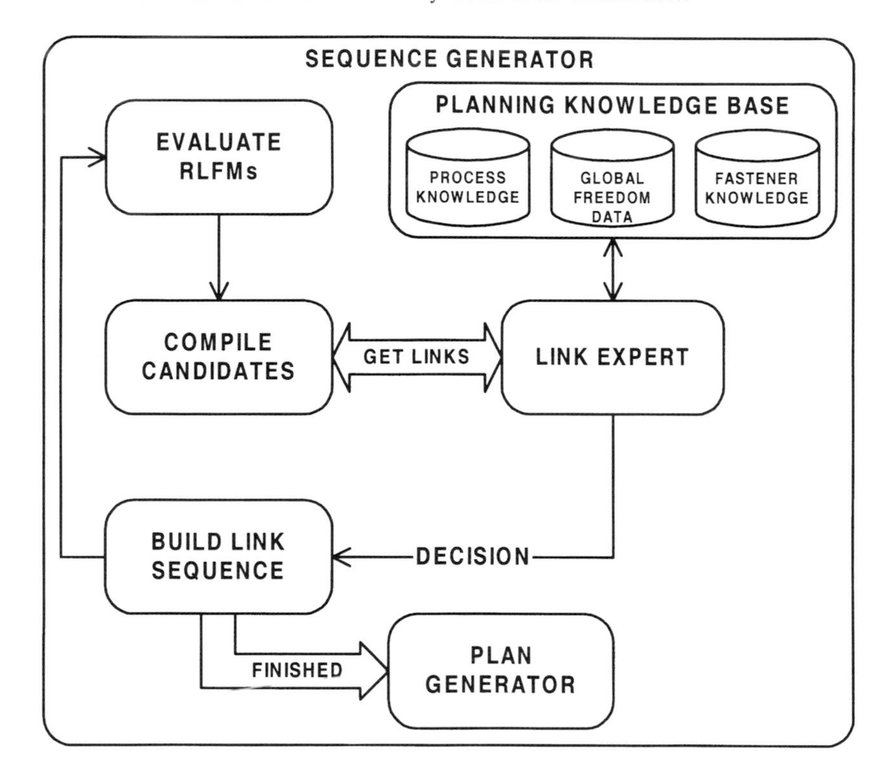

Figure 3: An overview of the sequence generator.

Once complete, the sequence lists are prepared for visualisation using a variation of the AND/OR graph (10) and the task details are extracted. Relevant data is easy to extract as the sequence is a list of links to be broken rather than components to be removed. The sequence graph provides an excellent method of plan navigation, the user can pull up task details from the links in the graph and view configuration status as a solid model.

CONCLUSIONS

This paper describes a novel method of assembly planning, based on the integration of CAAPP and data management tools to generate a selection of process sequences for assembly. No evaluation of the sequences is to be considered in this work, their goodness is governed by the quality of the knowledge used in the reasoning processes. A broad aim of this work is to produce an assembly plan that provides sufficient information for downstream applications to work on. Most assembly planning systems, particularly those providing automated sequence planning, do not provide information at a suitably high level to meet this aim

The system is still currently under development and relevant knowledge being compiled. A case study based on assembly structures obtained from the sponsor companies is to be completed post haste in order to evaluate the effectiveness of this approach.

ACKNOWLEDGEMENTS

This research is supported in part by EPSRC Research Grant GR/L66236. The support of Matra-British Aerospace Dynamics and Meritor Light Vehicle Systems is also greatly appreciated. Provision of Motiva DesignGroup by Photonic EDC is acknowledged with thanks. Individual thanks also go to Alan Crawford and Ray Muldoon of LSC Group and Andy Gordon of Photonic EDC.

REFERENCES

(1) Pham, D.T., & S.S. Dimov. "A system for automatic extraction of feature-based assembly information." *Journal of Engineering Manufacture*, 1999, Vol. 213, Part B, pp 97-101.
(2) Wang, E., & Y.S. Kim. "Feature-based assembly mating reasoning." *Journal of Manufacturing Systems*, 1999, Vol. 18, No. 3, pp 187-202.
(3) Ong, N.S., & Y.C. Wong. "Automatic subassembly detection from a product model for disassembly sequence generation." *International Journal of Advanced Manufacturing Technology*, 1999, Vol. 15, pp. 425-431.
(4) Dini, G, & M. Santochi. "Automated sequencing and subassembly detection in assembly planning." *Annals of the CIRP*, 1992, Vol. 41, No. 1, pp 1-4.
(5) You, C.-F, & C.-C. Chiu. "An automated assembly environment in feature-based design." *International Journal of Advanced Manufacturing Technology*, 1996, Vol. 12, pp. 280-287.
(6) Lee, K, & D.C. Gossard. "A hierarchical data structure for representing assemblies: part1." *Computer-Aided Design*, Vol.17, No. 1, pp 15-19.
(7) Tran, P., & S. Grewal. "A data model for an assembly planning software system." *Computer Integrated Manufacturing Systems*, 1998, Vol. 10, No. 4, pp 265-275.
(8) Bowland, N.W., J.X. Gao & R. Crawford. "Embedded knowledge-based functionality for process planning software." *Second International Workshop on Intelligent Manufacturing Systems 1999 (IMS 1999)*, September 1999.
(9) Sharma, R., J.X. Gao & N.W. Bowland. "A PDM-based progressive design and manufacturing evaluation system." To appear, *International Manufacturing Conference of China (IMCC 2000)*, August 2000.
(10) Homem de Mello, L., & A. Sanderson. "AND/OR represenation of assembly plans." *IEEE Transactions on Robotics and Automation*, 1990, Vol. 6, No. 2, pp 188-199.

Forming Processes

CAPE/022/2000

Computer-aided hot forging sequence design for non-axisymmetric products by eliminating form features on series of cross sections

T OHASHI and **M MOTOMURA**
Dept of Mechanical System Engineering, Gunma University, Japan

SYNOPSIS

The authors develop prototype computer-aided process planning system for non-axisymmetric hot forging products. The system manages the shape of a forging product in two ways, rough shape described by three dimensional primitive shape units and precise real shape described by series of cross sections. The system designs forging processes and transit shapes of the preforms required for the each process from the shape of the product to a raw material as the inverse of manufacturing. At first., the system designs rough process plan using primitive shape units. The system estimates basic strategy of forming, i.e. forming method, forming direction, volume loss and phase of forging. The second, system designs cross sections of the preform with regarding for the above strategy. The design of the cross section is done by 'feature elimination.' Thus, using two different knowledge-based programs applied for different shape expression cooperatively, the system designs process plan of three dimensional non-axisymmetric hot forging products.

1 INTRODUCTION

In this paper, authors report prototype computer-aided process planning system for three dimensional non-axisymmetric hot forging product. Computer-aided forging process planning systems have been studied and developed prosperously, however, most of them can treat only simple shape, for example axisymmetric products. Authors have developed the system for non-axisymmetric products which are expressed by combination of primitive shapes[1], however its shape representation method is rough in comparison with real forging products.

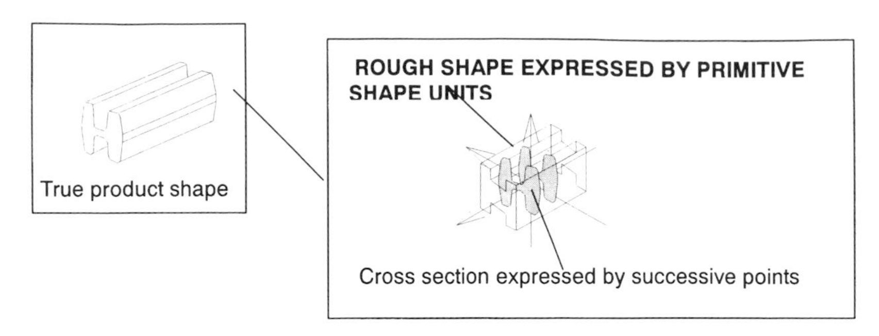

Fig.1 Two way shape representation

Improved system in this paper manages the shape of a forging product in two ways, rough shape described by three dimensional primitive shape units and precise real shape described by series of cross sections. In addition, authors develop two different knowledge-based sub programs applied for the above two different shape expressions. Using the two sub programs cooperatively, the system designs process plan of three dimensional non-axisymmetric hot forging products.

2 SHAPE REPRESENTATION FOR FORGING PRODUCTS

2.1 Two way shape representation

To treat forging shape by knowledge based techniques, we must symbolize it. However it is difficult to describe three dimensional shape freely only by symbols. So, it is efficient to find convenient way for knowledge-based approach, of which capability for shape expression is even a little limited. In this paper, the system manages the shape of a forging product in two ways, rough shape described by three dimensional primitive shape units and precise real shape described by series of cross sections. It is shown in **Fig.1**. Former way has been used in our developed system for non-axisymmetric hot forging products[1].It is convenient to deal three dimensional shape by symbols, but rough expression for real forging products. Latter way also has been used in our developed system for axisymmetric forging products[2]. It can express precise shape of each cross section using counter-clockwise series of points, but the system cannot treat the shape over sections. It is our aim of this paper to combine them.

2.2 Shape representation using primitive shapes

In the system, the rough shape of the product is represented by using primitive shapes such as shown in **Fig.2**. The primitive shape has its axis and two connecting points, start and end point, at both ends of the axis, which is prepared by the system designer. The product shape is built by tying multiple primitive shapes at the connecting points. Peculiar primitives that have no geometric substance are thus prepared as shown in **Fig.3**. Vector primitives are primitive shapes that have only an axis and two connecting points. They are used for moving connecting points. Connecting primitives are shapes having only multiple connecting points. They are used to connect multiple primitive shapes at the same connecting point. The user builds the shape of the forging product by using a dedicated shape editor for the system. The editor automatically connects primitive shapes while the user operates the system.

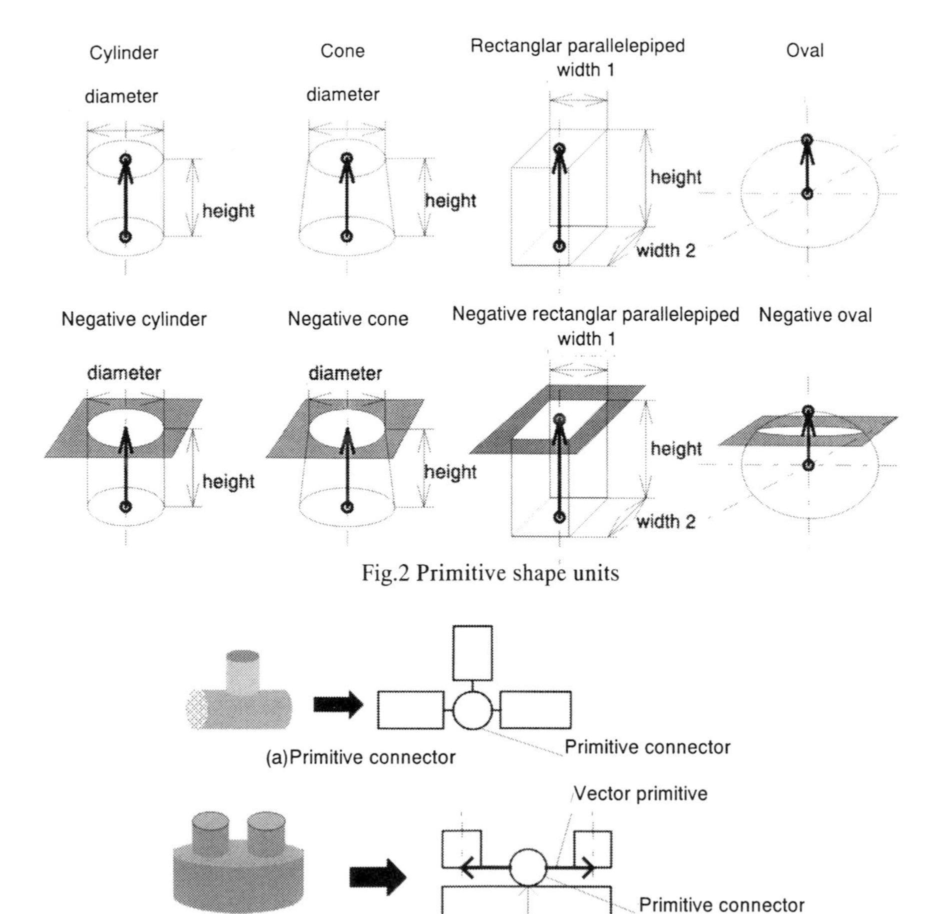

Fig.2 Primitive shape units

Fig.3 Peculiar primitive shapes

2.3 Shape representation using series of cross sections

A forging product is also given by series of the free curved cross sections drawn by counter-clockwise successive points (**Fig. 4**). The system then draws a vector line tying points with the neighbour. When the change of the angle of the vector from a neighbour exceeds 180 degrees, the system marks it with '-.' If the angle is equal to 180 degrees, the system marks with '0,' and if the angle is less than 180 degrees, the system marks it with '+.' In the above, the system produces an angle symbol line consisting of symbols '+,' '-,' and '0.' When the system finds consecutive identical symbols in the line, it combines them into one segment symbol (**Fig. 5**). These marks are prepared for feature extraction in process planning by 'feature elimination.'

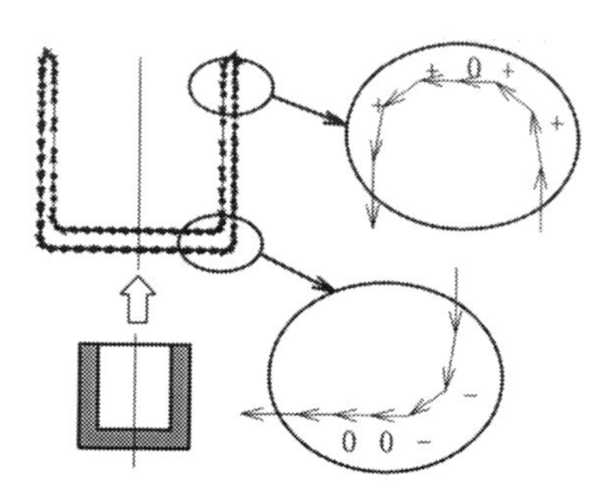

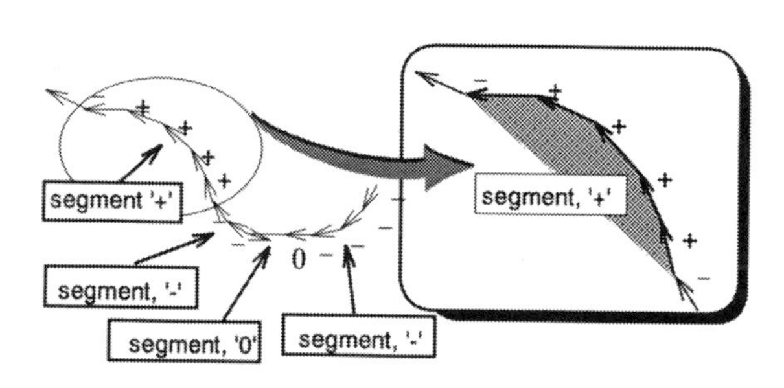

Fig.5 Example of the segment symbols and their line. In this case, there is four segments and their segment symbol line is expressed as (...,-,0,-,+,...).

3 TWO SUB PROGRAMS FOR PROCESS PLANNING

The system plans rough forming plan including shape of preform at first with using primitive shapes. Then it designs each cross section of the preform with using the result of rough planning and cross section data of postform. Each knowledge-based sub program of process planning employed for each shape representation method are developed . **Figure 6** shows summarise of cooperative action of the two sub programs. Each detail is described as followings.

3.1 Sub program for rough process planning using primitive shapes

The authors consider a forging process to be a complex of simple virtual processes in the CAPP system. This is illustrated in **Fig. 7**. In this case, a real bending process consists of two virtual processes. We can say that the virtual processes exist only in the forge expert's mind. Based on the virtual processes, the sub program designs a plan for the real process. The authors call database of virtual processes the "processing case base" and each case data the "processing case." The sub program uses the "processing case base" to design a virtual process plan such as shown in **Fig.8**. The program designs a rough preform from a rough

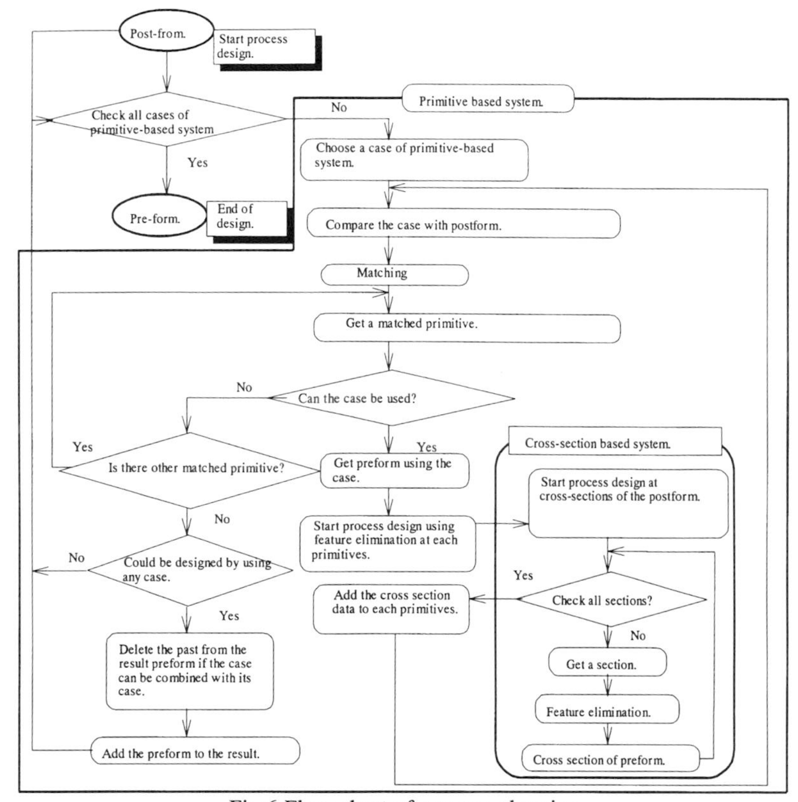

Fig.6 Flow chart of process planning

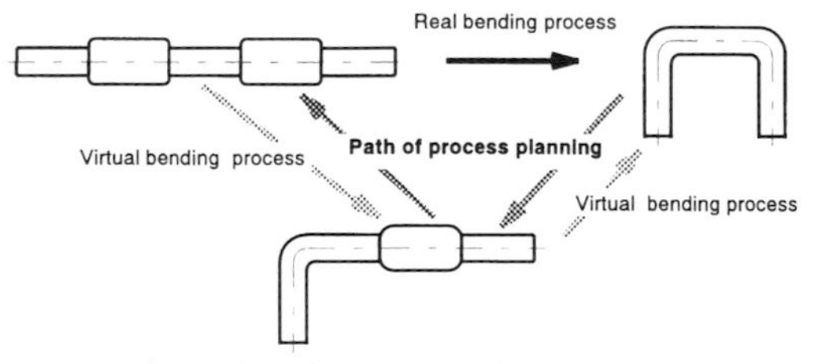

Fig.7 Virtual process and real process

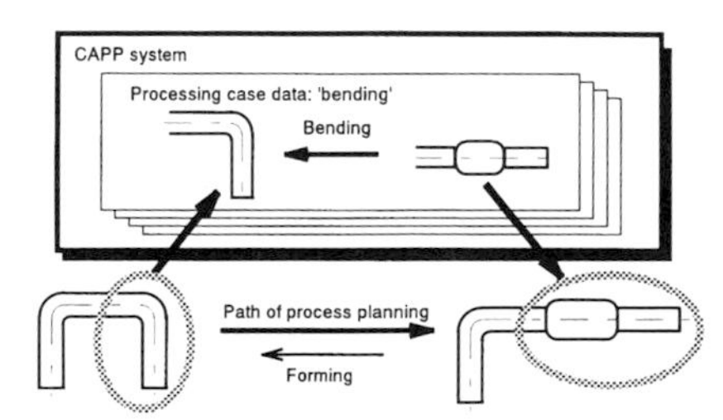

Fig.8 Process planning using processing case

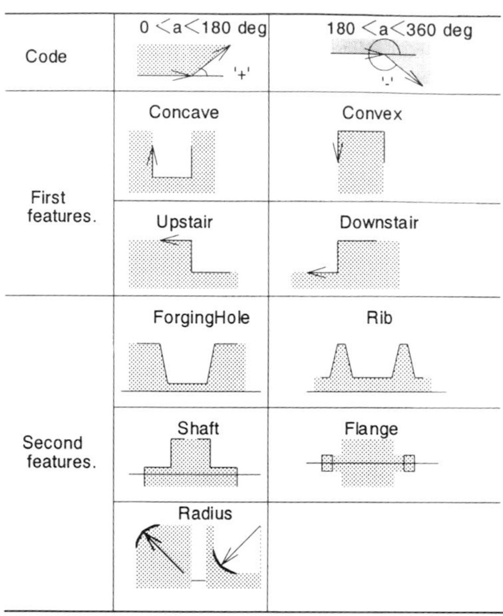

Fig.9 Features for cross section of connecting rods

postform in each virtual process using an inverse forming procedure. The processing case is implemented as object oriented data of smalltalk language, which contains the following data..

(1)Preform shape as a search key.
(2)Rules to design partial preform shape.
(3)Rules to check if the case can be applied or not.
(4)Rules to write constraint information to the rules of other virtual processes on the 'constraint blackboard.'

First, when the program starts the design process of rough process planning, it scans processing cases in the processing case base having the same preform shape as a part of the product's shape and fork thread at every hit.
Second, the program checks one of the cases by using the rules described above (3) referring to the "constraint blackboard." The "constraint blackboard" is a common working memory which can be accessed by the knowledge sources described in processing cases. Via constraint blackboard, virtual processes exchanges information and restrict each other to design a real forming process cooperatively. The rules (3) combine the limit of forming, geometric limit, and restrictions to other virtual processes in continual thread. Third, the program obtains a rough preform shape by replacing a part of the preform shape using the rules (2). Finally, the program adds information to the constraint blackboard using the rules (4). The information is represented by symbols, strings and numerals. Every rule is executed independently, and each virtual process is designed independently. Once one virtual process is designed, the planning thread is forked and the other virtual process is tried. The program designs several realizable plans concurrently. In a branch of planning, rules to design the other virtual process are restricted by the information on the constraint blackboard. The information on processing cases is created by the rules and used for planning other virtual processes. Thus, virtual

processes are designed cooperatively, and the processes which cannot be combined are automatically prevented from being planned in the same planning branch.
The program continues to assemble virtual processes into one real process until it fails to find a virtual process which is consistent with past fixed virtual processes in the branch.

3.2 Sub program for design of cross sections of preform using feature elimination

3.2.1 Feature extraction

After finishing planning of rough single process by the above sub program, the system designs cross section of preform using another sub program. At first, the sub program extract form features from data of the cross section. The program searches the feature database for a feature having the same segment-symbol line as a part of the symbol line of the section. If the system finds the data, it extracts a feature using it. Features that the system extracts are shown in **Fig.9**. The feature data set consists of following four parts:

(1) Symbol line composing the feature.
   This part of the data set is used as the searching key.
(2) Name of the feature.
   This just represents the name of the feature.
(3) Procedure to check whether or not it is able to apply itself.
   When the system finds a feature from the database using the symbol line, it checks whether it is able to apply the feature by running this procedure. This procedure checks conformity by using Euclid's distance[3] and checks for geometrical contradiction between the axis or other features.
(4) Procedure to fix the scope of the feature in the line
   This procedure is used to fix the scope of the feature in the product shape.

This data set is prepared in the form of object-oriented data, and all the above data and procedures are built into the set. The sub program first runs checking procedure (c). If this check is satisfactory, the program creates a new segment symbol list by replacing part of the segment symbol list with the detected feature name. Finally, it enters the information of the extracted feature (i.e., the feature's name and scope on the point list of the product) into the feature pool using data (b) and procedure (d). The program extracts features by repeating this process for all the candidate feature data and combinations of segment symbols. The system does not use only symbols (i.e. '+,' '-,' and '0') for extraction but also feature names. Features extracted from first generation symbol list are called "first features," and those from the second and third generations, "second features" and "third features."

3.2.2 Feature elimination by using database of processing cases

Features are eliminated using a database of processing cases. Note that process planning can not be performed by just eliminating features geometrically. The program must eliminate features from the manufactured shape by using an "actual manufacturing process" to plan a process. The program has a database of such "actual manufacturing processes," and uses it to eliminate features from a product until it finally obtains the raw material. We call these data 'processing (or eliminating) cases for feature elimination.' Its basic idea is same as the processing case in **chapter 3.1**. In this case, the processing case is the data set which ensures the elimination relates to the actual manufacturing process. It eliminates features in the

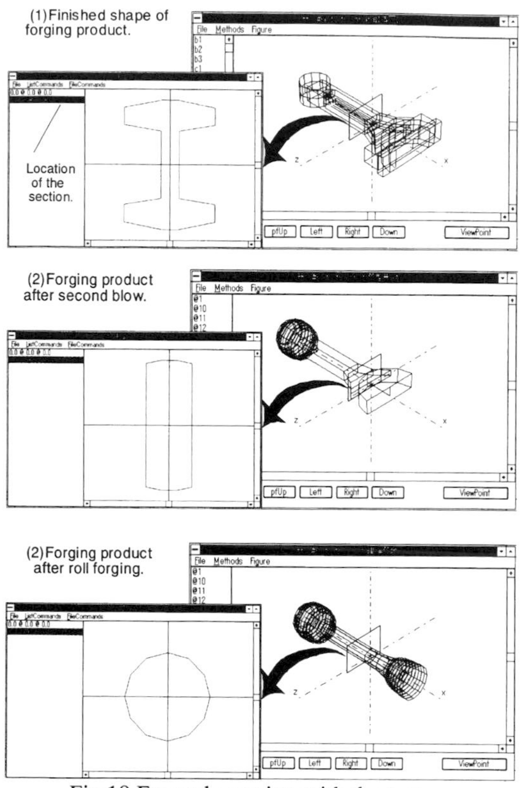

Fig.10 Example session with the system

manner of an actual forging process. The processing case for feature elimination also consists of the following four parts:

(1)Feature which can be eliminated by the case(; search key).
(2)Rules to eliminate feature.
(3)Rules to check if the case can be applied or not.
(4)Rules to write constraint information to the rules of other virtual processes on the 'constraint blackboard.'

In an actual manufacturing process, several features are produced in a time. This means that the system must combine the elimination of several features. We use the algorithm using a 'constraint-blackboard' for this purpose as same as **chapter 3.1**. Elimination of each feature is considered as the inverse of a virtual manufacturing process and corresponds to one 'processing case.' An actual manufacturing process is composed of these virtual processes.

4 EXAMPLE PROCESS PLANNING

Figure 10 shows example session of process planning for a connecting rod. In the figure, right windows shows result of rough process planning with using primitive shape and left windows

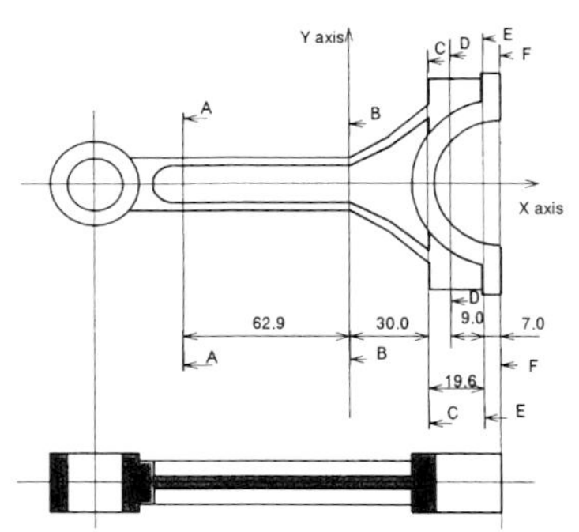

Fig.11 Example forging product

shows the result at a cross section. In this case, the cross section locates at I bar. List pane of the left window, i.e. '0.0@0.0@0.0,' indicates the location of cross section in the local coordinate of the primitive shape User clicks one of them to select and browse the cross section data. Selected one is inversed and the profile of corresponding cross section is displayed in a left pane of the window. **Figure 12** shows the result of process planning at each cross section for the forging product shown in **Fig.11**.

5 CONCLUSION

The authors develop prototype computer-aided process planning system for non-axisymmetric hot forging products. The system manages the shape of a forging product in two ways, rough shape described by three dimensional primitive shape units and precise real shape described by series of cross sections. The system designs forging processes and transit shapes of the preforms required for the each process from the shape of the product to a raw material as the inverse of manufacturing. At first., the system designs rough process plan using primitive shape units. The system estimates basic strategy of forming, i.e. forming method, forming direction and phase of forging, this time and predicts volume loss of material. The system has database of forming process of which search key is described by primitive shape units, and applies it to create the preform. The second, system designs cross sections of the preform with regarding for the above strategy. The design of the cross section is done by 'feature elimination.' The author think forging is the procedure adding form features to a raw material, and think process planning is the inverse procedure, eliminating form features from the product. The system has database of cases of the forming processes having search key described by the form features which the process can eliminate. Candidate cases are restricted by forming strategy determined at the rough process planning step. Dimension of preform is calculated by using volume constancy law and volume loss.

Thus, using two different knowledge-based programs applied for different shape expression cooperatively, the system designs process plan of three dimensional non-axisymmetric hot forging products.

REFERENCES

1. T.Ohashi, M.Motomura, Proceedings of 6th International Conference on Technology of Plasticity) , Germany, Nuremberg, 6-1(1999.9), pp.137-142.2.

2. T. Ohashi, S. Imamura, T. Shimizu and M. Motomura, Abstract book of 5th IURMS International Conference on Advanced Materials, P.R.China, Beijing, (1999.6), pp. 385.
3. G.J. Schmucker (Translated to Japanese by T.Onizawa), Fuzzy Set, Natural Language Computations, and Risk Analysis, 1990, Keigaku-Shuppan.

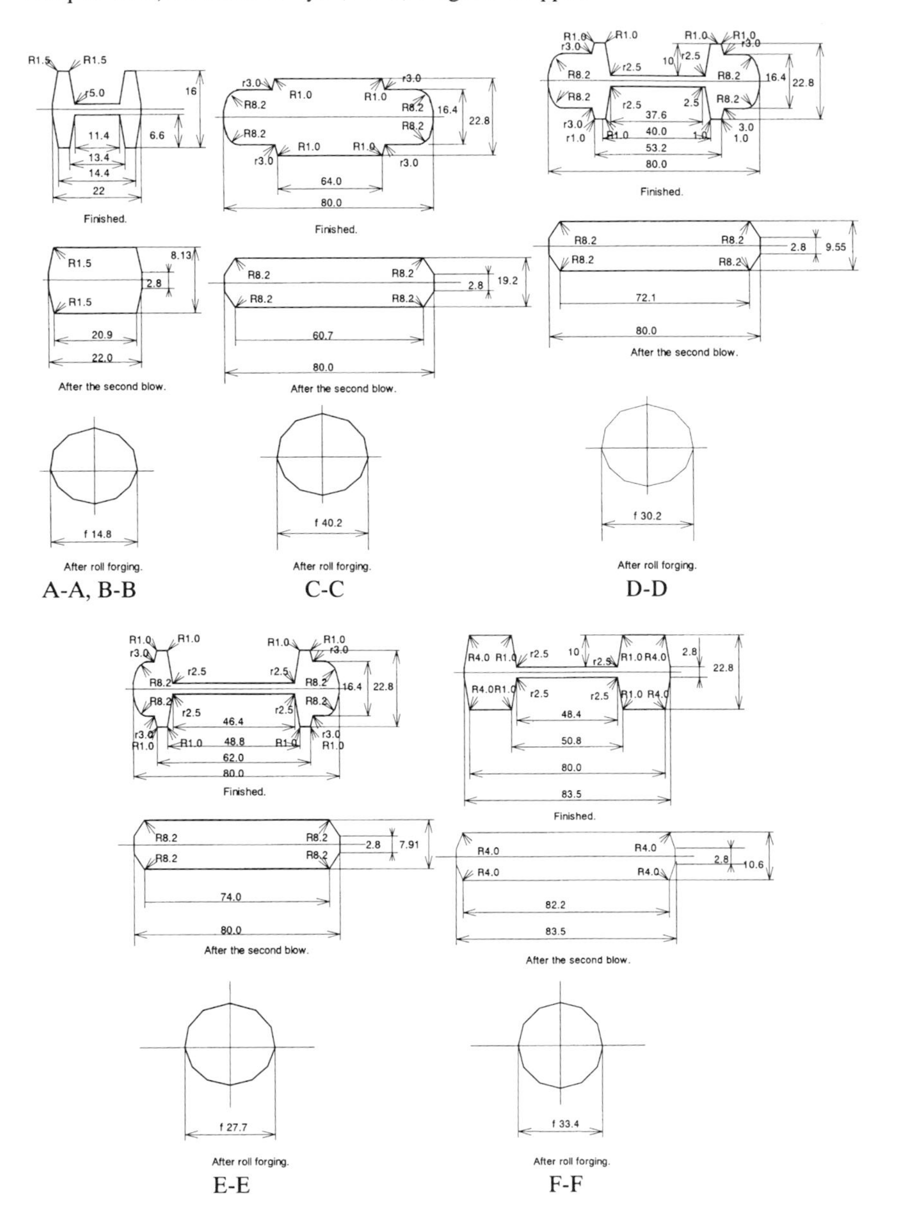

Fig.12 Example process plan produced by the system

CAPE/053/2000

Analysis of micro-plasto-hydrodynamic lubrication using strip drawing rig

R AHMED
Dept of Mechanical & Chemical Engineering, Heriot-Watt University, UK
M P F SUTCLIFFE
Department of Engineering, Cambridge University, UK

ABSTRACT

This paper addresses the evolution of surface topography and lubrication mechanisms in plane strain conditions using a strip drawing rig. Stainless steel strips containing artificial features such as Vickers and Brinell indentations were used in this study. Preliminary analytical investigations were made using Lo and Wilson model [10] to simulate micro-pool lubrication to investigate the influence of pit geometry on fluid film thickness and the evolution of surface roughness. Experimental and analytical results indicate that a transition from hydrostatic to hydrodynamic conditions caused by sliding between the tool and strip surface can assist elimination of individual pits. In addition, pit spacing and geometry affects the evolution of surface. The oil extraction, whilst beneficial for individual pit reduction however, may cause further problems of sheet conformance in the area surrounding the pits, especially for lubrication conditions in the mixed regime. Bite analysis indicates that most of the changes in surface features take place at the inlet to the bite.

1 INTRODUCTION

It is well established that the lubrication conditions within the roll bite during the cold rolling process controls the evolution of strip surface. Hydrodynamic entrainment of oil in the bite tends to keep the surfaces separated and prevents effective flattening of the asperities on the strip [1–3]. However, it also seems likely that a significant amount of lubricant is trapped within the isolated pits on the strip surface. Build-up of hydrostatic pressure in this trapped oil will tend to prevent these features being eliminated. Experimental measurements by Mizuno and Okamoto [4] with an artificial rough workpiece surface show how the trapped oil can be drawn out of the pits during the rolling process by the sliding action between the strip and the roll. An alternative approach considers the way in which oil is not completely trapped in the pits, but can flow out along channels in the rolling direction [5, 6]. Various experimental [7–9] and analytical [10] studies have considered these mechanisms, but the details of the lubrication mechanisms and the evolution of surface features remains unclear. This study thus addresses the evolution of surface roughness and the lubrication mechanisms to simulate the plane strain conditions using a strip drawing rig.

2 STRIP DRAWING and PLANE STRAIN COMPRESSION TESTS

2.1 Description of Strip Drawing Rig

Figure 1 shows a schematic of the strip drawing rig. It consists of a rigid rectangular frame, which houses the main parts of the rig. Two hydraulic rams R_i and R_d are attached to the outside of the frame. R_i is called the indentation ram and is capable of inserting a normal compressive load of 12 Ton on the surface of the specimen strip across the drawing direction. R_d is called the drawing ram and is also capable of inserting a load of 12 Ton on the surface of the specimen strip along the direction of draw. R_i holds a die D1 at its end via a die holder. Hence this die (D1) moves with the movement of the indentation ram. An identical die D2 is attached to the main frame of the rig via a die holder. The specimen strip is placed between the set of dies (D1, D2) and is held in a shackle at one end. This shackle is attached to the end of R_d (drawing ram).

The indentation is either measured by the load on the indentation ram or by a pair of strain gauge bridges, which are located in the die holder of the fixed die (D2). The push force during the strip drawing process is measured via a strain gauge attached to the shackle at the end of the drawing ram. The output from the bridge is sent to an oscilloscope and chart recorder via an amplifier.

2.2 Experimental Test Procedure

Stainless steel strips of 250 × 20 × 3 mm and 250 × 20 × 4.1 mm dimensions were used as test specimens. 3 mm strips were taken from the bright-annealed cold rolled sheet along the rolling direction. Artificial features were produced on the strip surface using the Vickers and Brinell indentations. 4.1 mm strips were taken along the rolling direction from the annealed and shot blasted hot rolled sheet (white hot-band). Two sets of dies made in EN31 material (2%Cr) and having a semi-wedge angle (β) of 4° and 6° respectively were used during this investigation. The variation in the wedge angle can be used to vary the hydrodynamic lubricant film thickness and also to change the percentage reduction in the strip thickness. Three different lubricants namely HVI-650 (gear oil), V-68 (base oil), and a commercial Sendzimir mill rolling oil were used during this study. The physical properties of these lubricants are shown in Table 1.

The strip drawing rig was also used to perform tests in plane strain compression mode. This was achieved by symmetrically indenting the strip by a set of flat face dies. These dies had a face width of 5mm and no pull force was applied in these tests.

3 MATHEMATICAL MODELLING

A two-dimensional surface evolution model developed by Lo & Wilson (1997) was used for the analytical investigations. The details of the mathematical derivations can be appreciated from [10] and, a brief description is reproduced in the Appendix for clarity. This model assumes a surface containing an array of triangular pits with an angle θ as shown in Figure 2. According to this model if *Lo* is the initial spacing of pits and *Ao* the corresponding area of contact ratio then, the oil pressure in the pit (Pb) and at the asperity (Pa), and also the non-dimensional lubricant film thickness can be evaluated from the solutions of the Newton-Raphson method at every time step (t). These solutions were then used to predict the variation in the area of contact ratio (*A*) and the inlet film thickness per unit sliding distance. The sliding speed (u) of 16 mm/s measured during the strip drawing tests was used in this analysis. Bulk strain rate ($\varepsilon^{\cdot}$) was evaluated from the following relation:

$$\varepsilon^{\cdot} = \frac{2\beta u}{I} \tag{1}$$

where I is the inlet strip thickness and β the die semi-wedge angle in radians.

EXPERIMENTAL TEST RESULTS

4.1 Plane Strain Compression Tests

Plane strain compression tests were carried out to establish the effect of lubrication entrapment on the elimination of surface features. These tests were carried out on the white-hot band as well as the 3 mm bright-annealed strip containing artificially developed features. Figure 3 shows the variations in RMS surface roughness (R_q) of the white-hot band strip surface at different bulk strain values for the lubricated and dry conditions. Similar tests on the bright-annealed material were performed by introducing Vickers and Brinell indentations at a load of 20 and 15 kg respectively. Table 2 shows the variation in the pit area and volume of indentations.

4.2 Strip Drawing Tests

Typical strip drawing test results for the artificially produced features on the bright-annealed surface are shown in Table 3. These tests were carried out at a bulk strain of 30% using the three different oils (Table 1). The lubrication regime (λ) shown in Table 3 was calculated as the ratio of the smooth film thickness (h_s), the values of which were evaluated from the Wilson and Walowit relation [1], to the average surface roughness R_q of the tool and strip surfaces. Figure 4 shows the optical view of the indentations after the draw and shows the oil extrusion from the indentations at different orientations. Table 3 also gives an approximate value of the area covered by MPHL and values of R_q in that area.

5 DISCUSSION

5.1 Influence of Oil Hydrostatics

Figure 3 indicates the importance of the hydrostatic oil entrapment in surface features to resist sheet conformance. The R_q of the sheet surface for HVI-650 lubricant is almost twice to that for dry conditions. The values for the rolling oil are in between the two cases and indicate that the lubricant properties such as viscosity and pressure viscosity coefficient also play an important role in hydrostatic conditions. These results indicate that the presence of oil resists sheet conformance due to hydrostatic pressure build-up in surface pits. Similar trends can be seen from strip drawing tests on indentations produced on the bright-annealed surface (Table 2). These indentations show less conformance in lubricated conditions in comparison to the dry case. It is to be noted from Table 2 that the area proportion increased for the Brinell indentations i.e. surface spread-out but decreased for the Vickers indentations for lubricated tests. However, volume decreased in both cases. This indicates that pit geometry also plays an important role in the evolution of surface.

5.2 Influence of Hydrodynamic Pressure Distribution

In the presence of sliding, the pressure distribution in the pits changes from hydrostatic to hydrodynamic. Depending upon this pressure distribution, it is possible that entrapped oil may be completely or partially released from surface pits, resulting in better conformance of sheet surface. This effect can be appreciated by comparing Tables 2 and 3, in which sliding of the sheet has caused an increase in the volume change of the indentations from 50% (for plane strain compression tests) to values in excess of 75% (for strip drawing tests). Although, this may seem advantageous in improving the surface quality in the first instance, it can have the drawback of causing excessive lubrication in the trailing area of the pits. This is essentially the effect shown in Figure 4. The oil released from the pits caused micropitting of the surface at the trailing edge of the pit. The extent of micropitting was not only dependent upon the lubricant properties e.g. reduced micropitting with the rolling oil, but also on the pit geometry. It is suggested that the longer 'comet-tails' in the case where the indentation faces are perpendicular to the draw direction was due to the earlier closure of pit at the inlet region allowing less back flow of oil, entrapping more oil in the pit and hence inducing more MPHL. Also the sliding distance over which the pit is closed by the die is longer in the case where the indentation face is perpendicular to the direction of draw. This can be explained from Figure 5, which shows that the sliding distance (d) over which MPHL can occur, can be approximated as:

$$d = D - x(1+\mathrm{r}) \tag{2}$$

where D is contact length, x is width of pit and r is the reduction in the pass. This analysis can be extended to estimate the reduction in strip over which MPHL can't occur as the pit is not completely enclosed at the entry and exit. Strip reduction over the start and end sections can be approximated for the strip drawing process using the following relation:

$$\xi = \textit{Proportion per pass with back (out) flow of oil} = 2\,x\,(2+\mathrm{r}/2)\,/\,\mathrm{l}\,(\mathrm{Tan}\,(\beta)) \tag{3}$$

For the tests on bright-annealed strip, x for the two cases of either the diagonal or the face running along the drawing direction was equal to 0.49 and 0.34 mm respectively. Hence the value of d under the given test conditions can be evaluated as 3.6 mm and 3.8 mm respectively. The length of comet tails in the two cases (Figure 4) were similar to these values, i.e. 2.7 and 3.5 mm respectively. Differences in the case of the indentation diagonal parallel to the drawing direction can be attributed to the differences in pit angle and also the fact that the asperity strain rate may be different from bulk strain rate. This difference between the asperity and bulk strain rate may be appreciated from the mathematical model shown in the Appendix. Similarly the proportion of reduction in which MPHL occurred was 20 % and 23 % of the total bulk reduction of 30%, for the two cases. This led to the small differences in the pit volume for the two cases as shown in Table 3.

5.3 Influence of Lubrication Regime

The effect of the lubrication regime can be seen from Table 3, which shows higher roughness outside the MPHL area for the HVI-650 and V-68 oil (R_q=0.27) than the rolling oil (R_q=0.14). The conformance of strip surface is constrained at higher λ values for the HVI-650 and V-68 oil, whereas sheet conformed well to the die surface (Rq =0.1 μm) for the rolling oil. This indicates that the rolling speed, which governs the smooth film thickness in the cold rolling process, can have a significant effect on sheet conformance in the later passes of rolling, where the strip surface is relatively smooth.

5.4 Analysis of Surface Features within the Bite

The evolution of surface features through the bite for V-68 oil is shown in Figure 6. This analysis was performed using the mathematical model for the identification of the strip surface features, the details of which can be seen in Ahmed and Sutcliffe [11]. The strip sample was taken from the bite during the strip drawing process for the 4.1 mm thick shotblasted surface. Surface profilometry measurements were made along the bite length using a three-dimensional interferometer (Zygo). It can be appreciated from Figure 6 that much of the changes in surface features occurred at the inlet to the bite. In this inlet region, the asperities were relatively easily crushed, leading to a rapid reduction in the deep pit areas. This could have been assisted by the backflow of oil from surface pits at the inlet to the bite as discussed in section 5.2. This rapid change at the inlet can be further confirmed from Figure 7, which shows the optical micrograph of the strip surface at the entry to the bite. Half of the figure shows the area just outside the bite while the remaining half is in the bite and shows a sharp smoothening of the strip surface as indicated in the early part of the bite length in Figure 6.

5.5 Predictions from Mathematical Model

The mathematical model was formulated in the Matlab programme environment to investigate different cases of pit geometry, lubricant properties, strip velocity. The initial area ratio (Ao) of shot blast surface was evaluated as 0.5. This was based upon the pit analysis of the white hot-band using three-dimensional interferometry [11]. AFM measurements of surface pits on the sheet surface in the later passes of rolling indicated that the pit angle (θ) was in the range of 10~35 degree. A typical result of AFM measurement, to represent pit geometry is shown in Figure 8. Lubricant was selected as V-68 and HVI-650. The sliding speed from strip drawing tests was calculated as 16 mm/s. Figure 9(a and b), shows the area of contact ratio and inlet film thickness for a pit angle of 35 degree. Table 4 shows a comparison of results in terms of area ratio and film thickness after a sliding distance of 4mm for two oils (V-68 and HVI-650) and various pit geometries

It can be established from Table 4 that it is more difficult to eliminate pits as the pit angle increases (i.e. pits become steep), as seen for the case of V-68 oil at pit angle of 35° and 17°. This is because of hydrodynamics of oil within the pit. The inlet film thickness increases as the pit angle decreases. Also, the film thickness falls rapidly to its minimum value as the pit angle decreases. This sharp decline in the film thickness results in faster elimination of small angle pits then the steep ones. Table 4 also shows the effect of pit spacing on the area ratio and film thickness for a pit angle of 35 degree. The decrease in pit spacing at fixed Ao represents the decrease in pit depth and hence its volume. It can be appreciated that the decrease in pit volume not only decreases the film thickness but also the bulk strain required to achieve similar area ratio. Similarly, results indicate that pit elimination is more dependent upon pit angle than pit volume. Comparison of the last two columns of Table 4 shows the effect of changing lubricant viscosity (HVI-650) on the elimination of pits. It can be established that as the lubricant viscosity increases, the film thickness increases and it is more difficult to achieve high area of contact ratios above a certain value. Hence the factors which dominate the elimination of surface pits during MPHL are pit angle, spacing and depth.

6 CONCLUSIONS

1) Plane strain compression tests indicate that the hydrostatic pressure build-up in the entrapped oil can significantly reduce sheet conformance.
2) A transition from hydrostatic to hydrodynamic conditions caused by sliding in the bite can assist elimination of individual pits. This may cause problems in sheet conformance in an area around the pits due to oil extrusion, specially at higher speeds and lower λ values.
3) Bite analysis indicates that most of the surface feature changes take place at the inlet to the bite.
4) Pit angle, spacing and depth can have a significant effect on the MPHL and consequently on the evolution of surface.

ACKNOWLEDGEMENTS

The authors are most grateful for the help of Dr. John Williams and Dr Huirong Le at Cambridge University Engineering Department and for the input from Dr Didier Farrugia, Mr Ken King and others at the collaborating companies (Corus plc/British Steel Ltd., Swinden Technical Centre and Avesta Sheffield). The financial support of the collaborating companies, the Engineering and Physical Sciences Research Council and the Isaac Newton Trust is gratefully acknowledged.

REFERENCES

[1] Wilson, W. R. D., and Walowit, J. A., 1972, "An isothermal hydrodynamic lubrication theory for strip with front and back tension", Tribology Convention, Institute of Mechanical Engineers, London, 164-172.

[2] Sutcliffe, M. P. F. and Johnson, K. L., 1990, "Experimental measurements of lubricant film thickness in cold strip rolling", *Proc. Instn Mech Engrs*, **204**, 263-273.

[3] Sheu, S. and Wilson, W. R. D., 1994, "Mixed lubrication of strip rolling", *STLE, Tribology Transactions*, **37**, 483-493.

[4] Mizuno, T., and Okamoto, M., 1982, "Effects of lubricant viscosity at pressure and sliding velocity on lubricating conditions in the compression friction test on sheet metals", *Journal of Lubrication Technology*, **104**, 53-59.

[5] Kihara, J., Kataoka, S., and Aizawat, T., 1992, "Quantitative evaluation of micro-pool lubrication mechanism", *Journal of Japan Society for Technology of Plasticity*, **33-376**, 556-561.

[6] Lin, H. S., Marsault, N. and Wilson, W. R. D., 1998, " A Mixed Lubrication Model for Cold Strip Rolling Part I: Theoretical", *Tribology Transaction*, **41**, 317-326.

[7] Fudanoki, F., 1997, "Development and evaluation of model for mechanism of formation of surface properties of cold-rolled stainless steel," First International Conference on Tribology in Manufacturing Processes, Gifu, Japan, 378-383.

[8] Wang, Z., Dohda, K., Yokoi, N., and Haruyama, Y., 1997, "Outflow behaviour of lubricant in micro pits in metal forming," First International Conference on Tribology in Manufacturing Processes, Gifu, Japan, 77-82.

[9] Wang, Z., Kondo, K., and Mori, T., 1995, "A consideration of optimum conditions for surface smoothing based on lubricating mechanisms in ironing process", *Journal of Engineering for Industry*, **117**, 351-356.

[10] Lo, S., and Wilson, W. R. D., 1997, "A theoretical model of micro-pool lubrication in metal forming," First International Conference on Tribology in Manufacturing Processes, Gifu, Japan, 83-90.

[11] Ahmed, R and Sutcliffe M. P. F., 1999, "Evolution of surface roughness within the roll bite during cold rolling of stainless steel", Modelling of metal rolling processes-III, London, ISBN 1-86125-105-X, 390-399.

[12] Wilson, W. R. D., & Sheu, S., 1988, "Real area of contact and boundary friction in metal forming", Int Jr of Mech: Sci., **30 (7)**, 475-489.

APPENDIX

This Appendix summarises the model of Lo and Wilson. Consider a surface containing an array of triangular pits with an angle θ as shown in Figure 2. Let *Lo* be the initial spacing of pits and *Ao* the corresponding area of contact. If ε and ε_a represent the bulk and asperity strain and $\varepsilon^{\cdot}$ and $\varepsilon^{\cdot}_a$ the corresponding strain rates, then from the definition of natural strain, it can be proved that:

$$A\,L = Ao\,Lo \exp(\varepsilon_a) \tag{A1}$$
$$L = Lo \exp(\varepsilon) \tag{A2}$$
$$A = Ao \exp(\varepsilon_a - \varepsilon) \tag{A3}$$

where A and L are the changes in asperity spacing and area ratio with bulk strain. If is assumed that pit angle remains unaltered during reduction, then for a given sliding speed of u, non-dimensional film thickness (H_1= Inlet film thickness / *Lo*) based upon the mass conservation can be evaluated as:

$$H_1 = \frac{-\theta C_1(1-A)(B-BaA)\exp(2\varepsilon)}{2+\theta A Ba \exp(\varepsilon)} \tag{A4}$$

where:

$$C_1 = 2\,\theta \tag{A5}$$

$$Ba = \frac{\varepsilon_a^{\cdot} Lo}{u\theta} \tag{A6}$$

$$B = \frac{\varepsilon^{\cdot} Lo}{u\theta} \tag{A7}$$

Based upon Reynolds equation, if oil pressure in the pit is assumed as Pb and at asperity as Pa, then non-dimensional film thickness can be evaluated from:

$$H_1 = \frac{CV_1}{\exp(-\Gamma Pb) - \exp(-\Gamma Pa)} \tag{A8}$$

$V_1=u_a/u$, which is the propagation speed of the lubricant. u_a is the average velocity of two surfaces. $\Gamma = \gamma k$, where γ is the pressure viscosity coefficient, k is the shear strength of strip material.

$$C = \frac{6\eta_o \gamma u}{\theta Lo} \tag{A9}$$

η_o is the lubricant viscosity at atmospheric pressure. Pa and Pb can be approximated from Wilson & Sheu's hardness analysis [12], as follows:

$$Pa = P^* + (1-A)\, Ha \tag{A10}$$

$$Pb = P^* - A\, Ha \tag{A11}$$

P* is the non-dimensional average tool pressure. Ha is the non-dimensional hardness which can be approximated from semi-empirical functions of A, (*f1*(a) and *f2*(A)) , i.e.

$$Ha = \frac{2}{f1(A)E + f2(A)} \tag{A12}$$

Then the value of Ba, which yields same H_1 in both equations (A4) and (A8), can be evaluated from Newton-Raphson method at every time step (t). The non-dimensional time (T) can be evaluated from following relation:

$$T = \frac{\theta u t}{Lo} \tag{A13}$$

Sliding distance can then be evaluated from the equation below

$$\text{Sliding distance} = \frac{TLo}{\theta} \tag{A14}$$

Table 1. Lubricant properties.

Parameter\Lubricant	**HVI-650**	**V-68**	**Rolling Oil**
η_o (Ns/m^2)	1.915	0.135	0.00713
α (m^2/N)	3.29E-08	2.2E-08	1.20E-08

Table 2. Plane strain compression test results for bright-annealed surface at bulk strain of 30 %.

		Area ($10^4 \mu m^2$)	Volume ($10^5 \mu m$)	Reduction in area (%)	Reduction in volume (%)
Vickers	Before test	12	27	-	-
	HVI-650	11.9	13.8	2	49
	V-68	12.7	13.6	-6	50
	Rolling oil	11.6	11.8	3	56
	Dry test	4.18	1.43	65	94
Brinell	Before test	9.6	8.8	-	-
	HVI-650	12.7	6.8	-31	23
	V-68	11.8	6.6	-22	25
	Rolling oil	11.6	5.5	-20	38
	Dry test	1.8	1.4	81	98

Table 3. Strip drawing results for bright-annealed surface at bulk reduction of 30%.

		Volume ($10^5 \mu m^3$)	Reduction in volume (%)	Area of MPHL ($10^6 \mu m^2$)	R_q of sheet (μm)	R_q in MPHL area (μm)	λ
Vickers (Draw along the face)	HVI-650	6.3	76	1.9	0.26	0.53	0.2
	V-68	5.9	78	2.1	0.27	0.55	0.01
	Rolling oil	6.3	76	0.62	0.14	0.32	0.0003
Vickers (Draw along the diagonal)	HVI-650	6.1	77	1.6	0.26	0.38	0.2
	V-68	4.8	82	1.5	0.27	0.45	0.01
	Rolling oil	4.8	82	0.62	0.14	0.31	0.0003
Brinell	HVI-650	0.25	97	1.4	0.26	0.54	0.2
	Rolling oil	0.7	92	0.6	0.14	0.24	0.0003

Table 4. Comparison of area ratio and minimum film thickness for various configurations after a sliding distance of 4mm & reduction of 16%.

Oil	V-68	V-68	V-68	HVI-650
θ	35°	17°	35°	35°
Lo	300 μm	300 μm	100 μm	100 μm
Area ratio	0.7	0.85	0.79	>0.85
H_{min} at start	0.9 μm	1.2 μm	0.23 μm	0.8 μm
H_{min} at exit	0.6 μm	0.3 μm	0.07 μm	0.1 μm

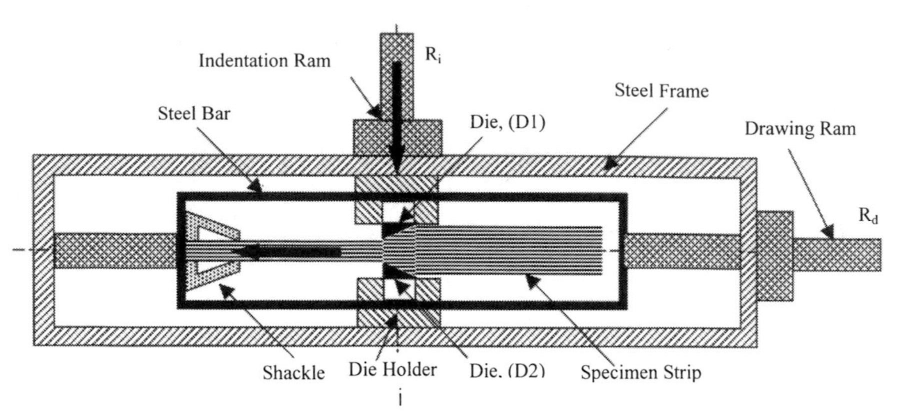

Figure 1. Schematic of the strip drawing rig

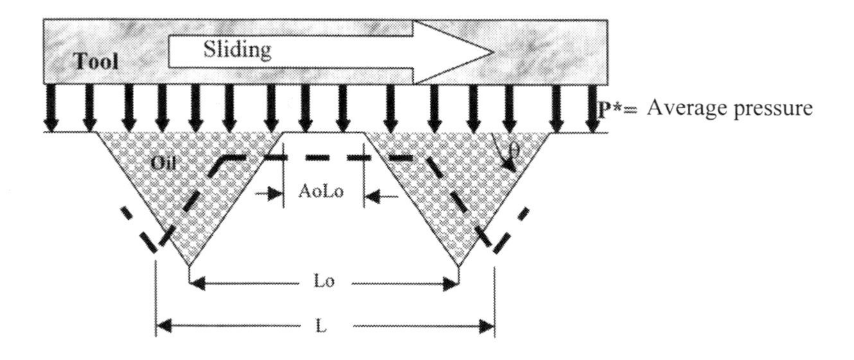

Figure 2. Pit spacing and area ratio for MPHL modelling

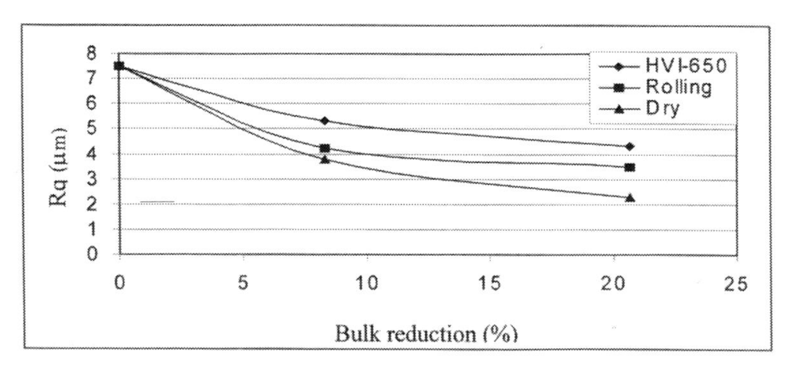

Figure 3. Variation in surface roughness of shot blast surface in plane strain compression tests.

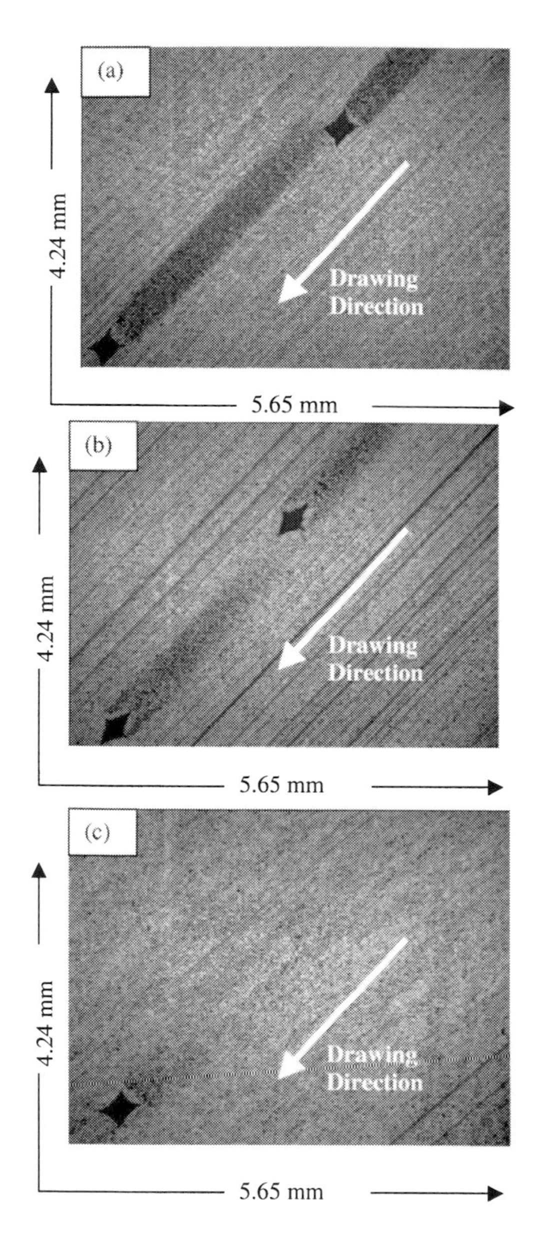

Figure 4. MPHL lubrication of Vickers indents on bright annealed strip after drawing with a 30% single-pass reduction: (a) Vitrea 68, indent side parallel to sliding direction, (b) Vitrea 68, indent diagonal parallel to sliding direction, (c) Rolling oil

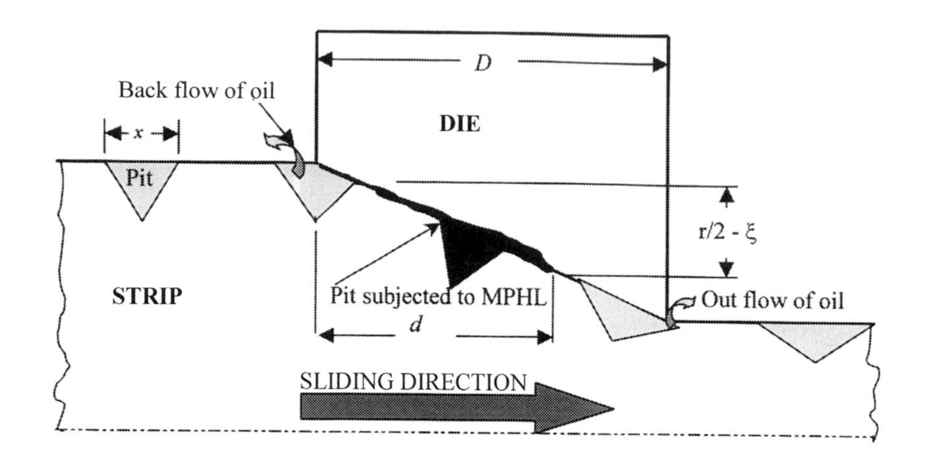

Figure 5. Behaviour of surface pit through the bite.

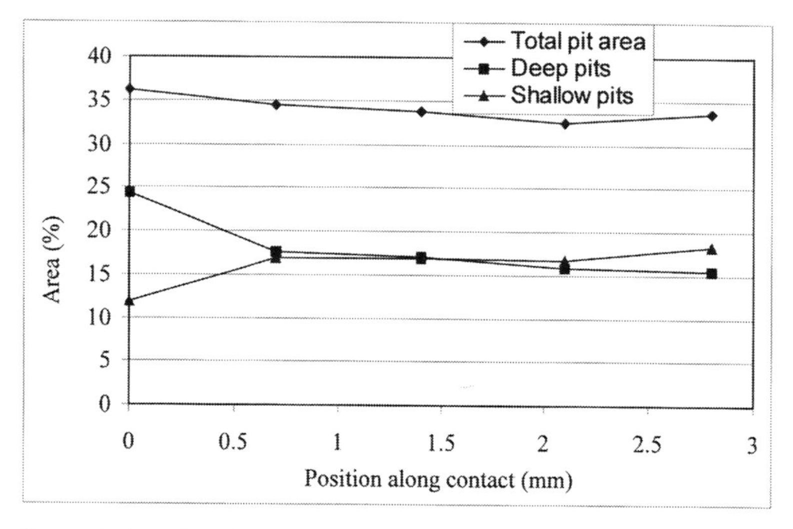

Figure 6. Area fraction of strip features through the bite for V-68 lubricant

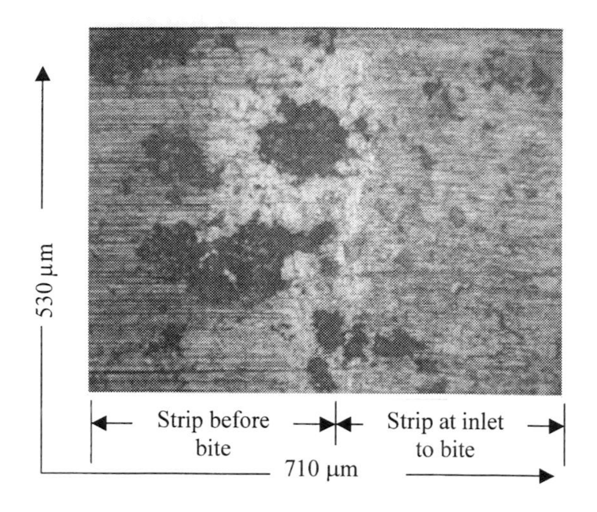

Figure 7. Optical micrograph of the strip surface at the entry to the bite.

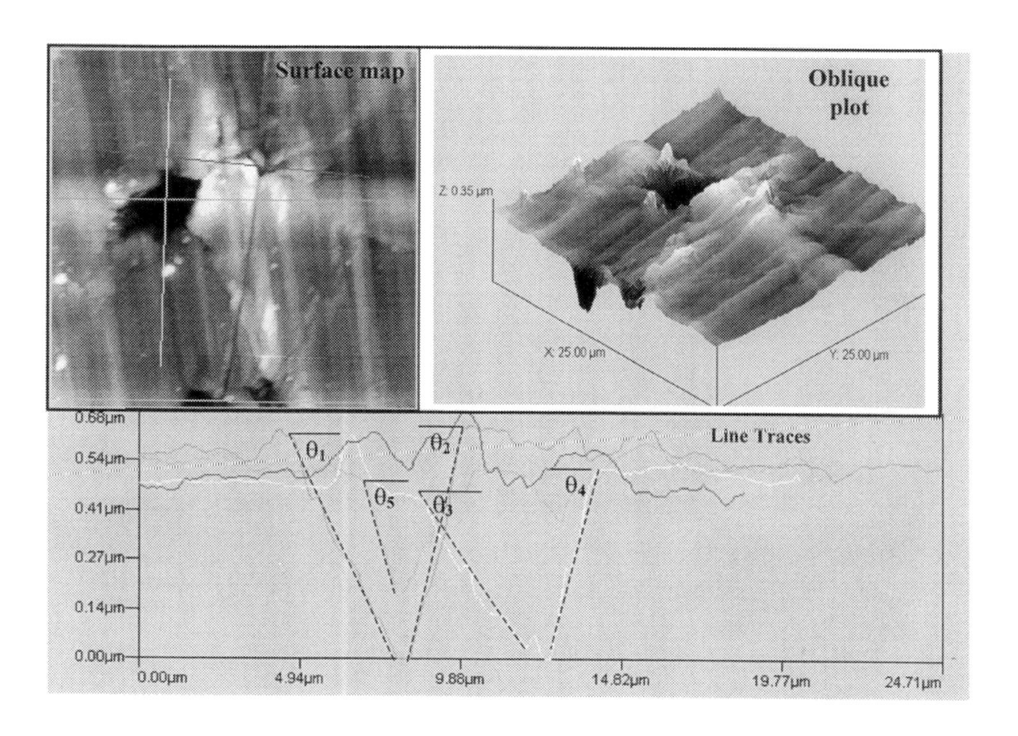

Figure 8. AFM measurement to reveal the geometry of remnant pits after an intermediate pass of rolling.

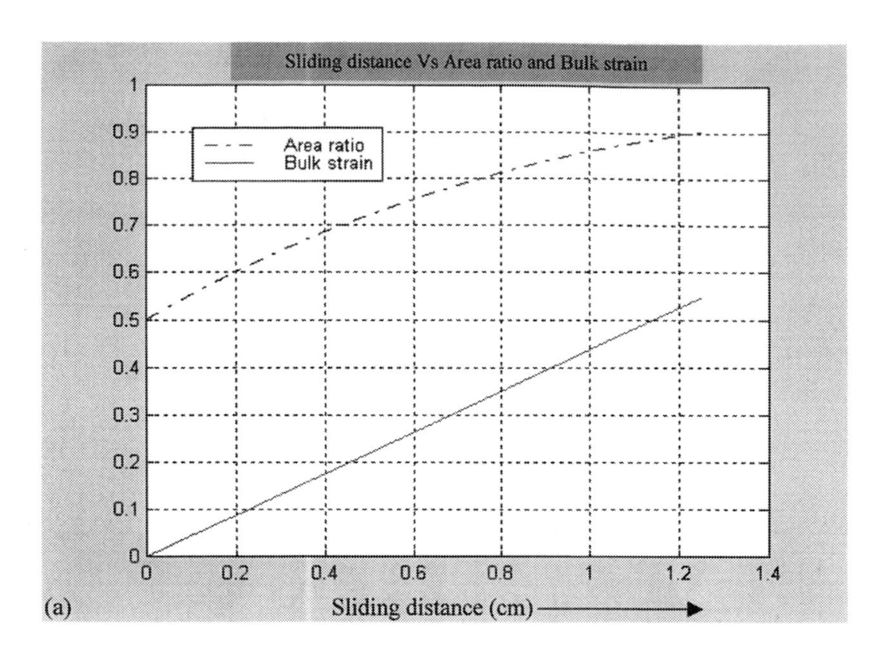

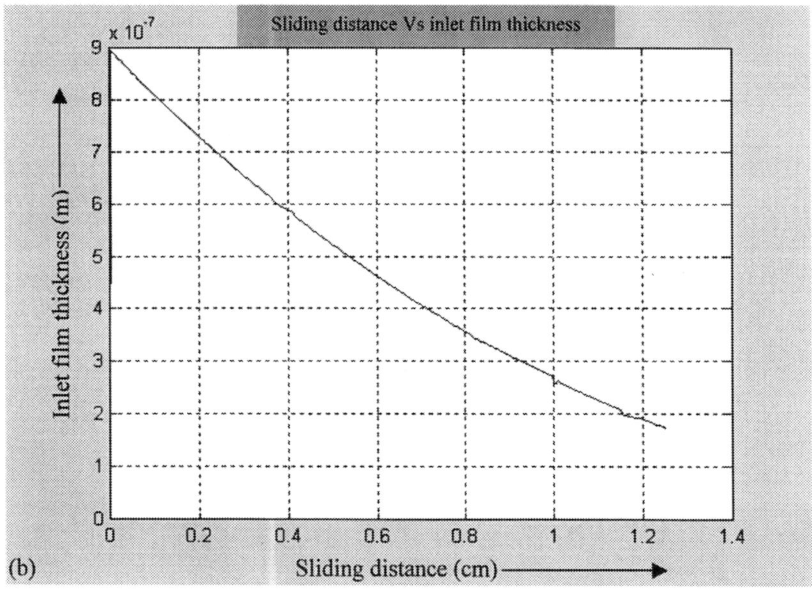

Figure 9. Area ratio and film thickness results for pit angle of 35 degree, V-68 oil (Ao = 0.5; Lo = 300μm; $\varepsilon^{\cdot}$ = 0.7; u=16 mm/s; B = 0.0219; C = 1.99E-06

CAPE/008/2000

Optimization of electrochemical grinding process for titanium alloys

A CLAPP and **J A McGEOUGH**
School of Mechanical Engineering, University of Edinburgh, UK
H A SENBEL
Faculty of Engineering, Ain Shams University, Cairo, Egypt
A DE SILVA
Department of Engineering, Glasgow Caledonian University, UK

Abstract
Titanium poses difficulties for conventional manufacturing owing to its thermal conductivity that can cause high thermal tool loads, and its high reaction affinity with atmospheric oxygen. An alternative technique, electrochemical grinding is discussed here, in which mechanical and electrochemical removal are combined. Optimisation of the process to secure appropriate machining variables for acceptable surface quality and rate of metal removal is investigated. Development of artificial intelligence techniques for a knowledge-based system arising from these results is discussed. A Manufacturing Target Diagram is used to provide a general overview of the results obtained.

1. INTRODUCTION

In aeronautical and medical engineering, lightweight items are often necessity, for which titanium is a preferred material. The significant characteristics of titanium are its low density, high strength, and in contrast to other light alloys, high corrosion- and heat- resistance [1]. Titanium is a light alloy; in the periodic table it sits in the fourth group of the fourth period. Owing to its four free valence electrons, titanium is electrically-conducting, it can be overstrained, and alloyed with other elements [2,3,4]. As the extraction of titanium is generally very expensive, the main way to achieve relatively low cost manufacture is by optimisation of the operating procedures. In particular, near net-shape procedures could reduce the costs of machining of titanium and raise material utilisation. Such procedures could include super-plastic forming, diffusion welding, laser machining and electrochemical removal. The manufacturing of titanium by electrochemical grinding (ECG) is investigated here. ECG is a process that links mechanical- and electrochemical removal. Depending on the choice of control variables, either kind of removal can predominate. Nonetheless, the control variables should be selected in a manner that guarantees an electrochemical removal of 90%.

In order to interpret the effects of ECG on machining, both procedures of removal must be understood. In this paper different cutting experiments reveal the influence of electrolyte concentration, current density, and spindle speed of the grinding wheel on machining. Optimisation of the process variables is therefore of direct industrial interest for securing low cost and high machining performance.

2. THE PASSIVITY OF TITANIUM

In electrochemical machining the removal rate can be increased by using a higher voltage. As a result the anode electrode potential rises. On reaching a certain "Flade" potential, at the anode the aqueous electrolyte solution will be oxygenated. On the basis of the high affinity of titanium for oxygen, the anode-surface becomes coated by a layer of titanium dioxide, which hinders the transfer of ions into the electrolyte-solution. The effect of such coatings is called "passivity". This condition decreases the efficiency of the process and should be avoided.

In electrochemical grinding (ECG), such coatings may be mechanically removed by the diamond grains, maintaining the process-efficiency by an unhindered transfer of alloy-ions into the aqueous electrolyte solution.

3. EFFECTS ON ECM/ECG

Effects on ECM/ECG can be classified into five groups, Figure 1. They are mostly affected by workpiece-material, electrolyte, tool-and workpiece-geometry, process-parameters and precision of the machine. In order to achieve repeatability of machining, these effects have to be maintained constant. This requirement demands homogeneity of materials, constant tool- and workpiece-geometry, constant properties of electrolyte and a highly accurate machine tool. In industrial practice, alloyed materials are often used. Their inhomogeneous grain structures lead to varying machining results even for the same process conditions. Also, the specific removal rate depends on the chemical compound of the alloys used [5,6]. The electrolyte characteristic required include high chemical resistance, in order to guarantee constant electrolyte conditions, no reactivity with machine components and a high physiological compatibility, so that health is not adversely affected.
A detailed account of the relationship between individual effects on machining is difficult to provide, owing to the complexity of the process and the number of factors.

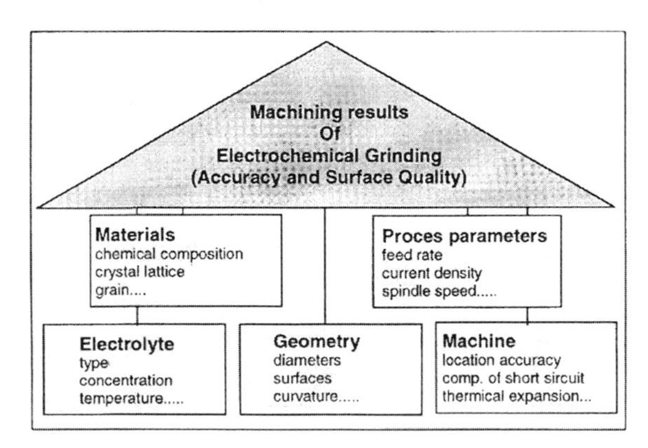

Figure 1 Some aspects affecting ECM/ECG

4. APPLICATION OF ARTIFICIAL INTELLIGENCE TECHNIQUES

The performance of ECM/ECG is greatly dependent on how the process variables are selected, in order to achieve high rate of metal removal with maximum surface accuracy. Information is not always readily obtainable as the bulk of the knowledge is based on practice, and is not conveniently contained or easily consulted in documents or databanks. To increase the rate of metal removal and to enhance high surface quality, a computerised fuzzy logic, artificial neural networks and knowledge-based system (KBS) are a promising approach to overcoming difficulties. Fuzzy controllers which use qualitative rules of discrete fuzzy classes, instead of continuous variables to optimise non-traditional machining performance have already been applied in the case of electro-discharge machining albeit with limited success, (7,8,9). Such approaches for non-traditional processes were likewise anticipated for ultraprecision, traditional methods (10); they emphasised the need for a machining strategy based on computer software to control the performances of all subsystems. Knowledge-based systems (KBS) have not yet been fully utilised, although some work has been reported for electrodischarge (EDM) and electrochemical machining (ECM) (11). The first stage in this investigation was to gather experimental data that could subsequently be included in a knowledge-base system. The flow chart in Figure 2 summarises the steps of optimising the machining parameters to achieve the best performance (surface quality) as well as maximise the metal removal rate.

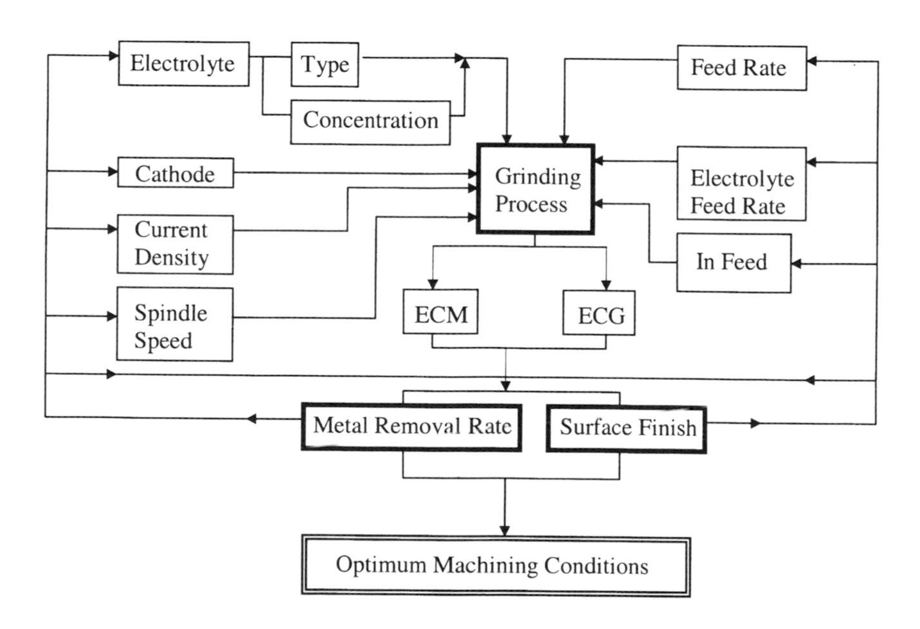

Figure 2 Flow chart of optimisation strategy

5. THE GRINDING PROCESS

In these experiments a metal bonded diamond grinding wheel is used for cylindrical surface grinding, Figure 2. The grinding wheel rotates with a circumferential speed V_s, and approaches the workpiece with an in-feed a_e at a feed-speed V_w. During this movement the grinding wheels active width b_s and the mesh a_p are of the same size.

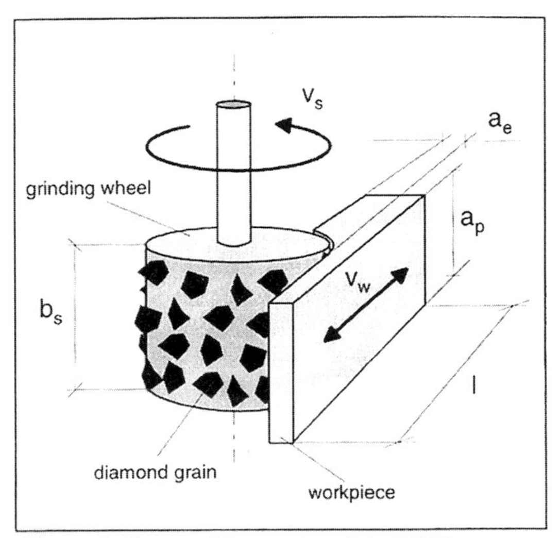

Figure 3 Contact kinematics of the ECG-process

A characteristic control variable of grinding process is the volume of metal removed by cutting V_w. As shown in the figure 3, V_w is calculated by the product of workpiece length, in-feed a_e and mesh a_p.

$$W_w = I.a_p \,.\, a_e \qquad [1]$$

In terms of the removal rate Q_w, a further control variable in grinding technology, the differential coefficient of removal rate V_w and time t is useful:

$$Q_w = dV_w / dt \qquad [2]$$

If the removal rate remains constant throughout the process, removal rate with time Q_w can be calculated from equation [1].

$$Q_w = (I \,.\, a_p \,.\, a_e) / t = (I / t) \,.\, a_p \,.\, a_e = V_w \,.\, a_p \,.\, a_e \qquad [3]$$

In order to compare grinding processes, with a variable width of grinding-wheels and constant other parameters, the time-removal-rate referred to a grinding wheel width of one millimetre $Q`_w$ has been introduced.

$$Q`_w = Q_w / b_s = (V_w \,.\, a_p \,.\, a_e) / b_s \qquad [4]$$

In the case of a grinding wheel width b_s equal to the mesh width a_p equation [4] changes to

$$Q`_w = V_w \,.\, a_e \qquad [5]$$

6. EXPERIMENTAL WORK

The electrochemical grinding equipment was adapted from a converted milling machine. The main characteristics are: the spindle speed, n, range (100-25000 rpm), table feed speed (X-axis) (30-20000 mm/min),vertical path (Y-axis) (700 mm), and spindle power (3kW). The ECG apparatus is indicated in Figure 4.

The titanium workpiece had a rectangular geometry (50x20x4)mm, and were be machined on all four edges. Aqueous Sodium Chloride, Sodium Bromide, Sodium Fluoride and Sodium Nitrate were employed in concentrations ranging from (5-20%).

In the ECG of titanium alloys, sodium chloride produces a good surface finish, although superior results can be expected by using a mixture of sodium chloride (14.14 kg/100 l), sodium bromide (4.7 kg/100 l) and sodium fluoride (0.188 kg/100 l) [12]. In the initial experiments, a pure copper cathode of diameter 15mm and length 20mm were used. After satisfactory process conditions had been achieved, the metal bonded diamond grinding wheel type (D 251 M45-B 25Z), with the same dimensions, was employed. These procedures were adapted in order to avoid damage to a costly grinding wheel, also to give a comparison between ECM and ECG.

The resulting surface finish was measured on a TALYSURF 4 instrument.

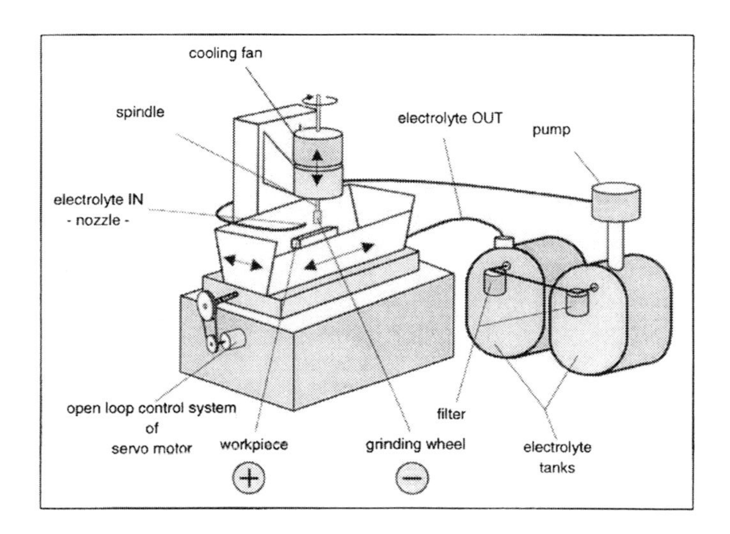

Figure 4 ECG-apparatus

7. EXPERIMENTAL RESULTS

The experimental results of ECM and ECG are presented graphically. They contain information on average surface roughness R_a and current density J for different spindle speeds and electrolyte concentrations for both copper and grinding wheel cathodes.

7.1 Influence of current density

To study the effects of current density on surface roughness, experiments with a rotating copper cathode, and a grinding wheel have been undertaken. In these experiments the spindle speed was approximately 19 m/s and the current density varied from 0.1 to 0.5 A/mm^2. Figure 5 shows the results of pure electrochemical machining with a rotating copper cathode. The best surface qualities were achieved with an electrolyte concentration of 5%. Increasing the current density up to 0.3 A/mm^2 improved the surface roughness from 2.23 μm to 2 μm R_a. The higher the current density the poorer became the surface quality.

Use of a grinding wheel as the cathode in electrochemical grinding leads to better surface qualities than pure ECM, as shown in Figure 6.

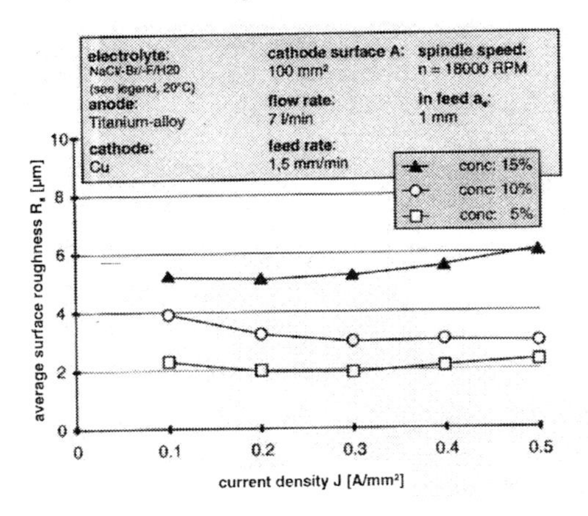

Figure 5 Influence of increasing current density on surface finish, by use of a rotating copper cathode

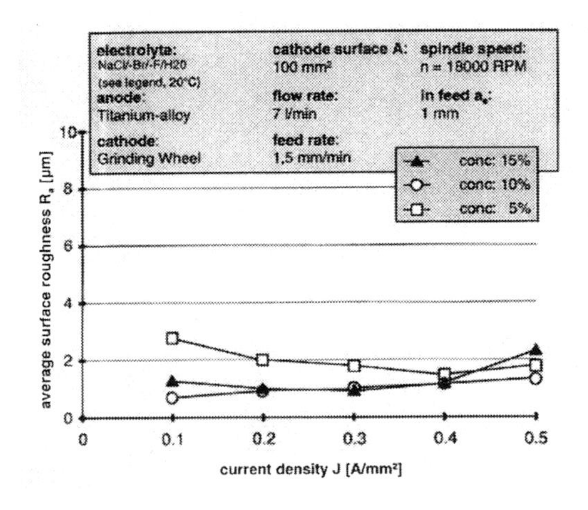

Figure 6 Influence of current density and electrolyte concentration on surface finish in ECG

This effect depends on the mechanical removal of oxide layers by grinding. Best results in surface qualities are about R_a 0.8-1μm. A current density of 0.1 A/mm^2 leads to the smallest width of the interelectrode gap. For that reason, the thickness of cut in grinding is high and leads to rough surfaces. When the current density increases, more ECM occurs, the gap width grows simultaneously and the thickness of cut decreases. Therefore the surface quality improves. High current density densities (0.5 A/mm^2) enlarge the gap width, so that grinding is eliminated, and the oxide layers cannot be removed. As a result the surface finish is poor.

7.2 Influence of electrolyte concentration

In order to establish the influence of electrolyte concentration on the surface finish, experiments with three different concentrations (15, 10 and 5%) were undertaken, as shown in Figures 5,6. It is clear that the gap width in the electrochemical process increases when a higher electrolyte concentration is used. This leads to a low thickness of cut due to mechanical abrasion, and hence to a better surface finish, as with a higher current density, where the grinding grains lose contact with the anode surface, and the oxide layers cannot be removed mechanically.

When an electrolyte concentration of 5% is used, and a low current density is applied, the gap width becomes very small, and the thickness of cut increases. These conditions lead to higher surface roughness. When current density is altered to higher values, the percentage of grinding in metal removal decreases, since the gap width become greater. The thickness of cut reduces and the surface is improved.

7.3 Influence of spindle speed

Figure 7 shows that an increase in spindle speed generally leads to a better surface finish. This effect depends on a higher electrolyte flow rate, caused by a higher spindle speed on a decreasing thickness of cut through grinding with a higher circumferential speed. When a spindle speed of 24000 rpm is used, the surface quality deteriorates, owing to increasingly turbulent electrolyte flow.

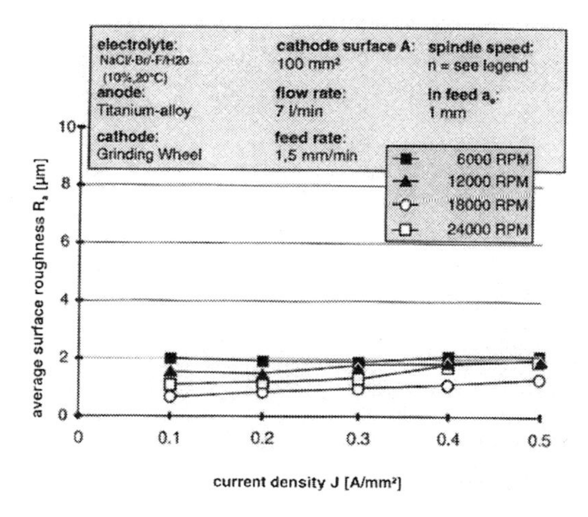

Figure 7 Influence of different spindle speeds on the surface finish in ECG

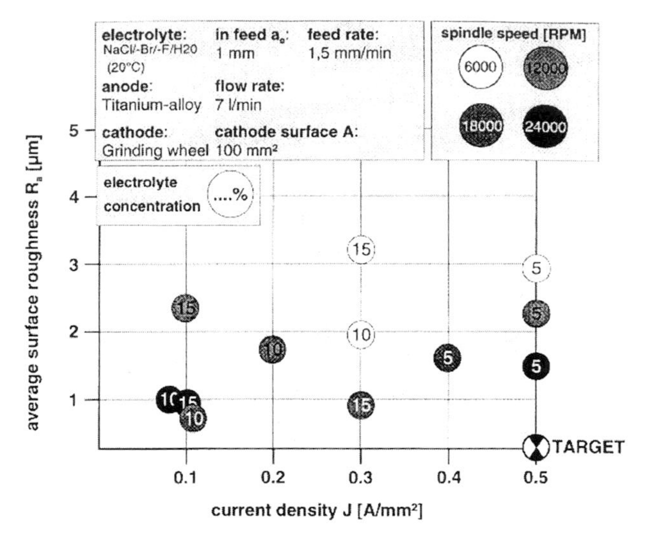

Figure 8 Manufacturing target diagram of ECG

8. CONCLUSIONS

The Manufacturing Target Diagram (MTD), Figure 8, provides a general overview of the optimum experimental results. The results can be compared with a best performance (surface quality) datum line, represented by the X-axis. The results are presented in the form of circles, the different shades stand for different spindle speeds. The circles include a number that indicates the electrolyte concentration used during experiments. The manufacturing target lies at the right hand side of the X-axis, in order to reach best surface roughness in combination with a high removal rate when a high current density is used. This leads to shorter production times whilst guaranteeing acceptable surface finishes.

A computer-aided process depending on a knowledge-based system (KBS) has been developed for optimising the machining variables to obtain the best surface quality as well as maximising the rate of metal removal. The best surface finish was achieved by use of an electrolyte concentration of 10 or 15% in combination with a spindle speed about 18000 to 24000 rpm. In contrast to ECM alone, where high speeds led to deterioration in surface roughness, a low thickness of cut in grinding leads to better surface qualities than that obtained from low spindle speeds.

ACKNOWLEDGEMENT

The authors would you like to acknowledge the contribution of Eng. Dirk Friedrich and Y.Tang with the experimental work.

REFERENCES

1. M. Peters, "Titanium alloys for aerospace applications", Titanium World 2, pp. 15-17, 1995.
2. J.M. Donachi, "Titanium, A technical guide", ASM International, Metal Park, OH, USA, 1988.
3. H.J. Ratzer-Scheibe, H.Buhl, "Repassivation of titanium and titanium alloys", Titanium Science and Technology AIME 2641,1984.
4. H.A. Lipsitt, "Titanium Aluminides – an overview"; in: "high-temperature ordered intermetallic alloys", MRS, Pittsburgh, PA, USA, pp 351-364, 1985.
5. J.A. McGeough, "Principles of Electrochemical Machining", Chapman and Hall, London, 1974.
6. J.A. McGeough, "Advanced Methods of Machining", Chapman and Hall, London, 1988.
7. Morita, A., et al, "Fuzzy Controller for EDM", Proc. ISEM-X, 1989, pp. 236-239.
8. Xiong, Y., "Fuzzy Pattern Recognition and Fuzzy Control in EDM", Proc. ISEM-X, 1989, pp.357-360.
9. Kruth, J.P., "Advances in Physical and Chemical Machining", Advancement in Intelligent Production, E. Usui Editor, Sept. 1994, pp. 123-138.
10. Ikawa, N., et al, "Ultraprecision Metal Cutting, The Past, The Present and The Future", Annals of the CIRP, Vol. 42/2, (1991), pp. 587-594.
11. Khairy, A.B., "A Knowledge-Based System for Electrochemical Machining Procedure", Journal of Materials Processing Technology, Vol. 58, (1996), pp. 121-130.
12. J.F. Wilson, "Practice and Theory of ECM", Malabar, Florida, Reprint Edition, 1982.

CAPE/064/2000

Effect of vibration on axisymmetric indentation of a model elasto-viscoplastic material

Z HUANG and **M LUCAS**
Department of Mechanical Engineering, University of Glasgow, UK
M J ADAMS
Unilever Rresearch Port Sunlight, Bebington, UK

SYNOPSIS

In this study, a series of axisymmetric indentation simulations was carried out for a model elasto-viscoplastic material, plasticine, using the finite element method in order to gain insight into the bulk mechanical flow of the material and the interface frictional characteristics for rigid spherical indenters. Experiments were conducted to obtain selected data as a basis for successful validations of the numerical simulations. Subsequently, a validated study of superimposed vibration loading on standard indentation measurements is described, which investigates the effect on interfacial conditions. The results show that the reduction in indentation load by superimposed oscillations may be explained by a combination of stress superposition and friction reduction.

1. INTRODUCTION

Superimposing oscillations on dies has been investigated in many areas of metal forming, such as wire and tube drawing, welding and cutting [1]. Several studies have demonstrated that vibratory metal working operations have the capacity to achieve improved surface qualities, a reduction in working loads and also a reduction in friction between the die and workpiece. These advantages depend on parameters such as process speed, the lubricant used, the mode of oscillation of the die and the material being deformed. In this study, the benefits of superimposed oscillations on the bulk mechanical flow and interface frictional characteristics for a model elasto-viscoplastic material are investigated by adopting indentation testing as a standard method which can be readily adapted for practical studies of the oscillation effects.

Indentation tests are a convenient means of assessing several mechanical properties of solids as they provide a simple, inexpensive, non-destructive and objective test method using small volume specimens. In the traditional hardness test, a diamond indenter is pressed into the surface of a solid specimen with a prescribed load, P, and either the depth of penetration, d, is measured or a characteristic length of the indent such as the contact radius, a. . Various shapes of indenter may be used in practice: the most common geometries are spherical (Brinell test): conical (Rockwell test) and a rectangular pyramid (Vickers test). In recent years, depth sensing indentation tests have also been used to characterise a wide variety of material properties, such as creep resistance [2], stress relaxation [3], Young's modulus [4] and subsurface damage [5]. In using indentation to measure the elastic and plastic properties of a material, it is often assumed that contact is frictionless between the indenter and specimen [4,6]. However, plastic deformation often produces a permanent indent in the form of a cavity, and friction between the face of the indenter and the material significantly influences the mode of deformation and the force of adhesion [7]. Evidence of the effect of friction induced by the wall roughness on the measurement of Young's modulus and other mechanical properties has been reported [8,9].

Computer simulation offers a fundamental understanding of the indentation process and the capability to predict a detailed material deformation history. In particular, a detailed knowledge of the deformed configurations at any instant during the indentation process can be derived through a combination of experimentation and finite element simulation. However, modelling indentation contact is a complex problem, since the constitutive equations are non-linear and a number of material parameters must be included to describe material behaviour. The aim of this paper is to present a study of the deformation of an incompressible elasto-viscoplastic solid, which is indented by a rigid spherical indenter with a lubricated wall. The effect of the application of an oscillation to the indenter during penetration has been investigated on the basis of measurements of the indentation load as a function of the depth. The emphasis is on the effects of oscillation on the friction induced by the interaction between the rigid spherical indenter and the specimen. A simple Coulombic wall boundary condition is assumed to be imposed along the wall with a friction coefficient, μ. The vibrational indentation problem was investigated by both finite element contact techniques using ABAQUS and a new indentation experiment developed to allow the application of a superimposed oscillatory load.

2. VIBRATIONAL INDENTATION TEST

The vibrational indentation technique is based on a spherical indentation. Typically, the experimental procedure involves moving the indenter into the specimen under a combination of static and vibrational loading, and measuring the forces and displacements associated with the indentation process.

2.1 Apparatus

The test apparatus is shown in figure 1. Four stainless steel spherical indenters with radii of 2, 3, 5 and 10 mm were used in the study. The static indentation force was provided by a LLOYD testing machine, which is capable of controlling the loading speed at 1mm/min and providing a maximum load of 100N. The mechanical vibration is generated by a signal generator in order to provide a specified vibration amplitude at the working surface of the spherical indenter. The vibrational load was therefore superimposed on the static load from the LLOYD test machine.

The load recorded by the force transducer on the crosshead is the mean indentation load since the response time of this beam transducer was not sufficiently short. For measuring the superimposed vibrational load, a piezo-electric force transducer was mounted on the vibration exciter. A non-contact laser Doppler vibrometer (LDV), which monitors the velocity of the target surface, was used to measure indenter vibration. The LDV signal was processed to extract vibration displacement, and post-processed using CADA-X3.2 (LMS International) vibration analysis software. The output of the amplitude measurement was fed back such that the driving power for the vibrator was controlled to maintain a constant vibrational amplitude.

2.2 Model Material

Plasticine has been used as a common model material to simulate the plastic flow of hot metal in fundamental experimental studies of extrusion, upsetting and rolling processes [10-12]. The rheological characteristics of plasticine, such as the strain rate sensitivity and the stress-strain behaviour, are very similar to that of steel at elevated temperatures. [11]. Plasticine generally exhibits marked elasto-viscoplastic behaviour, which is characterised by elastic deformation at low strains and rate dependent plastic flow when the yield criterion is satisfied. The processing of the material invariably involves relatively large strains and finite strain rate. Under these circumstances, the constitutive behaviour may be approximated by a Herschel-Bulkley relationship [13]. In the current study, the effect of oscillation on the material flow characteristics is examined using plasticine as the model material for the spherical indentation tests.

The plasticine used in this study is a highly concentrated dispersion of clay in a hydrocarbon liquid medium. It was homogenised in a Z-blade mixer and cylindrical specimens of the required thickness were prepared by compression between parallel platens on a LLOYD test machine and then cut to the required radius. The samples were allowed to thermally equilibrate for 24 hours in the laboratory at 20^{0}C. The specimens used in this study were all 40 mm in height and 60 mm in diameter, which is more than an order of magnitude greater than the characteristic indentation dimension which is the contact radius.

2.3 Experimental Procedure

The vibration-assisted experiments were conducted using two different loading arrangements as shown in figure 2. During each test, the indenter was loaded statically until a specified depth was reached. A vibrational load was then superimposed on the static load until a vibration amplitude A was reached. After maintaining a constant vibration amplitude of A for 30 s, the oscillation was removed. For case 1, the specimen continued to be deformed under static loading until the indentation reached a specified depth appropriate for the sphere radius, when the specimen is unloaded. For case 2, the specimen was unloaded when the oscillations were discontinued removed. A static indentation test was also conducted (case 3 in figure 2). Before each test, the surface of the plasticine disk and also indenter were cleaned and re-lubricated with a thin layer of silicon oil in order to maintain the lubricant condition. The indentation load-depth data were recorded using dedicated software.

The applied indentation loads and associated penetration depths were measured during the static indentation test (case 3, shown in figure 2d) with a nominally frictionless boundary condition. This test provided the data used to calculate the material flow parameters, which relied on a combination of elasticity, plasticity theories and semi-empirical relationships which govern material behaviour under indentation loading and unloading. The detailed analytical procedures for deriving the Young's modulus and yield stress have been described in a previous study [14]. The material properties for the spherical indenter and the plasticine are listed in table 1 assuming isotropic properties for the plasticine.

Table 1 Material properties of indenter and plasticine.

	E (MPa)	σ_0 (MPa)	ρ (Kg/m^3)	ν	n	k_e (MPaS^n)
Spherical indenter	193000	200	7833	0.30		
Plasticine	17.5	0.23	1878	0.49	0.34	0.039

3. FINITE ELEMENT SIMULATION

Finite element simulations using ABAQUS have been carried out to model the oscillatory indentation using the zoned mesh refinement shown in figure 3 and for sphere radii equal to those used in the experimental study. As a consequence of the axisymmetric conditions, only the right half cross-section of the axisymmetric plane was modelled. The surface of the indenter was defined using an analytic rigid surface definition since it may be regarded as much stiffer than the plasticine. The specimen was modelled using a set of 4-noded bilinear axisymmetric quadrilateral elements and an elasto-viscoplastic constitutive relationship (13). A fine mesh in the vicinity of the indenter and a gradually coarser mesh away from the indenter was used to ensure a high degree of numerical accuracy and a sufficiently accurate representation of a semi-infinite solid.

Between the potential contact surfaces, node-to-node type elements were located. The base of the specimen was fixed, while the load was introduced to the top of the rigid sphere segment. The static load was controlled by a constant velocity. Vibration of the indenter was introduced by an oscillatory load superimposed on the static load, and was specified by its sinusoidal displacement amplitude and frequency.

The contact algorithm used iterations to satisfy the stress and displacement contact conditions, as illustrated in figure 4. The displacement and stress contact conditions for a rigid sphere are:

$d_j = u_i + h_i;\ p_i > 0$ for points inside the contact area

$d_j = u_i + h_i;\ p_i \leq 0$ for points outside the contact area

where d_j is the sphere approach, u_i is the deformation of the surface points of the specimen, h_i is the initial gap between the bodies at point i and p_i represents the contact pressure at point i. The reaction forces at the nodes of the contact area are proportional to the contact pressure distribution. If the reaction forces at assumed boundary points of the contact area are positive (representing compression), these points are inside the contact area. If the reaction force is negative (representing tension) the point should be outside the contact area. When contact occurs, a Coulombic coefficient of friction equal to 0.2 was given for the lubricated boundary condition. In this way, the contact area was continuously modified during each iteration and the total normal load, P, was calculated.

4. RESULTS AND DISCUSSION

First, the results for the static indentations (case 3) will be described.The variation of the load with the indentation depth obtained from the numerical solution and experimental results are presented in figure 5. It may be seen that there is an excellent agreement between the numerical and experimental values. With the aid of the finite element simulations, it is possible to obtain detailed information at any instant of the deformation. The contours of the effective Mises stress and plastic strain are shown in figure 6, (a) and (b) corresponding to indentation depths of 0.15 and 1.5 mm for the spherical indenter having a radius of 5mm. The contour levels (in MPa) are also indicated in the figures. Comparison of figure 6(a) and 6(b) correctly shows a growth of the plastic zone due to additional indentation.

A reduction in the static load required to deform the material was observed and predicted numerically in all cases when there was a superimposed oscillation. The load-depth relationships during vibrational indentation for case 1 loading at different vibration amplitudes and a frequency of 4Hz are shown in figure 7. The FE results agree well with the experimental measurements during static loading but, in each case, the load reduction measured experimentally during superimposed vibration loading is greater than the FE prediction.

Vibration at higher frequencies was also investigated. A finite element solution for the load-depth relationship is shown in figure 8(a) using the procedure described previously for a specimen indented by a sphere of 5mm radius at a vibration amplitude of 0.04mm and frequency of 40 Hz. Figure 8(b) shows an expanded view of the loading history during the period of vibrational loading. Again, the measured reduction in force is greater than the FE prediction. The difference between the finite element simulation and experimental results may be explained by a further reduction in friction achieved under superimposed vibration, as a result of pumping and improving the 'activation' of the lubricant. The finite element model only simulates the superposition effect due to the vibration. Updating the finite element model by changing the coefficient of friction to 0.01 during the period of vibration loading allows much closer agreement with experimental data to be achieved, as shown in figure 8(b).

The hardening of the material during vibration loading, which can be seen as the over-stress during period A on the FE curve in figure 8(a), is due to the strain rate dependency of plasticine during plastic deformation.

For both the FE simulation and experimental study of vibrational indentation, the unloading curves remain unaffected for both loading cases 1 and 2 described in figure 2. Also, the elastic curve back to the plastic deformation curve on cessation of oscillation, and the following plastic deformation curve, for case 1 remains unaffected. Thus it can be deduced that there is no absolute change in the mechanical properties of plasticine following the application of vibration.

These results are consistent with the findings in a study of vibration effects on an upsetting process for plasticine[15].

5. CONCLUSIONS

A detailed finite element simulation and experimental study of spherical indentation has been conducted for the model elasto-viscoplastic material, plasticine. The finite element solution provides a details of the displacement, strain and stress history of the specimen during the indentation. These results are in good agreement with the experiments conducted under similar boundary and loading conditions.

Introducing vibration in an indentation test, results in a reduction in the forming force required to deform the material. The force reduction may be explained by a combination of stress superposition and friction reduction due to vibration. There was no absolute change observed in the mechanical properties of the material following the application of vibration. These results supported the findings of a previous study of an upsetting process.

REFERENCE

(1) Eaves, A.E., Smith, A.W., Waterhouse, W.J. and Sansome, D.H., Review of the application of ultrasonic vibrations to deforming metals, Ultrasonics, 162-170, July 1975.

(2) Bower, A.F., Fleck, N.A., Needleman, A. and Ogbonna, N., Indentation of a power law creeping solid, Proc. R. Soc. Lond. A., Vol. 44, 97-124, 1993.

(3) Lawrence, C.J., Adams, M.J., Briscoe, B.J. and Kothari, D.C., Wedge indentation and stress relaxation of a viscoelastic paste, World Congress on Particle Technology 3

(4) Oliver, W.C. and Pharr, G.M., An improved technique for determining hardness and elastic modulus using load and displacement sensing indentation experiments, J. Mater. Res., Vol. 7, No. 6, 1564-1583, 1992.

(5) Munawar, M., Subsurface deformation patterns around indentations in work-hardened mild steel, Philosophical Magazine Letter, Vol. 67, No. 2, 107-115, 1993.

(6) Briscoe, B.J. and Sebastian, K.S., The elastoplastic response of poly (methylmethacrylate) to indentation, Proc. R. Soc. Lond. A, Vol. 452, 439-457, 1996.

(7) Johnson, K.L., Contact mechanics, Cambridge University Press, 1985.

(8) Adams, M.J., Briscoe, B.J., Kothari, D.C. and Lawrence, C.J., Plane-strain wedge indentation of a soft plastic solid, The 1997 Jubilee Research Event, 317-320, 1997.

(9) Yang, F. and Li, J.C.M., Effect of friction and cavity depth on the elastic indentation of a cylindrical rod, Mechanics and Materials, Vol. 25, 163-172, 1997.

(10) Green, A.P., The use of plasticine to simulate the plastic flow of metals, Philosophical Magazine, 42, 365-373, 1951.

(11) Yagishita, K., Tsukamoto, H., Egawa, T., Oomori, S. and Ibushi, J., Study of simulative model test for metal forming using plasticine, Mitsubishi Heavy Industries, Mitsubishi Tech.Bull., 1-11, 1974.

(12) Sofuoglu, H. and Rasty, J., On the measurement of friction coefficient utilizing the ring compression test-II: effects of deformation speed, strain rate and barrelling, Engineering Systems design and Analysis, ASME, 3, 189-197, 1996.

(13) Adams, M.J., Edmondson, B., Caughey, D.G. and Yahya, R., An experimental and theoretical study of the squeeze-film deformation and flow of elastoplastic fluids, Journal of Non-Newtonian Fluid Mechanics, 51, 61-78, 1994.

(14) Huang, Z., Lucas, M and Adams, M.J., The elasto-viscoplastic response of plasticine to indentation, Modern Practice in Stress and Vibration Analysis - IV, Nottingham, 2000.

(15) Huang, Z., Lucas, M and Adams, M.J., A finite element study for optimising wall boundary conditions in an elasto-viscoplastic material forming process, Proceedings of the 15th international conference on CAPE, Part 1, 117-124, Durham, 1999.

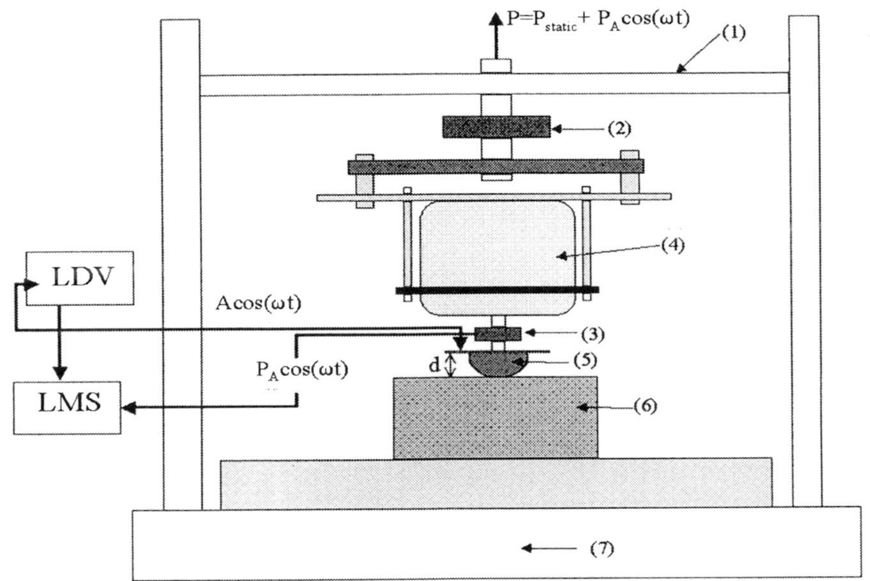

(1), crosshead; (2), load cell; (3), piezoelectric force transducer; (4), vibrator; (5), indenter; (6), Plasticine specimen; (7), LLOYD base.

Fig. 1 Apparatus for oscillatory indentation

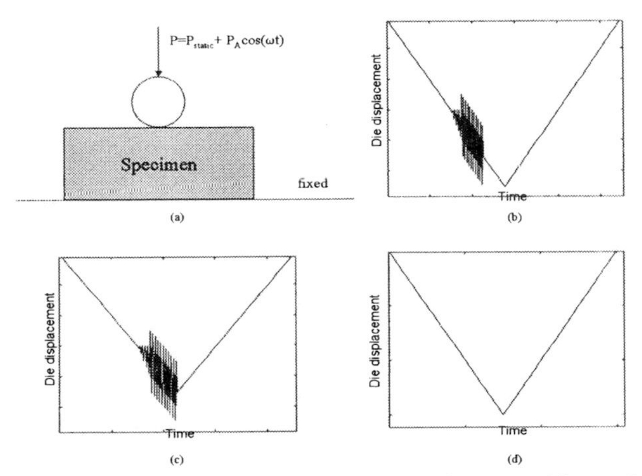

Fig. 2 (a) Spherical indentation with vibration. Displacement history for (b) case 1, (c) case 2 and (d) case 3.

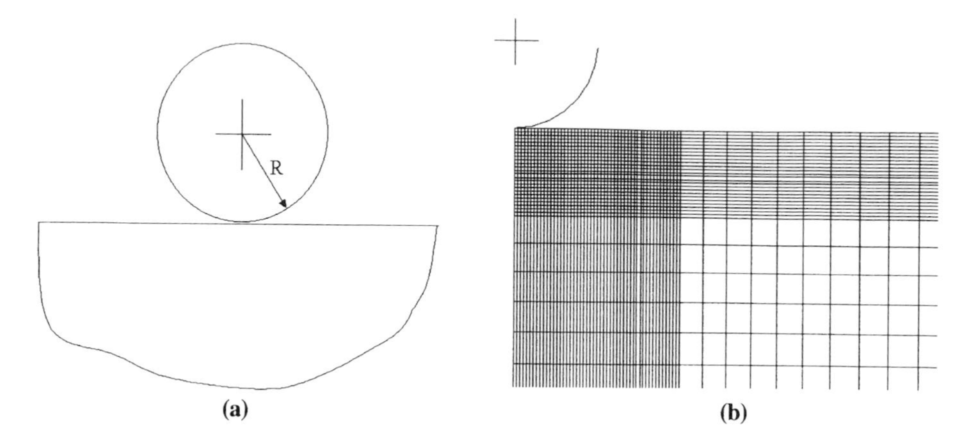

Fig. 3 (a) Geometry and (b) undeformed mesh of FE model for axisymmetric spherical indentation

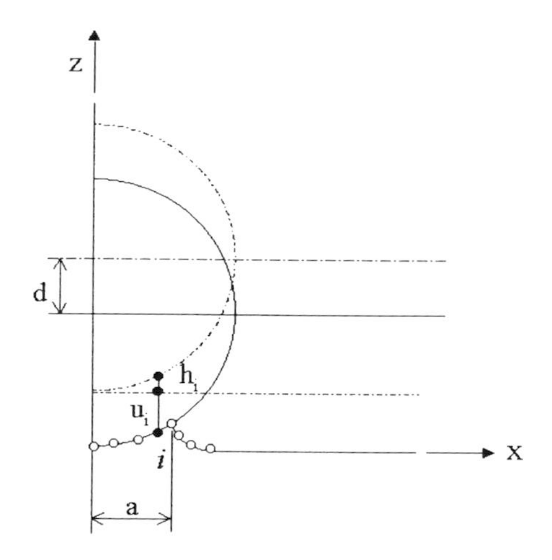

Fig. 4 Displacement contact condition

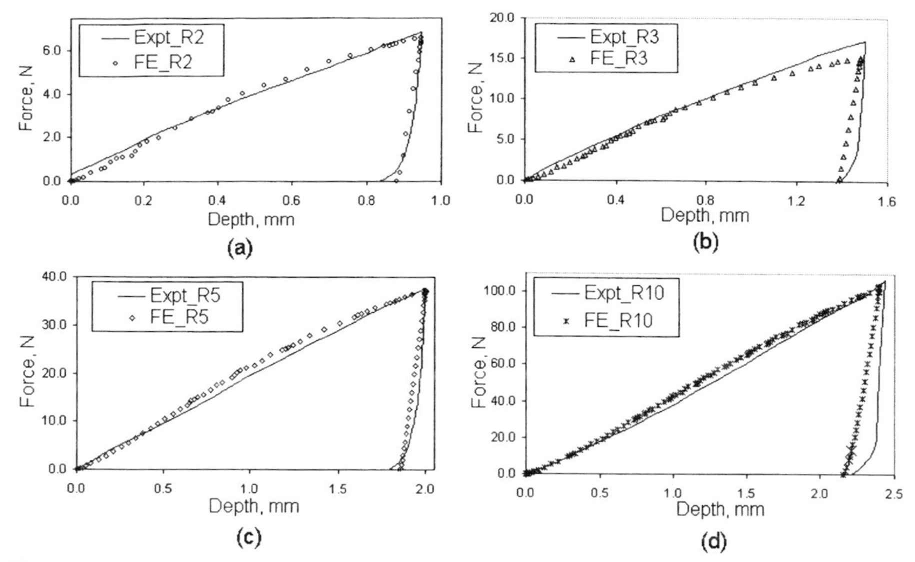

Fig. 5 Load as a function of indentation depth obtained numerically and experimentally for spherical indenter with radius (a) 2, (b) 3, (c) 5 and (d) 10 mm.

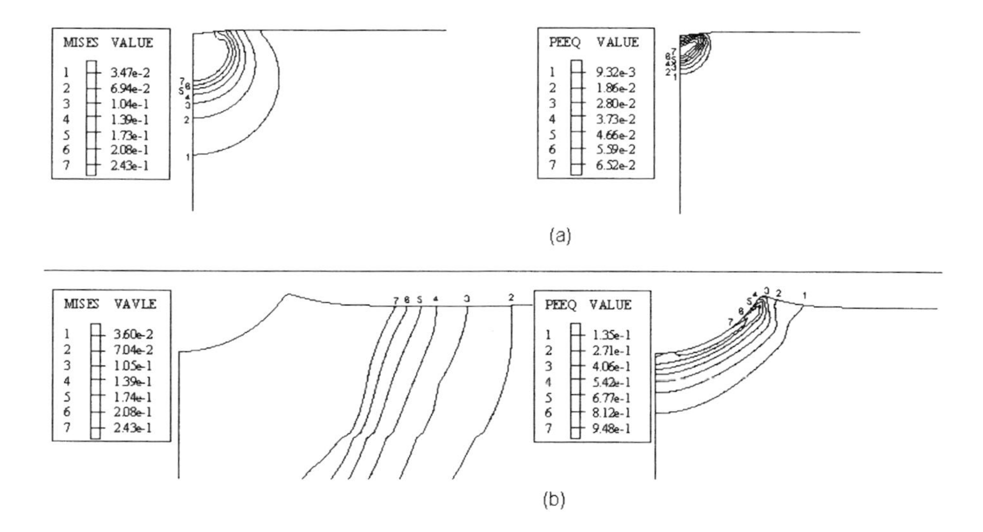

Fig. 6 Contours of equivalent Mises stress and plastic strain corresponding to indentation depth (a) 0.15 and (b) 1.5 mm for a spherical indenter of radius of 5mm.

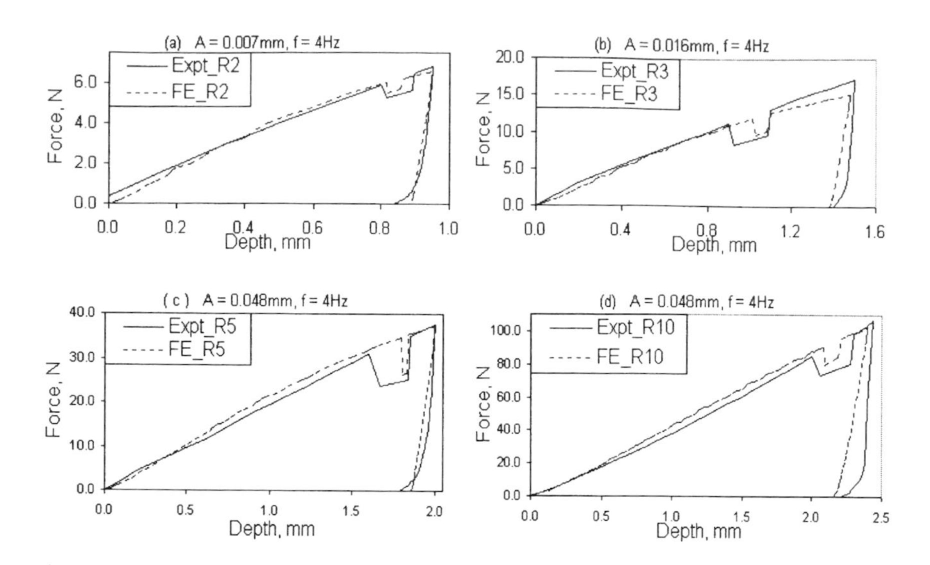

Fig. 7 Load as a function of indentation depth for a spherical indenter of radius (a) 2, (b) 3, (c) 5 and (d) 10 mm, during vibrational indentation test (case 1) with various vibration amplitudes and frequency of 4Hz.

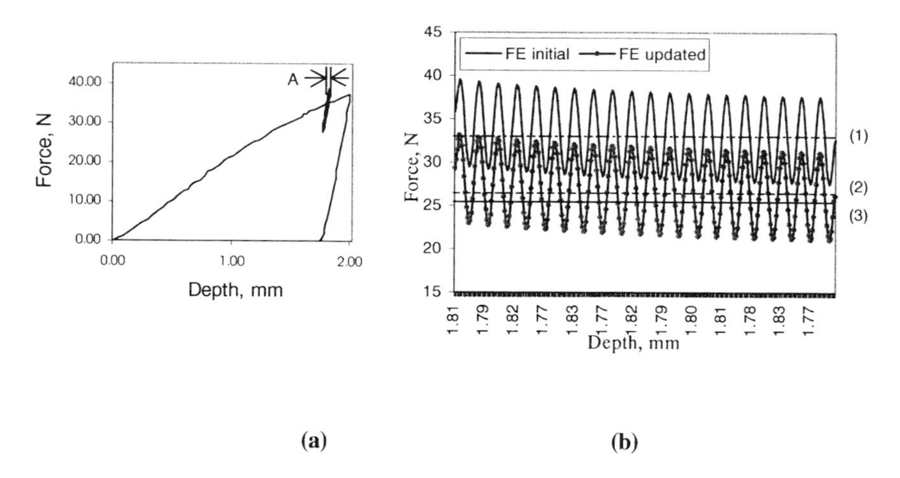

Fig. 8 (a) FE simulation of force–depth data for plasticine indented by a spherical indenter of radius 5mm, (b) expanded view during period A of vibrational loading at an amplitude of 0.04mm and frequency of 40Hz under a lubricated boundary condition, the horizontal straight lines represent the mean force value obtained from (1), initial FE; (2), updated FE and (3) experimental study.

Development of a methodology of analysis and compensation of component form-errors for high-precision forming

Y QIN and **R BALENDRA**
Department of Design, Manufacturing and Engineering, University of Strathclyde, UK

ABSTRACT

Previously, the forming of high-precision components were achieved, largely, by trials on machinery and tools. More effective design/analysis methods were developed, recently, for the qualification of component form-errors and design of the machinery and tools to effect high-precision forming. The design and analysis was conducted for individual processes and components. Research is, currently, being conducted to define a systematic methodology for compensation of component form-errors. The methodology would be used as a theoretical guide to support concurrent "component and process" design for high-precision forming of engineering components.

1 INTRODUCTION

The configuration for the forming of high-precision components was traditionally achieved through production trials. The machine may be enhanced to improve its stiffness and tool structures, and die-cavity geometries may be modified by iterations to minimise dimensional errors of the components. The trials invariably added the cost of manufacturing and resulted in the increased product-development cycle. Subsequent to identifying the influence of tool-deflection on component-forms, efforts were made either to reduce forming-pressures or to enhance tool-stiffness. These studies were assisted by application of conventional analysis to predict working-pressures on the tools. Until Finite Element techniques were developed for the analysis of forming processes, detailed information on stresses and strains in the work-material and tools were not available to the designer; regardless, production trials were necessary, particularly for manufacturing complex component-forms. An FE simulation on the generation of component form-errors during a nett-forming process showed that form-errors of the component could be generated at any stage of forming [1], and an accurate analysis would have to address the inter-relationship between the work-material and tools. The existence of mutual influences between the work-material and tools during loading, unloading and ejection of the component suggests that an analysis procedure for the qualification of component form-errors may not be based exclusively on tool-elasticity. Pressure contours on the die-surfaces may be established by physical modelling experiments for both, loading and unloading; elastic FE analysis may then be conducted, using measured pressure-values, to qualify deformations and stresses [2]. Such a procedure could only be used for preliminary estimates of component form-errors. An accurate analysis would have to be effected using an elastic-plastic modelling procedure [1, 3].

Another development in the design and analysis of high-precision forming was to qualify both, mechanical and thermal effects, on component-forms [4]. Investigations suggest that the thermal components could be more significant than mechanical components of error-contribution [5]. The component-forms also varied with manufacturing cycles because temperature in the tools changed with time [6-7]. The effort to qualify thermal parameters which could influence the analysis of forming processes [8-11] didn't address issues concerning the variation of component-forms with manufacturing cycles.

The approaches developed [1-11] may be used for the analysis of forming processes. A further issue to be addressed is the approach for the compensation of form-errors. Initial efforts were to design a ring-shaped die [12] and a truncate-shell die [13] to enable in-process compensation for forward extrusion of solid bars. Another attempt was to design shrink-fit structures to force the die to pre-contract, thereby reducing or eliminating component-errors which result from the die-deflections [14]. These efforts were, however, insufficient to enable definition of a systematic methodology of compensation for component form-errors.

Since industry needs a theoretical guide to support the design of processes, machines and tools, research was initiated to develop a systematic methodology for the analysis and compensation of component form-errors which could be generated during high-precision production.

2 COMPONENT FORM-ERRORS

In precision forming, component form-errors could be generated at any point in the forming cycle. Prior to the development of methods to reduce or completely eliminate these errors, the type of each error-component and corresponding generation-causes have to be defined. With this in mind, the following investigations were conducted:

- Investigating high-precision forming production procedures – from preparations of the billet, die manufacturing, machine design and uses, set-up of tooling, process control, production environment and the handling of components. The investigation is necessary because errors in component-form are influenced by factors which are associated with every stage of the production.

- Investigating material deformation, die-deflection, heat transfer at every stage of the forming, such as loading (forming), unloading, ejection and the cooling of the component. Variation of component-form with time was also investigated. The investigations were effected by examining engineering components provided by industry, experimental and FE simulation results.

Based on those investigations, component form-errors were defined as the following types:

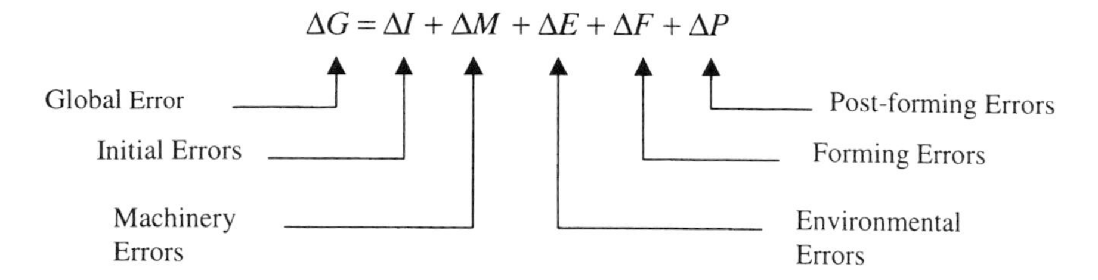

The global error of a component-form is the sum of the following errors:

- ΔI – initial errors, including manufacturing errors introduced into the die-cavity, geometrical errors of the billet, setting-up errors of the tooling and the errors due to die-wear.
- ΔM – errors in machine geometry and/or due to errors in guideways, and machine behaviour during operation.
- ΔE – environmental errors, including that generated due to changes of environmental temperature or use of auxiliary media.
- ΔF – forming errors, including that generated due to die-elasticity, material elasticity-plasticity, and thermal changes during forming.
- ΔP – post-forming errors, including that generated during the handling of the component after forming (delivery, storage), secondary processing, etc.

Forming errors are ones which were, previously, not defined accurately, due to deficiencies in analytical methods and experimental procedures which supported the definition. Through detailed analyses, major sources which contribute to component form-errors were defined as followings:

- Die-elasticity;
- Secondary yielding of the work-material;
- Material deflection upon ejection of the component;
- Temperature changes in tools;
- Temperature changes in the component.

The errors resulting from these sources have to be compensated for. To enable definition of an approach to effect compensation, factors which could influence the generation of each type of error was examined, these being listed in the Table 1.

3 CUMULATION OF INDIVIDUAL ERRORS

A variety of factors could influence the generation of component form-errors; an ideal algorithm may be one which encompasses all these factors.

Efficiency of defining final die-cavity specifications or constructing tools to compensate for component form-errors depends largely on the accuracy and efficiency of the approach to define global form-errors which may be generated during a particular forming-process and work-material. Theoretical approaches and procedures were developed, respectively, to address each individual issue concerning the generation of component form-errors. A general procedure is defined for the integration of individual analyses with a view to enabling cumulation of individual error-components.

Table 1 Forming-error components and factors influencing error-generations

Error-Components	Factors Influencing Error-Generations	
Die-elasticity dependent errors	*Forming pressure -* Magnitude and distribution; Loading characteristics (static and dynamic); Process (loading, unloading); Friction at workpiece/die interfaces; Machine-behaviour.	*Die-configuration -* Die-cavity geometry; Die-construction; Die-material.
Errors resulting from the secondary yielding of the material	*Material property -* Bauschinger effect; Isotropy and anisotropy; Rate-dependent; Temperature-dependent; Behaviours under reverse, loading-orientation-dependent.	*Loading path (history) -* Manufacturing process; Manufacturing cycle; Die-cavity geometry. *Interfacial -* Friction conditions
Errors upon ejection of the component	*Material property -* Elastic-plastic behaviours; Strain distribution after the forming; Residual stresses in the material after the forming. *Component-form -* Die-cavity form (constraints by the die).	*Sequence of component release -* Tool kinematics; Friction at workpiece/die interfaces; Temperature distributions in component and die; Handling of component after the ejection.
Errors resulting from temperature changes in the tools	*Tool/die material -* Physical properties of material (thermal conductivity, specific heat, thermal expansion coefficient). *Tool/die construction -* Single solid die/tool; Compound die/tool-sets; Tool/die-machine interfacial conditions.	*Workpiece/tool interface conditions -* Working pressure; Material/tool surface texture; Lubricant. *Manufacturing cycle -* Duration of the contact between the material and the tools; Number of manufacturing cycle. *Working environment -* Temperature, media.
Errors resulting from temperature changes in the component	*Work material -* Material mechanical property; Material physical property. *Material/tool interface conditions -* Working pressure; Material/tool surface texture; Lubricant/preparation of billet; Initial tool/die temperature. *Component-form -* Die-cavity geometry.	*Process parameters -* Strains to be achieved; Deformation rate (punch velocity); Forming cycle (loading, unloading, ejection of component). *Cooling of component -* Environmental temperature; Handling of component.

Two types of methods may be used for the cumulation of individual error-components generated (Fig. 1), i.e. analytical (conventional) and FE method. The conventional method is to conduct linear-elastic and thermal-elastic computation of expansion and contraction of the die and the work-piece at different stages of the forming cycle, by assuming forming-pressure and temperature values.

FE simulation has been proven to be most efficient means of analysing metal forming operations. To meet requirements on the simulation of high-precision forming, some particular theories and procedures have been developed at the University of Strathclyde; these enable consideration of the influences of elastic, plastic, thermal, and combined factors on the generation of component form-errors during high-precision forming.

Three models were identified as being suitable [15] for elastic and elastic/plastic analysis of high-precision forming. A more complete approach - an elastic/plastic model with cyclical heat-loading was developed for the combined analysis which included coupled thermo-mechanical analysis and heat transfer analysis. The analysis considered both, elastic, plastic, thermal, and the number of manufacturing cycles. In this model, tool kinematics was defined to simulate loading, unloading, and ejection of the component. To consider the tools as elastic bodies with heat transfer characteristics for elastic/plastic flow analysis of a complete forming cycle, particular measures were taken to effect numerical stability of the analysis.

A high-precision forming process could deliver a component to within a few microns. An accurate constitutive model is, therefore, required to characterise the property of the material for subjecting to complex loading cycles. Plastic deformation of the work-material which occurred during unloading of the tool [1] suggested that a simplified elastic-plastic constitutive model may not be sufficiently accurate to predict the material deformation during

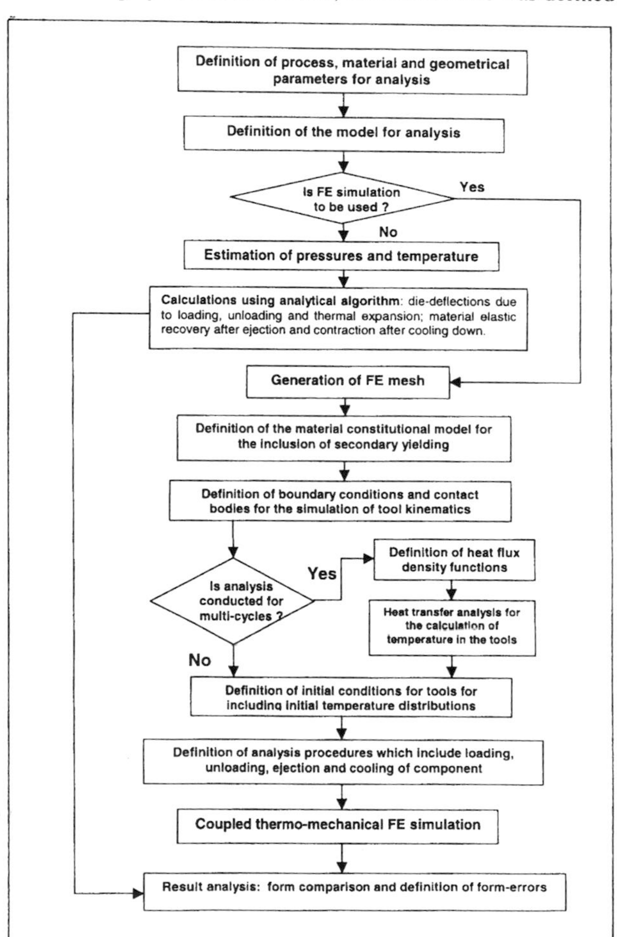

Fig. 1 Procedure for the cumulation of component form-errors

high-precision forming. Methods and experimental procedures were, therefore, developed to create new constitutive models which enabled the prediction of material deformation under complex loading and unloading. The models were linked to the FE code ABAQUS [16], to initiate a coupled thermo-mechanical FE simulation.

In FE simulation, the initial temperature of the billet and tools were defined with reference to the environment and manufacturing cycles. To consider variations of temperature in the tools during high-precision forming, an algorithm was developed for the simulation of temperature changes in the tools during multi-cycle thermal loading. The computed temperature-distributions at the end of each cycle was imported as the initial tool-temperature for the simulation of a subsequent cycle. Based on this algorithm, variations of component form-errors with manufacturing cycles can, now, be qualified.

Boundary conditions were defined by defining contacts and movements between the workpiece and tools; thermal contact conductivity, coefficient of friction and velocity of tool-movement were defined for each contact. FE simulation of a manufacturing cycle included several analysis-steps which were defined in a FE input file, i.e., analysis of billet stabilisation, loading, unloading, ejection and cooling of the component.

The coupled thermo-mechanical FE simulation was conducted for each step defined in the input file; material-deformations and die-deflections at the end of each step were then extracted for evaluation of form-errors. The errors derived from the final contour of the component after cooling were, theoretically, defined as form-compensation requirements.

4 ERROR-COMPENSATION

Based on the examination on the factors listed in the Table 1, a series of simulations were conducted to establish feasible concept of form-error compensation.

Material – mechanical and physical properties of the work- and tool-material significantly influence accuracy of the formed component. Component form-errors may be reduced by selecting an appropriate combination of the work- and tool-material, if the combination does not raise other concerns for high-precision forming of the component.

Geometrical – die-cavity geometries may be modified, with reference to the component form-errors identified. The application of this approach may be limited because the component errors, in most cases, occur in a complex form (nonlinear distribution). The manufacturing of complex of high accuracy could be prohibitive. Preform design can be used as a means of improving the accuracy of component-forms. The billet-geometry may be designed to enable better material flow during forming, consequently, forming-pressure requirements may be reduced, and more acceptable springback behaviour may be introduced.

Equipment – setting-up changes of tooling, construction of dies/tools (e.g. using shrink-fit structures, designing tools to enable in-process compensation [13]) can be used as means of compensation for component form-errors. A "flexible die" concept is being developed, in which the die (or die-sets) may be designed to effect a "self-adjustment" of die-deflection with changes in forming-pressure and temperature. Using such a design, the die-bore would contract or expand in response to the changes of forming pressure and temperature. As a result, variation of material-forms caused by the changes of forming pressure and temperature can be negated. Initial trials were conducted for forward extrusion; the design considerations are being extended to thermal factors. Further effort is required to develop "in-process" compensation configurations for other forming processes.

Operational – design of lubrication, control of deformation-rate, monitoring of die-wear, design of forming-sequence, etc. can also be used as means for form-error compensation. Friction at the die/workpiece interface influences forming-pressure requirements during

Table 2 List of processes, errors and recommended compensation-methods

Process	Illustration	Specifications on form-errors	Cumulation of errors	Recommended methods for compensation
Completely closed-die upsetting		Non-uniform dimensional-errors	Analytical algorithm and FE simulation procedure developed	Modification of die-cavity geometries, use of a shrink-fit structure for compensation
Forward extrusion		Different form-errors for extrudate and material contained in the die	Analytical algorithm and FE simulation procedure developed	Split die-structures, compensate errors respectively for two-parts
Back-can extrusion		Non-uniform dimensional-errors for can and solid block, thin-wall of the can tends to bend	FE simulation procedure developed	Modification of die-cavity and punch-nose, use of a shrink-fit structure
Double-can extrusion		As above	FE simulation procedure developed	As above
Forward combined with back-can extrusion		Different form-errors for extrudate and material contained in the die	Analytical algorithm and FE simulation procedure	Split die-structures, compensate errors respectively for two-parts
Flashless closed-die forging		Non-uniform errors across the sections, due to non-uniform material-flow and spring-back	FE simulation procedure developed	Preform design to improve material-flow, modification of die-cavity
Injection Forging		Different errors for the bar and the flange	FE simulation procedure developed	Split die-structures, preform design to reduce pressure requirements
Injection Forging combined with forward extrusion		Different errors for the bar and the can (flange)	As above	As above, and the method used for can extrusion
Injection Forging combined with back-can extrusion		Different errors for the bar, can and flange	As above	As above

loading and die-contraction during unloading; hence, the friction influences component form-errors. The deformation-rate (or production-rate) significantly influences heat generation during forming and temperature rise with manufacturing cycles. The forming-sequence determines distribution of plastic strains in the work-material, and hence, heat generation. All these may be considered as possible control-parameters for reducing component form-errors.

Environmental – controlling the temperature of the forming environment may also be used as a means for compensating for component-errors which result from temperature changes in the component and tools; handling of workpiece may be carefully planned to reduce non-uniformity of the cooling of the component.

In principle, the methods listed above can be used for any a high-precision forming process; recommendations for some typical processes are listed in Fig. 2. Other concerns about the selection of a compensation method are the cost of tool-manufacturing, equipment requirements for process control, production-rates, life of tools, etc. A synthesising algorithm is to be developed to encompass manufacturing and economic issues of error compensations.

5 THE PROCEDURE OF DESIGN AND MANUFACTURING

To achieve high precision during forming, issues other than analysis and compensation would also have to be addressed, because compensation for component form-errors may not be an isolated activity. It is necessary to define a general procedure which would enable integration of design and manufacturing for high-precision forming of engineering components.

An ideal system for design, analysis and manufacturing of high-precision forming should be one which is supported by a series of design, analysis and manufacturing modules (refer to the flow-diagram in Fig. 2). For given specifications on component-form, decisions have to be made to select a nett-forming process, such as forward extrusion, back-can extrusion, injection forging, closed-die forging or variations or combinations of these processes. This may be effected using a variant approach, based on the information on component-forms produced previously. A database containing information on previous products, which are associated with certain high-precision forming processes, is required. The same strategy may be applied to the selection of forming sequences, materials, lubrication, process parameters and tools. A link to a knowledge-based system for forming is required, with a view to providing an efficient support to the decision-making. Previous work in developing a decision-support system for nett-forming [17] will be extended to high-precision forming.

For a new component-form, analysis may be conducted to quantify the component form-errors, with reference to a selected process, materials and tools, either using analytical algorithms, or using FEM. The analysis may be effected by following the procedure defined for the cumulation of individual error-components (Fig. 1). This is followed by decision-making on the selection of an approach for compensating for the form-errors identified.

After tool-geometries, material and processing parameters have been determined, detailed CAD-descriptions of tools may be produced for use in NC programming for tool manufacture. Die-cavities may be formed by EDM. Components can, then, be manufactured by the nett-forming process selected. The components, finally, may be subjected to inspection.

Compared with other design and manufacturing procedures, the procedure for high-precision forming includes four distinguishing phases, i.e., design, analysis, compensation and manufacturing. The compensation is a necessary addition to the design and manufacture for nett-forming.

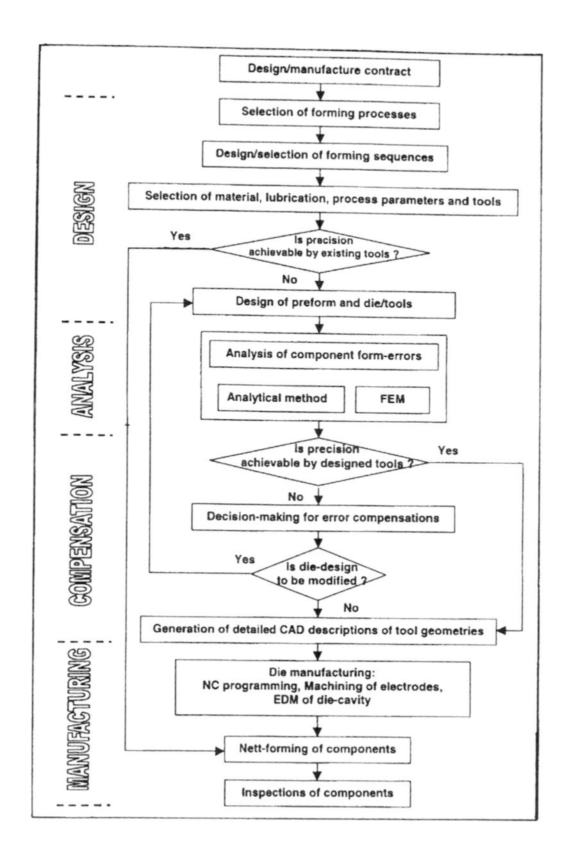

Fig. 2 Procedure of design and manufacturing for HPF

6 CONCLUSIONS

A methodology is being developed to enable concurrent design, analysis and compensation of high-precision forming of engineering components. For a new component-form, component form-errors may be qualified for a selected nett-forming process, material and tools, using the analytical procedures developed. Appropriate methods may then be selected for compensation of the error-components identified, which can be in material, geometrical, operational and environmental. The methods may be assessed, with reference to product-precision and manufacturing costs, to select an "optimal" means of compensation of component form-errors. A synthesis algorithm needs to be developed to effect this assessment efficiently.

REFERENCES

[1] Y. Qin and R. Balendra, "FE simulation of the influence of die-elasticity on component dimensions in forward extrusion", Int. J. of Mach. Tools Manuf., Vol. 37, No. 2, pp. 183-192, 1997.

[2] R. Balendra and Y. Qin, "Die-elasticity errors during nett-forming of engineering components", Advances in Manufacturing Technology IX, edited by D. Stockton and C. Wainwright, Taylor & Francis, London, pp. 204-208, 1995.

[3] R. Balendra, Y. Qin and X. Lu, "Development of approach and technology for die-cavity elasticity error analysis and compensation for nett-forming of components", Proc. of the

Int. Conf. on Challenges to Civil and Mechanical Engineering, Wroclaw, Poland, Vol. II, pp. 49-59, 1996.

[4] M H Sadeghi and T A Dean, "Analysis of dimensional accuracy of precision forged axisymmetrical components", Proc. of the Institution of Mechanical Engineers, Part B - J. of Engineering Manufact. Vol. 205, No. 3, pp. 171-178, 1991.

[5] H. Long and R. Balendra, "The influence of thermal and elastic effects of the accuracy of cold-formed components", Proc. 32nd Int. MATADOR Conf. Manchester, UK, pp. 361-366, 1997.

[6] Y. Qin and R. Balendra, “A method for the simulation of temperature stabilisation during multi-cycle cold forging operations”, Proc. of the 15th Int. Conf. on Computer-aided Production Engineering, April, Durham, UK, pp. 499-506, 1999.

[7] T. Kato, Y. Tozawa, K. Nakanishi and T. Kawabe, "Prediction of dimensional accuracy of cold extruded can during repeating operation", Annals of the CIRP, Vol. 35/1, 1986.

[8] J. G. Lenard and M.E. Davies, "The distribution of temperature in a hot/cold die set - the effect of the pressure, temperature, and material", Trans. of the ASME, J. of Engng Materials, Vol. 117, No. 2, pp. 220-227, 1995.

[9] S. Shamasundar, A.G. Marathe, S.K. Biswas, "Computer-aided prediction of temperature history in a upset forging operation", Proc. of the 25th Int. MATADOR Conf. Machester, UK, pp. 405-411, 1985.

[10] T. Kato, Y. Tozawa and T. Tanaka, "Thermal analysis for backward cold extrusion", Proc. 7th ICFC, pp. 397-403, 1985.

[11] T. Udagawa, E. Kropp and T. Altan, "Investigation of metal flow and temperatures by FEM in the extrusion of Ti-6Al-4V tubes", J. of Material Processing Technology, Vol. 33, No. 1-2, pp. 155-174, 1992.

[12] T. Wanheim, R. Balendra and Y. Qin, "An extrusion die for in-process compensation of component-errors due to die-elasticity", J. of Material Processing Technol. Vol. 72, No. 1, pp. 177-182, 1997.

[13] K. Osakada, "Precision cold forging with controlled elastic deformation of die", 27th ICFG Plenary Meeting, Sept. 19-23th, Padova, Italy, 1994.

[14] Aoa Ibhadode, "On a method for die container design in completely closed die forging", ASME Transactions, J. of Manufact. Sci. and Engng. Vol. 119, No. 3, pp. 438-440, 1997.

[15] X. Lu and R. Balendra, "Evaluation of FE models for the calculation of die-cavity compensation", J. of Material Processing Technol. Vol. 58, No. 2-3, pp. 212-216, 1996.

[16] Hibbitt, Karlsson and Sorensen, Inc. ABAQUS Manual, 1998.

[17] Research Reports, "Decision-support system for nett-forming (NETTFORM)", Contract No. BRE2-CT92-0304, Brite-Euram No. BE-5819.

Analysing the parameters in deep drawing using analytic hierarchy process

A W LABIB, J ATKINSON, and **S HINDUJA**
Department of Mechanical Engineering, University of Manchester Institute of Science and Technology, UK

SYNOPSIS

This paper describes the results obtained from a study to identify and prioritise the different parameters that could be incorporated in an intelligent controller for a sheet metal forming press. The initial step was to develop a model based on part defects, their causes, and remedial actions needed. A questionnaire based on this model was circulated to various experts in order to prioritise the defects and their causes. The analysis was performed using a multiple criteria decision-making tool called the Analytic Hierarchy Process. The results obtained identify the critical defects and a sensitivity analysis of the three most critical defects is presented.

1. INTRODUCTION

Sheet metal forming is a highly complex process to model and control due to the many parameters involved and the non-linear interaction between them. Consequently, one of the main dilemmas in designing a controller for this type of process lies in determining the parameters that influence the behaviour of the system and their relative importance.

Although significant advances have been made in sheet metal forming, twenty-five percent of the die manufacturing time is spent on a trial-and-error approach to re-shape and re-position the drawbeads and vary the forming parameters to obtain a good quality part (1). This time-consuming and inefficient process is due to the complexity of the forming problem, which mainly involves non-linear material, and lubricant behaviour and the development of instabilities during deformation. Parameters affecting the forming process have been studied by several authors (2-4). The present paper is concerned with establishing the relative importance of part quality defects, their causes and remedial actions using a multiple criteria group decision-making system. The results obtained from this work have been used to design and implement a controller for improving part quality in deep drawing (5).

2. THE ANALYTIC HIERARCHY PROCESS

In most organisations, decisions are made collectively, and it is sometimes difficult to achieve a consensus among group members. On the other hand, group decision-making generally produces better solutions to a particular problem than those made in isolation by individuals, as it conveys different points of view, interests and backgrounds from its participants. The Analytical Hierarchy Process (AHP), developed by Saaty (6) is a methodology that focuses on the choice phase of the decision-making process. It is designed to solve complex problems involving multiple criteria with respect to competing objectives. The AHP is a method for breaking down a complex, unstructured problem into its component parts. These parts, or variables, are arranged into a hierarchical order. Numerical values are then assigned to subjective judgements on the relative importance of each variable. Next, judgements are synthesised to determine which variables have the highest priority and should be acted upon to influence the outcome of the problem. The judgements are used in deriving ratio scale priorities for the decision criteria and alternatives. Finally, the output of AHP is a prioritised ranking indicating the overall preference of the decision alternatives. A by-product of the process is the provision of a measure of consistency. It also facilitates a sensitivity analysis. In summary, the three main steps in AHP, due to Saaty (6), are as follows:

(i) Decomposing the problem into a hierarchical structure.

(ii) Making comparative judgements in order to establish priorities for the elements of the hierarchy.

(iii) Performing a synthesis of the various competing elements.

2.1 Hierarchical structure

In the first step, the overall problem is decomposed into its basic components, thus enabling a hierarchical structure to be established. Several hierarchical structures are available, such as: goal, criteria, alternatives; goal, criteria, sub-criteria, alternatives; goal, scenarios, criteria, sub-criteria, alternatives, etc. In the present work, a hierarchical structure consisting of the overall goal, criteria (part defects), sub-criteria (main causes of the part defects) and alternatives (remedial actions for the causes of the problem) has been employed.

In this work, the overall goal is to establish a priority ranking of the part defects that can affect the sheet metal process, as identified by process experts, in terms of criticality.

The criteria are the various part defects that occur in sheet metal work, such as tearing and/or thinning, underform, geometric distortion, panting, springback, wrinkles, scoring, slugpick and burrs.

The sub-criteria are the causes that are responsible for the aforementioned defects (see Table 1). Many of these are common to more than one problem. Each of these defects can result from several causes. For example, wrinkling may be caused by one or more of the following: material being not to specifications (*MATNOT*); tool being not to specifications (*TOOLNOT*); blank holding and die cushion pressures, and cushion pin lengths not set correctly (*SETTING*); inappropriate lubrication (*LUBRICAT*); wear of cushion pads (*CUSHION*); and misalignment of tool (*MISLOC*).

In the case of sheet metal forming, the alternatives are the remedial actions for the causes identified. However, since there is a one-to-one relationship between a cause and the remedial action, a two-level hierarchical structure is sufficient, with the part defects at level one and their causes at level two.

2.2 Comparative judgements to establish priorities

Having established a hierarchical structure, the next step is to evaluate the relative importance of: (i) the various defects; and (ii) the causes with respect to a given defect. It was not necessary to evaluate the relative importance of the remedies with respect to each of the causes since there is a one-to-one relationship between a cause and its remedial action.

AHP uses pairwise comparisons to derive priorities instead of arbitrarily assigning values as is done in many score and weight methodologies. The derived priorities are easier to justify and are usually more accurate. In our work, the pairwise comparison has been made on a 9-point scale, which is a reasonable basis for discriminating between two items (7). For example, if alternative A is "extremely preferred" to alternative B, then a value of 9 is used. At the other end of the scale, if both alternatives are "equally preferred", then a value of 1 is used.

As a starting point, all the experts (press users, press manufacturers, researchers in sheet metal forming) involved, identified and agreed on the most important defects -and their related causes- that occur in the sheet metal forming process. Following this, a questionnaire was

DEFECTS → CAUSES ↓	Splitting and/or thinning	Underform	Geometric Distortion	Panting	Springback	Wrinkling	Scoring	Burrs	Slugpickup
Material not to specification	X	X	X	X	X	X	X	X	n/a
Tooling not to specification	X	X	X	X	X	X	X	X	X
Mis-location	X	n/a	n/a	n/a	n/a	X	n/a	X	n/a
Setting incorrect	X	X	X	X	X	X	n/a	n/a	X
Cushions pins pad worn	n/a	X	n/a	n/a	n/a	X	n/a	n/a	n/a
Lubrication incorrect	X	n/a	X	X	n/a	X	X	X	X
Press wear	X	X	X	X	n/a	n/a	n/a	X	n/a
Low material temperature	X	X	n/a	n/a	n/a	n/a	n/a	n/a	n/a

Table 1 Defects and their causes

prepared in which the participants were asked to do pair-wise comparisons among all the defects, and their respective causes. As an example, Figure 1 shows fragments of the questionnaire completed by one of the expert press users. In this questionnaire the decision-maker was asked to express the relative importance between part defects. For instance, it can be noticed that in line 4 the 'tearing' is considered to be eight times more critical than 'springback' , whilst in line 29 'burrs' is seven times more critical than 'springback'.

Compare the relative IMPORTANCE with respect to: GOAL

1=EQUAL 3=MODERATE 5=STRONG 7=VERY STRONG 9=EXTREME

1	TEARING	9	8	7	6	5	**4**	3	2	1	2	3	4	5	6	7	8	9	UNDERFORM
2	TEARING	9	8	7	6	**5**	4	3	2	1	2	3	4	5	6	7	8	9	G.DISTORTION
3	TEARING	9	8	**7**	6	5	4	3	2	1	2	3	4	5	6	7	8	9	PANTING
4	TEARING	9	**8**	7	6	5	4	3	2	1	2	3	4	5	6	7	8	9	SPRINGBACK
5	TEARING	9	8	7	6	5	4	**3**	2	1	2	3	4	5	6	7	8	9	WRINKLES
6	TEARING	9	8	7	6	**5**	4	3	2	1	2	3	4	5	6	7	8	9	SCORING
7	TEARING	9	8	7	**6**	5	4	3	2	1	2	3	4	5	6	7	8	9	BURRS
8	TEARING	9	**8**	7	6	5	4	3	2	1	2	3	4	5	6	7	8	9	SLUGPICK
9	UNDERFORM	9	8	7	6	5	**4**	3	2	1	2	3	4	5	6	7	8	9	G.DISTORTION

26	PANTING	9	8	7	6	5	4	3	2	1	2	3	4	5	**6**	7	8	9	SLUGPICK
27	SPRINBACK	9	8	7	6	5	4	3	2	1	2	3	4	**5**	6	7	8	9	WRINKLES
28	SPRINBACK	9	8	7	6	5	4	3	2	1	2	3	**4**	5	6	7	8	9	SCORING
29	SPRINBACK	9	8	7	6	5	4	3	2	1	2	3	4	5	6	**7**	8	9	BURRS
30	SPRINBACK	9	8	7	6	5	4	3	**2**	1	2	3	4	5	6	7	8	9	SLUGPICK
31	WRINKLES	9	8	7	6	5	4	**3**	2	1	2	3	4	5	6	7	8	9	SCORING
32	WRINKLES	9	8	7	6	5	4	**3**	2	1	2	3	4	5	6	7	8	9	BURRS
33	WRINKLES	9	8	7	**6**	5	4	3	2	1	2	3	4	5	6	7	8	9	SLUGPICK

Figure 1 Fragments of one completed pairwise comparison questionnaire
(Numbers in bold represent user's selection)

Category	Problem	Criticality Rating (%)
Highly Critical	Splitting and/or thinning	34,1
	Underform	19,7
	Ripples	14,2
Medium Critical	Burrs	9,7
	Geometric distortion	7,5
	Scoring	6,8
Low Critical	Slug pick up	3,1
	Spring back	2,8
	Panting	2,1

Table 2 Relative importance of part defects

3. RESULTS AND DISCUSSION

The responses from the different experts were synthesised and the resulting prioritisation of the defects in deep drawn parts is shown in Table 2. Note that in this table, a unanimous prioritised ranking is presented since all the experts were asked to agree amongst themselves on each of the judgements in the pairwise comparisons, rather than at the end having to perform an average of their prioritisations. In this table, the defects have been classified into three categories depending on their relative importance. Table 3 shows the prioritisation of the causes relative to the overall goal. From this table, it is clear that material and tool being not to specifications have the highest priority. *Setting* i.e incorrect setting of the blankholding and die cushion pressure and/or cushion pin lengths, tends to have the next highest priority. *LowTemp* i.e. low material temperature has the lowest priority.

Experts	HIGHEST ←			IMPORTANCE				→ LOWEST
Press User A	MatNot	ToolNot	MisLoc	Setting	Cushion	Lubricat	Press	LowTemp
Press Manufac-turer	MatNot	ToolNot	Setting	Lubricat	Cushion	Press	MisLoc	LowTemp
Press User B	ToolNot	MatNot	Setting	Cushion	Press	MisLoc	Lubricat	LowTemp
Research Expert	ToolNot	Setting	Lubricat	Cushion	MatNot	Press	MisLoc	LowTemp

Table 3. Relative importance of causes

From the prioritised ranking of the defects (Table 2), it is clear that by attending to the five most critical problems i.e. tearing/thinning, underform, wrinkling, burrs and geometric distortion, 85% of the defects can be accounted for. From Table 3, the main causes for these defects, arranged in groups of decreasing priority, are:

Highest Priority	Intermediate Priority	Medium Priority
Material specification	Setting	Press wear
Tooling specification	Wear of Cushion pads	Mislocation
		Lubrication

While it is desirable to control all the above factors, there are limitations on the actions or adjustments that the operator can actually perform during the deep drawing operation. If control over the blank holder force, cushion pressure and lubrication, as is conventionally done, then the process problems will not be fully addressed.

In order to get a better understanding of the decision analysis, a sensitivity analysis was carried out. In this type of analysis, the priority of a particular defect is changed and its effect on the priorities of the causes is examined. This sensitivity analysis has been carried out for only three critical defects; it was performed using the gradient sensitivity method as this provides all the information visually.

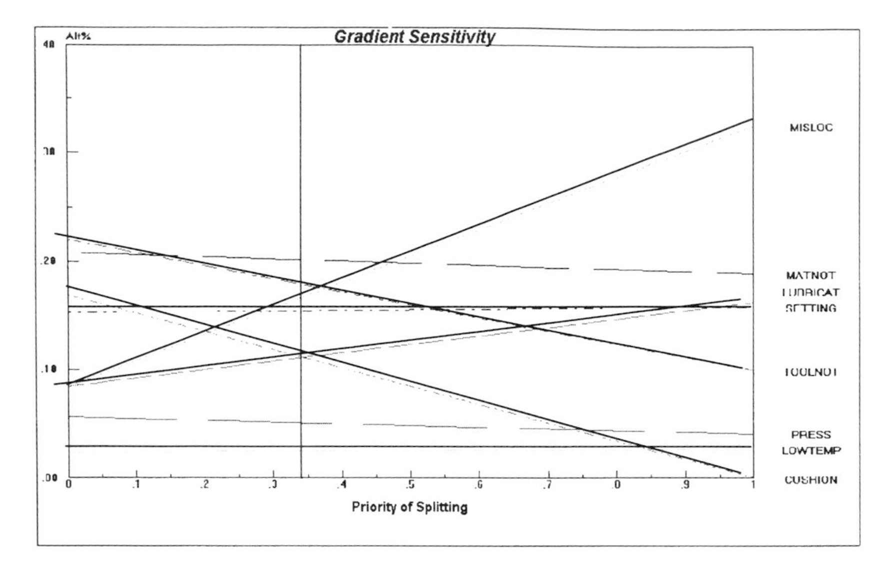

Figure 2 Gradient sensitivity analysis for tearing/thinning(splitting)

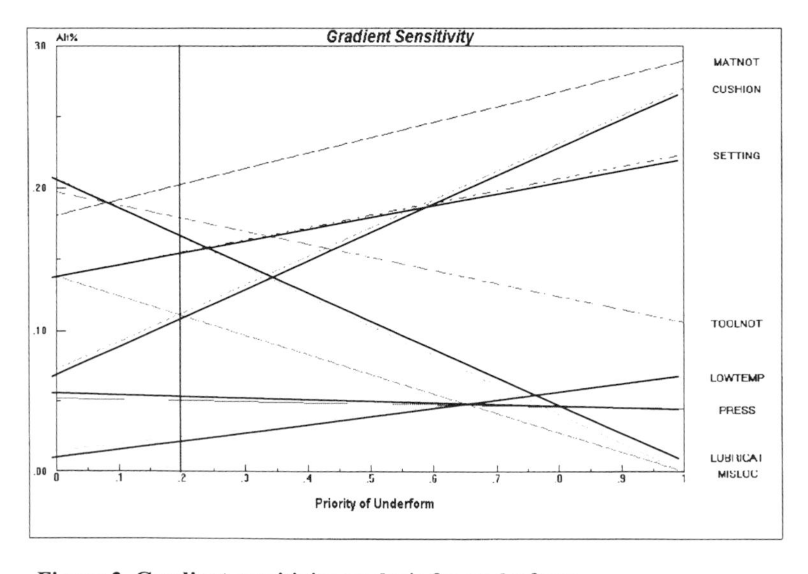

Figure 3 Gradient sensitivity analysis for underform

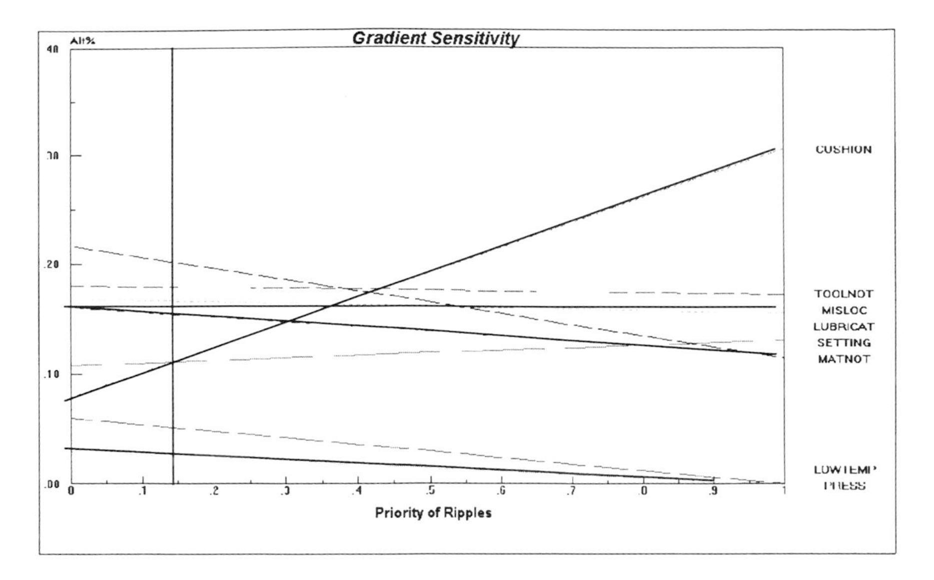

Figure 4 Gradient sensitivity analysis for wrinkling (Ripples)

Figure 2 shows the sensitivity analysis for tearing/thinning. The intersection of the vertical line with the horizontal scale shows the actual value of the priority of tearing/thinning i.e. 34%. The intersection of the other lines representing the causes with this vertical line, represents their priorities. Moving the vertical line to the left or to the right alters the priority of tearing/thinning. In Figure 2, only if the priority of thinning/tearing is increased from its present value of 34% to about 50% by moving the line to its right, will the main cause for this defect shift from material not to specification to mislocation. Conversely, if tearing/thinning is given a much lower priority, say 10%, by moving the vertical line to the left, then the most important cause would be tooling not to specification.

Figure 3 shows that underform is relatively less sensitive to its priority ranking than tearing/thinning. If its priority were increased, the three main causes would be MATNOT, CUSHION and SETTING.

Wrinkles, on the other hand, have four main causes (TOOLNOT, MISLOC, LUBRICAT and SETTING) which have local priorities between 16 and 21% (see Fig.4). If the priority ranking of wrinkles is increased or decreased by 50% from its present priority value, the order of the causes will not change and their magnitude is only marginally affected.

4. CONCLUSIONS

(i) A two-level hierarchical structure in AHP was adequate to represent the sheet metal problem.

(ii) According to the consensus of the experts consulted, tearing, underform, and wrinkles are the most critical defects. These critical defects together with burrs and geometric distortion account for more than 85% of the defects in deep drawing.

(iii) With reference to these three critical defects, the causes with high priority are material and tool not to specifications, cushion pressure, blankholding force and lubrication.

(iv) The sensitivity analysis has demonstrated its value by giving a complete picture of the problem.

ACKNOWLEGEMENTS

The authors would like to acknowledge the EEC for the Brite-Euram Grant. Also our partners: Fagor Arrasate, Spain; Robotiker, Spain; Gestamp Noury, France; and GDA, U.K. Thanks are also due to Manuel A. Pulido Gonzalez for his assistance with this research.

REFERENCES

(1) Coduti, P.L., "Topological behaviour of solid lubricant films on bare and coated sheet steel products", *SAE Technical paper 870648*, 1997.
(2) Hardt, D. and Fenn, R., "Real-Time Control of Sheet Stability During Forming", *Journal of Engineering Industry*, Vol. 115, August, pp 299-308, 1997.
(3) Siegert, K., "Advances in sheet metal forming processes", *Automotive Engineering*, Vol. 105, No 11, November, pp 98-101, 1991.
(4) Mejlessi, S. and Lee, D., "Deep Drawing of Square-Shaped Sheet Metal Parts, Part 2: Experimental Study", *Trans ASME J of Engg and Ind*, Vol. 115, February, pp 110-117, 1993.
(5) Hinduja, S., Atkinson, J., Lau, D.K.Y., Labib, A.W. and Agirrezabal, P., "An intelligent controller for improving the quality of deep drawn components", *Annals of the CIRP*, Vol 1, 2000.
(6) Saaty, T. L., *"The Analytic Hierarchy Process"*, McGraw-Hill, New York, 1980.
(7) Miller, G. "The magical number seven plus or minus two: some limits on our capacity of processing information", *Psychological Rev.*, Vol 63, pp 81-97, 1956.

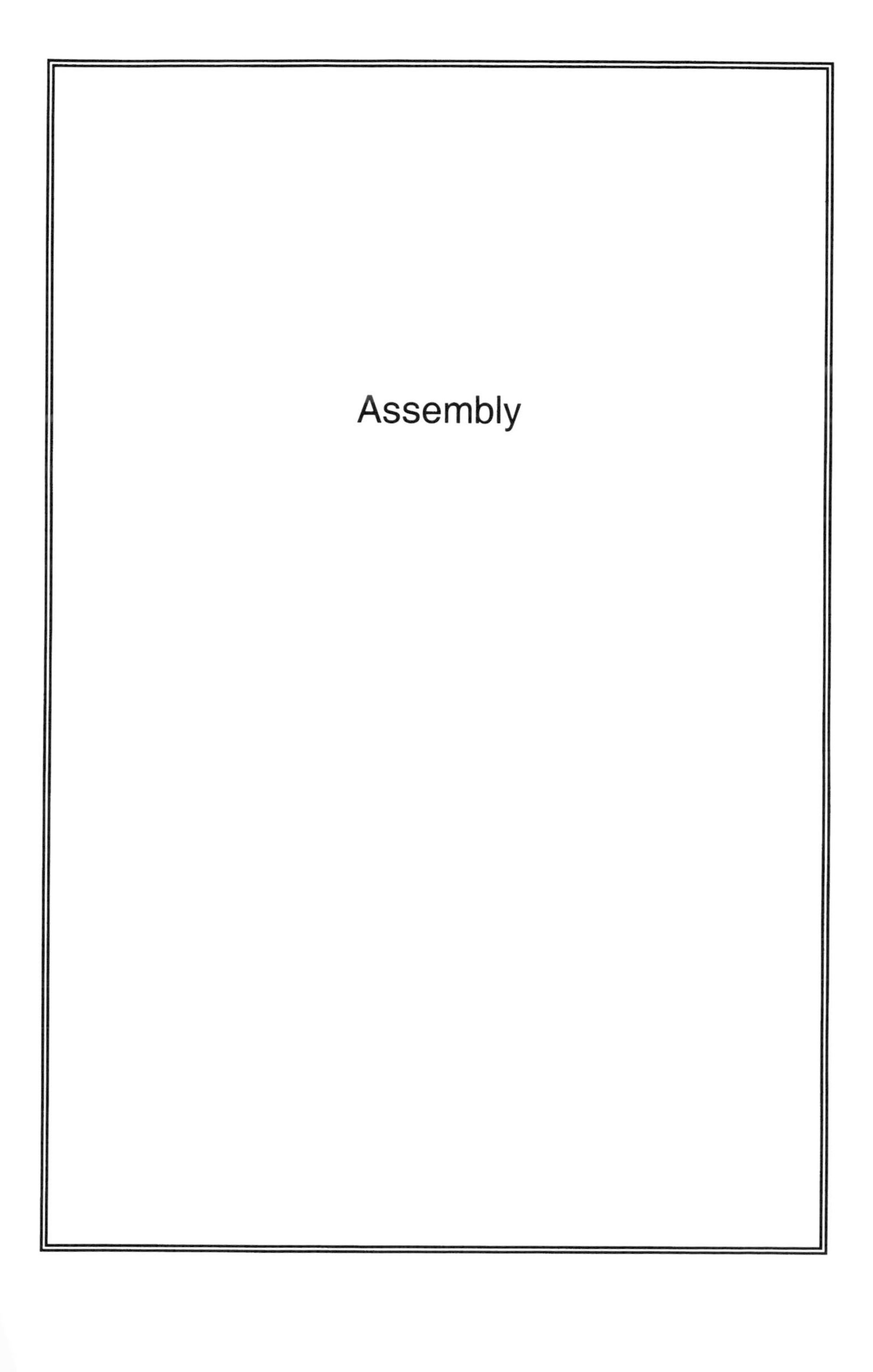

Assembly

CAPE/054/2000

Use of a continuum actuator as a form of controllable, remote centre compliance device

J B C DAVIES
Department of Mechanical & Chemical Engineering, Heriot-Watt University, Edinburgh, UK

INTRODUCTION

The applications of robots to assembly activities has been hampered by the jamming of components during insertion. Whitney (1) identified the problems associated with peg-in-hole insertion tasks and produced a reliable analysis relating frictional forces, insertion forces and peg and hole geometry. This analysis was applied to the design of his well known Remote Centre Compliance (RCC) device which relies upon the use of a compliant robotic wrist with a fixed virtual centre of rotation. This centre of rotation is located at the outer surface of the peg to be inserted and, for any specific RCC, can only accommodate a single component length, figure 1.

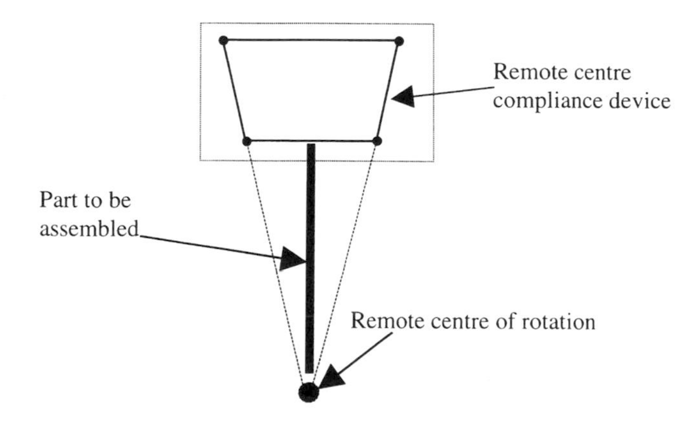

Figure 1: Whitney remote centre compliance device

A device such as the RCC is necessary because the jamming condition that can occur during the assembly process is exacerbated by the rotational characteristics of standard industrial robot joint arrangements. These joints allow rotations about centres that are located outside the component work space so that it is almost impossible to accommodate jamming during the peg - in hole insertion. Jamming is distinguished from wedging in that if the insertion force is removed, a jammed component is free to be moved or removed. On the other hand, if a component is wedged, removing the insertion force does not free the component for subsequent movement. In this paper the device described only addresses the jamming condition.

Following the initial steps made by Whitney, several other devices have been produced, but most still suffer from the limitation of only being useful for a single length of component.

Other researchers such as Jeong and Cho (2) have used a vibratory pneumatic wrist to attempt to overcome the same problem. whilst Hirzinger et al (3) initiated a force feed-back control approach that also been adopted by other workers in the field.

A new approach, based upon a new form of flexible motion generator, called a "continuum actuator" eliminates many of the difficulties associated with peg in hole insertion and provides techniques that are both controllable and predictable. The actuator is mounted either as a gripper or as a wrist with controlled deflections perpendicular to the plane of the object to be inserted. In both circumstances, a circular search pattern is employed and a passive force is employed to enable insertion to proceed when the correct geometric conditions obtain.

Actuator Description

The continuum actuator is an assembly of three elements whose design is such that increasing internal pressure within an element causes a significant change in the axial length of the element. In this context, a significant change is of the order of 10 - 20% of the original length. The elements are assembled with their longitudinal axes parallel and with each element mounted at the vertices of an equilateral triangle. By varying the pressures in each element, the deflections produced cause the actuator to adopt a curved position. The radius of curvature of the actuator and the magnitude of the tip deflection from the unloaded position is a function of the internal pressure differentials, figure 2.

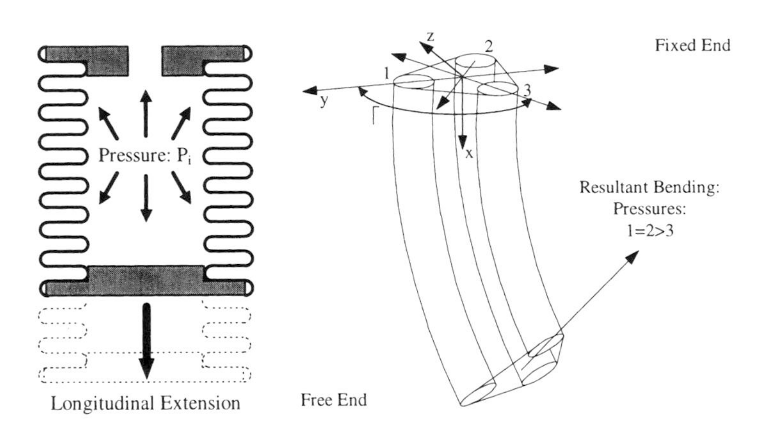

Figure 2.: Actuator operating principle.

The actuator itself may be used in a groups of three to form the basis of a dextrous gripper, figure 3, and has been used as such in several subsea applications, (4), (5), (6).

Figure 3: Subsea dextrous gripper

For assembly purposes, although the actuators are mounted in a disposition similar to that of a three fingered gripper, in this case the actuators are directly attached to the peg that is to be inserted, allowing the orientation of the peg to adjusted by varying the pressures in the relevant actuator elements. This removes any variables associated with the part grasping parameters, i.e. it eliminates the possibility of dropping the part during the insertion tests.

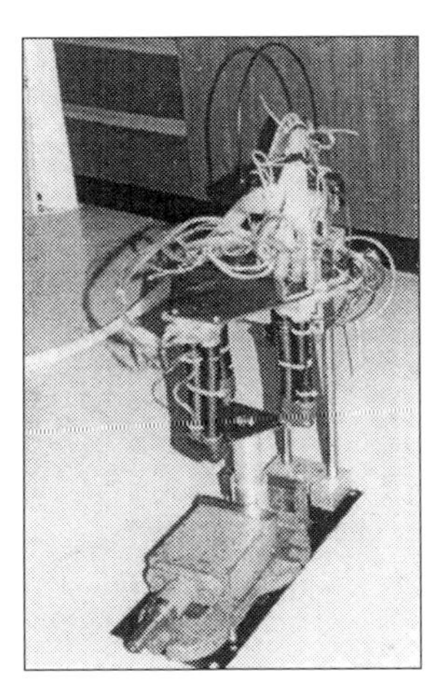

Figure 4: Actuator insertion test rig

To control the part orientation through the deflection of the actuator, the magnitude of the internal pressure in each individual element must be known and controlled. Figure 5 shows the co-ordinate system employed and the orientation, θ, of the plane of curvature of the actuator a for a known set of element pressures, P_1, P_2, P_3.

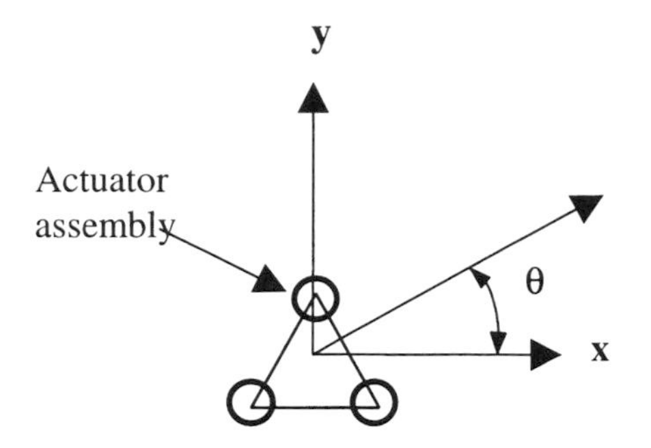

Figure 5: Actuator co-ordinate system

If the pressure in element i is given by P_ii, for an actuator consisting of three elements mounted at the vertices of an equilateral triangle, the orientation of the bending plane is given by equation {1}, derived by Davies (7)

$$\theta = \frac{\frac{1}{2} - \left(\frac{P_1}{2\sum_{i=1}^{3} P_i} + \frac{P_3}{\sum_{i=1}^{3} P_i} \right)}{\sin 60 \left(\frac{P_1}{\sum_{i=1}^{3} P_i} - \frac{1}{3} \right)} . \quad (1)$$

The magnitude of the tip deflection is a function of the bending moment generated by the internal pressures and the axial and bending stiffness of the actuator assembly. For any given actuator, the bending stiffness and axial stiffness maybe assumed constant over the pressure range considered and hence the tip deflection and the radius of the sircle described by the tip as θ varies is only a function of the applied pressures. The bending moment developed by the actuator is given by equatio {2} aslo derived by Daveis (7)

$$M_b = \frac{2h}{3} \sum_{i=1}^{3} P_i A \cos \phi_i \ . \quad (2)$$

Where

A = crossectional area of element i

h = altitude of the equilaterla triangle formed by the elements of the actuator

ϕ = orientation of element i relative to θ, the orientation of the bending plane.

Test rig description

A standard 35mm F6/g7 peg and hole assembly were used for the test with the female portion mounted in a machine vice. The male portion was attached to a triangular 3mm mild steel plate using two 5mm countersunk set screws. The entrance hole was chamfered 1mm x 45°

The plate was attached to each of the actuators such that the actuator assembly formed an equilateral triangle of side 120mm.

The actuators were similarly mounted at their upper end to a plate and bearing arrangement that allowed linear movement in the vertical plane. Masses were added to the upper plate such that the vertical force was constant at 30 newtons.

The internal pressure in each actuator element, nine elements in total, was provided from the 10 bar laboratory air line. The pressure in each element was varied using nine Watson Smith 101X proportional pressure control valves. These valves provide a 1% linearity over a 0 -10 bar range with 256 step resolution. The valves were controlled from a 64 channel transputer based control system instructed from a 486PC.

Maximum operating pressure during operation was limited to 6 bar.

The machine vice was attached to a small x - y table to provide lateral displacement.

Angular errors were introduced by mounting spacers between the actuator and the attachment plate and measured using angle slips.

Experimental Procedure

A range of assembly conditions were assessed with angular errors ranging from 0 ±2° and lateral errors from 0 ± 1.75mm.

The angular orientation of the curvature of each actuator and the radius of the circle described by the tip of the actuator were controlled through the internal pressure differentials using equations {1} and {2}.

A simple programme was written to enable the angle θ to be varied from 0 to 360 degrees with the maximum pressure differential increasing by 0.5 bar at the completion of each circuit.

The insertion force was fixed at 30 newtons for any given set up the peg was placed in position over the hole and the search strategy started as the peg was released. The force was fixed at 30 newtons through trial error. This vale represented close to the minimum for which consistent insertion was achieved with zero measurable misalignment.

For any particular combination of alignment conditions, the time to successful insertion was recorded to the nearest second.

The constant passive load attempts to complete the insertion process after the peg has been released. The angular and lateral mis-alignments were varied and the actuator used to achieve complete insertion. For any misalignment condition, the actuator element pressures were varied such that the portion of the component remote from the hole was caused to move in a circle.

In effect, the actuator is copying the action performed by a human being when inserting a component. Having initiated the insertion task, if insertion does not proceed smoothly, the hand is moved in a circular motion causing the remote end of the peg to travel in a circle relative to the hole. When the axis of peg and the axis of the hole are aligned, insertion proceeds. In the test rig, insertion occurs when the same geometric conditions are reached and the passive insertion force causes insertion to continue.

This is the basic modus operandi of the device, figure 6.

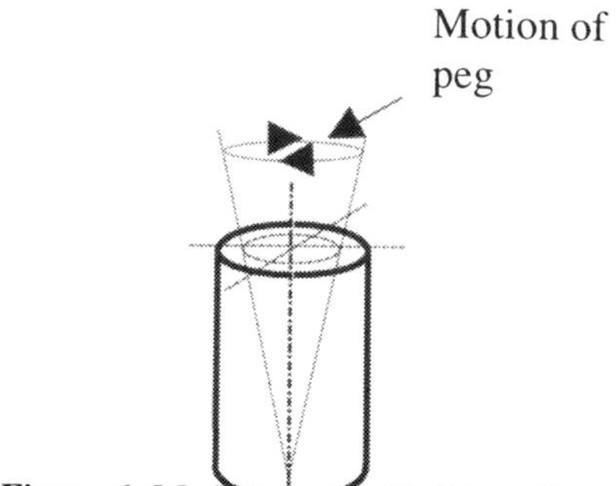

Figure 6: Modus operandi of insertion strategy

Successful insertions were achieved with lateral errors up to ±1.75mm and angular errors up ±2°. Each condition was repeated 10 times and figure 7 records the mean values from these results. At deflections out side those recorded, insertion failed and no time was recorded. The results obtained are summarised in figure 7.

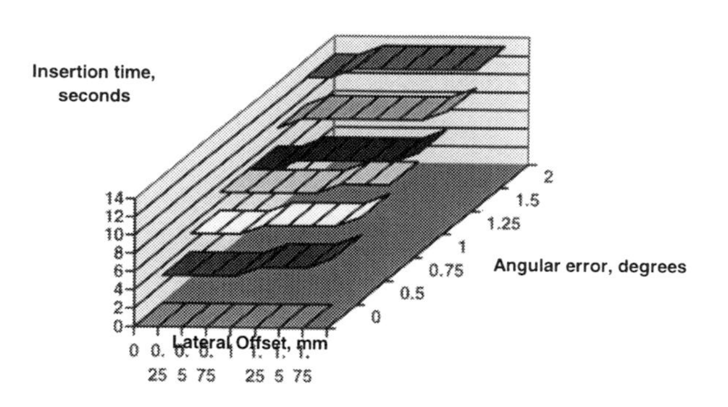

Figure 7: Lateral offset/angular error/time

Conclusions

The continuum actuator forms a useful basis for facilitating peg in hole insertion procedures. The use of a circular search strategy with an increasing radius can succeed in inserting components with a 35mm F6/g7 clearance assembly with a maximum angular displacement of ±2 degrees and a maximum lateral misalignment of ±1.75 mm. This is achieved with a hole entrance chamfer of 1.5mm. In these conditions, the search strategy took up to 15 seconds to achieve mating, but insertion was always successful. If the peg is placed within the hole, the magnitude of the chamfer is unimportant as the search/insertion procedure is an active process not dependent upon local contact forces for insertion success. In assembly conditions where an entrance chamfer is unacceptable, the unit could be integrated with a combination of visual and tactile sensors to begin the insertion process. In these conditions, the actuator may then perform the basic search strategy, ceasing when motion sensors indicate that insertion is in progress/been achieved.

References

1. Whitney, D., *"What is the remote centre compliance device?"* Proc. Intl. Symp. on Industrial Robots, 1979, pp 135 - 152
2. Jeong, J.W., Cho, H.S., *"Development of pneumatic vibratory wrist for robotic assembly"* Robotica Vol. 7 pp 156- 162, 1989
3. Hirzinger G., Brunet, U., *"Fast and self improving compliance using digital force control"* Proc. 4th Intl. Conference on Assembly Automation pp 269 281
4. Robinson, G.C., Davies, J.B.C. *"The Amadeus Project: An Overview"*, Industrial Robot Journal, vol 24, no 4, pp 290-296.
5. Davies, J.B.C., et al *"AMADEUS Dexterous hand, Design Modelling and Sensor Processing"*, IEEE J. Oceanic Engineering, Vol 24, No 1, Jan 1999, pp 96 - 112
6. Davies, J.B.C., *"Alternative to Rigid Mechanisms"* Industrial Robot J., 1991, Vol 18, No 4, pp29-31AM2
7. Davies, J.B.C., *"Flexible three dimensional motion generator"*, PhD Thesis, Heriot-Watt University 1996

CAPE/086/2000

A micro system tool for improved maintenance, quality assurance and recycling

W GRUDZIEN and **G SILIGER**
Institute for Machine Tools and Factory Management in Production Technology Centre Berlin, Technical University Berlin, Germany
A MIDDENDORF and **H REICHL**
Research Center for Microperipheric Technologies, Technical University Berlin, Germany

ABSTRACT

The publication presents the further development of the LCU concept published during the CAPE'99. The LCU allows an active data management of products so that processes such as disassembly and maintenance can be simplified. Starting from a description of the system of the LCU this paper will discuss present results and show the path for future research. Further, the field of application for the LCU is being laid out and illustrated by using partially implemented examples. (1)

1 THE CONCEPT OF THE LIFE CYCLE UNIT

1.1 Description

The LCU is a tool for the sensing, storing, processing, transmitting and application of product and process data before, during and after the usage phase. It simplifies the processes of adaptation on the basis of these data. The adaptation is used to adjust a product according to the demands and needs of the next usage phase. The product can be re-used or further-used. An adaptation process includes the processes disassembly, cleaning, reassembly, processing, component advection and removal, examination and sorting.

The LCU consisting of the four modules sensor, LCB, actuator and marking is being developed as a modular component system. The sensor acquires information, which are processed, stored and transferred by the LCB and applied by an actuator. Marking represents an information that can be sensed more easily. The modules of the LCU are represented in Figure 1 with the relevant functions.

1.2 Deterioration

In addition to different user preferences it is mainly the state of the product towards the end of an usage phase which decides about a products possible new life-cycle. The state of the product is subject to external and internal influences. External causes can be usage and/or

environmentally caused. Negative, value reducing modifications are referred to as wear. Both, wear of material modifications and changing user preferences are considered to be wear.
Usage- and environmentally caused stress results in physical modifications such as abrasion, aging, corrosion or breakage and can be reduced to mechanical, radiation physical, thermal, chemical or biological mechanisms. Product qualities such as material and fabrication, external influences like temperature and dust as well as usage conditions and intensity determine its extend. Varied demands can be the result of technical progress, law amendments, value change or fashion trends. They can be ascribed to environmental conditions, especially politics, society, economy and usage.

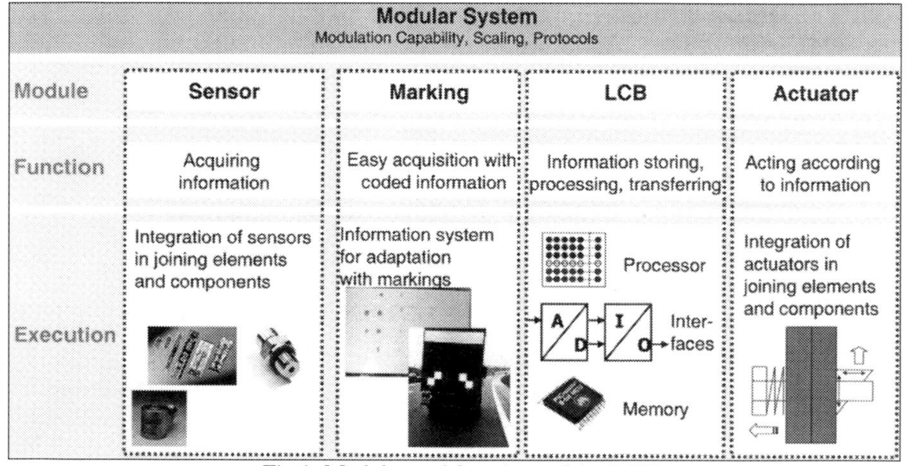

Fig 1: Modules and functions of the LCU

1.3 Adaptation

Information about type, amount, construction and state of the product as well as adaptation processes help to decrease expenditure for an adaptation and increase the net profits from the recovery of resources.
The basis for all adaptation processes is the provision of information as well as a determined reaction according to these information. Information need to be sensed, processed, stored and transmitted for any provision of data. A reaction is being realized by the application of the provided information. The spectrum of a further use can be increased if time and place of the last use as well as the product state are known and can be influenced by means of contractual arrangements. A secured access onto the products at the end of an usage phase improves the calculation of the returned volume and the availability of the resources. Great potentials for a secured access come from usage marketing i.e. the provider remains in the possession of the product, the customer only acquires the benefit of usage. The disposal rights are secured for the provider. By applying a data processing tool like the LCU to realize a continuous product surveillance one can provide data on the state of the product. Options on re-usage and further-usage could be determined more reliably. The product state can be influenced via adaptation processes like maintenance, repair and other measures to allow a renewed re-usage on a higher stage of production.

2 RESEARCH FIELDS

2.1 Preliminary works

For the individual modules of the LCU, different preliminary works have already been carried out. In such a way, sensor systems different for the sensor module were developed for the measurement of physical changes. Screws with integrated sensors are represented in Figure 2, each measuring the pre-tension and an environmental value like temperature, pH-value, UVB-intensity and humidity.
Several sample products were equipped with a sensor system and a LCB in order to record the life history and/or to monitor the product. At a washing machine, the life history as well as the disassembly plan are registered and stored. A system to monitor the passenger car cooling unit was developed (see Figure 2) allowing the determination of the temperature up to which the refrigerant remains in the fluid state, i.e. it does not freeze.
The most frequent failure cause of small fans (PC-cooler, processor-cooler) is bearing wear, that becomes apparent by temperature rise of the bearing and reduction of RPM. To detect those temperature and RPM fluctuations, a LCU was developed, providing the required sensors.

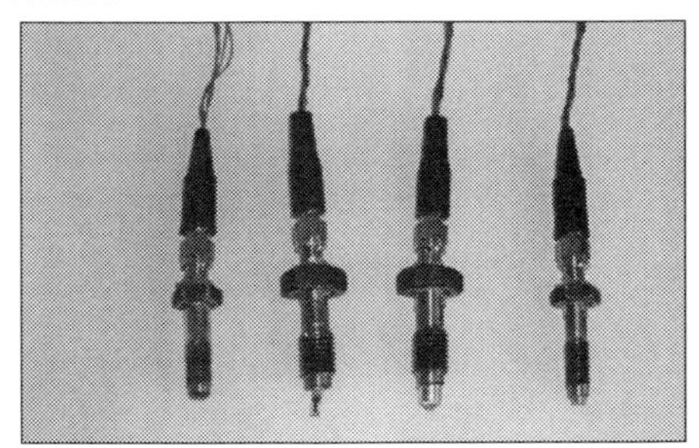

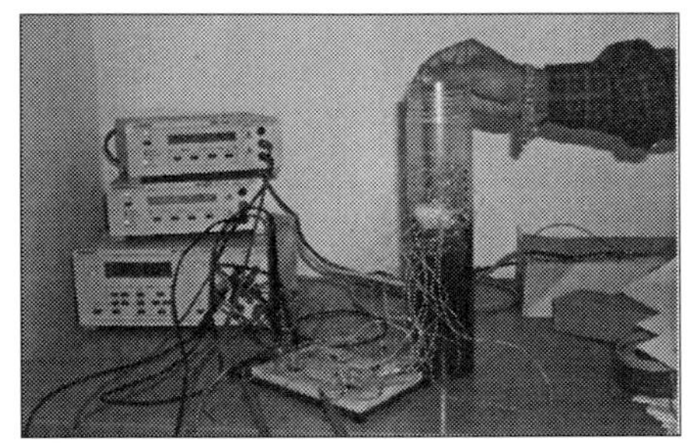

Fig. 2: Prototypes of LCUs (left: for screws; right: for car cooling unit)

On account of increasing circulation of accumulators, the question of a possible reuse ability wins significant importance. For this reason, a LCU was developed for the determination of the absolute loading capacity of accumulators consisting of a sensor system and a LCB. The LCB is based on microcontroller models 167 and 6225 from Siemens and/or SGS Thompson.
In the next development phase, the LCB will be implemented as a microsystem with standardized mechanical and fluidic interfaces. For this purpose, a modular microsystem component system including an Application and a Supplier Kit (see Figure 3) has been implemented.

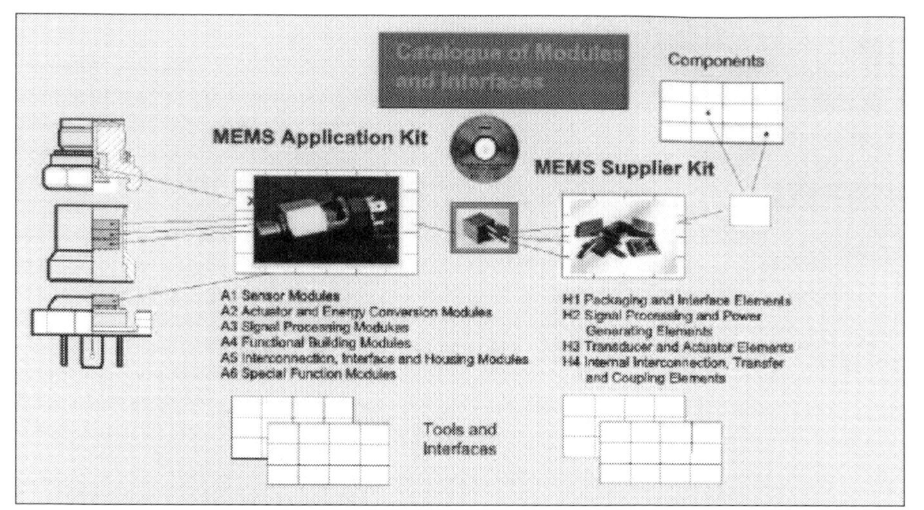

Fig. 3: Modular microsystem including an Application and a Supplier kit

From elements of the manufacturer component system, microsystem technical elements are being produced fulfilling defined functions and acting as modules of the manufacturer component system.
The first by means of Top-Bottom-BGA (TB-BGA) implemented elements are a controller element, a memory chip and an interface element. Microsystems built up modularly are constructed completely by assembling the modules of the customer component system. With the modular microsystem, the conventional sensor was upgraded to an intelligent sensor with a bus interface. The majority of the work on the concept of this framework for the design and production of modular microsystems was carried out in a research project. The funding for the project was provided by the German Ministry for Education and Research (BMB+F).
The modular design and production framework has the potential to overcome present restrictions of the industrial application of microsystems. On the one hand, the framework promotes the industrial application of modular microsystems by industries usually not familiar with microstructuring technologies. On the other hand, the modular approach enables the economically efficient design and medium-scale production of consumer-adapted smart microsystems. (2, 3)

The LCU-module marking allows a low-cost storage of disassembly relevant information such as cut outlines for cutting tools.
In order to achieve high quality marking satisfying the needs of disassembly some requirements must be met. The marker must be economical, long-lived, non-poisonous, easily recognizable, applicable in small amounts, stable and resistant against detergents. Varnish paints were employed for the first experiments to support disassembly by color marking. Determined geometrical forms and colors serve as marking. The combination of form and color offers an unambiguous interpretation for the disassembly process: As an example, the form "triangle" always refers to an additional memory chip in its middle. In this case, the color of the triangle determines the type of memory. In such a way, a red triangle means that the memory is a linear bar code. A green triangle refers to a two-dimensional bar code and a blue triangle to a transponder. The form "circle" always marks joining elements whereas the

type of the joining element is determined by the rings color: thus, red rings are screws, blue rings snap-on connections, and so forth. The form "oblong" always refers to an outline for guiding a tool. The color of the oblong indicates the model of the tool. In order to mark a closed face the corners of the face are marked with oblongs.
For this simple disassembly marking system, a color image processing system acquiring and valuing the markers was developed. The graphical data acquired in such a way is transmitted to a disassembly robot which then follows the outlines with a cutting tool. The marked sidewall of a washing machine and the recognized outlines are represented in Figure 4.

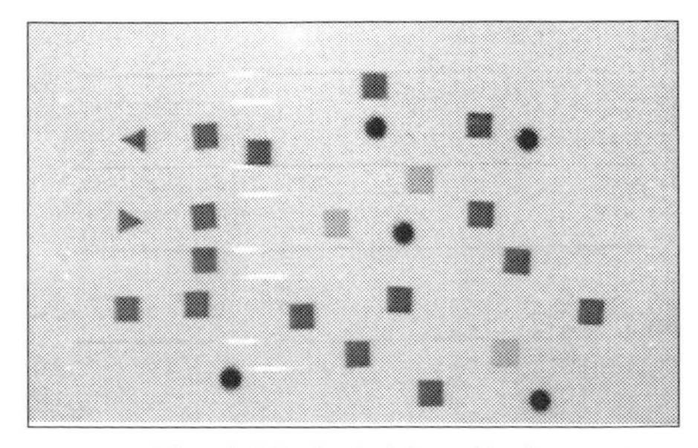
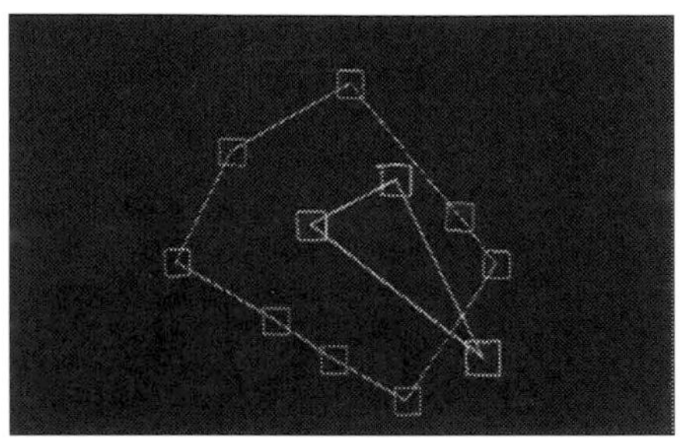

Fig. 4: Marked sidewall of a washing machine with the recognized outlines

In order to preventively oppose and/or to compensate the deterioration and support the disassembly process using joining elements with integrated actuators, actuators based systems are being classified according to their principles of effect. Analogously to sensors, the principles are based on the six possible energy forms of a measurement variable, with an actuator transforming a measurement variable into one of the six forms of energy. The classification of actuators into electronic, electro-magnetic, fluidic, piezoelectric, magnetostrictive, electro-rheological, magneto-rheological, thermo-bi-metallic and electro-chemical as well shape memory and expansion material actuators occurs corresponding to the energy forms.

2.2 Future works

Disassembly is supported by integration of actuators into joining elements. In such a way, an integrated magnetostrictive actuator e.g. can open and close a joint by applying a magnetic field. As a result, new possibilities like sequential, simultaneous and selective disassembly of products, components and parts become possible. Secondly new possibilities of construction and design arise as joining elements need to be no longer accessible to disassembly tools.
The LCB should be developed as a microsystem. It must fulfill the functions acquiring, storing and transmission of data (bus accessibility, where appropriate radio-controlled or with local display). Sub-functions are the processing of data (compressing and pre-evaluation) as well as energy supply (local or external with buffering) for the modules sensor and actuator. After the development of the hardware and software and/or firmware, the LCB is at first built up discreetly with SMD. After a revision of detailed specification on account of first test outputs, system design and layout the LCB is re-worked.
Based on a technology adjusted manufacturing preparation the LCB is manufactured as a single chip LCB, followed by the system integration of the hardware and software for the LCB.
On the one hand, this is required since the relevant devices have micro system components on account of their functionality (Fuzzy logic). On the other hand, the acquired data must if

possible be processes and prepared on site so that the spatial measurements must in particular be considered.
At the example of the circuit board deterioration parameters will be determined qualitatively and quantitatively. On this basis, the LCU can extend the usage time and the quality of circuit boards.

3 APPLICATION FIELDS

3.1 Adaptation

The field for application of the information provided by the LCU covers inspection, maintenance, quality protection during the usage, support of production and R&D, as well as the product adaptation actively guided re- and disassembly processes. Additionally, the product specific usage data can be applied in the field of product liability for legal safeguarding against product liability damages.
In the field of maintenance there are many thinkable applications where usage data can be applied to intensify the usage phase of products and components:

- calculation of the remaining service life,
- weak point analyses,
- appropriate replacement of wearing parts,
- appropriate disposition of repair parts and workshop stays,
- fast return from the workshop to use,
- simple error localization and subsystem assignment,
- damage reduction by early recognition,
- longer standby through maintenance extension of deadline,
- increase of availability by decrease in down-times,
- reduction of the maintenance costs by fast and precise failure localization and
- behavior influencing by stress-oriented use.

3.2 Quality Assurance

Through its enormous documentation possibilities the LCU allows to extend the quality assurance of products from the production phase into the usage phase and post-usage phases. Today, quality assurance instruments are only applied during the usage phase of products in the form of evaluation of field data and loss analyses. The quality assurance instruments applied in R&D and production aren't used during the usage phase of the product.
The five approaches being used today in order to protect quality in R&D and production have the potential to secure a lasting quality during the usage phase of products. These five approaches are Quality Function Deployment (QFD), Failure Mode and Effect Analysis (FMEA), Failure Tree Analysis (FTA), Design of Experiments (DOE) and Statistical Process Control (SPC). (4)

Today, the possibilities to influence a product by means of QA-measures during the usage phase sharply decrease along the added value chain. The usage phase holds only a small part of the total internal added value of the product, the phase of disposal an even lower share (see Figure 5). During the usage phase the added value is realized by maintenance measures, during the phase of disposal through the Dis- and Reassembly of products.

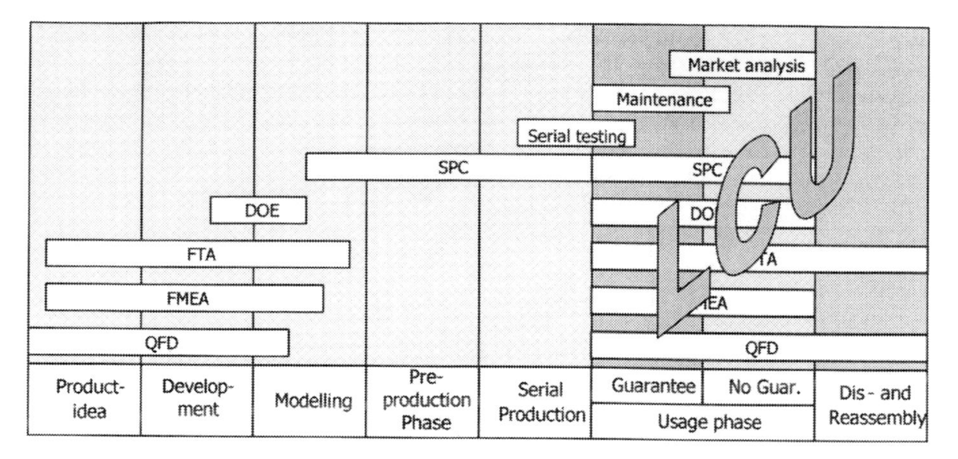

Fig. 5: QA-instruments during the product's life span

By the extension of the five known QA-approaches into the usage phase of a product, the spectrum of the QA-influence possibilities enlarges clearly. Thus, the weight of the internal added value of a product shifts noticeably towards its usage phase of the product.
The approaches derived from the known QA-instruments for the application during the usage phase of a product join into a comprehensive concept called FOCUS promoting the usage phase of the product as an integral element in its added value chain. (5, 6)

FOCUS consists of the five QA-instruments described above. The abbreviation stands for Flaw detection, Optimization, Customer-oriented and Safeguarding.

FTA - the surveillance of crucial places in failure trees allows the early warning i.e. the recognition of threats (flaw-detection) and in addition helps to avoid redundancies.

DOE - the extension of the statistical experimental planning permits a lastingly optimization of product setting during the usage phase of the product and adds to the realization of time and cost advantages for customers and manufacturers.

FMEA - the direct derivation of the particular customer profile from usage data and the individualized importance of known mistakes allow a customer-oriented maintenance strategy.

QFD - the consideration of customer requirements in the process of planning the maintenance and under the regard of usage data, supports a lasting customer-oriented maintenance strategy.

SPC - the provision of network data allows the protection of processes during the usage phase.

The common objective of the instruments identified here is to focus quality assurance towards the usage phase, which up to now is being strongly neglected in regard to the entire life-cycle of the product. In the presented model, all QA-instruments share the data which is provided by the Life Cycle Unit. In addition to quality assurance during the usage phase the previously introduced instruments also widen the framework of action following the completion of the usage phase. The QFD individualizes the estimation of the restorable deterioration stock of products in the case of the reassembly of products. By means of the LCU's memory capacity the FTA allows the precise localization of the mistakes occurring during the use of the product, in the course of dis- and reassembly. The emphasis on the usage phase as the span during the life-cycle of a product that is naturally the longest has a strong influence on the overall added value. The accentuation of the factors protection of the state, customer-oriented maintenance, optimization of the product and early failure recognition add to the weight of the added value during and after the use of the product.

3.3 Product liability

Whether and to what extend a product is to be judged dangerous in relation to its environment can be shown by a risk factor analysis.

By accompanying the usage phase the LCU allows deriving data to describe the place, the kind and the strength of usage for the purpose of carrying out a risk factor analysis. The objective is it to obtain a secure product, provide precise instructions for usage and maintenance, take specifically designed influence on the distribution and to insure the residual remaining risk. The LCU adds for the systematically aim-oriented reaching of these targeted states. Figure 6 shows the product liability areas influenced by the LCU. (7)

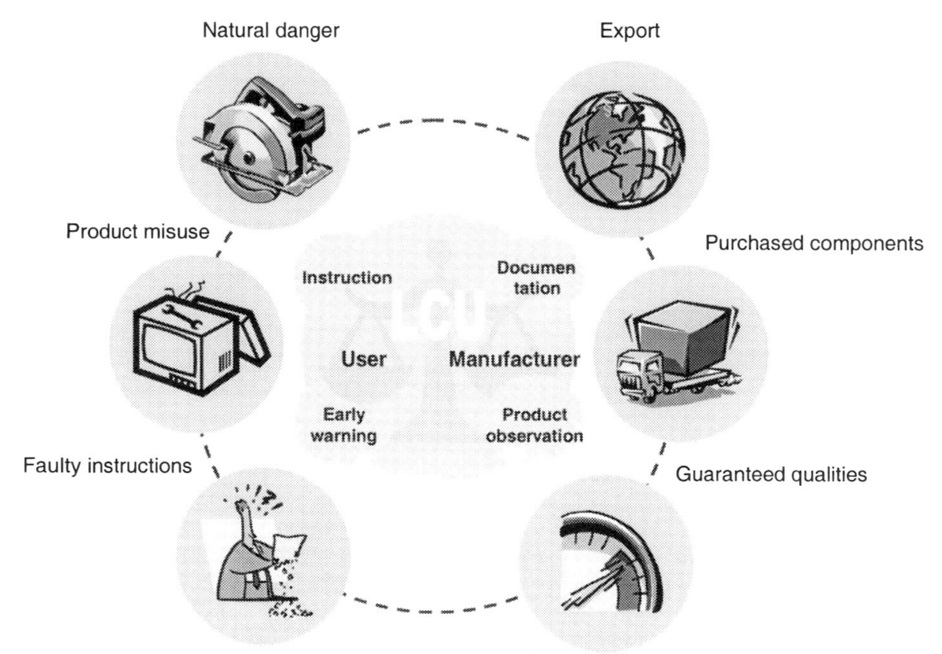

Fig. 6: Extended product liability with the LCU

REFERENCES

(1) Seliger, G.; Grudzien, W.; Zaidi, H.N.: The Life Cycle Unit Concept. In: Proceedings of the CAPE'99, Durham, England, 1999, pp. 223-230.

(2) Grosser, V.; Michel, B.; Leutenbauer, R.; Sommer, J.-P.: Reichl, H.; TB-BGA - Basic Elements for microsystem Packages. Proc. of Actuator´98, pp. 161-63.
(3) Leutenbauer, R.; Amiri Jam, K.; Sommer, J.-P.; Grosser, V.; Reichl, H.; Michel, B.: Thermo-mechanical Aspects of Modular Vertical Integration Techniques (TB-BGA). Proc. of MicroMat´97, pp. 1169-71.
(4) Wallisch, F.: Nutzen statt Selbstzweck, QZ 44 (1999) 4 pp. 462-468.
(5) Hoffmann, J., Huber, M., Knaus, T., 1999, QFD – den Kunden im Visier, QZ 44 (1999) 9 pp. 1127-1131.
(6) Dietzsch, M., Althaus, K., Brander, T., 1999, Fehler früh erkennen, QZ 44 (1999) 11 pp. 1394-1398.
(7) Bodenschatz, W., Fichna, G., Voth, D.: 1990, Produkthaftung, 4. Auflage, Maschinenbauverlag, VDMA.

CAPE/091/2000

Automatic generation of optimal assembly sequences using simulated annealing

P G MAROPOULOS and **A LAGUDA**
School of Engineering, University of Durham, UK

ABSTRACT

The automatic generation of assembly sequences is recognised as an important aspect of assembly planning and optimisation in order to streamline production and reduce lead times and costs. It also plays an important role in designing and planning the configuration of the assembly system. A methodology has been developed for the creation and selection of optimal assembly sequences using simulated annealing. This is a combinatorial optimisation problem and the following control parameters are used to obtain a set of optimal assembly sequences: (*i*) parallelism; (*ii*) number of re-orientations; (*iii*) stability of subassemblies; (*iv*) *globally* good sequences; and (*v*) *locally* good assembly sequences. The methods are based on an object-oriented assembly planning system and have been tested using products and factory data supplied by a large manufacturing company with good results.

1 INTRODUCTION

The highly competitive nature of today's global market coupled with the dynamic requirements of the consumers drives the need within the manufacturing industry to decrease the life-cycle of new and/or redesigned products. At the same time, these factors also propagate the increase in the complexity of the assembled products by increasing the customised features of manufactured products. In order to shorten the time required for the development of a product and its associated manufacturing processes in a concurrent engineering environment, it is desirable for the process planning activity to be automated. Furthermore, it is imperative that such a process planner should be available at the earlier stages of design, for the full potential in terms of cost savings to be realised.

Numerous research works have been published with regards to assembly including the automatic generation of assembly sequences. The majority of these seek to developed algorithms that generate assembly plans based on a single criterion evaluation scheme. Lin and Chang (1) proposed a framework for automated generation of assembly sequences that utilised a knowledge base system to reduce the number of feasible assembly sequences. A different and popular approach towards generating a feasible sequence of assembly operations is to inverse a feasible disassembly sequence of operations (2). This approach circumvents some of the inherent difficulties in generating an assembly sequence when starting with the base components. Other methods of interest include the generation of assembly sequences using observation as devised by Ikechi et al (3) and the works of Ko and Lee (4).

A closely related topic is that of assembly representation and/or modelling. Most of the work to date within this field can be grouped into three categories namely, network graphs, fastening method, and robot-programming language based. Bourjault (5) characterised an assembly by a graph of nodes and arcs where the nodes represent parts and the arcs between nodes represent the relations between parts, better know as *'liaison'*. Other network graph representation methodologies of interest include the works of Krogh and Sanderson (6), and Homen de Mello and Sanderson (7,8).

Only a few authors have considered the optimisation of assembly sequences. Of this group an even smaller number have dealt with the issue of multi criteria optimisation namely, De Mello and Sanderson (8) and Laperrière and ElMaraghy (9). This paper presents an aggregate product and assembly modelling process. A method for the multi criteria optimisation of feasible assembly sequences based on a simulated annealing approach is also presented.

2 ASSEMBLY MODELLING

In conceptual design, decisions are made between alternative structures, which could meet the functional specification of the product (10). This determines the basic list of components and their principal attributes. It may be neither possible nor desirable at this stage to produce a geometrical representation of the part, since this will depend on factors yet to be considered. At this stage, however, the developer should be able to make an assessment of the relative manufacturability of alternative conceptual design options in order to select the most appropriate design and process solutions. This is important since a large proportion of life-cycle cost is determined during conceptual design.

2.1 Aggregate Product Modelling

Assembly modelling is a procedure for describing the assembled state of a given product model in terms of its basic assembly connections. An aggregate assembly modelling and planning (AAMP) method and the corresponding computer based system has been developed at Durham University under the supervision of Prof. P.G. Maropoulos for the initial stages of product development (18). The starting point for the development of an aggregate product model is obtained from the bill of materials (interactive modelling via user interface) or a CAD product model. An aggregate product model is created by extracting assembly features from the components of the product. Assembly feature connections (AFCs) or mating relationships are subsequently established between the components' assembly features. The assembly feature connections emulate assembly processes or operations performed on the shop floor of manufacturing firms. The steps involved in the generation of an aggregate product model are shown in Figure 1. The classification of assembly feature connections considered in this research is shown in Table 1.

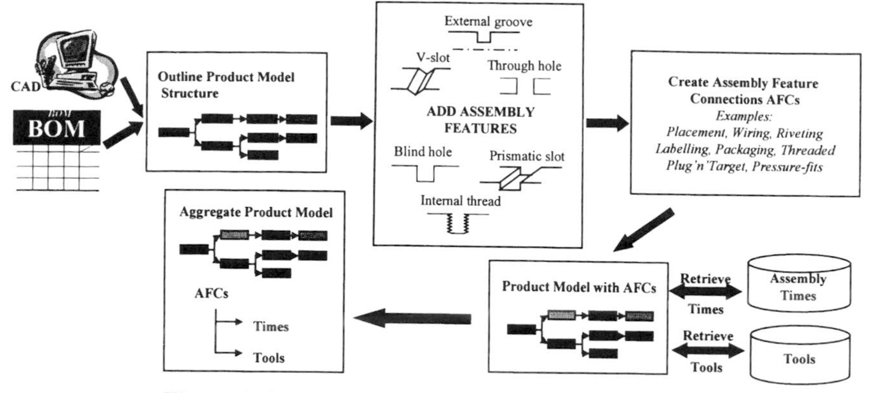

Figure 1: Generation of aggregate product model

AFCs Classifications	Assembly Feature Connections	Assembly Feature Connections Sub-Type
Standard Insertion Assembly Operations	Placement	
	Plug 'n' Target	Cylindrical
		Non-Cylindrical
Reversible Insertion Assembly operations	Packaging	
	Pressure Fits	
	Threaded	Screwing
		Bolting
	Wiring	Tag connectors
		Screw connectors
		Pressure-fit connectors
Permanent Insertion Assembly Operations	Adhesives	
	Riveted	
	Thermoplastic Welding	Ultrasonic welding
		Spin welding
		Hot plate welding
		Vibration welding

Table 1: Classification of AFCs

3 THE CONNECTIVITY MODEL AND SEQUENCING RULES

A sequence planning system requires a model of the logical relationships (surface contacts and attachments) as well as for non-geometric information (such as attachment forces), that is not related to part geometry but never the less affect assembly methods. A connectivity model is used to represent relations among the components of an assembly, using the aggregate product model as the starting-point of this relational model. The connectivity model uses the hierarchical representation of the product model as illustrated in Figure 2. When an AFC is created in the connectivity model, two subscript indices are attached to it (AFC_{ij}). The following notation is used.

i = subscript for the level at which the AFC is created, within the hierarchical product model
j = subscript for the ranking of the AFC within its level

Each AFC generated within the connectivity model has the following basic data attached to it; AFC type (such as threaded, placement, thermoplastic welding and wiring), mating components, mating features, dimensions of mating components and features, and maximum reorientation angles (18).

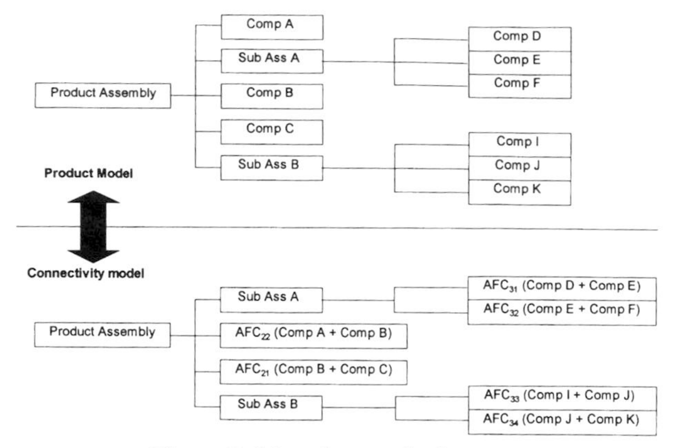

Figure 2: Mapping method

Three algorithms were employed for the generation of the connectivity model.

- Contact Constraint Algorithm

Contact constraints specify which parts are connected to other parts in terms of an assembly operation, denoted as the AFC's parent. For example, if the external thread (etd) of a screw is to mate with the internal thread (itd) of a nut to form a threaded AFC, the parents of the AFC are the screw and nut, denoted as screw_nut. Such a list is generated for each component of the product. A contact constraint algorithm is used to generate this list; the list is subsequently used as one of the inputs for the precedence constraint algorithm.

- Precedence Constraint Algorithm

Precedence constraints represent the fact that some components have to be assembled before others. They are imposed within the hierarchical levels of the product model as well as within subassemblies. A logical set of rules can be derived to establish an adequate ranking system for all AFCs considered. AFCs that are restricted to subassemblies within a given hierarchical level are called "fixed AFCs". An AFC allowed to move freely between hierarchical levels or subassemblies is deemed to be "floating". This algorithm establishes the relative priorities of the AFCs included in the connectivity model.

- Technological Constraint Algorithm

Technological constraints apply when multiple mating conditions occur in addition to the regular sequential assembly processes. For example, a component may be assembled with more than one component (multiple mating) or there may be a variety of AFCs involved including permanent or reversible AFCs. In such cases the technological constraints are applied to prioritise such operations. For example, permanent AFCs (such as welding and joining using adhesives) are performed after executing all reversible AFCs (such as placement and threaded) of the same level.

4 ASSEMBLY SEQUENCE OPTIMISATION: EVALUATION CRITERIA AND MATHEMATICAL MODELS

The minimisation of assembly times is achieved by using the following procedure, which results in defining the values of three assembly rating variables c_1, c_2 and c_3.

4.1 Minimisation of the number of reorientations (c_1)

Some intermediate assembly operations require reorientation of the sub-assembly and/or components. The objective here is to minimise the number of reorientations required for product assembly. In manual assembly it is more efficient to group parts according to their assembly direction. The maximum orientation values of the adjacent AFCs are compared to determine if reorientation is needed for a particular sequence. The function for reorientation is;

$$RE = \sum_{i=1}^{n-1} x_{re} \qquad [1]$$

where;

n is the total number of AFCs.

x_{re} is the reorientation index which is 1, 0.75 or 0 depending on the alpha and beta values (11). For example, for the sequence "$AFC_{21} \rightarrow AFC_{22}$", if the $(\alpha+\beta)$ value of the moving part in AFC_{22} is greater than or equal to $540°$, then reorientation is required and $x_{re} = 1$.

This algorithm minimises the reorientation function and c_1 = Min RE.

4.2 Maximisation of parallelism (c_2)

The selection of an assembly plan that allows parallelism, that is the concurrent execution of assembly operations leads to significant reduction in the total assembly time. A function for measuring parallelism was derived by Laperrière and ElMaraghy (9).

$$PA = \frac{\overline{diff}}{d} \qquad [2]$$

where;

$$\overline{diff} = \begin{cases} 0 & \text{if diff} = 0 \text{ and n is even} \\ 0 & \text{if diff} = 1 \text{ and n is odd} \\ diff & \text{otherwise} \end{cases}$$

$d = n - 2,$

d, is maximum difference between the number of components in the subassemblies created resulting from a spilt in the parent assembly.

n, is the total number of components in the parent assembly

This algorithm maximises the parallelism function and c_2 = Max PA.

4.3 Maximisation of the stability of intermediate subassemblies (c_3)

The choice of an assembly sequence that involves highly stable subassemblies improves the reliability of assembly operations. This reduces assembly cycle time by preventing errors during execution and avoiding unstable subassemblies when building a product. For the purpose of this research, a stability rating system has been derived for all AFC types (adhesives, plug'n'target, threaded, pressure-fits, riveted, placement, labelling, wiring, packaging and welding) following a set of logical rules. The function for stability is;

$$ST = \sum_{i=1}^{n-1} x_{st} \qquad [3]$$

where;

n is the total number of AFCs.

x_{st} is the stability index.

This algorithm maximises the stability function and c_3 = Max ST.

An overall assembly rating variable (c) can then be derived using the normalised variables c_1, c_2, c_3 and by applying weighting factors (w_{re}, w_{pa}, w_{st}). This allows the user to define the relative priorities for analysis.

$$c = (w_{re})\frac{X_{re}}{c_1} + (w_{pa}) * c_2 + (w_{st})\frac{c_3}{X_{st}} \qquad [4]$$

X_{re} and X_{st} are the maximum values for the reorientation and stability indices used to normalise c_1 and c_2 respectively.

Two optimisation methods are then used for the optimisation of assembly sequences. These are:

1. Globally good optimisation, if an assembly plan is globally good, then all functions described above are considered.
2. Locally good optimisation, if an assembly plan is locally good, then at least one of the functions described above have been selected.

5 SIMULATED ANNEALING (SA) ALGORITHM

The SA approach can be viewed as an enhanced version of local optimisation or an iterative improvement, in which an initial solution is repeatedly improved upon by making small local alterations until no such alteration yields a better solution. SA algorithms have been used successfully in solving various combinatorial optimisation problems, including VLSI design (13), scheduling (14) and assembly line balancing (15).

5.1 Simulated Annealing Parameters

The parameters of the simulated annealing assembly sequence algorithm that initiated the best results are as follows:

1. Initial assembly sequence is obtained form the connectivity model, via the precedence constraint algorithm. The sequence is encoded into a string of numbers representing the indexes of the AFCs. The encoded string has one number for each AFC. Positions of the numbers in the string correspond to the indexes of the AFC, and their values denote the priorities of the corresponding AFCs.
2. Initial temperature. The starting temperature for the annealing process is usually high. For the purpose of this research a typical value of 100 is used.
3. A random number for the probability of acceptance of simulated annealing is generated from a uniform distribution between 0 and 1.
4. A new solution is made by interchanging two random floating AFCs.
5. A maximum number of cooling schedules $cs = 5$ has been chosen for the purpose of this study. The cooling schedules represent a finite time implementation for the simulated annealing algorithm.
6. At a particular temperature, if the ratio of accepted solutions to the number of solutions generated is less than a predetermined number, that has been set to 0.1, the cooling schedule is increased by 1.
7. The cooling rate *(cr)*, is the rate of change of temperature with increasing number of cooling schedules. This is usually less than one; higher values of cooling rate correspond to a slower cooling process. A cooling rate of 0.95 is used herein.

6 AN ILLUSTRATIVE EXAMPLE

The performance of the method presented was tested using a number of products. The example presented herein is in relation to an outdoor, lightweight product. The product and connectivity model is shown in Figure 3. The contact, precedence and technological constraints algorithms were used to generate the corresponding connectivity model from the aggregate product model. The connectivity model was used to generate an initial assembly sequence. This acts as the initial assembly sequence for the simulated annealing program. The AFCs shaded grey are fixed AFCs and all other AFCs are regarded to be floating. Table 2 shows the sequence of AFCs generated from the precedence constraint algorithm; the mating components and the encoded strings are also shown in Table 2. The values of the random numbers generated for encoding (see Table 2) are linked to the priorities placed on each AFC. For example, an AFC with a higher priority is given a higher value random number.

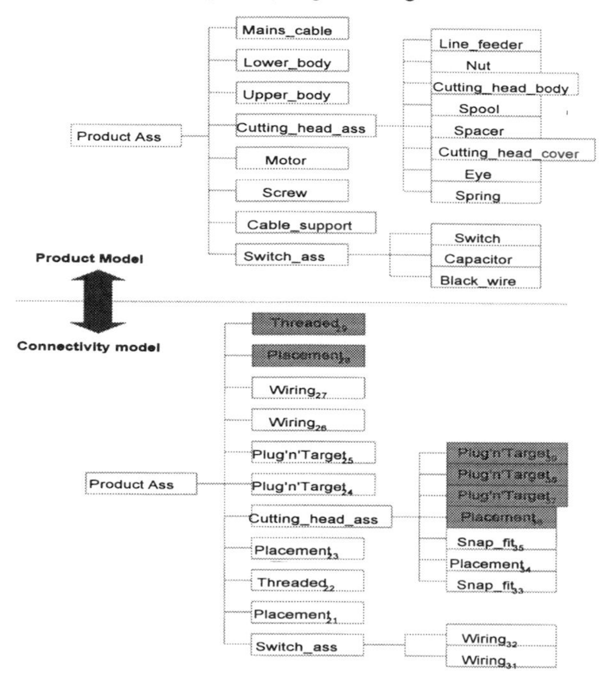

Figure 3: Example product model and generated connectivity model

Two optimisation procedures were performed; a locally good optimisation was performed prior to a globally good optimisation. A locally good optimisation is achieved by applying suitable weightings to the assembly rating functions. For example, if reorientation is to be ignored, the weighting factor for reorientation in equation 4 should be $w_{re} = 0$. Table 3 shows a locally good optimisation with; $w_{st} = 0.5$, $w_{re} = 0.0$, $w_{pa} = 0.5$.

In the case of a globally good optimisation, all three assembly rating functions are considered. Table 4 shows a globally good optimisation with; $w_{st} = 0.5$, $w_{re} = 0.5$, $w_{pa} = 0.5$. The result of this analysis is shown in Table 4.

Both locally and globally good optimisation methods produced good results. In the case of locally good optimisation, it can be seen from Table 3 that several AFCs have moved from level two to level three. Notably, the mains_cable and cable_support are now members of the switch assembly. Also, the motor has been included to the cutting_head_ass in order to satisfy the parallelism criterion.

No.	AFCs (afrtype)	Components within AFC	Encoded string
1	Wiring$_{31}$	Switch,capacitor	0.03
2	Wiring$_{32}$	Switch,black_wire	0.09
3	Plug'n'Target$_{37}$	line_feeder,spring	0.11
4	Placement$_{36}$	Cutting_head_body,line_feeder	0.12
5	Snap_fit$_{35}$	Cutting_head_body,eye	0.14
6	Plug'n'Target$_{39}$	Cutting_head_body,nut	0.15
7	Plug'n'Target$_{38}$	Cutting_head_body,spacer	0.25
8	Placement$_{34}$	Cutting_head_body,spool	0.36
9	Snap_fit$_{33}$	Cutting_head_body,cutting_head_cover	0.41
10	Placement$_{21}$	Lower_body,motor	0.45
11	Threaded$_{22}$	Cutting_head_ass,motor	0.46
12	Placement$_{23}$	Lower_body,switch_ass	0.48
13	Plug'n'Target$_{24}$	cable_support,mains_cable	0.49
14	Plug'n'Target$_{25}$	Lower_body,cable_support	0.65
15	Wiring$_{26}$	Switch_ass,mains_cable	0.68
16	Wiring$_{27}$	Motor,switch_ass	0.77
17	Placement$_{28}$	Lower_body,upper_body	0.81
18	Threaded$_{29}$	Upper_body,lower_body,screw	0.92

Table 2: Initial sequence generated

No.	Encoded string	AFC type	Components within AFC
1	0.03	Wiring$_{31}$	switch,capacitor
2	0.09	Wiring$_{32}$	switch,black_wire
3	0.49	Plug'n'Target$_{24}$	cable_support,mains_cable
4	0.65	Plug'n'Target$_{25}$	lower_body,cable_support
5	0.48	Placement$_{23}$	lower_body,switch_ass
6	0.68	Wiring$_{26}$	switch_ass,mains_cable
7	0.11	Plug'n'Target$_{37}$	line_feeder,spring
8	0.12	Placement$_{36}$	cutting_head_body,line_feeder
9	0.14	Snap_fit$_{35}$	cutting_head_body,eye
10	0.15	Plug'n'Target$_{39}$	cutting_head_body,nut
11	0.25	Plug'n'Target$_{38}$	cutting_head_body,spacer
12	0.36	Placement$_{34}$	cutting_head_body,spool
13	0.41	Snap_fit$_{33}$	cutting_head_body,cutting_head_cover
14	0.46	Threaded$_{22}$	cutting_head_ass,motor
15	0.45	Placement$_{21}$	lower_body,motor
16	0.77	Wiring$_{27}$	motor,switch_ass
17	0.81	Placement$_{28}$	lower_body,upper_body
18	0.92	Threaded$_{29}$	upper_body,lower_body,screw

Table 3: Results of locally good analysis

No.	Encoded string	AFC type	Components within AFC
1	0.15	Plug'n'Target$_{39}$	cutting_head_body,nut
2	0.25	Plug'n'Target$_{38}$	cutting_head_body,spacer
3	0.11	Plug'n'Target$_{37}$	line_feeder,spring
4	0.12	Placement$_{36}$	cutting_head_body,line_feeder
5	0.14	Snap_fit$_{35}$	cutting_head_body,eye
6	0.36	Placement$_{34}$	cutting_head_body,spool
7	0.41	Snap_fit$_{33}$	cutting_head_body,cutting_head_cover
8	0.46	Threaded$_{22}$	cutting_head_ass,motor
9	0.45	Placement$_{21}$	lower_body,motor
10	0.03	Wiring$_{31}$	switch,capacitor
11	0.09	Wiring$_{32}$	switch,black_wire
12	0.49	Plug'n'Target$_{24}$	cable_support,mains_cable
13	0.65	Plug'n'Target$_{25}$	lower_body,cable_support
14	0.48	Placement$_{23}$	lower_body,switch_ass
15	0.68	Wiring$_{26}$	switch_ass,mains_cable
16	0.77	Wiring$_{27}$	motor,switch_ass
17	0.81	Placement$_{28}$	lower_body,upper_body
18	0.92	Threaded$_{29}$	upper_body,lower_body,screw

Table 4: Results of globally good analysis

The globally good optimisation process introduced the reorientation criterion. The general trend of the assembly sequence generated follows that of the locally good solution, with AFCs migrating from lower levels to higher assembly levels, increasing stability and parallelism. The effect of introducing the reorientation index is evident by the new sequence generated for the assembly of the cutting head. The system also suggests assembling the "cutting head" assembly prior to the switch assembly. The optimisation processes have created what appears to be a completely top-down assembly direction for this subassembly. This has satisfied the reorientation criterion.

The assembly sequences generated for both locally and globally optimised scenarios were mapped to a predefined assembly line layout to establish assembly times for the assembly plans generated. The estimated difference in total assembly time between the locally and globally optimised scenarios is approximately 8%, with the global solution offering the lowest time.

DISCUSSION AND CONCLUSIONS

The main purpose of the research described herein, is to create a system suitable for the automatic generation of an optimal assembly sequence for a given product at the early stages of design. The method is based on the creation of an aggregate product model and the subsequent extraction of contact, precedence and technological relationships from the aggregate product model to create a connectivity model. The extraction of such relationships facilitates the generation of an initial rudimentary assembly plan, which reduces the search space. The generated assembly plan is then refined through a series of optimisation methods using simulated annealing. The simulated annealing algorithm seeks to optimise an assembly rating variable, which includes functions for reorientation, parallelism and stability.

The ability of the system developed to quickly generate and optimise assembly plans locally as well as globally when various criteria are enabled or their relative importance is changed makes it an effective tool for simultaneously considering several manufacturing considerations at the design stage. The results obtained using the method presented have been very encouraging. Indeed, a total of four industrial products have been modelled and optimal sequences generated, with good results.

It is important to note that the generation of an optimal assembly sequence does not in itself imply an optimal assembly plan. An optimal assembly plan can only be realised when the available resources, human and equipment, are taken into consideration. At the moment the assembly plan generated is independent of factory constraints. The generated optimal assembly sequence essentially provides a suitable input for optimising the line balancing and this is being investigated at present.

ACKNOWLEDGEMENTS

The authors wish to acknowledge the substantial contribution of Mr Michael Betteridge in the area of Assembly Modelling and Planning. We are also thankful to our industrial partners for providing access to product and factory data.

REFERENCES

1. Lin, A.C. and Chang, T.C., Three dimensional mechanical assembly planning systems. Journal of Manufacturing Systems, 1993, 12, pp. 437-456.
2. Lee, K., Disassembly Planning based on Subassembly Extraction. Proceedings of the Third ORSA/TIMS Conference on Flexible Manufacturing Systems, edited by K.E.Stecke and R.Suri, 1989, pp. 383-388.
3. Ikechi, K. and Suehiro, T. , Tanguy, P. and Wheeler, M., Assembly plan from Observation. Annual research Review, The Robotics Institute, Carnegie Mellon University, 1990, pp. 37-55.
4. Ko, H. and Lee, K., Automatic assembling procedure generation from mating conditions. Computer Aided Design, 1987, 19 (1), pp. 3-10.
5. Bourjault, A., Contribution to a methodology approach of automate assembly; automatic generation of assembly sequences, Ph.D. Thesis 1984 Universite de France-Comte, Besancon, France.
6. Krogh, B.H. and Sanderson, A.C., Modelling and Control of Assembly Task Systems, Technical report CMU-RI-TR-86-1, 1985 Robotics Institute, Carnegie Mellon University Pittsburgh, PA, USA.
7. Homen de Mello, L.S. and Sanderson A.C, Representation of mechanical assembly sequences. IEEE Transactions on Robotics and Automation, 1991a, 7(2), pp. 211-227.
8. Homen de Mello, L.S. and Sanderson A.C, A Correct and Complete algorithm for the generation of mechanical assembly sequences. IEEE Transactions on Robotics and Automation, 1991b, 7(2), pp. 228-240.
9. Laperrie, L. and ElMaraghy, H.A., Planning of products assembly and disassembly. Annals of the CIRP, 1992, 42,pp. 5-9.
10. Bradley, H.D. and Maropoulos, P.G., A Concurrent Engineering Support System for the Assessment of Manufacturing Options at Early Design Stages, Proceeding of the Thirty-first International Matador Conference, 1995, pp. 485-492.
11. Boothroyd, G. and Dewhurst, P., Product Design for Manufacture and Assembly, Marcel Bekker, Inc. 1996.
12. Delchambre, A. CAD Method for Industrial Assembly. John Wiley & Sons Ltd. 1996.
13. Gerez, S.H., Algorithms for VLSI Design Automation. John Wiley & Sons Ltd. 1999.
14. Kim, J.U. and Kim, Y.D., Simulated Annealing and Genetic Algorithms for Scheduling Products with Multi-level Product Structure. Computer and Operational Research, 1996, Vol.23, No. 9, pp. 857-868.
15. Suresh, G. and Sahu, S., Stochastic assembly line balancing using simulated annealing. International Journal of Production Research, 1994, 32, pp. 1801-1810.
16. Kirkpatrick, S., Gelatt, C.D., Jr and Vecchi, M.P., Optimisation by Simulated Annealing. Science, 1983, 220, pp. 671-680.
17. Bean, J.C., Genetic Algorithms and random keys for sequencing and optimisation. ORSA Journal of Computing, 1994, 6, pp. 154-160.
18. Betteridge, M. J., A methodology for Aggregate Assembly Modelling and Planning. PhD Thesis, 2000, University of Durham, Durham, United Kingdom.

CAPE/097/2000

A knowledge-based design approach for materials handling in flexible manufacturing assembly lines

A KHAN and **A J DAY**
Department of Mechanical & Medical Engineering, University of Bradford, UK

ABSTRACT

This paper presents a Knowledge Based Design Methodology for the selection of material handling technology in flexible manufacturing assembly lines, developed in Application Manager, a ruled-based Expert System Shell. Actual information about the requirements of general manufacturing, the requirements of individual components and detailed individual product operations is first collected, which enables each individual product to be clearly linked with their component parts and operational requirements. This information is used in Knowledge Based (KB) decision-making on major design issues such as material handling system selection, appropriate location for each operator, station design in terms of the complete bill of materials requirements, and selection of a layout for the line. The methodology enables the selection of a suitable material handling system from a number of alternative systems.

1 INTRODUCTION

Assembly is an essential part of the process for producing discrete manufactured items where production volumes are anything more than minimal, and especially in large volume manufacture, careful planning is required before installing an actual assembly process so as to meet the requirements of flexible demand and economical production. Product designers have determined that up to 75 % of manufacturing costs depend upon product design decisions (1). The planning and development stage of a product or a group of products for assembly contributes only 12 % of the total manufacturing cost which is responsible for 75 % of total manufacturing cost (2). Design For Assembly (DFA) principles are well established and should always be followed for minimising the number of parts being assembled (3).

The performance of such assembly systems always depends upon the production plan which must identify the requirements list of raw material, auxiliary parts and sub-assemblies for the final production in some specific time period (4). Based on the production plan, the materials handling activities are then monitored for balancing and post-balancing requirements so as to achieve the scheduled objectives accordingly. The balancing of lines is classified as a combinatorial optimisation problem which has received much attention from researchers in the past (5), (6). The post-balancing requirements such as efficient materials handling activities has been the subject of research which is spread over the literature relating to assembly lines. Such activities mainly influence the design strategy for the materials handling system, which side the operator works on the line, station design, station requirements, and the layout for the assembly line under consideration (2), (7).

This paper presents a Knowledge-Based Design Methodology (KBDM) which gathers critical design information for defining economical and effective post-balancing activities including the selection of an appropriate material handling system, the best location for each line-operator, station design based on parts input and supply conditions and a suitable line layout for the system under consideration. The collected information is stored in KB tables and lists which can be retrieved for reviewing, modifying, updating and / or amending so as to respond flexibly to market needs.

The paper mainly focuses on the selection of an appropriate material handling system. The overall structure of the KBDM is described and is used to highlight the importance of the KB production information. The applications of KB production information for the selection of an appropriate material handling system is then described. Finally an illustrated example of the methodology is presented which is then followed by conclusions and recommendations.

2 NOTATION

A = Angle of displacement (degrees)
AM = Asymmetric multiplier (1 – 0.00032A)
CM = Coupling multiplier (1 for good and 0.9 for poor)
D = Vertical distance moved between origin and destination (cm)
DM = Distance multiplier (0.82 + 4.5/D)
FM = Frequency multiplier (0.45)
H = Horizontal location of hands from the midpoint between the ankles (cm)
HM = Horizontal multiplier (25/H)
LC = Load constant (23 kg)
N = Number of workstations
RWL = Recommended Weight Limit (kg)
V = Vertical location of hands from the floor (cm)
VM = Vertical multiplier $(1 - 0.003|V - 75|)$
WC_{min} = Minimum work carrier required

3 KBDM STRUCTURE

The KBDM has been developed through the use of an Expert System Shell, called Application Manager, which is Windows based with advanced facilities of on-line help (8). It is a Ruled Based development tool: a set of KB production rule questions are prepared for each section, and the user is asked to answer each question, which is then compared by IF-THEN-ELSE production rules within the developed system for deriving an expert conclusion.

In order to develop the KB for material handling activities in flexible manufacturing assembly lines, information about production was collected and arranged in the form of KB tables which clearly link each individual product with component parts and operational requirements. The bill of material can be generated for individual products on a mixed-product line as well as for products on a multi-product line. This information (in the form of KB tables) was developed as the database for the KBDM for application in a systematic step by step approach to decision-making on the selection of the post-balancing activities defined in Section 1. Based on the design attributes of, and selection criteria for, assembly lines (2), (9), the KB production information was subdivided into (i) general manufacturing information, (ii) auxiliary part requirements, and (iii) operational information, and a data module was set up for each as follows.

(i) The general manufacturing information module contains data relating to the general manufacturing requirements for the system, based mainly on the weight, size, nature and material used in overall production. This enables decision-making for appropriate station design on or off-conveyor, suitable selection of a material handling system, and specification of the access required to operate the line. An option takes account of the type of manufacture (delicate or robust).

(ii) In the auxiliary part information module, the sub-assembly requirements for all individual components in any particular product is identified. The KBDM can handle a maximum of 26 various auxiliary parts and requires a clear linkage of any auxiliary part and quantity with the specific assembly operations for each product. Advice is provided on valid entry limits for the part number and maximum number of parts required in the assembly; the most important information is the part number, category, sub-category, weight, size and form which can then subsequently define how and when the part is supplied to the required workstation. As auxiliary parts are only needed for some main assembly operations, this KB information identifies only those main assembly operations. The remaining main assembly operations (if any) are identified in the operational information module.

(iii) The operational information module collects knowledge and factual information for the assembly operation category and sub-category type. The assembly line operations were divided into 4 categories of main assembly operations, secondary assembly operations, adjustment operations and inspection operations. For an economical assembly line, the number of main assembly operations must be greater than the total number of any other remaining operations (10). Each operation category was divided into a number of sub-category operations to clearly identify the nature of the operations on the line (2).

The KBDM was also provided with a random data generation and a "GO-BACK" facility, which allowed the user to change previously entered data, e.g. to analyse a system for a manufacturing assembly line under a range of test conditions. In each section, a number of KB questions are asked of the user, and on the completion of each section, the information entered is displayed to the system user. The user is asked to review the section and can change or re-enter the information if necessary. Once the user is satisfied, the analysis of the next KB section can be started. The collected production information is stored in the form of tables and lists (bill of materials) and covers material handling system selection (weight, dimensions, and work carrier requirements), station design (on or off-line and part supply condition), station requirements (list of parts required and selection of appropriate tools), and the location of the operator on the line (left, right or either side) (11).

4 SELECTION FOR MATERIAL HANDLING SYSTEM

In terms of the material handling system, the two main types of assembly lines in common use are "non-mechanical" and "belt" assembly lines (12). Non-mechanical lines are those where no mechanical or automated material handling system is used and materials are transported from one station to another by hand (12). The selection of an appropriate material handling system in this paper was based on the belt type of assembly line, where the workparts are transferred from station to station on the line by means of conveyor belts. The conveyor belt type of assembly lines is itself divided into a number of sub-categories which depend upon the nature of operations to be performed at each station on the line and the nature of production in the system (see Fig. 1).

A number of KB production rules are applied to trigger the appropriate selection of a material handling system. The information required by these KB production rules can be abstracted from the KB information gathered in the production module. A typical example of the selection of a suitable material handling system using KB manufacturing information is:

KB Rule

IF Products are heavy-weight
AND Sizes of products are large
AND Precise location is not required
AND Job is not stationary during operation
AND Production is rigid
THEN Arrange jobs on a moving belt and allow the operator to move with the job

OR Analyse other material handling systems

The necessary data from the production module for the KB decision are: weight of the finished product, size of the finished product, nature of the product and operational requirements in the system under consideration.

The KB production rules have been developed to differentiate between heavy-weight and light-weight jobs for the assembly line applications by considering a number of design issues. These included the distances moved in various directions (horizontal, vertical and distance from origin to destination), angle of displacement (from origin to destination), frequency (number of times weight lifted in one minute) and coupling (degree of care in holding the weight). The Recommended Weight Limit (RWL) between heavy-weight and light-weight jobs on assembly lines was identified by the relationship (13):

$$RWL = LC \times HM \times VM \times DM \times AM \times FM \times CM \qquad \text{(Equation 1)}$$

Applying this relationship and allowing standard limits for the distance moved, the recommended weight limit for a heavy-weight job on an assembly line application was determined as 4 kilograms. Products weighing more than this were considered to be "heavy-

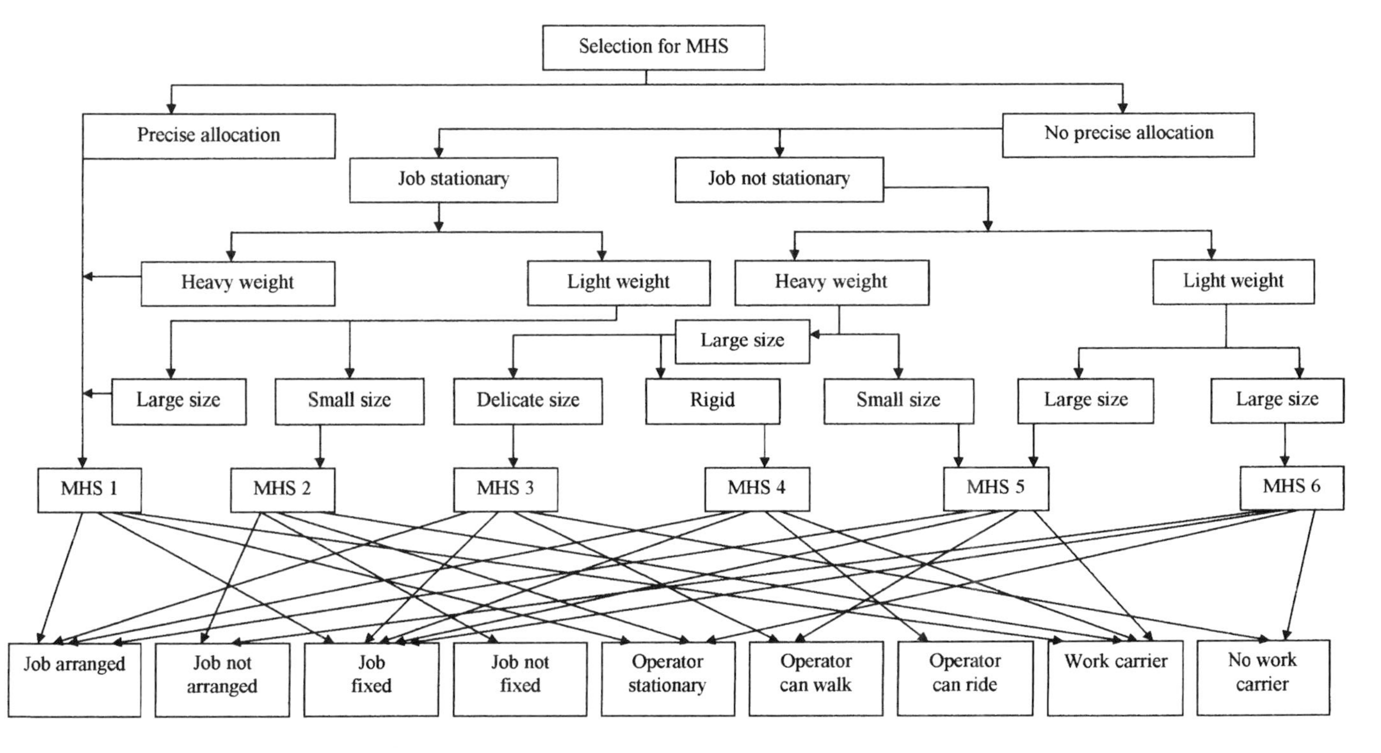

Fig. 1 KB selection for appropriate Material Handling System (MHS)

weight" products. The permissible limits for a maximum 4 kilogram weight are (13):

- Maximum limit for horizontal displacement is 35 cm,
- Limit for vertical height is between 60 to 100 cm,
- Maximum distance moved in the vertical direction is 40 cm, while the total height (summation of 2nd & 3rd assumption) should not exceed more than 100 cm,
- Maximum angle of rotation is 90 degrees,
- Maximum frequency per minute is less than or equal to 4,

Heavy-weight and large products are difficult to remove from the line or within the workstations, so a KB decision is required to select a type of material handling system where the stations are on the line and the jobs remain on line during assembly. When the production is heavy-weight, then it is necessary to ensure that the product arrives at the station in the required orientation which automatically necessitates the use of a work carrier on the line (2). For light-weight products a KB option allows the design of on- or off-conveyor workstations. Also if the nature of production is such that it requires precise location at a station on the line, then irrespective of the size and weight of the product, the work carrier is recommended (2). The overall minimum requirements of the work carrier is calculated for the system under consideration by the following formula:

$$WC_{min} = 5 \times N \quad \text{(Equation 2)}$$

The selection of a suitable material handling system is based on four decision parameters:

(i) Is the product launched onto the line in an arranged form or a non-arranged form?
(ii) Is the product launched fixed onto the line or can it be removed from the line?
(iii) Will operators on the line be stationary, can they walk or possibly ride on product?
(iv) Will the product require the use of work carriers on the line or not, and if required then how many?

The input information is analysed and organised for the selection of a suitable material handling system, which has solutions for all four-decision parameters. The input information of the production module which monitors the selection of an appropriate material handling system is as follows:

- Does the production require precise location at the workstation or not?
- Will the workpart be stationary or moving while the operation is performed?
- Is the product light-weight or heavy-weight?
- Is the product small or large?
- Is the overall production nature robust or delicate?

Based on this information, material handling systems were divided into six different categories and the appropriate selection was made from nine different decision parameters (Figure 1). The KBDM analyses various alternatives for the selection of a suitable material handling system and makes an appropriate selection for the line under consideration.

5 ILLUSTRATED EXAMPLE

The KBDM is a multi-attribute design tool which performs a systematic analysis on the input data and provides comprehensive advice for the complete design of an assembly system. As the design of each section depends on the others, it is not practicable to use the methodology for one module alone. However the input data provided for one of the example was analysed and the decision arrived on the selcction of an appropriate material handling system was as follows:

1. The weight of each product was 6 kg, which classified as "heavy-weight" based on Equation 1.
2. Product dimensions were more than 300 mm which fell in the category of large size.
3. The product was robust.
4. Precise location required.
5. 4 workstations were required for the proposed line (14)

The first two decisions confirmed that the product must be fixed on to the line and must arrive at each workstation in the required orientation. This necessitated work carriers on the line. The 3rd and 4th decisions considered that the product should be stationary at each workstation on the line. The last decision was that 20 work carriers (Equation 2) would be required for the proposed line. These decisions can be traced "MHS 1" (Fig. 1) as the suitable material handling system for the proposed assembly line while the final conclusion arrived can be seen in Fig. 2.

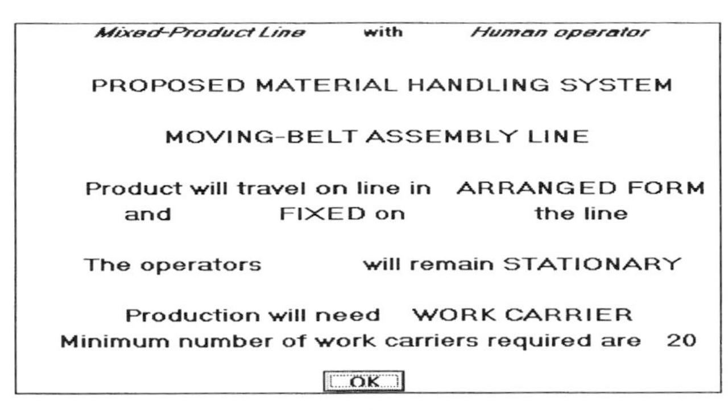

Fig. 2 Material handling system selection

6 CONCLUSIONS

The KBDM presented here uses production information about general manufacturing, auxiliary part / sub-assembly requirements and operational information to analyse the post balancing material handling activities in an assembly system. Each individual product in the system is clearly linked to operational requirements, auxiliary part requirements and tooling requirements. Each individual auxiliary part / sub-assembly (specified in terms of dimensions, weight, nature and supply conditions to stations on the line) could provide the information to prepare the bill of material requirements for one complete cycle on the line. Assembly operations were classified as main assembly operations and related assembly operations. For main assembly operations, the accessible position along with the dimensions and weight of the product can be used to identify which side an operator should be positioned on the line.

Gathering the manufacturing information for appropriate knowledge based decision-making in major facility design issues is key to the success of the methodology presented. For example, in workstation design, the operator's location on the line and the selection of a suitable material handling system are crucial design issues for which detailed manufacturing information is essential. Product weight, product dimensions, the nature of product and the operational requirements can be used to select a suitable material handling system by analysis. Work carriers are required for precise location of products on the assembly line. Once the use of work carriers has been justified, the total number of work carriers required for the system can also be determined.

At the end of using the software program, the complete database, the overall information for workstation design, the bill of materials requirements and the overall design achievements are displayed for the planned assembly system. The software program is comprehensive and easy to use and can handle 20 products, 26 auxiliary parts / sub-assemblies and 60 operations in one time. The conclusion is that the KBDM is a viable tool for the design of Material Handling in Flexible Manufacturing Assembly Lines.

REFERENCES

1. Boothroyd, G. (1988), Estimate Costs at an Early Stage, American Machinist, Vol.132, 54-57.

2. Lotter, B. (1989), Manufacturing Assembly Hand Book, Butterworths.

3. Boothroyd, G. and Alting, L. (1992), Design for Assembly and Disassembly, Annals of the CIRP Vol.41, No.2, 625-636.

4. Buffa, E. S. and Sarin, R. K. (1987), Modern Production / Operations Management, 8th Edition, John Wiley & Sons.

5. Ghosh, S. and Gagnon, R. J. (1989), A Comprehensive Literature Review and Analysis of the Design, Balancing and Scheduling of Assembly Systems, Int. J. Prod. Res., Vol.27, No.4, 637-670.

6. Baybars, I. (1985), A Survey of Inexact Algorithms for the Simple Assembly Line Balancing Problem, Published in Pittsburgh: Carregii Mellan University Graduate School of Industrial Administration.

7. Boothroyd, G. (1982), Economics of Assembly Systems, Journal of Manufacturing Systems Vol.1, No.1, 111-127.

8. AM for Windows, (1995), Reference Guide, Version 8.2. Intelligent Environment.

9. Besant, C. B. and Lui, C. W. K. (1986) Computer-Added Design and Manufacture, John Wiley & Sons New York.

10. Nof, S. Y., Wilbert, E. W and Hans, J. W. (1997) Industrial Assembly, Chapman & Hall, London.

11. Khan, A. (1998), A Knowledge-Based System for the Design and Analysis of Assembly Lines. Ph.D. Thesis, University of Bradford.

12. Wild, R. (1975) On the Selection of Mass Production Systems, Int. J. Prod. Res., Vol.13, No.5, 443-461.

13. Helander, M. (1995) A Guide to the Ergonomics of Manufacturing, Taylor & Francis London.

14. Khan, A. and Khan, M. K. (1998), A Knowledge Based Balancing for a Mixed-Product Assembly Line. Pacific Congress on Manufacturing Management, Australia.

AI Applications in Manufacturing

CAPE/035/2000

Product engineering patterns

R G DEWAR
Napier University Business School, Edinburgh, UK
J M RITCHIE and **I BLACK**
Department of Mechanical and Chemical Engineering, Heriot-Watt University, Edinburgh, UK

ABSTRACT

This is the second paper in a series which explores the use of patterns (descriptions of successful, reflective solutions to recurring problems in context) to manage the knowledge asset in a product engineering environment. The authors extend the notion beyond mere introduction and include patterns that have been seen in industry that relate specifically to manufacturing management. A prototype implementation of a web based pattern language is also described. It is intended that this prototype will be deployed in a company intranet to measure its impact on capturing, retaining and sharing knowledge.

1 INTRODUCTION

The value of knowledge as a strategic asset has been recognised for many years (1, 2). Indeed, manufacturing organisations have traditionally managed knowledge well through the use of paper documents for explicit knowledge (particularly emphasised during the push for quality accreditation in the 1980s and 90s) and apprenticeships for more tacit knowledge. The explicit knowledge of the business is also often encoded in computer software in an attempt to automate processes, standardise data, re-use information, record intent and improve communications. Examples demonstrating varying degrees of success can be seen in the application of product data management technologies (3). Even concurrent engineering, with its emphasis on collocation, improved communication and the synergy of the collective endeavour, is an instance of Knowledge Management (KM).

However, knowledge is increasingly being presented as an enabler for sustainable competitive advantage (4, 5, 6). This is coupled with the relentless pressure for faster innovations, which themselves rely on the retention and creation of knowledge. Add to this the rise and rise of the internet allowing new ways of working and providing new opportunities for communication (7, 8) and it is perhaps not surprising that companies in a variety of sectors are exploring how they might manage knowledge better.

Unfortunately, Lucier & Torsilieri (9) estimate that around one third of KM programmes have little business impact. They suggest the symptoms of such programmes are that they have poor objectives and focus, do not link organisational change with knowledge creation and use and suffer from a lack of top management sponsorship. The authors would also add that people's reluctance to share knowledge (10), the implicit nature of the most valuable expertise, the effort of maintaining the knowledge repository and information overload exacerbate this problem.

Since attempts at recording engineering expertise, such as the MIT Process Handbook (11), have had limited success and uptake, an opportunity remains to create a useful and accessible knowledge repository. With the intention of seizing this opportunity and taking account of the aforementioned difficulties of KM, this proposal considers the application of patterns (12, 13) to product engineering.

2 PATTERNS AND PATTERN LANGUAGES

The concept of the pattern was introduced by the architect and philosopher Christopher Alexander in the late 1970s (14, 15). He recognised that certain attributes in building and urban design frequently occurred throughout history and across cultures. In other words, he identified successful solutions to recurring problems in context and found a way of communicating these by standardising the format into what he called a pattern. He also recognised that each solution resolves the competing forces that the architect faces (for instance light, comfort, aesthetics, etc.), to a greater or lesser degree. This meant that the solution transformed the initial context in some way and generally had advantages and disadvantages. By being explicit about the nature of the resulting context, the solutions were not necessarily prescriptive and became reflective instead, forcing the pattern user to think for themselves rather than blindly following a procedure.

Furthermore, a pattern language is formed when patterns are collected together and cross-references are provided between those that provide mutual support. This connectivity creates synergy between the patterns and so enhances the resulting course of action that the reflective pattern user takes.

The form of the pattern is a subject of some debate (16) but a typical template is for them to have:

- a short evocative name (providing a lingua franca for engineers);
- a description of the context in which the pattern has arisen including the prevailing forces (such as time, cost, quality, availability and risk);
- a problem that recurs given the context and forces;
- the solution to that problem;
- the resultant context (or consequences) which describes the solution's pros and cons and states how well the forces called up in the context have been resolved;
- related patterns that may work well with, before, after or instead of this one;
- known uses (see next paragraph).

For patterns to attain a useful level of abstraction and for users to have some faith in them, they should be validated. The widely recognised heuristic applied by the patterns community is the *Rule of Three* (12) - the pattern can be said to be valid if it has been seen on at least three separate occasions. For this reason, the known uses called up in the pattern provide evidence of its validity, but they can also point the reader to the people, projects or resources where more information can be found – the so called *know-who* to augment the *know-how*.

Although patterns have become a powerful and increasingly popular KM tool in software engineering (17, 18), the manufacturing community has not yet exploited them. The only contribution found has been a tentative position paper from Arnold and Podehl (19) who were looking to help mechanical and software engineers communicate.

3 PATTERN EXAMPLES

Several patterns have already been identified from the authors' own experiences of product engineering in a mainly mechanical domain. These deal with: the use of prototypes (20); the perennial dilemma over whether to make or buy parts; parametric CAD functions; part libraries; product structures; project management per se; and design for assembly. Although the authors would hesitate to call their growing collection a pattern language at present, connections are beginning to appear between patterns and instances are certainly recurring. Unfortunately, space restrictions only permit the inclusion of one substantial example here (Stock Downsizing - see Figure 1), which deals with reducing inventory, along with two ancillary patterns tackling aspects of the ubiquitous problem of vendor management (War Room and Vendor Visit – see Figures 2 and 3, respectively).

Notice how the patterns contain underlining in the *Related Patterns* and *Known Uses* sections. These indicate hyperlinks in a web-based pattern catalogue. In this way, the patterns point users at other sources of expertise. These connections, in turn, provide synergy by allowing successful solutions to work together. They also reduce the maintenance overhead since knowledge is well bounded and generic and large volumes of explicit context do not need to be codified. Whilst these patterns are not company specific, the authors would expect that those developed for and by an organisation's practitioners will contain additional cross-references to resources such as procedures, standards, drawings, bills of material and other project/product documents.

Stock Downsizing (aka Deprecate Stock).	
Context:	The size of your stock catalogue has been steadily increasing over the years as new products and developments have demanded more variety. Designers have more choice and can optimise designs, but the expense of illiquid assets, stockholding, procurement, marshalling and planning is now burdensome.
Problem:	How can the size of the stock catalogue be reduced?
Solution:	***Outlaw low turnover stock items***. Task a senior manager with the ownership of downsizing stockholdings. Consider the turnover of each item within the catalogue and group the items under categories (e.g. fasteners, pipe, etc). Only allow designers to specify items for new developments from a category that has a relatively high turnover and prevent requests for new additions to stock. Special cases may provide exceptions. Flag all low turnover items so that specific authorisation is required instead of automatic reordering. Look for opportunities in one-off designs or short runs of production to use up the low turnover items and investigate their re-sale value. Direct product development efforts at redesigning product lines to only use high turnover components. If your company has a spares business some items may be required indefinitely, in which case match stock holding against historical demand and audit regularly.
Consequences:	Without a champion the process will fail. Gradually the amount of, and demand for, low turnover stock will decrease. Eventually these items can be removed from the catalogue as they fall into disuse. Volumes of the remaining stock will increase, providing more opportunities for purchasing discounts. On the shopfloor, the decreasing variety of components should produce higher quality as fewer errors will be made, for instance using similar looking components for the wrong duties. Inevitably there will be instances where designers are forced to significantly over engineer in order to use preferred stock items. Here, a case for an alternative may be justified.
Related Patterns:	Consider **Bonded Stock** for higher value or longer lead time items. For parts that may be required indefinitely as spares, consider **Badging** for items that are not part of your core manufacturing competency.
Known Uses:	**Rick Dewar** has seen this pattern within Rolls-Royce, but has witnessed parts of it within other organisations, such as reducing the choices available to designers at NCR. Some aspects of the approach are also called up within the design for assembly methodology.

Fig. 1 Stock Downsizing Pattern (underlined items imply hyperlinks)

War Room	
Context:	You are dealing with an uncooperative vendor. The project is suffering as a result.
Problem:	How do I get the vendor to tackle the outstanding issues?
Solution:	Call a war room with all interested external and internal parties. Prioritise issues and agree on actions and time scales.
Consequences:	By having all parties together, the scale of the problem can be seen. The collective pressure of the group forum can help to isolate the people inhibiting progress and coerce them to commit to actions. However, the promises of the guilty may still not be realised away from the war room.
Related Patterns:	To improve relationships, it may be worthwhile employing a **Work Shop**, although it may be too late for a **Vendor Visit**.
Known Uses:	**Rick Dewar** has seen this used in a variety of industries from heavy engineering to financial services.

Fig. 2 War Room Pattern (underlined items imply hyperlinks)

Vendor Visit	
Context:	You have not dealt with a potential vendor before. Their sales visits and promotional material seem professional, but you are unsure about their capabilities and future viability.
Problem:	How can you help assess a vendor?
Solution:	Hold one of your pre-contract award meetings at the vendor's premises.
Consequences:	By visiting their site you can at least see if they operate from their mother's garage or not. You can judge the professionalism of the organisation and see what sort of resources they have. Of course, you may only end up seeing what they want you to see and a visit can only supplement any evidence you get from reference sites and credit checks.
Related Patterns:	A **Vendor Prototype** may also help you assess the company's competencies.
Known Uses:	**Rick Dewar** has seen this used in a variety of industries from heavy engineering to financial services.

Fig. 3 Vendor Visit Pattern (underlined items imply hyperlinks)

4 WEB-BASED DELIVERY

In a parallel activity to identifying and verifying patterns, the authors are developing a computerised tool to help with the dissemination and management of the growing patterns catalogue. Mentzas & Apostolou (21) point out that specialist IT applications are becoming less critical to the success of KM initiatives as intranets become more widely available. In light of this and the growing familiarity of the general public with the browser interface, a web-based implementation of the catalogue was chosen. The wiki web model (22, 23) provides a useful open-source foundation from which to build the tool. Wiki webs provide editable web pages that are ideal for intranets. In addition they allow key phrases to become recognised and cross-referenced automatically throughout the wiki web. Figure 4 shows a typical screen from the evolving product engineering patterns wiki web. The patterns are listed towards the top of the screen and the functions that act upon the repository are at the bottom.

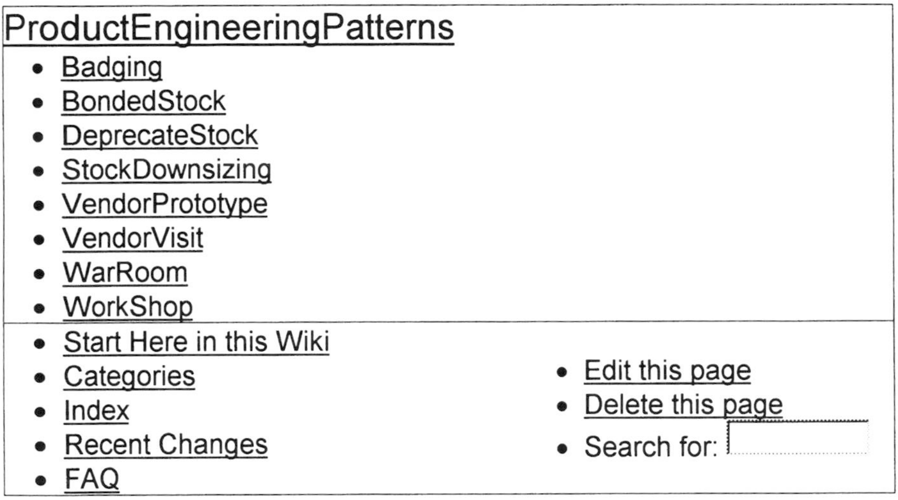

ProductEngineeringPatterns

- Badging
- BondedStock
- DeprecateStock
- StockDownsizing
- VendorPrototype
- VendorVisit
- WarRoom
- WorkShop

- Start Here in this Wiki
- Categories
- Index
- Recent Changes
- FAQ

- Edit this page
- Delete this page
- Search for:

Fig. 4 Typical Screen from the Product Engineering Patterns Wiki Web

Sellens & Wilson (24) confirm the success of intranets as dissemination medium and emphasise the importance of allowing people to have control over managing their knowledge needs. Indeed, self-determinism is at the heart of the authors' methodology and experts will be encouraged to contribute to and use the tool. Furthermore, if the users do manage their own knowledge needs, the repository becomes self-healing, avoids explicit maintenance overheads and allows the effort to be absorbed into normal working practices.

5 CONCLUSIONS

It could be argued that patterns can be used to better shape the presentation of expertise contained in "*best practice*" literature. However, Maffin *et al.* (25) point out that such literature is often too general in scope and too prescriptive in nature to accommodate the unique contexts that surround product developments. Indeed, the forms that patterns take when they evolve within a company are likely to be markedly distinct from those that occur in the literature. This divergence arises by virtue of there being much richer sources of context sensitive expertise from which they can draw, and more complex trade-offs that they must address, within the organisation than in the idealised and sanitised abstraction that so often constitutes "*best practice*".

The authors contend that a lack of practical approaches to KM is negatively impacting the competitiveness of manufacturing industry. However, the problem is not that expertise does not exist in industry - or that industry is somehow ignorant of best practice - it is that the knowledge is implicit, inconsistent, prone to disappearing as staff turn over and difficult for novices to acquire. Moreover, the volume and complexity of such knowledge makes it impractical to codify. A better approach is to capture generic solutions that are not necessarily prescriptive and which can point knowledge users at the people and resources that can provide further elucidation. In this way, the knowledge repository is accessible and maintainable.

The challenge and opportunity is to make tacit expertise not only explicit but to disseminate, refine, validate and evolve it in such a way as to support better and more consistent decision making as well as supporting reflective learning in novices. At the same time, individuals must be empowered to manage their own knowledge needs. This approach avoids bureaucracy and allows the KM effort to be absorbed into people's normal working practices.

The work seeks to capture engineering expertise and, where possible, present this using the pattern formalism. Already the authors have observed how the patterns are far from exclusively technical and in fact include references to sociological issues (for instance teams, politics and relationships), as well as prevailing strategic and competitive forces (specifically time, cost, quality, risk and flexibility). Furthermore, patterns provide a common language which practitioners can use to communicate expert solutions. However, until these patterns have been identified, written down, validated and disseminated, the knowledge they encapsulate only resides in the minds, notebooks and top drawers of a few individuals.

To date this work has been mainly targeted at the acquisition and formalisation of knowledge from the authors' own experiences. This activity acts as a pilot for future investigations, which is an oft cited indicator of success (9, 24, 26) for KM programmes. The ultimate aim is to take the evolving methodology into a manufacturing organisation. To this end, the organisations that have been approached by the authors have been enthusiastic about the work. They have recognised that, through the use of patterns, the solution need not be prescriptive and the potential consequences (positive and negative) of the solution are made clear to encourage reflection by practitioners. Too often they have seen documentation that is too process-oriented and inflexible being blindly followed with less then desirable results.

Having witnessed the success of patterns in software engineering (27, 28), there is no reason to suspect that they cannot have a similar impact on product engineering, particularly since it now possible to learn from the former community's experiences (29). At the same time, the prevalence of the internet, and the growing familiarity with it from the general public, bodes well for the acceptance of the chosen manifestation of the knowledge repository. The authors believe that now is an ideal time to exploit the web and realise the benefits of patterns in manufacturing industry. These benefits are expected to be realised through reduced training times, common errors being avoided, new knowledge being created to support innovation, reduced design stage lead times, greater consistency in decision making and improved productivity.

REFERENCES

1. Penrose E.T. (1959) The Theory of the Growth of the Firm, Basil Blackwell, Oxford.
2. Machlup F. (1980) Knowledge and Knowledge Production, Princeton University Press, NJ.
3. Hameri A.P., Nihtila J. (1998) Product data management - exploratory study on state-of-the-art in one-of-a-kind industry, Computers in Industry, vol.35, no.3, pp.195-206.
4. Boisot, M. and Griffiths, D. (1999) Possession is nine tenths of the law: managing a firm's knowledge base in a regime of weak appropriability. Journal of Technology Management vol.17, no.6, pp.662-676
5. Boisot, M. (1998) Knowledge Assets: securing competitive advantage in the information economy, Oxford University Press.
6. Teece, D.J. (1998) Capturing value from knowledge assets: the new economy, markets for know how and intangible assets. California Management Review vol.40, no.3, pp.55-79
7. Drucker P.F. (1999) Management Challenges for the 21st Century, Butterworth-Heinemann, ISBN: 0750644567
8. Alavi M., Leidner D.E. (1999) Knowledge Management Systems: issues, challenges, and benefits, Communications of the Association for Information Systems, vol. 1, article 7.
9. Lucier, C.E., Torsilieri, J.D. (1997) Why knowledge programs fail: a CEO's guide to managing learning, Strategy & Business, 9(Fourth Quarter), pp 14-28.
10. Dash J. (1998) Encouraging Employee Buy-In, Software Magazine (March).
11. MIT (1999) The MIT Process Handbook Project, http://ccs.mit.edu/ph/
12. Appleton, B. (1997) Patterns and Software: Essential Concepts and Terminology, http://www.enteract.com/~bradapp/docs/patterns-intro.html
13. Rising L. (1999) Patterns: A way to reuse expertise, IEEE Communications Magazine, vol.37, no.4, pp.34-36
14. Alexander C, Ishikawa S, Silverstein M (1977) A pattern language : towns, buildings, construction, Oxford University Press, New York.
15. Alexander C (1979) The timeless way of building, Oxford University Press, New York.
16. Meszaros G, Doble J (1997) A Pattern Language for Pattern Writing, http://hillside.net/patterns/Writing/pattern_index.html
17. Gamma E, Helm R, Johnson R, Vlissides J (1995) Design Patterns : Elements of Reusable Object-Oriented Software, Addison-Wesley, Reading, Mass.
18. Coplien J O, Schmidt D C (1995) Pattern Languages of Program Design, Addison Wesley Publishing Company, ISBN: 0201607344.
19. Arnold F, Podehl G (1999) Features and design patterns - a comparison, Swiss Conference on CAD/CAM, pp 140-148.

20. Dewar R.G., Ritchie J.M., Black I. (1999) Patterns - a novel approach to representing product engineering knowledge, 15th National Conference on Manufacturing Technology (Advances in Manufacturing Technology), Bath, UK, pp.351-355, ISBN 1 86058 227 3.
21. Mentzas G, Apostolou D (1998) Managing Corporate Knowledge: a comparative analysis of experiences in consulting firms, Proceedings of the 2nd International Conference on Practical Aspects of Knowledge Management (PAKM98), Basel, Switzerland.
22. Bower A. (2000) Dolphin Wiki, www.object-arts.com/wiki/html/Dolphin/FrontPage.htm, Object Arts Ltd.
23. Cunningham W. (2000) The WikiWikiWeb, c2.com/cgi-bin/wiki, Cunningham & Cunningham, Inc.
24. Sellens C, Wilson O L F (1998) The CMG Knowledge Intranet, Proceedings of the 2nd International Conference on Practical Aspects of Knowledge Management (PAKM98), Basel, Switzerland, October.
25. Maffin D, Thwaites A, Alderman N, Braiden P, Hills B (1997) Managing the product development process: Combining best practice with company and project contexts, Technology Analysis & Strategic Management, vol.9, no.1, pp.53-74.
26. Wigg K.M. (1998) Perspectives on introducing enterprise knowledge management, Proceedings of the 2nd International Conference on Practical Aspects of Knowledge Management (PAKM98), Basel, Switzerland.
27. Dewar R., Lloyd A.D, Pooley R., Stevens P. (1999) Identifying and communicating expertise in systems reengineering: a patterns approach, IEE Proceedings - Software, vol.146, no.3, pp.145-152.
28. Lloyd A.D., Dewar R., Pooley R. (1999) Legacy Information Systems and Business Process Change: A Patterns Perspective, Communications of the Association for Information Systems, (cais.isworld.org), vol.2, article 24.
29. Manns M.L. (1999) Evolving a patterns culture, Proceedings of PLoP.

CAPE/094/2000

An Internet-based tool selection system for turning operations

P G MAROPOULOS, M E VELÁSQUEZ, and **L A VELÁSQUEZ**
School of Engineering, University of Durham, UK

ABSTRACT

This paper presents the technical aspects and distributed functionality of ***Seltool***, an open tool selection system for turning operations. The system is able to deliver information about suitable selections of inserts and toolholders, for specific machining operations, workpiece material group and cutting type. Developed under an open and distributed philosophy, ***Seltool*** can be downloaded by remote users through any computer with Internet connection and using conventional Java enabled browsers. The benefits arising from using this system are based in its world-wide access capability which means a fast, updated and cheap information delivery to users. ***Seltool*** has been tested with encouraging results.

1. INTRODUCTION

Traditional cutting tool selection systems have oriented their solutions towards recommending different methodologies to choose the best performance solutions from a wide range of alternative options (1), achieve a higher utilisation of the selected tools (2), improve the efficiency of machining processes (3,4), and reduce tool inventory. However, these benefits are usually obtained on stand-alone environments (5), where the advantages of sharing valuable information and knowledge from varied and geographically dispersed locations are not fully exploited (6,7).

The Internet has significantly enhanced the interaction between companies and their customers, suppliers and partners, acting as an important platform to deploy open and distributed manufacturing solutions (8,9). The popularity of Internet-based applications, the functionality of which must be supported by Database operations, is growing considerably, increasing the applicability of Java-based development environments. The different applications are programmed using Java language and published in a Web-Server, where they are accessed from remote locations. The development of distributed manufacturing solutions using the Internet, involves the successful resolution of a range of technical issues such as connectivity, database management, security, user interfaces and fast access (10,11).

In the tooling sector, large databases storing information about the use of tools, machines, operations, materials and multiple trials are constantly being generated. Moreover, for large corporations, with world-wide client-portfolio, it is highly suitable to rely on a distributed infrastructure that can satisfy remote requirements (12,13), allowing staff and external users to access information sources in an efficient manner. Hence, in order to provide useful strategies and distributed solutions to these companies, research efforts have focused on open and distributed architectures to support technical and logistic functions (14,15). The following section describes the justification and functionality of ***Seltool*** and discusses its performance and evaluation using real tooling data.

2. *SELTOOL*, FUNCTIONALITY

Seltool was developed after analysing methods for implementing Internet-based systems.

2.1.Justification

The main reasons for developing the methods and prototype system described hereby were:
a) Satisfy tool manufacturers requirements: the system was developed for a tooling company as a pilot project in order to test the feasibility of future Internet-based tool selection methods.
b) Industrial usefulness of Seltool: because of its extensive tool selection functionality and the numerical and graphical capabilities this system is useful for the machine tool industry.
c) Availability of up-to-date tooling data and knowledge: technical tooling data and certain grouping criteria for the selection of specific tools is available from three sources: paper-based catalogues, PDF-based format (Compact Disc), and ACCESS Database-based format.
d) World-wide access: ***Seltool*** allows remote and distributed access (16), cheaper information exchange processes and direct interaction between users and company products.

2.2.Functionality of *Seltool*

The main functional features include:
a) Tool selection for external and internal Turning, Grooving and Threading operations.
b) Contains the following range of insert information: 10 grades, 7 types of chipbreakers for external and internal Turning and common standard profiles for Threading.
c) Contains Toolholder information for suitable inserts and approach angles.
d) Covers 19 material groups.
e) Includes 3 cutting types: finishing, medium roughing and roughing.
f) The system is easily expandable because to its modular structure.
g) The accessing and processing of data are distributed, while the storage of the information in the database is centralised.
h) The interface with the user is made through URL address in the way of an html document, which invokes the applet containing the operational structure. The system has been downloaded successfully using current versions of proprietary browsers.
i) Independent computational platform, because of using Java language and a 100% pure Java

driver for database access. That means it is possible to download the system from any computer connected to the Internet without using middle tiers software or hardware between clients and server sides (17,18). **Figure 1** shows the functional architecture of ***Seltool***.

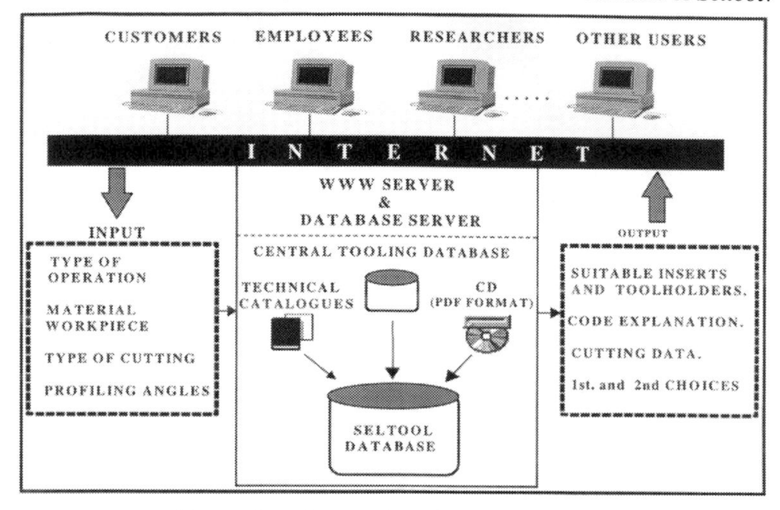

Figure 1 ***Seeltool*** **– Functional Architecture**

3. LOGIC ASSOCIATED WITH THE TOOL SELECTION PROCESS

This section describes the tooling knowledge implemented for turning operations regarding matching Inserts with Toolholders. It also describes the input and output user interface.

3.1 Logic of Matching Inserts and Toolholders with input parameters

The system requires 3 input parameters: type of operation, workpiece material group and type of cutting. *Seltool* searches the database for suitable inserts, matching the input parameters specified. For each insert found, the program searches for toolholders matching the insert selection.

If the user knows the workpiece shape to be machined, optional parameters, such as the Profile Out and In Angles, can also be introduced. **Figure 2** shows these angles and their relationship with the approach (κ) and trailing angles (ψ) of the tool. Depending on the profile angles submitted, the system searches for compatible toolholders with Approach angle and Trailing angle higher than the maximum workpiece angles given by the user.

The "Profile-OUT" angle is the profile angle generated by the cutting edge in the direction of the feed rate. Hence, this angle is compared with the approach angle (κ) of the tool as shown in **Figure 2.(a)** and **2.(c)**. The "Profile-IN" angle is compared with the trailing angle of the tool (ψ) during the generation of recesses. The user can input the maximum values of profile angle values and this acts as an additional constraint during the selection of toolholders.

The output information for inserts includes *first and second choices* together with, *shape, clearance angle, type, cutting edge length, thickness, radius, tolerance, grade, chipbreaker, cutting depth, cutting speed, feed rate* and *ordering number*. The information supplied for toolholders includes *hand of tool, tool length, locking system, weight, tool style, shank height, shank width* and *ordering number*.

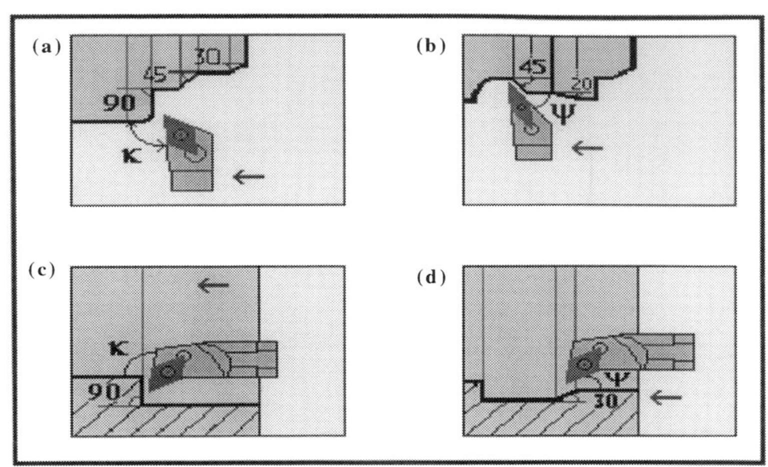

Figure 2. (a) Profile OUT and Approach angles - Ext.Turn., (b) Profile IN and Trailing angles - Ext.Turn., (c) Profile OUT and Approach angles - Int.Turn. and (d) Profile IN and Trailing angles, Int. Turn.

4. INTERNET CONNECTIVITY AND DATABASE CONSIDERATIONS

In order to establish the database connectivity through the Internet, a proprietary driver (19) was used, which is a 100% pure Java Driver and can be used by applets (20). One important benefit of using this driver is that the problem of asking the users for downloading and configuring the driver is eliminated.

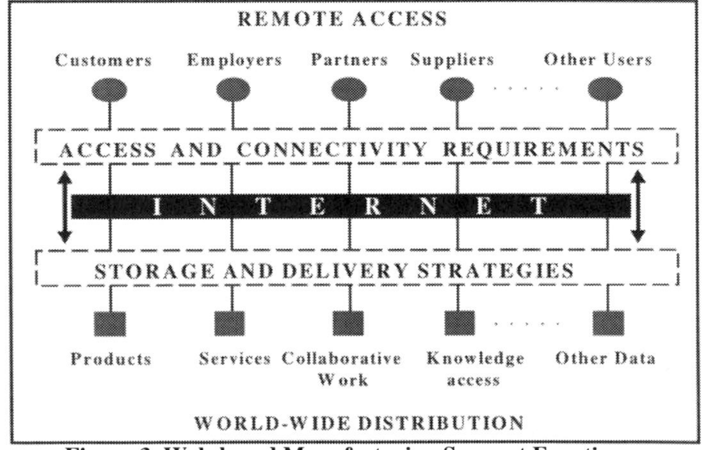

Figure 3. Web-based Manufacturing Support Functions.

The database of ***Seltool*** contains 13 relational tables storing tooling information. An SQL DataBase Management System was used for creating the database (21). In order to minimise human errors when the data is added, optimise update times of the information and provide a better choice for the maintenance of the database, two programs were developed: the Database Populate (DPS) and the Database Migratory System (DMS). The tooling data stored in a Compact Disc (PDF format) containing tooling information is used to create text files. DPS reads text files and after performing information filtering processes, the data is automatically written in the database. DMS converts useful data from an existing tooling database to the SQL database of ***Seltool***. **Figure 3** shows that access and connectivity functions are vital for performing manufacture support operations using the Internet (22, 23).

5. PERFORMANCE EVALUATION

In order to evaluate the functionality of the developed system and identify the completeness of the database to satisfy user-requests, a testing phase was conducted.

5.1. A Case Study

The first stage of the test phase was the creation of cases-studies to validate the results provided by the system against existing information in catalogues.

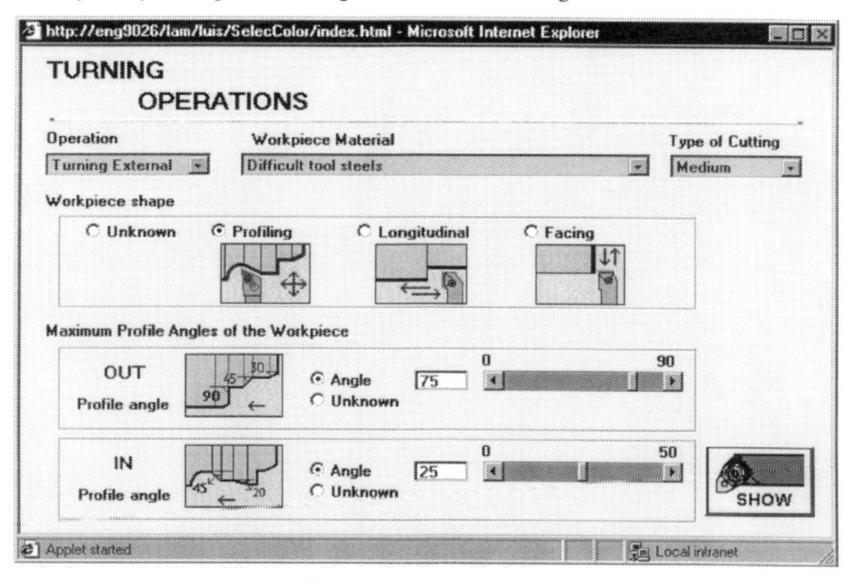

Figure 4. Input Screen.

To demonstrate the functionality of ***Seltool***, one of these studies will be discussed. The material of the workpiece to be machined is a "*Difficult tool steel*", the type of operation is "*External Turning*" and the type of cutting "*medium roughing*". It is also known that the shape of the workpiece is profiling with a "*maximum profiling OUT angle = 75°*" and *a* "*maximum profile IN angle = 28°*". The next step is to introduce these input values to the system and obtain the corresponding results.

The system creates two lists of suitable inserts, the "first choice" and the "second choice" inserts respectively. "First choice" inserts correspond to the usually recommended solution by the tool manufacturer, whilst the "second choice" list includes several other alternative inserts. **Figures 4** and **5** show the input and output screens of this example.

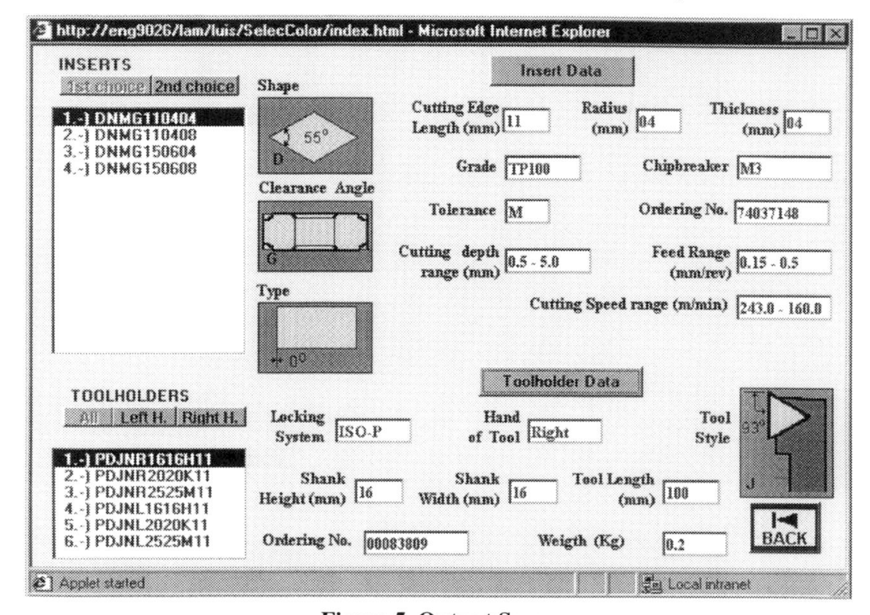

Figure 5. Output Screen.

Four "first choice" Inserts were obtained. The specification of the insert code and suitable cutting data range for these inserts are shown in the top-rigth part of the output screen. When any insert is selected from the Inserts list (**Figure 5 top-left**) the suitable toolholders list for the specified insert is shown (**Figure 5 Bottom-left**). For the first insert chosen, there are six possible toolholders with different type of hand (left and Right) and sizes in the Shank Height and Width. If only left hand toolholders are needed, the button **Left H.** can be pressed and just left hand toolholders will be presented in the list.

In the same way, second choice Inserts list is shown when the **second choice** button is pressed. The second choice inserts are those that can be considered as alternative options but without providing the best performance of the first choice inserts. The second option is very important when first choice results cannot be obtained. In the case-study presented, five inserts were found as second choice.

As can be seen from the screens, the graphical interface permits the user to submit the input parameters and obtain the results in a user friendly way. All the specifications of inserts, toolholders and cutting data are presented in one screen and aditional options for second choice inserts are provided. These characteristics allow a better visualisation, and a faster and interactive way of searching suitable tools than conventional representation schemes provided by catalogues.

5.2 Completeness of the database

The second stage of the test phase was the execution of experiments with different input parameters to find out the levels of completeness of the database.

Table 1. Inserts found for External and Internal Turning Operations.

Material Groups	External Turning						Internal Turning					
	Frist Choice			Second Choice			Frist Choice			Second Choice		
	Fin.	Med.	Rough	Fin.	Med.	Rough	Fin.	Med.	Rough	Fin.	Med.	Rough
Steel < 90 fg/mm^2	35	18	10	44	25	41	26	11	4	32	17	14
Steel > 90 fg/mm^2	26	14	10	30	18	25	17	8	4	20	11	8
Easy-cut. & moder. austenitic steels	31	15	12	66	42	24	22	9	4	45	25	8
Austenitic and duplex stainless steels	26	24	0	31	15	16	17	14	0	22	9	6
Cast iron	26	14	10	43	18	25	17	8	4	30	11	8
Aluminium & other non-ferrous alloys	22	0	0	9	0	0	13	0	0	9	0	0

Taking into consideration that the most used operations are external and internal turning, the experiment focused in searching for all possible solutions with these two operations. The workpiece material groups were used with finishing, medium and rough type of cutting, and both list of inserts, first and second choices, were taken into consideration for the test conclusions. **Table 1** shows the results for the combination of these input parameters.

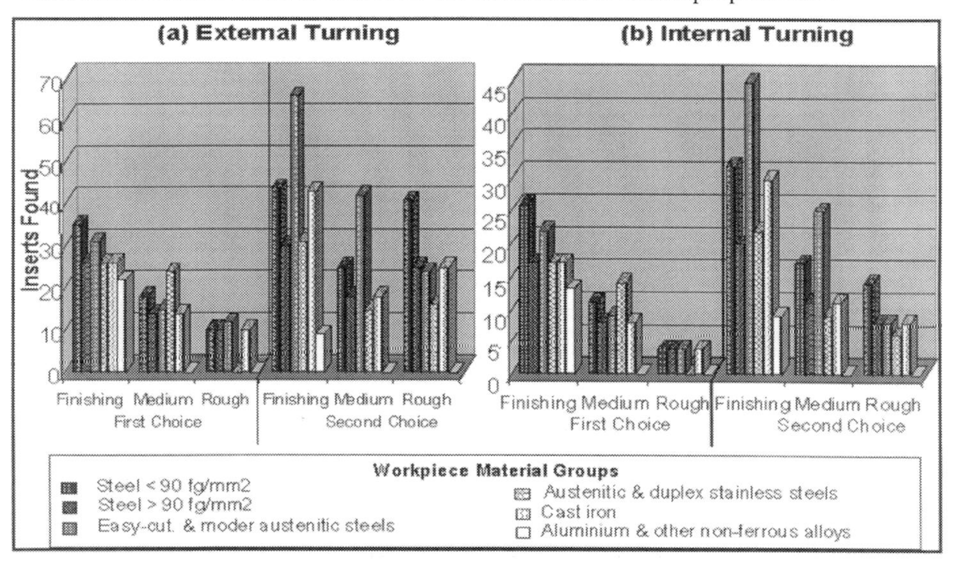

Figure 6. First and Second Choice Inserts for External and Internal Turning.

Figure 6 shows the results. From considerations of this figure it can be seen that the system is able to provide suitable first and second choice inserts for all specified material groups.

However for some type of material and type of cut there are no suitable inserts. For example, in the case of medium and rough type of cutting, the search for inserts for aluminium and non-ferrous alloys, results in no inserts.

In the same fashion, the system is not able to provide suitable first choice inserts for cast iron and rough type of cutting, but when the second choice option is used, the system supplies a list of alternative inserts.

The values observed in **Figure 6 (b)**, for Internal Operations, indicate the same trend observed in **Figure 6 (a)**, but the number of inserts found for each type of cutting is smaller than that found for external operations.

In order to verify the effectiveness of using the profile angles features to obtain more accurate results in the search for tools, a particular test was carried out. This test allows the comparison of unconstrained results, which do not have to satisfy particular profile shapes, with constrained results that meet the profile shape requirements in terms of profile angles.

For 10 cases study , inserts were selected, initially assuming that the geometry of the profile was unknown, as shown in **Table 2**. Once the number of inserts for each case was obtained, the next step involved changing the value of the profile angles to find out new sets of tools.

Table 2. Number of inserts found

No	Wpc. Shape *unknown*	Prof. OUT 40°	Prof. OUT 90°	Prof. IN 18°	Prof. IN 23°
1	35	31	25	15	13
2	24	24	22	12	9
3	26	22	17	9	7
4	4	4	4	2	0
5	8	8	8	4	2
6	31	27	26	15	13
7	10	10	8	4	2
8	15	15	13	7	4
9	22	22	21	12	11
10	11	11	11	5	3

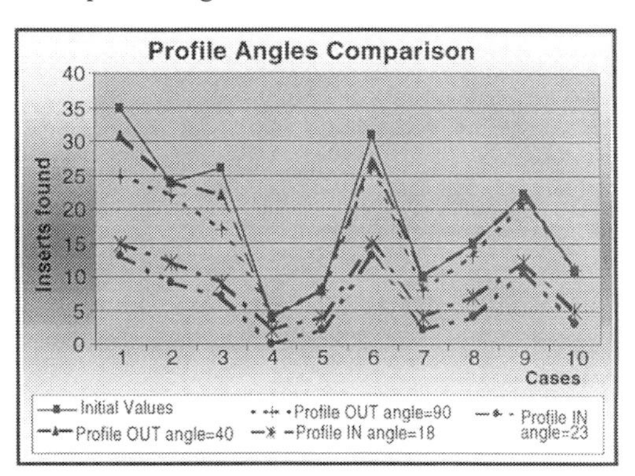

Figure 7. Inserts found using Profile Angles.

The data from **Table 2** is represented graphically in **Figure 7**. It can be seen that the number of inserts found diminishes when the profile angles values are increased. A particularly rapid reduction in the number of suitable tools can be seen by imposing a "profile-IN" constraint, which limits the selection of a large number of tools due to trailing angle unsuitability. For example, cases 1, 2, 3, 6 and 9 show a considerable reduction in the number of inserts found.

It is important to underline that in this last test only one profile angle (OUT or IN) was used at the time. When both angles are used at the same time, the number of matches found is even less, making easier the final selection by the user.

5.3 Discussion of Results

The system is able to recommend a suitable selection of inserts and toolholders for a specific operation, workpiece material group and cutting type, and show their code explanation together with the respective cutting data. The system is also able to recommend a second choice, providing in this way, alternative solutions.

Seltool finds only few matches for sometypes of materials such as austenitic & duplex stainless steels and aluminium & other non-ferrous alloys. This fact is only reflecting the need to increase even more the current size of the tooling database. The inclusion of new records will improve the capabilities of the system to satisfy a wider range of input requests. If the workpiece shape is known, the specification of profile angles provide a substantial reduction of suitable inserts found, which helps the user to make a better final decision.

6. CONCLUSIONS

In a very competitive world and increasing global markets, manufacturing companies are taking advantage from networking activities in which resources, learning, experience and knowledge can be shared. The use of the Internet as a vehicle to carry out these activities is seen as a good strategy to provide distributed solutions. It has become widely accepted that the future of manufacturing organisations will be information-oriented, knowledge driven and much of the daily operations will be automated around a global information network that connects multiple users. These facts and trends resulted in the creation of ***Seltool***, an Internet-Based Tool Selection System for Turning Operations, which was developed using an open and distributed philosophy.

The development of ***Seltool*** involved the analysis of technical criteria to establish appropriate selection of inserts, toolholders and cutting data for turning, threading and grooving operations, as well as the definition of mechanisms to access, store and retrieve information from remote locations. Hence, ***Seltool*** can be accessed through the Internet, which makes it suitable for downloading by remote users using their respective local computers and conventional browsers.

At the moment this research was conducted, commercially available browsers were able of downloading Java-applets managing databases, using only the Java Development Kit 1.02 edition. It is important to upgrade the compatibility of browsers with later editions of Java in order to access database applications efficiently. Further, the recent launch of faster processors allowing higher transfer speeds of video, audio and general data, open-up new possibilities to run Internet-based applications faster.

ACKNOWLEDGEMENTS

The authors wish to thank the co-operation given by the company Seco Tools (UK) Ltd, and the financial support provided by the following Venezuelan institutions: Com. Nac. de Investig. Cientificas y Tecnologicas (**CONICIT**) and Universidad Nac. de Guayana (**UNEG**).

REFERENCES

1 **Maropoulos, P G** 'Cutting Tool Selection: an Intelligent Methodology and its Interfaces with Technical and Planning Functions', Part B, Journal of Engineering Manufacture, 1992, Vol. 206, pp 49-60.

2 **Alamin, B** 'Tool Life Prediction and Management for and Integrated Tool selection System', PhD Thesis, 1996, University of Durham, Durham, UK.
3 **Maropoulos, P G** 'Intelligent Tool Selection for Machining Cylindrical Components. Part 1: Logic of the Knowledge-based Module', Proc. Inst. Mech. Eng., 1995, Vol. 209, pp 173-182.
4 **Maropoulos, P G** 'Intelligent Tool Selection for Machining Cylindrical Components. Part 2: Results from the Testing of the Knowledge-based Module', Proc. Inst. Mech. Eng., 1995, Vol. 209, pp 183-192.
5 **Wu, B.** 'Manufacturing Systems Design and Analysis', Chapman & Hall, 1992, Vol. 8 No. 1, (1995), pp 1-28.
6 **Smith C. and Wright P.,** 'CyberCut: A World Wide Web Based Design to Fabrication Tool', University of California, 1998.
7 **Leung, R, Leung, H and Hill, J** 'Multimedia/Hypermedia in CIM: state-of-the-art review and research implications (part I & II), Computer Integrated manufacturing Systems, Vol. 8 No. 4, (1995), pp 255-268.
8 **Park, H, Tenenbaum, J and Dove, R** 'Agile Infrastructure for Manufacturing Systems (AIMS) A Vision for Transforming the US Manufacturing Base, Technical Report on Internet (1993).
9 **Casavant T. and Singhal M.,** 'Distributed Computing Systems', IEEE Computer Society Press, 1994, pp 6-30,116-132.
10 **Denning D.,** 'Internet Besieged: Countering Cyberspace Scofflaws', Addison Wesley, 1998, pp 1-27.
11 **Stankovic J.,** 'Distributed Computing', Readings in Distributed Computing Systems, IEEE, Computer Society Press, 1994, pp 6-30.
12 **Tian G., Zhao Z. and Baines R.,** 'Agile Manufacturing Information based on Computer Network of the World-Wide-Web (WWW)', Advances in Manufacturing Technology XI, edited by D.K. Harrison, 1997, pp 571-576.
13 **Klein M.,** 'Capturing Geometry Rationale for Collaborative Design', Proceeding of the Sixth Workshops on Enabling Technologies: Infrastructure for Collaborative Enterprises, 1997, pp 22-28.
14 **Regli, W** 'Internet-enabled Computer-Aided Design', IEEE Internet Computing, January-February (1997), pp 39-50.
15 **Saha D. and Chandrakasan A.,** 'A Framework for Distributed Web-based Microsystems Design', Proceeding of the Sixth Workshops on Enabling Technologies: Infrastructure for Collaborative Enterprises, 1997, pp 69-74.
16 **McFarlan W.,** 'Harvard University Conference on the Internet and Society', Site Address: http://www.harvnet.harvard.edu/
17 **Hamilton G., Cattell R. and Fisher M.,** 'JDBC Database Access with Java', Addison Wesley, 1997, pp 33-89.
18 **Cornelius B.,** 'Developing Distributed Systems'. IT Services, University of Durham, Durham, UK, 1997.
19 **Sybase Inc.,** 'PowerJ, Programmer's Guide', Sybase Inc., 1997.
20 **Lemay, L, Perkins Ch** 'Java 1.1 in 21 Days', Sams.net Publishing, 1997.
21 **Sybase Inc.,** 'Sybase SQL Anhywhere', Sybase Inc., 1997.
22 **Mathews, J.** 'Organisational Foundations of Intelligent Manufacturing Systems, The Holonic Viewpoint', Computer Integrated manufacturing Systems, Vol. 8 No. 4, (1995), pp 237-243.
23 **Rembold, U, Nnaji, B and Storr, A** 'Computer Integrated Manufacturing and Engineering', Addison Wesley, 1993, pp 1-45, 371-401.

Petri net as a tool with application in modelling an Internet-based virtual manufacturing enterprise

M L YANG, K CHENG, G E TAYLOR, and **A DOW**
School of Engineering, Leeds Metropolitan University, UK

Abstract

This paper describes an on going research by the authors who are seeking methods to investigate the mechanism of Internet based Virtual Manufacturing Enterprises (IVMEs). The primary aim is to identify the different components of IVMEs and understand the way they operate. Petri net technique is used in this research to set up a hierarchical model for an IVME and simulation is also implemented based on that model. Modelling methodology and important modelling concepts used during this research are also discussed.

Key words: Internet based Virtual Manufacturing Enterprise, Petri nets, Enterprise Modelling, and Simulation

1. Introduction

For companies to be competitive in today's global manufacturing environment, it is critical that companies form strategic alliances with partners that can help them manufacture products in a timely and cost-effective manner. These partners, probably distributed throughout the world, are using information-based enterprise applications such as design tools, CNC manufacturing facilities, planning and scheduling software, etc [1]. Due to rapid advances in computer based technologies including microelectronics (especially microprocessors for PCs), software engineering (production of robust complex software modules; reusable, cross-platform software libraries or software chips), high speed intra-networking and inter-networking communications, it is more and more feasible to re-organise the business operation to meet new marketing demands. New manufacturing paradigms have successively emerged such as concurrent Engineering, agile manufacturing, and more recently the virtual manufacturing enterprise or extended enterprise as shown in Figure 1 [2].

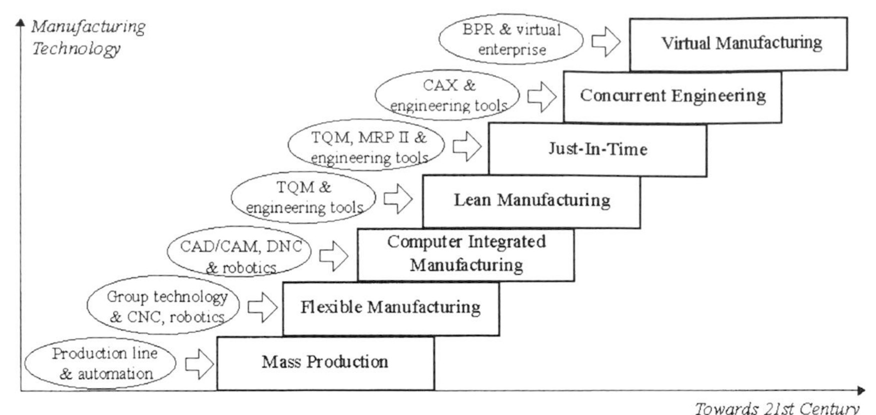

Figure 1 Development of manufacturing technology

Although the idea of virtual organizations is not newly proposed, recent developments in information technology allow the development of new implementations of virtual organizations that exploit the capabilities of the up-to date technologies [3].

The development of Internet especially the World Wide Web (WWW) has been successful in connecting of information resources without regard to physical locations. Easily communicating both within corporate Intranets and across the global Internet has given rise to new forms of manufacturing business. This leads to the emergence of Internet based Virtual Manufacturing Enterprises (IVMEs). An IVME is an organisation without the constraints of geographic location, and with a membership that often intersects multiple traditional organisations. An IVME can be formed within a large corporation (consisting of groups located at distribute sites), with multiple corporations as part of a business alliance or combined taskforce, and even involving individuals working independently of any corporate connections. In this situation efforts are concentrated on building relationship with those who have distinctive and complementary core competence. Thus, core competencies are selected from several partners and then synthesized into a single powerful entity. Figure 2 diagrammatically illustrates the concepts of an Internet based virtual manufacturing enterprise.

By nature, all enterprises are complex dynamic systems. An Internet based Virtual Manufacturing Enterprise can be perceived as a set of concurrent processes executed on the enterprise means (resources or functional entities) according to the enterprise objectives and subject to business or external constraints, based on the Internet as the medium. In order to understand IVME's operations clearly, an effective enterprise model with in-depth simulated analysis would be very helpful.

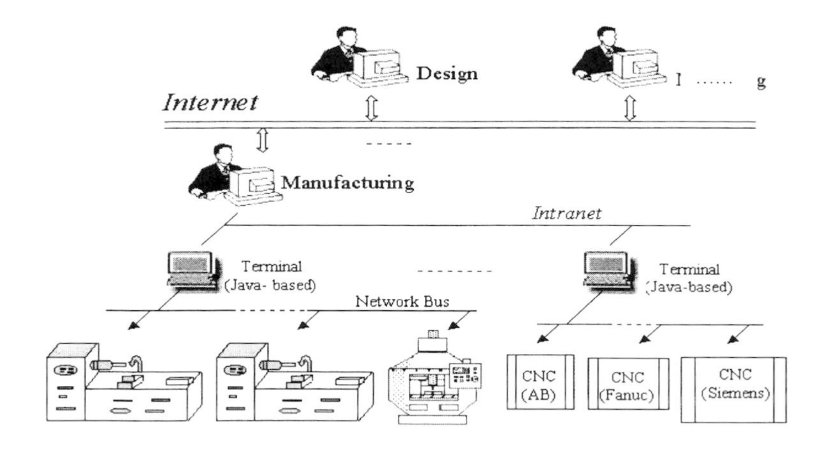

Figure 2 Concept of Virtual Manufacturing Enterprises

Petri nets (PNs) have recently emerged as a promising approach for modelling flexible and automated manufacturing systems [4]. It is a graphical and mathematical modelling technique that is becoming more and more popular and frequently used as powerful tools for the representation and analysis of systems which exhibit concurrency, parallelism, synchronization, and resource sharing features. The advantage of PNs over other analysis tools such as queuing networks and Markov chains is that PNs can deal with systems that have a large number of states[5]. This is especially important for modelling IVMEs because there are a large number of elements need to be represented at various levels across a manufacturing enterprise.

2. Petri net basics

There are many applications of Petri net or its modifications in the modelling of different systems such as workflow systems, information systems and project networks [6]. A Petri net is composed of four parts: a set of places, P; a set of transitions, T; an input function, I; and an output function, O [7]. The input and output functions relate transitions and places. The input function I is a mapping from a transition t_i to a collection of places $I(t_i)$ and the output function O maps a transition t_i to a collection of places $O(t_i)$. So a Petri net structure, C, is a four-touple, i.e. $C = (P, T, I, O)$.
Places represent passive system components that store "tokens" to indicate particular local system states. Graphically, places are represented by circles. Transitions represent the active system components that may produce, transport or consume "tokens". For each transition there is a set of input places and a set of output places. Graphically, transitions are represented by rectangles. Inputs and outputs are represented by arcs. They connect places with transitions

and stands for the relations between them. The direction of an arc indicates the flow of information/material (token flow) through the model. Arcs have a positive integer weight that indicates the number of tokens that can be consumed from an input place by a connected transition or produced by a transition on a connected output place.

In terms of modelling, the set of places is used to represent the system states and the set of transitions represents the set of actions which can be performed within the system. The state of the system is defined by the marking of places, i.e. the number of tokens in each place. The initial state of the system is defined by an initial marking.

The set of firing rules associated to the net specifies the way tokens are allowed to flow across the net, i.e. when and which transitions can be fired. When a transition is fired, it consumes tokens in its input places and produces tokens in its output places. The firing rules model the behaviour of the system.

A firing sequence of transitions $t_1, ..., t_n$ at a marking M produces a sequence of markings M' = $M_0, M_1, ... , M_n$ such that $M_{i-1}[t_i] \rightarrow M_i$ for i =1, ..., n. A marking M' is said to be reachable from a marking M if there is a sequence of intermediate markings (and transitions) leading from M to M'. The set of all reachable markings from an initial marking M0 is called reachability set and can be represented by the Reachability Tree. The Reachability Tree (RT) of a Petri net represents all possible firing sequences. It is a directed transition-labelled graph built according to the following inductive rules:
(1) The initial marking M_0 is a node of the RT
(2) If the marking M is a node of RT and t is an enabled transition at M such that M[t] -> M', the M' is also a node of RT and the arc (M, t, M') belongs to RT.
(3) No other nodes and arcs belong to RT.

A marking M is called a dead marking if no transition is enabled at M. A transition t is called dead at a marking M if no marking M' reachable from M exist such that t is enabled at M'.

3. Petri net modelling and manufacturing

Manufacturing modelling in the last decade was mostly implemented as a computer description of the process in a sequential way. Engineers make continuous efforts to cut development costs, to shorten product development cycles and to predict performance of a manufacturing process without actual manufacturing and measurements [8]. Petri nets have the capability of representing system characteristics such as contention, concurrency and synchronization which are realistic common in a manufacturing entity. For example, Petri net can represent system components that compete with each other in the form of two or more transitions connected by one common place. The vertical and horizontal properties of an enterprise such as an Internet based Virtual Manufacturing Enterprise (IVME) can be reflected by the nature of the Petri net models, the horizontal property of the system is represented by a contention form of Petri nets and a hierarchical (vertical) relationship among the components of the system is represented by several closed cycle form of Petri nets.

The main problem in modelling complex systems like IVMEs is that of managing the complexity of the resulting models. Despite the assistance of the graphical format, the complexity of network model will make it more difficult to be understood, thereby reducing its value to the analysts. Consequently, the first step in developing the model should be the design of an intelligible network; this design will not need the detail required for a more conventional model. As it may easily be modified at a later stage, the design therefore begins with the basic definition of rationale for the subsequent development of the model.

The approach that the authors used is based on the analysis of an IVME's system structure and properties. The fundamental components of this new approach are Petri net modelling and decomposition (sub-networking). The proposed modelling methodology consists of two basic phases:

- The first phase is enterprise level modelling. In order to create a model that could be easily understood. At this stage, the whole Internet based virtual manufacturing enterprise is modelled using Petri nets. Different enterprise components are recognized and represented. The relationship and communication among them are also set up and analysed at the enterprise level. The associated incidence matrix is generated for later mathematical based analysis, such as Reachability Tree analysis.
- The second phase is system decomposition (sub-networking). In this phase, focus will be moved onto individual business component, for example the manufacturer's sub-system. More manufacturing process related details will be analysed. One important concept that is always implemented during this modelling process is only the relevant business processes and enterprise objects concerned by the IVME are modelled. For instance, one big manufacturing company might be involved in the IVME. There is no need to build a model to represent the whole company. Only relevant shop floor or manufacturing cell is necessary to be modelled. Modelling scope is a very important concept that influences the efficiency of the IVME modelling greatly. Property analysis of the sub-systems is also implemented at this stage.

4. IVME modelling and simulation example

This study starts at an abstract enterprise level model of an Internet based virtual manufacturing enterprise (IVME). It consists of five individual components including the Co-ordinator, the Designer, the Manufacturer, the Deliver partner and the Maintenance partner. The Co-ordinator is mainly responsible for both internal (within the Virtual Enterprise) and external (with the Customer) communication. It also woks as an organiser, to adjust the time scale, to deal with inquiry and feedback, etc. The Designer's responsibility is to provide product design that meets the customer's demand according to the relevant timetable. It could be an individual designer or part of an organisation, e.g. a team within a big company's Design Department. The Manufacturer is responsible for manufacturing the product. Similar to the Designer, it could be part of an organisation (e.g. a manufacturing cell(s), workstation(s)) or an individual contractor such as a handmaker. They both mainly contact the Co-ordinator for co-ordination within the virtual manufacturing enterprise. But they also contact each other directly in some cases because of their specific function role in the enterprise. The Delivery and Maintenance partner respectively deal with product delivery and maintenance. They co-operate with the rest of the enterprise through the Co-ordinator. The abstract model of what being described is showed in Figure 3.

To apply the Petri net modelling technique to a manufacturing system, the main task is to match the system components with adequate Petri net touples. [9]

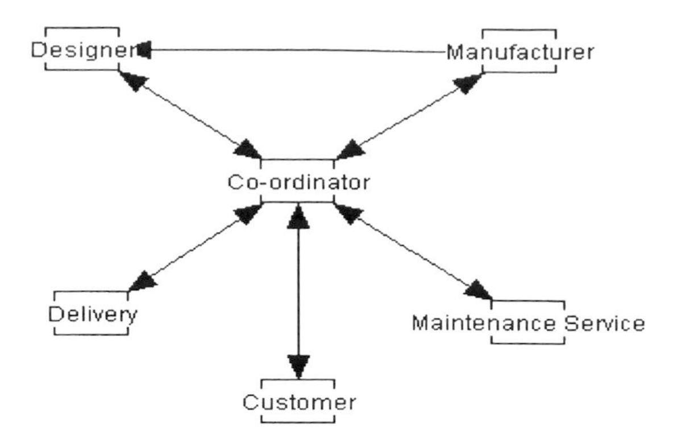

Figure 3 The abstract model of the IVME

The Petri Net model in Figure 4 shows how the previous diagram can be recast using Petri Net elements, transitions and places.

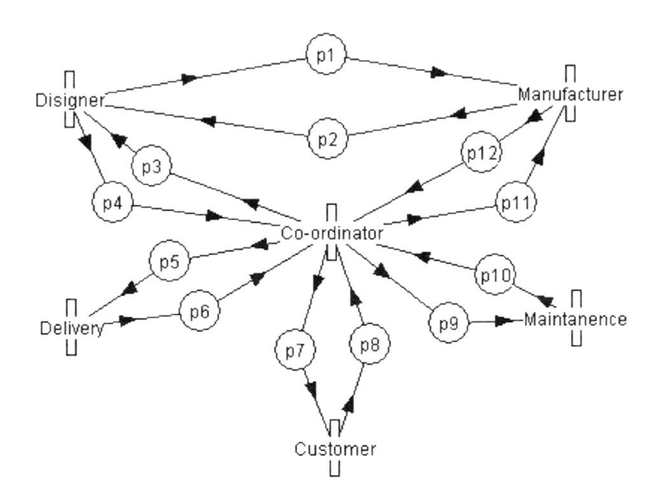

Figure 4 The Petri net model of an IVME at enterprise level

The physical system components are directly translated into transitions, while the system interconnections are expressed in the form of places and directed arcs. Every part of the business in presented and it forms the highest level of the model, its associated incidence matrix is as below for further analysis later.

$$
\begin{pmatrix}
0 & -1 & 0 & 0 & 1 & 0 \\
0 & 1 & 0 & 0 & -1 & 0 \\
0 & 0 & 0 & 1 & -1 & 0 \\
0 & 0 & 0 & -1 & 1 & 0 \\
0 & 0 & 0 & 0 & -1 & 1 \\
0 & 0 & 0 & 0 & 1 & -1 \\
0 & 0 & 1 & 0 & -1 & 0 \\
0 & 0 & -1 & 0 & 1 & 0 \\
-1 & 0 & 0 & 0 & 1 & 0 \\
1 & 0 & 0 & 0 & -1 & 0 \\
-1 & 1 & 0 & 0 & 0 & 0 \\
1 & -1 & 0 & 0 & 0 & 0
\end{pmatrix}
$$

One step further into the IVME model, the focus is moved onto the manufacturer's sub-nets. For instance two manufacturing cells are involved in the IVME. Each cell is equipped with one or more CNC milling machines(s), one robot, one CNC turning machine and one workstation. These two cells are designated as Cell 1 and Cell 2. There is a buffer area used to connect them. The layout of the manufacturing system is as shown in Figure 5.

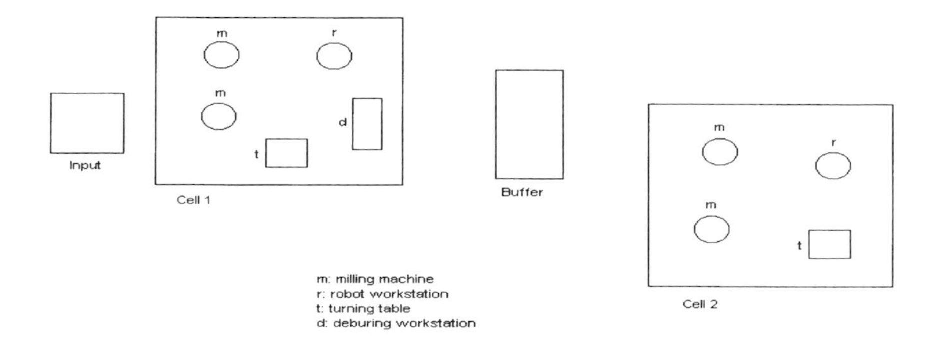

Figure 5 The layout of the manufacturer's sub-nets

Two different types of parts are processed in this manufacturing system. They are denoted as part A and part B. Part A needs to go through both Cell 1 and Cell 2. Part B only goes through Cell 2. Only one part is allowed in a cell at one time. For those parts that will go through Cell 2, they have to wait at the Buffer until Cell 2 is available. The processing path of the parts is demonstrated in Figure 6.

To model the manufacturing system described above using Petri Net, the system components need to be matched with proper Petri Net tuples. Here, cells and the Buffer are considered as resources. The status of resources is represented by places with or without a token(s) in the

Petri net model. The status of the system transforming from one state to another is reflected by transitions. Therefore, the manufacturing system can be represented as shown in Figure 6.

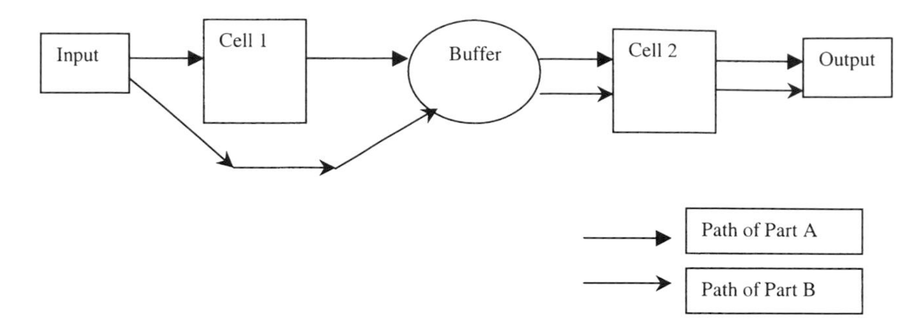

Figure 6 Parts processing path in the manufacturing system

Places Input "A" and Input "B" are used to represent the supply of the Part A and Part B to the system. Place C_i (i =1, 2) shows the availability of Cell i for a part. B as a place represents the availability of the Buffer; UB indicates the Buffer is occupied by a part; and PC_i (i=1, 2) represents the processing state of a Cell of one part. Finally, t_j (j=1, 2, ..., 6) are used as transitions from one state to another within the manufacturing system and the details as shown in Table 1. The Petri Net model of the system is shown in Figure 7.

Table 1 Function Descriptions of Transitions

Transitions	*Function Description*
t1	Cell 1 start processing Part A
t2	Cell 1 finish processing Part A, move it to the Buffer
t3	Cell 2 start processing Part B
t4	Cell 2 start processing Part A
t5	Cell 2 finish processing Part A and move it to the Output place
t6	Cell 2 finish processing Part B and move it to the Output place

In the Petri Net model, sharing of the system's resources between two parts is structured by places with two or more output transitions in contention, such as the Place B. The requirement of two resources for one operation is demonstrated as transitions connecting with two input places for synchronization, such as transition t1. The concurrency of the activities is symbolized by transitions linking with two or more output places, such as t3. The incidence matrix shown below clearly illustrates the contention, synchronizstion and concurrency properties within the system in a mathematical form.

$$
\begin{pmatrix}
0 & 0 & 0 & 0 & 1 & 1 \\
0 & 0 & -1 & 0 & 0 & 0 \\
-1 & 0 & 0 & 0 & 0 & 0 \\
0 & 0 & 1 & 1 & -1 & -1 \\
0 & 1 & 0 & -1 & 0 & 0 \\
0 & 0 & -1 & -1 & 1 & 1 \\
0 & -1 & -1 & 1 & 0 & 0 \\
1 & -1 & 0 & 0 & 0 & 0 \\
-1 & 1 & 0 & 0 & 0 & 0
\end{pmatrix}
$$

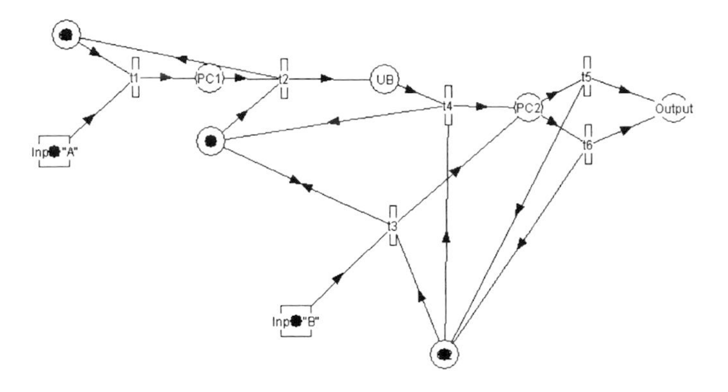

Figure 7 The Petri net model (with the 1st Initial marking) of the manufacturing system

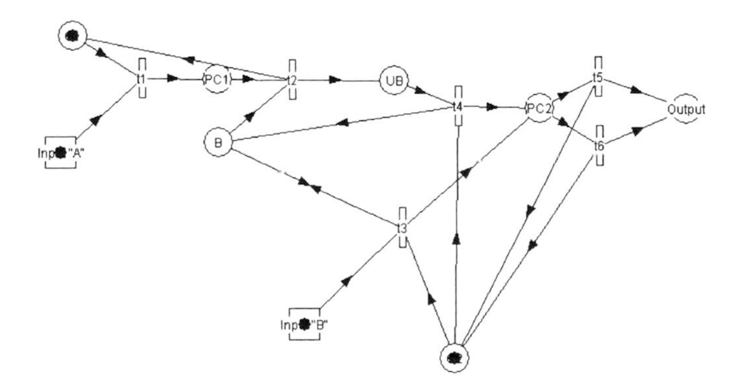

Figure 8 The Second Initial marking for the Petri net model

In order to analysis the performance feature of the manufacturing system, two different initial markings of the previous model is shown in Figure 7 and Figure 8. The free-choice firing rule is set for the current Petri Net Model. When they are executed, the first instance shown in Figure 7 runs continually, but the second instance stops as shown in Figure 9, from which we can see the availability of the Buffer is a key factor in this manufacturing system. It therefore has to be designed with the proper size to ensure the high efficiency in the manufacturing sub-system. The Reachability Tree of the Nets is also analysed in this research and the Figure 10 shows the shortest path for manufacturing Part A, for example. It indicates the chain that the IVME should focus on to improve the manufacturing performance for Part A.

With the graphical representation capability of Petri nets, we can intuitively see the physical flow of the product process. The different system states can also be simulated using different initial markings. It is very important for the business partners within the IVME to see the business operations of the enterprise simultaneously on the network, while it is the next goal of this research.

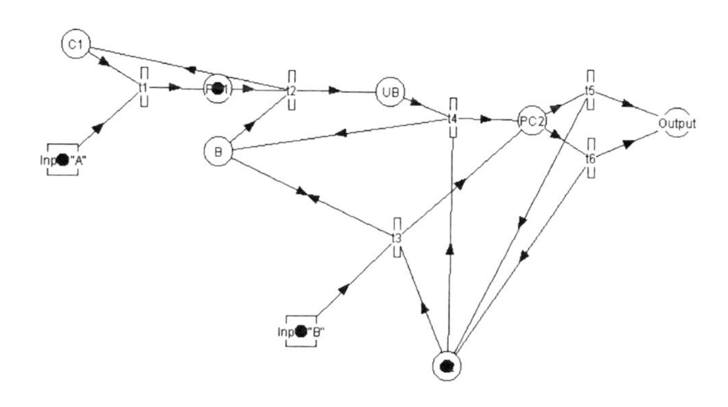

Figure 9 The 'stop' state of the Petri net model

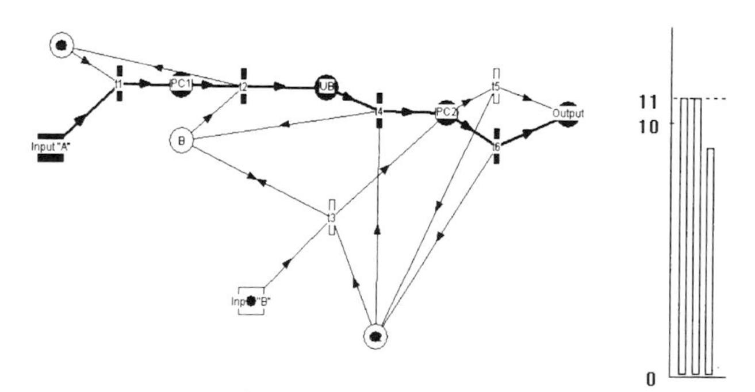

Figure 10 The analysis state of the Petri net model

5. Concluding remarks

In this paper, a new approach is presented to modelling and analysis of the Internet based virtual enterprises (IVMEs) using Petri Net. The capability of the Petri net modelling and system decomposition is further explore with a case study.

The proposed approach consists of Petri net modelling of the IVMEs, IVMEs' system decomposition and analysis. Quantifying the accuracy in the system simulation and the sensitivity in the analysis provides the basics to better understand to the mechanisms of the inter-enterprise business operation within Internet based virtual manufacturing enterprises.

As the project progressed, the research is based on a few well-understood types of Petri nets such as free-choice and pure Petri nets, etc. Further research will focus on better understanding connections and interrelations within the IVMEs, more sophisticated model based on advanced Petri net model such as Predicate Nets, and more detailed quantitative and qualitative analysis.

References

1. K. Cheng, D.K. Harrison and P.Y. Pan, *An Internet-based architecture of implementing design and manufacturing agility for rolling bearings.* Journal of Materials Processing Technology, 1998. **76**(1.3): pp. 96-101.
2. F.B. Vernadat, *Enterprise Modeling and Integration.* First Edition ed. 1996, London: Chapman & Hall. p. 513.
3. D.E. O'Leary, D. Kuokka and R. Plant, *Artificial Intelligence and Virtual Organizations.* COMMUNICATIONS OF THE ACM, 1997. **40**(1): pp. 52-59.
4. K.E. Moore, *Petri net models of flexible and automated manufacturing systems: a survey.* INT. J. PROD. RES., 1996. **34**(11): pp. 3001-3035.
5. A. Destochers, *Modeling and control of automated manufacturing systems.* 1990, Washington, DC: IEEE Computer Society Press.
6. H.M. Shih, *Management Petri net - a modelling tool for management systems.* INT. J. PROD. RES, 1997. **35**(6): pp. 1665-1680.
7. J. Peterson, *Petri net theory and the modeling of systems.* First edition. 1981, Englewood Cliffs, N.J. 07632: Prentice-Hall, Inc. p. 290.
8. L. Horvath, *Evaluation of Petri net process model representation as a tool of virtual manufacturing.* in *1998 IEEE International Conference on Systems, Man, and Cybernetics.* 1998. Hyatt Regency La Jolla, San Diego, California, USA: IEEE.
9. H. Seifoddini, *Application of simulation and Petri net modelling in manufacturing control systems.* INT. J. PROD. RES., 1996. **34**(1): pp. 191-207.

CAPE/023/2000

A welding process planning system based on the bead-on-plate database

T KOJIMA, H KOBAYASHI, S NAKAHARA, and **S OHTANI**
Dept of Applied Physics and Information Systems, Mechanical Engineering Laboratory, Japan

SYNOPSIS

In the complex welding operation, process variables are imperative designing parameters to accomplish requisite welding. A computer aided process planning system has been proposed which is of assistance for the complex welding along with combined welding method consisting of MIG and TIG. The system is based on the bead-on-plate experiment database and is also quite appropriate for those operations involved additional filler metal and magnetic control. It is WWW based and can be integrated with other metallurgical program and database on the Internet using XML. The system performance was practically examined using dissimilar metal weld of Cr-Mo steel and stainless steel, and the derived condition is confirmed practical by the associated welding procedure.

1 INTRODUCTION

Welding process plan is the process to obtain the adequate welding conditions when the base metal and the welding quality is given. Welding consists of a series of welding metallurgical processes associated with high rate of cooling and heating. The rate of cooling is detrimental factor to achieve the desired morphology of the microstructures. To achieve the optimal morphology of the microstructures, the process planning is considered as a most important factor.

Therefore, the engineering knowledge and data are integrated to be used in the planning (1) and the application software systems have been also developed (2). Most of which are classified into numerical modelling/analysis method and welding expert systems (3), (4).

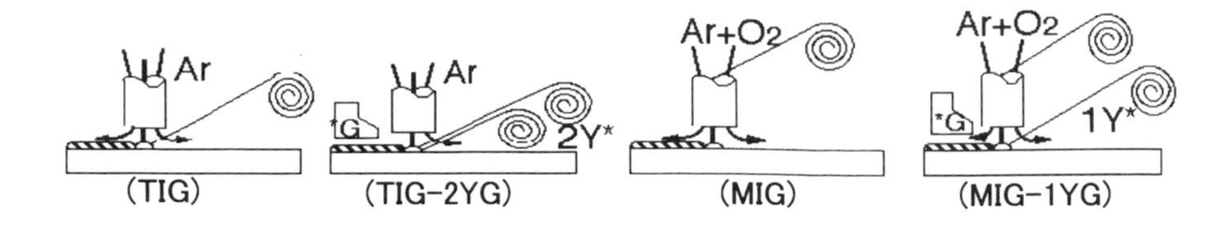

Fig. 1 Elements of complex arc welding

Nevertheless, the process planning is still largely based on the experiments (5) and experiences.
The high demand for the development of computer aided process planning system especially for the relatively difficult welding such as for the clad metals and dissimilar metals are necessary required in recent years. The complex arc welding (6) developed by the authors is an approach to be applied to the area. It consists of four element welding methods of MIG, TIG and those with additional filler metal and magnetic control concurrently, and they are combined and applied properly so as to make best use of their features. This paper proposes a computer aided process planning system based on the bead-on-plate experimental data. The system is implemented as a WWW based system and it can be integrated with other metallurgical program and database on the Internet using XML. The performance was practically examined using dissimilar metal weld and the derived condition is confirmed practical by the associated welding procedure.
At present, it covers butt joint, single-V-Groove with/without backing and flat welding. It has been observed that the system has known its worth to apply even for the plate with 6mm to 20mm thickness. The system can be used for ordinary TIG and MIG welding, as well.

2 WELDING PROCESS PLANING SYSTEM

2.1 Overview of the system

The welding process planning is a design process to select recommended wire and to determine the welding conditions by inputting the base metal specifications and expected performance of the welding. Several systems (3), (4) have been developed, but they do not cover relatively difficult welding such as for the clad metals and dissimilar metals. Besides, all of them are standalone systems and their functions are determined completely at the system construction phase.
In this paper, the complex arc welding (6) is introduced as base component of the welding process planning system. It is implemented as a WWW based welding process planning system to be used on the Internet. So, it can be integrated with other components of programs and database also on the Internet. The functional level of this type of the system is wholly dependent on the components to be organized (7). In the paper, it is aimed to construct the system in which base component of the process planning system plays the role to examine other components to obtain reliable solution. XML is used to integrate the components, which are discussed in chapter 3.

2.2 Complex arc welding method

The welding method discussed in the paper is a proper combination of four types of element welding shown in Figure 1. They are TIG (for high quality) and MIG (high efficiency) as basic methods, and MIG-1YG and TIG-2YG as additional methods. Here, "Y" indicates the number n of additional filler wire supply and "G" indicates magnetic control, respectively. We can summarize the performance of the element types from the experiments, as follows:

Efficiency: TIG < TIG-2YG <MIG < MIG-1YG
Quality: [MIG|MIG-1YG] < [TIG| TIG-2YG]
Welding heat input: [MIG] < [MIG-1YG] < [TIG] < [TIG-2YG]
Torch opening: [MIG-1YG|TIG-2YG] < [MIG|TIG] ... [1]

In the equation, a<b shows b is better than a, [a | b] shows the selection of a exclusive or b, respectively. There is no performance fall off with additional filler wire supply and magnetic control under the proper welding conditions. On the contrary, it is verified to be effective on the points of quality and cost. For example, MIG-1YG was 40% increased in efficiency as compared with MIG under the same condition. However, TIG-2YG is effective in lower dilution. Therefore, keeping in view the above remarks, we can obtain a sequence of proper element welding passes as follows:

Similar_metals: <root_pass><intermediate_layer> <finishing_layer>
Clad_metals:<root_pass><intermediate_layer><boundary_layer><finishing_layer>
Dissimilar_metals: <buttering><root_pass><intermediate_layer><finishing_layer> ... [2]

Here,

<root_pass>: [<with_backing> | <without_backing>]
<with_backing> : [TIG | MIG]
<without_backing>: TIG
<intermediate_layer>: [{TIG} | {MIG} | {MIG-1YG}]
<finishing_layer>: [{MIG} | [{TIG} | {TIG-2YG}]]
<clad_boundary>: TIG-2YG
<buttering>: [null | {MIG} | {TIG}]

In the equation, <> corresponds to the layer and multi-pass is represented as {} for each layer.

2.3 Bead-on-plate experiment database

The bead-on-plate experiment is widely used as test welding and the data include essential and practically valuable information as a whole. The database is constructed in conformity with the JIS standard (8). In the TIG welding, Tungsten electrode of 3.2mm diameter is utilized. Other common conditions are welding wire of 1.2mm diameter, electrode/wire extension of 4.5mm, fixed torch angle vertical to the welding direction. The base metal used in the experiment is SS400 (rolled steel for general structure) and the arc welding metal is Y308 of stainless steel type. Experiments using other base metals and weld metals are not directly included in the database. The bead-on-plate experiment data was obtained from the steps as follows:

(a) arc voltage for four different current values is controlled to get 3mm arc length,

(b) proper welding speed is obtained by evaluating the welding result as good, by deposition rate, reinforcement of weld, weld penetration (shape), rate of dilution and chemical analysis of molten pool.

The experiments were done for TIG, TIG-2YG, MIG and MIG-1YG, respectively. The database also includes the optional data items of preheating condition, inter-pass temperature, feeding speed of additional wire, frequency and magnetic flux density of magnetic control, etc.. The experiment results are shape and volume of reinforcement (deposition) and penetration of weld, and the components element analysis of molten pool. At present, 160 experiments are input. An example bead-on-plate data is shown in Figure 2 (9). The root pass conditions without backing in the database are obtained by another set of experiments to get fine reverse side bead for different gap lengths.

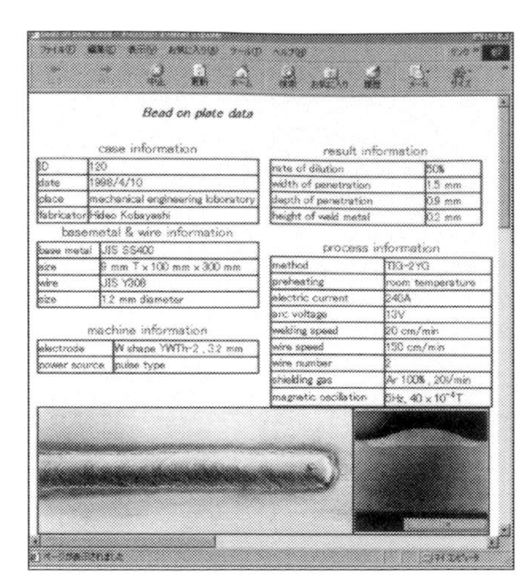

Fig. 2 Bead-on-plate database

The bead-on-plate data is different for base metal and welding metal. But, it is rather difficult to get the database for various combinations of them, practically. The total amount of experiments required will be huge if we consider only major parameters for each combination. So, we use the database above mentioned to get a set of initial predicted values and then they are examined by computational analyses to improve the precision of them. This combined approach will be discussed in the next section. In the practical welding procedures, test welding based on the welding process planning is done to confirm the plan if it is proper. The result of this test welding can be accumulated as example data in the database.

The process planning system with the bead-on-plate database is called basic component.

3 COOPERATION SYSTEM ON THE INTERNET

3.1 Base component

In the process planning system of base component, user inputs base material names, size, backing specification, and the required welding efficiency and quality. Then, the system select a candidate welding metal in accordance with JIS Z 3321-1985 (10) . The determination of element welding sequence of equation [1] is done by the decision rule of the system. The algorithm for equation [2] to determine the welding conditions of welding current, voltage, welding speed and other optional parameters is based on the computation how the V-groove is filled by welding passes using the bead-on-plate database.

3.2 Program and database on the Internet

3.2.1 Welding heat conduction simulator

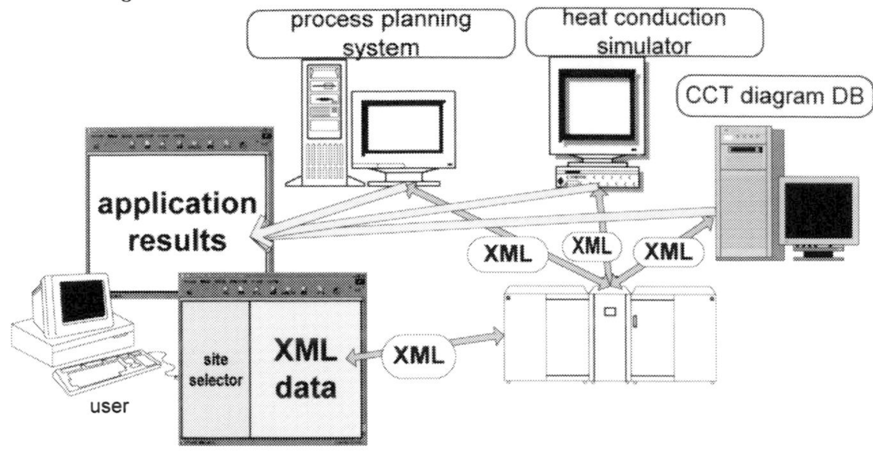

Fig. 3 System integration using XML

A program has been developed by NRIM (National Research Institute for Metals) to analyse heating/cooling pattern of moving heat source on the metal and it can be used on the Internet (11). When the physical properties and size of base metal and welding conditions of Welding current I(A), voltage E(V) and welding speed v(cm/min) is given, the cross section of molten pool and heat affected zone (HAZ) as well as the temperature history in a point by point manner are computed. This program is based on the thermal conduction equation. A simplification from the practical point of view, the coefficient of heat efficiency is set 0.7 to get the effective welding heat input (H) as H= 2.52x10**3xIxE/v (J/cm).

3.2.2 CCT diagram

CCT diagram (continuous cooling transformation diagram for welding) is an experimental diagram for steel and is used to estimate the change of structure and property by the transformation in HAZ and molten pool. As CCT diagram often shows different aspects for a small change of the composition, it can be used when the base metal is very close to the diagram and it is used to estimate the hardness of HAZ and to evaluate the tension at the joint. NRIM has collected around 100 diagrams and it can also be used on the Internet (12).

3.2.3 Schaeffler's diagram (13)

In the case for stainless steel, the quantity of molten pool is computed by the program discussed in 3.2.1, and we can obtain the rate of dilution rate. This can be used to evaluate Ferrite content and examine the possibility to obtain stable structure of welding metal by controlling the content less than 5%. When the content becomes over 10%, corrosion resistant property will drop.

3.3 Integration of components on the Internet using XML

The base component discussed in 3.1 is derived mainly from the welding experiments and experiences at the site. Other components shown in 3.2 are results from the knowledge of metallurgy. As they all can be used on the Internet using WWW, XML is used for the integration.

Major advantage introducing XML is the unification of the semantics of data and that of integration method, as well. The information can be interoperable by unifying and standardizing input/output data definition. Then by introducing XML as means to represent system input/output, the associated program execution can be proceeded uniformly. The integration image is shown in Figure 3. For this objective, each program/database site should be XML pages defining the functional summary, the name and the definition of its input and output in XML beforehand.
In the integration process, all the related welding information are defined in XML in accordance with the associated sites, and the system specifying the execution step of program/database at other sites is constructed in a server. The process starts by user input in XML and it is sent to the server. The server arranges the XML so as to fit for the step and receive the result also in XML. Then the result is returned to the site with input in XML. User can store the result and/or ask another step.

4 CASE STUDY, WELD OF Cr-Mo STEEL AND STAINLESS STEEL

4.1 Welding process planning

In the base component, user input base metals of Cr-Mo steel as JIS SCMV1 and stainless steel of JIS SUS304 with the specification for the welding grade of high efficiency and high quality. The input concerning the size of the material is 16mm thick and 100mm weld length. Penetration bead is specified. Then, the base component calculates the proper conditions based on the bead-on-plate database, as shown in Figure 4. As show in the figure, the welding sequence includes buttering process with thickness of 3mm, instead of welding dissimilar metals directly as defined in equation [2]. The succeeding processes can be regarded as similar metal weld and the conditions for the stainless steel is applied. A note concerning the buttering process is also output indicating the needs for precise control to keep the inter-pass temperature less than 423K. Actually, a welding operation was done based on the welding conditions obtained and the examinations of the subsequent procedures as shown in the following section from 4.2 to 4.4,

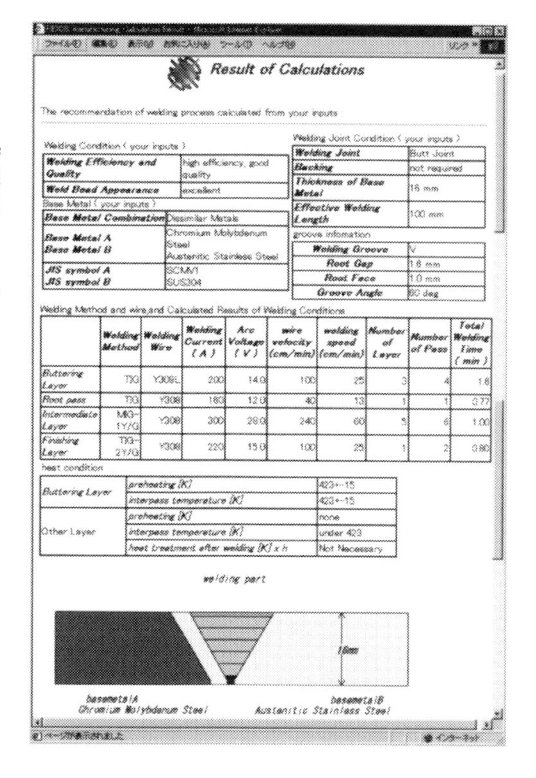

Fig.4. Output of the process planning system

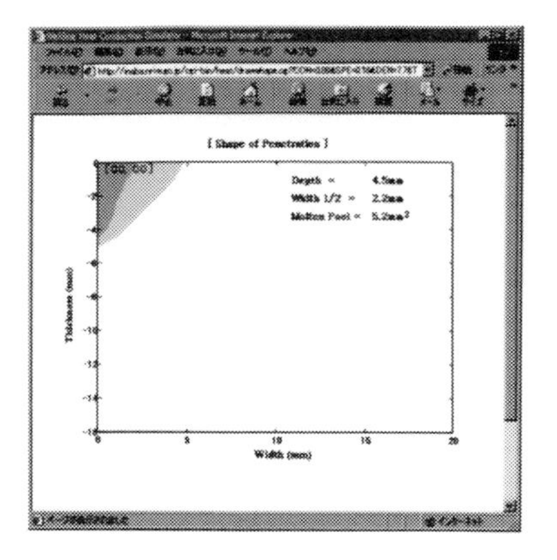

(a) Shape of the molten pool

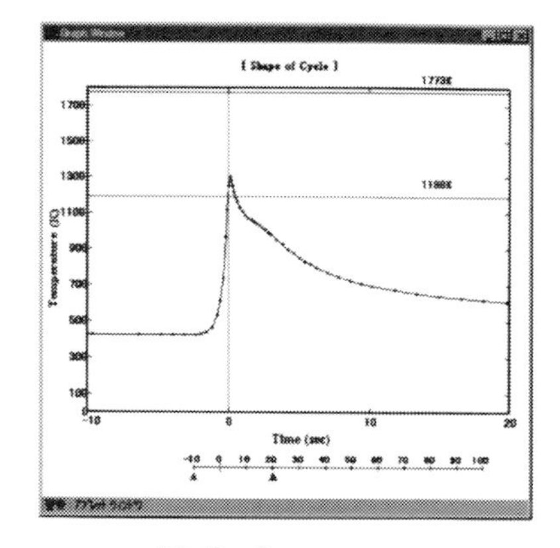

(b) Cooling curve

Fig. 5 simulation of heat conduction

and then evaluated. The bead appearance is normal and the result of tensile and bending tests can be acceptable.

4.2 Welding heat conduction simulator (part 1) (14)

When Martensite transformation occurred in HAZ, it may be the cause of defect by the change of mechanical properties. In the course of quality estimation of joint, it is useful to examine if such transformation occurs under the derived conditions. Data derived from the process planning for the buttering layer, is the welding conditions as follows: welding current of 200A, voltage of 14V, welding speed of 25cm/min, heat conductivity of Cr-Mo steel of 37.6J/(s,m,K) and other physical constants. Thickness of the base metal of 16mm and the preheating temperature of 423K is also input. . The shape of the molten pool is obtained as shown in Figure 5 (a). The time interval from 1188K (A3c point) to 773K in the cooling process is important for the transformation. When user indicates a point on the display, we can get the cooling curve of the position as shown in Figure 5 (b). We can read the cooling time of 5s from the graph.

4.3 Examination using CCT diagram

From the database of CCT diagram, "0.5Cr5Mo" is retrieved. The data is summarized on Table 1. It is the closest diagram to JIS SCMV1 based on the Carbon equivalents as shown in Figure 6. From the figure, hardness of 340Hv is read when the preheating condition is 423K. It is often said that HAZ may cause cold cracking if the hardness of the portion is over 350Hv. So, the welding conditions obtained are for the safety side. The value 340Hv is for the material with 0.12 wt-% of Carbon. As JIS SCMV1 denotes the material class up to 0.21wt-%

Table 1 Composition of Cr-Mo steel (wt-%)

	C	Si	Mn	P	S	Ni	Cr	Cu	Mo	Al
JIS SCMV1	< 0.21	< 0.40	0.55 - 0.80	< 0.030	< 0.030		0.50 - 0.80		0.45 - 0.60	
0.5Cr.5Mo	0.12	0.25	0.48			0.04	0.64		0.51	0.006

Carbon, the observation above cannot guaranteed for all the materials in the class. So, adjustment of the conditions derived from the process planning system is considered. As it is a recommendation for efficient and no defect welding operations, the change of preheating temperature is preferable compared with the change of welding current and voltage. The cooling time without preheating is read as 4s from the diagram and the estimated hardness comes up to 363Hv. This is closer to the dangerous conditions. The necessity and importance of preheating is concluded from the discussion using the CCT diagram.

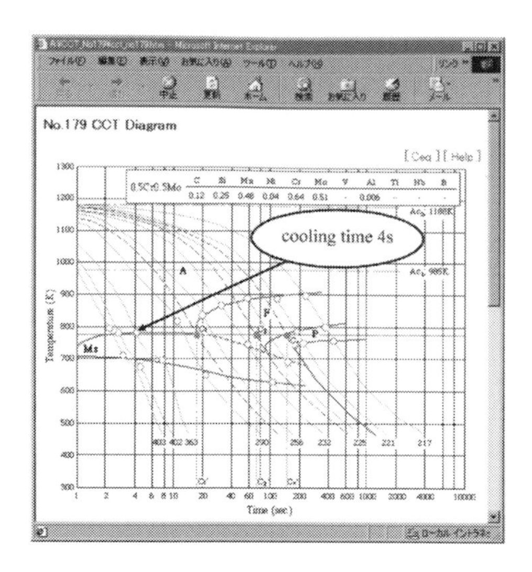

Fig.6 CCT diagram of a Cr-Mo steel 0.5Cr5Mo"

4.4 Welding heat conduction simulator (part 2)

When Austenite stainless steel is cooled gradually from 1072K to 823K, Cr-carbide deposits at the crystal interface and weld decay may occur. So, it is preferable for Austenite stainless steel to cool quickly so as to keep the corrosion resistant property. In the example, welding layer after buttering can be done on the assumption of similar material weld. The heat conductivity of stainless steel is 0.017J/(s,m,K) and they are quite different and the heat cycle during the layers may affect the heat history for Cr-Mo steel side. To investigate this, heat history of similar material weld is computed for both materials under the conditions of preheating and non preheating. The cooling time can be summarized as shown in Table 2.

Table 2 Cooling time of the base materials

	non preheating (293K)	preheating (423K)
Cr-Mo steel	4.1sec	7.2sec
stainless steel	8.9sec	18.3sec

In the dissimilar metal weld, the cooling time of Cr-Mo side is to relieve the cooling. So, we can consider the value as lower bound. Based on the similar reasoning, we can consider the value as upper bound for the stainless steel side, which is to increase the possibility of deposit of Cr-carbide at the crystal interface.

Under the condition without preheating, the cooling time is 5s for the Cr-Mo steel side, and 12s for the stainless steel side, respectively. The cooling time for the dissimilar metal weld will be longer for the Cr-Mo side, which is to keep down the hardness increase. Under the condition with preheating of 423K, which is optimal for the stainless steel side, the cooling time is 9s for the Cr-Mo steel side. But, that for the stainless steel side becomes 23s. So, in the buttering, preheating of stainless steel side should be avoided and cooling of the base metal is preferable during the process for quick cooling.

5 DISCUSSIONS

There remains many topics to be studied further. Here, we discuss them from the aspect of system integration.
In the integration process, the thesaurus in a broad sense is indispensable for the field. For example, base metal is input by JIS code name in the process planning system, in which the compositions are specified with range. But, in the CCT diagram database, it is specified by the compositions precisely. In some cases, it is not valid to put correspondence between them. In particular, each material has its history of invention and the thesaurus is preferable to include this inheritance relationship (15). In the welding metallurgy, multi-dimensional material characteristics is mapped into one dimensional value such as Ni equivalents , Cr equivalents, etc. and they are utilized effectively. This type of engineering knowledge should be modelled formally and included in the thesaurus. On the contrary, there are several terms with identical meaning. At present, this is processed individually.
At the moment, the number of verification tests using the recommended welding conditions from the process planning system is limited. But, their bead-on-plate data in the test welding are to be used effectively in keeping the quality of the system.

6 CONCLUSION

The result of the paper can be summarized as follows:
(a) A computer aided process planning system for complex welding has been developed. It is based on the bead-on-plate experiment database. The system is integrated with other metallurgical program and database to improve the solution, which shows the possibility to be used practically by case studies.
(b) The system is a WWW application system. The integration method on the Internet environment, is based on the use of XML and the method can be used for further integration in a unified manner.

Computer aided system for manufacturing, in general, is beneficial for the user, but it should not be used as a black box as it is difficult how far the system function is valid beforehand. Finally, it should be noted that different approaches to the welding process planning are also important such as to develop new welding metal and base material without special operational considerations.

ACKNOWLEDGEMENT

The authors would like to express their thanks to Dr. M. Fujita and Dr. J. Kinugawa of NRIM for their permission and advices in using the software discussed in the paper. They also express their thanks to Mr. H. Tsukui and Dr. K. Tsukui of Tsuruya Works Co., Ltd. for their valuable discussions during the research.

REFERENCES

(1)American Welding Society: WELDING HANDBOOK Eighth Edition, vol.1, (1991).
(2) Papers on numerical modelling and analysis of welding process in the Welding Research Supplement of the Welding Journal.

(3) J. Norrish & J. E. Strutt: Expert systems and computer software aids for welding engineers, Weld. Met. Fabr., 337/341, (1988).
(4) I.F.C. Smith & S.F. Bailey: The combination of independently developed expert systems, Proc. of AIENG 92, 305/315, (1992).
(5) W. G. Rippey & J. A. Falco: The NIST Automated Arc Welding Testbed, Proc. 7th Int. Conf. on Computer Technology in Welding, 1/8, (1997).
(6) H. Kobayashi & S. Nakahara: A complex welding process for clad steel, Welding International, 7-8, 81/87, (1993).
(7) Kojima, H.Sekiguchi, H. Kobayashi, S. Nakahara, S. Ohtani: An Expert System of Machining Operation Planning in Internet Environment, CAPE'99, 165/170, (1999).
(8) JIS Z 3605-1977, Recommended Practice for Semi-Automatic Arc Welding, Japanese Standard Association (JSA), (in Japanese), (1977).
(9) H. Kobayashi, et. al.: Welding Database on the World Wide Web, Quarterly J. of the JWS, 17-1,180/185, (1999). (http://www.aist.go.jp/RIODB/manufacturing/).
(10) JIS Z 3321-1985: Stainless Steel Welding Rods and Wires, JSA, (in Japanese), (1985).
(11) A. Okada, T. Kasugai and K. Hiraoka: Heat-Source Model in Arc Welding and Evaluation of Weld Heat-affected Zone, Trans. ISIJ, 28, 876/882, (1988).
(12) M. Fujita, Y. Kurihara, M. Shindo, N. Yokoyama, Y. Tachi, S. Kano, and S.Iwata, "A Distributed Database System for Mutual: Usage of Materials Information (Data-Free-Way) ", ASTM STP 1311, 249/260, (1997).
(13) A Martensite Boundary on the WRC-1992 Diagram, Welding Journal, 78-5, 180-s/192-s, (1999).
(14)M. Fujita, et. al.: Some Properties Prediction System for Welded Heat Affected Zone on Internet, Quarterly J. of the JWS, 17-1, 168/173, (1999).(http://inaba.nrim.go.jp/Weld/).
(15) http://mood.mech.hi-tech.ac.jp/home.html.

Using fuzzy logic in processing design specifications and its computerized implementation

P Y PAN
Department of Enigneering, Liverpool University, UK
K CHENG
School of Engineering, Leeds Metropolitan University, UK
D K HARRISON
Department of Engineering, Glasgow Caledonian University, UK

ABSTRACT

In order to effectively deal with the uncertainty or not well-defined imprecise inputs at the early stages of engineering design and manufacturing, researchers have carried out many approaches such as utility theory, probability method and Taguchi's method. But these approaches all have their pros and cons in some extent. In this paper a new approach based on fuzzy logic is proposed to cope with the inputs of imprecise information in design operations is proposed. This approach allows the designer or user to input imprecise parameters in 'region-wise' terms rather than 'point-wise' terms which are currently used in applying fuzzy logic. Then by means of a defuzzificaton method, such as the Centroid method or the Centre of Sums, to automatically convert the 'region-wise' input into a crisp numerical value. Compared with 'point-wise' input, the 'region-wise' input method does not require the designer to pinpoint crisp values for the input, it is therefore more suitable for dealing with imprecise descriptions at the early or conceptual stages of design operations. The advantage of this approach is further explored with a case study of applying this approach in bearing selection and the subsequent computerized system implementation.

1 INTRODUCTION

At the early stages of engineering design operations such as component selection, there are two types of inputs, certainty and uncertainty, for design specifications from a designer/user.

In order to deal with the uncertainty or not well-defined imprecise inputs, many methods such as utility theory, probability method and Taguchi's method are normally used. Compared with these methods, fuzzy logic seems to have significant promise and to be an effective approach to processing the uncertainty variables such as vague design specifications and not well-defined descriptive requirements, certainty and uncertainty, at the early or conceptual stage of an exercise in particular (1). There have been some researchers exploring the use of fuzzy logic in processing design specifications for a variety of applications. However, a practical approach is essential (2)(3) and much needed with particular reference to its implementation on computer based design support systems.

At the fuzzification stage of using a fuzzy logic method the way of dealing with uncertainty inputs does not amount to anything more than a look-up table or function evaluation. Thus an input is always a crisp numerical value (4), i.e. a 'point-wise' input limited to the universe of discourse of the input variable. It is not reasonable to ask the designer or user to commit themselves when they only have an imprecise or vague idea at this early and conceptual stage in particular. Therefore, this research is aimed at investigating a practical way of using fuzzy logic to represent imprecise or vague design requirement or descriptions and its association with computerized implementation in design support systems.

In this paper, a practical approach is proposed to use fuzzy logic in processing design specifications with particular reference to its application in an intelligent design support system. The proposed approach is verified with a case study of journal bearing selection. The work presented here is resulted from a research project being undertaken at Glasgow Caledonian University.

2 REVIEW OF THE 'POINT-WISE' INPUT IN A FUZZY MODEL

Fuzzy logic has been widely used for dealing with uncertainty. Fuzziness represents situations where membership in sets cannot be defined on a yes/no basis because the boundaries of sets are vague (5). The central concept of fuzzy-set theory is the membership function, which numerically represents the degree to which a parameter belongs to a set.

There are different kinds of membership functions, such as triangular, trapezoidal, gaussian, sigmoidal, etc., which are normally used to represent the fuzziness in a fuzzy set. Fuzziness is often described in linguistic terms such as high load, low pressure, fast speed and quick response, which are so common in processing design specifications. These descriptions are obviously easy to be understood by a designer but cannot be accepted directly by computers. Therefore, these descriptions need to be transformed into crisp values. That is the purpose of fuzzification in fuzzy logic.

In fact the membership function embodies the mathematical representation of membership in a set. If an element e.g. x in the universe X, is a member of a fuzzy set A, then the x can be mapped to a degree of membership as shown in Figure 1.

Where:
$\mu_A(x) = 1$, means x is totally in X.
$\mu_A(x) = 0$, means x is totally out of X.
$0< \mu_A(x) <1$, means x is partially in X.

Through this function evaluation all the elements belonging to the universe of discourse X can be mapped to corresponding degrees of membership. That means an element in a fuzzy set can be converted into crisp values via membership function. It is obvious that the input to a fuzzy model should be always a crisp numerical value in order to make the function evaluation. This method is called 'point-wise' input throughout this paper.

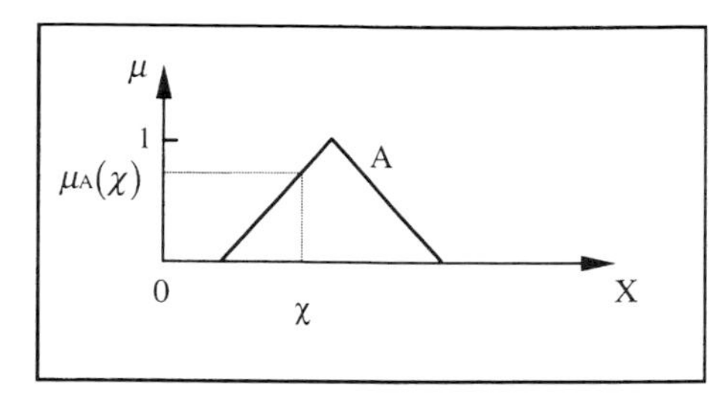

Figure 1 Membership function for fuzzy set A

3 A PROPOSED APPROACH: 'REGION-WISE' INPUT IN A FUZZY MODEL

3.1 Conception of 'region-wise' input

'Point-wise' terms are suitable and effective in applying fuzzy logic when the crisp inputs can be obtained, like in control systems where parameters can be measured by sensors, in some situations where actions and decisions implemented by humans or machines are crisp or binary.

At the early and conceptual stages of engineering design, however, the designer or user, may only have some imprecise or vague ideas on design parameters for them to make further decisions. For example, if the designer intends to design or select a voltage regulator, a few specifications such as cost, output of current and voltage, volume and appearance etc. will affect him/her to make an initial design decision. For instance, at such a stage the designer might only have the idea that the cost should be in the range of £50-70 rather than a specific crisp value like £55; the output of current could be in the range of 1-1.5A rather than an exact value such as 1.2A; the output of voltage should be within the range of 12-18V rather than just a crisp value of 13V. Therefore, it is not reasonable to ask him/her to commit himself/herself under such circumstances when using an intelligent design support system. If the system can accept 'region-wise' rather than 'point-wise' inputs, it will then make the design procedure more flexible and feasible at its early stages in particular. The 'region-wise' inputs are more commonly encountered in engineering applications.

3.2 Converting a 'region-wise' input into a crisp value

A 'region-wise' input is suitable for representing an engineering design requirement or specification. But it cannot be accepted directly by fuzzy models. It still needs to be converted into a crisp value. From Figure 2 it can be found that the region [a, b] covered from a 'region-wise' input looks like the output before defuzzification from a fuzzy logic system. The region really becomes a fuzzy quantity. However, mapping it into the corresponding degree of membership is essential.

Before converting the fuzzy quantity, i.e. the 'region-wise' input, into its corresponding degree of membership, it needs to transform it into a precise quantity. The methods used by a defuzzification procedure can be used here to find out the corresponding precise quantity of the region. There are different kinds of methods for defuzzifying fuzzy output functions, i.e. membership functions, such as Max-membership principle, Weighted average method, Mean-max membership, Centroid method and so on (6). In this research the Centroid method is focused and further explored.

(1) Centroid method
This method is also called 'center of area' or 'center of gravity' which is the most prevalent and physically appealing one of all the defuzzification methods. The defuzzified value c is given by the following expression:

$$c = \frac{\int \mu(\chi)\chi d\chi}{\int \mu(\chi) d\chi}$$ [1]

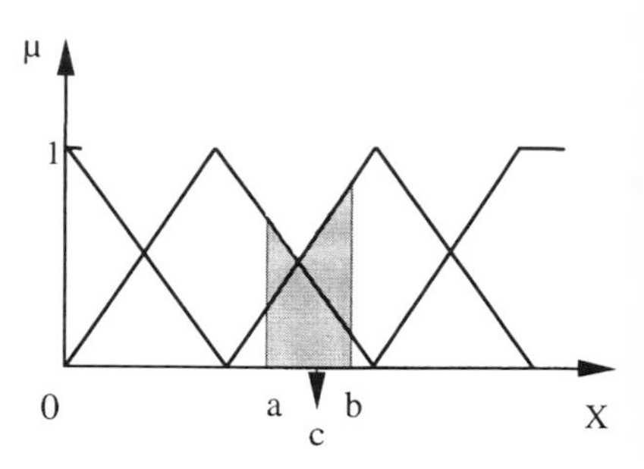

Figure 2 Region-wise input

(2) Center of sums
This method is faster than many defuzzification methods that are presently in use. This process involves the algebraic sum of individual output fuzzy sets, e.g. set A_1 and set A_2, instead of union. One drawback of this method is that the intersecting areas are added twice. The defuzzified value c is given by the following Equation:

$$c = \frac{\int \left[\chi \sum_{\kappa=1}^{n} \mu_{A\kappa}(\chi) \right] d\chi}{\int \left[\sum_{\kappa=1}^{n} \mu_{A\kappa}(\chi) \right] d\chi} \qquad [2]$$

where $\mu(\chi)$ in Equation [1] and $\mu_{A\kappa}(\chi)$ in Equation [2] are all membership functions and χ is the element of universe of discourse X.

As illustrated in Figure 3, the conversion from a 'region-wise' input to a crisp value is actually the preliminary process of fuzzification in a fuzzy logic application system. Other processes after the fuzzification such as fuzzy inference and defuzzification, are still the same as those in the 'point-wise' method.

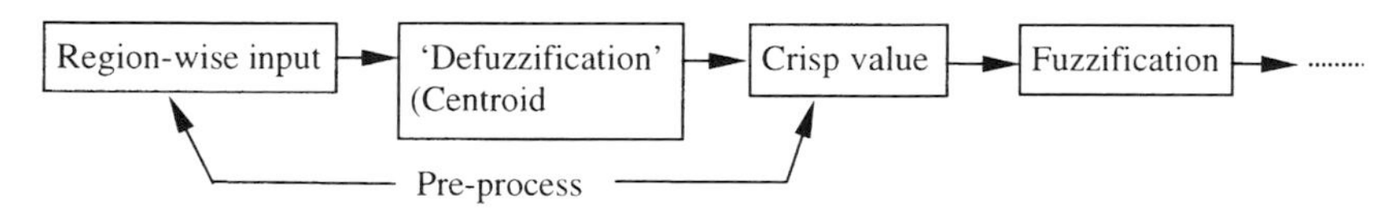

Figure 3 Process for 'region-wise' input before fuzzification

It is obvious that if the start point a and the end point b of the region [a, b] are equal then 'region-wise' method becomes 'point-wise' method. From this point of view, 'point-wise' method is just a special case of 'region-wise' method. Therefore, 'region-wise' input is more general with particular reference to engineering applications.

4 A CASE STUDY: SELECTION OF JOURNAL BEARINGS

In this case study, the proposed approach is used to process design specifications and requirements for the selection of journal bearings for designing a precision turning tools.

4.1 Journal bearings

Journal bearings are basic mechanical elements for supporting rotational and reciprocating motion in engineering products. Journal bearings are a massive family and are normally divided into seven main categories as shown in Figure 4. The bearings design includes the selection of bearings type, configuration and materials, bearing life calculation, mounting details, sealing devices, and lubrication specification. Therefore, the selection and design procedure of a journal bearings is very complicated and relies heavily on specialist knowledge and practical experience (7).

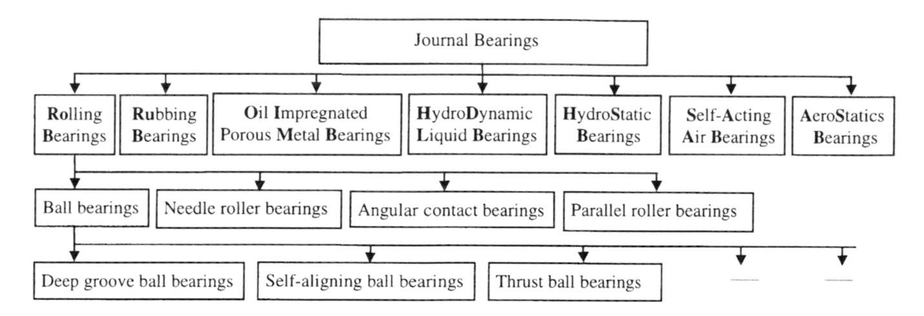

Figure 4 The family tree of journal bearings

Current methods of selection and design of journal bearings can be loosely classified as manual or computer aided. The manual method normally uses manufacturers' catalogues (book type) or electronic catalogues (CD-ROM media). In this method many complicated factors have to be considered by the designer in the decision making process toward the final selection and design outcome. Obviously it is very tedious, ineffective and even difficult to make the trade-off among these factors. Computer aided methods are usually implemented in algorithm-based systems or expert systems. These computer programs provide automated design procedures or rule based expertise. The paradigm is that of replacing human beings by intelligent computer programs. There have been some successful implementations but the systems also have their disadvantages. The most serious limitation of rule-based systems is an inability to learn from operating experience and the difficulty in building inference sequence with particular reference to the multi-disciplinary requirements from an engineering application.

4.2 A neural-fuzzy model for the selection of journal bearings

A neural-fuzzy based intelligent bearing selection system has been developed on a PC platform. In this system the kernel part for the selection is the neural-fuzzy model described in detail in (8). The structure of the model is illustrated in Figure 5. As shown in the figure, all the operating factors for the selection of journal bearings are described by membership functions. The input from the user will be fuzzified first and then the output from the fuzzification is used directly as the input to the neural networks. All the other processes in a fuzzy logic system such as fuzzy inference and defuzzification are replaced by the neural networks which are trained by specific design expertise widely collected.

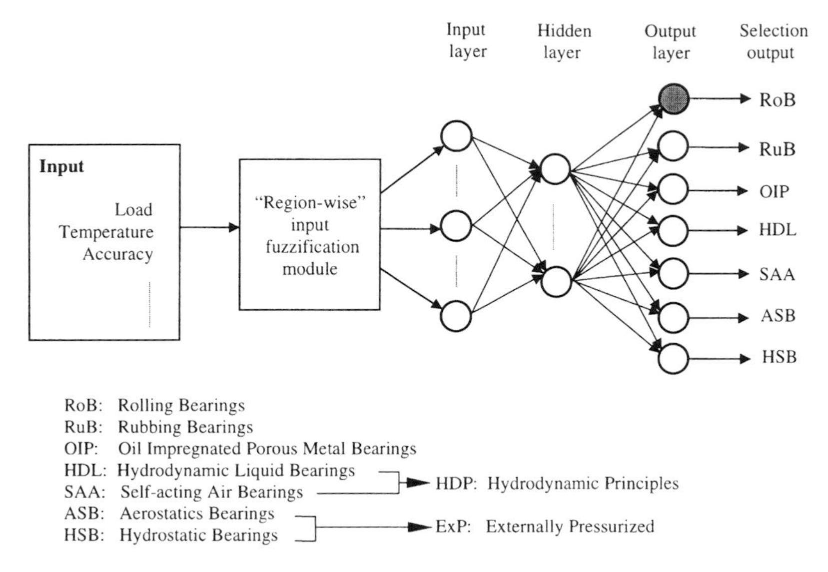

Figure 5 The structure of a neural-fuzzy based system for journal bearings selection

4.3 The procedure of pre-process and fuzzification

There are dozens of operating factors affecting the selection of journal bearings, such as:

High temperature	Limited space
Low temperature	High speed
Radial motion accuracy	Starting torque
Load capacity	Temperature rise
Wet and humid condition	Vacuum
Dirty or dusty condition	Simplicity of lubrication
External vibration	Availability of standard parts

Based on the world-wide expertise survey carried out by the authors, trapezoidal membership function as shown in Figure 6 is chosen for representing these operating factors. 'Region-wise' input method is used in the model of Figure 5 for processing the inputs from the user. For example, for load capacity, the description from the user may be like 'load capacity is high'. But this linguistic description is far from enough for a computer to deal with. The system will dialogue with the user to input the most ideal

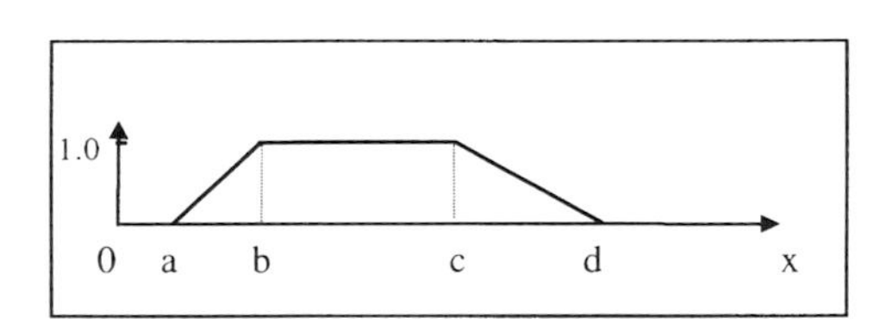

Figure 6 Membership function

working range for a specific application. Supposing the range of load capacity is 100 - 200 N, the four tuples of trapezoidal membership function of load capacity for rolling bearings is [50, 180, 500, 1000]. Before fuzzification, the 'region-wise' input should be pre-processed. The Centroid method is used for the processing which is based on Equation [1]. So there are:

$$c_1 = \int_{100}^{180} \frac{x-50}{130} \cdot x dx + \int_{180}^{200} x dx = 11882.05 \quad \text{and} \quad c_2 = \int_{100}^{180} \frac{x-50}{130} \cdot dx + \int_{180}^{200} dx = 75.4$$

Then the corresponding crisp number is $\frac{c_1}{c_2} = 157.6$, and the corresponding degree of membership of the crisp number is $\frac{(157.6-50)}{130} = 0.83$. For the other operating factors, such as temperature, radial motion accuracy, etc., their 'region-wise' inputs are supposed to be as follows:

temperature (°C): [25, 40]
radial motion accuracy (μm): [20, 30]
working speed (RPM): [400, 650]
low running torque (N): [35, 50]
limited space: [0.25,0.4]

The pre-processing and fuzzification for these inputs are the same as that for the input of load capacity. The results of the pre-processing and fuzzification are shown in Table 1.

Table 1. The results of the pre-processing and fuzzification for the 'region-wise' inputs

Operating Factors	Degrees of Membership Functions				
	RoB	RuB	OIP	HDP	ExP
Temperature (°C)	1	0.94	0.62	1	1
Radial Motion Accuracy (μm)	1	0.78	0.78	0.82	1
Load Capacity (N)	0.83	0.75	0.54	0.45	0.65
Working Speed (RPM)	1	0.31	0.38	1	1
Low Running Torque (N)	0.7	0.28	0.35	0.28	0.91
Limited Space	1	1	1	0.5	0.5

Table 1 clearly shows that for an operating factor x_i, there is a corresponding matching degree d_i which is associated with the corresponding bearings. For a specific application, each operating factor x_i is also associated with a weight rating k_i in the light of the importance of the factor for a specific application (9). In this example, according to domain experts' advises, weighting k_i =1 is assigned to load capacity, limited space and low running torque; weighting k_i =0.8 is assigned to the other three factors in Table 1. Using Equation [3] the average value D can be obtained for different kind of journal bearings.

$$D = \frac{\sum_{i=1}^{n}[k_i . d_i]}{n} \qquad [3]$$

After the pre-processing and fuzzification for the inputs associated with these operating factors, the data in Table 2 are obtained using Equation [3].

Table 2 'Region-wise' inputs and their corresponding matching degrees of journal bearings

tem	rma	lc	ws	lrt	ls	D_{RoB}	D_{RuB}	D_{OIP}	D_{HDP}	D_{ExP}
25, 40	20, 30	100, 200	400, 650	35, 50	0.25, 0.4	0.82	0.61	0.55	0.58	0.74

tem: temperature
rma: radial motion accuracy
lc: load capacity
ws: working speed
lrt: low running torque
ls: limited space

In Table 2, the data in the left most six columns are 'region-wise' inputs and the data in the other five right columns are corresponding matching degrees for different journal bearings. These corresponding degrees will be used as the input to the neural networks in the model.

4.4 Neural networks operations

The neural networks were trained by using plenty existing successful examples of design data which were obtained from domain experts and manufacturers' catalogues. After training the neural networks are ready for accepting the input from fuzzification module as shown in Figure 5.

5 COMPUTERISED IMPLEMENTATION FOR THE APPROACH

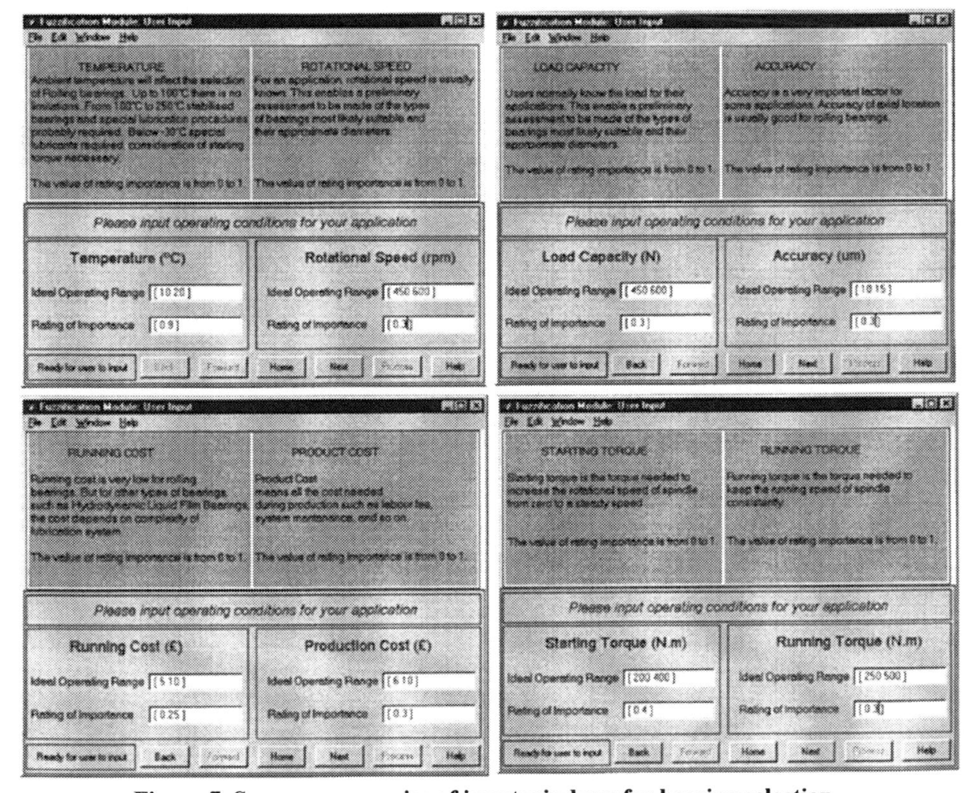

Figure 7 Some screen copies of input windows for bearing selection

This "region-wise" input fuzzy-neural approach was further developed for bearing design and selection. The system is programmed in C and C++. Figure 7 shows some screen copies of input windows from the developed system.

Table 3 Requirements for a bearing to support turning machine spindle

Application requirements		
The most ideal working range of		The rate of importance
Environmental temperature (°C)	10 – 20	0.9
Rotational speed (r.p.m.)	450 – 600	0.3
Load capacity (N)	450 – 600	0.3
Accuracy (μm)	10 – 15	0.3
Running cost (£)	5-10	0.25
Product cost (£)	6 – 10	0.3
Starting torque (N.m)	200-400	0.4
Running torque (N.m)	250-500	0.3

Table 3 illustrates an example of the application for this developed system. It includes the requirement of an assumed bearing application. All a user should do is just to input the region-wise values into the corresponding text boxes as shown in Figure 7. Then the developed system will automatically provide a recommended solution. Figure 8 is the output from the system. The result is based on the assumed application requirement in Table 3. As shown in Figure 8, a Y-bearing for the application is recommended.

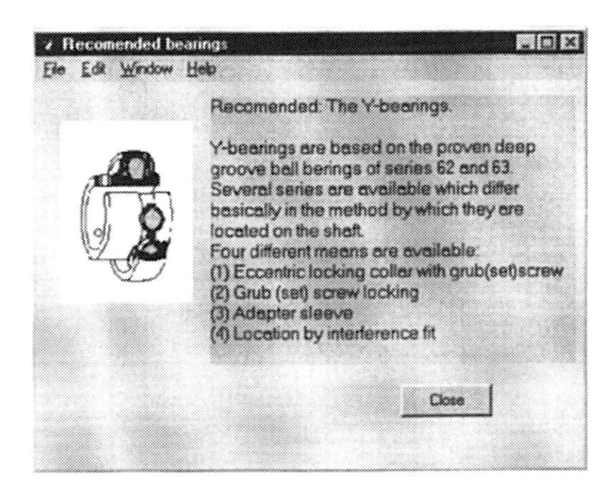

Figure 8 Recommended result from the system

6 CONCLUDING REMARKS

In this paper a new input method, 'region-wise' method, for fuzzy logic system is presented. The application of this method combined with neural-fuzzy based model for the selection of journal bearings is discussed in detail. It clearly shows that, compared with 'point-wise' method, 'region-wise' method is more suitable for engineering design purposes and more conforms with the thinking ways of designers or users of computer based design support systems.

REFERENCES

1. E.K. Antonsson and K.N. Otto, Imprecision in engineering design, Special 50th Anniversary Design Issue, Transactions of ASME, Vol. 117, June 1995, pp. 25-31.
2. A.R. Diaz; Fuzzy set based models in design optimization; Advances in Design Automation-1988, Vol.DE-14, pp 477-485, New York, ASME.
3. W.M. Dong and F.S. Wong; Fuzzy weighted averages and implementation of the extension principle; Fuzzy Sets and Systems, Vol. 21, No 2, 1987, pp. 183-199.
4. Roger Jang and N. Gulley; Fuzzy Logic Toolbox for Use with MATLAB; The Math Works Inc. 1995.
5. R. Bellman and L. Zadeh; Decision-making in a fuzzy environment, Management Science, Vol. 17, No. 4, 1970, pp. B-141-163.
6. T.J. Ross; Fuzzy Logic with Engineering Applications, McGraw-Hill, New York, 1995.
7. K. Cheng, D.K. Harrison and P.Y. Pan; Implementation of agile manufacturing - an AI and Internet based approach, Journal of Materials Processing Technology, Vol. 76, 1998, pp. 96-101.
8. P.Y. Pan, K. Cheng and D.K. Harrison; A neural-fuzzy approach to the selection of journal bearings; Proceedings of the Thirteenth National Conference on Manufacturing Research (NCMR), Glasgow, UK, 9-11 September 1997, pp. 427-432.
9. K. Cheng and W.B. Rowe; A selection strategy for the design of externally pressurized journal bearings; *Tribology International*, Vol. 28 No 7, 1995, pp. 465-474.

Neural network based process modelling and parameter selection for high-speed machining

A BOYLE and **A KALDOS**
School of Engineering & Technology Management, Liverpool John Moores University, UK
A NESTLER and **G SCHULTZ**
Institute of Production Engineering, Dresden University of Technology, Germany

ABSTRACT

High speed machining is an important new technology with considerable potential for increasing metal removal rates, for producing high quality surfaces with burr free edges, for providing good dimensional accuracy and for producing a virtually stress free component after machining. For high speed machining the selection of the optimum process parameters from a technological database is of considerable importance. The paper presents an examination of the potential for using neural network based modelling as an alternative to mathematical based modelling for the interactive selection of the machining parameters. This interactive system can be used to select the most suitable cutting parameters in various high-speed machining applications when cutting a number of different materials.

1 INTRODUCTION

The use of high cutting speeds when machining components is attractive from both the financial and the manufacturing time requirements. However the implementation of high cutting speeds in practice is difficult because of the considerable number of different factors that combine to limit the extent to which cutting speeds may be advantageously increased. These factors include the cutting temperature, the cutting tool material, the machine tool design, the geometry of the cutting tool, the cutting power, the surface integrity and the metallurgical condition of the workpiece after machining. Since the actual cutting speed achieved depends on the workpiece material, the type of cutting operation being employed and the cutting tool material used a full understanding of the technology requires a structured model to describe the process. A high speed machining process model is also necessary for the prediction of the process output parameters once the initial cutting conditions have been selected. The strategy most frequently adopted in developing models of the high-speed machining process is to provide a number of technological databases to allow actual cutting data to be recorded, interrogated and selected during the determination of the specific output parameters required.

Cutting values obtained from industrial experience or from detailed cutting experiments, are often considered to be more valuable and accurate than calculated cutting values. Hence the traditional method of determining cutting values is to make selections from machinability type tables, from cutting data handbooks and to use database information systems. Experience is also necessary and important for the selection of cutting values from a technology database. A number of different calculation methods may be employed for the determination of the output parameters from information selected from a database, including mathematical models and knowledge based methods, such as neural networks, fuzzy logic or generic algorithms. [1]. Unfortunately the complexity of the machining conditions and the interaction between the cutting parameters can not be fully described with only one of these methods. The present and future trends for the determination of machining parameters are moving towards hybrid, intelligent systems. However the application of mathematical models for technology data determination can give useful initial parameter values particularly when used in different technology software systems. In general, computer models, which allow for the prediction of the process output parameters, are essential for determining cutting parameter values in high speed machining [2].

A model of the high speed milling process is used to predict the effects of the independently selected input cutting values on the performance of the milling process [3]. In developing the model a number of databases are provided, which enables the input cutting data to be retrieved during a computer simulation of the process and the calculated results to be displayed either numerically or graphically. The computer model is designed with separate database sections for work materials, machine tool configurations, cutting conditions, cutting parameters and workpiece geometry. The separate database sections allow for the inclusion of interactive data input, for expanding the size of the database and for upgrading or changing the stored values. The process performance or output parameters of cutting temperatures, cutting forces and material removal rates are calculated with particular consideration being given to workpiece and cutting tool properties. The high speed milling process model provides a method of calculating the cutting parameters in an interactive way and gives a numerical indication of the process performance indicators.

2 EXPERIMENTAL WORK

Since one of the objects of the study is to verify the model against experimental results with particular reference to the milling of aluminium alloys suitable for aerospace components at high cutting speeds, aerospace aluminium alloy is employed as the workpiece material in a set of experiments. The workpiece material is received in the form of rectangular bar with a cross section measuring 45 mm by 300 mm and a length of 300 mm. The machine tool used is a Marwin three axis, vertical spindle, milling machine. The machine tool is equipped with a 30,000 min^{-1} high speed spindle with ceramic ball bearings and an operating power of 25 kW. It has a maximum feed rate in the X, Y and Z directions of 20 m min^{-1} and the execution time of the control system is 20 ms. The cutting tool material selected for the experiments is tungsten carbide brazed on to a high speed steel body. This material can be easily machined in its annealed state to the required shape, then heat treated to produce the desired mechanical properties for metal cutting. The cutters chosen are 50 mm and 25 mm diameter with two cutting points. The tool geometry is unique since it is designed and made for high speed milling of aluminium alloys. A three component Kistler dynamometer is used to measure the cutting forces in the X and Y directions in the horizontal plane. A 12-bit data acquisition card Lab PC Plus complete with Daqware software supplied by National Instruments UK performs the processing of the signals.

3 MATHEMATICAL MODELLING FOR CUTTING VALUE DETERMINATION

The strategy used is to create an extensive theoretical model of the high speed milling process based on mathematically determining the performance of the cutting process from a set of input cutting values. Application of such a model makes it possible to show the dependency of the selected cutting parameters and their effects on the mathematically simulated results [3][4]. By solving the equations simultaneously as a set, it is possible to see how the choice of input values selected by the user influences the high speed machining process. The computer model used consists of three major modules namely, the input data module, the numerical simulation module and the output module.

The input data module has an integrated database for the input parameter data sets associated with workpiece materials, tool materials, tool geometry, workpiece geometry, cutting parameters and machine tool parameters. Having received the input information, the numerical simulation module is designed to transform the geometric, mechanical and thermal properties information of the workpiece, the cutting tools, the machine tool and the cutting conditions into a suitable form for simulation. The module contains a numerical data controller with four sub-modules, which determines the cutting parameters, the cutting forces, the cutting energy and the cutting temperatures respectively [4]. Once the input data is entered, the numerical simulation module takes approximately 15-20 seconds to execute all its operations. The output module then generates both the numerical output data and a graphical representation of the cutting process.

The application of mathematical models for technological data determination can be used to give useful initial cutting parameter values for the desired application in different circumstances. Although the mathematical process model for the simulation of high speed milling has successfully been developed and used for generating cutting parameters, for cutting process energy analysis considerations and for process parameter verification it is necessary to verify the cutting values against practical tests. Using the input data from Brown [5] with a range of rake angle values from 6 degrees positive to 20 degrees negative, the predicted or theoretical cutting force Fc and the feed force Ff generated by the model are given in Figure 1 along with the set of experimental results for an aluminium 2017T4 workpiece material being machined at 1000 m min^{-1} at a depth of cut of 2.54 mm. The width of cut is 5 mm and the cutting tool diameter is 40 mm with 2 cutting points.

The results obtained suggest that higher negative rake angles do produce higher values for both the principal cutting force and the feed force. The predicted cutting forces plotted in Figure 1 compare well with the experimental work generally being 15% greater for the principal cutting force Fc and 9% greater for the feed force Ff. Similar decreases in force with increasing positive rake angles are reported by Armitage and Schmidt [6], von Turkovich [7], Fenton and Oxley [8] and Schulz [9]. Figure 2. shows the relationship between the experimentally measured principal cutting forces, the simulated or theoretical cutting forces and the cutting speed in both the lower speed range up to 2400 m min^{-1} and in the higher speed range from 2800 to 5000 m min^{-1}. The experimental cutting forces when plotted against the cutting speed generally correlates well with the theoretical results from the computer simulation of the mathematical modelling.

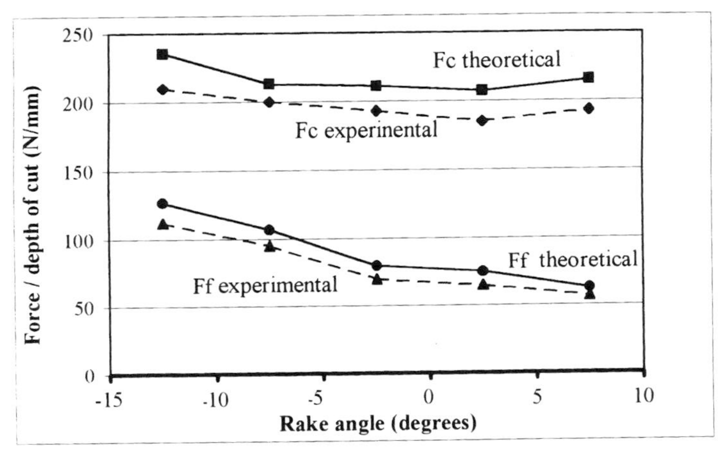

Figure 1 Rake angle and cutting forces

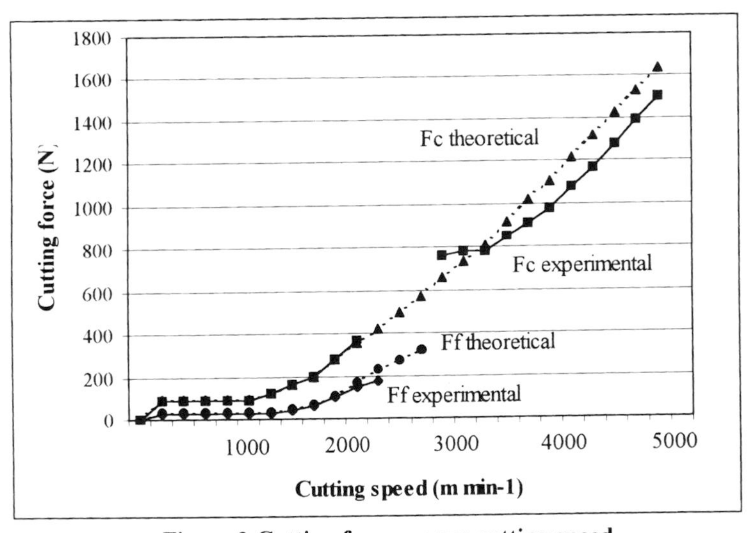

Figure 2 Cutting forces verses cutting speed

4 NEURAL NETWORKS FOR CUTTING VALUE DETERMINATION

For many technological problems, which have a large number of input and output variables with non linear interactions, it is very difficult to describe them by mathematical equations and it may be better to use suitable knowledge based methods for process modelling such as neural networks. Neural networks are able to integrate known cutting values from different sources to produce specific output values and the unknown relationships between the input and the output parameters can be learned by the neural network. For any technological situation it is necessary to have appropriate input and output parameters to describe the process and to select and provide a learning data set for the network. The provision of a learning data set means it is possible to calculate output values by training the network using known solutions. It is also feasible to apply different data sources for cutting process parameter selection [10]. A suitable cutting process parameter database has to be designed in support of the selected neural network configuration. The selection of the architecture of the neurons and the connections in the hidden layers largely depends on the specific application.

The arrangement of a suitable neural network structure is given in Figure 3. A cutting process parameter database complete with the relevant machining data is followed by the network configuration designed to suit the specific application [11]. The training of the network with a training database is needed for a thorough test of the trained network.

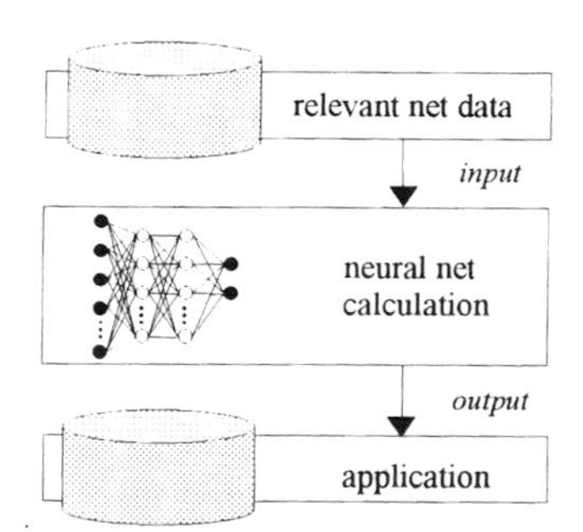

Figure 3 General processing step

In general, numerous experiments have shown that a feed-forward, multi-layer, back-propagation network with two hidden layers of 10 to 30 neurons each is appropriate for the modelling of high speed cutting processes [12]. The high flexibility and adaptability of neural networks allows a specific arrangement for the extremely complex cutting processes with a high number of input and output parameters to be created [13]. The network training is generally considered to be complete when the average error of the output is less than 10% while reproducing the training data. The influence of different net structures on system operation is examined by using measured values of a high speed cutting process for network training. Table 1 provides the neural network training data which gives workpiece temperature, chip temperature and chip/tool interface temperature as output parameters and cutting speed as the input parameter. The network structures and the simulated results for an increasing number of neurones and internal layers are shown in Figures 4, 5 and 6. The graphs in the figures illustrate only the chip/tool interface temperature for increasing cutting speed for the different trained nets. According to the increasing number of hidden neurons and connections, the networks have the capability to achieve a closer approximation of every single data point in the given data range.

Table 1 Cutting temperature data

Cutting speed [m min^{-1}]	Workpiece temperature [°C]	Chip temperature [°C]	Chip/Tool interf.ace temperature [°C]
60	37	459	680
120	36	459	897
180	36	459	860
240	36	456	920
300	35	456	985
360	35	456	1021

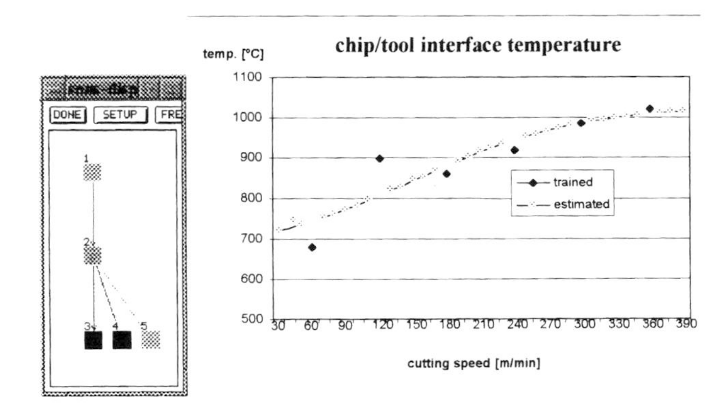

Figure 4 Network structure and simulated results for 5 neurons

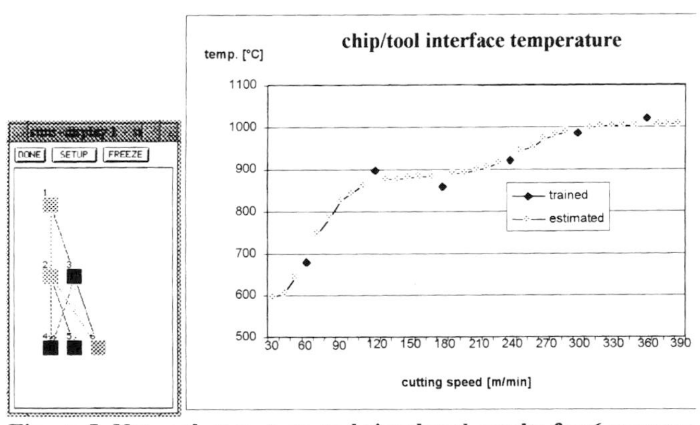

Figure 5 Network structure and simulated results for 6 neurons

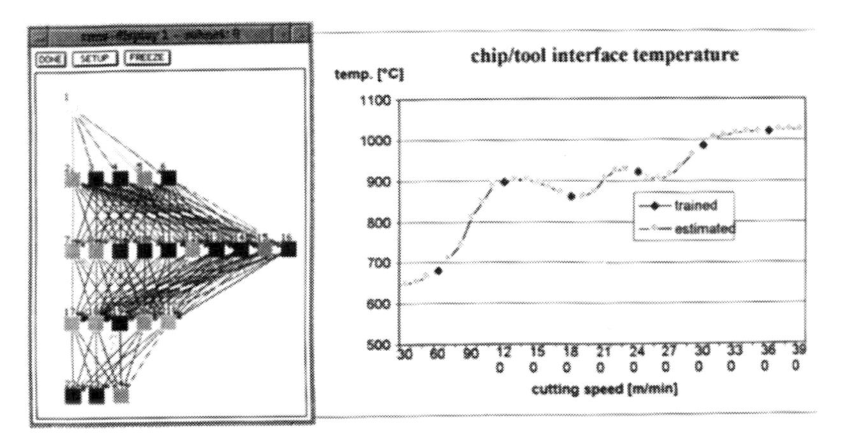

Figure 6 - Network structure and simulated results for 24 neurons

An examination of figures 4, 5 and 6 shows the capability of a neural network model to approximate non-linear functions describing high-speed cutting parameters for value prediction. The figures also indicate the process model has been verified by using neural network based models. It has been established that the neural network based models provides excellent simulation of the process provided the experimental data obtained is sufficiently accurate. This is demonstrated both experimentally and by means of the selected and trained neural network simulation models for cutting temperature prediction in the cutting speed range up to 300 m min^{-1}.

5 COMPARISON OF MODELLING METHODS

A set of technological parameters for high speed machining based upon mathematical modeling have been determined. The results calculated by this program are then used for training and testing several neural networks. The input data used for the calculations based upon mathematical modelling are given in Table 2.

Table 2 Input parameters and values for training the neural networks

No	Name	Unit	value
1	Width of Cut	mm	7,63
2	Depth of Cut	mm	31,75
3	Cutter diameter	mm	31,75
4	Cutting speed	m/min	500 - 10000
5	Rake Angle	degrees	-20 - +40
6	Feed Rate	Mm/min	90 - 4511
7	Material		Aluminium

For the development of the first neural network used to determine the cutting speed the mathematical based process model is used to generate nine sets of data. The data seta are divided into six sets for learning and three sets for training the network. A network structure of 27 neurones having three intermediate layers in a 1-3-5-8-10 arrangement produces an output error which is quite satisfactory. A second neural network to produce feed rate values is also created. Again the mathermatical model of the process is used to generate nine sets of data for the development of the neural network. These data seta are also divided into six sets for learning and three sets for training. A satisfactory network is built having 29 neurones on four intermediate layers in a 1-2-3-5-8-10 arrangement of neurones. The third neural network is created to deal with variations in rake angle. In this case the mathematical model is used to produce 13 data sets of which nine sets are used for learning and four sets are used for training. An accurate set of output values are produced using a neural network of 27 neurones aranged on four intermediate layers in a 1-2-3-5-7-9 arrangement as given in Figure 7. The output error generated by the neural network is given in Figure 8 and is seen to be quite satisfactory.

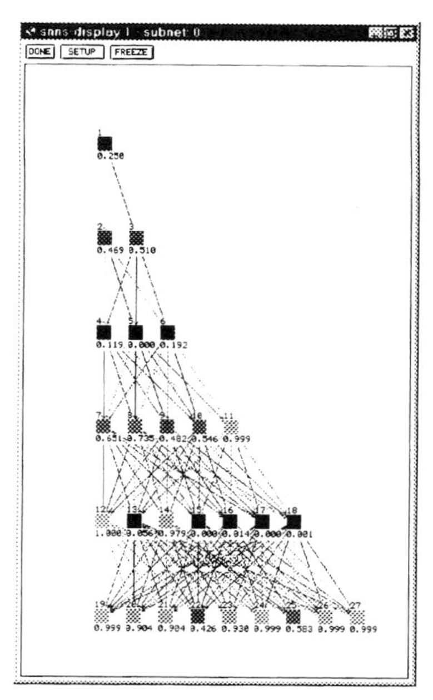

Fig. 7: Network structure for rake angle

The results of the comparison of the output data generated by the neural networks to the output data generated by the mathematical model is given in Figure 9. This figure shows the percentage difference between the values generated by the mathematical model and the neural network for all the output parameters.

The analysis of Figure. 7 shows the possibility to generate the data calculated with the simulation program with neural networks. The result is in the range of other examples for the use of neural networks.. An increase of the number of neurones would have a positive effect and it could be possible to generate even better networks. All structures of the neural networks are chosen from a lot of other structures, which were tested, and they supposed to be the best ones.

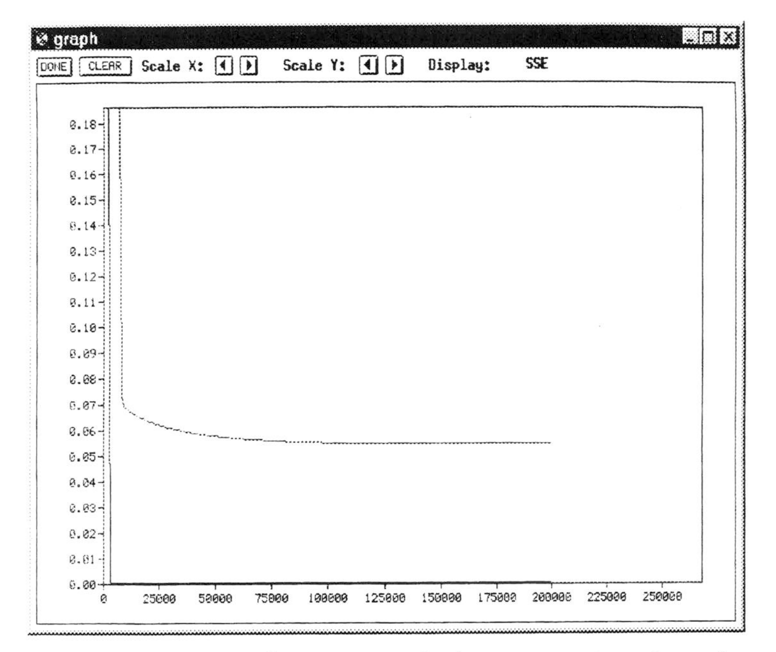

Fig. 8: Development of output error for input parameter rake angle

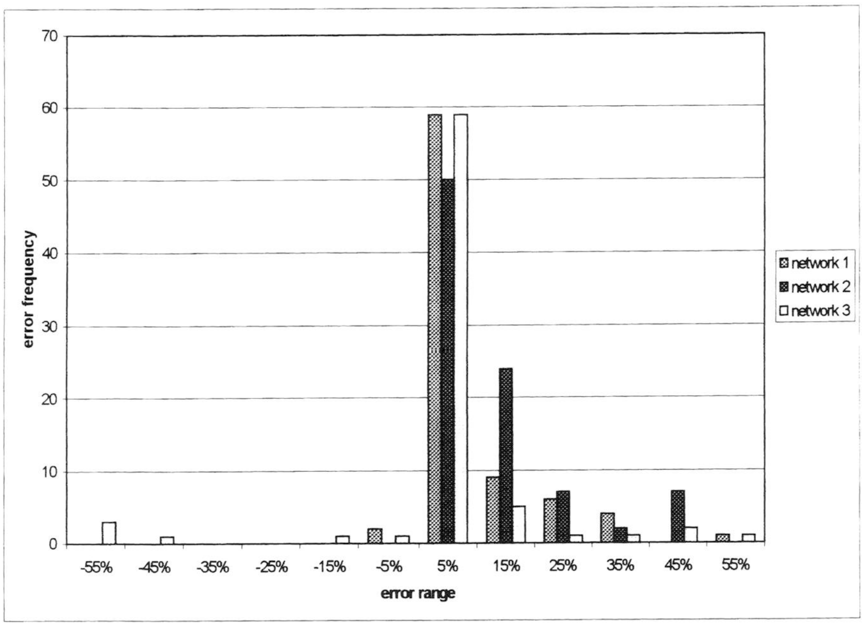

Figure 9 Output error frequency of the networks

6 CONCLUSIONS

The developed computer simulation model provides a sound foundation for process parameter selection and process optimisation for high speed milling primarily for the aerospace industry. It is envisaged that the model can also be used for predicting cutting parameters when cutting other materials such as hardened steels, which is of vital importance for the tool and die industry. The results obtained from experiments correlate well with the process parameters predicted by the computer model. In the study it is established that in high speed milling the cutting forces increase as the cutting speed increases. This is duly demonstrated experimentally and by means of the developed computer simulation model. This increase can be attributed to the effect of the momentum force which is in agreement with the findings of a number of published works. During the study it became clear that as the cutting speed increased the removal rate increased in the region up to 4700 m min^{-1}. As a result of computer model verification sufficient evidence has been gathered to justify the application of neural networks in system modelling and the initial neural system model is being trained at present.

7 ACKNOWLEDGEMENTS

The funding provided by the British Council and the Deutcher Akademischer Austausch-Dienst (DAAD) through the Academic Research Co-operation (ARC) programme in support of this project is greatly appreciated and acknowledged.

8 REFERENCES

1 J Balic, A Nestler and G Schulz. *Prediction and Optimisation of Cutting Values Using Neural Networks and Genetic Algorithms*. Proceedings of the 4th International Conference on Design for Manufacture in Modern Industry, DMMI'99. Maribor, Slovenia, pp 192-203, 1999.

2 A Kaldos, A Boyle, D Fichtner and I Dagiloke. *Development of an Integrated Process Parameter Selection Model for High Speed Machining*. Proceedings of the 14th National Conference on Manufacturing Research, University of Derby, Derby, UK, pp 125-130, 1998.

3 I.F. Dagiloke, A. Kaldos, S. Douglas and B. Mills. *Developing High Speed Cutting Simulation Suitable for Aerospace Materials*. Proceedings of Aerotech 94, IMechE, 1994.

4 A. Kaldos, I.F. Dagiloke and A. Boyle. *Computer Aided Cutting Process Parameter Selection for High Speed Milling*. Journal of Materials Processing Technology, 61, pp 219-224, 1996.

5 C.A. Brown. *A Practical Method for Estimating Machining Forces From Tool-chip Contact Length*. Annals of CIRP, 32/2, pp 91-96, 1983.

6 J.B. Armitage and A.O. Schmidt. *Radical Rake Angles in Face Milling*. Mechanical Engineering, New York, Vol. 66, p 403 and p 453, 1994.

7 B.F. von Turkovich. *Cutting Theory and Chip Morphology*, Handbook of High Speed Machining Technology, Chapman and Hall, London, pp 27-47, 1985.

8 R.G. Fenton and P.L.B. Oxley. *Predicting Cutting Forces at Super-High Cutting Speeds from Work Material Properties and Cutting Conditions*. Proceedings of the 8th

International Machine.Tool.Design and .Research Conference., pp 247-258, 1976.
9 H. Schulz. *High-Speed Machining*. Carl Hanser Verlag, Munchen, 1996.
10 D Fichtner, A Nestler and G Schulz. Wissensakquisition fur Schnittwerte beim Frasen unter Nutzung von neuronalen Netzen. Fortschr.-Berichte VDI Reihe 20, Nr. 304, Dusseldorf, VDI-Verlag,
11 D Fichtner et al. *Neue Perspektiven: Neuronale Netze ermitteln Schnittwerte*. Wt Produktion und Management, pp 101-105, 1997.
12 A Zell et al. *SNNS, Stuttgart Neural Network Simulator*, Software and User Manual, Version 4.0, Report No 6/95, 1`995.
13 G Schulz et al. *An Intelligent Tool for the Determination of Cutting Values Based on Neural Networks*. Proceedings of the Second World Congress on Intelligent Manufacturing processes and Systems, Budapest, pp 66-71, 1997.

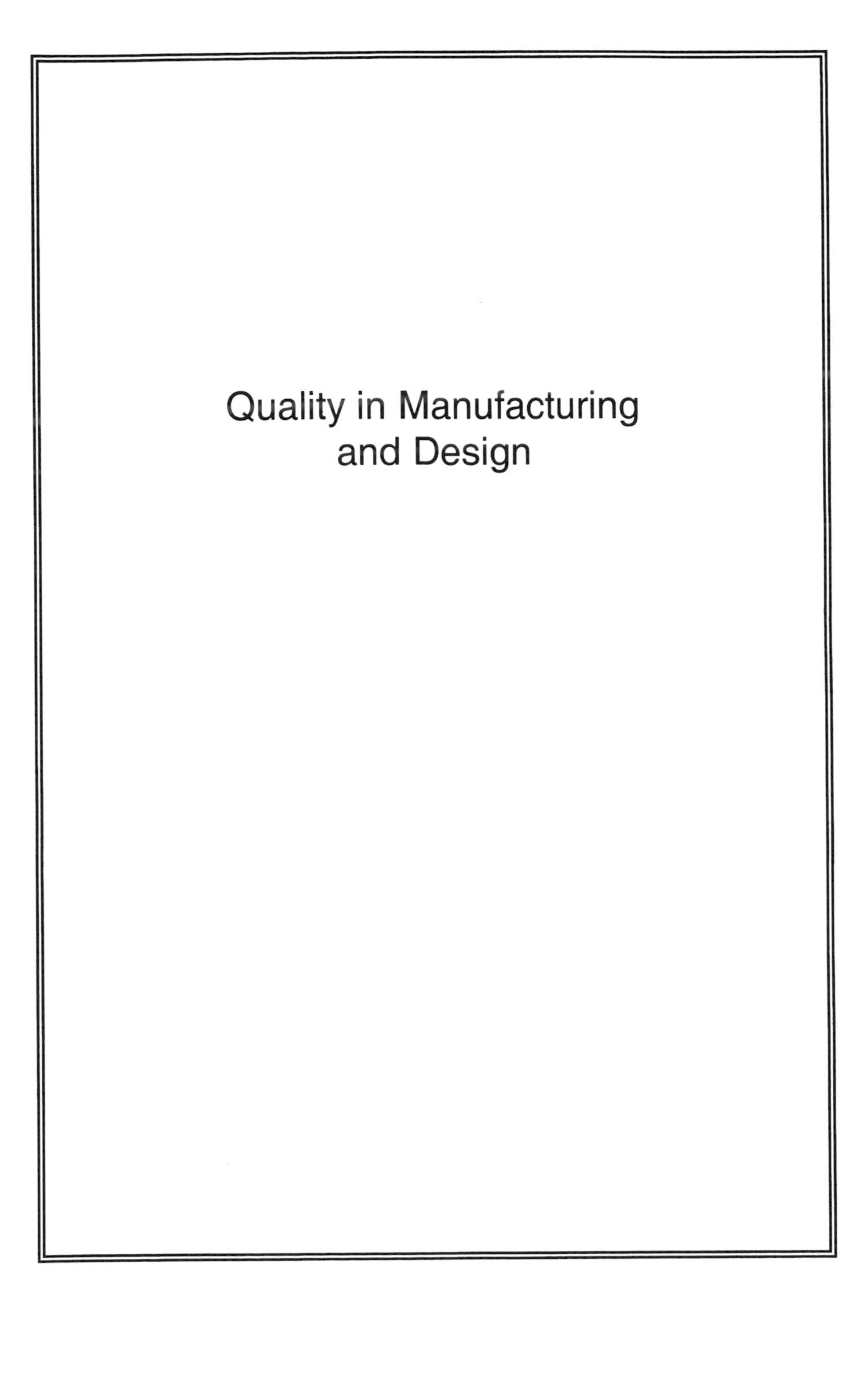
Quality in Manufacturing
and Design

CAPE/095/2000

Quality assurance through advanced sensor eddy current nondestructive evaluation

R TETI and **P BUONADONNA**
Department of Material and Production Engineering, University of Naples Federico II, Italy
G PELUSO
Istituto Nazionale per la Fsica della Materia (INFM), University of Naples, Italy

ABSTRACT

Eddy current (EC) nondestructive evaluation (NDE) consists in the use of electromagnetic techniques through which the inspection of the part is carried out by detecting magnetic anomalies in the material under examination. EC imaging techniques based on the detection of a voltage signal proportional to coil impedance variation or magnetic flux alteration were employed in this work. Measurements were performed through a conventional EC procedure, an advanced magnetometry technique (Fluxgate magnetometry) and an innovative approach based on superconductive material sensors (SQUID magnetometry). The experimental results obtained from conventional and innovative EC NDE methods were compared and critically assessed.

1 INTRODUCTION

Eddy current (EC) nondestructive evaluation (NDE) consists in the use of electromagnetic techniques through which the inspection of a part is carried out by detecting magnetic anomalies in the material under examination. A conventional and two innovative EC NDE methods, consisting of an advanced magnetometry technique (Fluxgate magnetometry) and a new approach based on superconductive material sensors (SQUID magnetometry), were used in this work to inspect conductive material samples containing artificial defects. The conventional EC method was based on probe impedance variations whereas both innovative techniques were based on magnetic flux changes. Measurement results, in the form of EC images obtained from EC scans of the conductive material samples, were compared and analysed to assess the capabilities of the new approaches, with particular reference to SQUID based EC NDE.

2 EXPERIMENTAL PROCEDURES AND MATERIALS

2.1 Conventional Eddy Current NDE

EC inspection is based on the principles of electromagnetic induction for inducing EC within a part placed within or adjacent to one or more induction coils. Changes in coupling between the induction coil and the part being inspected and changes in the electrical characteristics of the part cause variations in the loading and tuning of the generator (1).

The alternating current flowing in the induction coil, called the exciting current, causes EC to flow in the part as a result of electromagnetic induction. The magnetic field in the material region and its surroundings depends on both the exciting current in the coil and the EC flowing in closed loops within the part. EC magnitude and timing (or phase) depend on the original or primary field established by the exciting current, the electrical properties of the part, the presence or absence of flaws in the part, and the electromagnetic field established by EC flowing in the part.

The condition of the part can be monitored by observing the effect of the resulting magnetic field on the electrical impedance of the exciting coil or the induced voltage of either the exciting coil or other adjacent coil or coils. The main operating variables in EC testing include coil impedance, electrical conductivity, magnetic permeability, lift-off and fill factors, edge and skin effects (2).

The EC instruments available today are based on the bridge unbalance system and the induction bridge or driver pick-up configuration (2). The latter was used in this work.

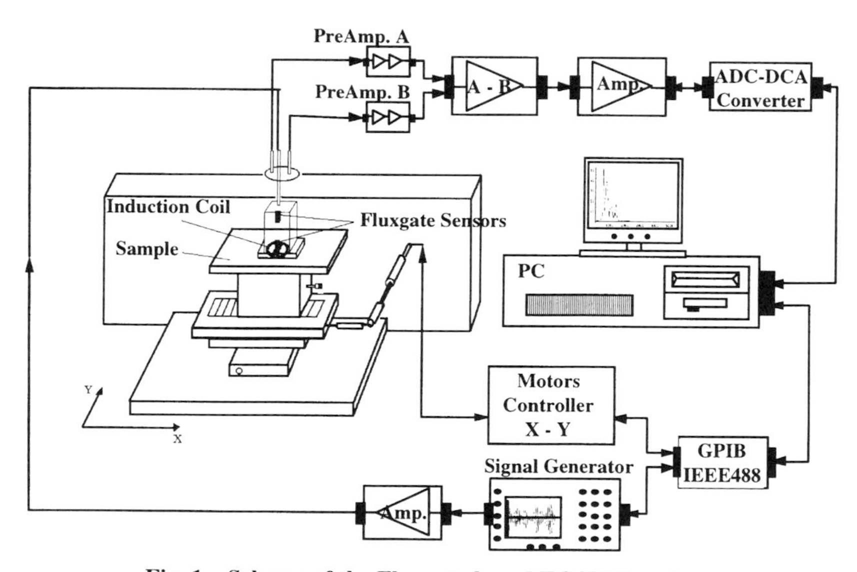

Fig. 1 – Scheme of the Fluxgate based EC NDE system.

An important part of an EC system is the readout device which can be an integral part of the system, a plug-in interchangeable module or an external computer (3). EC responses may be conveniently described by reference to the 'impedance plane', a graphical representation of the complex probe impedance where the abscissa (X-value) represents the resistance and the ordinate (Y-value) the inductive reactance.

In this paper, EC scanning of Al alloy samples containing artificial defects was carried out using a spot face differential probe (freq. 300 Hz - 100 kHz, diam. 11 mm) and a driver pick-up configuration EC instrument. Prior to scanning, the EC system was appropriately set up. The impedance plane diagram was rotated on the instrument display so that the response of the probe, as it moved over a defect, was given only by an amplitude variation along the Y axis (impedance modulus variation) whereas no appreciable phase change was observed.

During EC scanning, the instrument analog output proportional to the signal amplitude (Y component or impedance modulus) was digitized in a 2D numerical array. Each row in the array contained the numerical values of the detected signal for each material interrogation point during one scanning line with constant step. Details on this EC testing approach can be found in (4).

2.2 Fluxgate Magnetometry for Eddy Current NDE

Alternatively to coil impedance EC sensing, the resulting magnetic field can be detected through flux sensitive transducers such as Hall probes. An advanced version of this approach makes use of Fluxgate magnetometry for EC NDE. A Fluxgate magnetometer, known as core saturation magnetometer, is a solid-state device for measuring the magnitude and direction of the magnetic field vector in the sensitivity range between 10^{-10} T and 10^{-4} T. The main properties of Fluxgate sensors are: high linearity and stability, high sensitivity, simple realisability, robustness, low cost, and excellent sensitivity in a wide frequency range (from DC to 10 kHz) (5).

The Fluxgate EC NDE system used in this work is shown in Fig. 1. It consisted of an X-Y scanning system, a double D excitation coil to induce the EC (described in section 2.4), electronics for sensor signal control and conditioning, and two Fluxgate sensors in gradiometric configuration to reduce the noise caused by sources distant from the device. One Fluxgate sensor detected both signal and noise and the other only noise; their outputs were electronically subtracted by means of a vectorial amplifier (lock-in) which provided also for demodulation.

During Fluxgate EC scanning, the demodulated signal, representing the vertical component B_z of the resulting magnetic field due to EC flowing in the sample under test, was digitised in a 2D numerical array of B_z values detected with constant step at each material interrogation point.

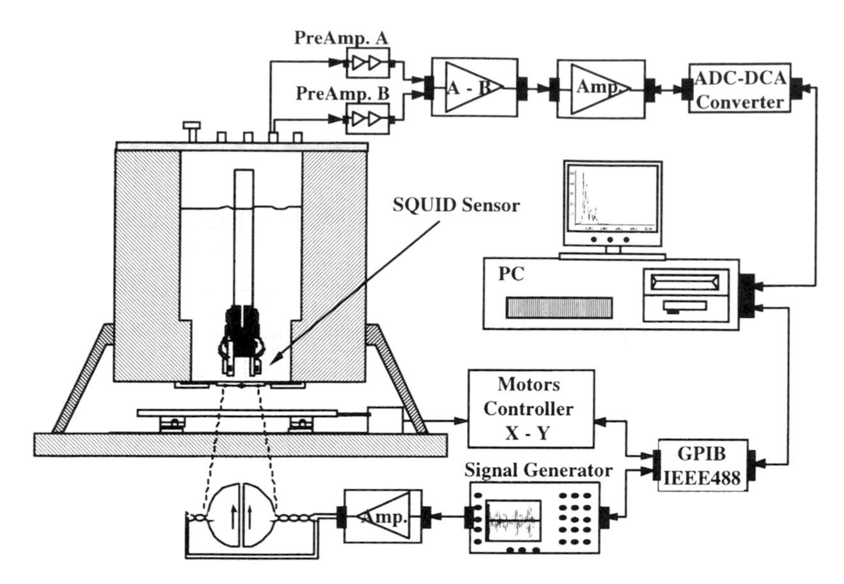

Fig. 2 – Scheme of the SQUID based EC NDE system

2.3 Eddy Current NDE Based on Superconducting Quantum Interface Device (SQUID)

Compared with traditional EC sensors, SQUID sensors offer higher spatial resolution and greater sensitivity in a wide frequency range (DC to 1 MHz) which make them particularly suited for detecting surface, sub-surface and deep defects in conducting materials (6). Rather bulky for certain applications, a portable system for defect detection in planar structures is a practical application of SQUID NDE. Active noise compensation techniques such as electronic gradiometry, total field compensation and individual flux compensation allow SQUID sensors to operate even in harsh environments. Moreover, due to flux-locked-loop electronics, these sensors work with a large dynamic range and high linearity. The advent of SQUIDs made of high critical temperature (HTc) superconductive materials working at the temperature of liquid N_2 (77 °K) allows to overcome the technical difficulties and high cost of liquid He cooling systems (7).

Fig. 2 shows the EC NDE system based on HTc SQUID sensors used in this work (8). The operative temperature of 77 °K was reached by liquid N_2 bath cooling in a dedicated fibreglass Dewar. The bottom of the Dewar cryostat had a 1 mm thick, 12 x 100 mm^2 window immediately under the SQUID. The system used two Tristan low-noise IMAG-3 HTc SQUIDs in gradiometric configuration for noise reduction; their outputs were electronically subtracted by a lock-in amplifier, demodulated and outputted as an AC signal with 6 Hz bandwidth. A dedicated X-Y scanning system, entirely in μ-metal material, with a sub-millimetre positioning accuracy, displaced the sample at 1 mm/s speed underneath the stationary Dewar cryostat. The motors were located 1.5 m away and coupled to the scanning system by plexiglass rods. During SQUID EC scanning, the demodulated signal (vertical component B_z of the resulting magnetic field) was digitised in a 2D numerical array of B_z values detected at each material interrogation point.

2.4 Innovative Eddy Current Systems Excitation Coil

The excitation coil in the Fluxgate and SQUID systems was a double D coil made of 60 turns, 25 mm diameter, 1 mm spacing between the Ds, and 2 mm total thickness (Fig. 3) (9). The coil AC source was a low-noise HP3245A waveform synthesiser. The currents in the double D central segments are equal in intensity and opposite in sign so that the field at the coil centre is zero.

EC flowing near a surface or subsurface defect tend to deviate, causing a change in the induced magnetic field and determining an unbalance between the global magnetic fields coupled with each D and between the current flows in each D. Thus, a non zero field is sensed by the Fluxgate or SQUID sensors positioned in gradiometric configuration coaxially with the double D coil.

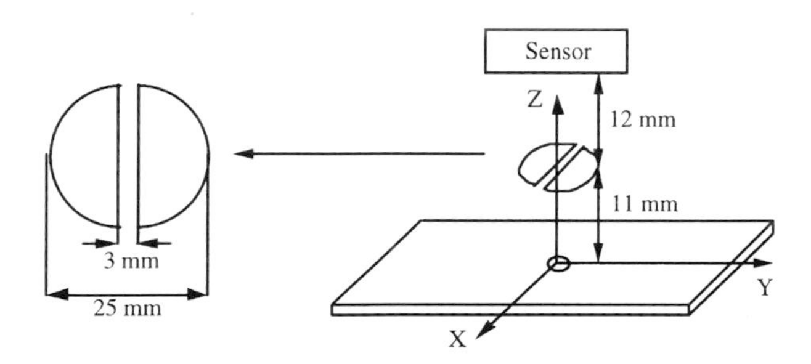

Fig. 3 – Double D shaped coil and distances between SQUID, coil and sample.

2.5 Testing Materials

Three 6061 Al alloy samples were tested with the conventional and innovative EC NDE systems. The first sample was obtained from the overlap of two 3 mm thick sheets, with the lower one containing a slot 1 mm deep, 2 mm wide, and 20 mm long. The two sheets were superimposed according to the two multilayer configurations (Figs. 4a and 4b). The second sample was a 3 mm sheet with a centre dead hole of diameter 6 mm and depth 2 mm (Fig. 4c). The third sample was a 15 mm thick plate with six EDM notches machined at regular intervals with width 0.5 mm, length 40 mm, and depth variable from 1 mm to 6 mm (Fig. 4d). EC scanning was carried out on the test samples by letting the conventional, Fluxgate or SQUID sensor scan the sample surface in an X-Y plane along successive parallel lines and interrogating the material at regular steps. This procedure allowed for the in-plane "radiographic" representation of the material structure. For each material interrogation point, either the impedance modulus (conventional EC sensor) or the vertical component B_z of the magnetic field (Fluxgate or SQUID sensor) was detected, digitised and stored. In Tab 1, the testing parameters for each experimental scan are summarised.

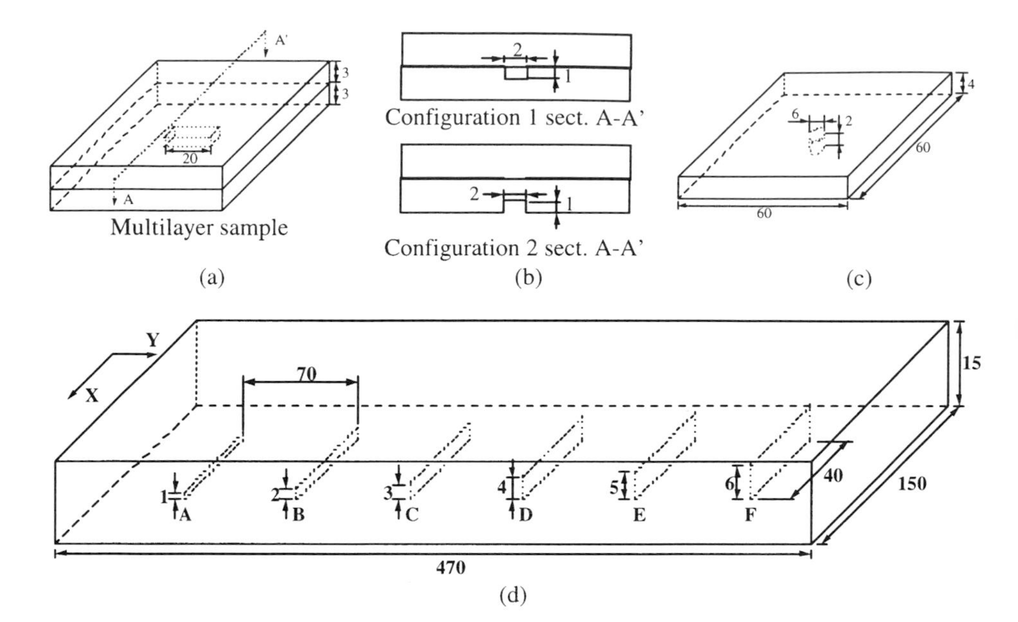

Fig. 4 – Schemes of the 6061 Al alloy test samples: (a) and (b) multilayer configurations, (c) sheet with centre dead hole, (d) thick plate with 6 notches of different depth. All dimensions in mm.

3 EDDY CURRENT IMAGING

A vast majority of the defect characterization schemes using EC NDE involve estimation of the size and shape of the defect based on a one dimensional signal obtained by scanning the surface of the test sample with a suitable probe (3). Recent years have witnessed increasing interest in the development of imaging techniques for characterizing defects. EC imaging methods involve a raster scan of the sample surface to obtain a 2D image (4). In this work, the instrument analog output during X-Y scanning of the test samples was digitised in a 2D numerical array. Each row in the array contained the values of the detected signal feature for each material interrogation point during one scan line with constant step. The set of scan lines made up the entire X-Y scan. In coil impedance EC scans, the elements of the 2D numerical array were the impedance modulus of the EC probe (10). In magnetic field sensing with Fluxgate or SQUID sensors, the elements of the 2D numerical array were the vertical component B_z of the resulting magnetic field.

A 2D image was obtained from the 2D numerical array by treating the detected signal parameter value as a grey level (11). Also 3D representations were realised by associating grey levels in the 2D array with Z-axis values in a 3D diagram. Image processing techniques can be applied to enhance 2D and 3D image readability and favour detail interpretation. Inverse techniques proposed to date rely largely on phenomenological models for analyzing the images to obtain estimates of the size and shape of the defect (12).

4 RESULTS AND DISCUSSION

4.1 Coil Impedance EC Scanning

Coil impedance EC scans were carried out on the two multilayer configurations with the sensor over the upper sheet and on the centre hole sheet with the sensor from the sheet side opposite to the defect (Tab. 1). Fig. 5 reports 2D EC images in 128 grey tones based on impedance modulus detection during EC scans of the two multilayer configurations and the centre hole sample (Tab.1). Multilayer configuration 1 allows for a good visualisation of the slot defect using an 800 Hz test frequency (Fig. 5a). The EC image obtained from multilayer configuration 2 using the same test frequency is less defined because of the higher depth of the slot. However, the slot location and size can still be clearly identified (Fig. 5b). Finally, the EC image from the centre hole sheet scan is very well outlined (Fig. 5c), due to the possibility of using a higher test frequency (8000 Hz) because of the lower material thickness over the dead hole area.

EC NDE allows to obtain 3D information on the inspected sample. The EC sensor signal, proportional to coil impedance variation due to the presence of a defect, is influenced by the defect position in the thickness direction.

Tab. 1 – EC test parameters. A_{eff} = effective area covered by sensor. Lift-off = 1 mm.

Scan id.	Sample	Sensor	Step	Scan Area	Test Parameters
1	Multilayer configuration 1 (control from the upper sheet)	EC differential probe Φ = 11 mm	0.5 mm	X = 33 mm Y = 50 mm	f = 800 Hz
2	Multilayer configuration 2 (control from the upper sheet)	EC differential probe Φ = 11 mm	0.5 mm	X = 30 mm Y = 30 mm	f = 800 Hz
3	Sheet with centre dead hole (control from the side opposite to the defect)	EC differential probe Φ = 11 mm	0.5 mm	X = 30 mm Y = 30 mm	f = 8000 Hz
4	Thick plate with six notches of different depths	EC differential probe Φ = 11 mm	1.0 mm	X = 70 mm Y = 460 mm	f = 300 Hz
5	Multilayer configuration 1 (control from the upper sheet)	Fluxgate A_{eff} • 30 mm^2	1.0 mm	X = 36 mm Y = 64 mm	I = 100 mA f = 377 Hz
6	Multilayer configuration 2 (control from the upper sheet)	Fluxgate A_{eff} • 30 mm^2	1.0 mm	X = 36 mm Y = 64 mm	I = 100 mA f = 377 Hz
7	Sheet with centre dead hole (control from the side opposite to the defect)	SQUID A_{eff} • 10 mm^2	1.0 mm	X = 40 mm Y = 40 mm	I = 5 mA f = 277 Hz
8	Thick plate with six notches: notches C, D, E (Fig. 4d)	SQUID A_{eff} • 10 mm^2	1.0 mm	X = 20 mm Y = 200 mm	I = 5 mA f = 277 Hz
9	Thick plate with six notches: notches D, E, F (Fig. 4d)	SQUID A_{eff} • 10 mm^2	1.0 mm	X = 20 mm Y = 200 mm	I = 5 mA f = 377 Hz

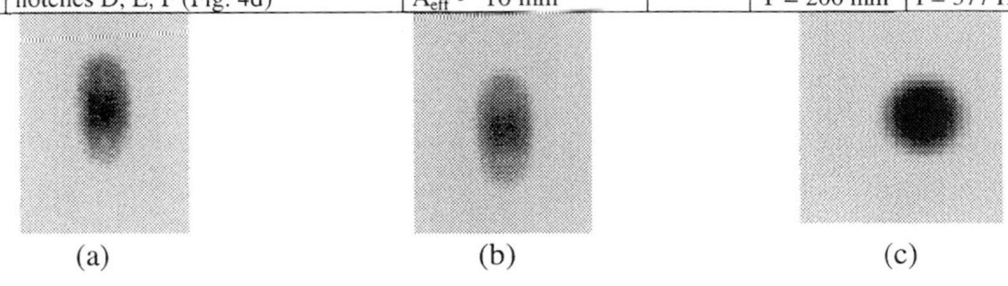

(a) (b) (c)

Fig. 5 – 2D EC images: (a) multilayer configuration 1, (b) multilayer configuration 2, (c) sheet with centre dead hole.

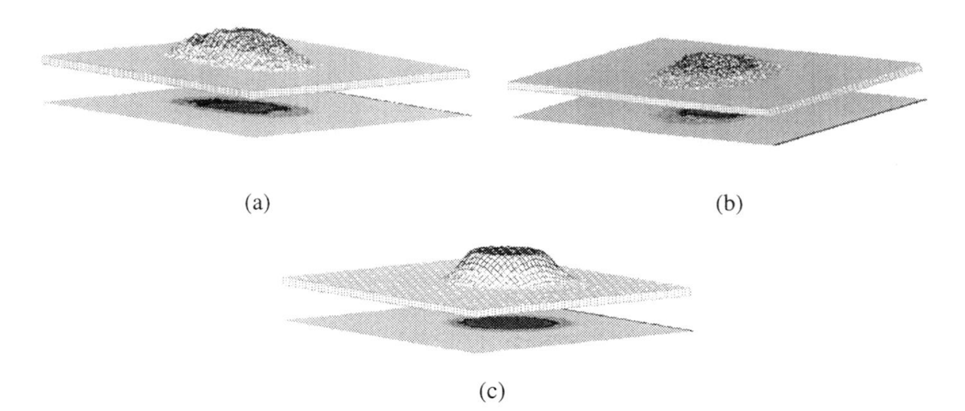

(a) (b)

(c)

Fig. 6 – 3D EC representations: (a) multilayer configuration 1, (b) multilayer configuration 2, (c) sheet with centre dead hole.

To obtain information on defect depth, 3D EC representations can be created by associating the values in the 2D numerical array to values on the Z-axis of a 3D diagram (4). The 3D representations in Fig. 6 characterise the defects through truncated conical or pseudo-conical projections of different height on a horizontal plane representing the conductive material surface.

Because of the detected sensor signal parameter, i.e. impedance modulus, the projection height is inversely proportional to defect depth. Fig. 6 shows that the projection corresponding to the slot in multilayer configuration 1 is higher than the one for the slot in multilayer configuration 2 by a ratio of approximately 5/3, proportional to the respective material thickness over the slot area.

To conclude the section on coil impedance EC testing, it is worth noting that EC scan # 4 carried out on the 15 mm Al alloy thick plate from the opposite side of the 6 notches with different depth (Tab. 1) did not allow for the EC imaging of any notch even when scanning the deepest one using the lowest test frequency (300 Hz). The material thickness between inspection surface and notch position (from 9 to 14 mm) was in all cases too high for the EC testing approach.

4.2 Fluxgate EC Scanning

Fluxgate EC scans were carried out on the two multilayer configurations with the sensor over the upper sheet (Tab. 1). The vertical component B_z of the resulting magnetic field was utilised for image creation. Fig. 7 reports the 2D images in 128 grey tones of the two multilayer configurations. In Fig. 8, the images are presented after application of contrast enhancement procedures.

The slot acts as a current dipole lying at a given depth, d, below the surface of the sample, which represents a conducting half-space, and producing a field that emerges from the material surface on one side of the slot and enters on the other side (13). The EC images allow for slot identification by means of two C shaped areas with opposite orientation and grey tone intensity. The lighter grey C shaped area represents the field emerging from the surface whereas the darker grey C shaped area the field entering the surface. The slot position in the plane is located between the C shaped areas, its length is related to the vertical extension of the C shaped areas, and its width to the horizontal distance between the two C shaped areas.

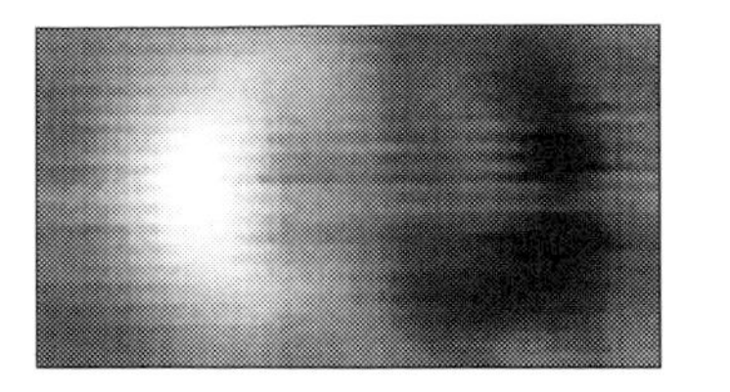
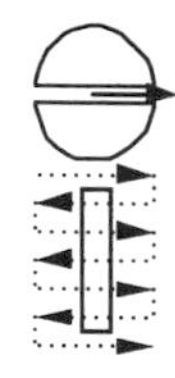
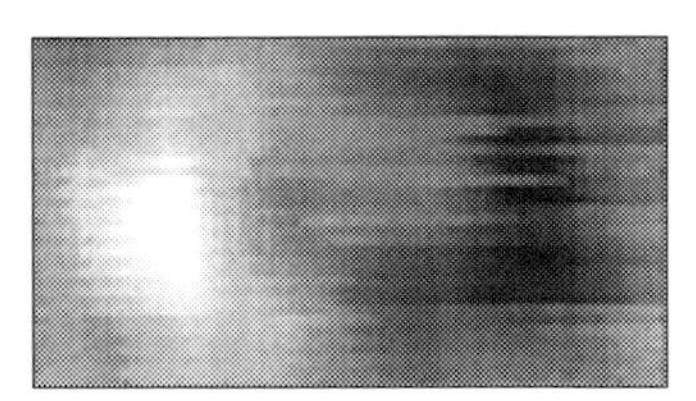

Configuration 1 Configuration 2

Fig. 7 – Multilayer sample: 2D EC images of configurations 1 and 2. Centre: excitation coil path.

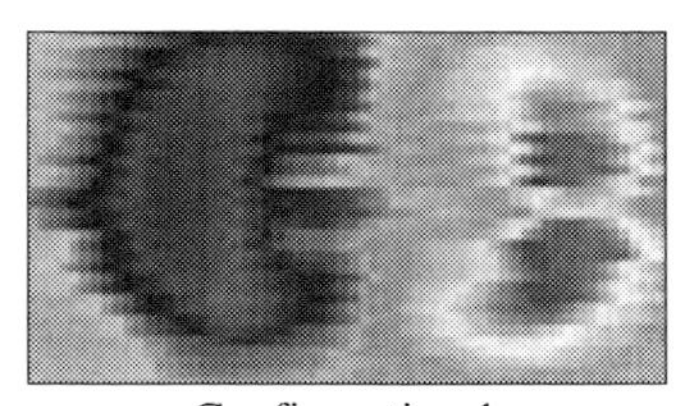
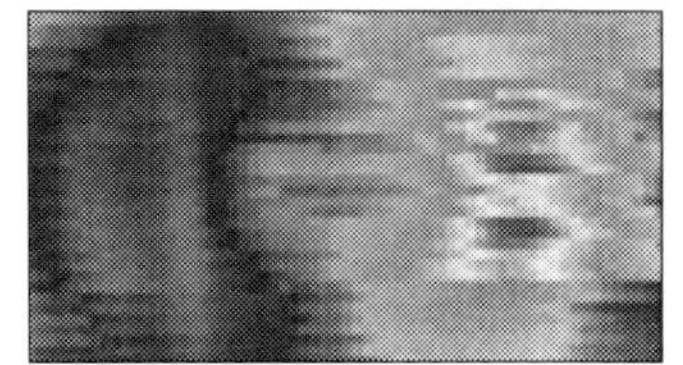

Configuration 1 Configuration 2

Fig. 8 – Multilayer sample: contrast enhanced 2D images.

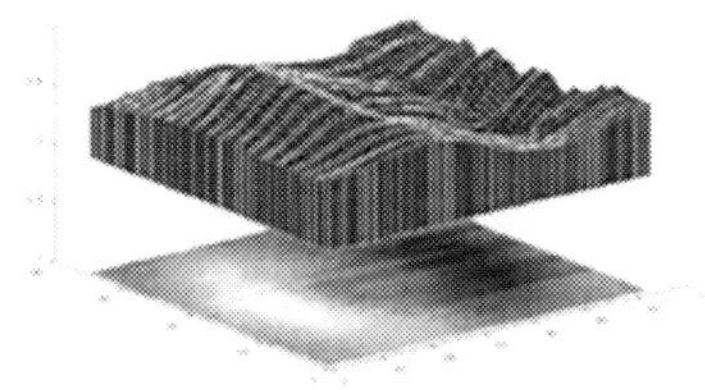

Configuration 1 Configuration 2

Fig. 9 – Multilayer sample: 3D representations combined with 2D grey tone images.

The slot depth, d, can be related to the difference between maximum and minimum grey levels in the C shaped areas: the higher the difference, the lower the slot depth. This difference is higher for multilayer configuration 1 (d = 3 mm) than for configuration 2 (d = 5 mm), as shown in the 3D representations of Fig. 9 obtained by reporting the B_z value on the Z axis of the 3D diagram.

4.3 SQUID EC Scanning

SQUID EC scanning was carried out on the centre dead hole sheet from the side opposite to the defect (Tab. 1). The vertical component B_z of the detected field was used for image creation.

In Fig. 10a, the 2D image in 128 grey tones from the scan of the dead hole area is shown with the indication of the excitation coil path. In Fig. 10b, the same image is presented after contrast enhancement. Also in this case, the dead hole acts as a current dipole lying at a given depth from the surface of the sample, which represents a conducting half-space, and producing a field that emerges from the material surface on one side of the dead hole and enters on the other (13).

The EC images allow for the identification of the dead hole by means of two semi-circular areas with opposite orientation and grey tone intensity. The semi-circular area with lighter grey tone values represents the vertical component B_z of the field emerging from the surface whereas the semi-circular area with darker grey tone values represents the field entering the surface.

The dead hole position in the sample plane is located at the centre of the two semi-circular areas. Its diameter is related to the diameter of the semi-circular areas. Its depth can be related to the difference between maximum and minimum grey tone values in the two semi-circular areas: the higher the difference, the lower the dead hole depth. This difference can be visualised in the 3D representations of Fig. 11 obtained by reporting the B_z value on the Z axis.

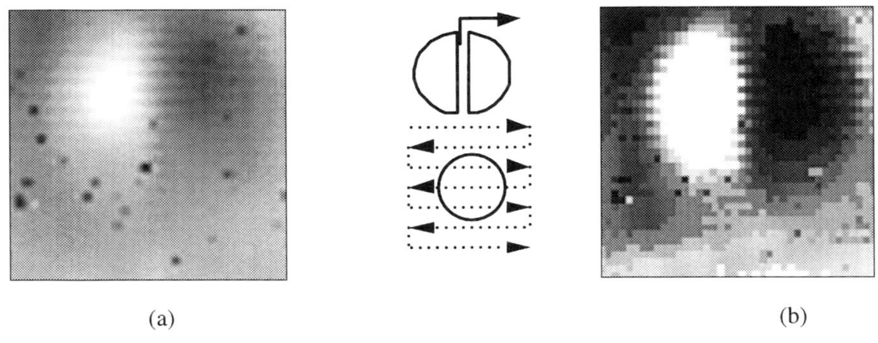

(a) (b)

Fig. 10 – Centre dead hole sheet: (a) normalised 2D grey tone image, (b) contrast enhanced 2D image. Centre: excitation coil path.

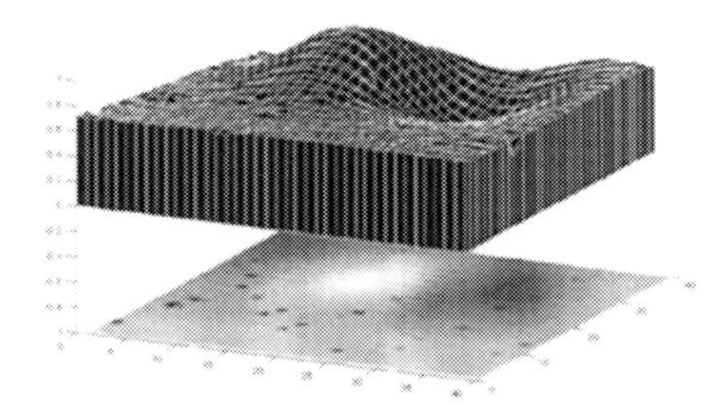

Fig. 11 – Centre dead hole sheet: 3D representation combined with 2D grey tone image.

The thick plate with six notches, which could not be successfully tested with the conventional EC system, was inspected with the SQUID system to exploit the SQUID sensor higher sensitivity.

Scan length (Y axis) was perpendicular to the notch direction and covered a maximum of 3 notches due to limitations in the dedicated SQUID scanning system; scan width (X axis) was kept smaller than notch width to avoid notch edge effects (Tab. 1, Fig. 4d). Figs. 12 to 14 report the results of scans # 8 and # 9 over notches C, D, E and D, E, F, respectively. Fig. 12 shows the sensor signal profiles detected during one scan line in the Y direction, Fig. 13 the 2D EC images of the X-Y scans, and Fig. 14 the 3D representations combined with the 2D images. The physical position of the notches in the thick plate along the scan length is indicated by vertical arrows.

The notch presence is identified by the wavy trend of the signal profile (Fig. 12) which, outside the notch area, has a constant trend. Like previous defects, also the notches act as current dipoles lying at different depths from the surface of the thick plate, which represents a conducting half-space. As the detected signal is proportional to the field that emerges from the surface on one side of each notch and enters on the other (13), the location of a notch is given by a horizontal '~' shaped portion of the signal profile. The upward part of the '~' shaped portion represents the vertical component B_z of the field emerging from the surface whereas the downward part of the '~' shaped portion represents the field entering the surface. Notch position is located at the centre of the '~' shape portion. In Figs. 12a and 12b, three subsequent '~' shaped signal profile portions can be identified, each corresponding to one notch in the thick plate.

The 2D images allow for notch identification by means of a transition from lighter to darker grey tone zones. The lighter grey tone zones represent the field emerging from the surface whereas the darker grey tone zones represent the field entering the surface. Both Figs. 13a and 13b show three transitions, roughly corresponding to the physical location of the notches in the thick plate plane indicated by vertical arrows. The '~' shaped portions of the signal profiles and the transitions of grey tone zones in the 2D images correspond to the 'waveforms' of the 3D representations of the SQUID scans (Fig. 14). Each waveform in the 3D surface corresponds to one notch, located approximately at mid-waveform, as indicated by the arrows in Fig. 14.

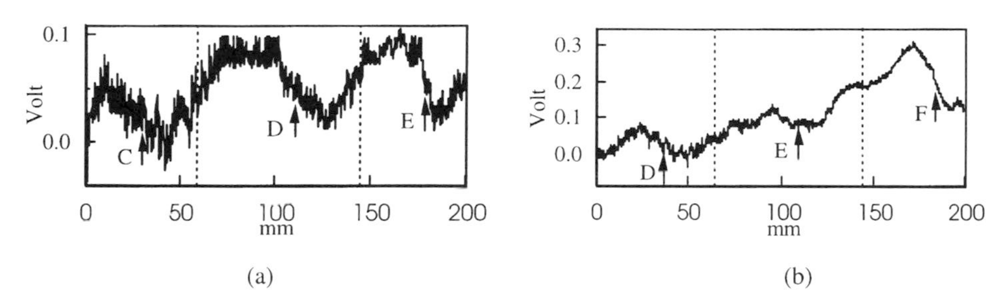

(a) (b)

Fig. 12 - Signal profile along one scan line in the Y direction: (a) scan # 8 on notches C, D, E; (b) scan # 9 on notches D, E, F. Vertical dotted lines delimit the '~' shaped signal portions.

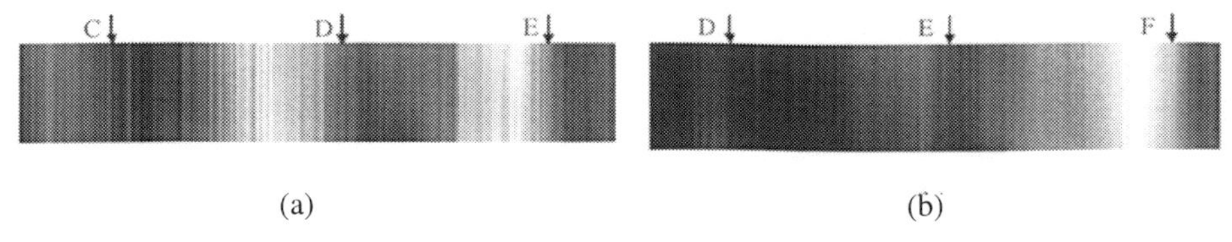

Fig. 13 - 2D grey tone image of the X-Y scan: (a) scan # 8 on notches C, D, E; (b) scan # 9 on notches D, E, F.

Fig. 14 - 3D representation combined with 2D grey tone image: (a) scan # 8 on notches C, D, E; (b) scan # 9 on notches D, E, F.

Information on notch depth can be obtained from the signal profiles (Fig. 12) by reporting the peak values of the '~' shaped portions versus notch depth: Fig. 15 shows that the peak value increases in both signal profiles from left to right as notch depth decreases from 12 mm for notch C to 9 mm for notch F, allowing for notch depth discrimination. This information is also available in the 3D representations of Fig. 14 which, however, display normalised information and, in this form, can only allow for depth discrimination within the same scan.

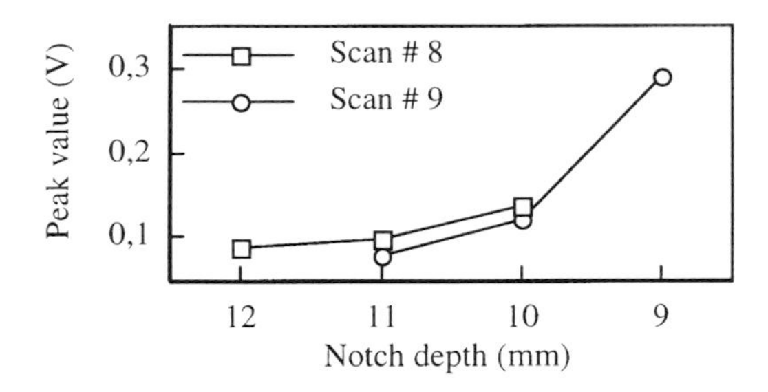

Fig. 15 – Peak value of signal profile '~' shaped portions vs. notch depth.

5 CONCLUSIONS

A traditional EC method, an advanced Fluxgate magnetometry technique and an innovative approach based on superconductive material SQUID sensors were used for EC NDE of Al alloy samples containing artificial defects. The conventional EC procedure was based on probe impedance variations whereas both advanced EC techniques were based on magnetic flux changes.

The experimental results were presented as 2D images or 3D representations obtained from X-Y scanning of the test samples. Coil impedance EC imaging allowed for the identification of defect presence and location by scanning the samples from the side opposite to the defects, except for the thick plate with EDM machined notches due to the high material thickness over the notches. The hole and slot defects could be identified with the advanced Fluxgate and SQUID EC imaging methods. Moreover, by taking advantage of the higher sensitivity of the SQUID sensor, also the presence and location of the notches in the thick plate could be successfully identified through SQUID NDE. The defect (hole, slot or notch) acts as a current dipole lying at a given depth below the surface of the sample, which represents a conducting half-space, and producing a field that emerges from the surface on one side of the defect and enters on the other side (13). Information on defect depth could be retrieved through analysis of the 3D representations obtained from X-Y scans by reporting the detected signal value on the Z-axis of a 3D diagram.

ACKNOWLEDGEMENTS

The present work was carried out with partial contribution from INFM Progetto Sud NDE. F. Finelli, U. Principio, A. Ruosi and M. Valentino are gratefully acknowledged for their support in the development of the EC NDE experimental program.

REFERENCES

(1) Cecco, V.S., Franklin, E.M., Houserman, H.E., Kincaid, T.G., Pellicer, J., Hagemaier, D., 1989, Eddy Current Inspection, in Nondestructive Evaluation and Quality Control, *Metals Handbook*, Vol. 17, ASM International, USA
(2) McMaster, R.C., McIntire, P., Mester, M.L., 1986, Electromagnetic Testing, *Nondestructive Testing Handbook*, Vol. 4, American Society for Nondestructive Testing (ASNT), USA
(3) Mix, P.E., 1987, *Introduction to Nondestructive Testing*, John Wiley & Sons, New York
(4) Teti, R., Buonadonna, P., 1999, Eddy Current NDE of CFRP Composite Laminates, *Advancing with Composites 2000*, Milan, May 9-11: 145-152
(5) Ripka, P., 1992, Review of Fluxgate Sensors, *Sensors and Actuators, A. 33*, Elsevier Sequoia: 129-141
(6) Wikswo, J.P., 1995, SQUID Magnetometers for Biomagnetism and Nondestructive Testing: Important Questions and Initial Answers, *IEEE Transactions on Applied Conductivity*, Vol. 5, no. 2: 74-120
(7) Pagano, S., et al., 1998, HTc SQUID for Nondestructive Evaluation, *Electromagnetic Nondestructive Evaluation (II)*, R. Albanese et al., Eds., IOS Press: 206-214
(8) Peluso, G., et al., 1997, A New Project on Nondestructive Evaluation with High Temperature SQUIDs, *XXIV Review of Progress in Quantitative NDE*, D.O. Thompson and D.E. Chimenti, Eds., Vol. 16A, Plenum Press, New York: 1083-1090
(9) Barone, A., et al., 1997, Design of a NDE Instrumentation Prototype with High Temperature SQUIDs, *Il Nuovo Cimento*, Vol. 19D, no. 8-9, Aug.-Sept.: 1495-1500
(10) Feil, J.M., 1981, in *Eddy Current Characterization of Structures and Materials*, G. Birnbaum and G. Free, Eds., ASTM SPT 722, Philadelphia: 449-463
(11) Buonadonna, P., 1999, Imaging for 3D Eddy Current Nondestructive Evaluation, *Research Reports of the LAPT - 2*, R. Teti, Ed., Dept. of Materials and Production Engineering, University of Naples Federico II: 59-66
(12) Udpa, L., Lord, W., 1987, in *XIV Review of Progress in Quantitative NDE*, D.O. Thompson and D.E. Chimenti, Eds., Vol. 6A, Plenum Press, New York: 899-906
(13) Romani, G.L., Williamson, S.J., Kaufman, L., 1982, Biomagnetic Instumentation, *Rev. Sci. Instrum.*, Vol. 53, no. 12: 1815-1844

CAPE/005/2000

Three-dimensional surface profile deviation evaluation based on CMM measured data

K CHENG, C H GAO, and **D WEBB**
School of Engineering, Leeds Metropolitan University, UK
D K HARRISON
Department of Enigneering, Galsgow Caledonian University, UK

ABSTRACT:

The deviations of machined surfaces of high precision mechanical components are one of the major error items that affect the precision and performance of the components. In this paper, a practical approach towards evaluating 3D surface profile deviation is proposed based on discrete co-ordinate measurement data. Spur gears and helical gears are used as inspection samples to test the approach developed. The trial results are very promising.

Keywords: Surface profile deviation, inspection, CMMs, genetic algorithms

1 INTRODUCTION

It is still many years away from the realization of taking a component model from a CAD system and automatically generating all the information required for down stream activities such as machining, assembly and inspection. The growing emphasis towards product quality has lead to an increasingly important role being played by inspection within manufacturing industry. An inspection system would ideally be able to measure the dimensional characteristics of randomly presented parts with virtually any configuration or complexity and to further provide real time feedback to the manufacturing process.

In the process of design of complex shaped parts or products in particular, reverse engineering is a useful approach for constructing a CAD model from a physical part/product that already exists. While in the process of manufacturing, the features of a part can be inspected by using a CMM. All of these processes start from digitizing the features of the part, i.e. the discrete points measured from the part surface. Ma and Kruth presented a method for the NURBS curve and surface fitting (1). The use of least squares fitting of inspection data is extensively used to nominal form features such as lines and circles, etc. Yau and Menq proposed an optimal match algorithm to eliminate the offset error between the measured data and design data by minimizing the shortest squared distance from each point to the nominal feature geometry (2). Jenning also presented that by skewing the density of the measured data points in a certain area, the fit would be skewed towards the more densely probed area. Ristic developed algorithms to minimize the time taken to best fit when dealing with a large number of data points resulting from data captured by laser scanning (3). Zhang and Gong presented the work on inspection of 3D surface form deviations using CMMs (4). However, the work described above is still far from satisfaction.

Based on an inspection model, using discrete measurement data on a measured surface to match the idea surface model such as the surface CAD design data and achieving the surface profile deviation is one of the major goals of this research. In this paper, preliminary results are presented focusing on establishing appropriate equations to fully describe a 3D free-form surface and its profile deviation evaluation. The authors also explore the measurement based on an optimal number of measurement points.

2 3D surface and its profile deviation

2.1 3D surface equations

For an ideal surface of a component as shown in Fig. 1, it can be expressed in Cartesian co-

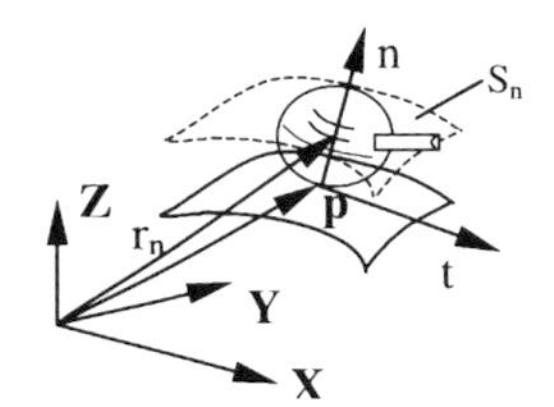

Fig. 1 3D surface

ordinates as:

$$f(x, y, z) = 0 \tag{1}$$

Its unit normal vector n on the surface can be represented as:

$$n = \cos\alpha \bullet i + \cos\beta \bullet j + \cos\gamma \bullet k \tag{2}$$

Where

$$\cos\alpha = f_x \Big/ \sqrt{f_x^2 + f_y^2 + f_z^2}$$
$$\cos\beta = f_y \Big/ \sqrt{f_x^2 + f_y^2 + f_z^2}$$
$$\cos\gamma = f_z \Big/ \sqrt{f_x^2 + f_y^2 + f_z^2}$$
$$f_x = \partial f / \partial x, \qquad f_y = \partial f / \partial y, \qquad f_z = \partial f / \partial z$$

The point vector p and spherical tangent vector t on the ideal surface is denoted as:

$$p = x \bullet i + y \bullet j + z \bullet k \tag{3}$$

$$t = p \times n = \begin{vmatrix} i & j & k \\ x & y & z \\ \cos\alpha & \cos\beta & \cos\gamma \end{vmatrix} \tag{4}$$

Considering the radius r_p of the probe tip, the inspection offset surface S_n will be:

$$r_n = p + r_p \bullet n \tag{5}$$

2.2 surface profile deviation

The surface profile deviation ∇f at point P of a measured surface (M_s) is the distance from P to the nominal surface (N_s) along the normal line, as shown in Fig. 2. It is namely,

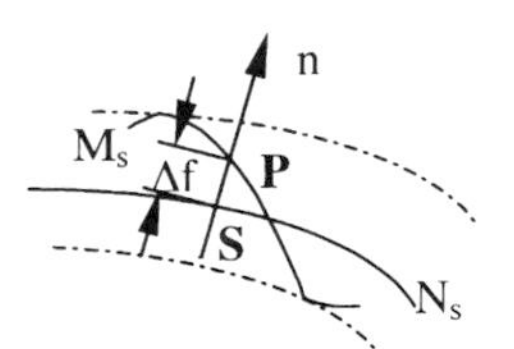

Fig. 2 Surface deviation

$$\nabla f = (P - S) \bullet n \tag{6}$$

Where, P and S are point vectors in the normal direction to the nominal surface N_s whose unit vector is n. The actual surface of the component to be measured may be specified by a set of discrete point p_i. The density and ordering of the point data is determined by the size and accuracy of the surface to be measured. It should be determined based on the distances of the points p_i to the ideal surface (or nominal surface) N_s.

For the co-ordinates of a random point p_j (j=1, 2, 3..., N) on the surface of a component measured using a CMM, it can be expressed in a vector format as:

$$p_j - x_j \bullet i + y_j \bullet j + z_j \bullet k \qquad (j = 1,\ 2,\ 3...,\ N) \tag{7}$$

The distance from point P_j to the ideal surface (N_s) and the surface deviation at this point can be represented as:

$$d_{p_j} = f(x, y, z) \big/ \sqrt{f_x^{\,2} + f_y^{\,2} + f_z^{\,2}} \tag{8}$$

$$\nabla f_{p_j} = d_{p_j} \bullet n \tag{9}$$

For the profile being measured over a range of the surface; the surface deviation will be defined as:

$$\nabla F = \min\{\max(\nabla f_{p_j}) - \min(\nabla f_{p_j})\} \qquad (j = 1, 2, 3, ..., N) \tag{10}$$

2.3 Calculation of co-ordinate transformation errors
As shown in Fig. 3, the co-ordinate system of the CMM is O_m-$x_m y_m z_m$, and the z_m axis is

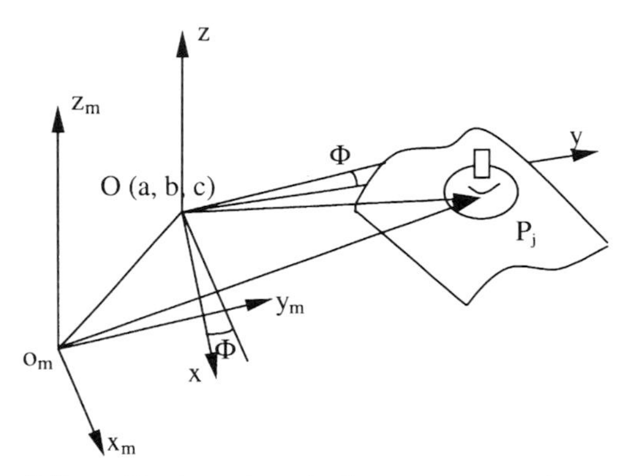

Fig. 3 Co-ordinate systems in the measurement process

vertical. The co-ordinate system of the work-piece is O-xyz, and the z-axis is vertical too. A point P_j (x_{mj}, y_{mj}, z_{mj}) is the centre of the probe. In the component co-ordinate system, the co-ordinates of the point P_j can be expressed using the transformation matrix as follows:

$$\begin{bmatrix} x_j \\ y_j \\ z_j \\ 1 \end{bmatrix} = \begin{bmatrix} \sin\Phi & -\cos\Phi & 0 & -a \\ \cos\Phi & \sin\Phi & 0 & -b \\ 0 & 0 & 1 & -c \\ 0 & 0 & 0 & 1 \end{bmatrix} \begin{bmatrix} x_{mj} \\ y_{mj} \\ z_{mj} \\ 1 \end{bmatrix} \tag{11}$$

The least squares and minimum zone methods are used for estimating the values of a, b, c, Φ and then the profile deviation Δf_{pj} (5)(6).

In reference (4), Zhang and Gong analysed the measurement errors for flat, concave and convex surface profiles. The measurement error for a concave profile as shown in Fig. 4 is

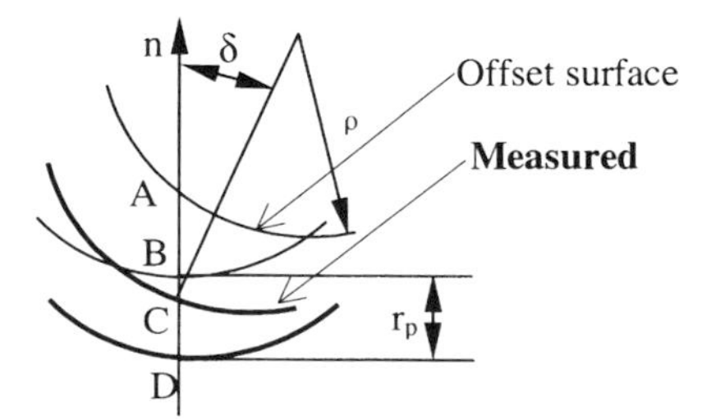

Fig. 4 Error analysis in a CMM

denoted as:

$$e_{rr} = (\rho + r_p)\cos\Phi - \sqrt{[(\rho + r_p)^2 \cos^2\Phi - (r_p^{\,2} + 2\bullet\rho\bullet r_p)]} \qquad (12)$$

Where, Φ denotes the slope deviation of the measured profile, r_p is the radius of the probe tip, ρ is the curvature radius. The measurement error e_{rr} will increase with the slope deviation Φ. In normal manufacturing circumstances, the slope deviation Φ of a part surface is very small. But in a measurement process, the co-ordinates system of a part itself and the measurement co-ordinates system are not the same. So to some extent, the measurement accuracy will be affected by the accuracy of their co-ordinates systems transformation.

3 3D surface profile deviation evaluation – a case study

The approach and associated modelling and algorithms developed were applied in evaluating gear tooth surface profile deviation. The major advantage of using a CMM for gears inspection is it can avoid using base discs that are essential in traditional gears metrology. Base discs are high precision mechanical components and thus being manufactured at high cost if just for inspecting a small batch of gears. The evaluation is based on the inspection data of gear tooth surfaces. It is important for a gear manufacturer to get full details of gear tooth surface profiles, which are essential for high quality mechanical transmission.

3.1 Gear surface equations

The equations for describing a helical gear tooth surface and its surface unit normal are (7)(8):

$$\begin{cases} x = r_b(\cos(\theta + \phi + \varphi) + \phi \bullet \sin(\theta + \phi + \varphi)) \\ y = \pm r_b(\sin(\theta + \phi + \varphi) - \phi \bullet \cos(\theta + \phi + \varphi)) \\ z = \pm r_b \varphi / \tan \beta_g \end{cases} \qquad (13)$$

$$n = \pm\cos\beta_g \bullet \sin(\theta + \phi + \varphi) \bullet i \mp \cos\beta_g \bullet \cos(\theta + \phi + \varphi) \bullet j \pm \sin\beta_g \bullet k \qquad (14)$$

Where r_b is the basic circle radius. β_g is the helical angle. The upper and lower signs in Equations (13) and (14) correspond to the right side tooth surface (its angles θ, φ and ϕ are measured counterclockwise) and left side tooth surface (its angles θ, φ and ϕ are measured clockwise) of the right-hand helical gear respecsively.

3.2 Gear tooth profile inspection and evaluation

The gear inspection was carried out based on some guidelines recommended in reference (9). The parameters of the gears measured are listed in Table 1. A table type co-ordinate measuring machine (Micromeasure™ III Brown & Sharpe) was used in the trials.

Table 1 Parameters of the gears used in the trials

Gears / Parameters	ZS1550	ZSH320	SH412	ZG320	ZSH2550
Number of teeth	50	20	12	20	50
Module (mm)	1.5	3	4	3	2.5
Helical angle (°)	17.45	17.45	17.45	Spur gear	17.45
Pressure angle (°)	20	20	20	20	20
Material	White Delrin	White Delrin	Steel 214 M15	White Delrin	White Delrin
Direction of spiral	Right hand	Right hand	Right hand	Spur gear	Right hand
Tooth width (mm)	15	36	48	20	20

The radius of the spherical probe used is 0.995 mm. The measured data were saved in text files that then being analyzed with the evaluation models built on MATLAB. The initial setting up values of a, b, c and Φ were obtained by least squares method and the minimum zone method.

Using above motioned method, on the jth measured point, the gear profile deviation will be:

$$\nabla f_{p_j} = d_{p_j} \bullet n = r \bullet \nabla \theta_j \bullet \cos \beta_g \qquad (j = 1,\ 2,\ 3,\ \ldots,\ N) \qquad (15)$$

For many scattered measurement data on the gear tooth surface, the gear profile deviation has to be obtained using equation (10) based on an optimization algorithms (10)(11)(12). The gear profile form deviations obtained are shown in Fig. 5.

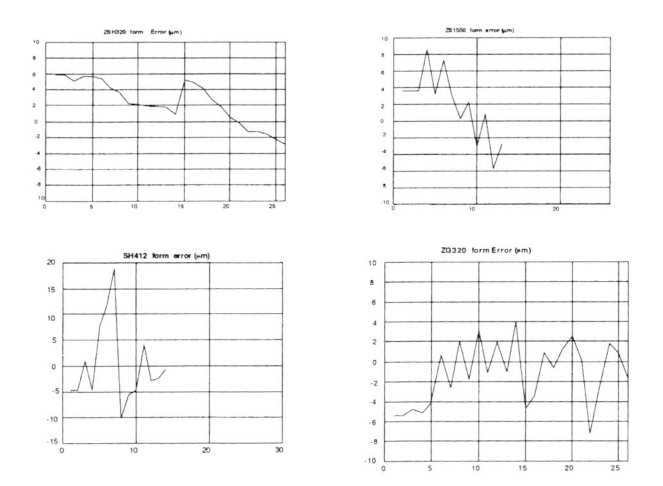

Fig. 5 Gear tooth surface profile deviations

4 OPTIMAL NUMBER OF MEASUREMENT POINTS

In theory, the more measurement points, the higher accuracy could be achieved. In the reality, however, a large number of measurement points means non-efficient and even probably unnecessary. So there must be an optimal number of measurement points as needed to precisely represent the surface real profile both efficiently and effectively.

For searching the optimal number of measurement points on a measured surface, genetic algorithms are used with heuristics based on a model which attempts to mimic the mechanics of natural biological evolution (13). The algorithms operate on a population of potential solutions stochastically choosing and using the fittest solutions to produce more potential solutions which it is hoped will yield better approximations to the final optimal solution.

The genetic search is implemented following the steps below:

(1) set a population of the measurement points on the measured surface;
(2) find the intensity error value for each measurement point (fitness value);
(3) stochastically choose a number of pairs of the measurement points;
(4) randomly combine the measurement points within each pair of measurement;
(5) replace measurement numbers in the previous population with newly created measurement number;
(6) repeat from step (2) until either the error goal or the maximum number of iterations (generations) is achieved.

5 CONCLUSIONS

For a 3D free-form surface, if its surface equations are known, the surface deviation can be inspected using the method as proposed above. In this investigation, the spur gears and helical gears were used as inspection samples. The gear tooth surfaces deviation can be precisely and quickly obtained. The paper also explores the idea of carrying out the inspection based on an optimal number of measurement points on a 3D surface measured, although only limited results are achieved as the project progressed so far.

6 REFERENCES

1. W. Ma and J.P. Kruth, NURBS curve and surface fitting for reverse engineering, International Journal of Advance Manufacturing Technology, 14(12), 918-927, 1998.
2. H.T. Yau and C.H. Menq, Automated CMM path planning for dimensional inspection of dies and molds having complex surfaces, International Journal of Machining Tools and Manufacturing, 35(6), 861-876, 1994.
3. D.I. Legge, Integration of design and inspection systems – a literature review, International Journal of Production Research, 34(5),1221-1241, 1995.
4. J. J. Zhang and D Gong, Precision inspection of deviation of machined surfaces, IMechE: Journal of Mechanical Engineering Science,211(8), 579-589, 1997.
5. T.S.R. Murthy and S.Z. Abdin, Minimum zone evaluation of surfaces, International Journal of Machine Tools and Research, 20 (2), 123-136, 1980.

6. M. Fukuda and A. Shimokohvbe, Algorithms for form error evaluation-methods of the minimum zone and the least squares, Proceedings of the International Symposium on Metrology for Quality Control on Production, Tokyo,197-202, 1984.
7. F.L. Litvin. Gear Geometry and Applied Theory, PTR Prentice Hall, 1994.
8. C. Gao, K. Cheng and D.K. Harrison, A novel approach to the inspection of gears with a coordiante measuring machine: theoretical aspects, Proceedings of the Fourth International Conference on Laser Metrology and Machine Performance, Newcastle, UK, July 13-15, 227-236, 1999.
9. D.J. Whitehouse, Handbook of Surface Metrology, Institute of Physics Publishing, Bristol, 1994.
10. X. Huang and P. Gu, CAD-model based inspection of sculptured surfaces with datum, International Journal of Production Research, 36(5), 1351-1367, 1998.
11. H. Qiu, K. Cheng, Y. Li, et al, An approach to form deviation evaluation for CMM measurement of 2D curve contours, Proceedings of the 15th International Conference on Computer Aided Production Engineering, Durham, 19-21 April, 307-314, 1999.
12. Y.L. Xiong, Computer aided measurement of profile error of complex surfaces and curves: theory and algorithm, International Journal of Machining Tools Manufacturing, 30(3), 339-357, 1990.
13. D.E. Goldberg, Genetic Algorithms in Search, Optimization and Machine Learning, Addison Wesley Longman Inc, 1989.

CAPE/006/2000

In search of quality in quality function deployment and Taguchi methods

E GENTILI
Dipto Ing. Meccanica, Università di Brescia, Italy
F GALETTO
Dipto Sistemi di Produzione ed Economia dell'Azienda, Politecnico di Torino, Turin, Italy

SYNOPSIS

The goal of "Quality Engineering" is to design Quality into every product and all the processes that build them. Design is the most important phase for Quality both of products and processes.
Two methodologies, originated in Japan, are claimed as very important for Quality design: Quality Function Deployment (QFD) and Taguchi Methods (TM). Often they are called upon as the best synergy a Company can do during development of products and processes. Taguchi Methods are wrongly considered better than the "classical" Design of Experiments (DOE).

1 INTRODUCTION

Design of Experiments (DOE) is a major element of the design activity and one of the most important methodologies to achieve Quality through intelligent testing of factors that influence Quality, during the design stage of product and processes. For processes DOE allows us to find the optimum setting of factors that provide the best Quality and economic yield, that is minimisation of "disquality cost". Quality Function Deployment (QFD) is considered by many people (generally indicated as "practitioners") as "*a tool which is able to ensure that the voice of the customer is deployed throughout the product planning and design stages*", via the use of various "houses of quality" that allow the collection and the organisation of information about quality, features, characteristics, parts, processes and goals relative to a product or a service. The literature on these matters is rapidly expanding. Therefore it seems important to stand-back a bit and meditate from a managerial point of view. The paper shows, using actual data that Logic and the Scientific Approach are able to provide the right route towards the good methods for Quality.

2 QUALITY FUNCTION DEPLOYMENT

Some people define QFD as "a method for structured product planning and development to specify clearly the customer's wants and needs, and then evaluate each product or service capability systematically in terms of its impact on meeting those needs. But QFD for somebody is not a tool. It is a planning process, as opposed to a tool for problem solving or analysis. Again, QFD can be a process (not a tool or a method or a technique). For other people QFD is a system (not a process or a method) for translating customer requirements into appropriate company requirements. The same authors use different definitions of QFD. It looks that QFD can be everything. Sometimes reading QFD papers and books the same authors claim that QFD is a tool, in some parts and process (not a tool) in other parts. QFD is said to be a planning process (methodology), but planning does not mean prevention. You can plan without doing any prevention, but QFD can help you in effective prevention. Some of the important points need to be highlighted: 1) What is the relationship between Quality and customers requirements? 2) What is the relationship between Quality and design characteristics? 3) QFD converts the consumer demands into "quality characteristics" and develops a design by systematically deploying the relationships between the demands and the characteristics. The information provided by consumers about the qualities that they want is Quality? 4) What are the methods suggested in QFD (books/papers)? 5) How are they related to other methods (reliability analyses and productions, tests, Design of Experiments, Control Charts)?

The Japanese pioneered the Quality Function Deployment (QFD) to meet the "customers' voice" throughout the design process and in the design of systems. QFD utilises market research of customer needs, engineering technical descriptions of that information, and competitive analysis from both customer and technical viewpoints. The information is presented to show engineering design standards, descriptions, and tests related to customer requirements for both existing and new competitive products. Therefore, QFD is a process, which integrates product requirements with product development. The information is also presented to show any interrelationships between engineering requirements and market tests. A customer might express a desire to own a car that is easy to start. The translation of the "voice of the customer" into technical language might be "car will start within 8 seconds of continuous cranking". QFD helps identify the important and difficult customer issues and design requirements that the team should focus on. Finally, it shows any design trade-offs that may be present. All of the information listed is summarised in a matrix called a house of quality. QFD converts the consumers' demands into "quality characteristics" (conversion of demanded qualities into measurable and engineering elements) and develops a design quality for the finished product by systematically deploying the relationships between the demands and the characteristics, starting with the quality of each functional component and extending the deployment to the quality of each part and process. The information provided by consumers about the qualities that they want is Quality (actual quality as demanded by the customer?) What are the methods suggested in QFD?

Authors on QFD are Akao (13), Dean (17), Sullivan (19), Bossert (22), Clausing, Hauser, Pugh, Eureka (24), Ryan (25), Cohen, Scollard, etc. Most of these people, who claim that QFD is very good, despite remarkable anomalies, also claim that "Taguchi Methods" are good: since there are many scientific proofs showing that the latter ones are wrong, it is almost impossible that they are right on QFD.

R.E. Reins, of Mack Trucks Corporation, wrote (1994) "The only scientific approach to customer satisfaction is quality function deployment". QFD must consider customer's needs and not just his satisfaction. Also cigarette makers want to achieve customer satisfaction: customer satisfaction is unfortunately different from customer's needs satisfaction. The health of smokers is not protected. W.M. Hancock of the University of Michigan told in 1988 "As a professor, I am always looking for concepts that are teachable. QFD is a teachable discipline." This is not enough to make QFD a valid and preventive method for Quality. For J.L. Bossert (22) "traditional experimental designs and Taguchi Techniques are tools utilised to understand the relationships in the body of the QFD matrix. There is much controversy among statisticians as to which tools do the job best... Use whatever works best for what you are trying to understand": the only problem is that Taguchi Methods don't work.

Before any analysis of further documents a definition of the term Quality is in order. F. Galetto (5,9) gave the following definition: Quality is set of characteristics that satisfy Customers, Users and Society Needs. This definition is broader than the one of Juran (11) and Feigenbaum (3). Prevention of any problem is the fundamental idea in this definition of the term Quality. We note that Customer Needs Satisfaction is very different from Customer Satisfaction. Moreover, Customer and Users mostly are different people with different needs, generally different from the needs of the Society. QFD "practitioners" misuses the term Quality. Actually the term "quality" in the acronym QFD stands for "qualities". This is the fundamental reason for the different meanings given by different authors to the acronym QFD. Professor Montgomery (26) gives a different definition of the term Quality. "We prefer a modern definition of quality: quality is inversely proportional to variability" He adds also "Note that this definition implies that if variability decreases, the quality of the product increases". Looking at the House of Quality found during the "design of a new undergraduate curricula" at the University of Wisconsin-Madison (27) we see that a "quality requirement" is "professor technical knowledge". Using Montgomery's definition one reaches the conclusion that "a professor who applies consistently a wrong method (e.g. Taguchi Methods) has more quality than a professor who applies the *methodical doubt* (accepting the variability as a rule of life". In the Day's Book (28) the definition of quality is the following: "It (quality) is an integral part of QFD. It is important, therefore, to have a clear understanding of how the term quality is used in this text". And he adds: "quality will be defined as follows: Goods and services that satisfied customers' requirements". Taguchi's definition of quality is not better: "Quality is the loss imparted to Society". Needs are left out from all these definitions of Quality.

Practitioners tell that QFD converts consumers' demands into quality characteristics: in effect they confuse quality with qualities. The quality characteristics become demanded qualities to convert into measurable and engineering elements. Bossert (22) admits: "Customer requirements are translated into manufacturer's terms". The example in Fig.1, derived from ASI (American Suppliers Institute), does not help in making Quality: Taguchi Methods are used instead of the Design of Experiments.

Prevention is always absent in the QFD literature. DOE and reliability help a lot in presenting problems. Managers have to learn DOE and Reliability to draw good decisions during product/process development act according Deming's "statistical thinking" and use the Scientific statistical approach to designs and experiments. Scientific statistic approach entails a) statistical design, b) correct statistical conduct, c) scientific statistical analysis of the data.

The absence of the scientific approach before, during and after any test gives relatively uninformative output of questionable real validity. QFD pragmatic books for practitioners lack Quality and scientific approach. According to Deming "It is a hazard to copy. It is necessary to understand the theory of what one wishes to do or to make. Without theory, experience has not meaning".

WHATS	HOWS: Implement simultaneous engineering partnership with suppliers	HOWS: Apply Taguchi methods and QFD at design stage	HOWS: Institute preventive antennae training for customers
Improved product quality and reliability			
Reduced delivery times and costs			

Fig. 1 A QFD matrix for quality development, according to ASI (num)

3 DESIGN OF EXPERIMENTS VERSUS TAGUCHI METHODS

On the basis of their actual experience (theoretical and practical in industrial companies) on Quality and teaching activity the authors want to highlight some important points of QFD and TM. The starting seed is the definition of the term Quality: Quality is the set of characteristics that satisfy the needs of the Customers, of the Users and of the Society. Using actual cases, the following points will be dealt with in the paper.

1. What is the relationship between Quality and customer requirements?
2. What is the relationship between Quality and design characteristics?

Any decision about Quality of product and processes is usually based on data, collected through tests on samples. There is always some risk of making a wrong decision because of the random variation always present. Managers want this risk to be acceptably small: the risk becomes smaller as the number of observations increases.

According to Deming's teaching if one really wants to make Quality he must acquire Statistical Thinking [2] that permits managers to understand the following fundamental principles:

1. A Scientific Statistical Approach is needed to draw logical conclusions and take sensible decisions, based on actual data, generated in designed experiments.

2. The absence of Statistical Approach before, during and after the experiments typically results in relatively uninformative output of questionable general validity.
3. A Scientific Statistical Approach of any experiment entails: statistical design, correct statistical conduct, scientific statistical analysis of the data. This is in accordance with Fisher's teaching (4).

It is clear that the more data we get, the higher the cost of testing is. It necessary to keep the cost of testing to a minimum consistent with the maximum risk of a wrong decision, that the manager is prepared to accept. DOE is a way to get the optimum testing effort: the least cost for the accepted risk. DOE (23) allows us to obtain more information for a smaller cost than can be obtained by traditional experimentation. According to [10] DOE there are three points to be considered: 1) the choice of the "dependent" (response) variable and factors, 2) factor levels, 3) collection and analysis of data. In order to design the experiments and correctly analyse the data we suggest a scientific method based on the Gauss-Markov Theorem called the G-Method.

Taguchi considers three stages in a product (or process) development: system design, parameter design, and tolerance design [1, 15, 18, 19]. The so called Taguchi Methods and related applications are generally shown for the stages of parameter design and of tolerance design, in order to find a "best" product or process, i.e. insensitive to environmental factors: such a product or process is called "robust". Since generally the number of test states and the replications are very large, according to certain rules a reduced design is carried out (fractional design). According to [20] it is obvious and un-managerial pretending that fractional designs provide you with the same information of complete designs. You can not estimate neither the factors effect, nor the interaction effect; they are inevitably entangled: you have generated the ALIAS Structure. Moreover, generally only one application is made of each treatment combination and therefore the estimator of the experimental error is not available. Taguchi pupils [6, 8, 9, 10, 12] hide this point.

3.1 A practical example

In order to compare the power of Taguchi Methods versus the G-Method we use an actual application to parameter design, prized as *best technical paper* by ASQC in 1986 [1] and in 1989 [19]. The data are in table 1. The experiment was carried out to find a method of assembling an elastomeric connector to a nylon tube such that the pull-off force (the "response variable") could be maximum. Four "controllable factors" (with 3 levels) and three uncontrollable "noise factors" (with 2 levels) were identified. Both kinds of factors were "controlled" during the test: they were "structural factors" in the terminology given above. Following the Taguchi parameter design methodology, onc experimental design was selected for the controllable factors A, B, C, D (inner array) and another for noise factors E, F, G (outer array). The factors were A=interference, B=connector wall thickness, C=insertion depth, D=percent adhesive in connector pre-dip, E=conditioning time, F=conditioning temperature, G=conditioning relative humidity. The inner array is the so-called "L9 orthogonal array", while the outer array is the so-called "L8 orthogonal array". The combined array comprised 9 x 8 = 72 test states; the pull-off force was observed for each state: the collected data on 9 x 8 = 72 test states and S/N (Signal to Noise ratio) are in the table 1 (the coding of the test is -1, 0, 1. We have not used the "standard" used by Taguchi: he uses 1, 2, 3 instead). Analysing the S/N, Byrne and Taguchi found the optimum state $A_2B_1C_2D_1$ (A at medium level, B at low level, C at medium level, D at low level), neglecting the interactions between the factors.

Since every phenomenon generates data that are not the same (due to some disturbing "random error"), in order to extract the "true behaviour" of the phenomenon we need to separate the random error from the collected data. This is precluded by the use of Taguchi-Methods (S/N ratios). To overcome this point Taguchi postulates that interactions are zero (are not present, have no effect) and then the SS (Sum of Squares) of interactions is used to estimate the experimental error. It is important to take into account that the residual error is not computed as a difference of the estimated factors and interactions from the corrected total sum of square SS, in the ANOVA tables 3 and 4. In the first analysis [20] we have shown that the significance of factors and interactions is hidden (if not forbidden) by the analysis of S/N. Moreover, firstly the noise factors E, F are much more important than the controlled factors A, B, C, D [9, 10, 20]. The so called "product array design" structure (product of the inner by the outer array) led to a very large experiment of 72 test states that did not permit the estimation of the interactions (so the authors were forced to neglect them).

Table 1 Data from (1)

			E	1	1	1	1	-1	-1	-1	-1	
			F	1	1	-1	-1	1	1	-1	-1	
			G	-1	1	-1	1	-1	1	-1	1	
Controlled				E, F, G "noise factors"								
A	B	C	D	Response								S/N
-1	-1	-1	-1	19.1	20.0	19.6	19.6	19.9	16.9	9.5	15.6	24.025
-1	0	0	0	21.9	24.2	19.8	19.7	19.6	19.4	16.2	15	25.522
-1	1	1	1	20.4	23.3	18.2	22.6	15.6	19.1	16.7	16.3	25.335
0	-1	0	1	24.7	23.2	18.9	21.0	18.6	18.9	17.4	18.3	25.904
0	0	1	-1	25.3	27.5	21.4	25.6	25.1	19.4	18.6	19.7	26.908
0	1	-1	0	24.7	22.5	19.6	14.7	19.8	20.0	16.3	16.2	25.326
1	-1	1	0	21.6	24.3	18.6	16.8	23.6	18.4	19.1	16.4	25.711
1	0	-1	1	24.2	23.2	19.6	17.8	16.8	15.1	15.6	14.2	24.832
1	1	0	-1	28.6	22.6	22.7	23.1	17.3	19.3	19.9	16.1	26.152

Table 2 Analysis of S/N data in Table 1, through ANOVA

source	df	SS	MS	Fc	F10%	significance
A	2	1.77	0.89	Does not exist	***does not exist***	**not assessable**
B	2	0.47	0.23	Does not exist	***does not exist***	**not assessable**
C	2	2.88	1.44	Does not exist	***does not exist***	**not assessable**
D	2	0.17	0.086	Does not exist	***does not exist***	**not assessable**
Residual	**0**	**0**	**does not exist**			

Table 3 Scientific analysis of the data, assuming that only noise factors E and F were used

Source	df	SS	MS	Fc	F10 %	Sign	source	df	SS	MS	Fc	F10%	Sign
A	2	50.58	25.29	6.17	2.46	*	***A*B*C***	8	132.55	16.57	4.04		*
B	2	13.38	6.69	1.63	2.46		***A*B*D***	8	87.63	10.95	2.67		*
C	2	68.59	34.30	8.36	2.46	*							
D	2	23.67	11.84	2.89	2.46	*							
A*B	4	92.27	23.07	5.63	2.11	*	**E**	1	275.7	275.7	66.77	2.86	*
A*C	4	37.06	9.26	2.26	2.11	*	**F**	1	161.7	161.7	39.44	2.86	*
A*D	4	81.97	20.49	5.00	2.11	*	***D*E***	2	21.75	10.87	2.65	2.46	*
B*C	4	74.25	18.56	4,53	2.11	*	*D*F*	2	15.45	7.73	1.88	2.46	
B*D	4	119.2	29.79	7.27	2.11	*							
C*D	4	63.96	15.99	4.64	2.11	*	Res	36	131.36	4.10			

Table 4 Scientific analysis of logarithmic transformation of the actual data, using G-method

sour	df	SS	MS	Fc	F10%	sig	Sour	df	SS	MS	Fc	F10%	sig
A	2	0.1381	0.069	25.65	4.32	*							
B	2	0.0336	0.0168	6.24	4.32	*							
C	2	0.2073	0.1036	38.39	4.32	*	**E**	1	0.7313	0.7313	271.61	7.71	*
D	2	0.0419	0.0209	7.78	4.32	*	**F**	1	0.4309	0.4309	160.00	7.71	*
A*B	4	0.2491	0.0623	23.13	4.11	*	***A*G***	2	0.0811	0.0406	15.07	4.32	*
A*C	4	0.0755	0.0189	7.01	4.11		***D*E***	2	0.0630	0.0315	11.70	4.32	*
A*D	4	0.2484	0.0602	22.36	4.11	*	***D*F***	2	0.0420	0.0210	7.80	4.32	*
B*C	4	0.1800	0.0450	16.71	4.11	*	***D*G***	2	0.0262	0.0131	4.86	4.32	*
B*D	4	0.2073	0.0518	19.24	4.11	*							
C*D	4	0.1717	0.0429	15.94	4.11	*	Res	4	0.0108	0.027			

If they had used the G-method they could have designed a "combined array" of the "structural factors" that would have been more likely to improve process understanding and decisions [10]. In [20] Galetto and Gentili, using the G-method, found that every factor was "entangled" with various interactions (the symbol **&** stands for the "entanglement relation"):

A**&**B*C**&**B*D**&**C*D; B**&**A*C**&**A*D**&**C*D; C**&**A*B**&**A*D**&**B*D; D**&**A*B**&**A*C**&**B*C

"Entanglement" is an "equivalence relation", in a logical sense. More precisely, it was stated that there was also the following ALIAS structure (the symbol @ stands for "equivalent to"), neglected by Byrne and Taguchi:

(A+B) @ C*D	(A+C) @ B*D	(A+D) @ B*C
(B+C) @ A*D	(B+D) @ A*C	(C+D) @ A*B

This means that changing "additively" any two factors is exactly the same as changing "interactively" the other two factors. As a consequence you can not choose the best levels of factors as though they were independent, "a magic feature of Taguchi orthogonal arrays": that was done in the *Best Technical Paper* [1] in which it is written "Taguchi teaches engineers to analyse their experiments without using ANOVA". In a recent book [19] on planned experiments, where the "*Best Technical Paper*" [1] is re-published with some corrections, the authors Byrne and Taguchi, to overcome the critic of "large experiment" [19], added a paragraph entitled "Alternative design layouts" to combat the criticisms that 72 test states were too many, suggesting to carry out 36 tests as a possible reduced experimental design. In particular the Authors proposed to use an L_4 outer array for the three noise factors informing the readers that "The primary purpose of the outer array is merely to create noise during the experimentation to aid in the selection of controllable factor levels. Specific interactions between controllable factors and noise factors could still be identified using this design".

Another problem arises: the alias structure becomes more complicated because more effects are now entangled. We decided to carry out two further analyses of the data in table 1.

- The first one considered a logarithmic transformation of the 72 data and the analysis of the transformed data: the results are in table 4. Since the transformation reduces the variability and stabilises the variance we got further significant factors and interactions, at 10% significant level: factors A, B, C, D are all significant and 10 interactions A*B, A*C, A*D, B*C, B*D, C*D, A*G, D*G, D*E, D*F are significant; the noise factors E, F, again appear to have more influence (they are more significant) than the controlled factors.
- Secondly we considered the experiment as though only 36 test states were carried out (since noise factor G was not significant!); if only 36 states are carried out, there is no residual to test the significance of factors and interactions. Since [20] factor G was not significant we decided to use table 1 as a fractional design of 36 states. In order to estimate the residual error we considered all the 72 data as 36 states replicated twice. The results are in table 3. It is important to note that now the Residual error is 4.10 (with 36 degrees of freedom) while previously it was 2.135 (with 4 degrees of freedom, estimated through 4 interactions). Since the last Residual error is higher, some effects become insignificant: the "merit" of that pertains to various significant interactions ("that must not exist", according to Taguchi, like A*G and D*G). At 10% significant level: factors A, C, D are significant and ten 1st order interactions A*B, A*C, A*D, B*C, B*D, C*D, D*E, are significant, and two 2nd order interactions A*B*C, A*B*D are significant; the noise factors E, F, again appear to have more influence (are more significant) than the controlled factors.

The conclusions are similar to [20]. We can conclude that the analyses of Byrne-Taguchi in [1] and [19] are not scientifically correct and that the alias structure is not provided. This is pure logic, not statistics. These further analyses confirm that the entanglement and the ALIAS structure are again

A&B*C&B*D&C*D; B&A*C&A*D&C*D; C&A*B&A*D&B*D; D&A*B&A*C&B*C

(A+B) @ C*D	(A+C) @ B*D	(A+D) @ B*C
(B+C) @ A*D	(B+D) @ A*C	(C+D) @ A*B

These further analyses confirm how it is a bad practice of the so-called "practitioners" to adopt a-scientific methods. You can show all this using the G-Method [6, 8, 10]; in chapter 9 of ref. 10 where a method allowing one to find the bias of the estimate of the parameters of a "reduced model" is mentioned. The same idea can be used for finding the alias structure. From this it can easily be seen that:

- When a full design is carried out and a reduced model is considered, the estimators are biased
- When a fractional design is carried out, only a reduced model β_1, ALIASED, can be estimated.

Using Statistics correctly for the Byrne-Taguchi case, the optimum point is therefore different from the one found by Byrne-Taguchi, due to interactions. Using Logic, a Rational Manager is not dazzled by "the robust design methodology, following the modern Total Quality philosophy, ... (where) Taguchi proposes to use different types of response, characterised by great simplicity ... today it is possible even for inexperienced people thanks to the diffusion of advanced statistical software ..."[14]. In the same paper the problem of the alias structure is hoaxed-missed again. The entanglement can be found by the G-method. Unfortunately, at least 90% of the papers on application of TM do not provide you with the alias structure.

4 CONCLUSIONS

There are many cases where interactions are important; therefore it is quite non-managerial pretending, before any test, to say (Taguchi) *"... when there is interaction, it is because insufficient research has been done on the characteristic values."*, or to say, after a test (Phadke), *"... if we observe that for a particular objective function the interactions among the control factors are strong, we should look for the possibility that the objective function may have been selected incorrectly"*. Many cases analysed show that facts and figures are useless, if not dangerous, without a sound theory. Interactions are really very important, according to the fundamental principles F1 and F2 of GIQA [see 6, 8, 9].

Prevention is different from improvement. Unfortunately in the ISO 9001 there are misunderstandings about the two things and the point 4.14 considers corrective and preventive action. Preventive actions must include (4.14.3) "the use of appropriate sources of information such as... service reports and customer complaints to detect, analyse and eliminate potential cause of nonconformities, etc." Prevention and improvement are not here two distinct things, but they are confounded in one, producing a dangerous misunderstanding in people not able to think with their brain, but only able to accept standards as they are, also when they wrong. Actions that must avoid actual defects are corrective actions. If we design a new product or service, but similar to precedent one, we take into account our experience, doing a good improvement action. Unfortunately this is not prevention and it is shame that this confusion is present in standards as important as ISO 9000, particularly because of the wide use in firms.

5 REFERENCES

(1) Byrne D, Taguchi S.; 1986, The Taguchi approach to parameters design, ASQC86 and Quality Progress Dec 87.

(2) Deming W.E.; 1986, "Out of the Crisis", Cambridge Press

(3) Feigenbaum A.V.; 1993, "Total Quality Control", 3a ed., McGraw-Hill.
(4) Fisher R. A.; 1935, "The Design of Experiments", Oliver and Boyd, Edinburg
(5) Galetto F.; 1988, Quality and Reliability. A must for Industry, 19th ISATA, Montecarlo
(6) Galetto F.; 1989, Quality of methods for quality is important, EOQC Conference, Vienna
(7) Galetto F.; 1990, Basic and managerial concerns on Taguchi Methods, ISATA, Florence.
(8) Galetto F., Levi R.; 1993, Planned Experiments-Key factors for product Quality, 3rd AMST 93, Udine
(9) Galetto F.; 1999, GIQA Golden Integral Quality Approach: from Management of Quality to Quality of Management, Total Quality Management, vol. 9
(10) Galetto F.; 1997, "Qualità. Alcuni metodi statistici da Manager", CUSL, Torino.
(11) Juran, Grina F. M.; 1993, "Quality Planning and Analysis", 3rd ed., McGraw-Hill, N.Y.
(12) Levi R.; 1991, Piani sperimentali e metodi Taguchi: luci e ombre, ATA, vol 44, n. 11, 777- 781
(13) Akao Y.; 1990, "Q.F.D. Integrating Customer Requirements into Product Design", Productivity Press.
(14) Lombardo A.; 1997, Product-Array and Combined Array for Robust Design, Statistica Applicata, vol 9, n. 1
(15) Phadke M.S. 1989, "Quality Engineering using Robust Design", Prentice Hall.
(16) Pistone G., Wynn H. P.; 1996 Generalised confounding with Groebner bases, Biometrica 83,1
(17) Dean E.B.; 1999, "Quality Function Deployment", Internet Paper.
(18) Taguchi G.; 1987, "System of Experimental Design", vol. 1 & 2, UNIPUB/Kraus Publications, N.Y.
(19) Bendell A., Disney J., Pridmore W.A.; 1989, "Taguchi Methods: Applications in World Industry, IFS Publications/Springer Verlag, London
(20) Galetto F., Gentili E.; 1999, The Need of Quality of the Methods Used for Making Quality, CAPE '99, Durham UK
(21) Sullivan L.; 1989, "Quality Function Deployment", Quality Progress, Prentice Hall.
(22) Bossert J.L.; 1991, "Quality Function Deployment", ASQC Quality Press, Milwaukee, USA
(23) Moen R.D., Nolan T.W., Provost, L.P.; 1991, "Improving Quality Through Planned Experimentation, McGraw-Hill, N.Y.
(24) Eureka W., Ryan N.E.; 1989, "The Customer Driven Company: Managerial Perspectives on QFD", ASI Press.
(25) Ryan N.E.; 1989, "I metodi Taguchi e il QFD", Franco Angeli (ELEA-Olivetti).
(26) Montgomery D.C.; 1997, "Introduction to Statistical Quality Control", John Wiley & Sons, Inc., N.Y.
(27) Ermer D.; 1995, "QFD Becomes an Educational Experience for Students and Faculty", Quality Press.
(28) Day R.G.; 1993, "Quality Function Deployment. Linking a Company with Customers, ASQC Quality Press.
(29) Gevirtz C.; 1994, "Developing New Products with TQM", Mc Graw-Hill, N.Y.

CAPE/011/2000

Ten stages for optimizing electronic manufacturing

L B NEWNES and **A R MILEHAM**
Department of Mechanical Engineering, University of Bath, UK
A DONIAVI
Faculty of Engineering, University of Urmia, Iran

SYNOPSIS

With the increasing competition in the area of electronic manufacturing the authors have focussed their research on optimising the electronic manufacturing processes. Optimising a manufacturing system is a complex problem where many attributes and goals need to be considered. The aim of this research has been to use a Computer Aided Systems Engineering (CASE) approach to optimising electronic manufacturing in terms of yield.

The research work that has been undertaken is based on the author's view that to truly optimise a manufacturing system, the system needs to be examined as a whole, enabling the whole system to be optimised not individual sub-systems.

This paper introduces a CASE framework, which has been developed, verified and validated in a number of electronic manufacturing companies. The framework consists of three phases. Stages one to two make up phase 1, which is the system-modelling phase where a model of the system to be optimised is created. Phase 2, which consists of stages three to seven, involves system analysis and control. In this phase the focus is on identifying areas of the manufacturing process where analysis and control needs to be undertaken. A collection of techniques can be used systematically to generate such information and provide an understanding of an individual process's capability. These techniques include process capability indices, measurement operation evaluation indices and process failure mode and effect critical analysis. The final phase is used to optimise the system using techniques such as experimental design and response surface models. This phase is represented by stages eight to ten.

This paper describes the proposed framework and examines its use within the three main areas of electronic manufacturing; printed circuit board manufacture, silicon device manufacture and electronic assembly.

1. INTRODUCTION

Electronics manufacturing (EM) includes three major process namely printed circuit board (PCB) manufacturing, semiconductor device (SD) manufacturing and electronics assembly (EA). PCB's allow electronic components to be electrically interconnected and mechanically supported. SD manufacturing is used to fabricate active components such as diodes, transistors, thyristors and Integrated Circuits and electronics assembly is used to bring a multitude of electronic components and the PCB together to provide a reliable and useful appliance. Electronic assembly involves various sub-processes such as component onsertion/insertion, soldering, testing and packing **(1,2,3)**. Much research has been carried out into various aspects of the EM process. Some research has looked in detail at particular process parameters. From the literature and industrial sources it is considered that no overall methodology exists through which a systems approach can be employed to a group of manufacturing processes.

2. SYSTEMS APPROACH METHODOLOGY

Many researchers within electronic manufacturing have focused on specific processes which is very relevant when attempting to overcome particular performance problems but does not offer an integrated approach for optimising the manufacturing process. The research described in this paper offers a framework for modelling such manufacturing systems. A model is a simplified representation of a system that can be used to predict, interpret, and understand the behaviour of a system as well as examine control strategies for a system and their effectiveness. A systems approach can be used to design and develop activities and has been recognised to be essential in the orderly evolution of generic models. It involves a series of steps accomplished in a logical manner and directed toward the development of an effective and efficient model. A systems approach is also an overall approach for solving manufacturing problems which optimises the whole of the system instead of optimising discrete sub-systems (**4**). The proposed methodology is aimed at improving the yield of the EM system as a whole i.e. the percentage of good products manufactured. It achieves this through a three phase, ten stage methodology, which is shown in fig. 1.

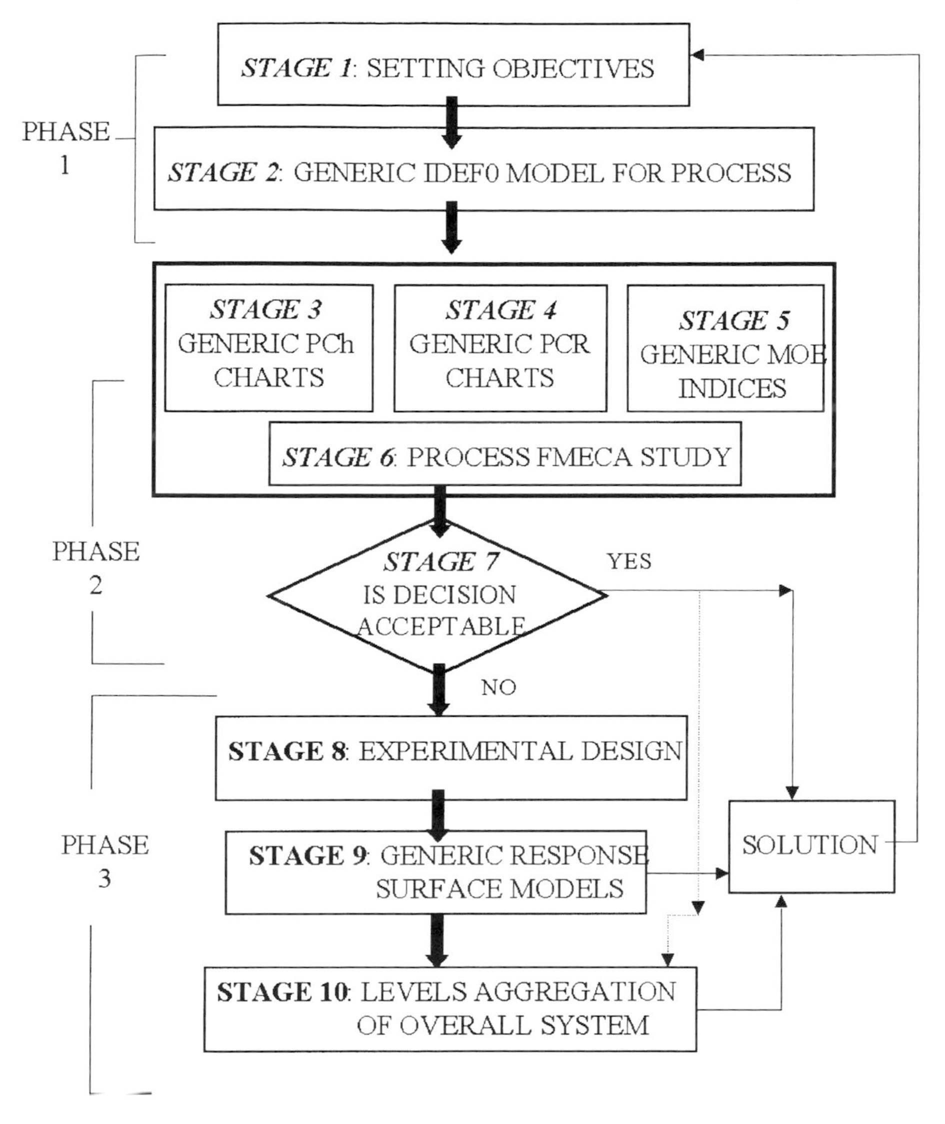

Fig. 1 Ten-Stage Methodology for Optimising Electronic Manufacturing

In phase 1 the aims and objectives of the system are set and its various elements, processes, outputs etc. are identified and verified. In phase 2 the existing quality levels of the various processes are determined. This involves identifying the critical process characteristics, determining their performance through the use of capability ratios etc and evaluating their influence on the process through FMECA (Failure Mode Effect Critical Analysis). At this point there is a possibility that incapable processes will be improved and made capable through the purchase of new equipment. However, optimisation of the process can be achieved in phase 3 through the application of Experimental Design (ED) and response surface optimisation techniques, after which optimal yield settings for each process are identified. Finally the yields for each sub-process and process are aggregated into an overall EM model. This model is analysed to produce yield statistics for all levels together with information of how this can be maximised.

Table 1: Methodology Application Areas

Phase	No. Stage	CASE stage	PCB Manufacturing	SD Manufacturing	Electronics Assembly
1	1	Setting objective	√	√	√
	2	Generic IDEF0	√	√	√
2	3	Generic PCh	√	√	√
	4	Generic PCR	√	√	√
	5	Generic MOE		√	
	6	FMECA study		√	
	7	Is the decision acceptable	√	√	
3	8	ED	√		
	9	RS	√		
	10	Levels aggregation	√	√	√

The above methodology has been applied in each of the key electronic manufacturing areas. Table 1 depicts each of the stages and the areas where they were verified and evaluated.

3. EXAMPLE STUDY

To illustrate the application of the methodology the following section will describe each of the stages and the reason for them. Stages 1, 2 and 4 will include their application within the semiconductor industry. The company used in the study manufactures products such as gate turn off devices.

3.1 Stage 1 - setting objectives

In this study the objective was to examine current manufacturing processes and ascertain any process problems. The overall aim being to improve the overall yield of semiconductor device manufacturing. A Gate Turn Off (GTO) device was used for the study. A GTO is a device, which is used, for electrical current control in power systems. Setting the objectives is an essential aim to ensure that after completion the process can be fed-back and a check can be undertaken to see whether or not the objectives are met.

3.2 Stage 2 - generic IDEF0 models of the process

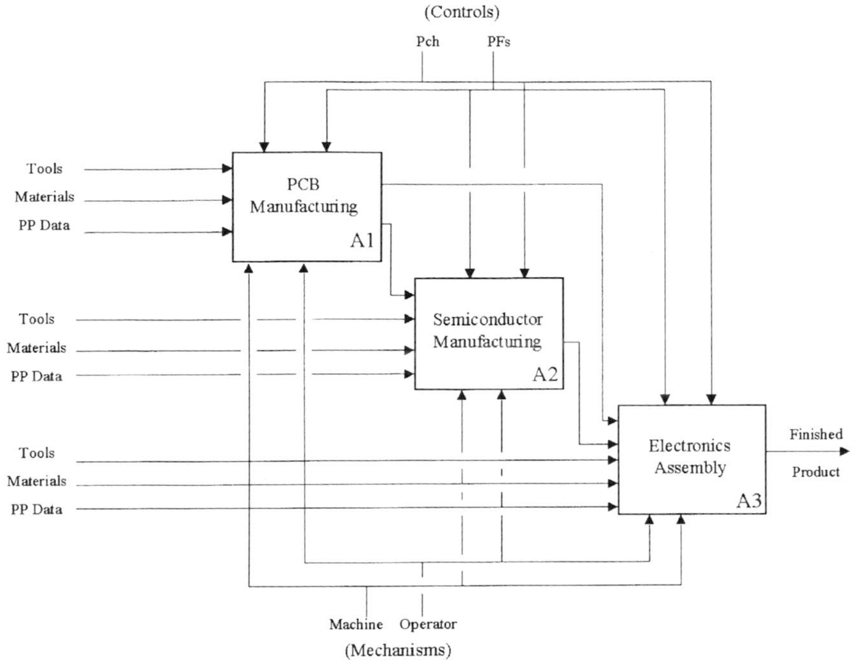

Fig. 2 IDEF0 Representation of the Three Key Electronic Manufacturing Areas

The authors selected IDEF as the mode for representing the relationship between processes. Any structured approach would be suitable for use depending on personal preference. The process stages for Silicon Device (SD) manufacturing are represented by A2 in Fig. 2. This box illustrates the top level for SD manufacture.

IDEF0 diagrams were produced for each of the key electronic manufacturing processes depicted in Fig. 2. At each stage the IDEF diagrams were verified by industrialists to ensure that the proposed methodology would be generic for electronic manufacturing. It is the authors' view that the approach could be adopted for other manufacturing methods.

Within SD manufacture a number of sub-processes are used such as; photolithography, impurity doping and etching. These processes are repeated several times. Fig. 3 shows the IDEF representation for GTO device manufacturing. The process steps in this case include; gallium (p-type) diffusion (A21), phosphorous (n-type) emitter diffusion (A22), n-type emitter definition (A23), metal deposition and definition (A24), life time control (A25), edge contour and passivation (A26) and finally, encapsulation and test.

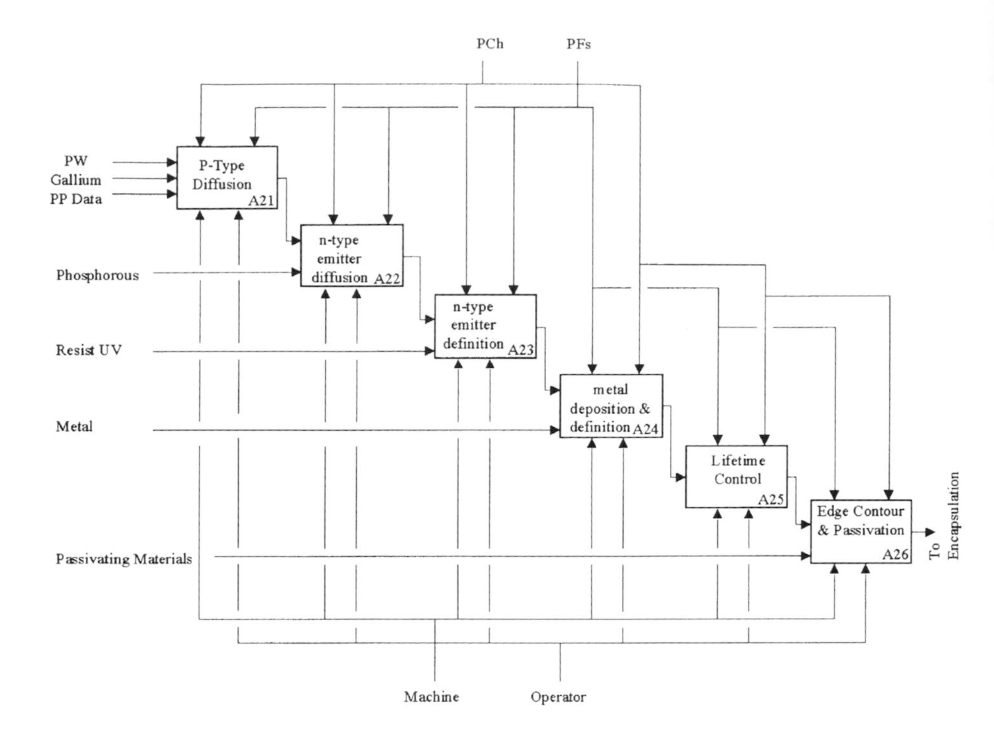

Fig. 3 - Level 2 IDEF0 Model of GTO Manufacturing Process

3.3 Stage 3 - process characteristic charts

At the third stage, it was necessary to analyse historical data of the various processes. Control chart techniques are generally used in industry to provide this information and are very effective for the analysis and control of a manufacturing process. Control charts such as Xbar-R, Xbar-S, I-MR, R, S, ARL and CUSUM charts are used to analyse variable data whilst, charts such as P, NP, C and U are used for attribute data. In this research, various charts and diagram were generated and used in the CASE environment.

3.4 Stage 4 - generic process capability indices

At the fourth stage, it is necessary to understand whether the process is capable and which process capability indices provide this information. Process capability ratio (PCR) analysis can be used to estimate process capability. This could be in the form of a probability distribution having a specified shape centre (mean), and spread (standard deviation). Capability indices are used to explain the relationship between the technical specification and the production capabilities. They are important criteria in comparing the process results and process specification. Various process capability indices are used in the electronics manufacturing environment such as Cp, Cpk, and Cpm. Each of these indices has specific characteristics e.g. Cpm uses the process mean. Cp, Cpk and Cpm indices were generated in this research for process control. The PCR indices are useful when comparing the efficiency of a manufacturing process with the process specification. After generating PCR indices, it is necessary to interpretate the process capability index. The interpreting is carried out in terms of the confidence interval and testing hypothesis methods.

For the GTO devices gallium restivity is used to control the maximum achievable breakthrough voltage, which defines the capability of the thyristor. Fig. 4 shows the results of the PCR analysis, which was used to understand whether the gallium restivity process was capable, or not. This analysis proved criteria, which was used in comparing the process results and process capability for gallium restivity.

The target specification was 25 ohms -cm, the upper limit being 30 and the lower limit 20. The number of observations taken was 45. This gave a mean of 25.30 and a standard deviation of 1.08.

Using confidence levels as defined by Montgomery **(5)** with a significance level of α=0.05, C_{pk} is calculated as;

$$\hat{C}_{pk}[1 - Z_{\alpha/2}\sqrt{\frac{1}{9n(C_{pk})^2} + \frac{1}{2(n-1)}}] \le C_{pk} \le \hat{C}_{pk}[1 + Z_{\alpha/2}\sqrt{\frac{1}{9n(C_{pk})^2} + \frac{1}{2(n-1)}}] \quad (1)$$

Where;

$$1.02 \le C_{pk} \le 1.85$$

Figure 4 shows that the value of C_{pk} equalled 1.44 which is greater than 1. A process is deemed to be capable if the C_{pk} is greater than 1. This needs to be calculated within confidence bounds.

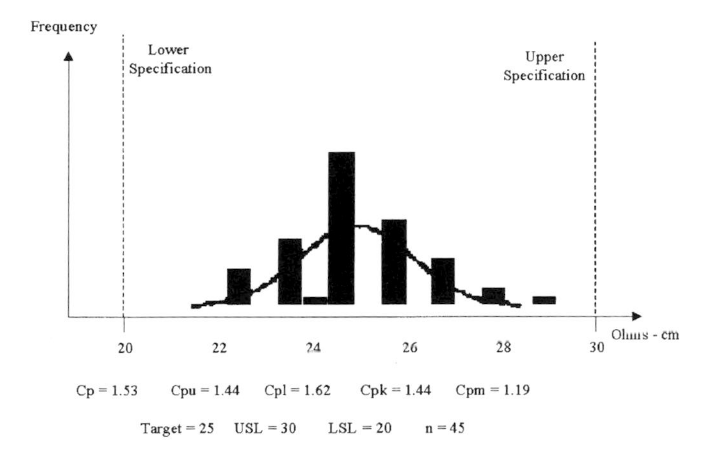

Fig. 4 Gallium Restivity

Various methods are used to test hypothesis for PCRs such as Kushler and Hurley **(6)**. Their research showed that the lower confidence bounds on C_{pk} could be established using:

$$L_{cpk,\alpha} \cong \hat{C}_{PK} - Z_{\alpha}\sqrt{\frac{1}{9n} + \frac{\hat{C}^2_{pk}}{2n-2}} \quad (2)$$

where,
$\hat{C}_{pk}$ = Estimated C_{pk}

The inference of the process capability using equation 2 shows that the lower confidence interval, L equals 1.19. This indicates that the null hypothesis can be rejected for C_{pk} less than 1 at the α=0.05 significance level. This indicates that the process is capable. In a similar example for testing of the blocking leakage current C_{pk} was calculated as 0.95 and the lower confidence limit 0.85. This indicated that the process was not capable.

3.5 Stage 5 - generic measurement operation evaluation capability indices

Stage five is required to determine whether the measurement system is capable. Measurement operation evaluation study is necessary to assess the capability of measurement operations; this includes various indices such as equipment variation, part variation, operator variation and the gauge index. The Gauge index (GRR) includes two components that assess the repeatability and the reproducibility. At this stage various measurement operation evaluation indices are generated for the process being investigated to determine the capability of the measurement operations. This assessment is necessary in order to understand the source of variation and avoid misinterpretation of whether the measurement tools or process is responsible.

3.6 Stage 6 - process failure mode and effect critical analysis

Process failure mode and effect critical analysis (FMECA) is used to identify and eliminate known or potential failures during manufacturing operations. It enables the magnitude of risk priority number (RPN) relative to process failure modes to be evaluated. In process FMECA, parameters such as process function, failure mode, failure effect, severity, occurrence, detection and risk priority number are considered. This technique offers the capability to recommend actions to eliminate the cause of failure and improve process quality. Process FMECA is very effective within the CASE environment and can identify which processes should be considered for further investigation using experimental design technique.

3.7 Stage 7 - decision stage

The decision stage is essential in determining whether the performance of the current manufacturing system is satisfactory or whether it needs to be optimised. Stage seven is a bridge stage between the analysis and optimisation of the manufacturing process. In this stage, the results from stages three to six are evaluated in terms of the desired targets for process characteristic control, process capability, measurement operations and risk analysis of the process. At this stage any obvious actions can be undertaken, such as purchasing new equipment. If the process needs more investigation then stages eight to ten are then recommended.

3.8 Stage 8 - design of experiments

An experimental design method is a formal plan for an experiment which includes the choice of the response, factors, levels, blocks and treatments aimed at achieving higher process yields, reduced variability, reduced development lead time, improve product quality and give greater customer satisfaction. Various experimental design methods are used in the design of manufacturing processes that can be classified in terms of levels and factors. The central composite design due to high capability was chosen for use in analysis and optimisation for this research. It was selected as it provided data information for second order response surface models. These models are used in the next stage.

3.9 Stage 9 - generic response surface models

An important aim of the RS model is to determine stationary points of process factors. Setting process factors in these points can be used to generate optimal response of the RS models. Second order RS models are used for determining these stationary points for the process factors. These points can be used by industry to set the process factors to improve the yield.

3.10 Stage 10 - levels aggregation

After stage nine, it is necessary to integrated the outputs (yield) of sub-levels and sub-systems to generate the output of overall yield of electronics manufacturing system. Levels aggregation of the CASE model is very important for on-line control of the EM process. This stage aggregates and integrates the sub-process outputs using a C program. This stage is a bottom up approach which starts from the lowest level and sub-process to generate experimental design and response surface models. This aggregation and integration enables the effect of sub-process outputs or process factors upon different levels of the overall EM model to be investigated.

4. CONCLUSION

Within this research paper the authors have introduced a methodology for optimising electronic manufacturing. The application of this methodology has been discussed with regard to SD manufacture. Phase 1 of the methodology examines the use of IDEF0 to provide process models of the manufacturing process being investigated. The authors used an industrial example within the semiconductor industry to illustrate this phase. Each of the IDEF models were verified by industrialists. Many of the sub-processes used within GTO device manufacture were analysed using PCR techniques which are used to explain the relationship between the technical specification and the the capability of the production process. In the area of gallium restivity the process was found to be capable. Other studies within the company indicated that not all their manufacturing processes were capable. Further details can be found in Doniavis thesis **(7)**.

The results obtained using the CASE methodology described have been used to provide significant improvements in product yield in the key processing areas. The total study investigated the use of IDEF0 techniques, process characteristics control, measurement operation evaluation and failure mode effect control analysis.

This paper has described some of this work by introducing the CASE methododlogy and the ten stages within this. It is the authors view that this methodology could be used to maximise manufacturing yield.

5. REFERENCES

(1) Brindley Keith, Newnes electronics assembly handbook, Techniques, Standards and quality assurance, Heinemann Newnes, 1990.

(2) Groover M. P.; Fundamental of modern manufacturing, material, processes, and systems, Prentice-Hall, 1996.

(3) Edwards P R, Manufacturing technology in the electronics industry, Chapman & Hall, 1991.

(4) Doniavi A., Mileham A. R., Newnes L. B., Computer integrated systems engineering in an electronics manufacturing environment, 13th National conference on manufacturing research, Glasgow, UK, September, 1997.

(5) Montgomery Douglas C., Introduction to Statistical Quality Control, , John Wiley & sons, 1996

(6) Kolarik William J., Creating quality, concepts, strategies, and tools, Texas Tech University, McGraw-Hill, Inc, 1995, chapter 18.

(7) Doniavi, A 'Computer Aided Systems Engineering Approach to Electronic Manufacturing Process Modelling', 1999, PhD thesis, University of Bath, UK.

CAPE/030/2000

On the application of quality function deployment in integrated supervisory process control

M BÄCKSTRÖM
Division of Production Engineering, Luleå University of Technology, Sweden
H WIKLUND
Division of Quality Technology and Statistics, Luleå University of Technology, Sweden

ABSTRACT

The paper outlines a comprehensive three-step approach incorporating module design, sensor design and model design, which deals with the development of advanced strategies for process control of a complex machining process. Well-known techniques from the quality movement, such as the seven MT-tools, Quality Function Deployment (QFD) and Conjoint Analysis (CA), are examples of TQM-tools that have been successfully utilized in the development process. The research framework is a developed concept for Integrated Supervisory Process Control (ISPC) which deals with multi-purpose control requirements utilizing individual advantages of sensors and modelling techniques together with varying pre-defined requirements on productivity and product quality.

KEYWORDS: QFD, Process Control, Integration.

1. INTRODUCTION

The growing complexity of manufacturing systems and produced parts together with new tools and work materials require that monitoring and control systems provide a reliable basis for decisions taken during the machining operations, see e.g. (1-6). This situation puts, in its turn, high requirements on the development of reliable, flexible and efficient systems for process monitoring and control. Difficulties in selecting among a high number of sensors available, developing adequate and advanced control modules and integrating developed monitoring applications and modelling techniques, are examples of central activities that have to be incorporated and integrated in such a systems solution. Recently the concept of integrated supervisory process control (ISPC) was presented as a systems approach dealing with multi-purpose control requirements by utilising the individual advantages of several modelling techniques, see (7-8). The ISPC concept was based on an integration of a variety of existing modelling techniques with necessary application modules and was aimed at improving the utilisation of available sensor systems and the performance of today's monitoring and control systems.

In order to develop the ISPC research platform a number of linked activities should be completed, all related to the selection and design of system components. In this paper the principles of quality function deployment (QFD), see e.g. (9) and (10), has been utilised in a three-step transformation process of pre-specified systems requirements down to the selection of sensors and design of modelling techniques and applications.

The aim of the QFD-study has been to systemise the design and selection of sensors and modelling approaches in order to improve the cost-efficiency of the developed research platform in terms of overall quality and productivity requirements and criteria's.

2. THE CONCEPT OF INTEGRATED SUPERVISORY PROCESS CONTROL

The ISPC concept is divided into three hierarchic levels, Figure 1. At the supervisory level, the actual systems requirements and goals of the machining process are co-ordinated and synthesised. Thereafter, the fused performance profile constitutes the basis for selection of needed functionality modules that trigger the selection and set-up of needed models and sensors at the set-up level.

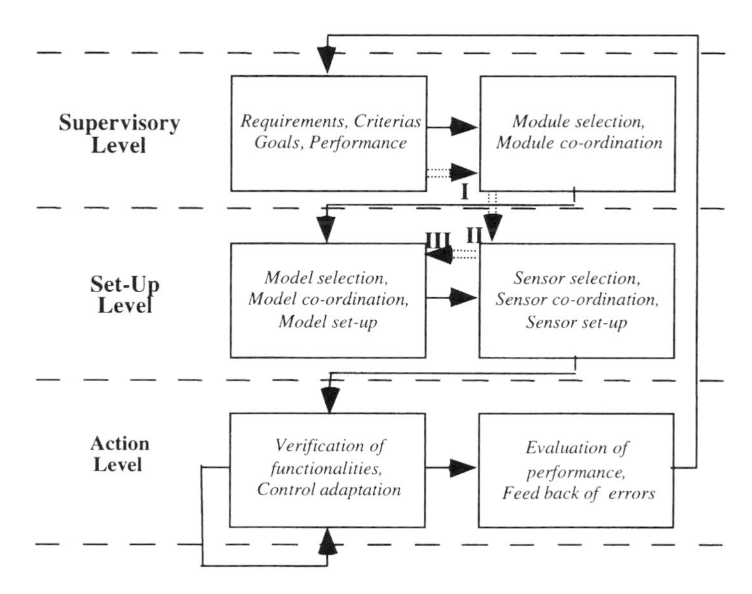

Figure 1. Different activity levels in the ISPC concept.

At the action level, the machining process is started and on-line data is starting to emerge. The decisions in the system have so far been based on accumulated knowledge and historical data. Verification is made as a safety precaution, first at the action level, to assure that the control actions carried out have the correct functionality.

The control of the machining process is adapted on the basis of the on-line signals and the previously established performance demands. At this level, the settings of model parameters are updated or altered but no changes to the model structure are allowed. Evaluation of the process performance is made continuously and any differences or errors are fed back to the supervisory level for eventual reconfiguration of the system.

The physical ISPC platform incorporates a five-axis Liechti Turbomill ST1200 machining centre, which represents the state of the art in machine tools, Figure 2. A prior research platform installation, see (11) and (7), has given valuable experiences and know-how when dealing with the requirements, design and implementation of a new research environment. The working space and speed are defined of the axis configurations as:

X-axis: 1400 [mm], 30 [m/min], 2.3 [m/s^2]	Y-axis: 500 [mm], 30 [m/min], 4.0 [m/s^2]	Z-axis: 500 [mm], 30 [m/min], 6.0 [m/s^2]
A-axis: -90-45 [°], 5760 [°/min]	B-axis: -50-50 [°], 10080 [°/min]	Spindle: 24000 [rpm], 30 [kW]

Figure 2. Liechti Turbomill ST1200 machining centre.

The control system is an Andron A 400 with the capability to perform spline interpolation. This, in combination with the digital Indramat drives and the SERCOS fibre optic servo coupling, renders the ability to perform motions with high speed and geometrical accuracy. In order to increase the available amount of process information an extensive sensor configuration is planned in the research platform (categorised as internal and external sensors) Figure 3. The internal information sources are originally fit in the machine and sensors denoted as external sensors are hardware sensors that have been retrofit to the machine.

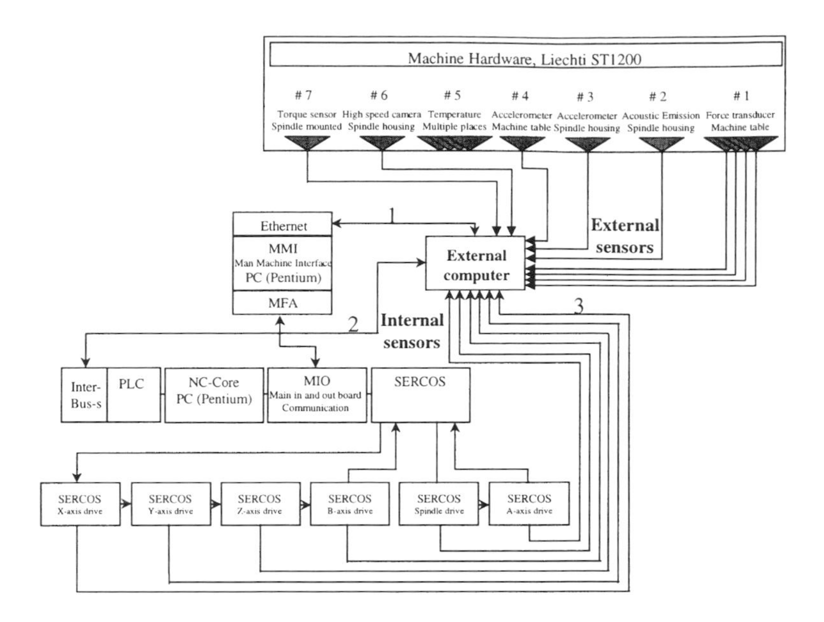

Figure 3. External and internal components in the research platform.

The internal information sources in Figure 3 are denoted 1-3, and constitute of: 1) Ethernet connection for NC-programs, tool pre-setting data, CAD-drawing etc.; 2) Interbus-S connection for PLC machine state register in- and output; 3) Digital drive connection for consumed power, axis position, torque, following error, axis speed and commanded position. Approximately 15 external sensors, such as force, acoustic emission, accelerometers, thermo elements and torque sensors have been discussed as possible sensors to be included in the platform.

3. THE DEPLOYMENT PROCESS

QFD is a structured process that establishes customer's value using the voice of the customer (here systems requirements) and transforms that value to design, production, and supportability process characteristics. In this paper, the transformation is made in three design steps, from module design to model design. The result is a system engineering process that prioritises and links the product development process so that it ensures product quality as defined by the systems requirements. In Figure 4 a flow chart illustrates the different activities in the conducted study. Figure 5 illustrates the so-called planning matrix (also called House of Quality, HoQ)

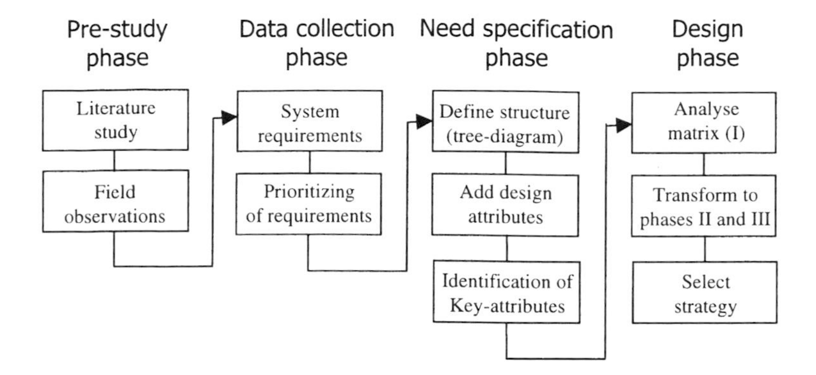

Figure 4. Activities in the study.

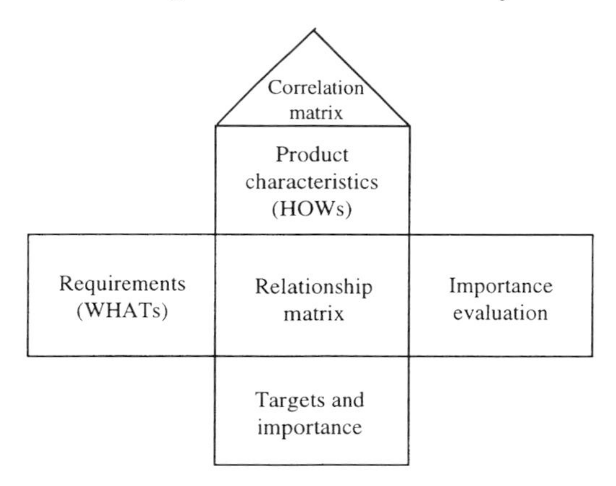

Figure 5. The planning matrix.

This process, based on the different phases of QFD, is aimed at transferring system requirements into system characteristics. The process follows the parts of the ISPC concept, Figure 1., and is divided into three different phases (I-III), see Figure 6.
In the first matrix (I) the system requirements are transferred to properties of system modules. In the next matrix (II) the system modules are transferred to the need of sensors and in the final matrix (III) the variety of process modelling techniques are designed. The first matrix (I) is illustrated in Figure 7 where the overall requirements on the ISPC system are specified.

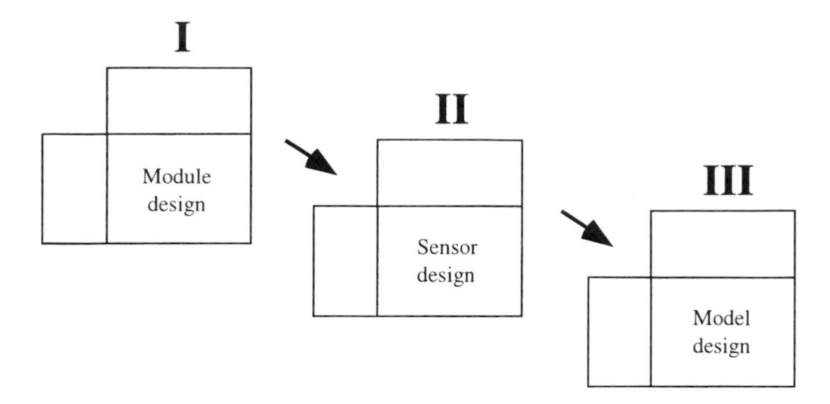

Figure 6. The deployment process.

Then, the relationships (correlations) between the requirements and the modules of interest are studied and marked with white, grey or black dots indicating the strength of relationship. A row with no black or grey dots indicates that there is no module that adequately handles the actual requirement and we therefore have to develop another module that satisfies the requirement. A column with no black or grey dots indicates that the actual module is redundant and may not be needed in the ISPC platform. For example, it is seen that the requirement "work piece protection" is satisfied in four or five of the modules while "high safety for the operator" is only satisfied in the "emergency stop"-module.

All of the dots in the different matrixes, Figure 7 have their own explanatory background and demand more space than available here to be fully described. However will some of the background and reasoning be outlined for the functionalities. Following the example of "high safety for the operator"- requirement and "emergency stop"-module into the second matrix (II), we can conclude that the sensors that are to be utilised to realise the functional module are primarily the cutting force, spindle power and torque.

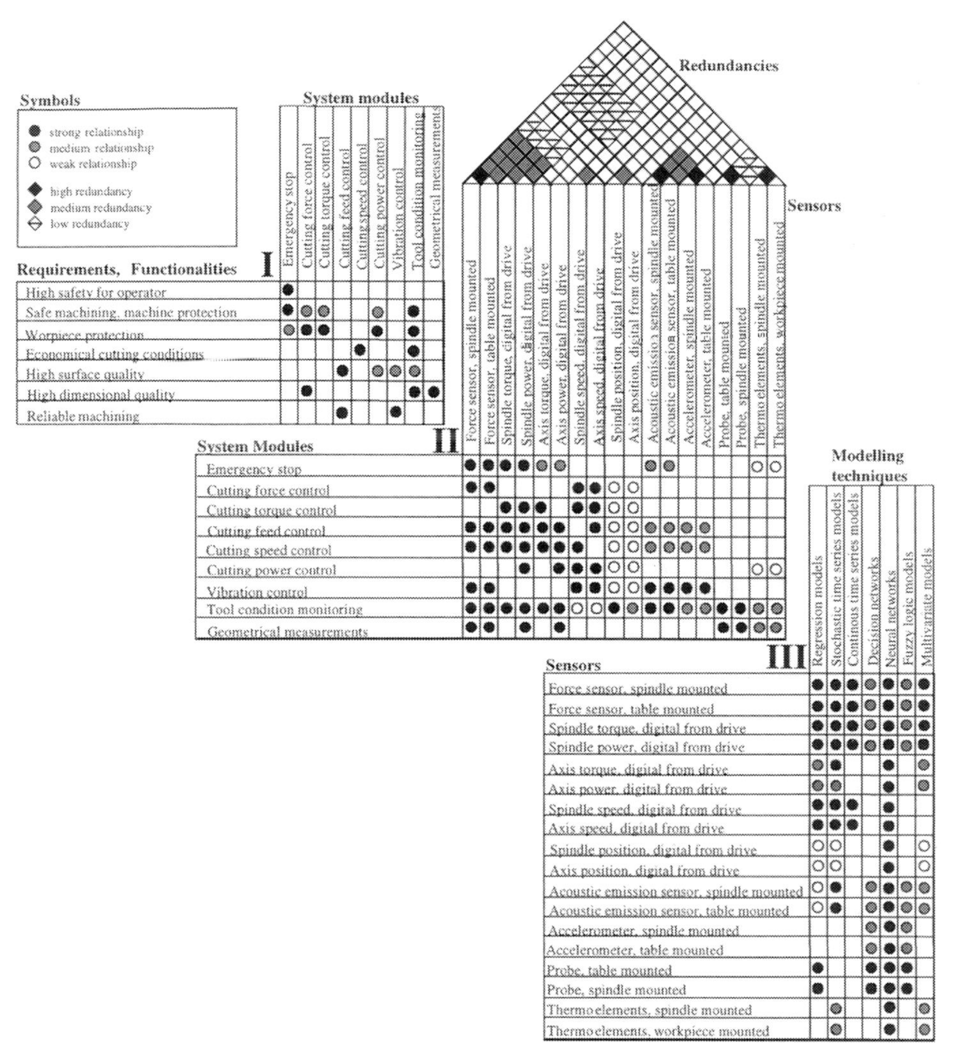

Figure 7. The three phases illustrating the design and deployment of planned configuration of the ISPC research platform.

Secondarily the axis power and torque should be used, and finally the thermo element sensors. Continuing this reasoning in matrix (III) the modelling techniques that are selected as most appropriate for the task are regression models, stochastic time series models and multivariate models. This is also possible to derive from matrix (III) by the strong correlations between the actual sensors and the modelling techniques. This exemplifies the deployment of system components originating from system requirements and resulting in suggestions of module, sensor and model design. The deployment technique also identifies the appearance of possible redundancies, which is illustrated at the top of matrix (II).

An example of the background reasoning for the functionality is as follows:

High safety for the operator: This is a basic requirement in machine systems when damage risks are obvious. The development towards higher cutting speeds and feed do also call upon a higher degree of active operator protection.

> Emergency stop: This module is the only one that possibly could serve as an operators protection where in fact all moving parts in the machine have to be brought to a stand still as quick as possible.
> Sensors: In this case do every sensor that could detect an overload or malfunction contribute in determining dangerous situations. Preferred are also sensors with a good resolution and sensitivity compared to the studied phenomena. Force sensors based on piezo-electric technique are good examples of fast and reliable sensor of dynamic forces and dangerous overloads can be detected and constitute an basis for machine shut down. Torque and power sensors are of value to detect abnormal levels in tooling, workpiece and machine elements. Acoustic emission sensors are favourable in detecting insert chipping and early warning in tool failure. Temperature sensors can assist in giving machine element status.
> Modelling: Basic requirement on the models are that they have to be fast in order to prevent operator accidents. The level of sophistication do not have to be high, mainly could simple threshold values be used to determine the action.

In the ISPC environment, several modeling techniques are developed and adapted for tool condition monitoring, see Figure 3.5. The different techniques, such as ANN and a number of statistical methods, are in practice associated with both advantages and drawbacks and are therefore handled by the principles of active data acquisition (ADA), see (13-14). Further, the methods are linked parts of different quality control (QC) activities where the developed methods and applications are integrated and post-process quality control is applied only as complement and for reference measurements.

All QC activities are supervised and fed with information from the ADA system. In Figure 8 the different methods are plotted versus model generality and process knowledge (how the methods stimulate increased process knowledge). The ambition has been to combine the advantages of the different methods and to facilitate increased process knowledge. Methods that adequately describe and predict the process and also increase process knowledge are given priority. If these fail, priority is given to more computer intensive (black-box) methods.

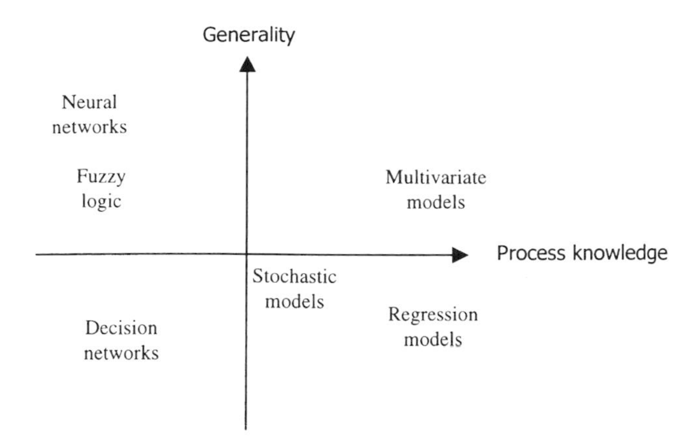

Figure 8. Some examples of used relationships between model generality and the process knowledge gained.

4. CONCLUSIONS

The ISPC concept and the research platform for its implementation are outlined. A "state of the art" high speed five axis machining centre equipped with an extensive sensor configuration constitutes the research equipment necessary for practical implementation.

In the paper, the principles of QFD have been utilised to develop the configuration of the ISPC concept and the system design is characterised by means of the deployment process. The deployment process includes the design and selection of modules, sensors and models, all derived from supervisory functional requirements of the system. The set-up of the system has been systemised and possible redundancies have been identified. The individual advantages of different sensors and modelling techniques are indicated and derived in relation to pre-determined requirements on productivity and product quality.

The total environment affecting the machining results are vast and relates to each other in complex ways and the QFD technique shows promising features in categorising and relating all necessary aspects in an integrated system.

The conducted deployment proccss indicate that the pre-specified requirements on the ISPC system, in terms of economy, quality and productivity expressed implicitly, are possible to satisfy by the developed system design and configuration.

The promising results obtained demonstrate a non-traditional approach to the deployment of quality and productivity requirements set by the system users, whereby the efficiency and systematization of the development of process control strategies can be significantly improved.

REFERENCES

1. Barschdorff, D., Monostori, L. (1991) Neural Networks-Their Applications and Perspectives in Intelligent Machining, *Computer in Industry (No.17)*, pp.101-119.
2. Dornfeld, D.A., DeVries, M.F. (1990) Neural Network Sensor Fusion for Tool Condition Monitoring. *Annals of the CIRP*, Vol. 39/1.
3. Malakooti, B., Zhou, Y. (1992) Applications of Adaptive Neural Networks for An In-Process Monitoring and Supervising System, *International Joint Conference on Neural Networks*, Vol.2 . IEEE, New York.
4. Matsushima, K., Sata, T. (1980) Development of Intelligent Machine Tools, *Journal of Faculty of Engineering, University of Tokyo, Vol.35, No.3*, pp.395-405.
5. Monostori, L., Nacsa, J. (1991) On the Application of Neural Nets in Real-Time Monitoring of Machining Processes. *Manufacturing Systems, Vol.20 (No.3)*, pp.223-230.
6. Monostori, L. (1993) A Step Towards Intelligent Manufacturing: Modelling and Monitoring of Manufacturing Processes through Artificial Neural Networks. *Annals of the CIRP*, Vol.42/1 (pp.485-488).
7. Bäckström, M., Wiklund, H. (1998) A Step towards Integrated Supervisory Control of Complex Machining Processes, *Changing the Ways We Work*, N. Mårtensson et.al. (Eds.), pp.769-778, IOS Press.
8. Bäckström, M., Wiklund, H. (1999) *An Approach to Integrated Process Monitoring and Quality Control of Advanced Machining*, Proceedings of the Advanced Summer Institute, ASI'99 September 22-24, Leuven, Belgium.
9. Akao, Y. (1990) History of Quality Function Deployment in Japan, *The Best on Quality*, pp.183-196. International Academy for Book Series, Vol.3, Hanser Publisher.
10. Mazur, G.H. (1994) *QFD Outside North America - Current Practice in Europe, The Pacific Rim, South Africa and Points Beyond*, Proceedings of the 6th Symposium on Quality Function Deployment, Novi, Michigan.
11. Bäckström, M. *Technologically Integrated Machining,* Proceedings of the National Conference on Technologically Integrated Manufacturing Engineering (TIME), pp.25-31, Linköping May 29, (1996).
12. Cai, D.Q., Xie, M, Goh, T.N., Wiklund, H., Bäckström, M. (1999) *A Study of Control Chart for Adjusted Processes,* Proceedings of the 26th International Conference on Computers and Industrial Engineering Vol. 1, pp.443-447, Melbourne, Australia.
13. Wiklund, H. (1999) A Statistical Approach to Real-Time Quality Control, *International Journal of Production Research*, Vol.37, No.18, pp.4141-4156.
14. Bäckström, M., Wiklund, H. (1998) Development of a Multi-Tooth Approach to Tool Condition Monitoring in Milling, *Insight*, Vol.40, No.8, pp.548-552.

CAPE/090/2000

Analysis of stresses and strains in the section roll when considering conditions of its regeneration

A ŚWIĄTONIOWSKI and **R GREGORCZYK**
Department of Technological Equipment and Environmental Protection, University of Mining and Metallurgy, Cracow, Poland

SYNOPSIS

The subject of this paper is the analysis of stress and strains in a regenerated section roll of a two-high rolling mill with the use of the Finite Element Method. Obtaining a description of the investigated phenomena more adequate when compared with the classical analytic methods is one of the conditions of appropriate design of working the roll during its regeneration.

1. ORIGINS OF THE RESEARCH

Rolled products nowadays are expected to meet higher and higher quality standards which aim at limiting allowance for their dimensions and shape discrepancies and tolerance for their surface condition. Such demand forces the producers to constant improvements both in technology and in tools applied in the production process. Coping with the requirements of competition is not possible without lowering the production costs on a constant basis, and a serious component of the costs is the so called roll wear. It is a factor expressed by a quotient of the roll weight (in kilograms) to the weight of products (in tons) obtained until its complete wear.

In rolling practice - excluding the immediate, originating from failure, roll damages, their cracking or breakage - rolls are changed as a result of damages to the barrel's working surface. This shows initially as a net of delicate super-fractures. It is then followed by some chipping, first small but growing bigger in time and causing defects in the rolled product surface.

The roll surface deterioration process is conditioned by stress and strain distribution in its sub-surface layer. These conditions are influenced by normal and tangential load forces accompanying the metal rolling, and also by thermal and internal stresses. To bring back the required smoothness to the working surface of the rolls and - which applies to section rolls, the subject of this paper - to restore the correct dimensions of roll passes, rolls need to be regenerated. Regeneration means a series of their redressing or regrinding, repeated in its working life (Table 1).

Thus resulting diminishing roll diameter is synonymous with its decreasing strength capacities, especially in the profiles exposed to notch operation. Time of rolls' operation between redressings depends on the position of the roll pass in the rolling process and on the rolling mill type, where the rolls of the first mill stands may as a rule work longer than the finishing stands.

Table 1 Average values of roll-off and wear-off of the rolls in section roll mills

Redressing or regrinding				
Rolling mill type	thickness of a single redressing	number of redressings or regrindings	total roll-off thickness (in % of the initial diameter)	Rolls wear-off
Billet mills	3 – 6 mm	8	10 - 13 %	0.2 - 0.3 kg/t
Large rail and section mills	3 – 6 mm	8 - 9	10 - 13 %	0.5 - 0.6 kg/t 0.8 - 1.0 kg/t
Medium-size and small section mills	1 – 5 mm	8 - 10	8 – 11 %	—

It should be mentioned here that for clear chill rolls or hardened steel rolls the total loss of thickness at their regeneration depends also, apart from the strength criteria, on the thickness of the hardened layer.

Stresses in the profiles of section rolls' roll passes subjected to a series of regeneration processes often reach the critical values. This fact is strictly connected with strong concentration of stress in the area of local roll pass shape changes, which generally present highly complicated geometry. As a consequence, the shape of a roll pass has to provide equally adequate conditions for the plastic strain of the metal and for meeting strength requirements. These two requirements are often contradictory, which provides for the difficulty of the problem. Traditional methods of stress and strain condition determination in the roll system are based on bringing it down to coplanar forces and using energetic methods. This requires assuming a series of simplifications which lower accuracy of such calculations. Much wider possibilities of analysis, and in a three-dimensional approach, is created by applying Computer Aided Engineering systems (CAE), here especially the ones based on FEM formulas.

2. DISCRETE MODEL OF A ROLL AND THE RESULTS OF THE CALCULATIONS

The object of the investigation were working rolls of rolling stands in a continuous rail and section mill.
They are cast rolls with a grey cast iron core and a hardened top layer of white inoculated cast iron.

The calculations have been carried in the linear range (elastic material), with assumption of a static nature of the loads operation (strip load pressure on the rolls). Values of the loads have been determined by experimental measurements.

Numerical simulation of the stress and strain distribution has been carried out basing on the MSC*/Nastran (7.0 version) software package.
At the first stage, CAD design section roll geometry has been directly transferred into MSC/Patran (7.6 version) database without any transaction or modification.
It should be noticed here that using graphic pre-processor of the MES system has proved more effective than an attempt to import geometric parameters of a roll model through the IGES file translator from the AutoCad R14 environment.
After that, a 9800 (Tet4 type) finite element model has been created and submitted for simulation and further visualisation of the simulated model's behaviour.
The HPK-XP computer with HP-UX 10.2 operating system has been used, which has enabled finishing calculations for a single roll in 14 minutes.

Evaluation of the results' accuracy has been performed by checking the balance conditions (consistence of the stresses with the load given) in the area of support reactions.
Increasing of the number of used elements up to 4,600 - at the subsequent stage of calculations - has not resulted in a significant change in so calculated distribution or values of stresses and strains of the rolls. The maximum discrepancies recorded amount to 7.4% for stress and 9.6% for strains at almost tripled calculation time.

Design of a roll pass subjected to investigation is presented in Fig. 1.
Calculation of stress values in particular cross-sections has been performed for a new roll and then after its regeneration which was redressing the roll passes in order to restore their proper dimensions and surface condition.
Contour lines of the reduced stresses acc. to von Mises hypothesis are shown in Fig. 2 and 3.
In figures 4 and 5 respectively, the course of stresses for the considered roll when new and after regeneration are presented.

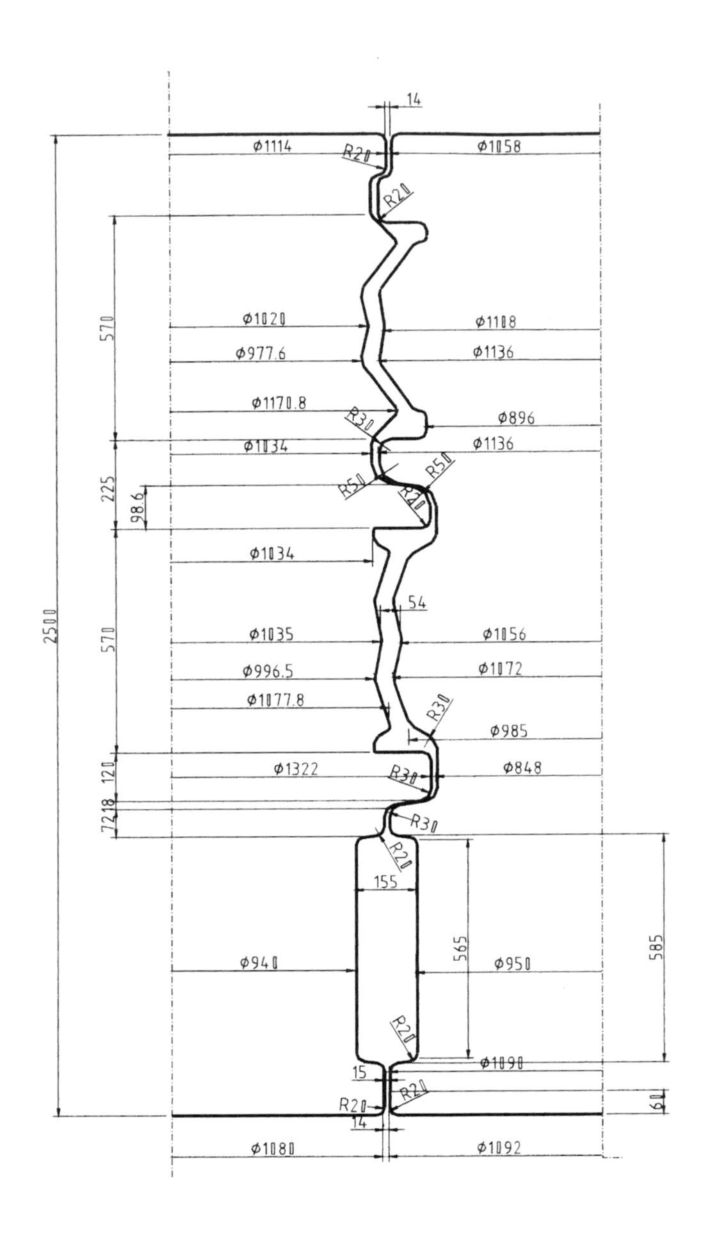

Fig. 1 Design of a roll pass subject to analysis

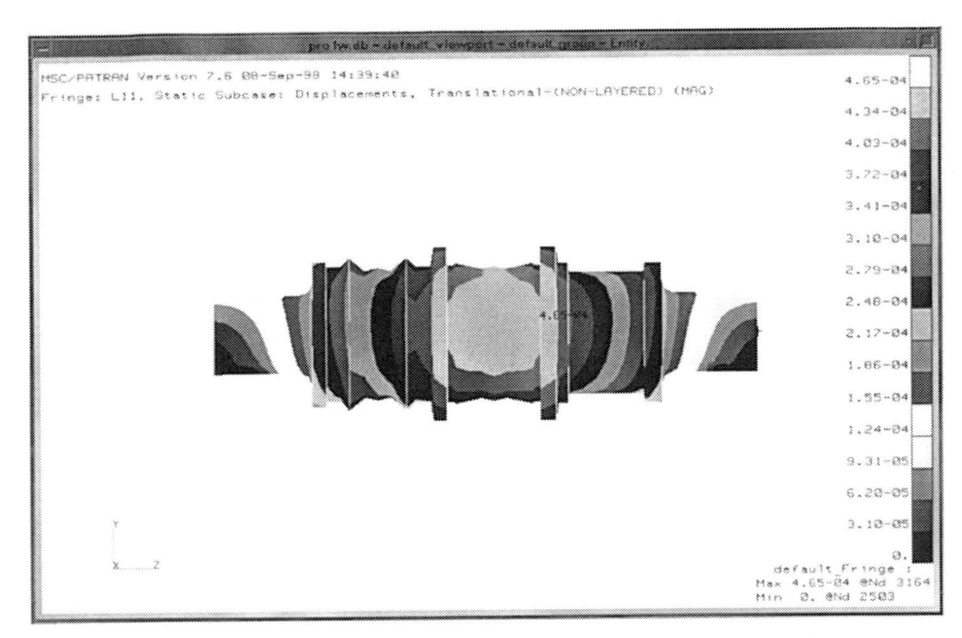

Fig. 2 Stress distribution in cross-sections of a new section roll

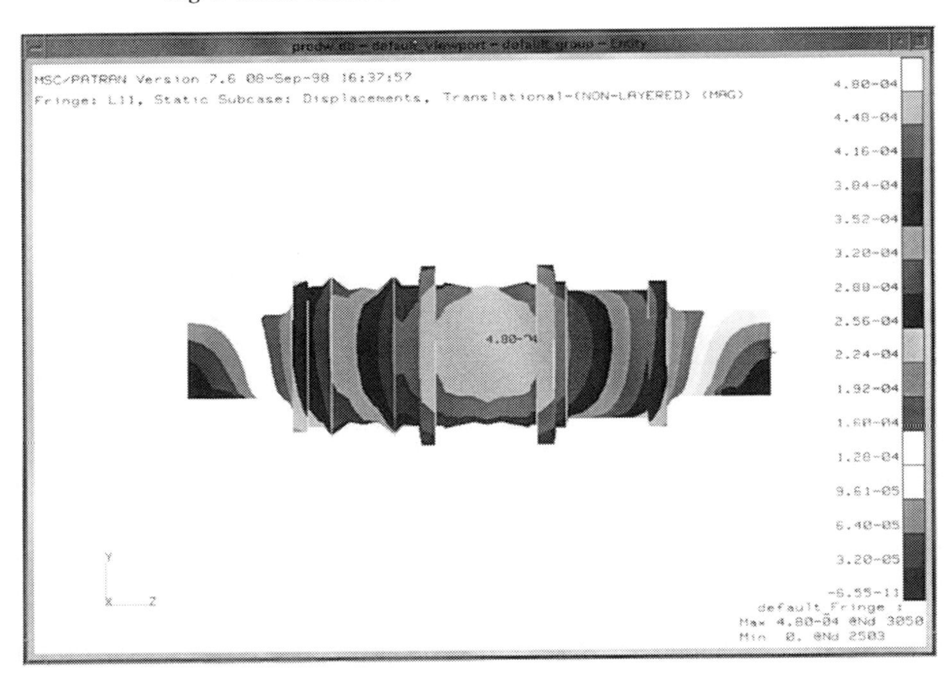

Fig. 3 Stress distribution in cross-sections of a section roll after its regeneration

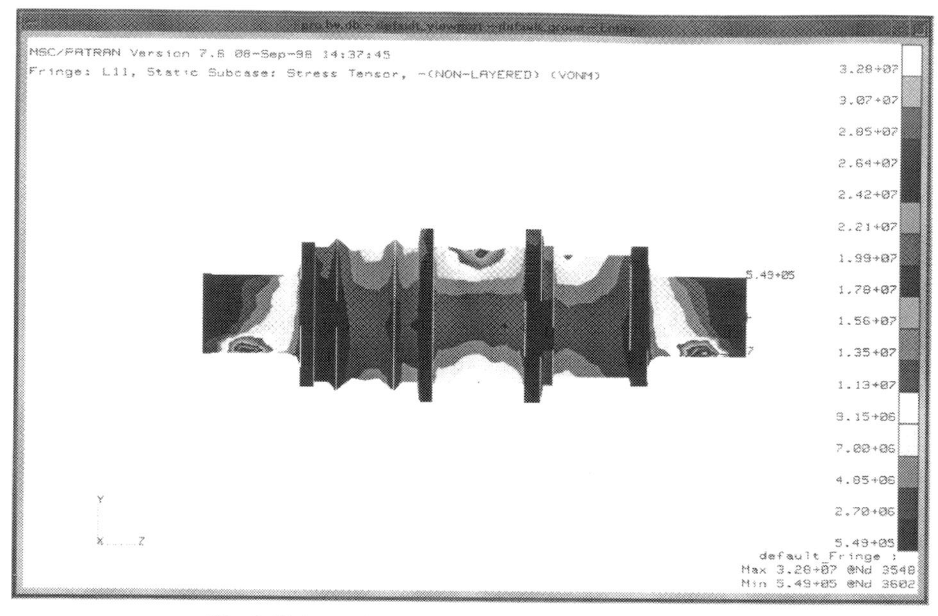

Fig.4 Values of elastic strains for a new section roll

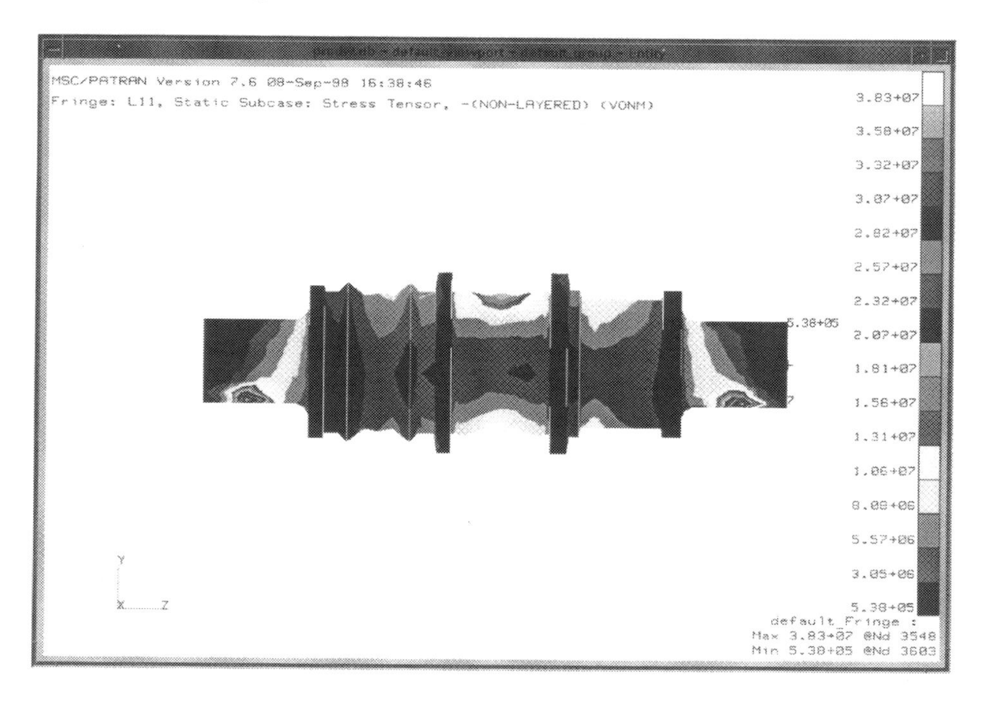

Fig. 5 Values of elastic strains for a section roll after its regeneration

3. CONCLUSIONS

Hitherto existing results obtained in rolling mill operation indicate that optimisation of the roll pass shape achieved as a result of calculation analysis has reduced the wear of the rolls by 7% on average at relatively small expenditure - when faced with the savings provided - connected with installation of the equipment platform and the software.

4. REFERENCES

(1) Zienkiewicz O.C., Taylor R.L., The finite element method Vol.1, Vol.2, McGraw - Hill Book Comp., London 1991

(2) MSC/NASTRAN (7.0 version), Theory Manual

Rapid Prototyping

CAPE/027/2000

Fractal scanning path for selective laser sintering

X ZHANG, H BIN, Z LIU, and **H GAO**
School of Mechanical Science and Engineering, HuaZhong University of Science and Technology, Peoples' Republic of China

Abstract

This paper reports on the study of scanning path in Selective Laser Sintering (SLS) process. SLS is a thermal process that uses a laser to sinter the powder layer to form solid dimensional components according to the contour of them. It is known that the scanning path is one of the important factors which affect the performances of final sintered components. So searching for certain paths which are suitable for SLS is of important. In this paper, a new kind of scanning path, the fractal scanning path is proposed for SLS. In order to compare the results of the different scanning paths, a 2-dimension scanning table without backlash is made and the CNC system is also developed. Here, attentions are focused on three main problems. First, the density of the components which are sintered with SLS is always low. Second, the sintering process may cause stress concentrations that will make some cracks in the components. Third, the shrinkage of the component after it was cooled down will greatly affect on the accuracy. Due to these problems, work is currently being undertaken. The influence of the scanning paths on the temperature field of the layer in SLS is researched. And experiments are done on the experimental device, several square parts made by resin and sand powders are sintered both by fractal scanning path respectively. Base on these, some conclusions are drown in the paper.

1 Introduction

Selective Laser Sintering (SLS) is a widely accepted rapid prototyping technique. It incorporates several technologies including laser, CNC machine, material and CAD/CAM. With SLS, it is feasible to directly produce special parts which can not be done with traditional tooling. However the parts sintered by SLS exhibit a stair-step surface texture and it is difficult to achieve the full density, the accuracy which can be obtained with traditional machining. In addition, the residual stress through thermal gradients during layer sintering can lead to warpage. In our experiences, the scanning path of SLS process is one of the important factors which influence on the final performance of the sintered parts, so an new scanning path called fractal scanning path for SLS process is proposed. Comparing with the traditional "S" scanning path, the fractal scanning path and its influence on the temperature field and residual stress of the layer in SLS process are described. Further, a equipment for fractal scanning path is set and examples are presented.

2 Fractal scanning path

The slice of the part to be produced possesses a 2-dimension outline. Usually, in SLS process the 2-dimensrion layer is made by laser following "S" line scanning path(see Figure 1). While the "S" line scanning path is easy for implementation with SLS, it remains some limitations which affect the performance of the sintered component. The long line sintering path results in the large shrinkage which affects on the accuracy of the sintered part, the poor combination between lines affects the mechanical strength. In addition, the residual stress through thermal gradients during laser sintering can lead to warpage and cracks. To address these issues, the fractal scanning path is introduced in SLS process. The fractal curve has its special characteristics, such as fine structure, iteration by itself, similarity itself (see Figure 2). Such fractal curve is developed from the unit R0 and iterated according to the rules:

$$L_{n+1} = +R_n F - L_n F L_n - F R_n +$$
$$R_{n+1} = -L_n F + R_n F R_n + F L_n -$$

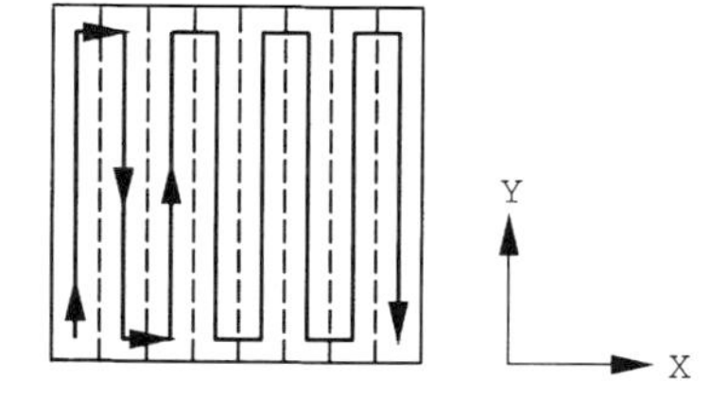

Figure 1: " S" scanning path

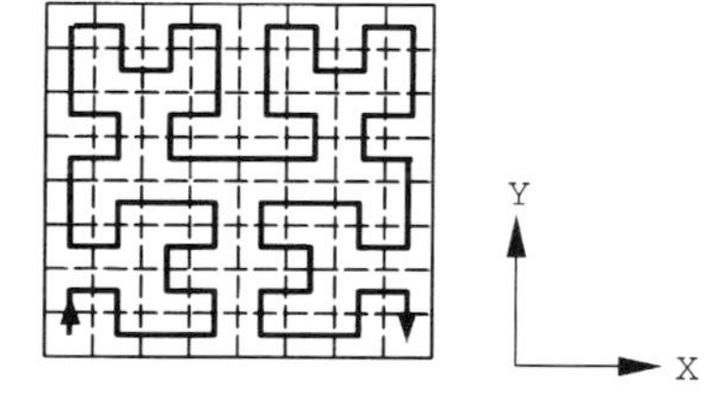

Figure 2: fractal path

Intuitively obvious, comparing with the "S" scanning path there is no clear separation between the lines in the fractal path, it is good to improve the density and mechanical strength. In addition, the fractal path is composed with short lines instead of long lines, as a result, the residual stress is decreased and the shrinkage of the sintered part is also decreased. Comparing the length of the lines in X direction with that in Y direction in the "S" path, it is obvious that the Y direction ones is longer than the X ones, so the shrinkage is not equal to each other in the two directions. While in the fractal path, the length of Y direction lines is almost equal to that of X direction one, the shrinkage in two direction are almost same, so the accuracy is improved.

3 Discussion of Thermal and Stress Issues

In SLS process, the powders are heated and melted by the laser beam, and combinated each other. After cooling down, the solid layer is present. Temperature gradients in SLS are always associated with internal stresses and distortions. Along the sintering line, some powder materials have cooled down, some are heating and some are melting. After all portions have cooled down and converged to the same temperature, in some area the powders are compressed and in other area extend. So that, a quantitative understanding of temperature gradient in layer manufacturing with SLS is an important issue to obtain layer's performance.

However, it is difficult to measure the temperature field in the SLS process, a computer simulation method is introduced. The temperature field of the layer in scanning sintering process is calculated with finite element method (FEM). Based on the following heat transfer model, some supposes are proposed for temperature field calculation in scanning sintering.

$$\frac{\partial}{\partial x}(\lambda\frac{\partial T}{\partial x})+\frac{\partial}{\partial y}(\lambda\frac{\partial T}{\partial y})+S=\rho c(\frac{\partial T}{\partial t})$$

Where "c" is the specific heat capacity, "•" is the coefficient of heat conduction, "S" is the heat source, "T" represents the temperature and "t" is the time. The calculated results of the layer's temperature field distribution under "S" line scanning path and fractal scanning path are shown by Figure 3 and Figure 4 respectively.

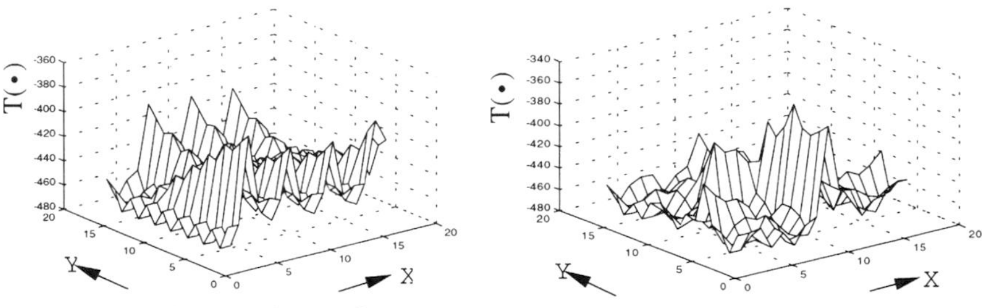

Figure 3:"S" scanning path temperature field

Figure 4: fractal scanning path temperature field

The results indicated that the temperature field under fractal scanning path is more smooth than that under "S" scanning path. It also means that the fractal scanning path can minimize temperature gradients during laser sintering process, and according to temperature gradients, the residual tension stresses are decreased.

4 Experimental

In order to compare the results of the two kinds of scanning path, several experiments are carried out on the self-developed experimental device which is designed and manufactured with high accuracy, small backlash, small weight and rapid response, and controlled by a 2D CNC system. This equipment is used to implement the SLS process in "S" scanning path and fractal scanning path. The material for sintering is the mixture of resin and sand powders. In the SLS process, the laser beam heats the resin and sand powders, and the resin powders are melted and then sand powders adhere to each other. After cooling down, the solid parts are produced. As the post processing, all these parts were heated under 250• condition and kept for a half hour. After that, the parts are tested on the compressive stress testing instrument individually. The instrument records the maximum of the compressive stress of each part before it is broken. Testing data of these parts compressive stress is shown in Table 1.

Table 1: Compressive stress results for the two scanning paths

Scanning path	Test 1 compressive stress [MPa]	Test 2 compressive stress [Mpa]	Test 3 compressive stress [MPa]	Average compressive stress [MPa]
"S" path	2.22	2.10	1.99	2.103
Fractal path	2.54	2.46	2.58	2.526

The average compressive stress of the parts sintered with fractal path is specified at 2.526 Mpa, hence 2.103 MPa of the parts with "S" path, it is higher about 20%.

5 Conclusions

The difference between the fractal scanning path and "S" line scanning path for SLS is discussed in this paper. Comparing with path structure, temperature gradient and residual stress, it is indicated that the fractal path is better than the "S" line path in SLS process. The computer simulation of the temperature gradient field has also got the same conclusion. A test facility for SLS process is developed and several parts have been sintered. The implementations of the facility and the sintered parts have demonstrated the feasibility of fractal scanning path. The compressive stress experiments show that the parts built with fractal path is stronger than the "S" path ones. However, several issues must be addressed to realize the high performance SLS parts. These issues include preheating the powders, a closed loop control system, laser power control and different fractal scanning paths. The influences of these issues must be further investigated.

References

(1) J.P.Kruth. etc. Material Increase Manufacturing by Rapid Prototyping Techniques, Annals of the CIRP, 1991

(2) Merz, etc. Shape Deposition Manufacturing, Proceeding of the Solid Freeform Fabrication symposium, The University of Texas at Austin, 1994

CAPE/087/2000

Computer technology in design of hip joint endoprostheses

K SKALSKI, M J HARABURDA, K KĘDZIOR, M BOSSAK, and **J DOMAŃSKI**
Institute of Mechanics and Design, Warsaw University of Technology, Poland

ABSTRACT:

This paper presents the design process in orthopaedic engineering using the example of an endoprosthesis of the hip joint. This computer-aided process is carried out utilising integrated techniques connected with the transformation of tomographic images (CT), the geometric modelling of bones and endoprostheses (CAD), and the verification of constructions using rapid prototyping technology (RP).

1. INTRODUCTION

In recent years, CAD/CAM computerised designing and manufacturing systems have become powerful and useful engineering tools. These systems, whose development and application were stimulated by the aeronautical and automotive industries, have made inroads into other disciplines as well, including that of bioengineering. The design and manufacture of endoprostheses (e.g. of the hip) is one such example of the application of these systems (6, 7, 10).
In orthopaedic engineering, the development of effective treatments for diseases and defects of the organs of locomotion has been achieved through alloplasty (4, 5). In particular, some problems are corrected through:

- the design of artificial joints such as hip, elbow, knee, ankle, etc.
- the development of materials ensuring the biological tolerance of implants

- engineering analysis of designed constructions with the aim of ensuring appropriate resistance and durability
- the development of manufacturing technology of endoprostheses of high functional utilitic

Anatomical endoprostheses adapted to individual patients' needs are geometrically complex structures. The designing and manufacturing of such structures in a computerised system requires a multi-stage design-manufacture process (3). In the first stage, normally undertaken after x-ray diagnosis using computer tomography (CT) (1), medical images are produced (having high resolution and precision of the reconstructed contours) and entered into a CAD system (9). In the second stage, after a 3D geometric model of the bones of the joint has been created, an endoprosthesis is designed to fit them. In the third stage, this construction is subjected to strength analysis in order to optimise its performance with regard to the applied loads and the durability of the materials used. In the fourth stage, the form of the designed enodprosthesis is evaluated and operating surgical procedures are pre-tested utilising the prototypes models of endoprosthesis and patient's femur which have been manufactured. Such prototypes , which can be quickly produced using stereolithographic technology, (2, 13) make it easier for the physician to make certain decisions, for example, where to make the resection of the essential bone parts such that fixation of the implant during surgery will be easy. In the fifth stage, a metal endoprosthesis is produced on a CNC (11, 12)machine tool according to the elaborated treatment program.

Following is a detailed look at the design process of a hip joint endoprosthesis, and at the manufacturing of a prototype, from computer tomography through computer-aided design to rapid prototyping, (Fig.1).

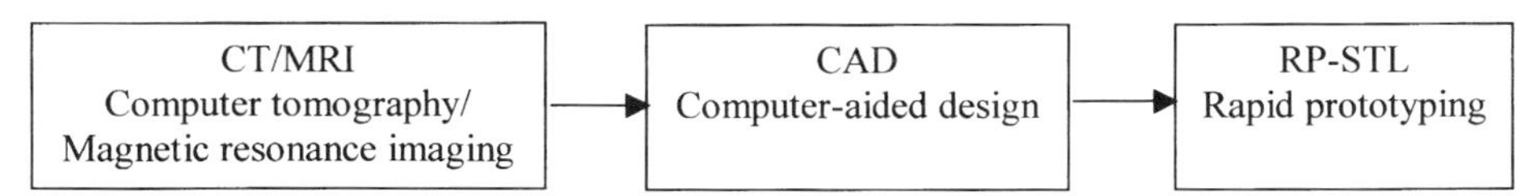

Fig. 1 Functional connection in the process of computer-aided design

2. IMAGE TRANSFORMATION

Inseparable from computer tomography is the problem of reconstructing images from the projection. Visualisation and interpretation of the graphic data obtained from the x-ray tomography (in our research a Siemens SOMATOM HiQ tomograph was used) is of critical importance, as it is the basis of further work towards the construction of a geometric model of the bone tissue (dense and porous) and soft tissue. In order to determine the relationship between tissue density and the value of the attenuation coefficient, the Hounsfield coefficient, C_T, was introduced. The value of the coefficient defines the relationship:

$$C_T = \frac{\mu_{obj} - \mu_{H_2 0}}{\mu_{H_2 0}} * 1000 \quad [HU];$$

Where: $\mu H_2 0$ is the attenuation coefficient of x-ray radiation for water, μObj the attenuation coefficient of the object under study, and C_T (Hu) the Hounsfield coefficient expressed as a number of units (Hu). The Hounsfield scale, composed of coefficients of tissues studied, contains 4096 degrees (units), from –1024 to 3071 (Hu). It is given that the attenuation coefficient of radiation for air is –1024, and for water, 0. The Hounsfield coefficient for

porous bone ranges from about 100 to 600, and for dense bone to 1000. For the visualisation of areas, which are cross-sections through the object of study, written in digital code in the form of a bitmap, the computer tomograph, depending on the value of the C_T coefficient, assigns to each bitmap pixel one of 256 shades of grey. Light shades indicate high Hounsfield numbers, dark shades low numbers.
Treatment of the tomographic data with the help of a TOMOCOMP system begins with the selection of those images which will serve in creating the model. Next, segmentation is carried out, during which bone tissue is separated out from the tomographic images according to C_T value (Fig. 2). Segmentation can be undertaken by manually specifying regions of interest, or automatically, through a previously established upper and lower limit of the C_T coefficient. Transverse bone sections of the femur are shown in Fig. 2a.

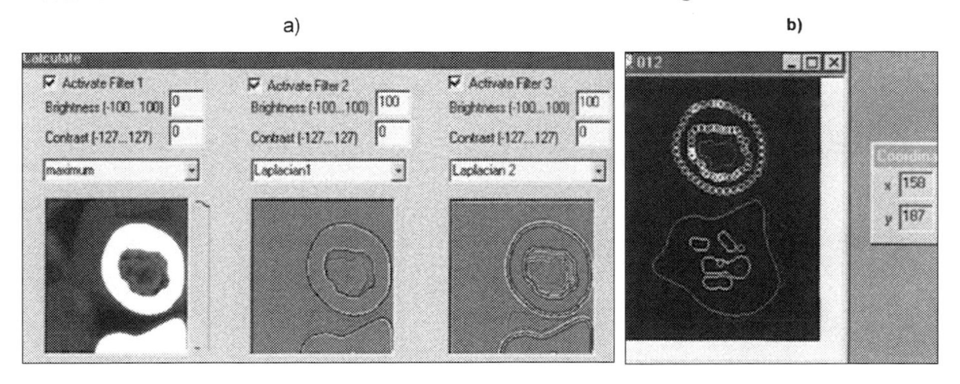

Fig. 2 Segmentation and contours detection (a) and coordinates definition (b).

In the next stage of tomographic image transformation, detection of edges in specific regions of the bone is carried out. It is necessary to determine the threshold value of the C_T coefficient. The range of C_T coefficient values for bone (dense and porous) is quite wide, and bone elements have varying density throughout their structure. The problem of selecting the threshold value is of particular importance in identifying the form of the marrow cavity of the femur, surrounded as this is by porous bone. An incorrect coefficient can lead to incorrect detection of the interior contours of the bone, resulting in a faulty endoprosthesis design. Fig. 3 presents two different tomographic sections of the femur, in the so-called distal region (Fig. 3a, the almost round contour) and in the proximal region (Fig. 3b, the roughly elliptical contour). These illustrations show the character of change in the value of the C_T (Hu) coefficient in areas of dense bone, porous bone, and soft tissue (the marrow cavity). The smaller the change in the attenuation coefficient value at the tissue border, the greater the influence of the threshold value on identification of the bone tissue. Changes in the coefficient value in regions of bone tissue have only a mild effect on the interior of the femoral marrow cavity. In the cross-section of the bone, changes in the C_T (Hu) coefficient with different degrees of flattening in the peak of the curve are clearly visible. This means that in these areas, such as at the edges of tissues, measurement error will be greatest.
The result of bone tissue detection is generation of the edges of the bones and of the marrow cavity. These contours are closed, and separately defined for each scan layer Z, as a set of X, Y coordinate points (Fig. 2b). The data set thus generated is then written in Common Language Interface (CLI) format, in which each scan layer is defined on the basis of the given level at which it is situated. The X, Y coordinates lying in a given layer are determined.

This format allows for two versions of the coordinates: as continuous text, or directly, in binary form.

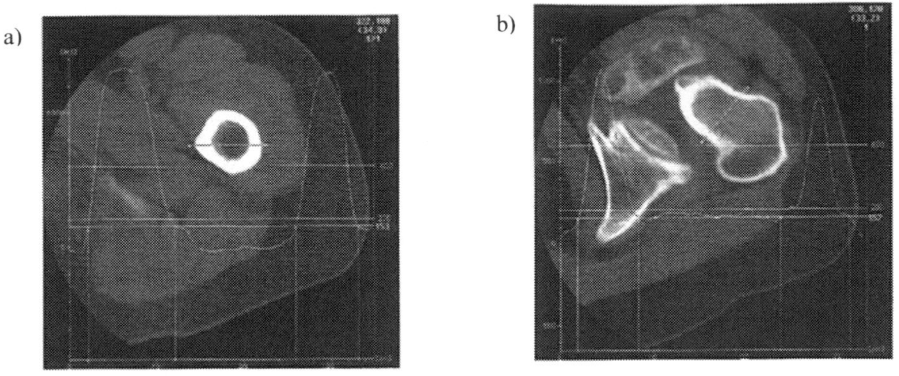

Fig. 3 Distribution of threshold values of C_T coefficient in the femur: a) distal, b) proximal

3. MODELLING IN A CAD SYSTEM

Currently, endoprostheses are designed in type-series for application in alloplasty of the hip joint. In the case of some patients, however, an individualised approach is necessary when choosing an implant. The criteria of selection should take into account the characteristic dimensional and formal features of the patient's joint, and any changes which have occurred in it as a result of disease, trauma or pathology. An example of such changes is shown in the tomographic images of Fig. 4.

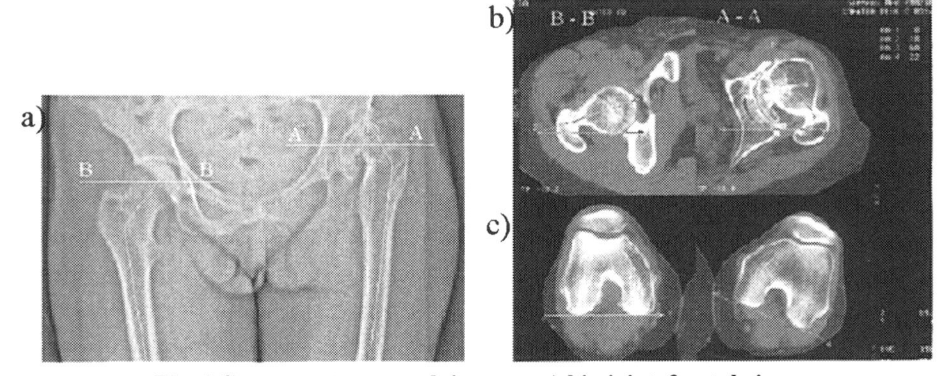

Fig. 4 Computer tomograph images, a) hip joint, frontal view
b) transverse sections of the hip joint A-A and B-B
c) section through both knees at the condyle

These images clearly show significant pathology of the patient's hip joint. The left joint, in comparison with the right, is shifted significantly upward (Fig. 4a). Also noticeable is a significant deformation of the head of the femur.

Fig. 4b shows cross-sections A-A and B-B of the joints (from Fig. 4a). The left joint, as well as having an altered form of both head and neck, displays rotation of the femur. A quantitative description of these changes, the so-called angle of anteversion, allows for a description of bone rotation (condyle) in the region of the knee joint, as presented in Fig. 4c.

3.1 Modelling the femur

The modelling process is carried out using a ProENGINEER or Unigraphics system. It reads out the characteristic points, then describes the dimensions and form of the interior and exterior surfaces of the bone. To produce a solid model in a CAD system (which will be the basis for producing a prototype), it is essential to have a description of the geometric elements of the model's curves and surfaces (8). These can be approximated through one of the selected parametric functions: Beziera, B-spline, NURBS.
In the TOMOCOMP system, data are obtained, as is known, as CLI text files. After reading out the data, it is essential to select the layers to be used in building the model. Designing on the basis of points from each and every layer is time-consuming and unnecessary. In practice, layers are used which are spaced about 9mm apart in the distal part of the bone, and about 6mm apart (or exceptionally 3mm) in the proximal femur. The sets of points obtained differ in their relation to the accepted threshold value of the Hounsfield coefficient (Fig. 5). In the distal part of the bone, changes of form and size of the curves produced from the points are small, and show that a higher coefficient causes reduction of the exterior contours of the bone, and an enlargement of the interior contours connected with the marrow cavity. In the proximal part, changes in form and dimension are significant, and result mainly from the difficulty in identifying the border between dense and porous bone tissue. It often happens that only by comparing the form of curves for various theshold values of the Hounsfield coefficient C_T(Hu) can sharp contour curves be generated.

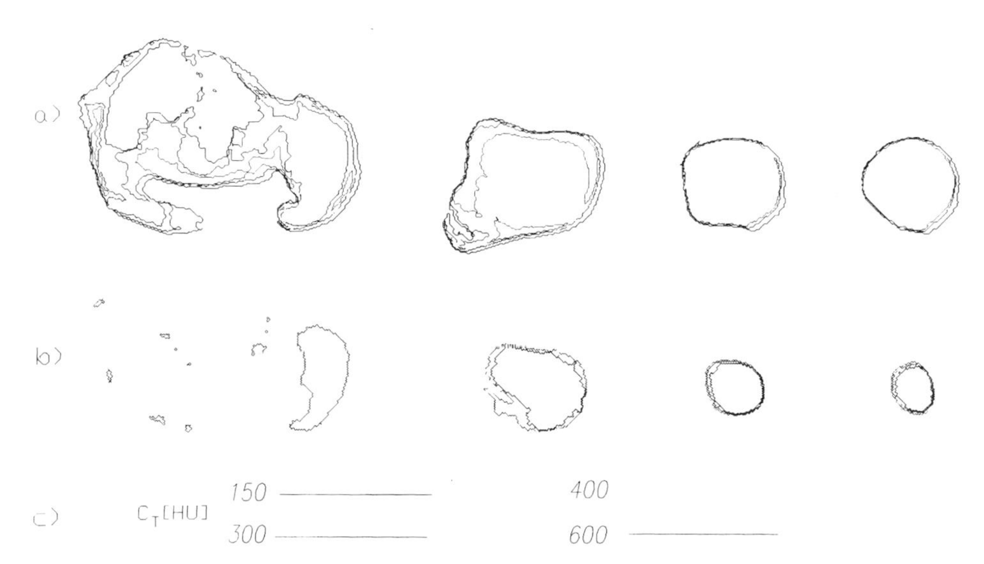

Fig. 5 Edge contours, a) exterior and, b) interior of the femur depending on chosen c) value of Hounsfield coefficient

Modelling the curves from the measurement points makes it possible to model the surface. The precision of surface models can be defined by assigning a parameter of deviation of the location of the surface from the measurement points. Surface models make it possible, in the next stage of the design process, to create solid models.

3.2 Designing of the Endoprosthesis

Beyond other requirements, prosthesis design should meet two main engineering criteria:

1. to model as faithfully as possible the form of the marrow cavity (enabling the bone to be filled to the maximum)
2. to enable the endoprosthesis to be inserted into the bone

Fulfillment of these criteria normally causes that in the parts of the bone nearer the trochanter minor, the endoprosthesis is fully fitted to the bone. In the distal part, the fit is less tight, and results from the accepted principle that each successively lower transverse section of the prosthesis must fit within the contour (Fig. 6) of the transverse section of the bone (viewed in the direction of insertion of the endoprosthesis).

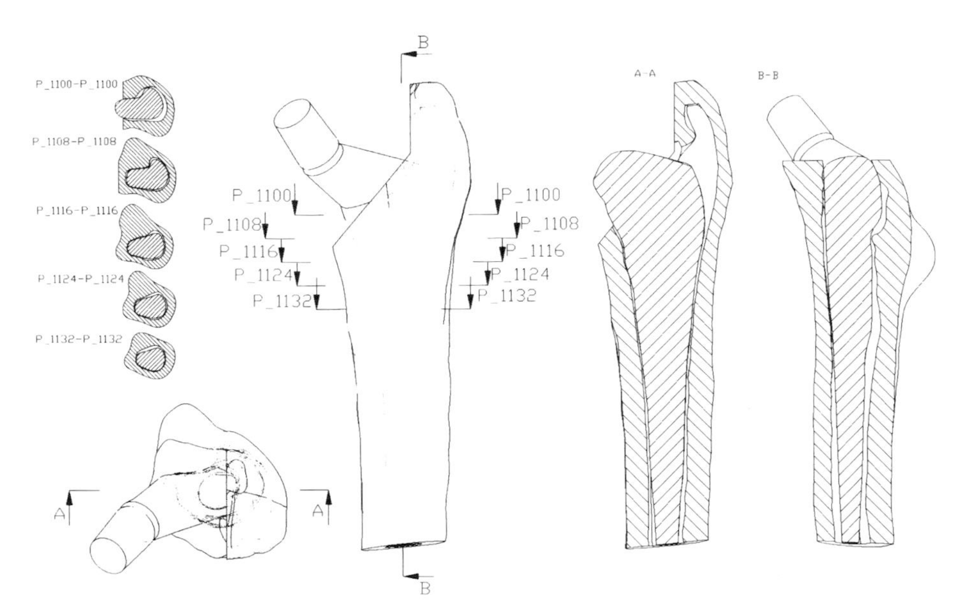

Fig. 6 Sections (A-A, B-B) through proximal-distal parts of the femur and the designed endoprosthesis. Sections P-P show filling of the bone by the prosthesis.

Design of the endoprosthesis is performed in successive steps by:

1. defining the curves describing the dimensions and form of the bone (after previously establishing the degree to which the marrow cavity is to be filled, the method of fixation, as well as the possibility of introducing the endoprosthesis into the marrow cavity)
2. filling out the created surface curves
3. defining the geometric dimensions and angles positioning the endoprosthesis neck axis, as well as designing the cone
4. creating a solid model (Figs. 7a,b,c)

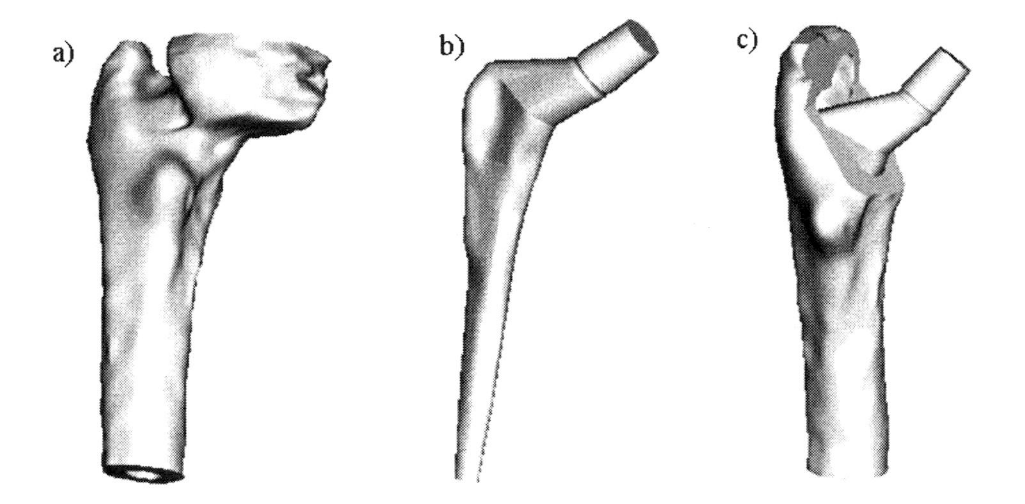

Fig. 7 Solid models: a) femur, b) hip endoprosthesis, c) implantation of the endoprosthesis in the bone

4. MANUFACTURING STEREOLITHOGRAPHIC PROTOTYPES USING THE RAPID PROTOTYPING TECHNOLOGY

Manufacturing a model prototype of the implant and of the femur – a solid of complex exterior and interior form – is time-consuming and expensive. These factors can be minimised by modern technologies developed for rapid production prototypes. One such technique is stereolithography (STL), belonging to the group of processes known as rapid prototyping (RP).

A stereolithographic model of the femur and a prototype of the endoprosthesis stem were created on a 3D Systems SLA250 stereolithographic machine. In the course of carrying out the stereolithographic process (Fig. 8), the model produced (1) is placed in a chamber (3) filled with a light-hardenable liquid. This liquid polymer changes its phase under the action of a laser beam. The model is placed on a platform (2) which moves vertically. In the stereolithogaphic process, objects are created layer by layer. Each individual layer is hardened by the light emitted from a helio-cadmium laser (4). The beam is guided by a mirror (5) controlled by the computer governing the operation of the stereolithographic machine. After execution of a single layer, the work table, along with the object, it moved downwards in an amount corresponding to the thickness of the next layer of the model. For an SLA250 series stereolithographic machine, the minimum layer thickness is 0.1mm. Stereolithographic prototypes of the femur and endoprosthesis are shown in Figs. 9a,b.

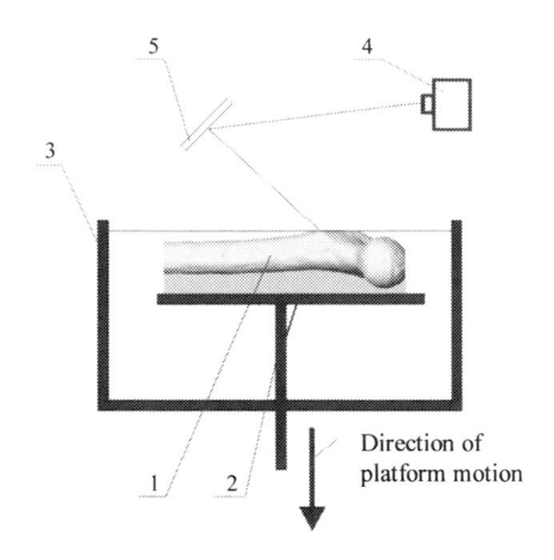

Fig. 8. Idea of stereolithography.

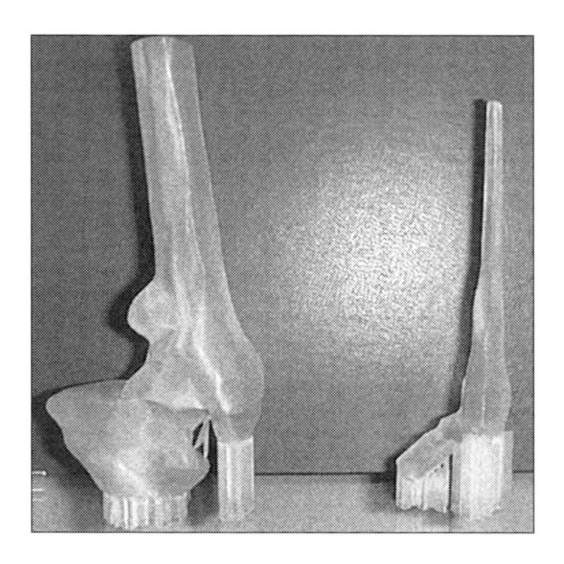

Fig. 9 Stereolithographic prototypes from synthetic resin femur and hip endoprosthesis

The process of manufacturing a prototype using the stereolithographic method can be divided into the following stages:

1. creation of a geometric model in a CAD system (as described above)
2. entry of the geometric model in .stl or .slc format

3. preparation of the model using the program MAESTRO for the purpose of machine manufacturing, comprising:
   a) verification of the correctness of data, and entry in .stl or.slc format
   b) orientation of the model in the program VIEW (part of MAESTRO) through:
      - definition of the minimum distance between the first layer of the model and its supporting base
      - minimalisation of closed volumes such as those areas in the model which will hold resin during movement of the platform
      - minimalisation of the number of oblique surfaces with attention to the so-called stepping effect
      - maximalisation of the number of complex sections in the X-Y plane with attention to greater distribution
      - ensuring that the model or models will fit into the chamber
   c) creation of the support (in the program VISTA) which will play the role of:
      - distinguishing the object from its base and enabling the model-prototype to be easily separated from the base
      - fixing the object to its base (platform)
      - supporting those parts of the object which require it
      - preventing deformations of the article during its creation
   d) generation of files guiding the work of the machine
4. production of a prototype of SL5170 epoxy resin
5. final treatment of the stereolithographic prototype:
   a) cleaning away fluid resin and removal of the supporting base
   b) final hardening in a separate chamber
   c) final touches such as grinding, polishing, painting, etc.

5. CONCLUSIONS

1. Computer-aided design of endoprostheses in orthopaedic engineering involves integrated techniques related to the transformation of medical images (CT), the geometric modelling of the femur and the endoprosthesis (CAD), and verification of the construction by means of rapid prototyping technology (RP). Integration of these systems requires the application of suitable formats which transferring large files of data between systems.
2. Analysis of computer tomography image files points to the critical influence of the threshold value of the Ct(Hu) coefficient on the resolution of contours, and thus on the correctness and precision of the designed endoprosthesis.
3. Among the many techniques available for describing curves in the modelling process of such complex objects as bones and endoprostheses, the NURBS method seems to be the most suitable.
4. RP technology is also effective, allowing for the creation of useful prototypes of bones, and of endoprostheses designed to fit them. Such prototypes enable a surgeon to plan an alloplastic operation by manually fitting the prototype before surgery. The design-manufacturing process requires close cooperation between engineers and physicians, in effect leading to integration of those communities.
5. A future goal of the methodology of design in orthopaedic engineering is to include other techniques of object identification: MRI (Fig. 1), the Coordinated Measurements Method (CMM), or laser scanning. Though not considered in this paper, one important problem seems to be the precision of stereolithographic prototypes obtained by the hardening of epoxy polymer resins.

REFERENCES

[1] Berman A.T., McGovern K.M., Paret R.S. Yanicko D.R.: The use of preoperative computed tomography , Clin. Orthop., No 222, 190, 1987.

[2] Bossak M., Skalski K., Święszkowski W., Werner A.: Application of rapid prototyping to the design process of endoprostheses., Conf. Proc. on Computer Methods and Systems in Scientific Researches and Engineering Design, Kraków, 1999; 289-294.

[3] Deitz, D.: Customer – Driven Product Delivery, Mechanical Engineering, v. 117, No12, 1995, p. 72-80.

[4] Huo M. H., Salvati E. A., Lieberman J. R., Burstein A. H., Wilson P. D: Custom – designed Femoral Prostheses in Total Hip Arthroplasty Done with Cement for Severe Dyplasia of the Hip, j. Bone Joint Surg., v.75-A, No10, 1993, p. 1497-1504.

[5] Murphy S. B., Kijewski P. K., Simon S. R., Chandler H.P., Peneberg B. I., Landy M. M.: Computer – aided simulation, analysis and design orthopaedic surgery, Orthop. Cli. North. Am., No17, 637, 1986.

[6] Reuben J. D., Chi-Chan C., Akin J. E., Lionberger D. R.: A Knowledge - Based Computer – Aided Design and manufacturing System for Total Hip Replacement, Clinical Orthop. Related Research, No 285, 1992, p. 48-56.

[7] Robertson D. D., Walker P.S., Granholm J.W., Nelson P.C., Weiss P.J., Fishman E.K., Magid D.: Design of custom hip stem prosthesis using 3-D modelling, J. Comp. Assist. Tomography, No 11, 804, 1987.

[8] Sarhar B., Meng C. H.: Smooth surface approximation and reserve engineering, Computer Aided Design, 1991.

[9] Smolik W., Brzeski P., Kędzior K., Skalski K., Szabatin R., Święszkowski W.: Tomographic Image Processing for Geometrical Modeling in CAD System., Mater. XVth World Congress on the Theory of Machine and Mechanisms, Oulu, Finland 1999: 1871-1876.

[10] Stulberg S.D., Stulberg B.N., Wixson R.L.: The rotationale , design characteristics and preliminary results of a primary custom total hip prosthesis, Clin. Orthop., 249, 79, 1989.

[11] Werner A., Skalski K., Kędzior K., Kukiełko M.: A method of the human hip joint endoprosthesis manufacture using numerically controlled machine tool, Proc. 15th Polish Conf. On the Theory of Machines and Mechanisms, Białystok – Białowierza, 1996, p. 522-530.

[12] Werner A., Skalski K., Piszczatowski S., Święszkowski W., Lechniak Z.: Reverse Engineering of Free – form Surfaces, J. Mater. Process. Techn., v. 76/1-3, 1988, p. 128-132.

[13] Werner A., Skalski K., Święszkowski W., Plewicki J.: Application of stereolithography and designing and manufacturing of the custom-made endoprostheses of the hip joint., Acta of Bioengineering and Biomechanics 1999; Vol.1, Supl.1: 547-550.

CAPE/089/2000

The effect of mould heating on the accuracy of rapid tooling

P DUNNE and **G BYRNE**
Department of Mechanical Engineering, University College Dublin, Ireland

ABSTRACT

The inherent technical limitations of rapid prototyping systems have lead to the development of rapid tooling techniques. These techniques allow manufacture of prototypes in final materials by the final manufacturing process. Enhanced silicone moulding is one such process that provides both a rapid and low cost route to the manufacture of prototype injection moulded plastic components. During the moulding cycle, the mould deforms due to the high pressures and temperatures experienced. This results in the manufacture of dimensionally inaccurate components. The main contributing factors have been identified as global expansion due to heating and local deformation due to injection pressure. This paper reports on the development of a model to predict the effects of mould heating. Two routes are reported, one using conventional analytical techniques and one using an approximated finite element technique. The validity of the models are assessed by comparison with experimental results.

1 INTRODUCTION

Advances in materials and processing methods within the injection moulding industry have allowed components of increasing complexity to be designed and manufactured (1). Associated with such components are equally complex moulds whose manufacture is both time consuming and expensive. It is therefore vital that thorough design evaluation is undertaken before commitment is made to expensive production tooling. Typically the component design is evaluated for performance (e.g. form, fit, function) and manufacturability (e.g. draft angles, split line). This evaluation is often best achieved through use of physical prototypes.

While the application of machining and rapid prototyping techniques to prototype manufacture is established, the prototypes produced do not have the same materials and process characteristics as moulded components (2). In order to overcome these limitations rapid tooling techniques are applied. These techniques involve manufacture of a low cost mould from which prototypes are produced.

The most widely used rapid tooling technique is vacuum casting as it is capable of providing low cost replicas in relatively short time periods. The process involves the manufacture of a mould by casting silicone rubber against a master pattern. This mould is then used in the manufacture of prototypes by vacuum casting polyurethane materials.

2 ENHANCED SILICONE MOULDING PROCESS

Despite being widely employed as a rapid tooling technique, vacuum casting does not provide a complete solution to the materials and process limitations of rapid prototyping and machining techniques (3). Low mould material stiffness restricts the range of materials that can be processed to those which can be gravity cast, polyurethanes for example. The pressures required to inject engineering thermoplastics would deform the silicone mould (4). Consequently the process is only capable of producing prototypes in replica materials by a low pressure casting rather than an injection moulding process.

To overcome these limitations an enhanced silicone moulding process is used (5). The component geometry is first used to manufacture a master pattern by rapid prototyping. The CNC machining of a master pattern is also feasible. During the mould manufacture stage metal powder filler is added to the silicone rubber which is then cast against the master pattern. The metal powder results in improved material stiffness. Several powder types have been used including aluminium oxide and iron powder with typical constituent ratios in the order of 50% wt. rubber and 50% wt. powder. Compression tests carried out on silicone rubber impregnated with 50% wt. 10 μm iron powder in accordance with BS 903 Part A4 have shown an average increase in stiffness of 38% across the full range of strains. Hardness tests showed an increase in hardness from 39 Shore A to 63 Shore A for the same material (6). Finally the mould is used in an injection moulding process to produce prototype components. The main stages in the mould manufacture process are shown in figure 1.

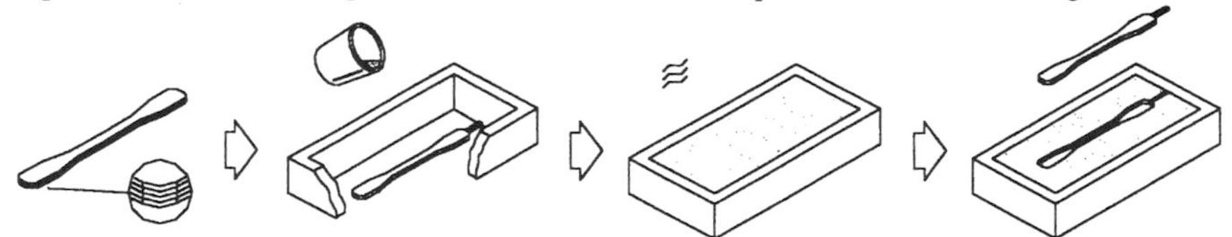

Fig. 1 Mould manufacture stages, (L-R) rapid prototype master pattern, material casting, material curing and pattern removal

While early process trials demonstrated that the injection moulding of thermoplastic prototypes using enhanced silicone moulds was possible, further investigation highlighted limitations regarding the dimensional accuracy attainable. Consequently an investigation of process accuracy issues has been undertaken.

3 PROCESS ACCURACY

3.1 Factors Affecting Dimensional Accuracy

The dimensional accuracy of any moulded component is dependant on the accuracy of the mould and the moulding process. In the enhanced silicone moulding process the mould is produced by casting against a master pattern. Therefore, the accuracy of the master pattern is critical. If the pattern is produced by rapid prototyping then pattern accuracy can be significant. For example, the selective laser sintering and stereolithography processes have claimed accuracies of ±0.05mm and ±0.1mm respectively (3). The effects of material shrinkage and swelling during the material curing stage can also impact on mould accuracy.

When the mould is used to produce components further inaccuracies will arise. Perhaps the most common is shrinkage of the injected material (8). Shrinkage, which can be in the order of 15% for some materials, can be compensated for through manufacture of an oversized cavity (7). Other common sources of dimensional inaccuracy which arise are warpage, bowing and ejection distortion. These effects can be largely controlled through variation of pressure, temperature and time parameters. In the enhanced silicone moulding process further sources of error arise as a result of limiting material properties, in particular, thermal expansion and stiffness.

Enhanced silicone rubber has a higher expansion rate than other commonly used mould materials such as steel. The coefficient of linear expansion of enhanced silicone rubber is 298×10^{-6} /K compared with 11×10^{-6} /K for steel (9). When the mould is heated, as is required to control the rate of cooling of the injected melt, the enhanced rubber insert expands at a greater rate than that of the metal mould frame within which it sits. This results in a bulging effect and the inevitable deformation of the cavity geometry. When the mould is closed during the clamping stage further deformation results. The amount of distortion experienced depends on many factors including mould temperature, cavity geometry and clamping pressure.

During the next stage of the moulding cycle, injection, the mould is subjected to the high pressures which are required to force the molten plastic into the mould. These pressures can range from 3.5 MPa for free flowing materials to 138 MPa for highly viscous ones (7). The mould material must therefore posses high stiffness properties. For metal moulds this is rarely a problem, however, for some rapid tooling solutions, including enhanced silicone moulding, material stiffness can be inadequate. The pressure of the injected melt results in compression of the rubber mould insert and production of components which are thicker than intended. This is referred to as thickening. The degree of thickening within a component depends on the melt flow index of the injected material, mould configuration and the component geometry. These two deformation modes are illustrated in figure 2.

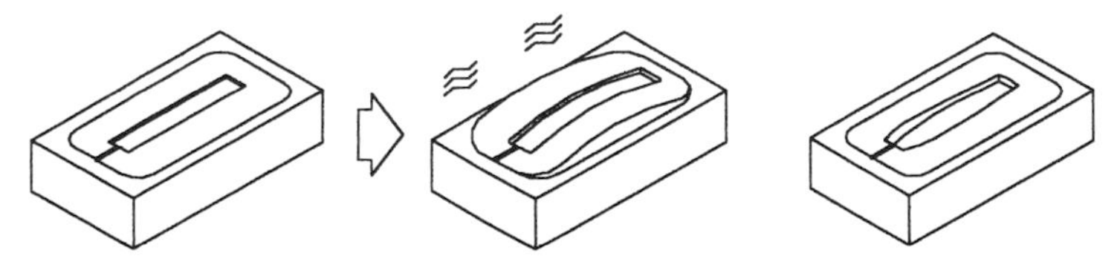

Fig. 2 Deformation due to thermal expansion (middle) and injection pressure (right)

3.2 Process Trials

In an effort to understand and identify the dominant factors which contribute to process inaccuracy, a series of process trials have been undertaken. Two moulds were manufactured, one with a tensile test cavity and one with a simple rectangular cavity. The first mould, made of 50% wt. silicone rubber and 50% wt. aluminium oxide powder (400 μm) was used to injection mould acrylonitrile-butadiene-styrene (ABS) components. The second mould was produced using 50% wt. iron powder (10 μm) as filler in the mould and low density polyethylene (LDPE) as the injected material.

Prior to mould manufacture the master patterns were measured for dimensional accuracy. The subsequent mould cavity produced was also measured. The moulds were then used to injection mould prototype components. Five of the components produced in each mould were then measured for dimensional accuracy. In all cases the components were found to be oversized, areas close to the gate were observed to be worst affected.

In both cases the inaccuracies incurred up to the moulding cycle (i.e. pattern and mould manufacture) are insignificant when compared to those incurred during the moulding cycle. For the tensile test mould for example, deviations from nominal in the order of -0.46% to +8.00% were measured on the mould cavity. These compare with maximum deviations on the ABS mouldings of +30.46% and +40.67% for component width (nominal 6.5 mm) and depth (nominal 3 mm) respectively. In real terms these figures represent thickening of between 1 mm and 2 mm. For the rectangular LDPE components, maximum deviations of approximately +0.4% for component width (nominal 25.4 mm) and +19.5% for component depth (nominal 3.175 mm) were recorded (6). These improved accuracies are primarily due to the use of a different injection material. Low density polyethylene is crystalline and has a higher melt flow index than amorphous ABS. Consequently, the injection pressure required is lower, this leads to reduced deformation of the mould and production of more accurate components (7). Furthermore the recommended mould temperature for LDPE is lower than that for ABS, 27°C against 80°C. This has two implications, (i) less deformation of the mould due to thermal expansion but also (ii) improved stiffness characteristics, the stiffness of rubber is temperature dependant, the higher the temperature the lower the stiffness (10).

Ultimately the process trials have indicated that overall process accuracy is most affected by phenomenon that occur during the moulding cycle. These include common moulding effects such as material shrinkage, and process specific effects such as mould expansion under heating and compression under injection. These less well understood latter effects are currently under investigation in the research. For the purposes of this paper an investigation into the effect of mould heating is undertaken.

4 MOULD HEATING

4.1 Approach

Two routes are reported, one using conventional analytic techniques and one using an approximated finite element technique. A simple mould configuration is modelled, this comprises an enhanced silicone rubber mould insert which sits in a metal mould frame. In order to simplify the analysis a 2D section of a blank mould (no cavity) is considered. This is shown in figure 3.

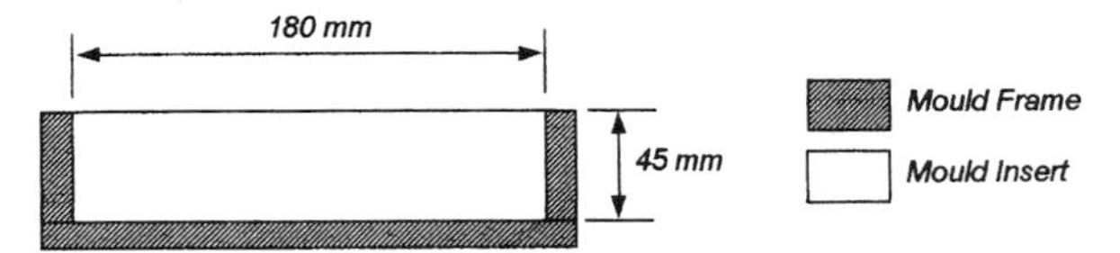

Fig. 3 Mould configuration

The mould heating is represented by a uniform temperature rise in both models. The materials properties required, namely coefficient of linear expansion, for these theoretical models were obtained in materials tests which are presented in section 4.2. The development of the analytic and finite element models are presented in section 4.3 and 4.4 respectively while section 4.5 outlines the experimental approach used to provide results for verification of the theoretical models.

4.2 Materials Tests

4.2.1 Material Preparation

The material used in the enhanced silicone moulding process is a moulding grade silicone rubber impregnated with iron powder. The material is prepared through addition of 10% wt. CAT 1300 catalyst to MCP1300T silicone rubber followed by thorough mixing of an equivalent weight (50% wt.) iron powder (10 μm). Processing tests have indicated that a minimum of 4 minutes manual mixing is required to ensure both complete wetting of the powder particles by the rubber material and even particle distribution. The material is then placed in a vacuum to remove air which becomes trapped during the mixing stage. A two stage vacuum strategy is employed. The material is first degassed in the mixing vessel at 740mm Hg for 20 minutes, the material is cast to final form and then subjected to a secondary degassing period at 740mm Hg for a further 20 minutes. The material is then left to cure at standard temperature and pressure for a period of 24 hours. The process by which the material cures is that of vulcanisation (10).

4.2.2 Thermal Expansion

The aims of these materials tests are twofold (i) to assess whether the material displays isotropic or anisotropic expansion characteristics and (ii) to determine the value of the coefficient of linear expansion α (/K). This testing is facilitated through use of rectangular block samples (100 x 50 x 25 mm) whose expansion in all three directions can be easily measured. The samples were heated using a hot plate device for a three hour period following which sample length, width and depth were recorded. The precise temperature within the samples at the time of measurement was recorded using a thermocouple embedded within the sample. In total the expansions at three temperatures, 20°C, 40°C and 60°C, were recorded. The results of the tests are shown in figure 4 which plots the coefficient of linear expansion in all directions for samples produced in two batches.

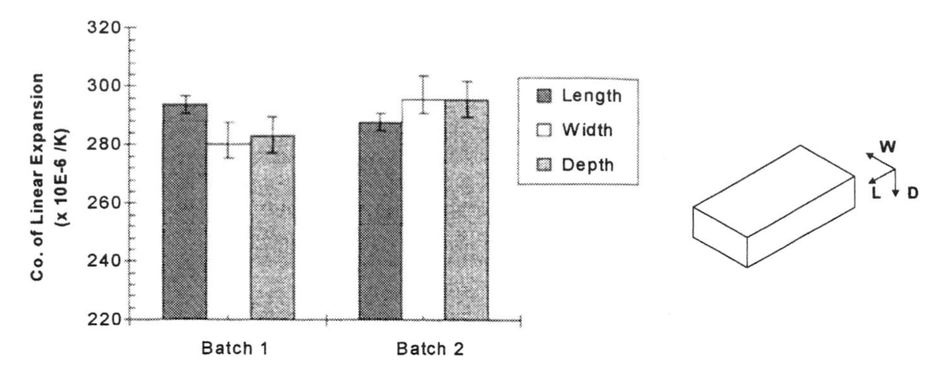

Fig. 4 Coefficient of thermal expansion results, sample geometry (inset)

These results clearly indicate that the material behaves in an isotropic manner with good correlation between results in the length, width and depth directions. This isotropic nature is because rubber consists of long chain molecules whose distribution in a stress free state is random (10)(11). The average coefficients of linear expansion are 285.6×10^{-6} /K and 293.0×10^{-6} /K for batch 1 and 2 samples respectively. The scatter of results indicated by maximum and minimum error bars is in the order of 2.0% – 4.3% (overall standard deviations 7.1×10^{-6} /K and 4.5×10^{-6} /K for batch 1 and 2 respectively). The coefficient of linear expansion for future reference is taken as the overall average, 289.3×10^{-6} /K.

4.3 Analytical Model

The development of the analytic model is based on calculation of the superficial (area) expansion of an isotropic material. This expanded area is then assumed to take up a certain geometry whose exact form can be obtained by equating the calculated expansion (materials based) with the area of the geometry (mathematics based).

The linear expansion dL (m) of an isotropic material, original length L (m), thermal expansion coefficient α (/K) under a change in temperature dT (K) is given by

$$dL = \alpha \, dT \, L \qquad [1]$$

For homogeneous isotropic materials, equation [1] can be extended to the two dimensional or superficial case, this is given by

$$dA = 2\alpha \, dT \, A \qquad [2]$$

where A is the original area (m^2) and dA the change in area (m^2) (12). This superficial expansion results in a change in mould section geometry. Preliminary experimental results have indicated that the expansion takes the form of bulging of the cavity. By assuming this bulge takes a standard geometric form (e.g. elliptical or parabolic) it is possible to predict the actual profile of the deformed cavity. For example, assume the bulge takes the form of a semi-ellipse. The area of a semi-ellipse dA (m^2), overall width 2a (m) and overall height b (m) is given by (9)

$$dA = \pi \, a \, b / 2 \qquad [3]$$

Assuming (i) the material to be incompressible, (ii) the mould section to be completely restricted by the frame and top mould plate (iii) the cavity deforms in the vertical direction only and (iv) ignoring the expansion of the steel frame, then equation [3] can be equated to the expansion area given by equation [2]. The resulting equation [4] is written in terms of the unknown dimension b which can be considered the maximum deformation.

$$b = 4\alpha \, dT \, A / \pi a \qquad [4]$$

The linear coefficient of expansion α is known from the materials tests, the change in temperature dT is based on the manufacturers recommended mould temperature for a particular polymer while the initial area A and overall cavity width given by 2a can be easily obtained from the particular moulding configuration. This provides all the information required to plot the deformed cavity profile. This technique can be repeated for other standard geometric forms including parabolic and exponential.

4.4 Finite Element Model

The deformation resulting from thermal expansion can be given by the thermal strain vector

$$\{\varepsilon^{th}\} = \Delta T \, [\, \alpha_x \;\; \alpha_y \;\; \alpha_z \;\; 0 \;\; 0 \;\; 0 \,]^T \qquad [5]$$

where ΔT is the change in temperature and α_x , α_y , α_z are the thermal coefficients of expansion in the x, y and z directions (13). This approach is applied using a finite element model of the silicone mould to predict the deformation due to mould heating. The development of model was undertaken in ABAQUS.

The geometry used for the analytic model was employed for the finite element model. The mould was described using 4 node continuum elements with an average size of 5 x 5 mm. Contact surfaces were defined as appropriate between the mould insert and mould frame. The materials used in the model were steel for the mould frame and top mould plate, and enhanced silicone rubber for the mould insert. The non linear rubber material was described using a first order polynomial function and using the thermal expansion results from the materials tests. The boundary conditions applied were simple displacement restraints in both horizontal and vertical (1,2) directions on both the mould frame and top mould plate. An initial temperature of 20°C was also specified throughout. Finally a simplified thermal loading was applied to the model. This comprised a uniform temperature rise to 50°C throughout.

4.5 Experimental Verification

In order to assess the accuracy of the theoretical models a mould heating apparatus was manufactured. The apparatus comprises an enhanced silicone rubber mould and a heating plate. The rubber mould measures 180 x 80 x 45 mm. The mould was heated to 50° C for a three hour period and the deformation measured using a co-ordinate measuring machine. The deformation was measured along the length of the mould at three different widths; along the centre line and 10 mm to either side of the centre line. Three thermocouples were embedded in the mould to allow mould temperature to be monitored.

5 RESULTS

The finite element results are presented in figure 5 which shows the deformed shape. Figure 6 shows the theoretical and experimental results, as the problem is symmetric only half of the results are shown. The graph plots the vertical deformation U2 (mm) against cavity position d (mm).

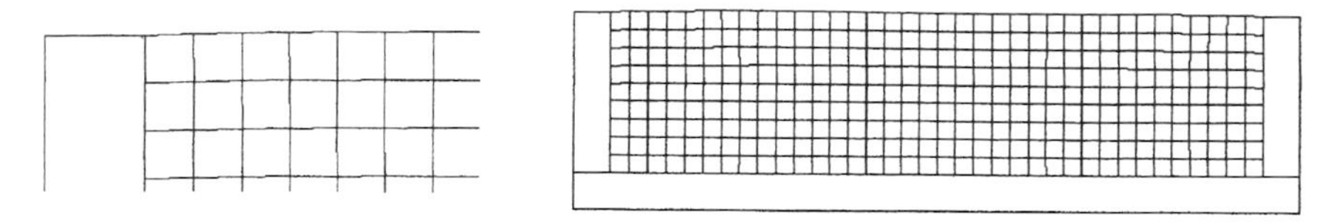

Fig. 5 Finite element deformation plot; edge detail (left) and full geometry (right)

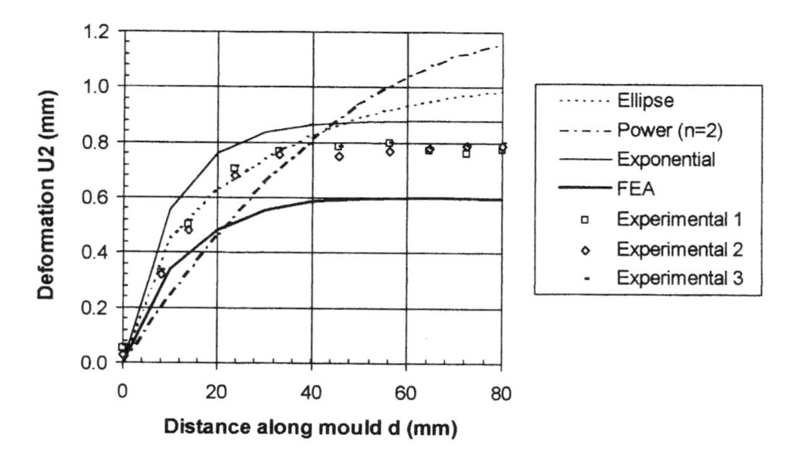

Fig. 6 Combined analytic, finite element and experimental results

6 DISCUSSION

6.1 General Observation

The experimental results shown in figure 6 indicate that the deformed shape comprises two regions; the boundary region where the deformation rises in a curved manner followed by a plateau region of constant deformation. For the specific geometry investigated in this work (length 180 mm) the plateau starts approximately 55 mm from the edge of the mould. The deformation over the plateau region is approximately 0.8 mm.

6.2 Analytical Analysis

The analytic results presented are based on three assumed geometries; elliptical, second order power and exponential. With regards form, it is clear that the second order power is least representative of the experimental results. This is because the geometry does not exhibit any approximate plateau region. This is also evident for the elliptical results, albeit to a lesser extend. Ultimately the exponential form provides the most representative form with a defined rise region followed by a near horizontal plateau region. It is possible to vary the distance over which the first rise region occurs through adjustment of the parameter c in the Cartesian equation

$$y = a\,(1 - e^{cx}) \qquad [6]$$

To date the value of c has been determined through trial and error, for this mould configuration a value of -0.1 has been found satisfactory. In terms of magnitude the analytical results predict greater deformation than actually observed. In the exponential case an average deformation of 0.88 mm over the plateau is predicted, this is +11% larger than the experimental result of 0.79 mm. This may be attributed to limitations of the analytic model. For example the material is assumed incompressible, to date compressibility tests have not been undertaken to quantify the materials bulk modulus. This assumption would lead to a higher predicted results than experimentally observed. The model also ignores the thermal expansion of the mould frame. If this were incorporated then the total calculated expansion of the rubber mould would be divided between the deformed bulge area and the expansion area of the mould frame, this would have the effect of reducing the magnitude of the deformation U2.

6.3 Finite Element Analysis

Figure 5 shows the deformation plot obtained from the finite element model. The results show the deformation U2 to rise from 0 mm at the edge to a maximum value of approximately 0.59 mm over the plateau region. The form of the deformation plot matches the experimental results well with similar rise and plateau regions, the plateau region for the experimental results starting 55 mm along the mould compared with 50 mm for the finite element results. The magnitude of the finite element results are however significantly lower then the experimental results with a deformation U2 in the order of 0.6 mm over the central region of the mould. This is approximately 24 % lower than the experimental result of 0.79 mm over the same region. The difference may be attributed to many factors. The finite element model represents an ideal 2D situation in which the materials properties, boundary conditions and loading obey exact rules. Any deviation from these in the experimental tests will result in an error. The rubber material used exhibits non-linear characteristics, within the finite element method such behaviour is described approximately using strain energy functions such as the Mooney-Rivlin and Ogden forms. The boundary conditions specified in the model also differ from the real case where constant sliding friction is used to describe the interaction between the rubber mould and the metal walls of the mould frame. The literature however suggests that the friction varies in a non-linear manner with temperature and is affected by surface roughness and sliding velocity (14). Finally the uniform thermal loading applied to the model may require improvement. In summary, the finite element results are reasonable. The trend matches that obtained experimentally while the difference in magnitudes can be addressed through refinement of the model.

7 CONCLUSIONS

This paper reports the results of a investigation into the effects of mould heating on the deformation of an enhanced silicone rubber mould. Two techniques have been investigated, one using analytical techniques and one using the finite element method. The results of the analytic technique have been promising, in particular where an exponential form is assumed. Ultimately this approach to deformation prediction is impractical as extension of the technique to consideration of 3D moulds with complicated cavities would be highly time consuming. The results of the finite element analysis are also promising. By comparison with the analytic technique, extension to a 3D analysis of a mould with complicated cavities is feasible. This is particularly true with the continued improvements being made to finite element software. Future work will involve development of this finite element model to 3D and inclusion of more precise materials, boundary conditions and loadings. Ultimately it is hoped that a combined model will be developed incorporating mould heating, mould clamping and injection pressure effects.

ACKNOWLEDGEMENTS

The authors wish to acknowledge the support of the Advanced Manufacturing Technologies (AMT) Ireland Programme for Advanced Technology (PAT) and the Science and Innovation Directorate at Enterprise Ireland.

REFERENCES

(1) Stokes, V.J.; Thermoplastics as Engineering Materials: The Mechanics, Materials, Design, Processing Link; In: Transactions of the ASME, pp. 448-455; Vol. 117.
(2) Barlow, J.W.; Beaman, J.F.; Balasubramanian, B.; A rapid mould-making system: material properties and design considerations; In: Rapid Prototyping Journal, pp. 4-15; V 2; N 3.
(3) Pham, D.; Dimov, S.; Lacen, F.; Techniques for firm tooling using rapid prototyping; In: Journal of Engineering Manufacture; pp. 269-277; Vol. 212; No. B4; 1998.
(4) Silicone – deformation
(5) Venus, A.D.; van de Crommert, S.J.; Rapid SLS Tools for Injection Moulding; In: Proc. 13th Conference of the Irish Manufacturing Committee, pp. 837-845; Limerick.
(6) Dunne P.; Young. P.; Byrne, G.; Dimensional Stability in Rapid Tooling Processes; In: Proc. of the 15th Conference of the Irish Manufacturing Committee; pp. 485-494; 1998.
(7) Bryce, D.; Injection Moulding... material selection and product design fundamentals; SME Publication; USA; ISBN 0-87263-488-4.
(8) Walker, J.; Martin, E.; Injection Moulding of Plastics; London Iliffe Books Ltd.; UK; 1966.
(9) Howatson, A.; Lund P.; Todd, J.; Engineering Tables and Data; Chapman and Hall; UK; SBN 412 11550 6; 1972.
(10) Mark, J.; Erman, B.; Eirich, F.; Science and Technology of Rubber; Academic Press, Inc.; USA; ISBN 0-12-472525-2; 1994.
(11) Treloar, L.; The Physics of Rubber Elasticity; Oxford Press; UK; ISBN 0 19 851355 0; 1975.
(12) Ohanian, H.; Physics; Second Edition Expanded; Norton & Company Ltd.; ISBN 0-393-95750-0; 1989.
(13) Kohnke, P.; ANSYS Theory Reference; Release 5.3; 7th Edition; DN-000656.
(14) Persson, B.; On the Theory of Rubber Friction; Surface Science; pp. 445-454; Vol. 401; 1998.

Manufacturing and Supply Chain Management

CAPE/047/2000

Computer-aided container handling system

J SZPYTKO
Faculty of Mechanical Engineering and Robotics, University of Mining and Metallurgy, Cracow, Poland
M CHMURAWA
R&D Centre Detrans, Bytom, Poland

ABSTRACT

The paper is describing a computer-aided container handling system in a logistic centre, which has been developed. The train serves the logistic centre. The train wagons are loaded and unloaded by cranes. Each container with a known goods specification and receiver/ sender address is stored at the warehouse. As a result of a mail order for container, the crane operator must identify and transport the requested container from point A to point B. The optimisation criteria are time and crane operation unit cost. The container transport process planning is assisted by the computer-aided system. The system has in-built: decision making strategy, container data base, warehouse and unloading/ loading ground virtual map. The above are linked via Internet and should be a part of a concurrent enterprise.

1. INTRODUCTION

Material handling systems need to be more effective and to overcome the present negative image of low productivity resulting from lower stack heights, lower moving speed of equipment, etc. For the modern container terminals the introduction of information technology and intelligent handling system based on cranes is crucial in order to achieve higher productivity and cost saving.

The container centre productivity is possible to increase throughout introduction following systems into the logistic centre: planning, operating and automated handling devices (Table 1). Increasing the level of planning and operating systems can be reached by using computer simulation analysis and implementation cranes with in-build intelligence into the real environmental.

Table 1. Results of implementation modern modules into warehouse.

SYSTEM	MODULES	TECHNOLOGY	MAIN EFFECT
planning system	berth assignment, train and truck planning, container terminal planning, etc.	optimisation models	time reduction, space maximising, equipment usage maximising
operating systems	container terminal control, equipment control, etc.	identification technologies, optimisation models	space maximising, equipment usage maximising, increasing productivity
automated handling system	crane, AGV, etc.	identification technologies, control technologies	labour saving, operation and maintenance costs saving

The paper is describing a computer-aided container handling system as a part of the logistic centre, which has been developed. The optimisation criteria are time and crane operation unit cost.

2. CONTAINER CENTRE PRODUCTIVITY

The logistic centre based on containers is served mostly by train, as well as by trucks (called container transport devices). The train wagons are loaded and unloaded by cranes. Each container with a known goods specification and receiver/ sender address is stored at the warehouse. Resulting from a mail order for container, the crane operator must identify and transport the requested container between the warehouse and container transport devices. The block scheme of the computer aided container-handling system is presented at Figure 1. Increasing the container's centre productivity is possible by improving crane automation level and introducing better quality crane management operation.

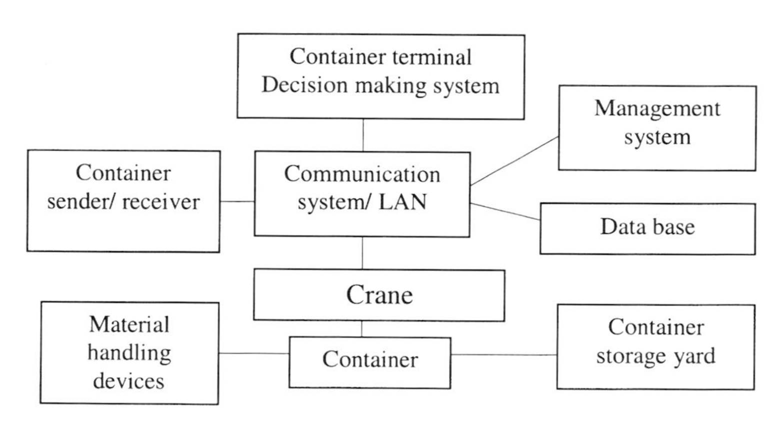

Figure 1. Computer aided container-handling system

3. CRANE AUTOMATION

Increasing the level of technology progress with decreasing cost's level in connection with globalisation aspects and evaluation the manufacture have the key influence to the material handling devices and system's updating. Evaluation of handling devices and systems with respect to the target of operation and to characteristics of the terminals is presented at Figure 2 (1).

The overall productivity of a container crane is determined by its automation and information functions (2). An important role is still played by the drive and control systems. The integrated crane subsystems, which support the device movement trajectory, include:

1. crane vision (AI eyes) system: profile scanning system, landing control sequence, crane alignment system, container positioning,
2. position unit identification of the: crane manipulator and others operating subsystems (travelling crab, device construction) in the 3D operation space, train (or truck) with containers,

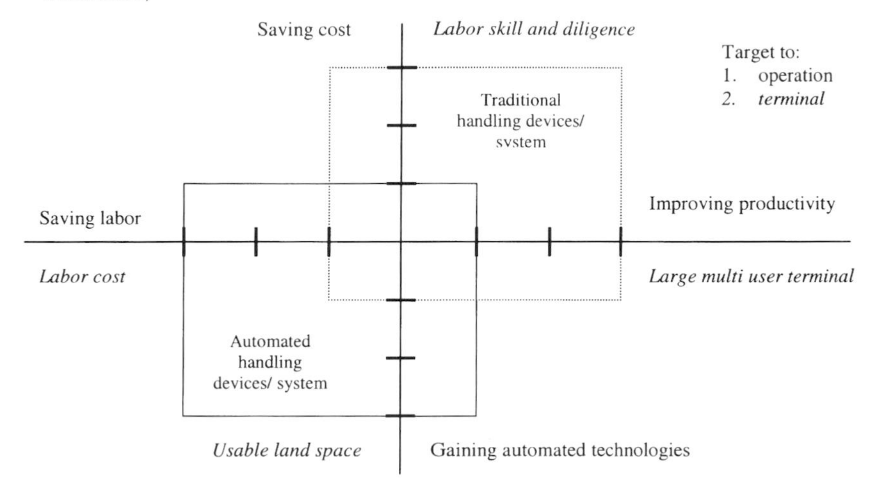

Figure 2. Material handling devices and systems evaluation

3. crane self-operation decision making unit based on crane technical state assessment unit,
4. drive mechanism which optimise path control unit,
5. crane safety units: skew control, anti-collision, anti-sway control,
6. crane monitoring and maintenance system,
7. remote communication system.

It has been established that the container handling system is based on the following transportation assumptions:

1. movement trajectory of the crane manipulator length minimisation,
2. operation time of the crane manipulator minimisation (its assumed that two crane drive mechanisms are operated simultaneously),
3. containers are placed evenly at the warehouse (to avoid the local pile up of the containers, to obtain better access to the containers).

4. CRANE MANAGEMENT OPERATION

The crane operation algorithm contains the following steps (Figure 1):

1. tasks of cranes and operation conditions (decision making system, remote communication system),
2. studying the operation procedures (management system, data base, communication system),
3. realisation the expressed work task by the crane (its movement mechanisms) according to established and accepted operations modules,
4. expression crane's work tasks and operation conditions (decision making system, remote communication system).

The key position is played by the management system, which is based on developed action modules:
Z1 – crane movement in order to collect a container from the transport device,
Z2 – crane movement in order to collect a container from the container storage yard,
Z3 – crane movement in order to store a container at the transport device,
Z4 – crane movement in order to store a container at the container storage yard,
Z5 - crane movement to the appointed point and awaiting the new orders,
and data base.

Action modules descriptions:

Z1 – crane movement in order to collect a container from the transport device:

1. identification of the A location/ point co-ordinates of the crane manipulator,
2. identification of the B location/ point co-ordinates of the container unloaded from the train carriage (or from truck/ tracks queue),
3. identification of the accessible and safety operation space of the crane manipulator (based on the 3D ground virtual map),
4. determination the crane manipulator movement trajectory between points A and B, simultaneously minimising the AB distance,
5. conversion the appointed movement trajectory of the manipulator into the operations of the crane executive mechanisms (based on the Programme Logical Controllers PLC types),
6. optimisation of the executive crane's mechanisms operations through their combination (crane operation time is limited by his mechanism with longer operation time: crane bridge travelling mechanism or travelling crab mechanism),

Z2 – crane movement in order to collect a container from the container storage yard:

1. identification of the A location/ point co-ordinates of the crane manipulator,
2. identification of the C location/ point co-ordinates of the selected container at the container storage yard,
3. identification of the accessible and safety operation space of the crane manipulator ,
4. determination the crane manipulator movement trajectory between points A and C, simultaneously minimising the AC distance,
5. conversion the appointed movement trajectory of the manipulator into the operations of the crane executive mechanisms,
6. optimisation of the executive crane's mechanisms operations through their combination,

Z3 – crane movement in order to store a container at the transport device:
1. the container owner (shipper) identification, as well as container code,
2. identification of the C location/ point co-ordinates of the crane manipulator,
3. identification of the B location/ point co-ordinates of the container at the loaded train carriage (or at the truck/ tracks queue),
4. identification of the accessible and safety operation space of the crane manipulator,
5. determination the crane manipulator movement trajectory between points C and B, simultaneously minimising the CB distance,
6. conversion the appointed movement trajectory of the manipulator into the operations of the crane executive mechanisms,
7. optimisation of the executive crane's mechanisms operations through their combination,

Z4 – crane movement in order to store a container at the container storage yard:
1. the container owner (shipper) identification, as well as container code,
2. identification of the B location/ point co-ordinates of the crane manipulator,
3. identification of the location/ locations of the selected shipper at the container storage yard,
4. identification of the possible C location/ point co-ordinates of the selected container at the nearest shipper yard and lowest store layer,
5. identification of the accessible and safety operation space of the crane manipulator,
6. determination the crane manipulator movement trajectory between points B and C, simultaneously minimising the BC distance,
7. conversion the appointed movement trajectory of the manipulator into the operations of the crane executive mechanisms,
8. optimisation of the executive crane's mechanisms operations through their combination,

Z5 - crane movement to the appointed point and awaiting for the new orders:
1. identification of the B/C point co-ordinates of the crane manipulator,
2. identification of the A awaiting point co-ordinates of the crane manipulator,
3. identification of the accessible and safety operation space of the crane manipulator,
4. determination the crane manipulator movement trajectory from point B/C to point A, simultaneously minimising the B/C-A distance,
5. conversion the appointed movement trajectory of the manipulator into the operations of the crane executive mechanisms,
6. optimisation of the executive crane's mechanisms operations through their put together.

Data base:
1. virtual map of the possible crane operation space in 3D, including locations of: train tracks, container trucks awaiting road, shipper storage yards, transportation corridors,
2. used containers type and their dimensions,
3. used container's transportation devices (rail carriages, truck-tractors) and their dimensions,
4. the quantitative and qualitative structure of the train set with containers (how many, what, where),
5. the quantitative and qualitative structure of the shipper containers yards set (how many, what, where, accessible).

5. CONTAINER HANDLING SYSTEM SIMULATION ANALYSIS

Computer simulation is based on the self-developed software called *Container unloaded and loaded optimisation.* Computer simulation is based on one or two gantry cranes, which are operated at the container logistic centre. When the train (trucks) is loaded and unloaded by two gantry cranes, the container storage yard and served train carriages are flexible divided per two areas which are directly controlled by each crane. It is assumed that the container loading and unloading operations are without any collision. Programme contain several modules (Figure 3):

1. way of loading and unloading: by use of one or two cranes,
2. train or trucks queue with container configurations: new, change for new, review,
3. container storage yard configuration: review, graphic presentation,
4. search out of the selected container: at the transport unit, at the yard.

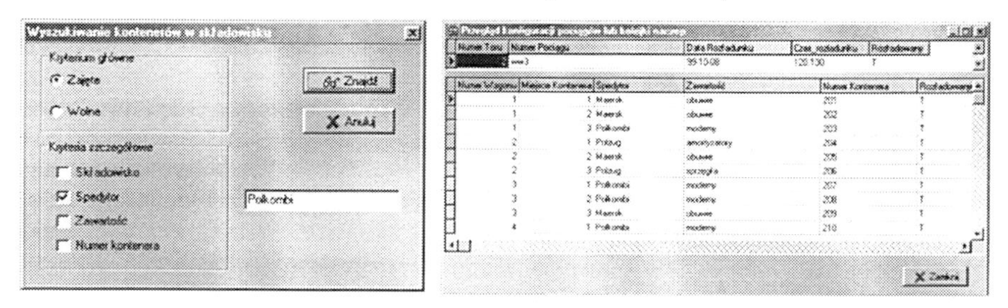

Figure 3. Programme selected modules: a) train with container configurations, b) search out of the selected container at the yard

A container terminal simulation model has the following capabilities: terminal layout, storage modes, traffics layout, loading specifications, equipment performance. Simulation model is used to establish required: cranes, container storage capacity, advanced technology systems for container terminal.

Container handling system simulation has been done using the following data:

1. container storage yard is at the open expanse and is based on the integrated cargo unit: yard dimension 37,4 x 18,6 = 696 [m^2], container are accumulate into 4 layers (owing to safety conditions) – Figure 4, 5,
2. integrated cargo unit is based on containers 1C type (dimensions H x L x S : 2590 x 6058 x 2438 [mm]),
3. integrated transport units is based on the: rail carriages 412Z type (with capacity till 3 piece of container 1C type) and truck-tractors with semitrailer NK32.122 type (with capacity till 2 piece of container 1C type),
4. trains/ trucks are loaded and unloaded by the gantry crane with capacity Q = 40 [t] and span L = 25 + 12 [m], speeds: bridge V_{js}= 46 [m/min], crab V_{jw}=58 [m/min], lifting V_{op}/V_p=24/16 [m/min]- Figure 5,
5. containers shipper: Maersk, Polkombi, Polzug, others.

6. CONCLUSIONS

Optimising container centre productivity using computer simulation analysis is crucial to optimising the layout and equipping the container handling system. Computer simulation of the container centre productivity helps:

1. to work out (depended on time) the scheduling programme of the crane operation (containers loading and unloading at/ from transport devices) based on guidelines conditions and administer reloading structure (which is known before container reloading process),
2. to support the crane operator activity or improve crane automation operation based on the device executive subsystems movements schedule,
3. optimisation the reloading structure, including: container storage yard geometry and infrastructure, technical operational data of used material handling devices, productivity and storage buffer requirement, number of container shipper, container turnover quantity,
4. to estimate the minimal stopover time of the container transportation devices (train, truck) in the container storage yard, waiting to be unloaded or loaded by cranes.

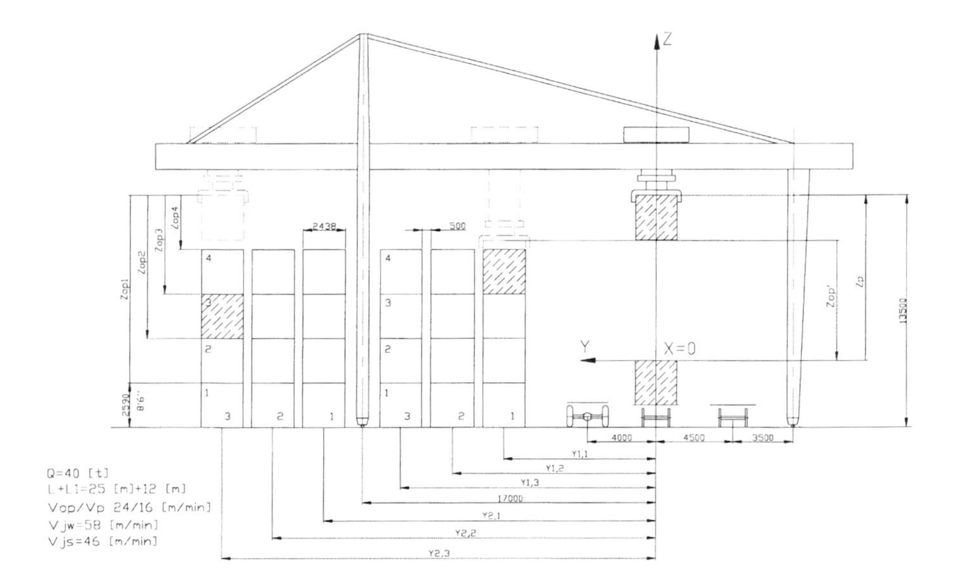

Figure 4. The container storage yard draft served by crane, train and trucks

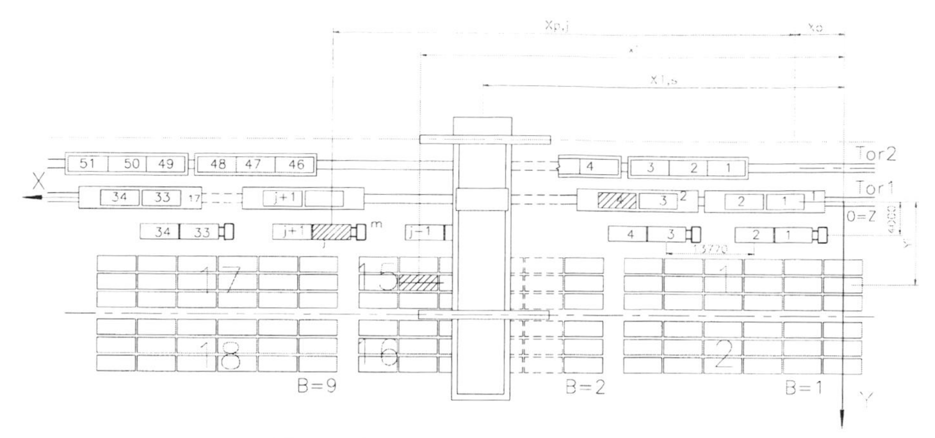

Figure 5. The crane position at the yard

The simulation package effectiveness depends on the container storage yard database. On the basis of existing data base it is possible:

1. fast identification of the given container at the storage yard,
2. quick identification of the fill level of the container storage yard,
3. better container management and labour cost reduction.

Computer simulation has been done on the example of the logistic centre served by train. Basing on the results of computer simulation analysis it has been affirmed that:

1. replacing one crane with two cranes reduce the container train reloading time approximately 38 % only,
2. earlier container grouping by each shipper at the train have no influence on the crane unloading time, increase crane unloading time when the yard is rather full,
3. the type of reloading container transportation device (train, truck) have no influence at crane unloading time,
4. when the yard is full on the level of 70 %, the crane unloading time maybe increased up to 50 %.

The position of cranes, as a transport device in modern manufacture and reloading/ logistics centres, is increasing. Business is looking for cranes with more intelligence in-build, both as a new designed and after modernisation. The crane productivity is determined by its automation (technique in-build), information functions based on technology and management.

ACKNOWLEDGEMENTS

Authors acknowledge with thanks to the KBN for the financially support under the RobCrane Eureka Programme.

REFERENCES

1. Nam K.,C. (1998): *Determination of handling systems at Pusan.* New Port. PTI, no 8, p.85-87
2. Szpytko J. (2000): *RobCrane ΣReports.* UMM, Cracow

CAPE/062/2000

An open tender approach to collaboration in parts supply system

T ITO and **M R SALLEH**
Department of Mechanical Engineering, The University of Tokushima, Japan

ABSTRACT

This paper proposes an open tender approach to enhance collaboration in parts supply system, which plays an important role in supply chain system. The idea of open tender provides an equal opportunity to all candidate suppliers for information sharing, and promotes an open competition to rank up the suppliers under open environment. The open competition among suppliers has two advantageous features. On one hand, it helps manufactures to find out one of the most appropriate suppliers from a number of candidate suppliers. On the other hand, even a supplier is not selected as a result of open competition, it gives some hints to the supplier in order to improve competency and to become more competitive in the future tenders. The selection mechanism in the open tender concept employs the idea of bulletin-board based negotiation, which is carried out by collaboration among intelligent agents dedicated to manufactures and suppliers.

1 INTRODUCTION

Manufacturers are required to develop and produce their products faster, better and cheaper in order to remain competitive in the market. Basic activities such as stable raw materials supply, avoidance of unnecessary inventory, or efficient parts distribution support manufacturers to comply with those requirements. Integration of these activities has been regarded as one of the important issues to become competitive in the market. Strenuous effort has been taken to realize the integration and several systems have been developed in the name of supply chain management system (1).

Parts supply system is a part of supply chain system, and plays an important role to ensure that the materials should be smoothly delivered to the manufacturers. Cooperation and collaboration between suppliers and manufacturers supports the success of the system, but integration of these activities is not achieved only with an information sharing approach.

This paper describes the development of methods and a computer based system to enhance collaboration in parts supply system and to provide supplier selection mechanism through open tender concept [2]. The system is called PASSOT (PArts Supply System through Open Tender concept) and is designed to provide a selection mechanism of suppliers in a collaborative manner by way of open and free competition among suppliers.

The open competition among suppliers has two advantageous features. On one hand, it helps manufactures to find out one of the most appropriate suppliers from a number of candidate suppliers. On the other hand, even a supplier is not selected as a result of open competition, it gives the supplier some hints in order to improve competency and some helps to become more competitive in the future tenders. The selection mechanism in the open tender concept employs the idea of bulletin-board based negotiation, which is carried out by collaboration among intelligent agents dedicated to manufactures and suppliers.

Furthermore, dynamic situations in parts order request are also considered. Order requests are given without any notice sometimes, when supply position may be very hard for some suppliers [3]. Even after a supply contract has been established between a manufacturer and a supplier, the manufacture may have to ask for an earlier delivery, or the supplier may have to ask for some delay in delivery. Under these circumstances, further collaboration may be required to manage some coordination, to which PASSOT works as a mediator and provides some solution.

2 FLOW OF OPEN TENDER PROCESS IN PASSOT

This section describes a flow of open tender process in PASSOT. Fig. 1 shows the overview of basic flow of open tender process to determine an appropriate supplier until a final agreement is reached between a manufacture and a supplier.

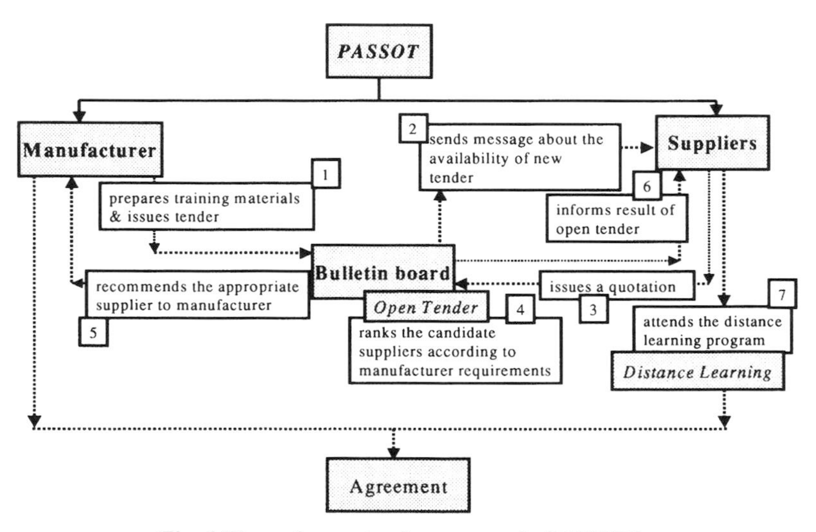

Fig. 1 Flow of open tender process in PASSOT

For the first step of the process, a manufacturer issues a tender and publishes it to candidate suppliers. Manufacturer gives its quotation selection criteria guideline, for example quote price, unit price, duration, location, delivery cost and other information required for bidding exercise, so that each candidate supplier can prepare a quotation according to manufacturer's requirement. The manufacture also prepares training materials for distance learning.

The tender information is given to Bulletin Board (BB) and, at the same time, BB sends a message to candidate suppliers, and informs them of the availability of new tender. If a supplier regards the tender as applicable, the supplier issues a quotation and sends it to PASSOT. After quotations are collected and displayed on BB for a certain period of time, these quotations are ranked up based on the selection criteria, and an appropriate supplier is recommended to the manufacture. When the tender process is finished, information about the tender results is fed back to all of the suppliers. In the mean time, distance learning training by the manufacture is conducted to the selected supplier to assure the quality of materials/parts handling. After the supplier is evaluated as appropriate, a mutual agreement is established between them.

3 TASKS OF COLLABORATION AGENTS IN OPEN TENDER PROCEDURE

Collaboration among agents is fundamental in e-commerce [4]. In open tender procedure, a manufacturer finds an appropriate supplier in each occasion in a dynamic manner under the support of collaborative agents. The collaborative agents in PASSOT are composed of inventory stock control agent (ISCA), manufacturer agent (MA), supplier agent (SA), supplier stock control agent (SSCA) and bulletin board control agent (BBCA). Tasks of these agents are to identify the critical information from incoming resources, to monitor the information, and to trigger appropriate actions based on the contents of information [5][6][7].

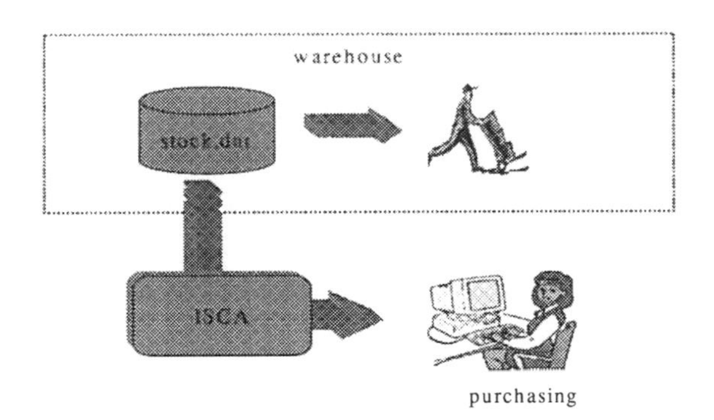

Fig. 2 Stock replenishment operation

The process of collaborative agents in open tender procedure is started with a stock replenishment activity. ISCA autonomously monitors the materials/parts stock level, and ensures that the level should keep at least the minimum. It also tries to make sure that the replenishment quantities are kept at the optimum level. The re-order level is set to the stock to keep it at a safety level. When the stock is reached at the minimum level, ISCA sends a signal to the purchasing department for order replacement. Fig. 2 shows the activity of ISCA for replenishing materials/parts stock.

BBCA controls BB, and manages the tenders and quotations. A manufacturer can obtain the recent quotation list as a reference to select an appropriate supplier. The interaction between a manufacturer and Bulletin board (BB) is taken care of by MA. The function of MA is to verify any tender and to notify a purchasing staff of the status of tender. The purchasing staff obtains the latest information about this tender, and scores the quotations submitted to PASSOT. Sometimes the purchasing staff checks the details of quotation. While MA behaves as a mediator between the purchasing department and BB, SA performs a task as a mediator between candidate suppliers and BB. SA notifies the candidate suppliers of the available tender on BB, and urges the candidate suppliers to prepare a quotation to participate in the bidding exercise. SA also informs the candidate suppliers of the open tender results, and asks the selected supplier to attend a training course of distance learning.

On the candidate suppliers side, SSCA agent monitors inventory stock of materials/parts, and informs the sales department of the available stock. Considering the stock condition, the sales department prepares a quotation to any available tender. The interaction of MA, BBCA and SA is illustrated in Fig. 3. Interaction of collaborative agents strongly supports an open tender process in PASSOT.

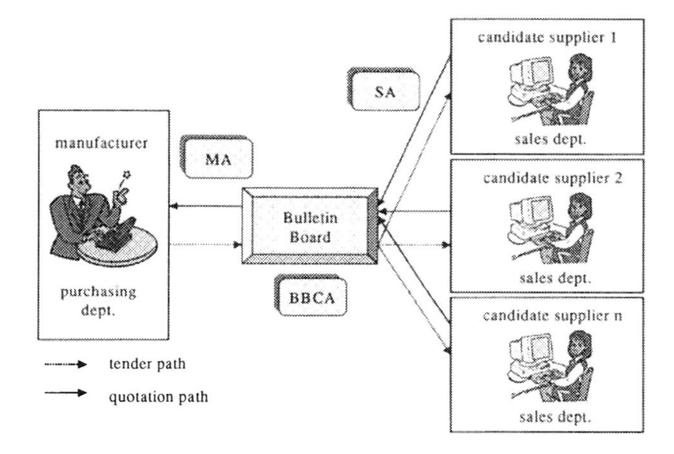

Fig. 3 Roles of agents in ordering and purchasing operation

4 CASE STUDY

This section shows an example of operations in materials/parts supply for car stereo products in PASSOT.

Tokugawa Electric Manufacturing is looking for a supplier to procure its stock of materials/parts. The stock of a screw with a part number of 123-321-4567 is running short at this time. In order to find an appropriate supplier to comply with the shortage of parts, Tokugawa Electric Manufacturing completes and submits a tender form as shown in Fig. 4 to

PASSOT. The form contains some required conditions including the unit price, duration, location and delivery cost. Tokugawa Electric also gives the guideline of selection criteria to rank up candidate suppliers.

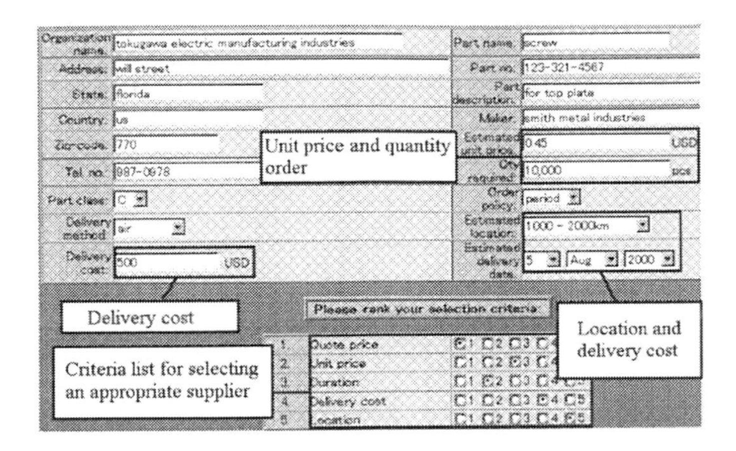

Fig. 4 A tender form for parts procurement process

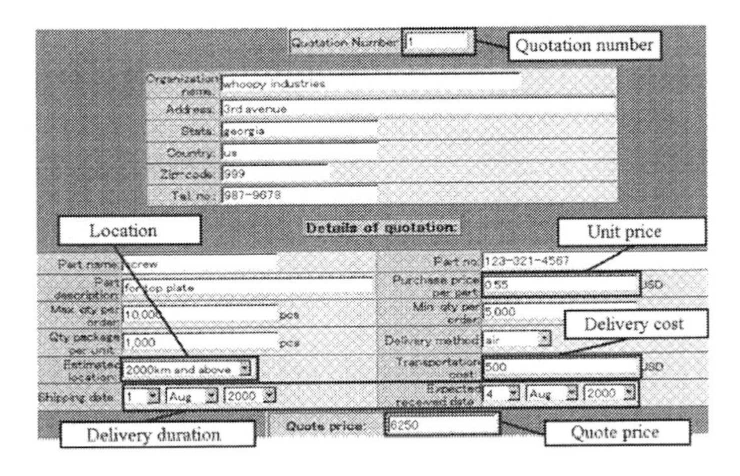

Fig. 5 A quotation form from a candidate supplier

Fig. 5 shows an example of quotation form, that is completed by a supplier in the bidding exercise, and that is processed by PASSOT.

As illustrated in Fig. 6, BB shows the names of manufacturer and candidate suppliers, and the number of quotations participated in the bidding exercise. In this example, BB receives four quotations from four different candidate suppliers. The search button at the bottom of BB window displays tender details for preparing a quotation. BB also displays the details of quotations, of which review is limited to the manufacturer, which means that the candidate suppliers are unable to view their competitors quotation.

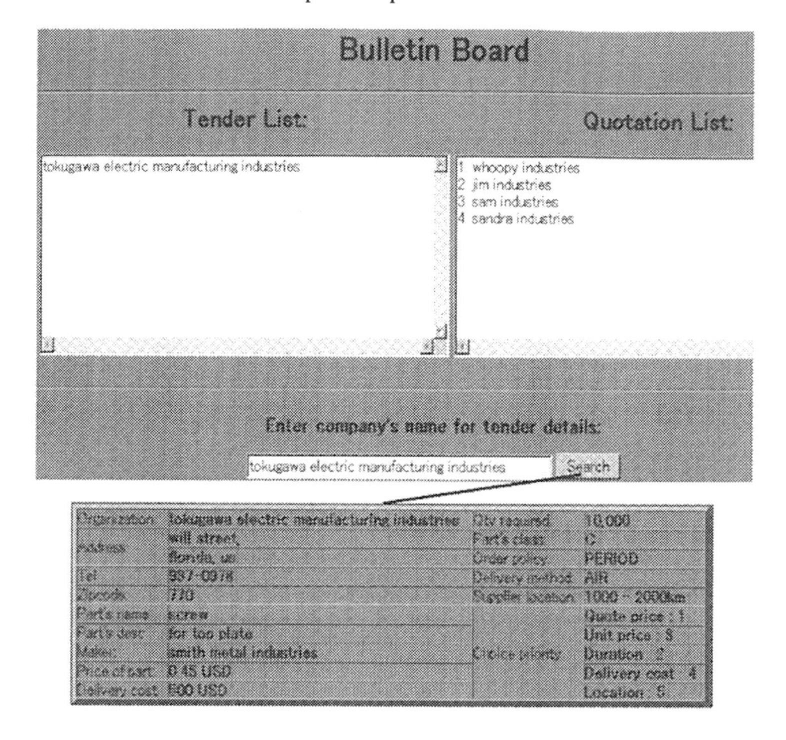

Fig. 6 A tender/quotation list and tender details for suppliers review

The manufacturer has two options for determining the supplier, or individual criteria and multiple criteria. For individual criteria, the manufacturer is provided with five lists of sorting results, those are ranked under the criteria of quote price, unit price, duration, location and delivery cost. In this bidding exercise, the manufacturer has selected Sam Industries as the supplier. This decision is made based on unit price basis. Sam Industries has quoted its unit price of a screw at 0.45 USD per piece. While Sandra Industries, Whoopy Industries and Jim Industries have offered 0.50 USD, 0.55 USD and 0.65 USD for each piece of screw, respectively. Fig. 7 shows a number of sorting lists of suppliers that are prepared in the bidding exercise. The manufacturer also considers all criteria for selecting an appropriate supplier.

The final decision on the selection of supplier from the list is up to the manufacturer. Although the manufacturer has disclosed the candidate suppliers with the selection criteria as a reference, the manufacturer sometimes may change the criteria without any notice to the suppliers. Fig. 7 shows an example of suppliers ranking list.

PASSOT also provides a function of distance learning [8], which is designed for a manufacturer to educate a supplier in order to enhance the quality of products. If any problems were identified in the previous transaction, those are analyzed by the manufacturer, solutions or countermeasures are proposed, and training materials are prepared. When an appropriate supplier is selected, distance learning takes place using the training materials to the supplier. After the training, some quizzes are given to the supplier for evaluation.

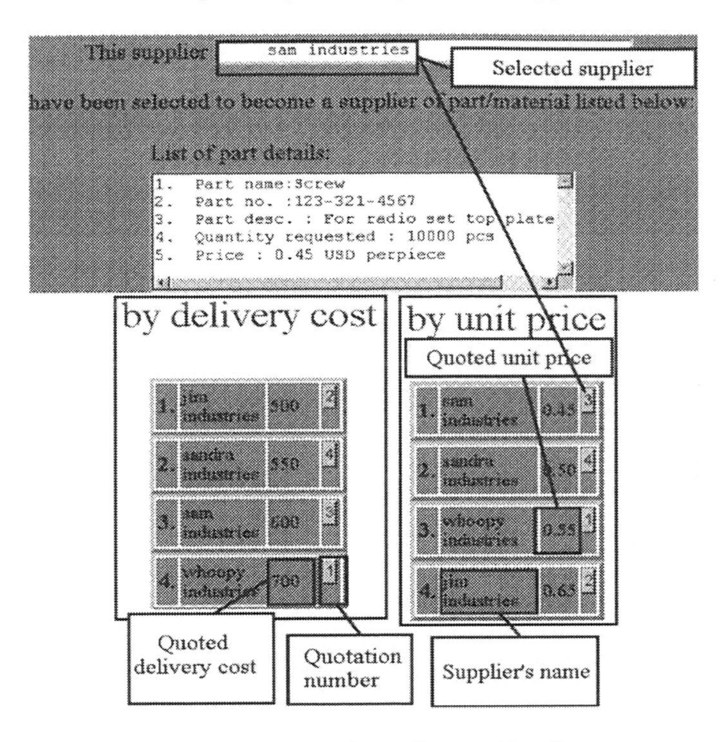

Fig. 7 Example of suppliers ranking list

5 CONCLUDING REMARKS

The paper presented the approach of open tender to enhance collaboration in parts supply system. The idea of open tender encourages an open competition in the supply process of materials/parts to achieve an efficient material flow and to shorten the production lead-time. Collaboration agents carry out the various tasks such as coordinating interactions, generating resolutions, and deriving agreements to satisfy the requirements.

Cooperation and collaboration between manufacturers and suppliers contribute to the success of PASSOT. They work together to reduce supply uncertainty, inventory stock and cost.

The advantageous features of the open tender approach in parts supply system are as follow. (i) An open environment to sound sharing of information, (ii) equal opportunity to candidate suppliers for open competition, (iii) quality improvement of products through distance learning, (iv) enhancement of productivity and customer satisfaction. For the future study, we would like to introduce an advanced negotiation process between a manufacturer and supplier.

In this negotiation process, both parties are allowed to further negotiate the quote price, unit price, quantity, and so.

REFERENCES

(1) Pillep, R. et al: Supply Chain Management Software – Market Overview, Benefits, Innovative Solutions, for Small and Medium Enterprises, CAPE'99, Durham UK, 1999, pp.637-642.

(2) Ito, T. and M. R. Salleh: A Blackboard-based Negotiation for Collaborative Supply Chain System, CAPE'99, Durham UK, 1999, pp. 631-636.

(3) Haan, J. and M. Yamamoto: Zero Inventory Management: Facts or Fiction? Lessons from Japan, International Journal Production Economics 59 , 1999, pp. 65-75.

(4) Oliver, J. R.: An Automated Negotiation and Electronic Commerce , The Wharton School, University of Pennsylvania. (http://opim.wharton.upenn.edu/~oliver27/ html/ diss. overview. Html)

(5) Parunak, H. V. D.: Practical and Industrial Application of Agent-based Systems, Industrial Technology Institute, 1998.

(6) Chavez, A. et al: Kasbah : An Agent Market Place for Buying and Selling Goods, Proceedings of the First International Conference on the Practical Application of Intelligent Agents and Multi-Agent Technology, London, UK, April 1996.

(7) Rhodes, B. J.: Remembrance Agent: A Continuously Running Automated Information Retrieval System, Proceedings of The First International Conference on The Practical Application Of Intelligent Agents and Multi Agent Technology, London, UK, 1996, pp. 487-495.

(8) Ito, T. and M.R. Salleh: Quality Improvement through Distance Learning in Supply Chain Management, SMC'99, Tokyo, October 1999, smc_2617.

CAPE/019/2000

Electronic commerce and its impact on the buyer-supplier interface

P HUMPHREYS and **G HUANG**
Department of Industrial and Manufacturing Systems Engineering, University of Hong Kong, Hong Kong
R McIVOR
School of Management, University of Ulster, UK

SYNOPSIS

The aim of this article is to show how electronic commerce can fundamentally change the inter-organisational processes at the interface between the buyer and supplier. The paper will indicate how electronic commerce is not only enabling the redesign of internal organisational processes but is also extended into both the buyer and supplier organisations. Two manufacturing case studies are presented outlining how various electronic commerce technologies have been implemented in a number of buyer-supplier environments. These case studies illustrate the benefits that organisations can achieve through the effective implementation of electronic commerce technologies such as electronic data interchange (EDI) and the Internet. The case studies will present electronic commerce technologies associated with MRPII and the product development process. It will be shown how electronic commerce is radically changing the way in which organisations have traditionally traded by: reshaping buyer-supplier relationships; improving core business processes; providing electronic inter-mediation; reaching new segments and markets; changing job roles and responsibilities.

1 INTRODUCTION

Electronic commerce is the process of doing business electronically and involves the automation of various business-to-business and business-to-consumer transactions (1). By reducing clerical procedures and eliminating paper handling, electronic commerce can accelerate ordering, delivery, and payment for goods and services while reducing operating and inventory costs. Electronic commerce is not just a single technology but a combination of technologies, applications, processes, business strategies and practises necessary to do business electronically.

Marketplaces that connect their customers and suppliers are prevalent in many product categories and are creating value by making trading more efficient. The experience of early participants suggests that an electronic marketplace can capture savings of 10 to 20 per cent and deliver lower prices for buyers (2). Electronic commerce technologies such as the Internet have been most prevalent in business-to-consumer type trading exchanges. For example, internet trading exchanges are most often experienced as consumer-oriented web stores selling everything from CDs to vintage wine. However, it is argued that the greatest potential for the application of electronic commerce technologies lies with business-to-business transactions (3).

This article explores the enabling role of electronic commerce in changing the traditional nature of relationships between customers and their suppliers. Two case studies are presented outlining how various electronic commerce technologies have been implemented in a number of buyer supplier environments. These case studies illustrate the benefits that organisations can achieve through the effective implementation of electronic commerce technologies such as electronic data interchange (EDI) and the Internet. As well as impacting the external trading arrangements between buyers and suppliers, electronic commerce is also affecting the traditional roles of the functions involved in managing the buyer supplier relationship. For example, electronic commerce is enabling the role of the purchasing professional to move from being involved in clerical type activities, such as invoice processing and expediting, to include activities such as integrating suppliers into their new product development processes and joint involvement in total cost analysis.

2 METHODOLOGY

Management research is characterised as being soft, applied and divergent and is undertaken in complex organisations which exist in a dynamic environment. As a consequence, a great deal of research in the management of organisations makes use of the inductive case study approach (4). The case study approach provides great richness and multiple perspectives of the many managers involved with regard to the data collected and is thus largely qualitative in nature. Unlike positivist research, however, the analysis of case study data is essentially interpretative and inductive. From the qualitative data, narratives or stories are developed which are examined for patterns. From these patterns inferences are drawn which yield propositions and can lead to specific hypotheses. Such hypotheses can then be tested in other situations and indeed if sufficiently specific can be tested via the more traditional survey methods of social science.

Gathering the large amounts of data associated with case study research involves a great deal of interaction between the researchers and the staff of the organisation being studied. With regard to the three organisations described in this paper, the authors have been extensively involved in examining procurement practices over a four year period. The research was concerned with: strategic purchasing decisions and their impact on SMEs; the make or buy decision making process; supplier development strategies. One important issue which emerged during the analysis of the case studies was the importance of electronic commerce in enabling the redesign of both the internal and external organisational processes in both the customer and supplier organisations. It is this theme which will now be explored in the two case studies presented.

3 CASE STUDIES

3.1 Case One

This case study outlines the implementation of a trading arrangement between an original equipment manufacturer (OEM) and an electronic component distributor. The company manufactures complex printed circuit board (PCB) assemblies. Due to the nature of the assembly process the company purchases a wide variety of electronic components from a large number of suppliers. For example, one type of PCB assembly has more than 2,500 electronic components with varying lead times and unit costs. In the past, the buyers in the purchasing function spent a considerable amount of their time in transaction type activities such as order planning, expediting and resolving supply problems for many of these components with a lot of this time being spent on low value commodity type components. Although the company had implemented EDI successfully with its key suppliers, it was not running EDI effectively with the majority of its commodity type suppliers. For example, in many cases the delivery forecast was sent by fax. However, firm orders were being sent by EDI with the buyer in the OEM also sending a fax of the same order as well. In the supplier organisation inefficiencies were also occurring with the supplier not dynamically loading the EDI message onto their manufacturing systems but printing the message out and manually entering the data. In addition, the communication channels within the OEM were affected by functional boundaries. For example, the buyers in the OEM were responsible for managing the suppliers while personnel in goods inwards managed the transfer of delivered items to stock.

The purchasing manager wanted to move to a situation where the buyers were involved in more value adding type activities such as the new product introduction (NPI) process. Therefore, the company decided to use an electronic component distributor to manage the logistics process between fifty of their commodity type suppliers. The objective was to ensure comparable service to that of the purchasing function, while increasing efficiency and reducing transaction costs with these commodity suppliers. In effect, the distributor would act as the interface between the OEM site and each component supplier maintaining and managing a store of items on the site. The operation of the entire system depends upon EDI with limited intervention in the OEM. The distributor uses its distribution centre as the 'hub' for the items it purchases from the suppliers. The initial running of the system begins with the distributor receiving a forecast of usage for each component for the proceeding six months. Consequently, each week the distributor receives a revised forecast of usage for the next four weeks from the OEM's manufacturing system. Based on this forecast the distributor must ensure that the distributor's on-site store has a stockholding equivalent to this forecasted four week usage. On receipt of the weekly forecast the distributor guarantees 24 hour delivery to ensure necessary replenishment of the on-site store. The distributor still 'owns' the stock while it is in the on-site store. When the OEM's manufacturing systems uses the stock from the store it then becomes the property of the OEM. Fig. 1 shows the set up of the system between the OEM, the distributor and the component suppliers.

The OEM is operating a consigned stock arrangement with the distributor. A consigned stock arrangement involves the supplier maintaining a stock in the customer's facility, under the customer's control. At the OEM site, the distributor maintains a stock equivalent to four weeks usage based on a weekly forecast transmitted via EDI. At the distribution centre, the distributor must provide additional stock and purchase order coverage in order to meet the manufacturing requirements of the OEM. Inventory levels will be at a level equating to a minimum 10% of forecasted annual usage. In relation to payment and component issue, manufacturing in the OEM draws components as required from the distributor's store and automatically updates the stock balance held on the Material Requirements Planning (MRP) system. Each week the OEM electronically notifies the distributor of all components drawn from the distributor carousel on the OEM site by part number, value and the date drawn. At the end of each month the distributor issues an invoice electronically giving a breakdown of the usage of each component for that month of issue to the OEM with payment due in thirty days. This effectively is 'self-billing' driven by usage, with weekly matching and monthly payment. Price is the purchase price from the original supplier of the components and includes a percentage mark-up agreed between the OEM and the distributor. In implementing this arrangement the OEM has drastically altered the business processes associated with purchasing to the point where it is no longer carrying out its traditional role. The major advantages for the OEM under this arrangement are as follows:

- Rationalisation of the supply base and a reduction in transaction costs;
- The distributor provides 'local' inventory with the OEM paying on use;
- The OEM obtains reduced lead times and quality inspection for the components has become the responsibility of the distributor;
- The implementation of EDI requires less purchasing resource with the buyers no longer being a link between the company and suppliers;
- EDI makes the processing of requisitions, order acknowledgements and invoices redundant;
- The buyers affected by this arrangement are able to focus on more value-adding tasks in the company such as implementing early supplier involvement in new product development activities;
- The OEM was uses an external source that provides greater levels of expertise and service in managing the logistics process than the OEM would achieve internally.

3.2 Case Two

During the last two decades there has been a major trend for firms and public organisations to externalise a wide range of functions that previously might have been carried out in-house. Increasingly business organisations are concentrating on core activities and outsourcing other functions to external suppliers. This ranges from major manufacturers increasing the proportion of components and sub-assemblies designed by suppliers to the contracting out of functions such as computer services, R&D and accountancy. There are a number of reasons for this trend, including rising global competition, more rapid technical change and the need for the faster development of products with higher quality and reliability. It is virtually impossible for any one firm to possess all the technical expertise needed to develop a complex product. This means that organisations have to focus on their core competencies and for other activities to draw on the best expertise available world-wide. Thus the old pattern of the large, vertically integrated business, is being replaced by one consisting of complex networks of collaborating organisations, and chains of buyers and suppliers (5).

In this new industrial structure, the design and development of complex engineering products is one of the activities that is being devolved back along the supply chain. The extent to which this occurs varies, with some manufacturers devolving most engineering design and development work to external suppliers. In other cases, there is often a mixed situation, in which the design of sub-assemblies and components are devolved to suppliers, or where in-house designers work closely with their suppliers to ensure that components of the required performance and quality are developed. It is therefore apparent that in this new structure, design and development not only has to be managed within one large organisation, but it also involves managing relationships between many companies in an extensive chain of buyers and suppliers.

A multinational electronics company has embarked, over the last five years, on an ambitious strategy to encourage its suppliers to take a more active role in the design of new products. The company has a number of key suppliers that are based in geographically dispersed locations across the Pacific Rim. During this period, the customer and designers at the suppliers' sites have had to deal with a number of problems related to information transfer during the product development process. The initial approach to managing product development involved face-to-face communication, the transfer of design documentation in paper and electronic format, and the use of other medium, such as fax, memos and e-mail. Due to the iterative nature of product design, these communication techniques were found, in some cases, to be costly and led to delays in the completion of the project.

In order to deal with these communication problems, the company has now introduced Intranet technology to assist in the management of the design process. Intranets are Web applications that are internal to organisations. An Intranet is not defined by a physical boundary or by geographical constraints, but by whom has access to the information. In this case, the boundaries are the design team from the customer and supplier organisations. The initial part of the project has involved one supplier site and has examined the application of the Morphological Chart Analysis method. The technique has been included in a number of major textbooks on Engineering Design as one of the most effective methods for concept design (6). Design concept specifies the form, function and overall purpose of the product, process or system under development and the benefits it will provide.

Fig. 2 outlines the morphological chart analysis method developed for the Intranet. The three Web pages correspond to the three major stages in concept design: functional analysis, concept generation and concept evaluation. The functional analysis stage is mainly concerned with establishing the functional structure of the product, process or system under development. The concept generation stage deals with identifying potential solutions that are able to meet the functional requirements. Finally, the solutions are assessed for their merits at the concept evaluation stage.

On the right hand side of Fig. 2, there are another two main components. They are the Web page for the concept browser/editor and the concept base. The homepage provides a point of entry to the system. The concept browser/editor allows the geographically distributed team to design and develop the structure and contents of the concept base. The concept base is the central database that is linked to all the other modules. It contains generic functional requirements expressed as goals, potential solution principles (means) and their relationships. It also serves as a repository for intermediate and final results. As an illustration of the terminology, consider the design of assembly systems for electrical plugs. The goals represent the desired functions of an assembly system, including feeding, transportation and handling. Means often used to achieve these functions include vibrating feeders, robotic handling and conveyors.

The Intranet module briefly outlined above has been beneficial in a number of areas to the designers in the customer and supplier organisations:

- *Developing specifications* - assists in changing, revising and modifying tolerances, features and specifications;
- *Interchangeable parts* - provision of parts and components that have commercially available specifications that can be used interchangeably to produce the same product without impairing the intended product utility or function;
- *Part standardisation and simplification* - identifying components which could be standardised and hence the greater availability and abundance of supply sources has led to a reduction in production lead time, resulting in lower product and inventory costs;
- *Part exclusions* - Bringing attention to specified items which had a long or unstable lead-time. Any shortages of such items can seriously hamper line balancing, resulting in costly delays and inefficient use of resources.

Finally, from a management perspective the Web-based design tool has improved the communication linkages between the customer and the supplier. Feedback between the customer and supplier on ongoing designs can occur on a continuous basis. In addition, within the electronics industry product development times are measured in months and companies need to find innovative ways of compressing the time-to-market in order to enhance their speed of response to the final customer. The system outlined above can assist product development by bringing the suppliers organisation into the project at an earlier stage. This should help in minimising design changes (and consequently reduce costs) later on in the process. At the same time, the close communication between designers in both companies as well as other members of the multi-functional procurement team, has enhanced the relationship between the two organisations. In addition, the supplier can now view his contribution as playing an important role in the larger value-adding network of the purchasing company with the potential impact and value that the supply network can add in the delivery of products/services to the final customer can now be recognised.

4 DISCUSSION OF FINDINGS

The key findings that can be drawn from the three case studies presented in this paper are as follows :

- Electronic commerce technologies are eliminating activities which, historically, have been carried out by the relevant participants in both the customer and supplier organisations. Such changes pose an immense challenge to the role of the functions involved in managing the inter-organisational interactions. For example, *CASE 1* illustrates that the traditional roles carried out by the procurement function are being automated or have become the responsibility of the supplier organisation.
- The application of electronic commerce technologies is 'blurring' the traditional boundaries in the value chain between suppliers, manufacturers and end customers. *CASE 1* has shown how the distributor is carrying out activities traditionally performed by the customer including managing an on-site store and component obsolescence for the customer. *CASE 2* illustrates how the supplier designers are taking on a more prominent role in the new product development activities of their customers.

- Innovations in electronic commerce are providing an opportunity for suppliers to add value to their customer's business. *CASE 1* illustrates how outsourcing part of the business to a distributor is reducing the inventory holding and stock holding costs for the customer. In *CASE 2* designers in the supplier organisation are developing solutions that improve the quality, ease of assembly and reduce the cost structure of the end product delivered to the customer. These changes have enabled the suppliers described to move from that of having a passive role to that of being a strategic resource for the customer.
- With electronic commerce technologies automating the transaction type activities, the participants at both the buyer and supplier interface are now able to focus on 'value adding activities'. In *CASE 1* electronic commerce enabled the role of the purchasing professional to move from being involved in clerical type activities, such as invoice processing and expediting, to include activities such as integrating suppliers into their new product development processes and joint involvement in total cost analysis. In this way, it is possible for the purchasing function to make the transition from being a transaction-oriented operation to one that has a strategic focus.
- Electronic commerce enabled suppliers to increase their leverage with customers as illustrated in the following areas :
    - In *CASE 1* the distributor was an important resource to the OEM in terms of the volume of business and service it was providing;
    - *CASE 2* has shown how the supplier has increased its role in the design process of its customer, which consequently should lead to the development of a longer term relationship;
- It has been shown how electronic commerce technologies have changed the traditional roles of the functions involved in managing the inter-organisational interactions. Clearly, such changes have considerable behavioural implications. For example, with the participants at both the buyer and supplier interface performing a more value-adding role there are training and development implications. Related to this is the fact that the participants are being 'empowered' to make more informed decisions through access to information. For example, *CASE 2* has shown how electronic commerce is empowering designers in the supplier organisation to make decisions on specifications.

5 CONCLUSION

The case studies provide evidence that electronic commerce is changing the competitive environment in a number of ways, by:

- reshaping buyer-supplier relationships;
- improving core business processes;
- providing electronic intermediation.

It has been shown that electronic commerce can provide support for relationships with strategic through to commodity suppliers. At one end of the continuum electronic commerce can create joint, inter-organisational processes at the interface between value-added stages, while at the other end increasing efficiency and reducing transaction costs with commodity suppliers. The development of close business partnerships to optimise inter-organisational processes remains one of the most difficult aspects given not only the technology issues but the strategic, cultural and organisational implications. The level of adaptation and co-operation which is becoming necessary in the supply chain means that electronic commerce takes on an increasingly critical role. An understanding of how electronic commerce can be deployed by firms to exchange information and to maintain and build relationships is important as it may impact on their ability to participate in a particular supply chain. Therefore, electronic commerce needs to be viewed in the context of its wider impact in enabling business process redesign, the opportunities it offers for exploiting information, the challenge of integration with internal systems and its implementation through supporting technologies and applications.

REFERENCES

(1) Laudon K.C. and Laudon, J.P. (1997) "Essentials of Management Information Systems - Organisations and Technology", Prentice-Hall.

(2) Harrington, L., Layton-Rodin, D. and Rerolle, V. (1998) "Electronic Commerce : Three Emerging Strategies", The McKinsey Quarterly, Winter, pp152-160.

(3) Nairn, G. (1997) "From Purchasing to Invoicing, Businesses are Linking Up", Financial Times, August 27, p20.

(4) Yin, R. (1984) *Case Study Research: Design and Methods*, Sage Publications.

(5) Rosseger, G. (1991) "Successful Industrial Innovation: Critical Factors for Success", R&D Management, Vol. 22, No. 3, pp. 221-239.

(6) Ulrich, K. and Eppinger, S. (1995) "Product Design and Development", McGraw-Hill, New York.

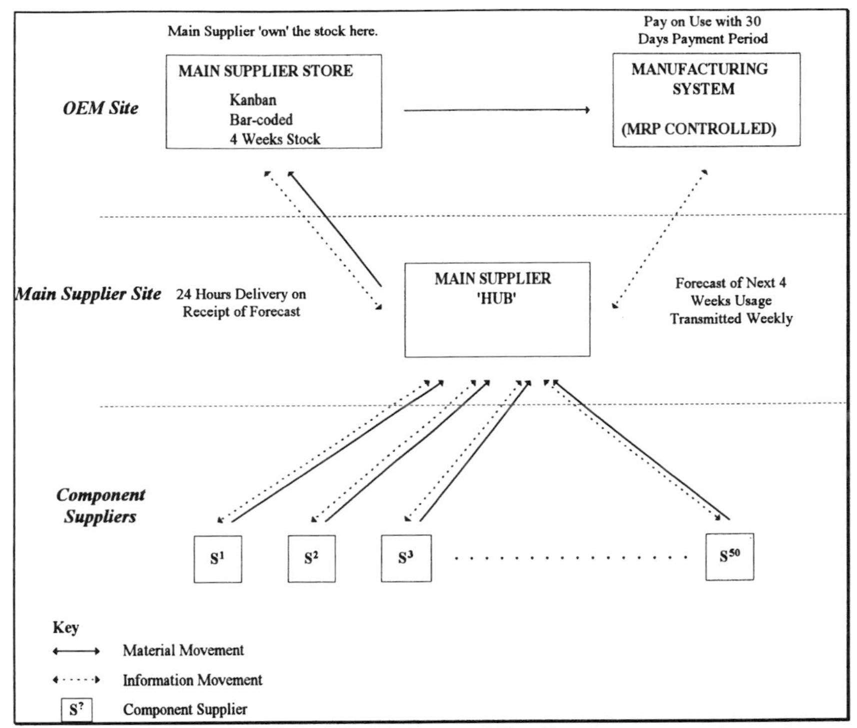

Fig. 1 Structure of Re-Engineered Trading Arrangement

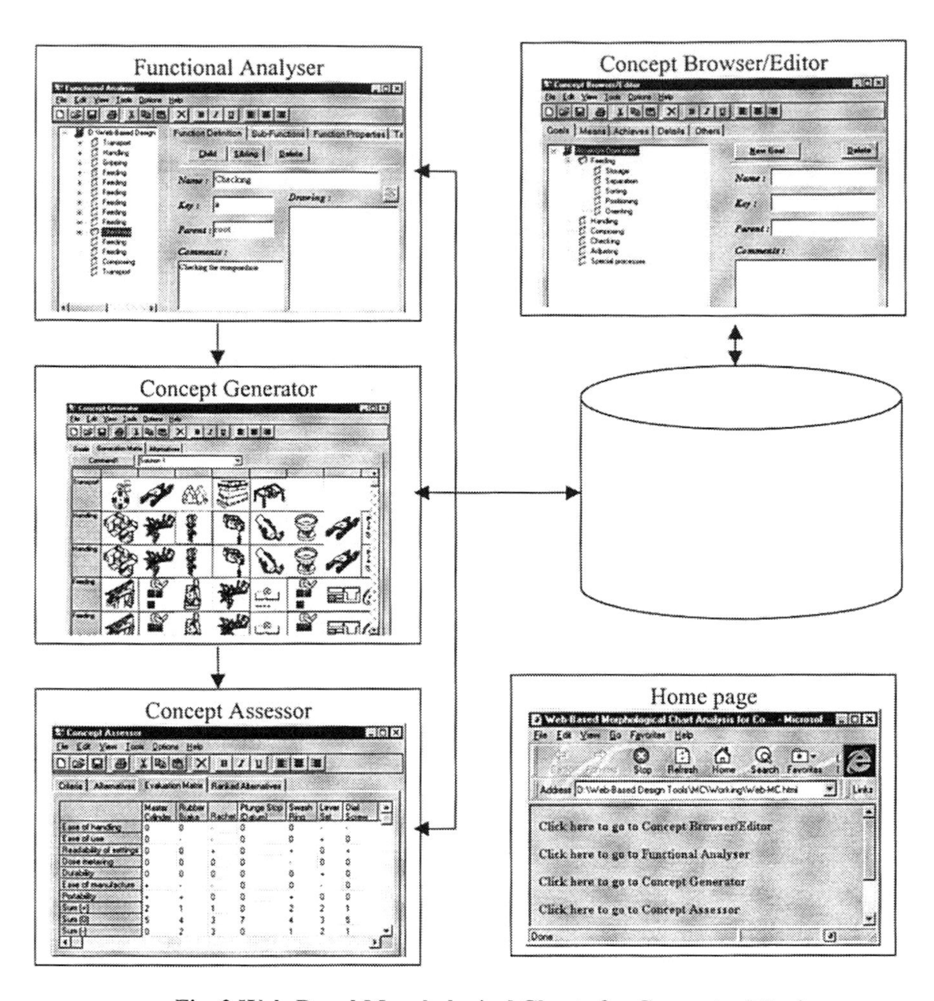

Fig. 2 Web-Based Morphological Charts for Conceptual Design

CAPE/074/2000

A methodology for engineering change impact analysis

G Q HUANG, V LOW, W Y YEE, and **K L MAK**
Department of Industrial and Manufacturing Systems Engineering, University of Hong Kong, Hong Kong

SYNOPSIS

This paper proposes a methodology for evaluating the impacts of Engineering Changes (ECs) or Design Changes on the various aspects of the business across a manufacturing company so that better decisions can be reached regarding the implementation of the proposed changes. The methodology includes a comprehensive worksheet for centralising and displaying the EC data, and a systematic procedure for identifying the causes and effects, with numerical ratings to indicate the occurrence of the cause and severity of the effect respectively. Both the occurrence and severity are used to calculate the priority. The analytical results are made available to the Engineering Change Committee or Board (ECB) who approve or disapprove the changes for further consideration and implementation.

1 INTRODUCTION

Engineering Changes (ECs) are changes and/or modifications in forms, fits, functions, materials, dimensions, etc, of products and constituent components after the design is released. An EC normally induces a series of downstream impacts on the associated products and operational processes in various disciplines across a company. Multiple disciplines are therefore involved in managing ECs. The research and practice literature on ECM is to some extent limited. After a thorough literature search, only about two dozens of relevant research papers have been found (1-20), together with less than half dozen of monographs and various standards (21-24). Basically, the messages from the literature are two folded. On the one hand, ECs are a serious problem within many manufacturing organisation and rigorous actions must be taken to address the problem. On the other hand, the existing system or practice in ECM varies from organisation to organisation. There are international and sectoral standards on ECM or configuration management (CM). Furthermore, there have been a variety of computerised ECM systems (e.g. product data management). However, the issue exists regarding the balance between efficiency and effectiveness. If a rigid procedure is followed as suggested in standards, the efficiency is limited. If ad-hoc ECM is adopted, effectiveness cannot be maintained. Many researchers have attempted to establish general guidelines for ECM, striking a balance between the efficiency and effectiveness. The authors (19, 20) have also addressed an ECM reference framework including steps, activities, approaches, and issues, for dealing with ECs.

However, one of the main issues, Engineering Change Impact Analysis (ECIA), has not yet been discussed in sufficient detail, though its importance and necessity are widely recognised in the literature. This has caused significant difficulties in Engineering Changes. Without evaluating the impacts of a proposed EC, it becomes difficult to classify and prioritise thus decide if this EC is worth making, if so, how best the change should be implemented. This paper contributes to the ECM literature to fill this gap.

ENGINEERING CHANGE IMPACT ANALYSIS WORKSHEET

Name of Analyst:	Name of Product:	Date of Analysis:

1	2	3	4	5	6	7	8	9
EC Title	**Cause**	**Importance Value (I)**	**Impacts**	**Severity Value (S)**	**Priority (PV =I×S)**	**Approval Effectivity**	**Action**	**Comment**

ECM Documents: | EC Request | EC Evaluation | EC Notice

Steps of ECIA Procedure: | 1 | 2 | 3 | 4 | 5 | 6 | 7

Figure 1 ECIA worksheet.

2 ENGINEERING CHANGE IMPACT ANALYSIS: WORKSHEET METHODOLOGY

The proposed ECIA methodology contains two components. One is a comprehensive worksheet that can clearly centralise and display the relevant data of the proposed EC. The other is a systematic procedure that can be easily followed to analyse the impact of the proposed EC thoroughly.

Figure 1 presents an overview of the ECIA worksheet. It is primarily concerned with collating the EC data relevant to the evaluation of the requested ECs in a proper format. It is made up of two parts: the heading and the main body. The heading provides the general information about ECIA projects. Information in the heading includes several basic items, for example, the name of product, the name of analyst, the date of analysis, and so on.

The main body provides specific EC data relevant to the entire ECIA process. The main body consists of a number of rows and columns. Each row corresponds to a proposed EC. One part in the product structure may have multiple ECs. Each column contains data pertaining to different ECM documents and steps of the ECIA procedure. The columns are described as follows:

- *Column 1: EC Title.* It is a short statement that briefly describes the proposed EC.
- *Column 2: Cause.* The reason(s) why a request for the EC is applied is listed in this

column against the proposed EC.

- *Column 3: Importance Value (I).* It is a numeric value or a descriptive qualifier that shows how important or urgent the proposed EC (or its cause) is.
- *Column 4: Impacts.* Potential impacts induced by the proposed EC are identified and listed in this column. The same proposed EC may have different impacts on different business processes or operations.
- *Column 5: Severity Value (S).* It is a numeric value or a descriptive qualifier that shows how severe the proposed EC (or its impact) is from a specific point of view.
- *Column 6: Priority Value (P).* For each proposed EC, a priority value is evaluated based on the importance and severity values identified in columns 3 and 5. That is, P = I * S.
- *Column 7: Effectivity.* It is a date by which the proposed EC should be implemented should it be approved.
- *Column 8: Actions.* Actions can be suggested or identified for the implementation of the proposed EC.
- *Column 9: Comment.* Any supplementary notes can be made in this column.

3 ENGINEERING CHANGE IMPACT ANALYSIS: SYSTEMATIC PROCEDURE

The ECIA methodology is characterised by a comprehensive worksheet which is logical to display the data and a systematic procedure which is easy to follow. This section discusses the steps embedded in the ECIA procedure. The ECIA procedure includes seven steps that basically correspond to the major columns in the ECIA worksheet. This procedure can be followed to analyse a proposed EC by identifying its causes and impacts, quantify the identified causes and impacts with the aid of pre-defined rating references, prioritise and authorise the requested EC, and plan the actions necessary for implementing the approved EC. In practice, it is unnecessary neither to include every step of the ECIA procedure nor follow the order of steps presented. Instead, there is, should be, flexible enough to be extended or tailored to suit the individual needs of companies. In the following, it is attempted to discuss the purpose of each step.

Step 1 - Identifying the causes

This step is concerned with identifying the reasons why the EC is proposed. This is basically an extension of the process of requesting an EC. That is, the originator or proposer of the EC should specify the reasons for proposing the EC in the first place. In fact, this section is usually grouped within the EC request form. The reasons identified at this step will be used in subsequent steps for evaluating and analysing the EC.

Step 2 - Determining the importance

At any time, there may be more than one EC proposed within an organisation. Different ECs may have differing degrees of importance, and therefore require differing degrees of urgency in their processing. Such degrees of importance are quantified by numeric ratings (e.g. between 1-5) or descriptive qualifiers (e.g. very important, important, routine, etc.).

Two issues need to be discussed regarding importance ratings. The first issue is related to the subjectivity or objectivity of the importance ratings. This is affected by the rating criteria. This is a complicated issue because different organisations or industrial sectors may apply different criteria. It is beyond the scope of this work to produce rating criteria for all the sectors. Nevertheless, some efforts have been made to outline a set of general guidelines in Table 1.

Table 1 Importance rating guidelines

Rating	Description
1	The product can work and be made without change. However, the change improves some aspect.
2	The product can still work and be made without change, but at unacceptable cost and/or quality.
3	The product can still work and be made without change, but at **extremely** unacceptable cost and/or quality.
4	The product may not work and/or be made without change.
5	The product neither works nor is made without the change

The second issue is whose importance should be rated: the EC itself or its causes. Obviously, the importance of the EC is derived from the importance of its causes. In theory, the causes or reasons should be rated in terms of their importance and then the EC importance is aggregated. Each EC may have multiple reasons. This implies more efforts for the analyst to rate their importance. In practice, the originator may propose an importance rating for the EC directly without analysing its causes. Both should be accommodated in the methodology.

In cases where each EC has multiple causes and they are individually rated, the issue is how individual ratings should be aggregated to form the overall rating for the EC. One method is to calculate the average of the individual ratings as the overall rating. Another method is to use the maximum cause rating as the overall EC rating. Other possible methods may include calculating the sum or products of the individual ratings as the overall rating.

Several activities may be involved in deriving the importance rating for ECs, for example,

- For each proposed EC, start with the first cause;
- Allocate the importance rating for the chosen cause;
- Repeat until all the causes are rated for the chosen EC;
- Repeat until all the ECs are analysed;
- Calculate the overall importance ratings for all the ECs.

Step 3 - Investigating the impacts

A proposed EC usually has impacts across the operation of the organisation. A series of downstream activities on products, processes, systems, and workflows of various business processes may be induced by an EC, whether the change is as simple as documentary change or complicated as entire product re-structuring. During this step, all these potential impacts are identified and evaluated.

Naturally, the impacts of an EC should be evaluated by concerned functions. In this proposed methodology, each function is invited to identify the potential effects the EC might have on its operation. All the effects are then collated in the ECIA worksheet for overall evaluation. The impact analysis involves several activities:

- For each proposed EC, invite different functions to evaluate the impacts;
- For each function, evaluate all the potential impacts of the EC on its operation;
- Collect all the impacts from all the individual functions and record them on the worksheet;

One problem in impact analysis is that different functions use different methods. For example, the Design Engineering function is mainly concerned with the impacts on the design drawings/documents, the impacts on the designs of other products or components, etc. The Store function is mainly concerned with if any materials in the store need to be scraped or re-purchased. The Manufacturing function is interested in the impacts on the tooling, routing, etc. On the one hand, the proposed ECIA methodology relies on individual functions to exercise their expertise in identifying effects. On the other hand, some efforts have been made to develop some general workbooks for different functions.

Finally, there are positive impacts and negative impacts. The analysts concerned should identify both of them.

Step 4: Determining the severity

The severity of an EC is defined as the degree of the impacts, either positive or negative. Such degree is evaluated in quantitative, qualitative, or semi-qualitative terms. Preferably, the impacts of an EC should be evaluated in numerical measurements. For example, the cost implications or cost savings of making an EC should ideally be calculated so that better informed decisions can be reached. In practice, such absolute quantitative measurement is not possible or the efforts spending on this cannot be justified. In such case, a compromise must be made. One alternative is to use numeric ratings (e.g. between 1-5) or descriptive qualifiers (e.g. very severe, severe, noticeable, negligible, etc.), just like the way that importance is rated for EC causes. The severity rating may involves several activities:

- For each proposed EC, start with an impact,
- Allocate the severity rating for the chosen impact;
- Repeat until all the impacts are rated for the chosen EC;
- Repeat until all the ECs are analysed;
- Calculate the overall severity ratings for all the ECs.

The rating approach is always afflicted by the issue of subjectivity or objectivity. This is affected by the rating criteria. This is a complicated issue because different organisations or industrial sectors may apply different criteria. Even more complicated is that the rating criteria used by the different functions of the same organisation may be different. It is beyond the scope of this work to produce rating criteria for all the sectors. Nevertheless, some efforts have been made to outline a set of general guidelines in Table 2.

A proposed EC usually has multiple impacts and they are individually rated. These individual ratings must be aggregated to form the overall severity ratings for the ECs. One method of aggregation is to calculate the average of the individual ratings as the overall rating. Another method is to use the maximum cause rating as the overall EC rating. Other possible methods may include calculating the sum or products of the individual ratings as the overall rating.

Table 2 Severity rating guidelines

Rating	Description
1	Mandatory disruption to the normal workflow of the business process.
2	Major disruption to the normal workflow of the business process.
3	Moderate disruption to the normal workflow of the business process.
4	Minor disruption to the normal workflow of the business process
5	No/Negligible disruption to the normal workflow of the business process:

Step 5 – Prioritising ECs

The priority of an EC can be established on the basis of (a) the importance of its reasons, (b) the severity of its impacts (positive and/or negative), and (c) a combination of (a) and (b). The ECIA methodology proposes to calculate the priority using the importance and severity ratings. Mathematically, the priority, P is evaluated as follows:

$$P = I \times S$$

where I is Importance Rating and S is Severity Rating

The more important (the causes of) an EC and the more severe (the impacts) of an EC, the higher priority the EC has to be processed and implemented. Otherwise, the EC receives lower priority for treatment. This is a valid rationale regardless of the impacts are positive or negative. The higher the positive impacts, the more beneficial the EC is (e.g. cost savings and quality improvement). The implementation of the EC is justified. On the other hand, higher negative impacts (in terms of absolute values) of an EC imply loss (of human lives or profits). Therefore, the change is also justified.

Based on the priority values, ECs can be rank ordered from the highest to the lowest. The prioritised ECs are submitted to the ECB (Engineering Change Board) meetings for approval or rejection.

For none technical managers, verbal descriptions such as "Routine", "Expedite", "Emergency", and "Mandatory" may be preferred to numeric priority values. If this is preferred by the ECB management, then a scheme is needed to convert the priority values into verbal descriptions. One method of doing this is to divide the priority into several regions, say 10-20 as Expedite, etc. Another method is to divide the priority values by the lowest priority value and then map the results onto several regions in a way similar to the first method. Table 3 exemplifies some possibilities.

Several activities may be involved in prioritising ECs, for example,

- For each proposed EC,
- Calculate its overall importance rating (I),
- Calculate its overall severity rating (S),
- Calculate its priority value $P = I \times S$,
- Repeat until all ECs are evaluated,
- Rank order ECs in terms of their priorities,
- If necessary, work out verbal descriptions about the priority of the ECs.

Table 3 Prioritisation reference.

Criteria	Implication
A: Routine ($PV \le 6.25$)	*Authorising at convenient time*: • If possible, the EC can be authorised at the coming usual EC meeting; or • If there are too many ECs queuing for the authorisation in the coming usual EC meeting, the EC will wait and be authorised at the next usual EC meeting. *Implementing at the most economical and least disturbed time*: • The EC must be implemented within six months. It is expected to implement the subject EC at the time that is the most economical expense and causes the least disruption to the normal workflow of the business. Open negotiation for the date is highly encouraged.
B: Expedite ($6.25 < PV \le 12.5$)	*Authorising as usual*: • If possible, the EC can be authorised at the coming usual EC meeting; or • If an urgent EC meeting is held before the coming usual EC meeting, the EC can be authorised at this coming urgent EC meeting. *Implementing at an economical and little disturbed time*: • The EC must be implemented within a month. It is expected to implement the subject EC at the time that is economical and causes little disruption to the normal workflow of the business. Some flexibility is necessary for the negotiation of the date.
C: Emergency ($12.5 < PV \le 18.75$)	*Authorising as soon as possible*: • If the coming usual EC meeting will be held within a week, the EC will be authorised at this coming usual EC meeting; • If there is an urgent EC meeting that is held for another ECs before the coming usual EC meeting within a week, the EC can be authorised at this coming urgent EC meeting; or • If neither the coming usual EC meeting nor an urgent EC meeting will be held within a week, an urgent EC meeting for the subject EC must be called and held within a week for authorising the EC. *Implementing as soon as possible*: • The EC must be implemented within two weeks. It is expected to implement the subject EC as soon as possible that may cost some certain amount of expense and cause certain degree of disruption to the normal workflow of the business. Little flexibility is possible for the negotiation of the date.
D: Mandatory ($18.75 < PV \le 25$)	*Authorising immediately*: • If the usual EC meeting will be held within two days, the EC will be authorised at this coming usual EC meeting; or • If the usual EC meeting will be held after two days, an urgent EC meeting must be called and held within two days for authorising the EC. *Implementing immediately*: • The EC must be implemented within a week. It is expected to implement the subject EC immediately without unnecessary delay, regardless of how much the expense is and how serious the disruption to the normal workflow of the business is. Seldom flexibility is allowed for the negotiation of the date.

Step 6: Authorising and approving EC proposals

Authorisation must be obtained in some form before a proposed EC is implemented. That is, the decision on whether the proposed EC should be implemented or not must be reached, if so, by what deadline (known as effectivity). For crucial ECs, the associated implementation plan should also be assessed and approved.

In organisations where formal ECM procedures/systems exist, an EC board assisted by an EC coordinator is responsible for approving the ECs. One usual form of approval process is to hold ECB meetings which are attended by delegates from various functions or units.

At the ECB meetings, the packages of the proposed ECs are examined and reviewed in the order of priorities established in previous steps. ECs with lower priorities may be approved or disapproved without in-depth investigations.

Step 7: Planning for implementation actions

Authorisation and approval discussed in the preceding step determines if a proposed EC should be made. This step is mainly concerned with how the EC should be implemented if approved. The input to this step includes the whole package about the EC, including the request, evaluation results, approval decisions (effectivity, etc.), and any relevant supporting documents. The output from this step is an action plan that describes clearly what should be done by whom, where and when. The action plan will form the main contents of the EC notice that is circulated to the parties or functions concerned for taking implementation actions.

One crucial issue in planning for EC implementation is the effectivity that may be specified by dates, product batches, or production schedules. Major actions must be planned and thereafter taken before the effectivity with some actions after the effectivity.

It is unlikely or unreasonable that the EC board should be tasked with the action planning at the EC board meetings. One reason is that the meeting schedule would be too tight for action planning. Another reason is that the delegates in the EC board may not be directly involved in the EC implementation and therefore cannot decide what exactly should be done in detail.

There are two possible points where actions can be formulated. One is before the approval by the EC board when the proposed EC is being evaluated in terms of causes and effects. The other is after the approval. Logically, the person who is involved in assessing the impacts and causes of the proposed EC is the best person to propose possible actions for implementing the EC. An early commitment approach is suitable to accommodate this option. Based on the proposed action plan, the EC board is able to approve the EC and its associated implementation plan. Alternatively, the board may suggest further improvement of the implementation plan or defer the planning when the EC notice is circulated.

4 CONCLUSIONS

This paper has proposed a new methodology for processing engineering changes from the request stage through impact evaluation stage to the approval and implementation stage. The proposed methodology is generic enough to be easily tailored or extended to suit specific requirements of manufacturing companies. The ECIA worksheet is comprehensive in the sense that it is used for centralising and displaying the EC data related to the ECIA. The ECIA procedure is systematic with a step-by-step approach.

The proposed ECIA method is being incorporated into a web-based prototype software system for engineering change management. It is expected to overcome many limitations of the paper-based ECM environment so that the sufficient effectiveness and efficiency can be achieved. Two prototypes have been developed, one with "fat-clients" and the other with "thin-clients". The design, development and implementation of the systems will be discussed separately in the near future.

REFERENCES

1. Balcerak, K.J., Dale, B.G. (1992) "Engineering Change Administration: The Key Issues", *Computer-Integrated Manufacturing Systems*, Vol. 5, No. 2, 125-132.
2. Boznak, R.G. (1993) *Competitive Product Development*, Milwaukee, WI: Business One Irwin/Quality Press.
3. British Standard 6488: 1984 "Configuration Management of Computer-based Systems", 1984, BSI.
4. Choi, F.C., Chan, L. (1997) "Business Process Reengineering: Evocation, Elucidation and Exploration", *Business Process Management Journal*, Vol. 3, 39-63.
5. Dale, B.G. (1982) "The Management of Engineering Change Procedure", *Engineering Management International*, Vol. 1, 201-208.
6. Diprima, M. (1982) "Engineering Change Control and Implementation Considerations", *Production and Inventory Management Journal*, Vol. 23, Part 1, 81-87.
7. Gianpaolo, C., Alfonso, F., Rierfrancesco, F., Gresse, W.C., Luigi, L., Serena, M., Rosaria, M., Guenther, R., Roberto, S. (1997) "A Case Study of Evaluation Configuration Management Practice with Goal-oriented Measurement", *International Software Metrics Symposium*, November, 5-7.
8. Harhalakis, G. (1986) "Engineering Changes for Made-to-order Products: How an MRP II System Should Handle Them", *Engineering Management International*, Vol. 4, 19-36.
9. Harris, S.B. (1996) "Business Strategy and the Role of Engineering Product Data Management: A Literature Review and Summary of the Emerging Research Questions", *Journal of Engineering Manufacture*, Vol. 210, 207-219.
10. Hegde, G.G., Kekre, S.H., Su, H. (1992) "Engineering Changes and Time Delays: A Field Investigation", *International Journal of Production Economics*, Vol. 28, 341-352.
11. Leech, D.J. Turner, B.T. (1985) *Engineering Design for Profit*, Ellis Horwood Limited, Chichester, England (Chapter 12).
12. Maull, R., Hughes, D., Bennett, J. (1992) "The Role of the Bill-of-Materials as a CAD/CAPM Interface and the Key Importance of Engineering Change Control", *Computing & Control Engineering Journal*, March 1992, 63-70.
13. Nichols, K. (1990) "Getting Engineering Changes Under Control", *Journal of Engineering Design*, Vol. 1, No 1, 1-6.
14. Reidelbach, M.A. (1991) "Engineering Change Management in Long-lead-time Environments", *Production and Inventory Management Journal*, Vol. 32, No. 2, 84-88.
15. Saeed, I., Bowen, D.M., Sohoni, V.S. (1993) "Avoiding Engineering Changes Through Focused Manufacturing Knowledge", *IEEE Transactions on Engineering Management*,

40 (1), 54-58.

16. Terwiesch, C., Loch, C. H. (1999) "Managing the Process of Engineering Change Orders: The Case of the Climate Control System in Automobile Development" *Journal of Product Innovation Management 1999*, Vol. 16, 160-172.
17. Watts, F. (1984) "Engineering Changes: A Case Study", *Production and Inventory Management Journal*, Vol. 25, Part 4, 55-62.
18. Wright, I.S. (1997) "A Review of Research into Engineering Change Management: Implications for Product Design", *Design Studies*, Vol. 18, No. 1, pp 33-42.
19. Huang, G.Q., Mak, K.L. (1997b) "Engineering Change Management: A Survey within UK Manufacturing Industries", In: *Proceedings of International Conference on Managing Enterprises*, Loughborough University, July 1997.
20. Yee, W.Y., Huang, G.Q., Mak, K.L. (1998) "Towards a Framework for Engineering Change Management", In: *Proceedings of 3rd Annual International Conference on Industrial Engineering Theories, Applications and Practice*, Hong Kong, December 1988.
21. ISO 9000: "Quality Management and Quality Assurance Standards", 1994.
22. ISO 10007: "Quality Management - Guidelines for Configuration Management", 1995.
23. Military Standard: :Configuration Management 973", 1992.
24. Monahan, R.E. (1995) *Engineering Documentation Control Practices and Procedures*, Marcel Dekker, Inc, New York, USA.

Concurrent Engineering and
Design for Manufacture

CAPE/052/2000

Linking product design and manufacturing capability through a manufacturing strategy representation

W M CHEUNG, J ZHAO, J M DORADOR, and **R I M YOUNG**
Department of Manufacturing Engineering, Loughborough University, UK

ABSTRACT

This paper presents a global manufacturing data model that can provide information structures to capture the manufacturing capability information in a global enterprise. In particular it focuses on a manufacturing strategy representation which links shape production to manufacturing processes and resources.

An object oriented manufacturing model based on the machining process, designed using UML and implemented using ObjectStore is discussed. A range of machining strategies for pockets, holes and planar faces are represented and their links to manufacturing processes and resources is described. The relationship between this model and manufacturing features within a product model is highlighted.

1 INTRODUCTION

The traditional feature based design approaches that have been the subject of integrated CADCAM research for many years are typically concerned with the incorporation of manufacturing features into CAD models (1). These features are generally geometric descriptions with some added, but limited, manufacturing information. Although these approaches include the manufacturing information required by the product, they keep only one possible method of manufacturing in one factory with specific facilities, making them inadequate for supporting products in global enterprises (1). Further, CAD systems focus heavily on the geometric description of a product and hence relevant manufacturing information known to the designer is lost because it can not be stored in these models (2).

The provision of relevant manufacturing information to support design decisions is important. Some component parts of a product may be costly simply because the designer did not understand the capabilities and constraints of the production process. For example, a designer may specify a small internal corner radius or excessively tight tolerances on a machined part without realizing that physically creating such a shape requires an expensive machining operation (3). The provision of such information is crucial so that the constraints of a process can be concisely communicated to designers.

In global enterprises products can be manufactured in different factories that have different resources and processes. In order to provide a flexible representation of manufacturing information that can support a range of alternative manufacturing methods there is a need for a new approach that goes beyond traditional machining features. This paper proposes such an approach based on the use of both product and manufacturing models.

Manufacturing models aim to capture manufacturing capability information in terms of manufacturing processes, manufacturing resources, and the manufacturing strategies that constrain their relationship (4). The focus of this paper is on the definition of machining strategies within a manufacturing model and how these can be flexibly linked to feature descriptions within a product model.

Machining Strategies can be defined as methods that utilise resources and processes to reduce manufacturing cost and increase manufacturing productivity. Defining sets of machining strategies involves defining sets of machining operations along with the constraints upon their use. The availability of tools, machine capability, surface conditions of the workpiece, component shape and dimensions are among the constraints to be considered. This paper explores how a manufacturing model can capture these alternatives and their constraints, relate them to available processes and resources in the enterprise and use this combined set of information to support design decision making. In addition to defining the necessary data structures, this paper goes on to explain the development and implementation of an Object-Oriented experimental environment. A simple example is illustrated to highlight the value of the approach taken.

2 MANUFACTURING INFORMATION MODELS

'Manufacturing Information Models' is the title of an EPSRC research grant (5) concerned with understanding of the roles of the product and manufacturing models and the enhanced data structures they require in order that they can support the generation of manufacturing information. The Product Model captures the information related to a product throughout its life cycle, whilst the Manufacturing Model captures the information of manufacturing facility (4). Both models have been defined as the central elements of a Model Oriented Simultaneous Engineering System (MOSES) (6) to support design and manufacturing functions in the product realization process (4).

The high level structure of a Manufacturing Data Model (7) captures manufacturing facility information from a general enterprise level through factory, shop and cell levels and down to individual station level manufacturing information. It is at the station level that this paper is concerned.

An example of a set of manufacturing resources and processes for machining is illustrated in Figure 1. Thus, resources include cutting tool descriptions and machine descriptions. Processes for machining in this case are defined as Machining Operations such as EndMilling, Drilling and FaceMilling etc. Strategies for machining capture the alternative methods by which machining features, such as pockets, holes and planar faces can be produced.

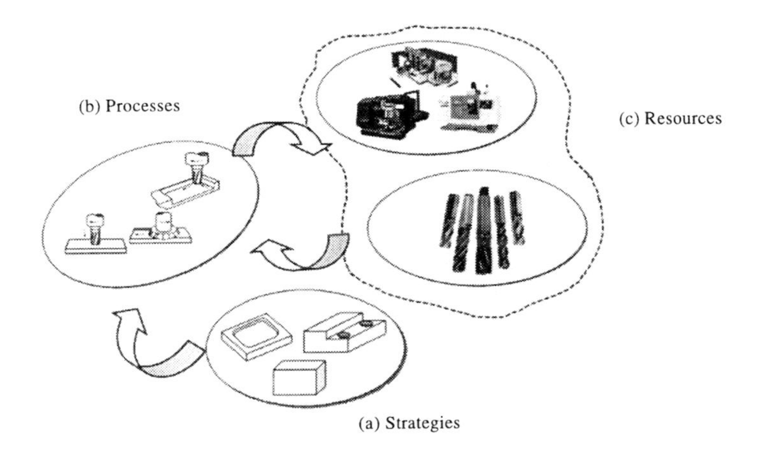

Figure 1 Shape production methods decisions

3 LOW LEVEL MANUFACTURING STRATEGIES IN A MANUFACTURING DATA MODEL

The role of Manufacturing Strategy for machining at the station level of a manufacturing model is concerned with the definition of sets of methods which can be used to produce particular shapes. Strategies have been represented in the Manufacturing Model for the typical machined shapes of ThroughPocket, ClosedPocket, RoundHole and PlanarFace. These are illustrated as a set of shapes in Figure 1(a), but it should be registered that the strategies capture the range of ways in which these shapes can be produced. These machining strategies are termed as ThroughPocketStrategy, ClosedPocketStrategy, RoundHoleStrategy and PlanarFaceStrategy in the Manufacturing Data Model. An example of producing a cylindrical hole is used to describe the development of RoundHoleStrategy.

There are a number of processes and process relationships associate with producing a cylindrical through or blind hole such as drilling, slot drilling, reaming and boring. Figure 2 illustrates these processes and some of their relationships. How these are used is dependent on the limits require on the hole in terms of dimensional accuracy, positional accuracy, size and surface finish. Centre Drilling is used for positional accuracy by ensuring the drill does not drift off the centreline. Improvements in finish and accuracy of drilled holes can be made by reaming. Boring offers the maximum in accuracy, roundness, alignment, straightness and finish. The range of methods to produce a hole defines the set of RoundHoleStrategies as illustrated in Figure 3. The data in the illustration can be captured in the Manufacturing Model along with the limitations and constraints on their use.

illustrated in Figure 3. The data in the illustration can be captured in the Manufacturing Model along with the limitations and constraints on their use.

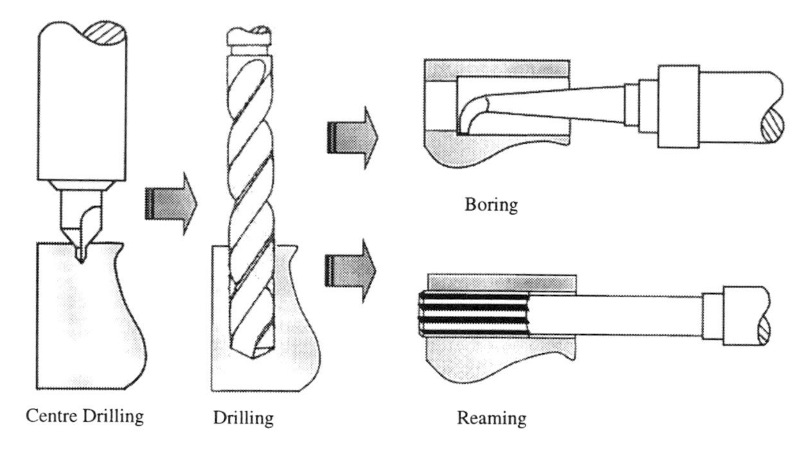

Figure 2 Hole producing processes and their relationships

The strategies are governed and restraint by the rules that indicates the operations needed, surface roughness requirements, dimensional and positional tolerances, the minimum and maximum hole diameter range etc. These rules typically relate to the condition of the workpiece before or after a process. The rules have therefore been captured as ‘Pre’ and ‘Post’ conditions of the operations in the manufacturing model. The availability of these options in the Manufacturing Data Model offers information to directly support design for manufacture decisions.

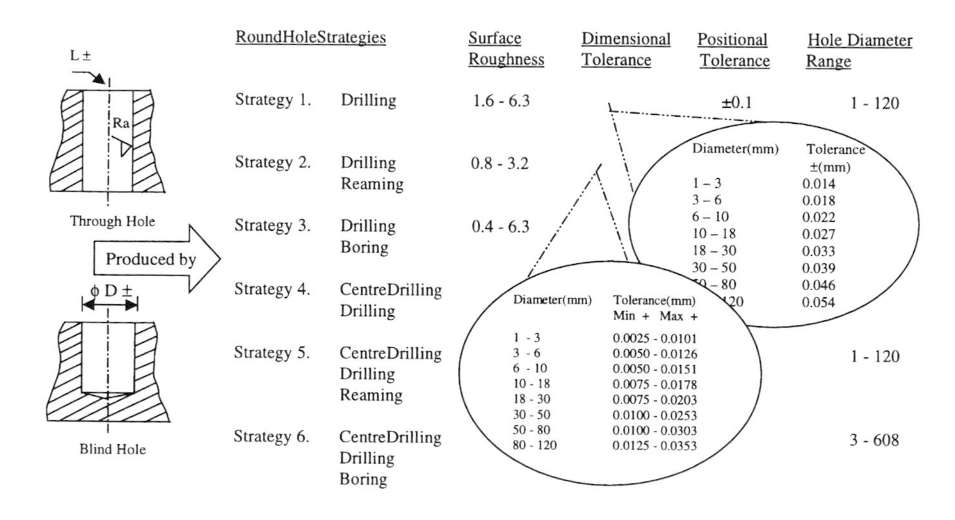

Figure 3 Different strategies for hole making

Feature based systems typically link manufacturing methods directly to product features. In this work the product feature parameters can be used to search the Manufacturing Model for appropriate manufacturing strategies. Figure 4 uses a simple cylindrical through hole to highlight the relationship of machining strategies to Manufacturing Processes and Resources. A set of basic parameters from a Product Model will be extracted and used to perform queries on the Manufacturing Model. The RoundHoleStrategies to be selected depends upon the information and rules stored in 'Pre' and 'Post' conditions of individual machining processes in the Manufacturing Model. Thus, an appropriate sequence of operations and resources associate with that particular strategy will be output to produce the shape.

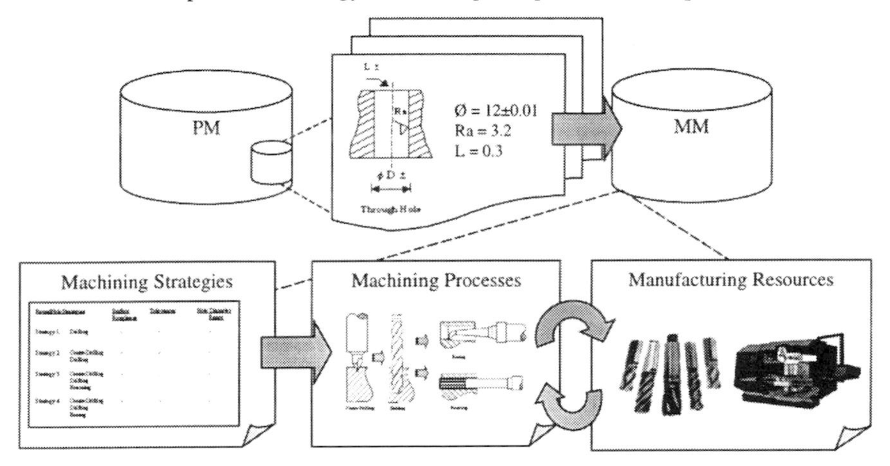

Figure 4 Relationships between product and manufacturing information

4 MANUFACTURING STRATEGIES LINK TO 'PRE' AND 'POST' CONDITIONS

Figure 5 is an example to illustrate how the Manufacturing Model is used to capture the relevant manufacturing capabilities. The example emphasizes particularly the relationship of MachiningStrategy with MachiningOperation 'Pre' and 'Post' conditions.

The example is focused on two strategies, *'Drilling'* and *'CentreDrilling / Drilling / Reaming'* respectively. The first strategy *'Drilling'* supports the production of a cylindrical hole with low requirements on surface roughness (1.6 – 6.3) μm and positional tolerance ± 0.1 etc. Vice versa, if better quality requires for the finishing product then a second alternative such as *'CentreDrilling / Drilling / Reaming'* will be used. In order to select the appropriate RoundHoleStrategies the Product Model is checked to identify the limitations specified by the product designer. The manufacturing model can then check against surface roughness, tolerances etc in the MachiningOperation 'Post' conditions and to establish the possible finishing operations. The 'Pre' conditions will establish the operation sequence needed.

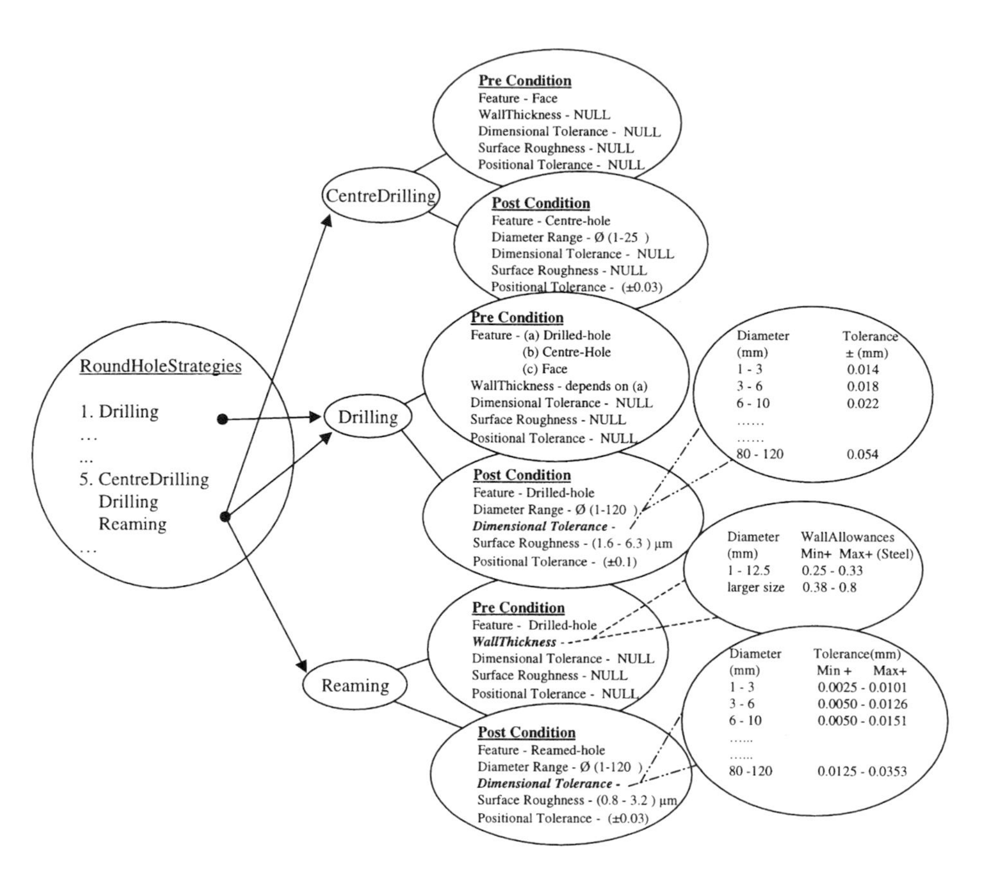

Figure 5 Example of manufacturing strategies link to 'Pre' and 'Post' conditions.

5 AN UML OBJECT ORIENTED MANUFACTURING DATA MODEL

Figure 6 illustrates a single conceptual manufacturing model and represents the high-level UML diagram. This is the core element of the complete MDM infrastructure which consists of main classes and relationship for a manufacturing enterprise. Details and descriptions of individual class and their relationship can be found in reference (7).

The low-level representation of this MDM has been developed based on the Resources, Processes and Strategies super-classes. The low-level structure has focused on the capture of manufacturing facility information at the station level. Manufacturing Resource is defined as the base class of Resources and consists of RawMaterial, MachineCentre and Tooling respectively. Tooling consists of several sub-classes such as CuttingTool and CuttingToolType. Process at this level is described as MachiningOperation including CentreDrilling, Drilling, Reaming, Boring, SlotDrilling, RoughEndMilling etc. Each operation has a unique set of 'Pre' and 'Post' conditions which dictates the process capabilities for that operation.

Machining Strategy is defined as the *base class* of Strategies and including *sub-classes* of throughpocket, closedpocket, roundhole and planarface strategies. The fundamental aspect of Machining Strategy class is to provide all the methods for manufacturing that shape. These methodologies are sequences of machining operations, where each operation is defined in the MachiningOperation class. Thus, Machining Strategies is the core element in Manufacturing Model to support information sharing and interact with Product Model.

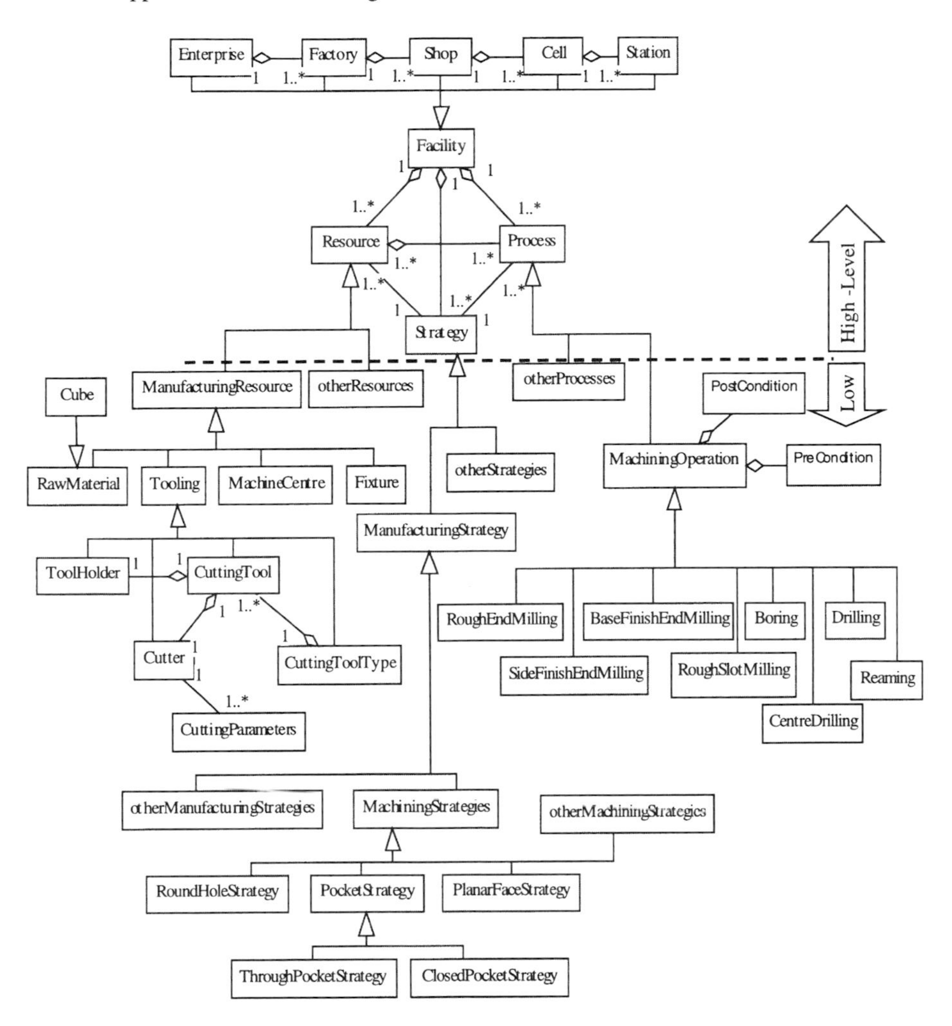

Figure 6 Manufacturing data model in UML representation

DISCUSSION/CONCLUSION

This paper has defined the relationship between manufacturing features in a product model and manufacturing facility information through the use of manufacturing strategies. In this way the information contained in two distinct, but related, models can be shared. One of the advantages of information sharing in this way is that product designers can check manufacturing capabilities to enhance and assist the design and subsequently identify product manufacturability. This is a significant advance on traditional feature based approaches in terms of the flexibility of manufacturing representation.

The machining strategies being implemented in the experimental system are based on the basic shapes of pockets, roundholes and plannarfaces. The Manufacturing Data Model is being constructed using the ObjectStore Object Oriented Database Management System and Visual C++ programming language.

ACKNOWLEDGEMENT

The authors would like to acknowledge the support of the EPSRC and the industrial collaborators. This paper is presented as part of the Manufacturing Information Models research project funded under EPSRC grant GR/L41493. The research undertaken by J.M.Dorador is funded by UNAM and CONACYT, Mexico.

REFERENCES

(1) Allada V., Anand S., 1995, Feature-Based Modelling Approaches For Integrated Manufacturing: State Of The Art Survey And Future Research Directions, International Journal Computer Integrated Manufacturing, 8 (6), pp. 411-440

(2) Lim, S.S., Lee, I.V.H., Lim, L.E.N., Ngoi, B.K.A., 1995, "Computer-Aided concurrent design of product and assembly processes: a literature review", Journal of Design and Manufacturing (5), pp. 67-88

(3) K.T. Ulrich, S.D. Eppinger, Product Design and Development, McGraw-Hill, Inc, 1995

(4) A. Molina, T.I.A. Ellis, R.I.M. Young and R. Bell, "Modelling Manufacturing Capabilities to support Concurrent Engineering." Concurrent Engineering, Research and Application Journal, 3(1) March 1995.

(5) R.I.M. Young and R. Bell, EPSRC Proposal, "Manufacturing Information Models", Department of Manufacturing Engineering, Loughborough University. 1996.

(6) T.I.A. Ellis, A. Molina, R.I.M. Young and R. Bell, "The Development of an Information Sharing Platform for Concurrent Engineering." Presented at IMSE '94, Grenoble, France, December 1994, pp.12-14.

(7) J. Zhao, W.M. Cheung, R.I.M. Young, R. Bell, " An Object Oriented Manufacturing Data Model for a Global Enterprise" 15th International Conference on Computer-Aided Production Engineering (CAPE'99), Durham, UK, 19-21st April 1999, pp.582-588.

Manufacturing analysis of conceptual and embodiment aerospace designs – an aggregate product model specification

K R McKAY, D G BRAMALL, P G COLQUHOUN, and **P G MAROPOULOS**
School of Engineering, University of Durham, UK

SYNOPSIS

In conjunction with a major European aerospace company a knowledge enriched aggregate process planning system, CAPABLE Space, is being developed to allow integrated design and manufacturing decisions to occur at an enterprise wide level. The aim of the system is to allow rapid manufacturability assessment of alternative designs and calculate the impact on existing plans from changes to the design, process and resource selections. Reported here is the first step in attaining this goal, namely an aggregate product model, which provides a representation of the design in terms of manufacturable features. Initially a review of the role of features within a product model is presented, relating the current research to those preceding, specifically to the previous CAPABLE system. An overview of the CAPABLE Space system is given and the object oriented aggregate product model is presented along with an explanation of its functionality and examples of the model applied to a satellite panel.

1 INTRODUCTION

Through the ever-increasing use and spread of the Internet to all sections of industry, an opportunity exists to combine a considerable number of supply network companies into a convenient portal site that systematically represents and assesses their manufacturing capabilities. A portal site is a one stop location that performs a variety of tasks, providing users with email and home pages and on a macro level supplies search engines, chat rooms, perhaps news and reviews and online shopping facilities.

Applying this to a manufacturing context requires that some individual companies are now users of the portal with their home pages being supplemented with areas that contain product, process and resource models that which (like html) are universally understood by the portal system. These models (or part thereof) are required to be; widely available to all users of the portal, discretely available to a selection of users or available to a specific user through delivery or automatic request by the system. An integrated process planning tool capable of inputting these distributed models and generating a plan that, like the above models, can be made available to a variety of users is a prerequisite to its successful application. With such a quantity of information available, the final requirement of the portal is to provide product data management (PDM) utilities to store versions of models and plans and act as the knowledge repository.

The aggregate process planning system CAPABLE Space, is being developed by the Design and Manufacturing Research Group (DMRG) at the University of Durham. Its aim is to quickly investigate the potential costs, time to market, manufacturing quality and design alternatives in order to maximise any potential improvements from the available design or process alternatives. With models in place the system also aims to be able to provide assessments regarding changes made to any of the models that from part of a process planning domain. This is part of the ongoing research and is a successor to the first version of CAPABLE (Concurrent Assembly and Process Assessment Blocks for Engineering manufacture) [1]. CAPABLE Space extends the original areas of heavy steel fabrication and welding to high technology assembly, joining and fabrication as used by the aerospace sector.

Figure 1 shows an overview of CAPABLE Space with a main company acting as the administrator of the portal and uses a secure intranet connection for internal operation of the software. Externally, the network of suppliers is connected via an internet connection. All of this is hidden from the users who deal only with the CAPABLE desktop utilities and from this interface can access the repository, create and modify products, resources and processes and perform a process plan.

In order to perform the planning at an early stage in the design, process planning tasks are clustered into three levels, namely, *aggregate*, *management* and *detailed* [2,3]. Each level is defined by the granularity of process modelling considerations with respect to the requirements of design and production control and the time cycle of concurrent activities in the design process. This paper introduces an Aggregate Product Model (APM) which concentrates on those elements most prevalent within the aerospace sector. Such an APM is necessary due to the lack of existing models which have sufficient functionality as is necessary here. A related paper [4] (presented at this conference) discusses the aggregate process model and its functionality within the system.

2 THE ROLE & REQUIREMENTS OF FEATURES

Feature based design comes in two distinct flavours for integrating the design and manufacturing tasks. With the first method, products are modelled via the use of a taxonomy of pre-defined features. The second allows the designer to create the product with the use of a

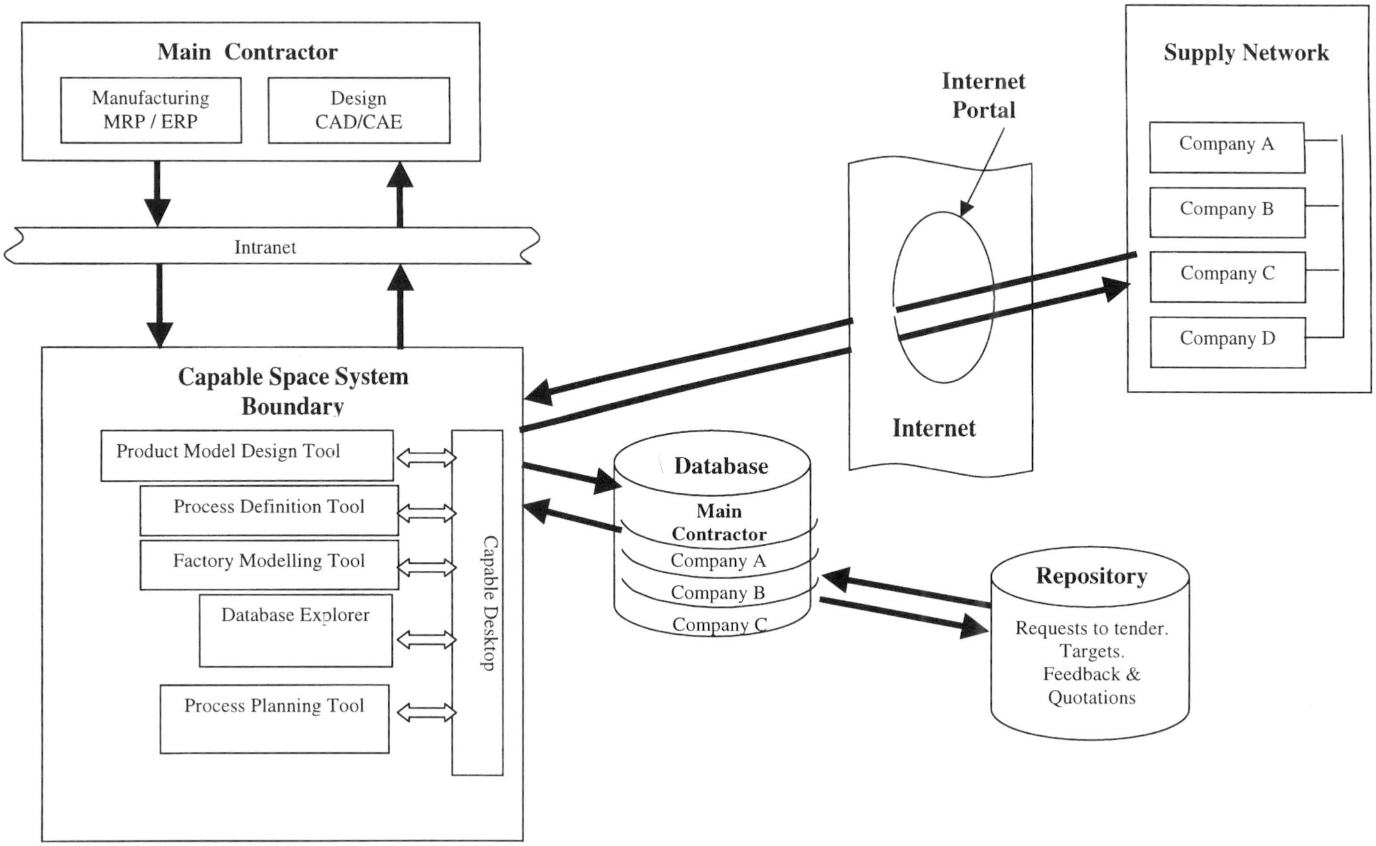

Figure 1. Overview of the distributed CAPABLE Space system.

geometric modeller, this is then followed by a post analysis to obtain features via a recognition algorithm [5]. The APM reported here uses the first of these two methods.

The basic building block of a product model is the feature, which has a requirement to perform three main tasks. The first is to be an encapsulation of knowledge [6]. Since the features exist within a taxonomy, a great deal of information is inferred from the classification and this can be increased via modifiers in the from of specific dimensions (such as radii, length etc). More complex features can include tolerances, end face constraints [7] and tool profiles [8].

Secondly, in order to represent products, the features must support complex inter-relationships [9] and this interaction forms the basis of the product model. Features constantly and inevitably interact with one another [10] from pose interactions which can be largely overcome via the application of tolerance entities to complex "Assembly feature connections"[11] or "Weld features"[12], which are specific instances of a feature, that provide high level information regarding assembly and joining methods. Large degrees of interaction occur simply from the requirement of one feature to align with the corresponding joining features.

To support both the design and manufacturing elements is the third task. A feature requires to link the geometric data to the information required by manufacturing. The concept of perspective [5] therefore exists that at its simplest recognises that a product model composed of *design features* would differ from a product model composed of *manufacturing features* when applied to the same model. For example a boss is more useful to a designer than to a manufacturer who would be more interested in the negative features that may be cut from around the boss in order to make it.

The two primary inputs to the APM were the existing product model of the previous CAPABLE system [13] which concentrated upon manufacturing features and elements of the Standard for the Exchange of Product model data (STEP) international standard [7]. Overlap does occur for many features and for compatibility issues the original CAPABLE features are used, whereas the STEP features introduce tolerance entities into the APM.

3 OVERVIEW OF CAPABLE SPACE

A high level description of the functionality of the CAPABLE Space system is to map features to processes, processes to resources and by applying routing optimisation to create the process plan. This entire process can be broken down into the following six steps:

1. Product model generation. The product model is the initial starting point for use of the system and is discussed in greater detail below.
2. Process option generation. The process option generation algorithm operates on each feature, querying a knowledge base of feature process combinations to assign process alternatives to features. The process options are assessed against the constraints (geometrical, quality) defined by the product model to eliminate any unsuitable processes. Each mapped process contains the procedural knowledge required to instantiate ancillary process objects required for pre/post processing.

3. Process Selection. Initial process selection is carried out on the basis of cost. The cycle time for processing each feature is calculated, using data aggregated over the resource type.
4. Machine Option Generation. A second mapping allocates a list of suitable resources for each process in the process set. Again, as with process selection, constraints are applied to eliminate resources which are unsuitable for reasons of volumetric or quality capability.
5. Job creation. The optimisation algorithm then creates an initial and valid process plan, which is a linear routing of jobs. A job is defined as a feature having a process occurring on it at a specified resource. In order to allow for a basic capacity checking each resource is booked out on its own diary.
6. Sequencing. Using a simulated annealing algorithm the initial process plan is optimised using user defined criteria, which are based upon a function of quality, cost and delivery.

4 THE AGGREGATE PRODUCT MODEL

The requirements of the APM are similar to the previous CAPABLE system but with the addition of management functions that allow for the engineering change evaluation to occur. This can be summarised via four principle objectives:

- The aggregate product model must support the transition of the design from the uncertain conceptual design through to the detailed design.
- The aggregate product model must represent the design in a format that is suitable for integration with the aggregate process planning system.
- The aggregate product model should have the capability for "engineering change management" of the design.
- The aggregate product model should provide a confidence on its accuracy.

4.1 Structure of the aggregate product model.

Figure 2 depicts a Universal Modelling Language (UML) description of the top level classes that comprise the product model. The complete taxonomy of all the features is too large to be shown here. Each feature specifies a form that has intrinsic meaning and this from is controlled via a few dimensional variables. Referencing figure 2, each of the pertinent features is discussed below.

The top most class within the product model is the *PMComponent* (Product Model Component) class. *PMComponents* are used to link *positive* features, *negative* features, *joint* features and child *PMComponents* in a meaningful sequence. The product model then takes on the form of a tree, with growth occurring with the addition of features and components in parent-child relationships. Child *PMComponents* provide clarity to the model and allow the systematic decomposition of the model by the process planning engine. The only inheriting class from *PMComponent* is the *Wire Harness Assembly* class and as the name implies is used as a specific instance to create a product model of a wire harness. Such features are simplified versions of those reported in [14].

The *positive feature* is a mechanical part with a simple or complex geometry that is composed of a single piece of material. In figure 2 only the very top *positive features* are shown. By dividing the features between prismatic and axi-symmetric, the process planning algorithm is

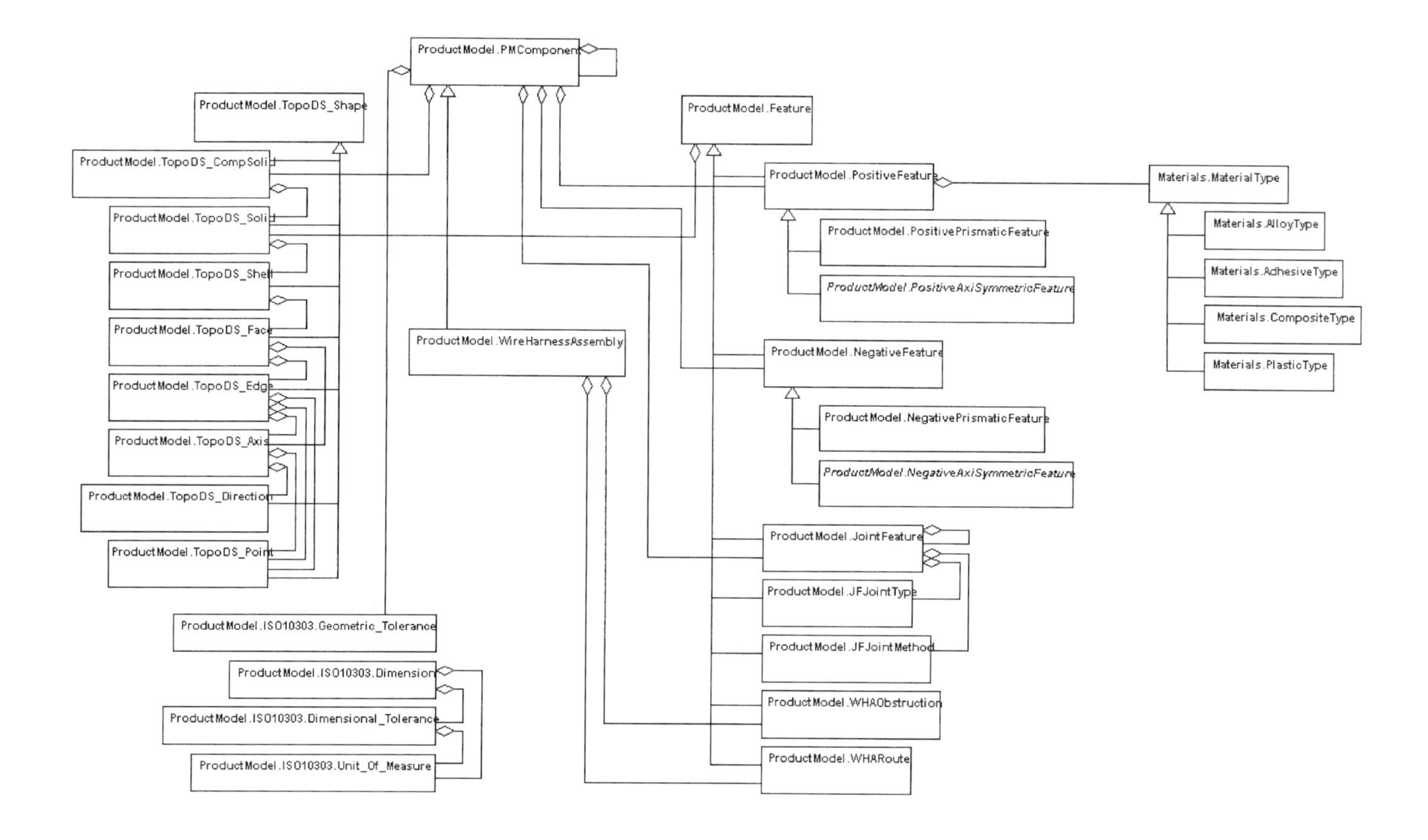

Figure 2. A simple UML description of the product model and its main classes.

able to distinguish when to apply turning as a process. The division continues by splitting the prismatic features into sheets, solids and formed parts and further still until sheet blanks, solid L-frames and cast parts are reached. It is these bottom level classes that are added to the product tree.

It should be noted however, that some *positive features* could be re-evaluated as *PMComponents*. For instance, a tube could be implemented either as a *positive feature* or as a new *PMComponent* via a *joint feature* and a *positive sheet feature*. With the *positive feature* classes largely defining geometry, the material that they are made from is provided via a database of the *material type* class. Each of the child *material types,* such as the *alloy type* or *adhesive type* is instantiated with sufficient data to specify material that fall into such groupings. This instantiated material object is then used to define the type of material that the positive feature is composed of.

Negative features relate to the material removal processes and like the *positive features* are split between axi-symmetric and prismatic types. The taxonomy continues with child *negative features* defining a more specific instance of a *negative feature*, for example the *negative prismatic feature* has holes, slots, edge cuts and face features as its children. Holes are further specified into through and blind holes, through and blind pockets, counter sunk and counter bored entities. These bottom level features are added to the product tree and by referencing the *positive features* that they interact with the process planning algorithm is able to obtain processes suitable to the cut material.

Joint features define the shape and physical method for the joining of two or more components/positive features. The *joint feature* consists of a *joint type* and a *joint method*. The type is used to define the shape of the joint such as tee, butt and lap, whilst the method defines the actual process implementation that should be used to make the connection, for example riveted or bonded. Joint features, like *PMComponents,* can have child *joint features* and this allows complex joints to be created with each child describing one part of the complex joint by its type and method.

The *topology* classes relate the features to topological entities that occur within a CAD system thus linking the manufacturing features to geometry; the primary mover for this is the TopoDS_Solid which occurs for each feature. Two exceptions to this are the *PMComponent* which instead has TopoDS_CompSolid and the *joint features* which have no topology. The TopoDS_Solid is representative of the solid representation that occurs within the Cas.Cade geometric modelling system. Cas.Cade is a package that provides all the functionality of a CAD system via a set of C++ libraries.

Tolerances are obtained via the STEP protocol [8]. Geometric tolerances can be applied directly to the features and components; for example, a through hole can have cylindricity and perpendicularity tolerances applied to it. The *dimension* entity is used to control the dimensions of all the features and this is achieved through the use of a *dimensional tolerance*. It should be noted that for clarity all the aggregation associations that connect the *dimension* entity are not shown. For example a *TopoDS_Point* is specified by (x,y,z) coordinates of a Cartesian axis, each coordinate is itself a dimension and thus has a value, a *positional tolerance* and a unit of measure.

4.2 Design using the aggregate product model

Depending on how much and what information is available the APM is designed to allow an increasing level of detail to be specified. It is in fact possible to create a product model that has sufficient information to allow an aggregate process plan to occur but is insufficient to allow a geometric modeller to generate the 3D solid from it. For example, a through hole can be specified by its radius and depth whilst additional information can be added later regarding positioning and tolerances, when they are determined, The lack of positioning information precludes the ability to generate the 3D solid.

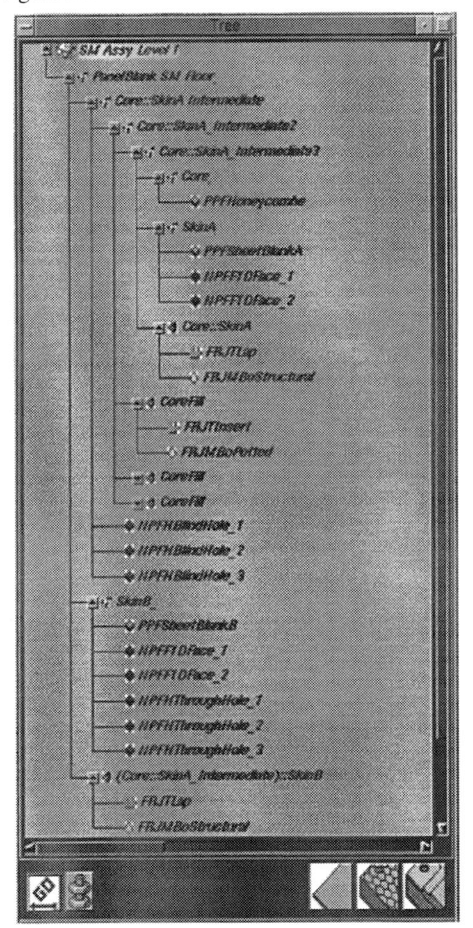

Figure 3. A featured panel blank.

An example is shown in figure 3 where the product model for a "panel blank" is given, the panel blank is a sandwich panel consisting of an aluminium honeycomb core sandwiched between two paper thin aluminium sheets. A stress weight optimisation results in the reduction in thickness over an area of both of the sheets and to ensure the correct alignment of these reduced areas a number of tooling holes are used.

A new product is created via the instantiation of the top level *PMComponent* and an identification tag is given to it, in this case "SM Assy Level 1". The top level *PMComponent* is used for manipulation of the entire product since by definition, all the subsequent children are appended to it. In order for the product model to be validly deconstructed by the process planning engine, the following simple rules need to be observed:

1. If a component has a child object, then all features of that child object must be made before those of the current component.
2. The negative features on a component must be made before the joint features
3. Processes at the same feature level are sequenced using a process anteriority.

With the construction of the panel there are two key elements that need to be observed. The first is that with the honeycomb core being very delicate, it first requires to be bonded to one of the sheets prior to any machining occurring. Secondly, the tooling holes are not cut directly into the honeycomb core but into a filler material previously set into specific cells of the core. With the above rules in mind, *PMComponents* are used to provide additional depth and structure to the panel blank's product model. The "Core::SkinA_Intermediate3" *PMComponent* is used to ensure that the panel and honeycomb are initially structurally bonded prior to the tooling holes and the "Core::SkinA_Intermediate2" *PMComponent* that the potting operations occur prior to the cutting of the three tooling holes ("NPFHBlindHoles" in the model). Finally the "Core::SkinA_Intermediate" component ensures that the tooling holes themselves are cut prior to its joining with the other skin. This final component is, according to the above rules, not entirely necessary for the correct understanding of the process planning engine. This however points a useful function of the *PMComponent,* which is to allow a meaningful representation of a physical entity.

5. FUTURE DEVELOPMENT

The present work is focusing upon the process planning section and once this has been completed to satisfaction, two primary capabilities are required to be added to the APM section of CAPABLE Space.

The first is engineering change management. This covers the requirement to automatically evaluate the impact of changes made to the models and is key to the systems' further development and successful use. This is particularly pertinent when reusing an existing design and/or designing during the conceptual stages where change is a constant occurrence. To enable this the "relative validity index" and "feature elasticity" concepts have been identified as possible solutions. The feature elasticity can be thought of as meta-tolerances; users can apply a degree of elasticity in the model to allow for change and like a spring the further the change deviates from the original designation of the feature/component the more work is required to apply that change.

The second requirement is to provide a workable connection to a commercial CAD package. This will enable the successful use of CAPABLE Space since it is necessary to blend the system into the existing methods of product design.

6. ACKNOWLEDGEMENTS

The authors are grateful to the collaborating company Matra Marconi Space (UK) Ltd. and to the Engineering and Physical Sciences Research Council (Grant GR/L98572) for their support of this work.

7. REFERENCES

1. Maropolous,P.G. (1997). CAPABLE: An aggregate process modelling techniques for process planning applications. *Proc. 13th Int. Conf. Computer-aided Production Eng.*, 49- 56
2. Maropolous,P.G. (1995). A novel process planning architecture for product-based manufacture. *J. Eng. Manufacture 209(4),* 267-276.
3. Maropolous,P.G. (1995). Aggregate process modelling techniques for process planning applications. *ImechE Conf. Trans. 1995-3. The 11th Int. Conf. Computer-aided Production Eng,.* 179-184.
4. Bramall,D.G., McKay,K.R., Colquhoun,P.G. & Maropoulos,P.G. (2000). Manufacturing analysis of conceptual and embodiment aerospace designs: An aggregate process model specification. *The 16th Int. Conf. Computer Aided Production Engineering* – CAPE2000.
5. Yeun,C.F. & Venuvinod,P.K. (1999). Geometric feature recognition: coping with the complexity and infinite variety of features. *Int. J. Computer Integrated Manufacturing, 1999, vol 12, No5*, 439-452
6. Hvam,L. (1999). A procedure for building product models. *Robotics and Computer-Integarated Manufacturing*. Vol.15, pp77-87
7. ISO10303. (1999). Mechanical product definition for process plans using machining features. part 224.
8. Gaines,D.M. & Hayes,C.C. (1999). CUSTOM-CUT: a customisable feature recogniser. *Computer-Aided Design*. vol.31, pp85-100
9. Case,K. & Wan Huran,W.A. (1999). A single representation to support assembly and process planning in feature based design machined parts. *ImechE, Proc. Instn. Mech Engrs. Vol213, PartB*, 143-155.
10. Hounsell,M.S. & Case,K. (1997) Structured multi-level feature interaction identification. *Proceedings of the Thirty-second Int. MATADOR Conf.* 10th-11th July 1997 UMIST, UK.
11. Laguda,A. & Maropoulos,P.G. (2000) Automatic generation of optimal assembly operation sequences using simulated annealing. *The 16th Int. Conf. Computer Aided Production Engineering* – CAPE2000.
12. Yao,Z. Bradley,H.G. & Maropolous,P.G. (1998). An aggregate weld product model for the early design stages. *Artificial Intel. for Eng. Des., Analysis and Manufacturing* 1998-12, 447-461.
13. Bradley,H.D. & Maropolous,P.G. (1997). A relation based product model for computer supported early design assessment . *Proc. 13th Int. Conf. Computer-aided Production Eng.* pp57-64.
14. Boothroyd,G.G., Dewhurst, P. & Knight, W. (1994). Product design for manufacture and assembly *NewYork: M Dekker,* ISBN:0824791762 m.

CAPE/093/2000

Manufacturing analysis of conceptual and embodiment aerospace designs – an aggregate process model specification

D G BRAMALL, K R McKAY, P G COLQUHOUN, and **P G MAROPOULOS**
School of Engineering, University of Durham, UK

ABSTRACT

There is a general lack of integration between the early stages of engineering design and manufacturing systems. A knowledge enriched aggregate process planning system is being developed to support early design work with analysis of possible manufacturing scenarios. This would allow the rapid assessment and analysis of alternative designs and facilitate evaluation of processes and manufacturing options within the extended enterprise. The creation of an aggregate process model to support this work is presented in this paper. Process models form a key part of the aggregate planning system, providing the basis for selection of suitable processes, equipment and the generation of routes for a given company.

1 INTRODUCTION

Concurrent Engineering (CE) promotes the reduction of the "time to market" through a simultaneous approach to product and process design (1), the two main activities required to realize a product. The manufacturing cost and performance of a product are greatly influenced by the decisions made at the early stages of design, yet it is not until later in the design cycle that most companies carry out process planning to determine which process will be used to manufacture the product. Computer-aided process planning (CAPP) is the selection of production methods based upon capability and production economics through the automatic interpretation of design data (2). The current generation prototype and commercial CAPP systems use manufacturing features from solid models to generate tool paths and perform tooling selection for NC applications. The level of detail required before these systems can be run their "micro process planning" algorithms is high, as the model must be fully-specified in terms of fully-toleranced geometry, materials and production volumes. These requirements make current CAPP systems unsuitable for use at the concept design stage.

Design for manufacture methods have long been recognised as critical to concurrent engineering strategy, promoting production oriented manufacturability assessment through the use of simple rules, such as, part-count reduction and poka-yoke principles. The Boothroyd and Dewhurst (3) Design for Assembly (DFA) methodology is widely used to predict manufacturing cycle times for assembly optimisation and highlight the manufacturability of a design. This method of assembly-based planning is not, however, well integrated with design tools, although research is underway to develop computer tools which link design characteristics to assembly models (4). Liebers et al. (5) also argue that DFA does not provide any feedback on design cost or resource availability.

Although the principles of CE are sound, the natural cycle of design means that, paradoxically, process planning has a significant impact upon product's performance yet cannot be accounted for until the majority of design information has been decided upon. In fact, it is recognised by Lenau (6) that in design for manufacture most effort is concentrated on the design of methodologies, rather than the synthesis of process models for the practical application of CE. This research sets out to overcome the incompatibilities in data granularity between concept design and process planning through the use of process models which predict manufacturing performance under conditions of partial design information. This aim is put into an industrial perspective by the results of a survey into responsive manufacturing in the UK aerospace industry (7) which showed that integrated product and process development and producability analysis are some of the most significant elements in achieving manufacturing responsiveness.

2 AGGREGATE PROCESS PLANNING AND INTEGRATED PRODUCT DEVELOPMENT

A novel Process Planning architecture has been proposed by Maropoulos (8). The Durham-developed methodology splits process planning into three levels of granularity; aggregate, management and detailed (AMD). The aggregate level envisages the sharing of partially complete design data between design functions thus, manufacturing performance estimates can guide designer as the design progresses from concept to embodiment. Very often, significant improvements in efficiency can be made by considering alternatives to traditional methods of production. Evaluating options using virtual process models greatly reduces the need to make expensive engineering changes as a consequence of breaching production constraints.

2.1 Theoretical Background: Aggregate Process Models for Early Manufacturability Analysis

An aggregate process model captures the fundamental aspects of a process' behaviour, complete with a selected set of process parameters and manufacturing knowledge. The requirements for process planning can be broken down into process selection and process performance modelling. Process selection ensures that design attributes fall within the capability of the process (9). Swift (10) has devised a method of manual process selection which is based upon broad process characteristics rather than detailed models. Gao and Huang (11) demonstrate the link between the component and feature representation of a product and the shape producing capabilities of a process. Process performance modelling is the estimation of the quality, cost, delivery and knowledge (QCD, K) behaviour of a process in a given situation. Aggregate process models link the key characteristics of the product model to production-related process variables in order to forecast performance. Maropoulos and Bradley (12) suggest the following theoretical guidelines for the definition of aggregate process models.

(i) Controlled simplification of detailed process models
Through simplification of detailed process models, aggregate models should operate with the limited amount of product information available at the concept design stage. The most significant feature characteristics relating to process performance should be used to drive process models.

(ii) Limited input data requirements
In order to limit the amount of data required, models shall be configured to function using minimal amounts of data. Basic elements of a design, such as structure and overall dimensions should exist but exact geometry, material choice and tolerance levels need not be defined until the embodiment stages of design.

(iii) Utilise company specific product and process knowledge
Where possible company specific knowledge should be used to increase the accuracy of models. Many process parameters are specified from experience and limited experimentation, such as the application of Taguchi methods.

(iv) Perform core technological checks concerning processes
Core capability checks should be made to confirm the applicability of a possible solution to the chosen resources and processes. The first stage of any automated process planning strategy is based upon selection of processes compatible with the type of feature. Initial checks should be made on specific feature properties to eliminate infeasible combinations due to mismatches in geometrical or quality limits.

(v) Incorporate only the basic, overall geometry of parts and features
Only the basic geometrical descriptions of parts or features should be sufficient for aggregate planning.

(vi) Function-driven operation and conformance with the team-based approach
The new planning methods should provide decision support based upon multiple-aspects of product development and factory modelling and must allow interaction between design teams, production teams and even members of external suppliers. Internet/intranet technologies should be taken advantage of to allow team members access to appropriate areas of the system.

(vii) Conformance with standards
Whenever possible the new modelling methods should be compatible with existing standards.

2.2 An Architecture for Integrated Product Development

Aggregate Process Planning recognizes the importance of a three elements in the production cycle, namely, the configuration of a design, the nature of the process and the capability of the production resources, in influencing the success or otherwise of a product. Three inter-linked object-oriented models are thus required to represent these three entities and to interface with the relevant development groups within a distributed product and process development team.

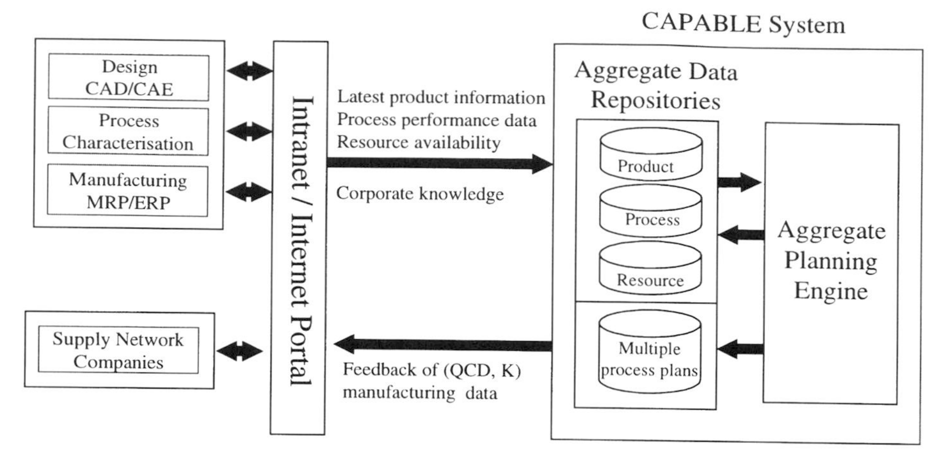

Figure 1. The architecture and its interfaces with dispersed members of the CE team

A Java-based technology demonstrator called CAPABLE Space is being developed to validate the aggregate planning theory for the aerospace industry, a business sector which is in need of responsive planning methods to cope with increasing demand for networks of high volume, low-cost satellites – a result of the current revolution in communications technology. Figure 1 shows the main data flows through the planning engine, and demonstrates the integration of CAPABLE Space with the internet/intranet functions of a company. The functions and aggregate data repository are accesses via the internet portal and can be used for internal manufacturing assessment as well as support the technical assessment of sub-contracting make-or-buy decision making using quality, cost, delivery and knowledge ranking criteria. The capture and use of company-specific knowledge enhances the aggregate models by allowing subjective criteria to be incorporated into the selection of processes and resources.

3 A GENERIC AGGREGATE PROCESS CLASSIFICATION

In order to carry out automated process selection the CAPABLE system requires a body of process models able to predict manufacturing performance, with acceptable levels of accuracy, across a wide range of manufacturing activities. The models must support the principles of aggregate planning and comply with the architecture shown in Figure 1. The aggregate process taxonomy follows, to a large extent, previous classifications but it is defined with the predominant requirements of aggregate models as defined in Section 2.1 and the need to rapidly incorporate new state-of-the-art processes. This is important, since the methods need to be applied for technology assessment, rather than post design process evaluation. If a team of designers wish to evaluate a production method for which no process model, or, a poorly-defined process model exists, a process class which captures user-defined standard time data is available to enable new processes to be included in the system.

At the highest level of granularity, the process model is divided into 5 core "packages". Manufacturing is characterised by two major activities; discrete parts production and the subsequent assembly of these parts (and bought-in materials) to generate the finished product. Assembly operations are further sub-divided into two categories; (a) part handling, which defines the orientation and placement of parts and (b) insertion, fastening and joining which represent the processes used to physically make the joint/assembly.

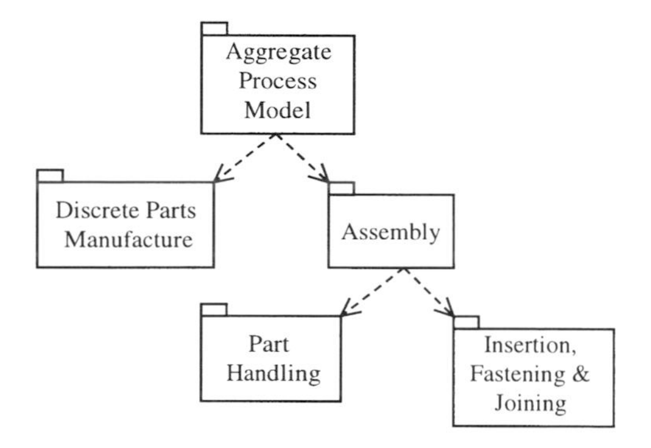

Figure 2. Process Model Package

3.1 Discrete Parts Manufacturing

The hierarchy of process models for discrete parts manufacture broadly follow the classification of Allen & Alting (13), grouping processes according to their morphological characteristics, as shown in Figure 3. Two broad categories of processes exist within this classification, shape-changing and non-shape changing. In particular, non-traditional machining processes are considered within the scope of the CAPABLE Space system, because they represent valid process alternatives which design teams may wish to consider. Other important process groups such as surface treatment and surface preparation for bonding are also encapsulated within the model.

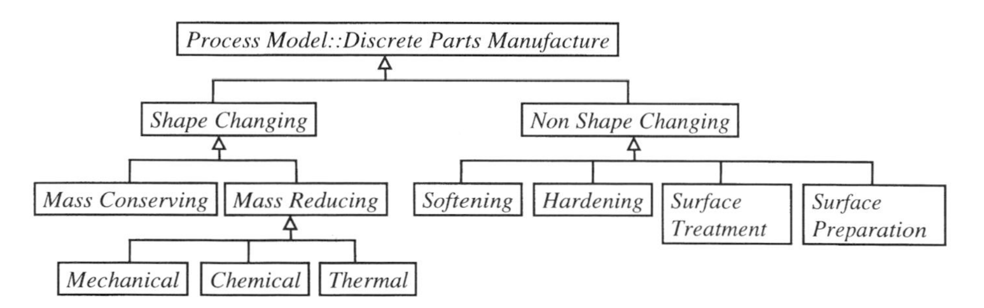

Figure 3. UML class diagram for the Discrete Parts Manufacture package

3.2 Process Models for Assembly

Assembly is fundamental to the manufacture of most products. Assembly operations model the bringing together of parts for the formation of a new structure/sub-assembly. Assemblies are represented within the product model by joint-feature connections which link two or more items in the bill of materials. The manufacture of this joint requires a combination of part handling (alignment of the parts) and the physical process of making the joint. These types of operations are modelled within two branches of the assembly tree. Assembly models for the CAPABLE Space are based upon Laguda's extension of Boothroyd & Dewhurst's work on Design for Assembly (3), (4).

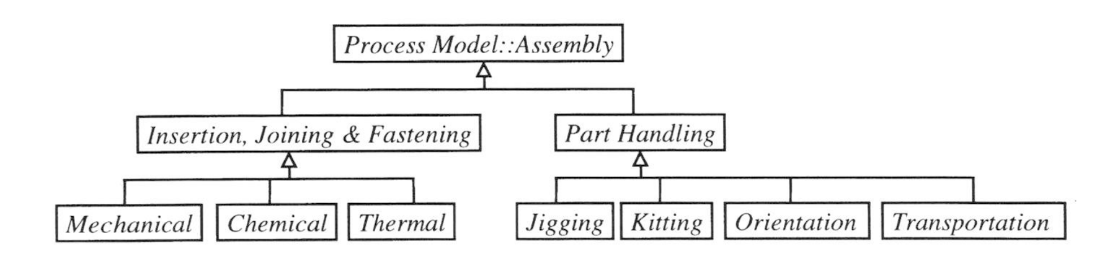

Figure 4. UML class diagram for the Assembly package

Joining operations provide a good example of some of the technological constraints applied via the process model. Conceptually, a designer may visualise a structural bond joining two components without deciding the exact type of joint until other considerations such as stress analysis techniques have been applied. Therefore, in the initial applications of the process planning algorithms all methods of structural bonding are considered. As more information about the joint configuration becomes known, invalid solutions are identified in the constraint checking method of each process, as shown in Figure 5, and eliminated from the solution space.

Joining method	Joint configuration: Butt	Corner	Lap	Tee	Filled edge closure	Folded edge closure	Taped edge closure	Insert (1Part)	Insert (2Part)
Core fill					■				
Foam bonding	■	■	■						
Insert potting								■	■
Film bonding		■	■						
Edge taping							■		
Paste bonding		■	■						

Figure 5. Joint configuration/joining method compatibilities

4 PREDICTION OF MANUFACTURING PERFORMANCE

A key task of process planning is the formation of a logical sequence of processing steps which transform raw material into a finished product. The manufacture of a complex product is governed by the production of sub-assemblies which have precedence relationships defined in the product model. The aggregate planning function uses simple heuristics, developed from production expertise, to sequence process operations on individual components. Each process also maintains information about ancillary pre- and post-processes which are required. These are instantiated in the correct sequence by the process planning engine.

The rules used for sequencing operations are;

1. If a component has a child object, then all features of that child object must be made before those of the current component.
2. The negative features on a component must be made before the joint features
3. Processes at the same feature level are sequenced using a process anteriority.
4. At the component level pre-process alternatives must occur before the seed process.
5. Post-process alternatives on a component must occur after the seed process.

In aggregate process planning the initial selection of process alternatives is made on the basis of the feature's geometrical type. A look-up table is then used to return the process objects which are suitable for manufacturing a given feature type. The compatibility of the selected process and the feature will also depend on many other factors such as materials suitability and workpiece geometry. Thus, the first stage of process planning is a simple constraint checking routine which checks each process, feature and resource combination against a set of pre-determined constraints such as materials suitability, workpiece geometry, feature size and machine availability.

A simulated annealing algorithm is used to find near optimal solutions from the large number of possible combinations of feature, process and resource. The algorithm automatically generates routes using the above rules and evaluates the fitness of the solutions. The key to ensuring that a product will be commercially viable is ensuring that the product can be manufactured with the available resources at high quality and lowest cost [14]. A simple objective function consisting of a weighted sum of the QCD values obtained from the process model is used. Aggregate process models store descriptions of the capabilities, requirements and parameters of manufacturing processes which enable the system to simulate the production of a component in order to forecast the probable performance using mathematical descriptions created from both theory and practical knowledge of the process.

4.1 Manufacturing Cost and Delivery

Delivery is a measure of the time required to manufacture a component. This acts as the input to the calculation of process cost and is also used to calculate the critical path for a process plan. Delivery has been defined to comprise four easily measurable aspects of manufacturing time, lead time, processing cycle time, part set-up time and batch set-up time. An objective function (Eq.1) has been defined for a simulated annealing algorithm to generate the optimal process plan for minimum cycle time from the large amount of available solutions. Manufacturing cycle times are derived from the controlled simplification of detailed process models, or, through the empirical measurement of process times.

$$\text{Total time} = \sum_{\text{processes}} \text{lead time} + \text{cycle time} + \text{part setup time} + \sum_{\text{resources}} \text{batch setup time} \qquad (1)$$

One of the most significant production metrics is the total cost of manufacture of a component. This comprises materials and consumables costs, labour costs, machine costs (maintenance, depreciation) and overheads. There are many ways to account for the manufacturing costs, however the one chosen for the first version of CAPABLE Space is Activity Based Costing (ABC). By averaging labour, overheads and machine costs within the factory to derive a cost rate for each activity undertaken by an employee a simple cost formula can be used to calculate cost (Eq.2). Labour and machine costs derived from the process time models are added to the cost of materials to obtain an estimate for the total cost of manufacture.

$$\text{Total cost} = \text{Raw material cost} + \sum_{\text{resources}} \text{ABC cost rate} \times (\text{cycle time} + \text{setup time}) \qquad (2)$$

4.2 Quality Measurement

Quality is a measure of the capability of the manufacturing system in terms of meeting the design tolerances and customer requirements. Selecting feature tolerances without reference to processes leads to escalating manufacturing cost (15). The reporting and measurement of useful quality information (that is to say, information reflecting the observed production quality) is an area which is particularly weak in existing CAPP systems, a view expressed by Marri et al in their state-of–the-art review of CAPP systems (16). In line with common aerospace practice, CAPABLE Space utilises Six-Sigma principles to reward those processes which are robust and can consistently produce parts within a given tolerance. The number of defects per million opportunities (DMPO) is used as the quality metric. DPMO is related to the process capability, C_p, and the tolerances defined by the product model (Eq. 3 & 4).

$$C_p = \frac{\text{USL - LSL}}{6\sigma} \quad (3), \qquad p = \frac{\text{DPMO}}{10^6} \quad (4)$$

where

C_p = Simple process capability

USL/LSL = Upper/Lower specification limit from the product model

σ = Estimate of process' standard deviation

p = Probability

A good example of the aggregate planning operation is in relation to the honeycomb-cored sandwich panel which is one of the major components in a satellite structure. This is constructed from two thin aluminium skins adhesively bonded to a lightweight core material. The UML representation of an aggregate process model for the film adhesive bonding process is shown in Figure 6. The process model for film bonding is responsible for extracting information about the bond from the joint feature in the product model, including the type of joint (JT_Lap, which is verified as being applicable to this process), the bond area of the joint and a cure cycle defined by the adhesive's materials specification. The process model also uses process capability information from resource model to generate a predicted value for the DPMO for the current feature limits. Estimates of the manufacturing performance are stored as objects which are available to the process planning engine. Pre and post process requirements are stored as a list which can be consulted by the process planner. In this case a preparatory process is required to degrease the bonding surfaces and after the adhesive has been laid up the adhesive must undergo at least one curing cycle.

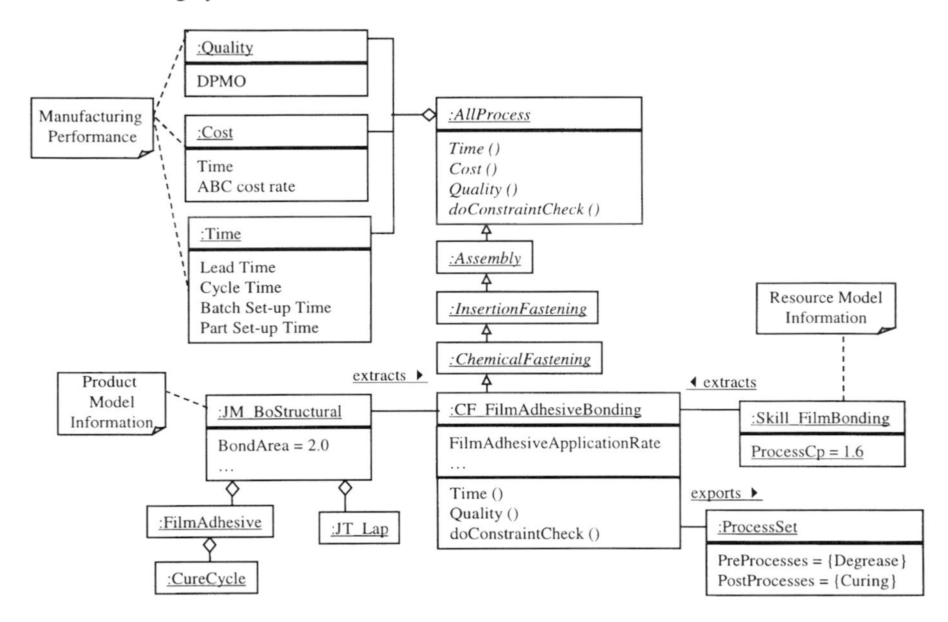

Figure 6. UML object diagram for Film Adhesive Bonding

5 CONCLUDING REMARKS

The research described herein forms the core of a computer-based system for the analysis and assessment of manufacturing alternatives at the earliest stages of design. It is based upon the re-definition of traditional process models to support an aggregate view of process planning, enabling production routes, complete with estimated performance, to be established for partially specified products. Early versions of CAPABLE Space are able to automatically select processes and resources and generate optimal production routes based upon QCD criteria. An extensive testing phase is planned to validate the work using real components and factory layouts.

Quality, cost and delivery indicators are widely understood within the manufacturing environment to show the customer perception of process performance. However, the need to gain competitive advantage in the ever-aggressive manufacturing sector has led to the realisation that knowledge is an asset and can be used as a driver of best practice. The future work will concentrate on the specification of new techniques for enriching process planning with in house product and process knowledge.

ACKNOWLEDGEMENTS

The authors are grateful to the Engineering and Physical Sciences Research Council for financially supporting this research (GR/L98572). We also acknowledge the support and contribution of the main collaborating company, Matra Marconi Space (UK) Ltd.

REFERENCES

(1) Pham DT and Dimov SS (1998), 'An approach to concurrent engineering', Proc Instn Mech Engrs, Vol 212 Part B, 13-27

(2) Alting L , Zhang H (1989) 'Computer Aided Process Planning: the state-of-the-art survey' Int. J. Prod. Res Vol27 No 4 553-585

(3) Boothroyd G, Dewhurst P, Knight W (1994) 'Product Design for Manufacture and Assembly' Marcel Dekker, Inc.

(4) Laguda A, Maropoulos PG (1999) 'Assembly modelling and Planning for Concurrent Engineering' Proceedings of 6th European conference on Concurrent Engineering 104-108

(5) Liebers A, Kals HJJ (1997) 'Cost Decision Support in Product Design' Annals of the CIRP Vol 46 No 1 107-112

(6) Lenau T (1996) The missing element in Design for Manufacture' Annals of the CIRP Vol 45 No 1 105-108

(7) Gindy NNZ (1999) 'Responsive manufacturing in UK aerospace industry' Proceedings of the 15th International Conference on Computer-Aided Production Engineering

(8) Maropoulos, PG (1995) 'A novel process planning architecture for product-based manufacture' Proc. Instn. Mech. Engrs. Vol 209 267-276

(9) Esawi AMK, Ashby, MF 'The development and use of a software tool for selecting manufacturing processes at the early stages of design' Proceedings of the 3rd Biennial Conference on Integrated Design and Process Technology (IDPT)

(10) Swift, KG & Booker JD (1997) 'Process Selection from Design to Manufacture' Wiley

(11) Gao JX & Huang XX (1996) 'Product and Manufacturing Capability Modelling in an integrated CAD/Process Planning Environment' Int. J. Adv. Manuf. Technol. (1996)11:43-51

(12) Maropoulos PG, Bradley HD, Yao Z (1997) 'CAPABLE: an aggregate process planning system for integrated product development' Journal of Materials Processing Technology No 76 16-22

(13) Alting, L & Allen DK (1986) 'Manufacturing Processes Student Manual' Brigham Young Univeristy, Utah, USA (ISBN 0-8424-22239-5)

(14) Abdalla HS and Knight J (1994), 'An expert system for concurrent product and process design of mechanical parts', Proc Instn Mech Engrs, Vol 208, 167-172

(15) Lee YC, Wei CC (1998) 'Process Capability-Based Tolerance Design to Minimise Manufacturing Loss' Int. J. Adv. Manuf. Technol. (1998)14:33-37

(16) Marri HB, Gunasekaran A, Grieve, RJ (1998) 'Computer-Aided Process Planning: A State-of-Art' Int. J. Adv. Manuf. Technol 14:261-268

CAPE/084/2000

Modelling of virtual cells for dynamic manufacturing environs – an overview

M M BAJIC and **K BAINES**
Department of Mechanical Engineering, The University of Adelaide, Australia

SYNOPSIS

This paper addresses a computer aided analytical approach which concurrently integrates cellular manufacturing *(CM)* systems with the material flow configurations. Specifically, to provide a methodology to derive virtual cell designs that are effective to operate. Firstly, a graph theory is applied to obtain machine groups, then the intercell material flow network is minimised and reoriented. Finally, a simulated annealing *(SA)* algorithm for integrating the material flow configurations optimises the shop layout. The proposed method is validated by using several comparative examples from the literature. This innovative layout approach considers simultaneously most major decisions involved in *CM* shop design.

1. INTRODUCTION

Existing approaches to cell formation concentrate on simultaneous part family formation and machine sharing, and do not use flow data. A fundamental problem faced when part families and independent cells are desired is that two or more cells may share a machine type. Hence, they cannot solve together the subproblems of machine grouping, machine sharing, intracell layout, intercell layout and material handling. Thus this work considers designing the virtual cellular manufacturing system *(VCMS)* layout concurrently with its material flow configuration in a manner that minimises the material handling within the system. In particular, this approach seeks to provide a computer programmed methodology to model and sustain economic cell designs that can be virtually reconfigured for any product variety scenario. The model is decomposed into subproblems and formatted in *MATLAB*. In the first subproblem a maximal weighted directed rooted spanning tree *(MWDRST)* is obtained which yields machine groups. The second subproblem concerns the reorientation of the material flow network to minimise the lengths of the intercell material flows. Finally, the *VCM* shop layout is developed in a manner that minimises resources and material handling, and incorporates machine capacity, cost factors, set-up times, machine utilisation, as well as production rates and batch sizes. A *SA* algorithm is employed which accounts fully for all physical constraints. Once the size and shapes of the resources are known the resulting *VCM* shop layout is determined and can be integrated with *CAPP*.

1.1. Analysis of Cellular Manufacturing Approaches

Modern factory cells may be designed to be autonomous, based on the current mix of parts and of demand volumes. Numerous factors which precipitate intercell flows between these cells are: machine failures, the need to keep expensive one-off type machines loaded, changes in part mix or production quantities for part families, non-integer machine requirements of bottleneck machines demanded by two or more part families, feasibility of using handling systems to induce intercell moves between adjacent cells, alternative routing for parts when identical machines become duplicated in cells,...etc. The traditional approach (5) to *CM* requires that an independent cell be designed for each part family. This creates a problem in deciding the integer number of each shared machine which must be assigned to each cell. Usually, each machine in a cell experiences either overloading or underloading. Thus, it is difficult to develop a method which has the ability to integrate problems of machine grouping, intracell layout, intercell layout, machine sharing and an approximate material handling system configuration. The interactions between these problems in *CM* often make it difficult to identify permanent machine-part compositions for the cells. Additionally, using the standard machine-part matrix clustering representation for cell design, it is possible to get the incorrect impression that cells can be created only through duplication of the shared machine types. The sizes, shapes and adjacencies of the different machine groups are lost in the pure linear arrangement for all the machines in the matrix output solution. From a layout perspective the cells can be placed around a common facility cell, containing all shared machine types. Based on Gallagher (8), *Figure 1* demonstrates the concept of a *VCM* layout that combines the properties of the standard functional layout with the cellular layout. The flow lines have been permutated in order to bring the identical machines next to each other. The machines remain dedicated to their part families. However, they are retained in functional layouts to allow flexibility in machine reassignments when machines fail or the part mix and/or demand changes. Suitable aisle configurations and handling capabilities are assumed which would allow permitted random flows.

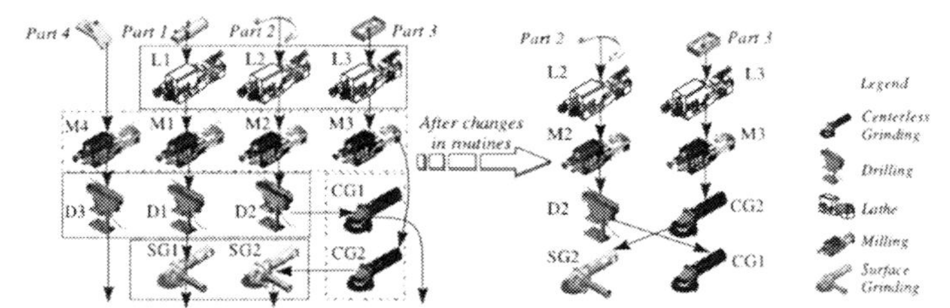

Figure 1. Virtual cellular layouts

2. INTEGRATED *VCM* SYSTEM DESIGN

The concept of a *VCM* layout is suggested as a result of the analysis of *CM* design approaches, which constitute the cell design problem and the associated interactions. These interactions suggest that Part Family *(PF)* formation and machine capacity calculations alone cannot accomplish the design of innovative *CM* cells to provide material flow flexibility. The objective here is to develop integrated layout design procedures which will facilitate the design of *VCMS* with improved material flow handling efficiencies. In a *VCMS* there are two types of material flow: the first is the flow between cells (intercell flow), and the second type is the flow between the machines within cells (intracell flow). The amount of intercell flow is affected by the cell formation. Intracell flow movements cannot be avoided and hence affect cell formation. Some of the factors which affect material flow efficiency are material flow controls, the layout of machines and cells, and the material handling systems. Thus, the required methodology relates to the manufacturing facilities layout (*Figure 2*). Traditionally, if more than one process plan is considered for each part, standard part/machine grouping techniques cannot be applied. Similarly, when multiple routings are considered, techniques that rely on knowing the exact flow between facilities cannot be used either. *Figure 2* defines the relationship between different design modules and outlines the tasks in each module.

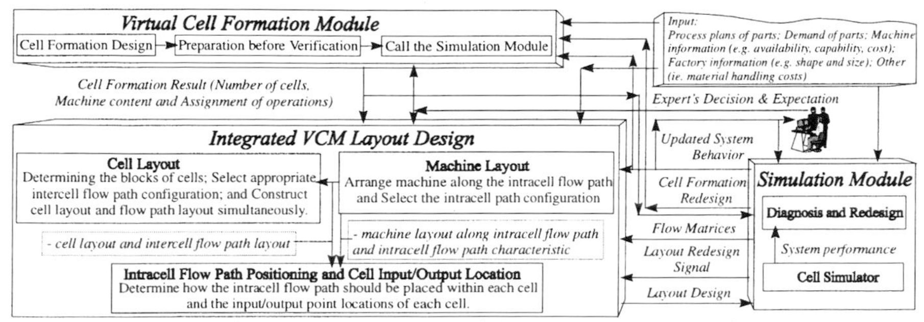

Figure 2. Modular structure of the *VCM* design procedure

2.1. Mathematical Modelling of Virtual Cells

This section describes proposed mathematical modelling approaches for flow decomposition of machine groupings and layout design of *VCM* cells. The theory of networks and graphs is applied to model this problem and some basic relevant definitions and notations from network analysis and graph theory texts are used. Development of the *MWDRST* arcs classification is in order to apply the concepts of Carrie (6 & 7) on unidirectional flow line design to intercell layout design. The initial design of undirectional flow lines is necessary because only then can backtrack flows be identified. Furthermore Carrie considered a single cell, and classified the flows along the flow line as in-sequence, bypass and backtrack. With respect to intracell layout, his classification of the flows is sufficient. Thus, connecting the machines in the flow line can provide for easy and rapid in-sequence and bypass flows. Backtrack flows require the services of a material handler for which queuing might be involved, and thus would take more time than in-sequence moves. However, when applied to intercell layout (6 & 7), an additional arc (crisscrossing) is necessary in order to classify the flows which may occur between any pair of flow lines. It is important to minimise the intercell flows (since these will have even higher queuing delays), and thus intracell machine duplication with throughput delays then becomes secondary.

2.2. Mathematical Formulation of the Model

The development of a single mathematical programming model for the combined machine grouping, intercell and intracell layout design problem has proved difficult. For intracell flow line construction no single model has been able to incorporate all aspects of machine grouping which consider: generation of all flow lines in the form of *MWDRST's*, interaction between intracell flow line configurations (flow distances for forward and backward arcs) and intercell layouts (flow distances for crisscrossing arcs), machine sharing for nonadjacent crisscrossing arcs, and optimal left to right order of the flow lines. Thus it is proposed that the required model be a combination of cluster analysis, *MWDRST*, permutation generation and two-dimensional facility location problems. A review of existing *NP*-complete problems (2) showed that each of these problems, previously listed in section 1.1, is itself a separate *NP*-

problem. There are no algorithms known for these problems, whose complexity is a polynomial function of the size of the input. Thus (2) showed that networks and graph theory were sufficient to structure the mathematical formulation and models for the proposed *VCM* layout design methodology, shown in *Figure 3*.

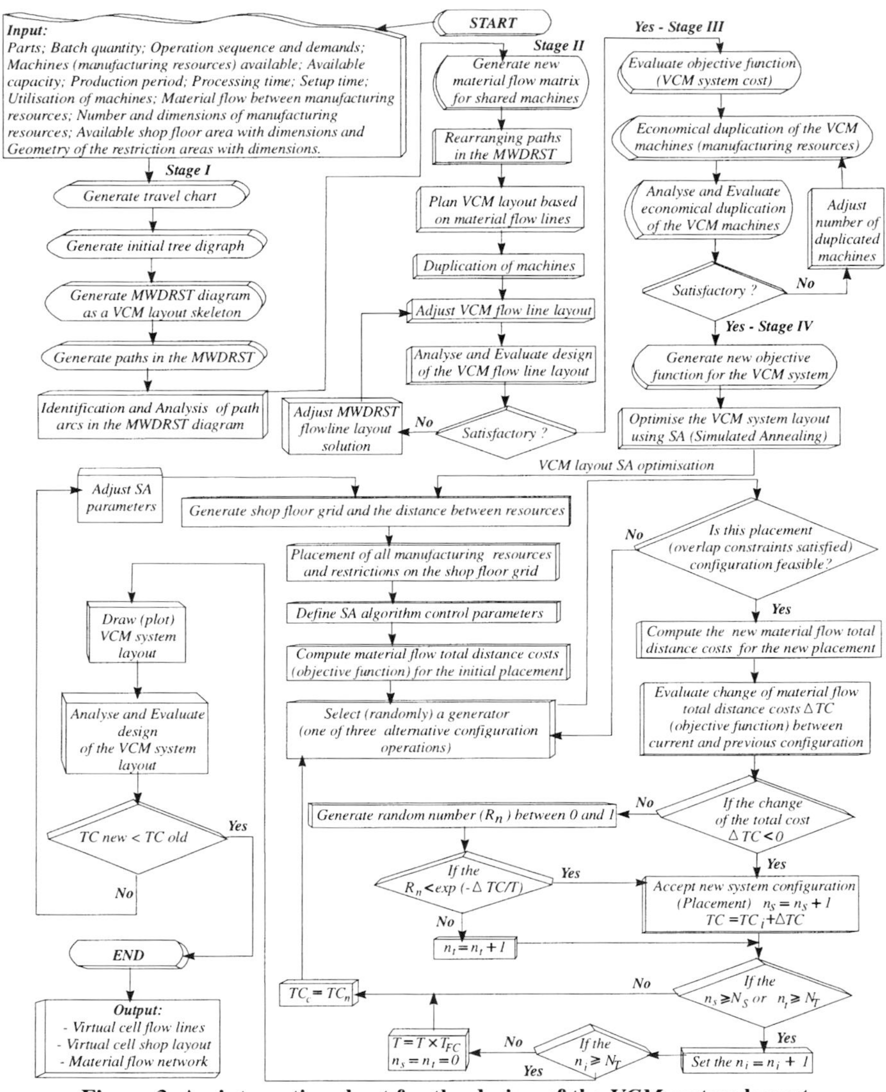

Figure 3. An interactive chart for the design of the *VCM* system layout

2.2.1. *Stages I and II Flow Line Directionality*

The feature of this section is to capture the directionality embedded in the operation sequences of a variety of parts produced in a facility for the *VCM* layout design. Also, to consider how to generate machine groups by identifying a flow line layout for each group, thus indicating which flow lines must be placed adjacent to each other to minimise intercell flow distances, and furthermore to determine an approximate configuration of the aisles. The approach is a unique combination of network analysis, graph theory and mathematical programming concepts. Its fundamental assumption is that machine sharing on a part family basis creates load-balancing problems. Thus, shared machines can be retained in functional layouts as long as machine sharing is within the cell or between adjacent cells.

<u>Stage I</u>: is a mathematical programming model to generate a *MWDRST*. This implicitly minimises travel distances for forward and backward arcs. <u>Stage II:</u> is a similar programming model derived from stage I. It finds the optimal orientation of the *MWDRT* in order to minimise travel distances and machine duplication for the crisscrossing arcs. With the crisscrossing arcs only, a matrix showing the intercell flows amongst all pairs of paths in the *MWDRST* can then be developed. This matrix permutes the branches of the *MWDRST*, without dividing any of the branching nodes (3).

2.2.2. Stage III Economical Machine Duplication

In order to eliminate the intercell machine sharing problems created at stage II, only those parts whose operational sequences contain predecessor and successor arcs need to be considered for capacity calculations, process planning or value analysis. Here additional system/machine variables (setup time, duration of the production period, machine capacity, utilisation of each machine type, flow volume, production volume, operation time, non-operation time, average time between failures and servicing time of the machine type, and availability of the machine type) are included for finding the required number of the flow line machines. To duplicate machines in several flow lines the mathematical model must compute the total capacity per machine of each type for the entire production period (2). This duplication is necessary because the machine type processing time may exceed the flow line machine's availability in the *VCM* cells. Thus, if the machine is a bottleneck machine then it can be duplicated in the flow lines, provided the corresponding part volumes are known.

2.2.3. Stage IV Shopfloor Machine Placement - Layout Optimisation

Stages I, II and III groups the production equipment of the manufacturing facility into virtual cells and forms a set of part families, which are processed within these cells. It does not, however, consider the relative placement of the machines within cells, or the relative placement of the cells and machines on the shop floor. Both these issues are crucial in terms of the total material flow within the shop and have been addressed by (2). To formulate the shop floor layout design problem, first of all a geometrical model *(Figure 4)* of the shop floor and of the manufacturing resources is developed. Subsequently, decision variables are introduced to model a discrete choice in the construction of the flow network and the location of the resources on the shop floor. Finally, the assumptions guiding the design are stated and an integer programming formulation of the layout problem is developed.

<u>*Definition of the Dynamic Cell Discrete Layout Model*</u> - The general model framework for the manufacturing shop, upon which this approach is based, considers an orthogonal unit imposed on its area *(Figure 4)*. This discrete model and solution algorithm approach identifies both the manufacturing resources to be located, and the shop floor enclosure in which they have to be placed, into a number of unit blocks. The blocks of a single resource are located on the shop floor, and allow the size and location of the resources to vary incrementally within the shop floor area. This approach must define in advance the geometrical shapes of the resources.

Geometrical Attributes Representation - The manufacturing resources which are to be placed on the shop floor form the set which is decomposed into unit square building blocks. Although each building block is treated as a distinct entity, mathematical relationships are established between adjacent blocks of the same resources in order to retain the size of the resources in the final layout solution. Each intersection of the grid *(Figure 4)* represents a node of the underlying graph, and graph arcs are candidates for the material flow paths (aisles). This discrete representation of the manufacturing shop and resources facilitates the introduction of decision variables to model the design problem to overcome overlapping area conflicts and complex flow path modelling.

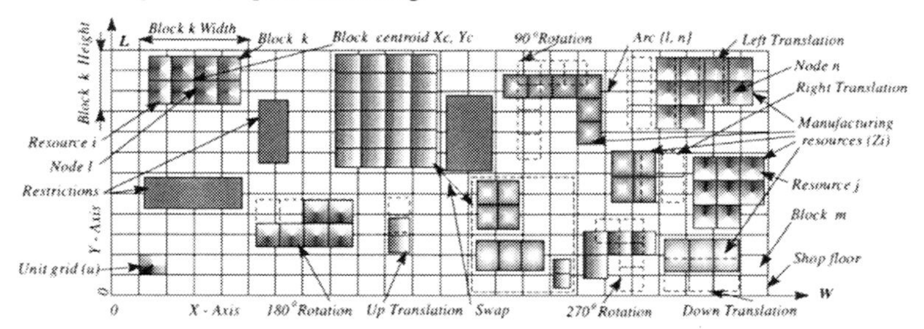

Figure 4. Manufacturing resources representation

Assumptions - The manufacturing resources of the *VCM* system are enclosed by rectangular work areas. Restrictions on the shop floor are also enclosed by rectangular areas *(Figure 4)*. The material flow between each resource pair represents the volume of interactions between them, and is calculated from the part routing's demands over the planning period and the batch sizes. The levels of material flow are assembled in a matrix which is known as the material flow matrix. A cost associated with each move is defined as the travel cost from/to resources.

Definition, Decision Variables and Parameters for VCM Layout Design - The layout problem consists of assigning the unit squares to the unit locations, whilst observing a number of layout constraints and objectives. Several decision variables (assignment of blocks of resources to the nodes, determines arcs activity in the network) are introduced which address important attributes of the shop design to model the discrete choices. Thus, a new formulation is proposed which attempts to avoid unacceptable resource shapes by introducing additional shape constraints. Based on the input parameters the feasibility constraints (resource area and their number of unit squares) can be immediately tested (2) *(Figure 4)*. Necessary information inputs for determining the shop floor layout method include: the set of manufacturing resources, the *SA* algorithm input parameters, the set of area restrictions, the minimum width of material handling corridors, the set of parts, and the traffic between resources.

Mathematical Formulation of the Virtual Cell Layout Approach - The main objective in stage IV is to design a conceptual block layout for a number of manufacturing resources. The manufacturing resources, which have material flow affinities must also satisfy a shape ratio constraint. The objective is to minimise the affinity weighted centroid-to-centroid rectilinear material flow cost distance rating. The placement problem consists of determining the relative positions of the manufacturing resources in the set, which might be either the set of machines belonging to a cell, or the set of manufacturing cells within the shop. The co-ordinates of the centroid of each department are linked to the assigned constraints. Since the coordinate space considered is discrete and finite, each point in the space is assigned a unique position number in order to simplify the analysis. The problem thus consists of determining the position number corresponding to each manufacturing resource in order to minimise the objective function (total cost). Thus, the minimisation problem is subject to a set of constraints, which states that each position can be occupied by not more than one manufacturing resource. The set of area overlapping constraints ensures that one and only one resource block is assigned to each node *(Figure 4)*, and that each resource block is assigned to a grid node. Adjacency constraints force adjacent blocks of each resource to occupy adjacent grid points between the locations occupied by these blocks and to be equal to the grid length. Non-overlapping conditions are also provided.

2.2.4. Simulated Annealing (SA) Algorithm Solution

The major issue for the layout of dynamic manufacturing systems is the placement of manufacturing resources (cells and machines) within the available area of the shop floor. Since the shop layout problem is an assignment problem, local improvement solution approaches are prone to converge to a local minimum, especially in the cases of shop design in which the objective function comprises several conflicting factors. The *SA* method has been previously shown (2) to converge to global near optimal solutions for such problems. This is the main reason it has been utilised in this proposed research methodology for stage IV. The superiority of *SA* based algorithms, with respect to other facility layout methods, has also been conclusively reported by many researchers. The *SA* based layout method minimises the objective function, which models the assignment of the manufacturing resources to the nodes of the grid. This method provides values to these variables, evaluates the resulting feasible layouts, and repeats this process until a near optimal shop layout is reached (2). The initial distance between all pairs of manufacturing resources is determined within stages II and III of the proposed methodology. The definition of all parameters, which is needed for an adequate problem set-up, together with the basic elements of the Metropolis (10) and Aarts (1) algorithms, is as follows:

Solution Space (Configuration) is given by assignment of the manufacturing resources to the candidate locations. *Description of System Configuration Changes* (Neighbourhood), consists of those configurations that result from the interchange of the manufacturing resource locations. Here, new system layout configurations *(Figure 4)* can be generated by applying the following operations: Translate - move of a manufacturing resource from its current position to another, previously unoccupied position; Swap – exchange of the positions of any two manufacturing resources; and Rotate – rotation of a manufacturing resources by 0^0, 90^0, 180^0 or 270^0. The computer implementation was tested using illustrative examples (2). Furthermore, computational simulations were carried out to validate the performance of *SA*.

3. Case Studies

In this section the proposed *MWDRST* research model and the final *VCMS* layout will be briefly explained using a rudimentary example Vakharia (12) *(Figure 5.a)* together with, for comparison, two other *CM* design methods (*MST* and *Cut Tree*). Combinatorial optimisation algorithms are also employed to solve the *NP*-problems, which are programmed in *MATLAB*.

3.1. Stage I

For this stage, the operation sequence and batch quantity of the parts is the only data. The travel chart obtained from this data is the input to stage I model to obtain the *MWDRST*. Thus, utilising the *MWDRST*, the forward and backward arcs in the original digraph can be eliminated. The example's initial number of machines available of each type is given. Using *MATLAB*, the stage I research model is solved to obtain the *MWDRST*, which generates the *MWDRST* material flow line paths as shown in *Figure 5.b*.

3.2. Stage II

In this stage II, permutations of the *MWDRST* paths are considered. Using only the matrix of intercell flows, the stage II result will yield the *MWDRST* shown in *Figure 5.c*. Finally, to provide aisles for the intercell flows, and by inspection of the crisscrossing arcs *(Figure 5.c)*, additional machines (types 4, 6, 7, 10 and 11) are identified as indicated in *Figure 6.a*. Their location must be based on flow directions without losing the functional layout of each machine type. The machines included in each crisscrossing arc then determines the adjacencies.

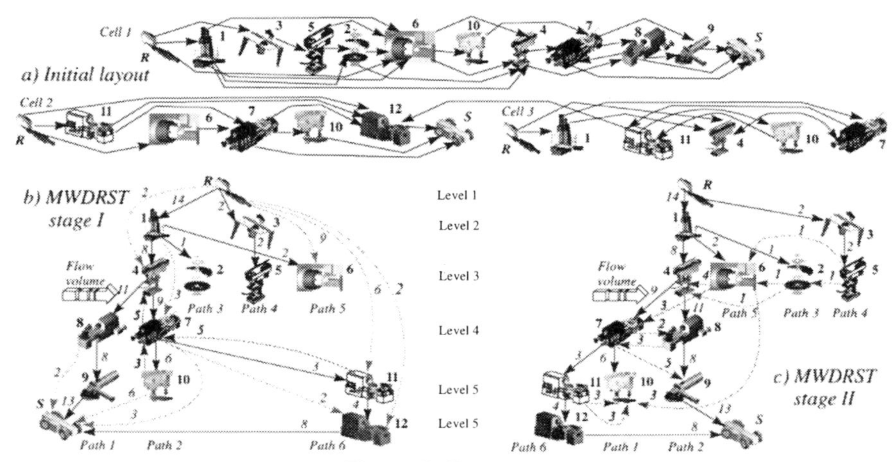

Figure 5. Case study

3.3. Stage III Economical Duplication

The result *(Figure 5.c)* from stage II of the proposed research methodology, together with the From – To machine material flow matrix, presents the minimum input data requirement for stage III. The first step is to consider the economic duplication of machines in several material flow line paths. To find the number of machines of each type in each module, it is assumed that each machine type must be an integer number, shared between material flow lines *(Figure 6.a)* and assigned to only one material flow line path. Again, the data from (12) is utilised. Thus, the economical numbers of each type of machine required in the different flow lines are calculated. However, in certain cases the integer requirements of machines summed over all the flow lines exceeds the available numbers of machines of each type. This is an easy problem to solve if extra machines of these types can be requisitioned for the flow lines (or if some of the flow lines can be partially replaced by multifunction machines).

3.4. Stage IV Economical Optimisation of VCM Layout

Evaluation of the material flowline networks is utilised because of having to change the assignment of the flowlines resulting from introducing a duplication of the machines. A new From – To machine chart is then developed as the input for this stage IV. Here the objective function is to minimise the total system cost, which is made up of the material handling and duplication costs.

Equal resource dimensions – These problems have been addressed by many researchers to demonstrate the effectiveness of different algorithms, including *SA*, and this aspect is not considered in this paper (the proposed methodology showed results (2) equal to Heragu (9), the best known methodology).

Unequal resource dimensions - In order to accommodate the approximate size of various resources, each entity is considered to be composed of an integer number of square building blocks (grid units). The one example considered is for twenty five resource types, the facilities test data again taken from (12). Here the initial From - To chart, together with the resource and restrictions data are the input for the *VCM* layout algorithm. A random initial placement of the resources is then applied to the shop size of a 12×10 grid, the algorithm resulted in a material flow cost distance of *260* units, with a *CPU* time of *65* seconds *(Figure 6.b)*. The next step in this comparative analysis is to analyse the change of the configuration (perturbation) and the generation factor (swap, translate and rotate), which is one of the essential components in applying the *SA* algorithm. Four classical Nugent (11) problems were used (2) to investigate the generation scheme property. The sizes were eight, twelve, fifteen and twenty facilities for the *QAP* problem. Bajic (2) showed that the best configuration changing factor was swapping of the facilities.

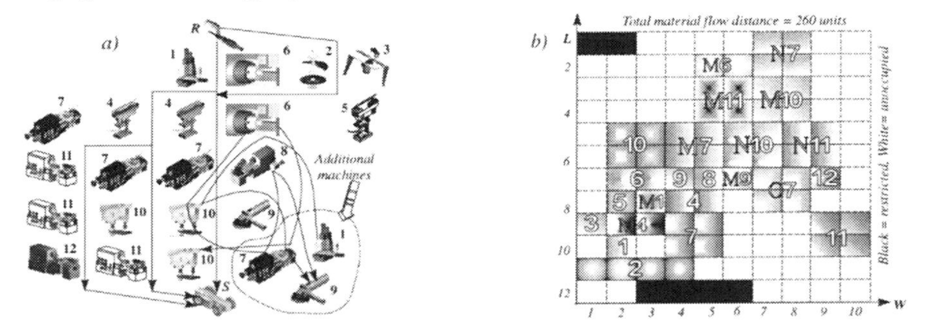

Figure 6. a) Economical (Stage III) duplication, and b) Final *VCM* layout (using SA)

4. Discussion

Stages I, II and III results show that the proposed method realistically models the immediate adjacencies of machines and the virtual cells to which they can be assigned. This enables flow distances and machine locations to be subsequently determined. Thus, this work can quickly evaluate a finite number of *VCM* layout alternatives which conform to the machine grouping, flow line layout and intercell machine adjacencies given by the second stage of the *MWDRST*. This helps to integrate the effects of reducing the lengths of individual material flow lines by increasing the width of the shop (converting a rectangle to a square with the same area). Furthermore the total weighted travel distances given by the *MST, Cut Tree* and *MWDRST* approaches (2), showed that the *MWDRST* gave better total backtrack and crisscross flows relative to the *MST* and *Cut Tree* analyses. The use of the proposed *MWDRST* is a feasible flow line design skeleton for generating a *VCM* layout for a dynamic-product facility, and unlike existing design skeletons, embeds directed material flow paths in the layout. Comparing the average utilisation of the resources shows promising results; in the *VCM* cell it is *77%* (system utilisation) which is higher than the average utilisation *51%* from (12), justifying the first three stages of the proposed methodology. Stage IV method is an advanced version of current *SA* based algorithms, since it considers: the size of manufacturing resources, restricted areas, and material flow paths. The proposed *SA* method showed a *15%* improvement in the material flow cost distance result, when compared with the results obtained by (4).These results show that *SA* can produce very efficient solutions to combinatorial optimisation problems, and that the best change of configuration scheme is swapping of the facilities, or locations. It should be noted that most of the past research work concentrates on layout facilities of equal sizes and shapes, and do not consider the material handling cost. However, the proposed research methodology advances current work by considering unequal shapes, the restrictions within a shop floor size, as well as material handling costs.

5. Conclusions

The manufacturing shop design problem is particularly complex, and here new methods for both the virtual cell formation and the manufacturing system layout problems have been presented. These methods address a host of practical issues, some of which have either been ignored or inadequately treated by past methodologies. This encourages the simultaneous design of a *VCM* layout to identify the machine groups, economical machine duplications, and intracell and intercell configurations. One of the most important contributions of this methodology is its practicality. The methodology for *VCM* design *(Figure 3)* may be used for the redesign of existing manufacturing shops and the design of planned facilities with known production period product demands and defined process routing's.

REFERENCE

1. Aarts E. and Korst J., *(1998),* Simulated annealing and Boltzman machines. *John Wiley.*
2. Bajic, M. M., *(2000),* Design of the dynamic cellular manufacturing system. PhD Thesis, To be published.
3. Bajic M. M., Baines K., Cresswell C., *(1997),* A cellular manufacturing approach towards *VM.* World Congress Manufacturing Technology Towards 2000, Cairns, Australia.
4. Bazaraa M.S., *(1975),* Computerised layout design: a branch and bound approach. *AIIE Transactions, Vol. 7, No. 4, pp. 432-437.*
5. Burbidge J.L., *(1995),* Back to production management. *Manufac. Engineering, pp. 66-71*
6. Carrie A.S. and Mannion J., *(1976),* Layout design and simulation of group cells. *Proc. 16th Inter. Conf. on Machine Tool Design and Research, Macmillan, London, pp. 99-105.*
7. Carrie A.S., Moore J., Roczniak M. and Seppanen J.J., *(1978),* Graph theory and computer aided facilities design. *Omega, Vol. 6, No4, pp. 353-364.*
8. Gallaghar C.C. and Knight W.A., *(1973),* Group Technology. *London Butterworth.*
9. Heragu S.S. and Kusiak A., *(1992),* Efficient models for the facility layout problem. *European Journal of Operational Research, Vol. 53, pp. 1-13.*
10. Metropolis N., Rosenbluth A., Teller A. and Teller E., *(1953),* Equation of state calculations by fast computing machines. *Jour. Chem. Phys., Vol. 21 pp. 1087-1099.*
11. Nugent C.E. and Vollman T.E., *(1968),* An experimental comparation of techniques for the assignment of facilities to locations. *Operation Research, Vol. 16, pp. 150-173.*
12. Vakharia A.J., *(1990),* Designing a cellular manufacturing systems: A materials flow approach based on operation sequences. *IIE Trans., Vol. 22, No. 1, pp. 84-97.*

Integration of solid modelling with manufacturing for the development of scroll compressors

Z JIANG and **D K HARRISON**
Department of Engineering, Glasgow Caledonian University, UK
K CHENG
School of Engineering, Leeds Metropolitan University, UK

ABSTRACT

Scroll type compressors are being widely used in the refrigeration and air conditioning industries. A concurrent engineering based integration approach to the design and manufacturing of these types of compressors is essential for achieving high product quality at low cost in the shortest delivery time. In this paper Pro/ENGINEER and Visual C^{++} design tools were used to implement the proposed CE based integration approach and associated design and manufacturing development. A visualised solid model of the compressor was developed first. Then finite element analysis was used to study the strain and stress distribution in the modelling. The deformation of the scroll wall was evaluated for maintaining the assembly clearance. The paper concludes with a discussion on the potential and application of the presented approach and development in mechanical product design.

Key words: Solid modelling, integration approach, concurrent engineering, and scroll compressors.

1 INTRODUCTION

The conventional design of scroll compressors has tended to use a serial engineering approach, which hampers the design and manufacturing engineers' ability to deliver the products effectively and efficiently in the increasingly competitive global marketplace. Concurrent Engineering (CE) is a promising approach to side step such difficulties for designers and manufacturing engineers (1,2). A literature survey shows there is much research on using CE. Sturges and Kilani (3) presented towards an integrated evaluation and reasoning system.

The system built on existing solid-modelling packages. The aim was the evaluation of the assemblability of the design, and the recommendation of design modifications that decrease the assembly time and the level of difficulty. The paper discussed the issues related to the development of an integrated design. It highlighted the need to develop models that extend beyond current geometric-modelling packages. These models would describe the component, system and process, and would lead to the fundamental data structures on which the evaluators of the assemblability factors can be built. Concurrent engineering criteria for effective implementation was studied in 1994 (4). The CE has developed into a significant management technique for orchestrating the multifunctional process of today's complex technical projects. Cross-function co-operation and effective teamwork are some of the crucial ingredients for making these processes work. People from many departments must collaborate over the product life cycle to ensure that it reflects customer needs and desires. All parties work together from the outset to anticipate challenges and bottlenecks and to eliminate them early on. In the process, they avoid organisational interface and project integration problems, which cause schedule delays, costly rework and problems in manufacturing, installation and service. Because of these more effective organisational processes, The CE offers more than just rapid implementation. Lee et al (5) presented a research on the development of a computer-based framework that supports concurrent mould manufacturing process planning. Vosnoiakos (6) presented an intelligent software system for the automatic generation of NC programs from wireframe models of 2-1/2D mechanical parts. In the paper an IGES post-processor was developed to the interface, then the system was converted to any CAD system for obtaining solid modelling geometry.

Though many researchers presented papers to introduce significant applications on CE, few papers (7) are related to the integration or the CE approach of CAD/CAM used in scroll compressors. It is necessary to develop an advanced approach to develop scroll compressors. The authors have developed a design approach that can cope with simultaneous design activities concurrently within one single environment. The approach is implemented using Pro/ENGINEER and Visual C^{++} programming. The work presented is focused on the integration of Constructive Solid Geometry (CSG) solid modelling and CAM to effect the design of scroll compressors in particular.

A typical refrigeration scroll compressor is shown schematically in Figure 1 (8). Its principal components include a fixed scroll, an orbiting scroll, a drive-shaft, frames, a counterweight and an electric motor. Obviously, the manufacturing accuracy of the scroll parts must be very high for their high qualities. It is difficult to implement the designed products. Firstly, because the two scrolls may contact each other in the radial direction at several points. The larger the machined clearances and assembly clearances of the two scrolls, the much higher the leakage of the compressed gas will be. Secondly, the working surfaces of the scrolls are inner surfaces, which are more difficult to machine than outside ones. Thus, this results in larger machining tolerance. As the scroll groove is an involute curve, the calculation for thermodynamic and aerodynamic properties is difficult. Thirdly, the deformation of the scroll wall will effect the properties of the compressors. These deformations might come from working pressure or from cutting forces or both of them. To determine the exact deformation is not easy. Therefore, it is necessary to use a powerful computer aided engineering tool to support the full development process of the scroll components, e.g. including their design, manufacture and inspection.

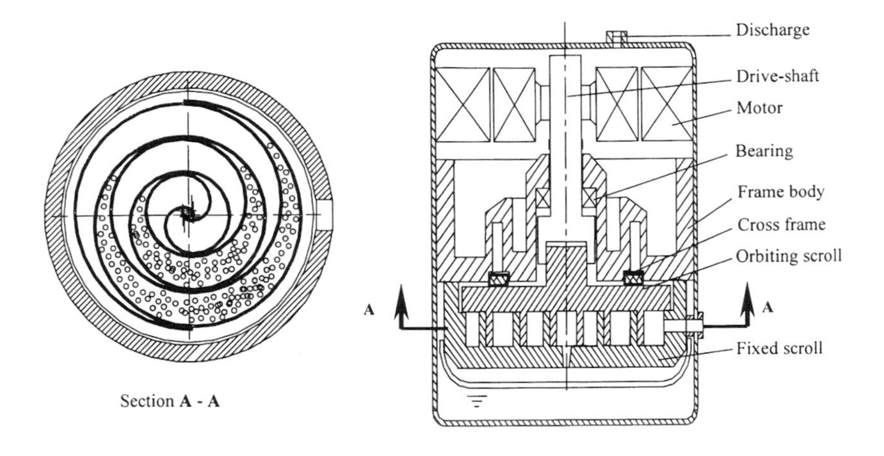

Fig. 1 The structure of a refrigeration scroll compressor

2 INTEGRATION OF SOLID MODELLING AND MANUFACTURING

As described above, scroll components will deform when they are working under high temperature and pressure. If the deformation is too great the compressor will fail and the component has to be redesigned and manufactured. In this approach, the deformation and other problems will hopefully be foreseen and overcome in advance of being in production. The integration of solid modelling and CAM is the kernel of the approach. It enables the design output to be optimal, robust and efficient and so achieve high quality products at low cost in the shortest delivery time.

Figure 2 shows the integration system of CAD and CAM. The main features of the scroll parts are created by designers in the CAD environment. These features have direct effects on the compressors performance. Therefore they should be carefully analysed in the environment. The created model including the properties of the parts is forwarded to FEM environment to analyse the deformation and stresses for the product quality and reliability. The results are analysed and compared with related design standards and criteria regarding manufacturing. The results will be continually fed back to the model until a final optimised model can be achieved. In the FEM environment optimisation is carried out to help the designer to optimise the height and thickness of the wall and other geometries.

When the workpiece is machined in the Computer Numerical Control (CNC) machine, the cutting force and heat will result in the deformation of the parts. To calculate this deformation is not easy because the deformation is affected by many factors such as the cutting depth of material, diameter of the tool, clipping force, heat and so on. On the other hand the cutting surface is a local plastic deformation combined with an elastic deformation, which is more complex. The situation can be simplified in a statistical way. The maximum deformation can be analysed out in FEM environment, when the related factors selected such as tools and cutting depth and size. The stresses and strains during manufacturing can be foreseen. If the results of FEM analysis in the manufacturing process are not satisfied the redesign is performed in the CAD environment as well. In this way CAM is integrated with the FEM and CAD. In this approach designers and manufacturing engineers can work concurrently. Many designers can be cross-function, co-operation and effective teamwork. They can transfer the model to a manufacturing engineer for process planning etc. The manufacturing engineer can also advise the designer if some revisions are needed. Even the model can be evaluated through the Internet. Many experts can check the design and manufacturing. Thus, solid modelling and machine cutting path simulation can bridge the gap between design and manufacturing which is essential for concurrent engineering.

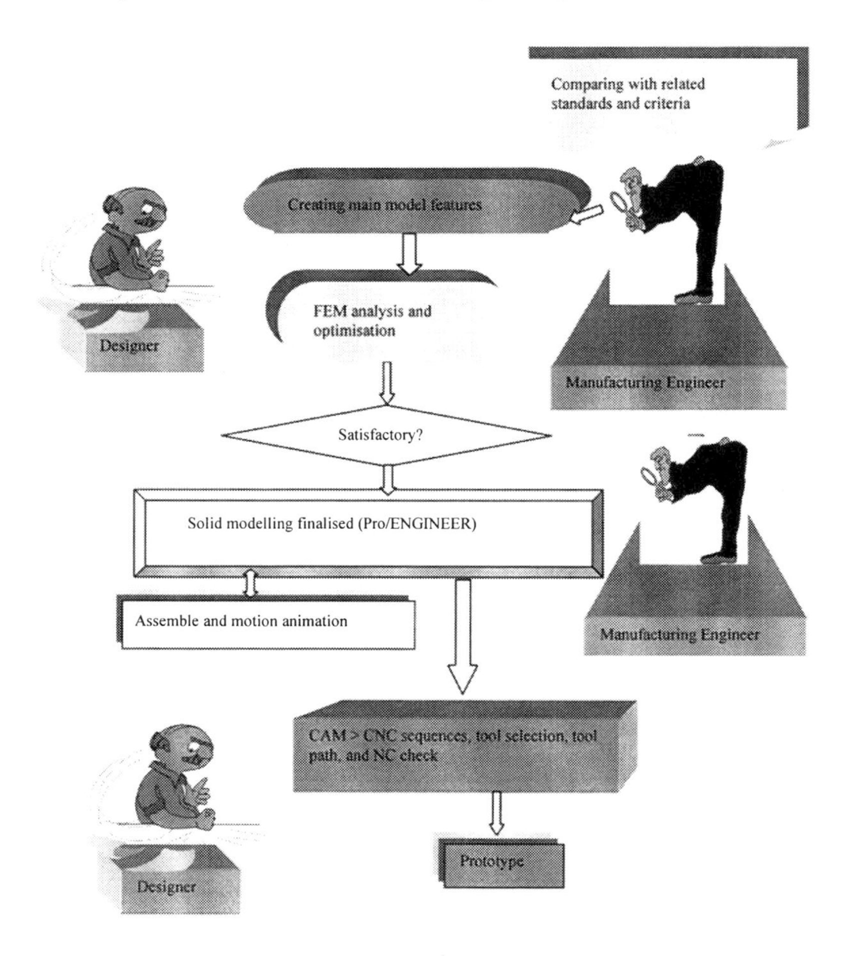

Fig. 2 The integration of CAD and CAM

3 SOLID MODELLING AND FEM FOR THE SCROLL COMPONENT

Solid modelling embodies information about the shape of an object as well as the material that it is made of and its mass properties. It is generally regarded as a superior form to wire frame modelling for the following reasons. Solid modelling is complete and unambiguous, which can reflect the full range of physical properties of designed objects. Solid modelling can provide superior visualisation of the object being designed since it is a true representation of the object. Therefore, solid modelling is significant for concurrent engineering because designers and manufacturing engineers should work at nearly the same time.

The essential element of initiating a CSG solid model of a scroll compressor design is to clarify the design task and design specifications such as the capacity of refrigeration, the working conditions and refrigerant type, etc. These specifications should be precisely converted and associated to the compressor's geometrical data and thermodynamic parameters during the early design stages. The calculation of the design is performed in Visual C++ program. The derived data is saved to a data file If the parameters are correct Pro/ENGINEER program will run.

3.1 Determination of Scroll Parameters in the Visual C++ Environment

The calculation of the design is performed in the Visual C++ environment. The required performance of the refrigeration compressor is determined by the customer. In this case study the performance eg. refrigeration capacity, condensing temperature and so on are assumed. The Visual C++ Program was created to perform the thermodynamic and dynamic calculations as shown in Figure 3. In the program an instruction window guides the users through the software. A dialogue window appears in the system and the designer of the compressor can fill in the data interactively. Running this program the necessary parameters of the scroll component will be calculated and displayed on the screen. If the designer needs to create the solid model of the scroll parts then it is just necessary to click the Run_ProENG button which is on the screen.

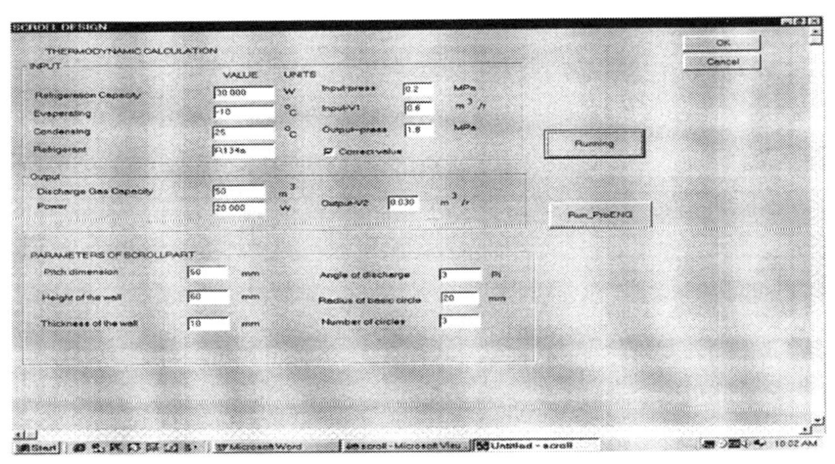

Fig. 3 The Visual C++ dialogue window for the calculation of the scroll part

3.2 Creation of the Solid Model of Scroll Component

The creation of the geometrical models and design features is implemented in the Pro/ENGINEER CAD environment, which are important for FEM analysis. The involute scroll curves are created from equations. The brief procedures are in the following ways.

Select **FEATURE** under **PART** environment. Then,
CREATE_DATUM_COORD SYS,
CREATE_DATUM_PLANE_DEFAULT,
CREATE_DATUM_CURVE_FROM EQUATION|DONE
GET COORDS|PICK_SET CSYS TYP_CARTESIAN.
Edit and enter the equations:

$$X_o = r[\cos(t\omega_o - \alpha) + t\omega_o \sin(t\omega_o - \alpha)]$$
$$Y_o = r[\sin(t\omega_o - \alpha) - t\omega_o \cos(t\omega_o - \alpha)]$$
$$Z_o = 0.0$$

This equation is used to create the outside involute curve of the scroll. The curve is a trajectory of the scroll wall. In the equations, r is the radius of the basic circle, t is a variable from 0 to 1, ω_0 is the total rotating angle of the scroll wall, and α is the initial angle of the involute curve. After the trajectory is created, the **SWEEP** method was used to generate features. The scroll feature created is shown in Figure 4. The fixed scroll shell features can be easily developed by the use of **PROTRUSION** and **REVOLVE**. They are shown in Figure 5. Many features are not included in these simplified models for FEM analysis, such as round corners, small holes and chamfers.

Fig. 4 A scroll feature created

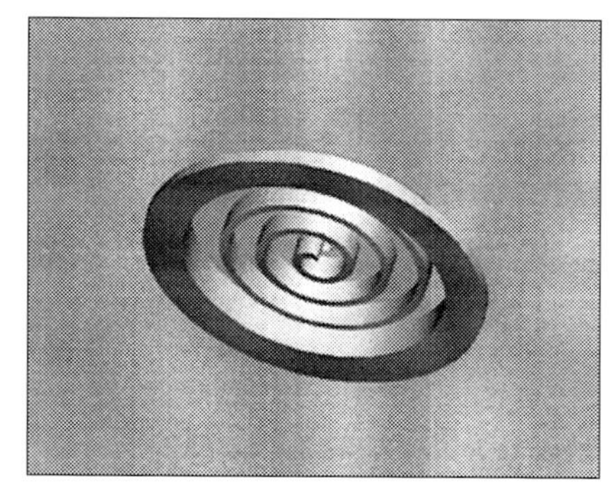

Fig. 5 Model of a fixed scroll

3.3 FEM Analysis on the Scroll component

As the scroll wall will deform under the gas pressure, the deformation will influence the performances of the compressor, it is necessary to precisely quantify the deformation. FEM analysis is a tool to evaluate such a feature of the scroll.

In the FEM environment the solid model was modified for FEM analysis. Several kinds of elements such as bar elements, tetrahedron element or mass element can be selected to mesh the model. The element size is entered. This size should be much smaller than that of the solid model. This element size is just a limitation of the element but not the exact size. For example global maximum, local maximum, global minimum and local minimum are the selected limitation of the size. The system can make out different size elements automatically.

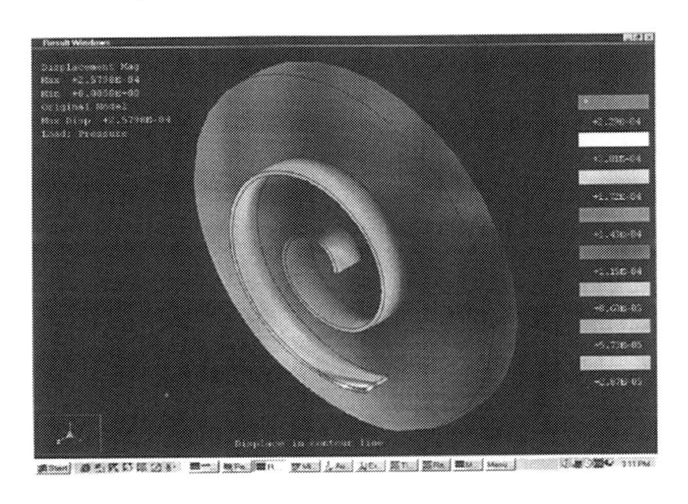

Fig. 6 Displacement of the scroll wall

Pro/MECHANICA is used to solve FEM problems in which there are three categories: **structure, motion and thermal**. **Structure** focuses on the structural aspects of design, which helps the designer to analyse stresses, strains, deformations and vibrations of the model as well as structure optimisation. **Thermal** focuses on the thermal aspects of design, which helps the designer to analyse heat transfer and thermal design optimisation, while **motion** focuses on the motion of components, which helps the designer to verify the mechanism's correct motion. The designer can simulate the motion and check the position, velocity and acceleration of the components without building and testing prototypes. As deformation and stress analysis are the main research purposes on the solid modelling, the structure is selected in this study.

The displacement, stress and mass are analysed. The displacement results of the scroll wall were shown in Figure 6. In the Figure the displacement increases gradually along the scroll wall from the inside to the outside and also from the bottom to the top. The reason is that the radius of the scroll wall increases, as the force increases in one direction. Beside this the bend moment is increased as well. This result is very important for scroll compressor assembly clearances. If this displacement exceed the allowed clearance the compressor will not run. Therefore the displacement or deformation should be known at the design stage. The maximum stress is at the bottom of the end of the scroll wall. It should be correct theoretically because there are concentrated stresses at this place. These stresses will change when the compressor is sucking and discharging the refrigerant gas periodically. The variable stresses will influence the life of the scroll wall.

4 MANUFACTURING SEQUENCES AND SIMULATIONS

The manufacturing operations for the scroll parts i.e. orbiting scroll and fixed scroll, utilise two machine tools the lathe and the milling machine. The parts should be machined in the lathe first because they have many cylindrical surfaces and circular disc surfaces. These surfaces are the datums of the scroll when these are machined. The parameters of machined scroll parts should be transferred to the CNC milling machine. Manufacturing the scroll component parts is the key procedure to develop the scroll compressors. Manufacturing sequences and tool path codes can be created in Pro/MANUFACTURE, which is the same CAD/CAM environment as that solid model created. MANUFACTURE comprises mainly **MFG model**, **MFG set up** and **Machining** as shown in Figure 7.

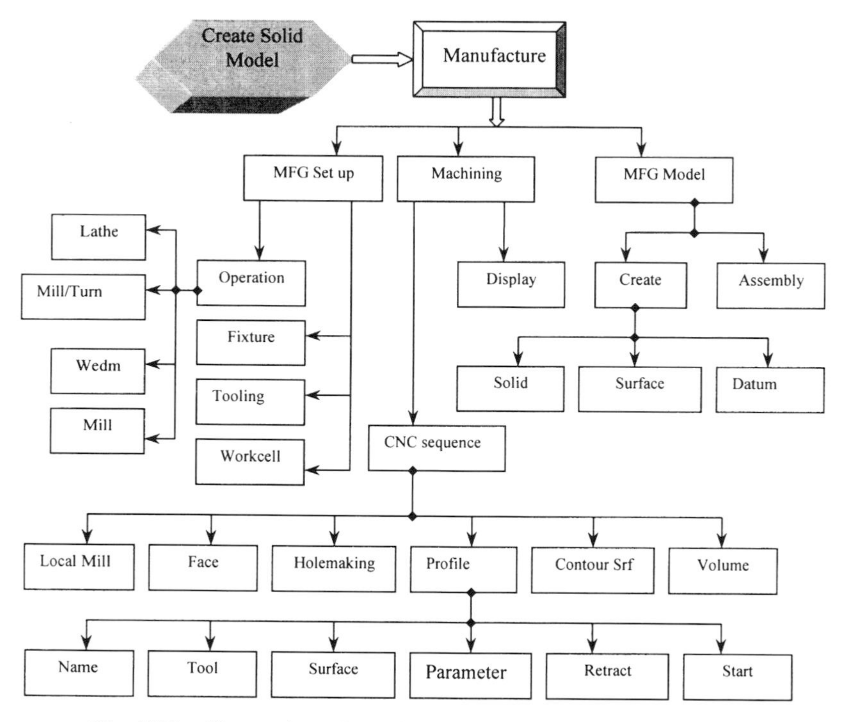

Fig. 7 The illustration of the CAM process for scroll parts

(1) **MFG Model**. The manufacturing model can be created or assembled in this environment. A workpiece is built independently first and then assembled with the solid model in the assembly environment. For this scroll part, the procedure of creating exact

workpieces is to build the surfaces that are the same as that on the model first. Then the surfaces built are merged into one quilt. Finally, the quilt is used to create a protrusion solid which is used as a workpiece scroll model.

(2) **MFG set up**. In this environment, the operation is used to determine the machine tools such as lathe, mill/turn, mill and welding machine. Once the machine tool is fixed the numbers of the axial should be selected. The fixture is used for complex shaped components or there is no suitable datum surface or there is a special cutting requirement. Fixture can be created here on the base of solid model.

(3) **Machining**. The CNC sequence can be created in this environment base on the former two environments i.e. MFG model and MFG set up. The machine ways are selected first such as hole-making, profile, contour surface, volume and so on. In each of the items the name, tool, parameters, surface, retract and so on should be chosen for CNC program The parameters, for example cut feed, spindle speed of the tool, step depth of each cutting must be determined for CNC path codes. To choose the surface to be machined is simple but care should be taken because many surfaces exist in the system. It is better to use query selection to select the surfaces to be machined.

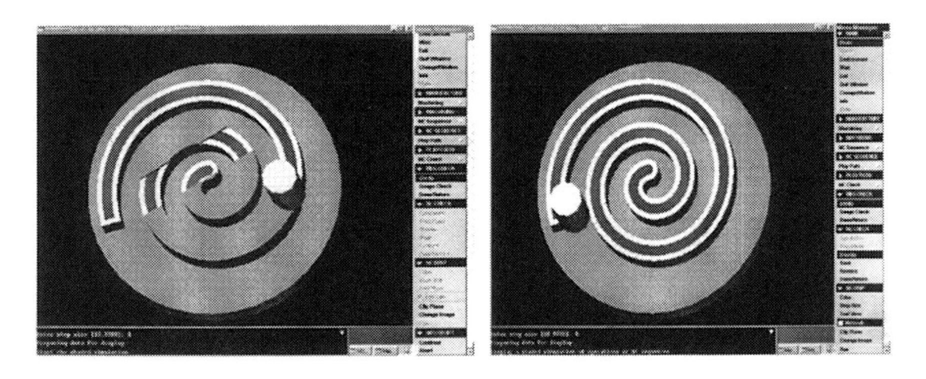

Fig. 8 Simulation of machining the top surface of scroll model

The CNC sequences created can be displayed on the system separately or serially as shown in Figure 8 which is the simulation of the CNC sequence to machine the top surface of the scroll wall. The left diagram shows that the tool is just at the half way point of the sequence and the right diagram is at the end of the sequence.

The offset area clearance is displayed in the system, which is useful to raster roughing. It removes material by machining a series of contours, producing a profile around the model and offsetting this initial path until the block limit is reached. The CNC tool path codes include rough and finish machining cuts. Rough machining allows a much more consistent depth of material to be left on the job before finishing is undertaken. This means that finishing can be undertaken more efficiently and at higher feed rates, with far less risk of tool damage or breakage.

5 CONCLUDING REMARKS

An integrated approach to create solid modelling of a scroll compressor and its manufacturing with the incorporation of FEM analysis in the Pro/ENGINEER and Pro/MECHANICA environment is presented in this paper. With this approach a correct and precise solid model can be achieved, which lead to an optimal design in terms of quality, delivery time and cost. The computer aided manufacturing approach and developed tool path codes are introduced. This work is essential for the manufacturing of the scroll components and their inspection on Co-ordinate Measuring Machines during manufacture. The approach and methodology have potential application for other mechanical components.

REFERENCES

(1) Turino, J., 1992, Managing Concurrent Engineering, Van Nostrand Reinhold Press, New York, pp. 1-4.

(2) Cheng, K., 1996, A designer based approach to product design in a concurrent engineering environment, Proceedings of the Second International Conference on Managing Integrated Manufacturing: Strategic, Organisation and Social Change, Leicester University, Leicestershire, UK, pp. 125-130.

(3) Sturges, R.H. and Kilani, M.I., 1992, Towards an integrated design for an assembly evaluation and reasoning system, Computer Aided Design, Vol. 24, No. 2, pp. 67-79.

(4) Thamhain, H.J., 1994, Concurrent engineering: criteria for effective implementation, Concurrent Engineering, No. 11-12, pp 29-32.

(5) Lee, R.S. Chen, Y.M. Cheng, H.Y. and Kuo, M.D., 1998, A frame work of a concurrent process planning system for mold manufacturing, Computer Integrated Manufacturing System, Vol. 11, No. 3, pp171-190.

(6) Vosniakos, G., 1998, An intelligent software system for the automatic generation of NC programs from wireframe models of 2-1/2D mechanical parts, Computer Integrated Manufacturing Systems, Vol. 11, No. 1-2, pp. 53-65.

(7) Jiang, Z.C., Cheng, K. and Harrison, D.K., 1999, Integration of solid modelling and FEM analysis for the design of scroll compressors, DETC99/DAC-8685, Proceedings of the 1999 ASME Design Engineering Technical Conferences, Las Vegas, Nevada, USA, CD-ROM.

(8) Li, W.L., Zhou, R. and Zhao, C., 1998, Rotary Type of Refrigeration Compressor (in Chinese), Mechanical Industry Press.

Authors' Index

T

V

W

Y

Z